Animal Diversity

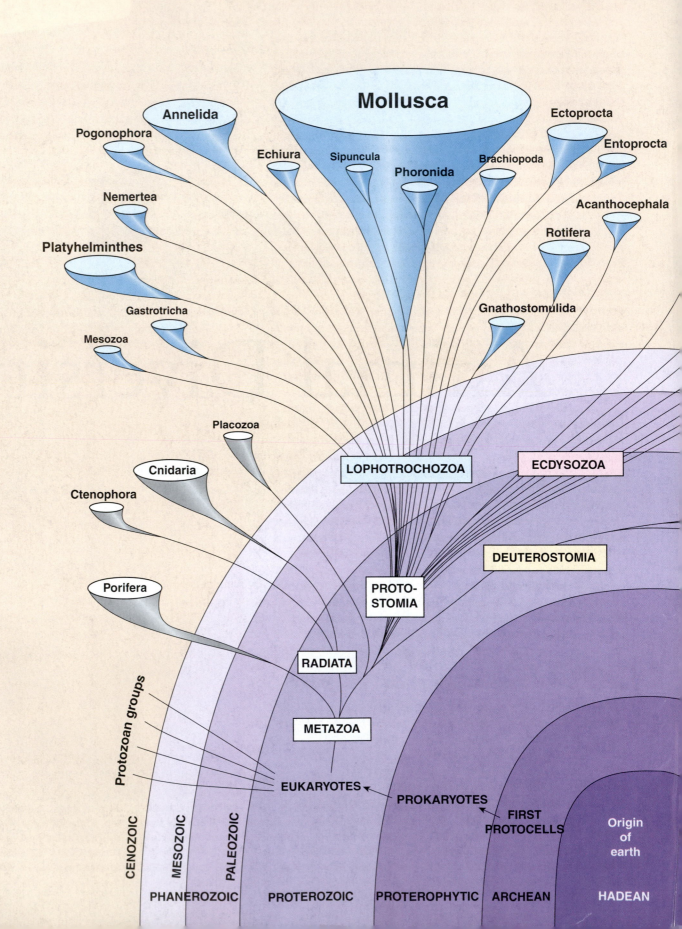

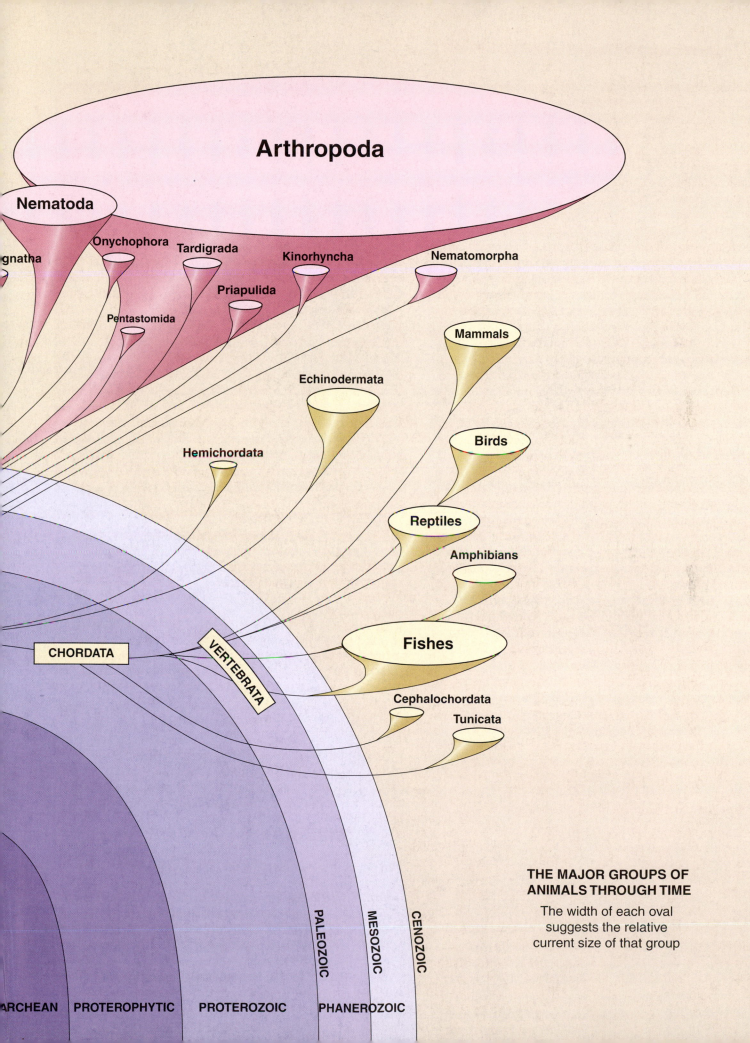

Arthropoda

Nematoda

gnatha

Onychophora

Tardigrada

Kinorhyncha

Nematomorpha

Priapulida

Pentastomida

Mammals

Echinodermata

Hemichordata

Birds

Reptiles

Amphibians

Fishes

CHORDATA

VERTEBRATA

Cephalochordata

Tunicata

**THE MAJOR GROUPS OF
ANIMALS THROUGH TIME**

The width of each oval
suggests the relative
current size of that group

PALEOZOIC

MESOZOIC

CENOZOIC

ARCHEAN

PROTEROPHYTIC

PROTEROZOIC

PHANEROZOIC

Animal Diversity

FOURTH EDITION

Cleveland P. Hickman, Jr.

Washington and Lee University

Larry S. Roberts

Florida International University

Susan L. Keen

University of California, Davis

Allan Larson

Washington University

David J. Eisenhour

Morehead State University

Original Artwork by

William C. Ober, M.D. and Claire W. Garrison, R.N.

Higher Education

Boston   Burr Ridge, IL   Dubuque, IA   Madison, WI   New York   San Francisco   St. Louis
Bangkok   Bogotá   Caracas   Kuala Lumpur   Lisbon   London   Madrid   Mexico City
Milan   Montreal   New Delhi   Santiago   Seoul   Singapore   Sydney   Taipei   Toronto

Higher Education

ANIMAL DIVERSITY, FOURTH EDITION

Published by McGraw-Hill, a business unit of The McGraw-Hill Companies, Inc., 1221 Avenue of the Americas, New York, NY 10020. Copyright © 2007 by The McGraw-Hill Companies, Inc. All rights reserved. No part of this publication may be reproduced or distributed in any form or by any means, or stored in a database or retrieval system, without the prior written consent of The McGraw-Hill Companies, Inc., including, but not limited to, in any network or other electronic storage or transmission, or broadcast for distance learning.

Some ancillaries, including electronic and print components, may not be available to customers outside the United States.

This book is printed on acid-free paper.

2 3 4 5 6 7 8 9 0 VNH/VNH 0 9 8 7

ISBN-13 978-0-07-252844-2
ISBN-10 0-07-252844-3

Publisher: *Margaret J. Kemp*
Developmental Editor: *Debra A. Henricks*
Marketing Manager: *Tami Petsche*
Project Manager: *April R. Southwood*
Senior Production Supervisor: *Kara Kudronowicz*
Senior Media Project Manager: *Tammy Juran*
Senior Designer: *David W. Hash*
Cover/Interior Designer: *Kaye Farmer*
Senior Photo Research Coordinator: *John C. Leland*
Photo Research: *Roberta Spieckerman*
Compositor: *Carlisle Publishing Services*
Typeface: *10/12 Garamond Book*
Printer: *Von Hoffmann Corporation*

Cover photos: 1st row: Caracal, ©Getty Images; Red Damselfly, ©F. Rauschenbach/zefa/Corbis; Bald Eagle, ©Stockbyte; Atlantic Oval Squid, ©Stephen Frink/Corbis. 2nd row: Iguana, ©Bryan F. Peterson/Corbis. 3rd row: Hornet, ©F. Rauschenbach/zefa/Corbis; Horsefly, ©F. Rauschenbach/zefa/Corbis; Squirrelfish, ©Digital Vision/Getty Images; Mantis Shrimp, ©Royalty-Free/Corbis. 4th row: Green Tree Python, ©Brand X Pictures/PunchStock; Buffalo, ©Getty Images; Cape Eagle Owl, ©Digital Vision/PunchStock; Indonesian Tarsier, ©Gary Bell/zefa/Corbis. 5th row: Red-Eyed Tree Frog, ©Creatas/PunchStock; Tarantula, ©Frank Krahmer/zefa/Corbis; Australian Pelican, ©Theo Allofs/zefa/Corbis; Mandrill, ©Creatas/PunchStock. 6th row: Nile Crocodile, ©Digital Vision Ltd.; Longlure Frogfish, ©Digital Vision/Getty Images; Bibron's Gecko, ©Royalty-Free/Corbis; Parrotfish, ©Bill Varie/Corbis.

Library of Congress Cataloging-in-Publication Data

Animal diversity / Cleveland P. Hickman, Jr. ... [et al.].— 4th ed.
   p. cm.
Rev. ed. of: Animal diversity / Cleveland P. Hickman, Jr., Larry S. Roberts, Allan Larson. 3rd. ed. 2003.
Includes bibliographical references and index.
ISBN 978-0-07-252844-2 — ISBN 0-07-252844-3 (hard copy : alk. paper)
1. Zoology. 2. Animal diversity. I. Hickman, Cleveland P. II. Hickman, Cleveland P. Animal diversity.

QL47.2.H527   2007
590—dc22   2005031218
                                                              CIP

www.mhhe.com

Brief Contents

Contents

About the Authors

Cleveland P. Hickman, Jr.

Cleveland P. Hickman, Jr., Professor Emeritus of Biology at Washington and Lee University in Lexington, Virginia, has taught zoology and animal physiology for more than 30 years. He received his Ph.D. in comparative physiology from the University of British Columbia, Vancouver, B.C., in 1958 and taught animal physiology at the University of Alberta before moving to Washington and Lee University in 1967. He has published numerous articles and research papers in fish physiology, in addition to co-authoring these highly successful texts: *Integrated Principles of Zoology, Biology of Animals, Animal Diversity, Laboratory Studies in Animal Diversity,* and *Laboratory Studies in Integrated Principles of Zoology.*

Over the years Dr. Hickman has led many field trips to the Galápagos Islands. His current research is on intertidal zonation and marine invertebrate systematics in the Galápagos. He has published three field guides in the Galápagos Marine Life Series for the identification of echinoderms, marine molluscs, and marine crustaceans. (To read more about these field guides, visit http://www.galapagosmarine.com.)

His interests include scuba diving, woodworking, and participating in chamber music ensembles.

Dr. Hickman can be contacted at:
hickmanc@wlu.edu.

Larry S. Roberts

Larry S. Roberts, Professor Emeritus of Biology at Texas Tech University and an adjunct professor at Florida International University, has extensive experience teaching invertebrate zoology, marine biology, parasitology, and developmental biology. He received his Sc.D. in parasitology at the Johns Hopkins University and is the lead author of Schmidt and Roberts' *Foundations of Parasitology,* sixth edition. Dr. Roberts is also co-author of *Integrated Principles of Zoology, Biology of Animals,* and *Animal Diversity,* and is author of *The Underwater World of Sport Diving.*

Dr. Roberts has published many research articles and reviews. He has served as President of the American Society of Parasitologists, Southwestern Association of Parasitologists, and Southeastern Society of Parasitologists, and is a member of numerous other professional societies. Dr. Roberts also serves on the Editorial Board of the journal, *Parasitology Research.* His hobbies include scuba diving, underwater photography, and tropical horticulture.

Dr. Roberts can be contacted at:
Lroberts1@compuserve.com.

Susan Keen

Susan Keen is a lecturer in the Section of Evolution and Ecology at the University of California at Davis. She received her Ph.D. in zoology from the University of California at Davis, following a M.Sc. from the University of Michigan at Ann Arbor. She is a native of Canada and obtained her undergraduate education at the University of British Columbia in Vancouver.

Dr. Keen is an invertebrate zoologist fascinated with jellyfish life histories. She has a particular interest in life cycles where both asexual and sexual phases of organisms are present, as they are in most jellyfishes. Her other research has included work on sessile marine invertebrate communities, spider populations, and Andean potato evolution.

Dr. Keen has been teaching evolution and animal diversity within the Introductory Biology series for 13 years. She enjoys all facets of the teaching process, from lectures and discussions to the design of effective laboratory exercises. In addition to her work with introductory biology, she offers seminars for the Davis Honors Challenge program, and for undergraduate and graduate students interested in teaching methods for biology. She was given an Excellence in Education Award from the Associated Students group at Davis in 2004. Her interests include weight training, horseback riding, gardening, travel, and mystery novels.

Dr. Keen can be contacted at:
slkeen@ucdavis.edu

Allan Larson

Allan Larson is a professor at Washington University, St. Louis, MO. He received his Ph.D. in genetics at the University of California, Berkeley. His fields of specialization include evolutionary biology, molecular population genetics and systematics, and amphibian systematics. He teaches courses in zoology genetics, macroevolution, molecular evolution, and the history of evolutionary theory, and has organized and taught a special course in evolutionary biology for high-school teachers.

Dr. Larson has an active research laboratory that uses DNA sequences to examine evolutionary relationships among vertebrate species, especially in salamanders and lizards. The students in Dr. Larson's laboratory have participated in zoological field studies around the world, including projects in Africa, Asia, Australia, Madagascar, North America, South America, and the Caribbean Islands. Dr. Larson has authored numerous scientific publications, and has edited for the journals *The American Naturalist, Evolution, Journal of Experimental Zoology, Molecular Phylogenetics and Evolution,* and *Systematic Biology.* Dr. Larson serves as an academic advisor to undergraduate students and supervises the undergraduate biology curriculum at Washington University.

Dr. Larson can be contacted at:
larson@wustl.edu.

David J. Eisenhour

David J. Eisenhour is an associate professor of biology at Morehead State University in Morehead, Kentucky. He received his Ph.D. in zoology from Southern Illinois University, Carbondale. He teaches courses in environmental science, human anatomy, general zoology, comparative anatomy, ichthyology, and vertebrate zoology. David has an active research program that focuses on systematics, conservation biology, and natural history of North American freshwater fishes. He has a particular interest in the diversity of Kentucky's fishes and is writing a book about that subject. He and his graduate students have authored several publications. David serves as an academic advisor to prepharmacy students.

His interests include fishing, landscaping, home remodeling, and entertaining his three young children, who, along with his wife, are enthusiastic participants in fieldwork.

Dr. Eisenhour can be contacted at:
d.eisenhour@morehead-st.edu

Preface

Animal Diversity is tailored for the restrictive requirements of a one-semester or one-quarter course in zoology, and is appropriate for both non-science and science majors of varying backgrounds. This fourth edition of *Animal Diversity* presents a survey of the animal kingdom with emphasis on diversity, evolutionary relationships, functional adaptations, environmental interactions, and certainly not least, readability.

We are fortunate to recruit as coauthors Susan L. Keen, who supervised this revision, and David J. Eisenhour. They bring fresh perspectives and the most current coverage from their research areas.

Organization and Coverage

The sixteen survey chapters of animal diversity are prefaced by four chapters presenting the principles of evolution, ecology, classification, and animal architecture. Throughout this revision we updated references and worked to simplify and streamline the writing.

Chapter 1 begins with a brief explanation of the scientific method—what science is (and what it is not)—and then moves to a discussion of evolutionary principles. Following an historical account of Charles Darwin's life and discoveries, the five major components of Darwin's evolutionary theory are presented, together with important challenges and revisions to his theory and an assessment of its current scientific status. This approach reflects our understanding that Darwinism is not a single, simple statement easily confirmed or refuted. It also prepares the student to dismiss the arguments of creationists who misconstrue scientific challenges to Darwinism as contradictions to the validity of organic evolution. The chapter ends with discussion of micro- and macroevolution.

Chapter 2 explains the principles of ecology, with emphasis on populations, community ecology, and variations in life history strategies of natural populations. The treatment includes discussions of niche, population growth and its regulation, limits to growth, competition, energy flow, nutrient cycles, and extinction.

Chapter 3 on animal architecture is a short but important chapter that describes the organization and development of body plans distinguishing major groups of animals. This chapter includes a picture essay of tissue types and a section explaining important developmental features associated with the evolutionary diversification of the bilateral metazoa.

Chapter 4 treats classification and phylogeny of animals. We present a brief history of how animal diversity has been organized for systematic study, emphasizing current use of Darwin's theory of common descent as the major principle underlying animal taxonomy. Continuing controversies over concepts of species and higher taxa are presented, including a discussion of how alternative taxonomic philosophies affect our study of evolution. Special attention is given to phylogenetic systematics (cladistics) and the interpretation of cladograms. Chapter 4 also emphasizes that current issues in ecology, evolution, and conservation biology all depend upon our taxonomic system.

The sixteen survey chapters are a comprehensive, modern, and thoroughly researched coverage of the animal phyla. We emphasize the unifying phylogenetic, architectural, and functional themes of each group. Structure and function of representative forms for major taxonomic groups are described, together with their ecological, behavioral, and evolutionary relationships. Each chapter includes succinct statements of "Position in the Animal Kingdom" and "Biological Contributions." Students find these highlights, a distinctive feature of this text, helpful in organizing their knowledge of animal diversity.

The Linnean classifications in each chapter are positioned following other coverage of a particular group, in most cases immediately preceding the summary at the end of the chapter. Discussions of phylogenetic relationships are written from a cladistic viewpoint, and cladograms are presented to show the structure of each group's history and the origin of the principal shared derived characters. Phylogenetic trees have been drawn to agree with cladistic analyses as closely as possible.

Changes in the Fourth Edition

Major revisions for the fourth edition include:

Scientific method—expanded explanation in Chapter 1.

Evolutionary mechanisms and theory—greater explanation of the rejection of teleology by Darwinian theory, sorting versus natural selection, population bottlenecks, roles of homeobox genes and mutations of large effect in evolution, and modes of species formation (Chapter 1).

Political controversies—updated coverage of controversies surrounding animal rights and "intelligent design" creationism (Chapter 1), and environmental issues (Chapter 2).

Metapopulation dynamics—added coverage of metapopulation dynamics in ecology (Chapter 2).

Life-history ecology—expanded coverage of life-history ecology, including general concepts (iteroparity versus semelparity, Chapter 2), evolution of cnidarian life cycles (Chapter 7), basic life histories of eels and hagfishes (Chapter 16), evidence for parental care in dinosaurs (Chapter 18), and social behavior of reproduction in birds (Chapter 19).

Community ecology—topics with expanded coverage include mutualism, Batesian and Müllerian mimicry, newly discovered hydrothermal-vent communities, and nutrient pools (Chapter 2).

Physiological ecology—expanded coverage of physiological ecology of many groups, particularly hemichordates (Chapter 14), echinoderms (locomotion and feeding, Chapter 14), tunicates (Chapter 15), fishes (swim bladder, fins, and osmotic regulation, Chapter 16), snakes (prey-capture strategies, Chapter 18), and mammals (feeding, Chapter 20).

Biodiversity and extinction—added discussion of animal diversity and extinction in the context of geological time (Chapter 2).

Epidemiology—new coverage of environmental epidemiological topics, including the fish-killing dinoflagellate, *Pfiesteria piscicida* (Chapter 5), the mosquito-borne West Nile virus (Chapter 12), incidence of snakebite in humans (Chapter 18), and declining amphibian populations (Chapter 17).

Body plans—expanded comparisons of major body plans, including formation of body plans and body cavities (Chapter 3) and implications of a possible sister-taxon relationship between annelids and molluscs for evolution of segmentation (Chapter 11).

Systematic concepts and theory—expanded explanation of systematic concepts, including species concepts, polytypic species, and the new taxonomic system PhyloCode as an alternative to Linnean taxonomy (Chapter 4).

Phylogenetic methodology—expanded explanation of molecular phylogenetic procedures (Chapter 4) and why evolutionary relationship of some taxa, such as chaetognaths, are difficult to discern (Chapter 13).

Phylogeny and classification of animals—updated phylogenies and taxonomies based largely on new comparative molecular studies (Chapters 4–20). Major cases include (1) new hypotheses for major prokaryotic and eukaryotic lineages and the concept of taxonomic domains above the kingdom level (Chapters 4–5); (2) phylogenetic position of acoelomorph flatworms outside all other bilaterians (Chapter 8); (3) pogonophorans subsumed into annelid class Polychaeta as clade Siboglinidae (Chapters 11 and 13); (4) paraphyly of annelid classes Polychaeta and Oligochaeta (Chapter 11); (5) arthropod taxon Uniramia abandoned in favor of four extant subphyla: Chelicerata, Myriapoda, Crustacea, and Hexapoda (Chapter 12); (6) arthropod subphylum Hexapoda revised to contain classes Entognatha and Insecta (Chapter 12); (7) revised phylogenetic relationships among many pseudocoelomate and lesser protostome phyla and their grouping into taxa Lophotrochozoa versus Ecdysozoa (Chapters 8 and 13); (8) sea daisies (formerly Concentricycloidea) subsumed into Asteroidea; (9) greatly revised cladograms and/or classifications for Mollusca (Chapter 10), Hemichordata (Chapter 14), Aves (Chapter 19), Mammalia (Chapter 20), and anthropoid apes (Chapters 4 and 20); and (10) priority of the name Urodela now given to salamanders (Chapter 14).

New taxa—addition of some newly described taxa, including a group of carnivorous sponges that lack choanocytes (Chapter 6) and the pseudocoelomate group Micrognathozoa (Chapter 9).

Vertebrate origins—expanded coverage of vertebrate origins, including ecological physiology, role of *Hox* genes and developmental changes (Chapter 15).

Paleontology—updated fossil discoveries, especially those pertaining to vertebrate origins (Chapter 15), sharing of derived characters between dinosaurs and birds (Chapter 19), and cynodont mammals and human ancestry (Chapter 20).

Group characteristics—extensive revision of group characteristics (Chapters 5–20).

Readability—reorganization of many topics to improve ease of reading; major cases include presentation of systematic principles (Chapter 4) and of phyla Cnidaria (Chapter 7), Platyhelminthes (Chapter 8), and Mollusca (Chapter 10).

These revisions include redrawing of many figures and enlargement of photos to improve clarity of presentation.

An extensively revised glossary includes all bolded key terms with the exception of those found in the "classification" boxes of their respective chapters. Additional unbolded terms that are useful in understanding terms in other definitions are also included.

Teaching and Learning Aids

Vocabulary Development

Key words are boldfaced and derivations of generic names of animals are given where they first appear in the text. In addition, derivations of many technical and zoological terms are provided in the text; in this way students gradually become familiar with the more common roots that recur in many technical terms. Updated for the fourth edition, the extensive glossary provides pronunciation, derivation, and definition of each term.

Chapter Prologues

A distinctive feature of this text is an opening essay placed in a panel at the beginning of each chapter. Each essay presents a theme or topic relating to the subject of the chapter to stimulate interest. Some present biological, particularly evolutionary, principles; others illuminate distinguishing characteristics of the group treated in the chapter.

Chapter Notes

Chapter notes, which appear throughout the book, augment the text material and offer interesting sidelights without interrupting the narrative.

For Review

Each chapter ends with a concise summary, a list of review questions, and annotated selected references. The review questions enable students to test themselves for retention and understanding of the more important chapter material.

Art Program

The appearance and usefulness of this text are much enhanced by numerous full color paintings by William C. Ober and Claire W. Garrison. Bill's artistic skills, knowledge of biology, and experience gained from an earlier career as a practicing physician, have enriched the authors' zoology texts through several editions. Claire practiced pediatric and obstetric nursing before turning to scientific illustration as a full-time career. Texts illustrated by Bill and Claire have received national recognition and won awards from the Association of Medical Illustrators, American Institute of Graphic Arts, Chicago Book Clinic, Printing Industries of America, and Bookbuilders West. Bill and Claire also are recipients of the Art Directors Award.

Web Pages

At the end of each survey chapter is a selection of related internet links dealing with the chapter's topics. The URLs for the pages are found in the text's Online Learning Center at *www.mhhe.com/hickmanad4e*.

Supplements

Instructor's Manual

The Instructor's Manual provides a chapter outline, commentary and lesson plan, and a listing of resource references for each chapter. We trust that this material will be particularly helpful for first-time users of the text, although experienced teachers also may find much of value. The Instructor's Manual is available on this text's Online Learning Center at *www.mhhe.com/hickmanad4e*.

Digital Content Manager

Created to accompany *Integrated Principles of Zoology,* thirteenth edition, by Hickman et al., this helpful CD-ROM can also be used with *Animal Diversity,* fourth edition. Instructors will find all of the illustrations, photos, and tables from *Integrated Principles of Zoology,* plus 200 additional animal diversity photos and 25 full-color animations illustrating key biological processes. All illustrations, photos, and tables have been pre-inserted into PowerPoint slides and can be easily modified for use with *Animal Diversity.*

Animal Diversity Online Learning Center

www.mhhe.com/hickmanad4e
This convenient website takes studying to a whole new level. **Students** will find chapter and animations quizzing, key term flashcards, interactive web links, and more. What a great way to get a better grade!

Instructors will appreciate a password-protected Instructor's Manual and test questions, PowerPoint lecture slides with images and species/origin information, a guide to teaching animal molecular phylogenetics, and more!

The *Animal Diversity* Online Learning Center is also home to the **Zoology Essential Study Partner.** This unique learning tool allows students to test their understanding of important zoological concepts through the use of animations, learning activities, quizzing, and interactive diagrams.

Animal Diversity Laboratory Manual

The laboratory manual by Cleveland P. Hickman Jr. and Lee B. Kats, *Laboratory Studies in Animal Diversity,* is designed specifically for a survey course in zoology.

Digital Zoology

Digital Zoology Interactive CD-ROM by Jon Houseman is an interactive guide to the specimens and materials covered in zoology laboratory and lecture sessions. Laboratory modules contain illustrations, photographs, annotations of the major structures of organisms, interactive quizzes, and video clips. Interactive cladograms within lab modules provide links to interactive synapomorphies of the various animal groups. Key terms throughout the program link to an interactive glossary. This CD-ROM is the perfect student study tool to promote learning both in and outside of the zoology laboratory, and also comes with an accompanying student workbook and website to provide additional study tips, exercises, and phyla characteristics.

Study Aid/Poster: Chief Taxonomic Subdivisions & Organ Systems of the Animal Phyla

This 30″ × 36″ poster is a great reference/study tool for students!

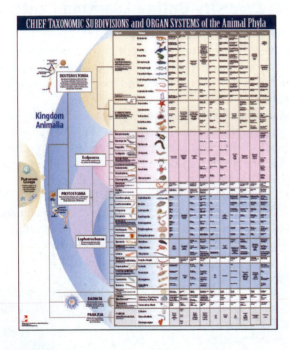

NEW! McGraw-Hill: *Biology Digitized Video Clips*

McGraw-Hill is pleased to offer digitized biology video clips on DVD! Licensed from some of the highest-quality science video producers in the world, these brief segments range from about five seconds to just under three minutes in length and cover all areas of general biology from cells to ecosystems. Engaging and informative, McGraw-Hill's digitized biology videos will help capture students' interest while illustrating key biological concepts and processes. Topics include: mitosis, amoeba locomotion, rainforest diversity, Darwin's finches, tarantula defense, nematodes, bird/water buffalo mutualism, poison dart frogs, echinoderms, and much more! ISBN-13: 978-0-07-312155-0 (ISBN-10: 0-07-312155-X)

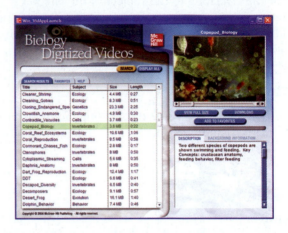

Acknowledgments

We wish to thank the following individuals for reviewing the third edition of this text:

Annalisa Berta, *San Diego State University*
Dana Brown Haine, *Central Piedmont Community College*
Walter M. Goldberg, *Florida International University*
Harold Heatwole, *North Carolina State University*
Stéphane Hourdez, *Pennsylvania State University*
Jeffrey Ihara, *MiraCosta College*
Robert D. Johnson, Jr., *Pierce College*
Richard N. Mariscal, *Florida State University*
Michael J. Shaughnessy Jr., *Morehead State University*
R. Pat Randolph and M. Frances Keller (*University of California, Davis*) made valuable contributions to the arthropod chapter

The authors express their gratitude to the able and conscientious staff of McGraw-Hill who brought this book to its present form. We extend special thanks to publisher Marge Kemp, developmental editor Debra Henricks, project manager April Southwood, photo researcher John Leland, and designer David Hash. All played essential roles in shaping the fourth edition.

Science of Zoology and Evolution of Animal Diversity

A Legacy of Change

Life's history is a legacy of perpetual change. Despite an apparent permanence of the natural world, change characterizes all things on earth and in the universe. Countless kinds of animals and plants have flourished and disappeared, leaving behind an imperfect fossil record of their existence. Many, but not all, have left living descendants that bear a partial resemblance to them.

We observe and measure life's changes in many ways. On a short evolutionary timescale, we see changes in the frequencies of different genetic traits within populations. Evolutionary changes in relative frequencies of light- and dark-colored moths were observed within a single human lifetime in polluted areas of industrial England. Formation of new species and dramatic changes in appearances of organisms, as illustrated by evolutionary diversification of Hawaiian birds, require longer timescales covering 100,000 to 1 million years. Major evolutionary trends and periodic mass extinctions occur on even larger timescales covering tens of millions of years. The fossil record of horses through the past 50 million years shows a series of different species replacing older ones through time and ending with the familiar horses that we know today. The fossil record of marine invertebrates shows a series of mass extinctions separated by intervals of approximately 26 million years.

Organic evolution is the irreversible, historical change that we observe in living populations and in the earth's fossil record. Because every feature of life is a product of evolutionary processes, biologists consider organic evolution the keystone of all biological knowledge.

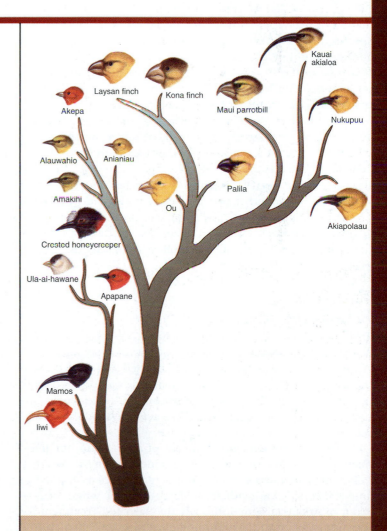

Evolutionary diversification of Hawaiian honeycreepers.

Z oology (Gr. *zōon*, animal, + *logos*, discourse on, study of) is the scientific study of animals. It is part of biology (Gr. *bios*, life, + *logos*, discourse on, study of), the study of all life. The panorama of animal diversity—how animals function, live, reproduce, and interact—is exciting, fascinating, and awe inspiring. A complete understanding of all phenomena included in zoology is beyond the ability of any single person, perhaps of all humanity, but we strive to know as much as possible.

To understand the diversity of animal life, we must study its long history, which began more than 600 million years ago. From the earliest animals to the millions of animal species living today, this history demonstrates extensive and ongoing change, which we call **evolution.** We depict the history of animal life as a branching genealogical tree, called a **phylogeny.** We place the earliest species ancestral to all animals at the trunk; then all living animal species fall at the growing tips of the branches. Each successive branching event represents formation of new species from an ancestral one. Newly formed species inherit many characteristics from their immediate ancestor, but they also evolve new features that appear for the first time in the history of life. Each branch therefore has its own unique combination of characteristics and contributes a new dimension to the spectrum of animal diversity.

Scientific study of animal diversity has two major goals. The first is to reconstruct a phylogeny of animal life and to find where in evolutionary history we can locate origins of multicellularity, a **coelom, spiral cleavage,** vertebrae, **homeothermy,** and all other dimensions of animal diversity as we know it. The second major goal is to understand historical processes that generate and maintain diverse species and adaptations throughout evolutionary history. Darwin's theory of evolution makes possible the application of scientific principles to attain both goals.

Principles of Science

A basic understanding of zoology requires understanding what science is, what it is not, and how knowledge is gained using the scientific method. In this section we examine methodology that zoology shares with science as a whole. These features distinguish sciences from those activities, such as art and religion, that we exclude from science.

Despite the enormous impact of science on our lives, many people have only a minimal understanding of science. Public misunderstanding of scientific principles as applied to animal diversity was evident on March 19, 1981, when the governor of Arkansas signed into law the Balanced Treatment for Creation-Science and Evolution-Science Act (Act 590 of 1981). This act falsely presented creation-science as a valid scientific endeavor. Creation-science is a religious position advocated by a minority of America's religious community and does not qualify as science.

Enactment of this law led to a historic lawsuit tried in December 1981 in the court of Judge William R. Overton, U.S.

District Court, Eastern District of Arkansas. The suit was brought by the American Civil Liberties Union on behalf of 23 plaintiffs, including religious leaders and groups representing several denominations, individual parents, and educational associations. Plaintiffs contended that this law violated the First Amendment to the U.S. Constitution, which prohibits establishment of religion by government. This amendment prohibits passing a law that would aid one religion or prefer one religion over another. On January 5, 1982, Judge Overton permanently prohibited Arkansas from enforcing Act 590.

Considerable testimony during the trial addressed the nature of science. On the basis of testimony by scientists, Judge Overton stated explicitly these essential characteristics of science:

1. It is guided by natural law.
2. It has to be explanatory by reference to natural law.
3. It is testable against the empirical world.
4. Its conclusions are tentative and not necessarily the final word.
5. It is falsifiable.

Pursuit of scientific knowledge must be guided by physical and chemical laws that govern the state of existence. Scientific knowledge must explain what is observed by reference to natural law without requiring intervention of any supernatural being or force. We must be able to make observations directly or indirectly to test hypotheses about nature. We must be ready to discard or to modify any conclusion if it is contradicted by further observations. As Judge Overton stated, "While anybody is free to approach a scientific inquiry in any fashion they choose, they cannot properly describe the methodology used as scientific, if they start with a conclusion and refuse to change it regardless of the evidence developed during the course of the investigation." Science is outside religion, and scientific knowledge does not favor one religious position over another.

Unfortunately, the religious position formerly called "creation science" has reappeared in American politics with the name "intelligent design theory." We are forced once again to defend the teaching of science against this scientifically meaningless doctrine.

Scientific Method

These essential criteria of science form the **hypothetico-deductive method.** The first step of this method is to generate **hypotheses,** or potential answers to a question being asked. These hypotheses are usually based on prior observations of nature (figure 1.1) or derived from theories based on such observations. Scientific hypotheses often constitute general statements that may explain a large number of diverse observations about nature. Natural selection, for example, explains our observations that many different species have properties that make them adapted to their environments. Based on a hypothesis, a scientist must say, "If my hypothesis is a valid explanation of past observations, then future observations ought to have certain characteristics."

figure 1.1

A few of the many dimensions of zoological research. **A,** Observing coral in the Caribbean Sea. **B,** Studying insect larvae collected from an arctic pond on Canada's Baffin Island. **C,** Separating growth stages of crab larvae at a marine laboratory. **D,** Observing nematocyst discharge (**E**) from hydrozoan tentacles (see p. 129).

The scientific method may be expressed as a series of steps:

1. Observation
2. Question
3. Hypothesis
4. Empirical test
5. Conclusions
6. Publication.

Observations illustrated in figure 1.1A–C form a critical first step in evaluating the life histories of natural populations. For example, observations of crab larvae shown in figure 1.1C might cause the observer to question whether rate of larval growth is higher in undisturbed populations than in ones exposed to a chemical pollutant. A null hypothesis is then generated to permit an empirical test. The null hypothesis is worded in a way that would permit data to reject it if it is false. The null hypothesis pertaining to larval growth rates of crabs in undisturbed versus polluted conditions is that the growth rates are equal in both kinds of populations. The investigator then performs an empirical test by gathering data on larval growth rates in a set of undisturbed crab populations and a set of populations subjected to the chemical pollutant. Ideally, the undisturbed populations and the chemically treated populations are equivalent for all conditions except presence of the chemical in question. If measurements show consistent differences in growth rate between the two sets of populations, the null hypothesis is rejected. One then concludes that the chemical pollutant does alter larval growth rates. A statistical test is usually needed to ensure that the differences between the two groups are greater than would be expected from chance fluctuations alone. If the null hypothesis cannot be rejected, one concludes that the data do not show any effect of the chemical treatment. The results of the study are then published to communicate findings to other researchers, who may repeat the results, perhaps using additional populations of the same or a different species. Conclusions of the initial study then serve as the observations for further questions and hypotheses to reiterate the scientific process.

Note that a null hypothesis cannot be proved correct using the scientific method. If the available data are compatible with it, the hypothesis serves as a guide for collecting additional data that potentially might reject it. Our most successful hypotheses are the ones that make specific predictions confirmed by large numbers of empirical tests.

If a hypothesis is very powerful in explaining a large variety of related phenomena, it attains the status of a **theory.** Evolution by natural selection is a good example; it provides a potential explanation for the occurrence of many different traits observed among animal species. Each of these instances constitutes a specific hypothesis generated from the theory of evolution by natural selection. The most useful theories are those that can explain the largest array of different natural phenomena.

We emphasize that the word "theory," when used by scientists, is not arbitrary speculation as often implied by nonscientific usage. Failure to make this distinction is prominent in criticism of evolution by creationists, who have called evolution "only a theory" to imply that it is little better than a random guess. In fact, evolutionary theory is supported by such massive evidence that most biologists view repudiation of evolution as tantamount to repudiation of reality. Nonetheless, evolution, like all other theories in science, is not proved by mathematical logic, but is testable, tentative, and falsifiable.

Experimental and Evolutionary Sciences

The many questions asked about animal life since Aristotle's time can be grouped into two major categories. The first category seeks to understand **proximate causes** (also called immediate causes) that underlie functioning of biological systems at all levels of complexity. It includes problems of explaining how animals perform their metabolic, physiological, and behavioral functions at molecular, cellular, organismal, and even population levels. For example, how is genetic information expressed to guide synthesis of proteins? What causes cells to divide to produce new cells? How does population density affect physiology and behavior of organisms?

Biological sciences that investigate proximate causes are called **experimental sciences** because they use the **experimental method.** This method consists of three steps: (1) predicting how a system being studied would respond to a disturbance, (2) making the disturbance, and then (3) comparing observed results to predicted ones. Experimental conditions are repeated to eliminate chance occurrences that might produce errors. **Controls** (repetitions of an experimental procedure that lack the disturbance) are established to eliminate any unperceived factors that may bias an experiment's outcome.

Processes by which animals maintain a body temperature under different environmental conditions, digest food, migrate to new habitats, or store energy are additional examples of phenomena studied using experimental methodology. Subfields of biology that constitute experimental sciences include molecular biology, cell biology, endocrinology, immunology, physiology, developmental biology, and community ecology.

In contrast to proximate causes, **evolutionary sciences** address questions of **ultimate causes** that have generated biological systems and their properties through evolutionary time. For example, what evolutionary factors have caused some birds to acquire complex patterns of seasonal migration between temperate and tropical regions? Why do different species of animals have different numbers of chromosomes in their cells? Why do some animal species maintain complex social systems, whereas animals of other species remain largely solitary?

A scientist's use of the phrase "ultimate cause," unlike Aristotle's usage, does not imply a preconceived goal for natural phenomena. An argument that nature has a predetermined goal, such as evolution of the human mind, is termed teleological. Teleology is the mistaken notion that evolution of living organisms is guided by purpose toward an optimal design. A major success of Darwinian evolutionary theory is its rejection of teleology in explaining biological diversification.

Evolutionary sciences proceed largely using the **comparative method** rather than experimentation. Characteristics of molecular biology, cell biology, organismal structure, development, and ecology are compared among related species to identify patterns of variation. Patterns of similarity and dissimilarity then can be used to test hypotheses of relatedness and thereby to reconstruct the evolutionary tree that relates the species being compared. Comparative studies also serve to test hypotheses of evolutionary processes that have generated animal diversity. Clearly, evolutionary sciences use results of experimental sciences as a starting point. Evolutionary sciences include comparative biochemistry, molecular evolution, comparative cell biology, comparative anatomy, comparative physiology, and phylogenetic systematics.

Origins of Darwinian Evolutionary Theory

Charles Robert Darwin and Alfred Russel Wallace (figure 1.2) were the first to establish evolution as a powerful scientific theory. Today organic evolution can be denied only by abandoning reason. As the noted English biologist Sir Julian Huxley wrote, "Charles Darwin effected the greatest of all revolutions in human thought, greater than Einstein's or Freud's or even Newton's, by simultaneously establishing the fact and discovering the mechanism of organic evolution." Darwinian theory allows us to understand both the genetics of populations and long-term trends in the fossil record. Darwin and Wallace were not the first, however, to consider the basic idea of organic evolution, which has an ancient history. We review the history of evolutionary thinking as it led to Darwin's theory and then discuss evidence supporting it.

Pre-Darwinian Evolutionary Ideas

Before the eighteenth century, speculation on the origin of species rested on myth and superstition, not on anything resembling a testable scientific hypothesis. Creation myths viewed the world as a constant entity that did not change after its creation. Nevertheless, some thinkers approached the idea that nature has a long history of perpetual and irreversible change.

Early Greek philosophers, notably Xenophanes, Empedocles, and Aristotle, developed a primitive idea of evolutionary change. They recognized fossils as evidence for former life, which they believed had been destroyed by natural catastrophe. Despite their spirit of intellectual inquiry, ancient Greeks failed to establish an evolutionary concept, and the issue declined well before the rise of Christianity. Opportunities for evolutionary thinking became even more restricted as the biblical account of the earth's creation became accepted as a tenet of faith. The year 4004 B.C. was fixed by Archbishop James Ussher (mid-seventeenth century) as the time of life's creation. Evolutionary views were considered rebellious and heretical. Still, some speculation continued. The French naturalist

A   **B**

figure 1.2

Founders of the theory of evolution by natural selection. **A,** Charles Robert Darwin (1809–1882), as he appeared in 1881, the year before his death. **B,** Alfred Russel Wallace (1823–1913) in 1895. Darwin and Wallace independently developed the same theory. A letter and essay from Wallace written to Darwin in 1858 spurred Darwin into writing *On the Origin of Species,* published in 1859.

Georges Louis Buffon (1707–1788) stressed environmental influences on modifications of animal type. He also extended earth's age to 70,000 years.

Lamarckism: The First Scientific Explanation of Evolution

The first complete explanation of evolution was authored by the French biologist Jean Baptiste de Lamarck (1744–1829) (figure 1.3) in 1809, the year of Darwin's birth. He made the first convincing case for the idea that fossils were remains of extinct animals. Lamarck's evolutionary mechanism, **inheritance of acquired characteristics,** was engagingly simple: organisms, by striving to meet demands of their environments, acquire adaptations and pass them by heredity to their offspring. According to Lamarck, giraffes evolved a long neck because their ancestors lengthened their necks by stretching to obtain food and then passed the lengthened neck to their offspring. Over many generations, these changes accumulated to produce the long neck of modern giraffes.

We call Lamarck's concept of evolution *transformational,* because it claims that individual organisms transform their characteristics to produce evolution. We now reject transformational theories because genetic studies show that traits acquired by an organism during its lifetime, such as strengthened muscles, are not inherited by offspring. Darwin's evolutionary theory differs from Lamarck's in being a *variational* theory. Evolutionary change is caused by differential survival and reproduction among organisms that differ in hereditary traits, not by inheritance of acquired characteristics.

The debate surrounding use of animals to serve human needs continues. Most controversial of all is animal use in biomedical and behavioral research and in testing commercial products.

Congress has passed a series of amendments to the Federal Animal Welfare Act, a body of laws covering animal care in laboratories and other facilities. These amendments are known as the three R's: *Reduction* in number of animals needed for research; *Refinement* of techniques that might cause stress or suffering; *Replacement* of live animals with simulations or cell cultures whenever possible. As a result, the total number of animals used each year in research and in testing of commercial products has declined. Developments in cellular and molecular biology also have contributed to decreased use of animals for research and testing. An animal rights movement has created awareness of needs of animals used in research and has stimulated researchers to discover cheaper, more efficient, and more humane alternatives.

Computers and culturing of cells can substitute for experiments on animals only when the basic principles involved are well known.

When principles themselves are being scrutinized and tested, computer modeling is not sufficient. A recent report by the National Research Council concedes that although a search for alternatives to animals in research and testing will continue, "the chance that alternatives will completely replace animals in the foreseeable future is nil." Realistic immediate goals, however, include reduction in number of animals used, replacement of mammals with other vertebrates, and refinement of experimental procedures to reduce discomfort of animals being tested.

Medical and veterinary progress depends on research using animals. Every drug and vaccine developed to improve human health has been tested first on animals. Research using animals has enabled medical science to eliminate smallpox and polio, and to immunize against diseases previously common and often deadly, including diphtheria, mumps, and rubella. It also has helped to create treatments for cancer, diabetes, heart disease, and manic-depressive psychoses, and to develop surgical procedures including heart surgery, blood transfusions, and cataract removal. AIDS research is wholly dependent on studies using animals. The similarity of simian AIDS, identified in rhesus monkeys, to human AIDS has permitted simian AIDS to serve as a model for human AIDS. Recent work indicates that cats, too, may be useful models for developing an AIDS vaccine. Skin grafting experiments, first done with cattle and later with other animals, opened a new era in immunological research with vast ramifications for treatment of disease in humans and other animals.

Research using animals also has benefited *other animals* through the development of veterinary cures. Vaccines for feline leukemia and canine parvovirus were first introduced to other cats and dogs. Many other vaccinations for serious diseases of animals were developed through research on animals: for example, rabies, distemper, anthrax, hepatitis, and tetanus. No endangered species is used in general research (except to protect that species from total extinction). Thus, research using animals has provided enormous benefits to humans and other animals. Still, much remains to be learned about treatment of diseases such as cancer, AIDS, diabetes, and heart disease, and research with animals will be required.

Despite the remarkable benefits produced by research on animals, advocates of animal rights often present a negative and emotionally distorted picture of this research. An ultimate goal of many animal rights activists, who have focused specifically on use of animals in science rather than on treatment of animals in all contexts, remains total abolition of all forms of research using animals. Our scientific community is deeply concerned about the impact of these attacks on our ability to conduct important experiments that will benefit people and animals. If we are justified in using animals for food and fiber and as pets, why are we not justified in experimentation to benefit human welfare when these studies are conducted humanely and ethically?

The Association for Assessment and Accreditation of Laboratory Animal Care International supports the use of animals to advance medicine and science when nonanimal alternatives are not available and when animals are treated in an ethical and humane way. Accreditation by this organization allows research institutions to demonstrate excellence in their standards of animal care. Nearly all major institutions receiving funding from the National Institutes of Health have sought and received this accreditation. See the website at **http://www.aaalac.org**

According to the U.S. Department of Health and Human Services, animal research has helped extend our life expectancy by 20.8 years.

for more information on accreditation of laboratory animal care.

References on Animal Rights Controversy

Commission on Life Sciences, National Research Council. 1988. Use of laboratory animals in biomedical and behavioral research. Washington, D.C., National Academy Press. *Statement of national policy on guidelines for use of animals in biomedical research. Includes a chapter on benefits derived from use of animals.*

Groves, J. M. 1997. Hearts and minds: the controversy over laboratory animals. Philadelphia, Pennsylvania, Temple University Press. *Thoughtful review of the controversy by an activist who conducted extensive interviews with activists and animal-research supporters.*

Paul, E. F., and J. Paul, eds. 2001. Why animal experimentation matters: the use of animals in medical research. New Brunswick, New Jersey, Social Philosophy and Policy Foundation, and Transaction Publishers. *Essays by scientists, historians, and philosophers that express a defense of animal experimentation, demonstrating its moral acceptability and historical importance.*

Charles Lyell and Uniformitarianism

The geologist Sir Charles Lyell (1797–1875) (figure 1.4) established in his *Principles of Geology* (1830–1833) the principle of **uniformitarianism.** Uniformitarianism encompasses two important principles that guide scientific study of the history of nature. These principles are (1) that laws of physics and chemistry remain consistent throughout earth's history, and (2) that past geological events occurred by natural processes similar to those that we observe in action today. Lyell showed that natural forces, acting over long periods of time, could explain formation of fossil-bearing rocks. Lyell's geological studies led him to conclude that earth's age must be reckoned in millions of years. These principles were important for discrediting miraculous and supernatural explanations of the history of nature and replacing them with scientific explanations. Lyell also stressed the gradual nature of geological changes that occur through time, and he argued further that such changes have no inherent directionality. Both of these claims left important marks on Darwin's evolutionary theory.

Darwin's Great Voyage of Discovery

"After having been twice driven back by heavy southwestern gales, Her Majesty's ship *Beagle,* a ten-gun brig, under the command of Captain Robert FitzRoy, R.N., sailed from Devonport on the 27th of December, 1831." Thus began Charles Darwin's account of the historic five-year voyage of the *Beagle* around the world (figure 1.5). Darwin, not quite 23 years old, had been asked to accompany Captain FitzRoy on the *Beagle,* a small vessel only 90 feet in length, which was about to depart on an

figure 1.3

Jean Baptiste de Lamarck (1744–1829), French naturalist who offered the first scientific explanation of evolution. Lamarck's hypothesis that evolution proceeds by inheritance of acquired characteristics was rejected by genetic research.

figure 1.4

Sir Charles Lyell (1797–1875), English geologist and friend of Darwin. His book *Principles of Geology* greatly influenced Darwin during Darwin's formative period. This photograph was made about 1856.

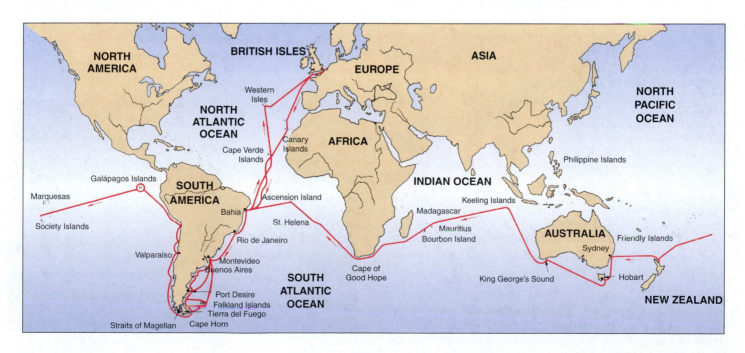

figure 1.5

Five-year voyage of H.M.S. *Beagle.*

extensive surveying voyage to South America and the Pacific (figure 1.6). It was the beginning of one of the most important voyages of the nineteenth century.

During this voyage (1831–1836), Darwin endured seasickness and erratic companionship from the authoritarian Captain FitzRoy. But Darwin's youthful physical strength and early training as a naturalist equipped him for his work. The *Beagle* made many stops along the coasts of South America and adjacent regions. Darwin made extensive collections and observations on the faunas and floras of these regions. He unearthed numerous fossils of animals long extinct and noted a resemblance between fossils of South American pampas and known fossils of North America. In the Andes he encountered seashells embedded in rocks at 13,000 feet. He experienced a severe earthquake and watched mountain torrents that relent-

lessly wore away the earth. These observations strengthened his conviction that natural forces were responsible for geological features of the earth.

In mid-September of 1835, the *Beagle* arrived at the Galápagos Islands, a volcanic archipelago straddling the equator 600 miles west of Ecuador (figure 1.7). The fame of these islands stems from their infinite strangeness. They are unlike any other islands on Earth. Some visitors today are struck with awe and wonder, others with a sense of depression and dejection. Circled by capricious currents, surrounded by shores of twisted lava, bearing skeletal brushwood baked by equatorial sunshine, almost devoid of vegetation, inhabited by strange reptiles and by convicts stranded by the Ecuadorian government, these islands indeed had few admirers among mariners. By the middle of the seventeenth century, the islands were already known to Spaniards as "Las Islas Galápagos"—the tortoise islands. The giant tortoises, used for food first by buccaneers and later by American and British whalers, sealers, and ships of war, were the islands' principal attraction. At the time of Darwin's visit, these tortoises already were heavily exploited.

During the *Beagle's* five-week visit to the Galápagos, Darwin began to develop his views of the evolution of life. His original observations of giant tortoises, marine iguanas, mockingbirds, and ground finches, all contributed to a turning point in Darwin's thinking.

Darwin was struck by the fact that, although the Galápagos Islands and Cape Verde Islands (visited earlier in this voyage) were similar in climate and topography, their fauna and flora were entirely different. He recognized that Galápagos plants and animals were related to those of the South American mainland, yet differed from them in curious ways. Each island often contained a unique species related to forms on other islands. In short, Galápagos life must have originated in continental South America and then undergone modification in various environmental conditions of different islands. He concluded that living forms were neither divinely created nor

A

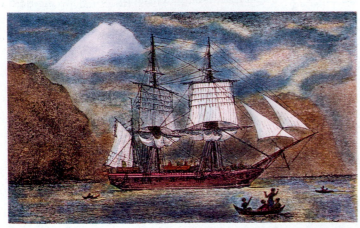

B

figure 1.6

Charles Darwin and H.M.S. *Beagle*. **A,** Darwin in 1840, four years after the *Beagle* returned to England, and a year after his marriage to his cousin, Emma Wedgwood. **B,** H.M.S. *Beagle* sails in Beagle Channel, Tierra del Fuego, on the southern tip of South America in 1833. This watercolor was painted by Conrad Martens, one of two official artists during the voyage of the *Beagle*.

figure 1.7

The Galápagos Islands viewed from the rim of a volcano.

immutable; they were, in fact, products of evolution. Although Darwin devoted only a few pages to Galápagos animals and plants in his monumental *On the Origin of Species*, published more than two decades later, his observations on the unique character of their animals and plants were, in his own words, the "origin of all my views."

"Whenever I have found that I have blundered, or that my work has been imperfect, and when I have been contemptuously criticized, and even when I have been overpraised, so that I have felt mortified, it has been my greatest comfort to say hundreds of times to myself that 'I have worked as hard and as well as I could, and no man can do more than this.' "

—*Charles Darwin, in his autobiography, 1876.*

On October 2, 1836, the *Beagle* returned to England, where Darwin conducted the remainder of his scientific work (figure 1.8). Most of Darwin's extensive collections had preceded him there, as had most of his notebooks and diaries kept during the cruise. Darwin's journal was published three years after the *Beagle's* return to England. It was an instant success and required two additional printings within its first year. In later versions, Darwin made extensive changes and titled his book *The Voyage of the Beagle.* The fascinating account of his observations written in a simple, appealing style has made this book one of the most lasting and popular travel books of all time.

Curiously, the main product of Darwin's voyage, his theory of evolution, did not appear in print for more than 20 years after the *Beagle's* return. In 1838, he "happened to read for amusement" an essay on populations by T. R. Malthus (1766–1834), who stated that animal and plant populations, including human populations, tend to increase beyond the capacity of their environment to support them. Darwin already had been gathering information on artificial selection of ani-

mals under domestication. After reading Malthus's article, Darwin realized that a process of selection in nature, a "struggle for existence" because of overpopulation, could be a powerful force for evolution of wild species.

He allowed the idea to develop in his own mind until it was presented in 1844 in a still-unpublished essay. Finally in 1856 he began to assemble his voluminous data into a work on origins of species. He expected to write four volumes, a very big book, "as perfect as I can make it." However, his plans were to take an unexpected turn.

In 1858, he received a manuscript from Alfred Russel Wallace (1823–1913), an English naturalist in Malaya with whom he corresponded. Darwin was stunned to find that in a few pages, Wallace summarized the main points of the natural selection theory on which Darwin had worked for two decades. Rather than withhold his own work in favor of Wallace as he was inclined to do, Darwin was persuaded by two close friends, Lyell and a botanist, Hooker, to publish his views in a brief statement that would appear together with Wallace's paper in *Journal of the Linnean Society.* Portions of both papers were read before an unimpressed audience on July 1, 1858.

For the next year, Darwin worked urgently to prepare an "abstract" of the planned four-volume work. This book was published in November 1859, with the title *On the Origin of Species by Means of Natural Selection, or the Preservation of Favoured Races in the Struggle for Life.* The 1250 copies of the first printing were sold the first day! The book instantly generated a storm that has never completely abated. Darwin's views were to have extraordinary consequences on scientific and religious beliefs and remain among the greatest intellectual achievements of all time.

Once Darwin's caution had been swept away by publication of *On the Origin of Species,* he entered an incredibly productive period of evolutionary thinking for the next 23 years, producing book after book. He died on April 19, 1882, and was buried in Westminster Abbey. The little *Beagle* had already disappeared, having been retired in 1870 and presumably sold for scrap.

Darwin's Theory of Evolution

Darwin's theory of evolution is now over 140 years old. Biologists frequently are asked, "What is Darwinism?" and "Do biologists still accept Darwin's theory of evolution?" These questions cannot be given simple answers because Darwinism encompasses several different, although mutually connected, theories. Professor Ernst Mayr of Harvard University argued that Darwinism could be viewed as five major theories. These five theories have somewhat different origins and fates and cannot be treated as only a single hypothesis. The theories are (1) **perpetual change,** (2) **common descent,** (3) **multiplication of species,** (4) **gradualism,** and (5) **natural selection.** The first three theories are generally accepted as having universal application throughout the living world. Gradualism and natural selection remain somewhat controversial among evolutionists. Gradualism

figure 1.8
Darwin's study at Down House in Kent, England, is preserved today much as it was when Darwin wrote *On the Origin of Species*.

and natural selection are clearly important evolutionary processes, but they might not be as pervasive as Darwin thought. Legitimate controversies regarding gradualism and natural selection often are misrepresented by creationists as challenges to the first three theories, whose validity is strongly supported by all relevant facts.

1. **Perpetual change.** This is the basic theory of evolution on which the others are based. It states that the living world is neither constant nor perpetually cycling, but is always changing. Characteristics of organisms undergo modification across generations throughout time. This theory originated in antiquity but did not gain widespread acceptance until Darwin advocated it in the context of his other four theories. "Perpetual change" is documented by the fossil record, which clearly refutes creationists' claims for a recent origin of all living forms. Because it has withstood repeated testing and is supported by an overwhelming number of observations, we now regard "perpetual change" as fact.

2. **Common descent.** The second Darwinian theory, "common descent," states that all forms of life propagated from a common ancestor through a branching of lineages (figure 1.9). An opposing argument, that different forms of life arose independently and descended to the present in linear, unbranched genealogies, has been refuted by comparative studies of organismal form, cell structure, and macromolecular structures (including those of the genetic material, DNA). All of these studies confirm the theory that life's history has the structure of a branching evolutionary tree, known as a phylogeny. Species that share relatively recent common ancestry have more similar features at all levels than do species whose most recent common ancestor is an ancient one. Much current research is guided by Darwin's theory of common descent to reconstruct life's phylogeny using patterns of similarity and dissimilarity observed among species. The resulting phylogeny provides the basis for our taxonomic classification of animals (see Chapter 4).

3. **Multiplication of species.** Darwin's third theory states that evolution produces new species by splitting and transforming older ones. Species are now generally viewed as reproductively distinct populations of organisms that usually but not always differ from each other in organismal form. Once species are fully formed, they propagate as separate evolutionary lineages, and interbreeding does not occur freely among members of different species, or the resulting hybrid offspring do not persist. Evolutionists generally agree that splitting and transformation of lineages produce new species, although much controversy remains concerning details of this process and precise meaning of the term "species" (see Chapter 4). Biologists are actively studying evolutionary processes that generate new species.

4. **Gradualism.** Darwin's theory of gradualism states that large differences in anatomical traits that characterize different species originate by accumulation of many small incremental changes over very long periods of time. This theory opposes the notion that large anatomi-

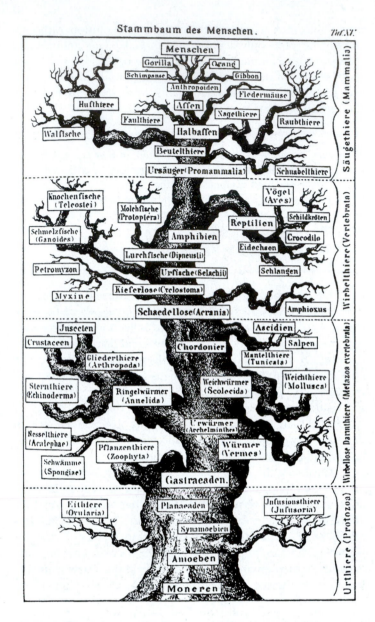

figure 1.9

An early tree of life drawn in 1874 by the German biologist, Ernst Haeckel, who was strongly influenced by Darwin's theory of common descent. Many phylogenetic hypotheses shown in this tree, including a unilateral progression of evolution toward humans (= Menschen, *top*), subsequently have been refuted.

cal differences arise by sudden genetic changes. This theory is important because genetic changes having very large effects on organismal form are usually harmful to an organism. It is possible, however, that some genetic variants that have large effects on an organism are nonetheless sufficiently beneficial to be favored by natural selection. Therefore, although gradual evolution is known to occur, it may not explain the origins of all structural differences observed among species. Scientists are studying this question actively.

5. **Natural selection.** Natural selection explains why organisms are constructed to meet demands of their

Thomas Henry Huxley (1825–1895), one of England's greatest zoologists, on first reading the convincing evidence of natural selection in Darwin's *On the Origin of Species* is said to have exclaimed, "How extremely stupid not to have thought of that!" He became Darwin's foremost advocate and engaged in often bitter debates with Darwin's critics. Darwin, who disliked publicly defending his own work, was glad to leave such encounters to his "bulldog," as Huxley called himself.

environments, a phenomenon called **adaptation.** This theory describes a natural process by which populations accumulate favorable characteristics throughout long periods of evolutionary time. Adaptation formerly was viewed as strong evidence against evolution. Darwin's theory of natural selection was therefore important for convincing people that a natural process, capable of being studied scientifically, could produce new adaptations and new species. Demonstration that natural processes could produce adaptation was important to the eventual acceptance of all five Darwinian theories. Darwin developed his theory of natural selection as a series of five observations and three inferences from them:

Observation 1—Organisms have great potential fertility. All populations produce large numbers of gametes and potentially large numbers of offspring each generation. Population size would increase exponentially at an enormous rate if all individuals produced each generation survived and reproduced. Darwin calculated that, even for slow-breeding organisms such as elephants, a single pair breeding from age 30 to 90 and having only six offspring could produce 19 million descendants in 750 years.

Observation 2—Natural populations normally remain constant in size, except for minor fluctuations. Natural populations fluctuate in size across generations and sometimes go extinct, but no natural populations show the

continued exponential growth that their reproductive capacity theoretically could sustain.

Observation 3—Natural resources are limited. Exponential growth of a natural population would require unlimited natural resources to provide food and habitat for an expanding population, but natural resources are finite.

Inference 1—A continuing *struggle for existence* occurs among members of a population. Survivors represent only a part, often a very small part, of all individuals produced each generation. Darwin wrote in *On the Origin of Species* that "it is the doctrine of Malthus applied with manifold force to the whole animal and vegetable kingdoms." Struggle for food, shelter, and space becomes increasingly severe as overpopulation develops.

Observation 4—All organisms show *variation.* No two individuals are exactly alike. They differ in size, color, physiology, behavior, and many other ways.

Observation 5—Variation is heritable. Darwin noted that offspring tend to resemble their parents, although he did not understand how. The hereditary mechanism discovered by Gregor Mendel would be applied to Darwin's theory many years later.

Inference 2—*Differential survival and reproduction* occur among varying organisms in a population. Survival in a struggle for existence is not random with respect to hereditary variation present in a population. Some traits give their possessors an advantage in using their environmental resources for effective survival and reproduction. Survivors transmit their favored traits to offspring, thereby causing those traits to accumulate in the population.

Inference 3—Over many generations, differential survival and reproduction generate new adaptations and new species. Differential reproduction of varying organisms gradually transforms species and results in their long-term "improvement." Darwin knew that people often use hereditary variation to produce useful new breeds of livestock and plants. *Natural* selection acting over millions of years should be even more effective in producing new types than *artificial* selection imposed during a human lifetime. Natural selection acting independently on geographically separated populations would cause them to diverge from each other, thereby generating reproductive barriers that lead to speciation.

Natural selection may be considered a two-step process with a random component and a nonrandom component. Production of variation among organisms is the random part. Mutational processes have no inherent tendency to generate traits that are favorable to an organism; if anything, the reverse is probably true. The nonrandom part is differential persistence of different traits, determined by effectiveness of traits in permitting their possessors to use environmental resources to survive and to reproduce. Differential survival and reproduction among varying organisms is

Darwin's Explanatory Model of Evolution by Natural Selection

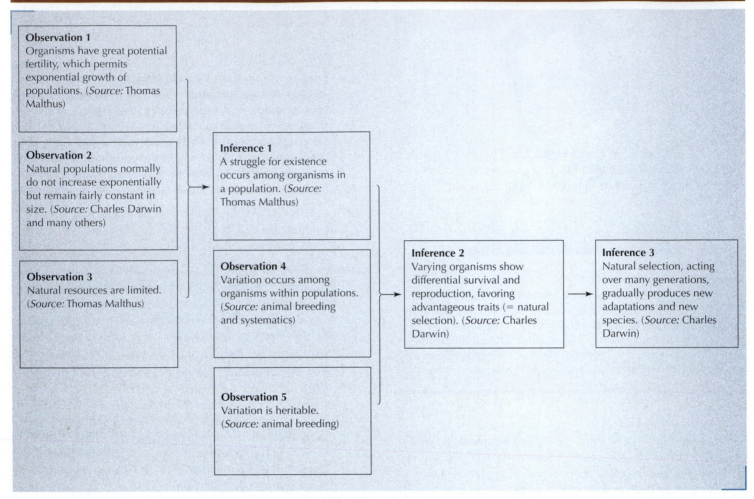

Observation 1
Organisms have great potential fertility, which permits exponential growth of populations. (*Source:* Thomas Malthus)

Observation 2
Natural populations normally do not increase exponentially but remain fairly constant in size. (*Source:* Charles Darwin and many others)

Observation 3
Natural resources are limited. (*Source:* Thomas Malthus)

Inference 1
A struggle for existence occurs among organisms in a population. (*Source:* Thomas Malthus)

Observation 4
Variation occurs among organisms within populations. (*Source:* animal breeding and systematics)

Observation 5
Variation is heritable. (*Source:* animal breeding)

Inference 2
Varying organisms show differential survival and reproduction, favoring advantageous traits (= natural selection). (*Source:* Charles Darwin)

Inference 3
Natural selection, acting over many generations, gradually produces new adaptations and new species. (*Source:* Charles Darwin)

Source: E. Mayr, One Long Argument, 1991, Harvard University Press, Cambridge, MA.

called **sorting** and should not be equated with natural selection. We now know that even random processes (genetic drift, p. 27) can produce sorting. If a garden planted with equal numbers of red- and white-flowered plants suffers severe damage from a hurricane, it is unlikely that equal numbers of red- and white-flowered plants will survive to produce seeds. If red-flowered plants constitute 70% of the survivors, sorting has occurred in favor of the red-flowered plants. In this case, flower color provided no advantage in withstanding the hurricane damage. Most likely, a larger number of red-flowered plants happened to be placed in better-protected locations, permitting their differential survival. This sorting therefore cannot be attributed to natural selection because the character being sorted had no causal influence on the outcome. If in the same garden, white-flowered plants produced more seeds and offspring because they were more visible to a nocturnal moth pollinator, we would observe sorting favoring the white flowers and could attribute this sorting to selection; white flower color in this case provided a reproductive advantage over red color, leading the white-flowered plants to increase in frequency in the next generation. Darwin's theory of natural selection states that sorting occurs *because certain traits give their possessors advantages in survival and reproduction* relative to others that lack those traits. Therefore, selection is one specific cause of sorting.

The popular phrase "survival of the fittest" was not originated by Darwin but was coined a few years earlier by British philosopher Herbert Spencer, who anticipated some of Darwin's principles of evolution. Unfortunately this phrase later came to be coupled with unbridled aggression and violence in a bloody, competitive world. In fact, natural selection operates through many other characteristics of living organisms. The fittest animal may be the most helpful or most caring. Fighting prowess is only one of several means toward successful reproductive advantage.

Evidence for Darwin's Five Theories of Evolution

Perpetual Change

Perpetual change in form and diversity of animal life throughout its 600- to 700-million-year history is seen most directly in the fossil record. A **fossil** is a remnant of past life uncovered from the earth's crust (figure 1.10). Some fossils constitute complete remains (insects in amber and mammoths), actual hard parts (teeth and bones), or petrified skeletal parts infiltrated with silica or other minerals (ostracoderms and molluscs). Other fossils include molds, casts, impressions, and fossil excrement (coprolites). In addition to documenting organismal evolution, fossils reveal profound changes in the earth's environments, including major changes in distributions of lands and seas. Because many organisms left no fossils, a complete record of past history is always beyond our reach; nonetheless, discovery of new fossils and reinterpretation of existing ones expand our knowledge of how forms and diversity of animals changed through geological time.

Fossil remains may on rare occasions include soft tissues preserved so well that recognizable cellular organelles can be viewed by electron microscopy! Insects are frequently found entombed in amber, fossilized resin of trees. One study of a fly entombed in 40-million-year-old amber revealed structures corresponding to muscle fibers, nuclei, ribosomes, lipid droplets, endoplasmic reticulum, and mitochondria (figure 1.10D). This extreme case of mummification probably occurred because chemicals in the plant sap diffused into the embalmed insect's tissues. A fictional extraction and cloning of DNA from embalmed insects that had bitten and then sucked blood of dinosaurs was the technical basis of Michael Crichton's best-seller *Jurassic Park*.

Interpreting the Fossil Record

The fossil record is biased because preservation is selective. Vertebrate skeletal parts and invertebrates with shells and other hard structures left the best record (figure 1.10). Soft-bodied

A

B

C

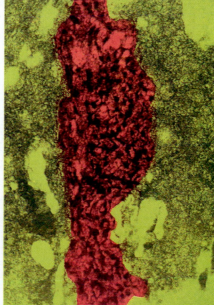

D

figure 1.10

Four examples of fossil material. **A,** Stalked crinoids (sea lilies, class Crinoidea, phylum Echinodermata p. 276) from 85-million-year-old Cretaceous rocks. The fossil record shows that these echinoderms reached their greatest diversity millions of years earlier and began a slow decline to the present. **B,** The fossil of an insect that got stuck in the resin of a tree 40 million years ago and the resin hardened into amber. **C,** Fish fossil from rocks of the Green River Formation, Wyoming. Such fish swam here during the Eocene epoch, approximately 55 million years ago. **D,** Electron micrograph of tissue from a fly fossilized as shown in **B;** the nucleus of a cell is marked in red.

figure 1.11

A, Fossil trilobites (p. 214) visible at the Burgess Shale Quarry, British Columbia. **B,** Animals of the Cambrian period, approximately 580 million years ago, as reconstructed from fossils preserved in the Burgess Shale of British Columbia, Canada. New body plans appeared rather abruptly at this time, including major body plans of animals alive today. **C,** Key to Burgess Shale drawing. *Amiskwia* (1), from an extinct phylum; *Odontogriphus* (2), from an extinct phylum; *Eldonia* (3), a possible echinoderm (p. 266); *Halichondrites* (4), a sponge (p. 111); *Anomalocaris canadensis* (5), from an extinct phylum; *Pikaia* (6), an early chordate (p. 286); *Canadia* (7), a polychaete; *Marrella splendens* (8), a unique arthropod; *Opabinia* (9), from an extinct phylum; *Ottoia* (10), a priapulid (p. 166); *Wiwaxia* (11), either a polychaete or from an extinct phylum; *Yohoia* (12), a unique arthropod; *Xianguangia* (13), an anemone-like animal; *Aysheaia* (14), an onychophoran (p. 260) or extinct phylum; *Sidneyia* (15), a unique arthropod (p. 213); *Dinomischus* (16), from an extinct phylum; *Hallucigenia* (17), from an extinct phylum.

A

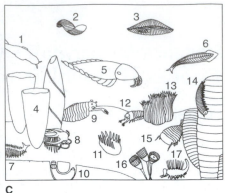

C

B

animals, including jellyfishes and most worms, are fossilized only under very unusual circumstances such as those that formed the Burgess Shale of British Columbia (figure 1.11). Exceptionally favorable conditions for fossilization produced a Precambrian fossil bed in South Australia, tar pits of Rancho La Brea (Hancock Park, Los Angeles), great dinosaur beds (Alberta, Canada, and Jensen, Utah; figure 1.12), the Olduvai Gorge of Tanzania, and the early Cambrian Chengjiang beds in China.

Fossils are deposited in stratified layers with new deposits forming above older ones. If left undisturbed, which is rare, a sequence is preserved with ages of fossils being directly proportional to their depth in stratified layers. Characteristic fossils often serve to identify particular layers. Certain widespread marine invertebrate fossils, including various foraminiferans (p. 105) and echinoderms (p. 274), are such good indicators of specific geological periods that they are called "index," or "guide," fossils. Unfortunately layers are usually tilted or folded or show faults (cracks). Old deposits exposed by erosion may be covered with new deposits in a different plane. When exposed to tremendous pressures or heat, stratified sedimentary rock metamorphoses into crystalline quartzite, slate, or marble, thereby destroying fossils.

Geological Time

Long before the earth's age was known, geologists divided its history into a table of succeeding events based on ordered layers of sedimentary rock. The "law of stratigraphy" produced a relative dating with oldest layers at the bottom and youngest at the top of a sequence. Time was divided into eons, eras, periods, and epochs as shown on endpapers inside the back cover of this book. Time during the last eon (Phanerozoic) is expressed in eras (for example, Cenozoic), periods (for example, Cambrian), epochs (for example, Paleocene), and sometimes smaller divisions of an epoch.

In the late 1940s, radiometric dating methods were developed for determining absolute ages in years of rock formations. Several independent methods are now used, all based on radioactive decay of naturally occurring elements into other elements. These "radioactive clocks" are independent of pressure and temperature changes and therefore not affected by often violent earth-building activities.

One method, potassium-argon dating, uses the decay of potassium-40 (^{40}K) to argon-40 (^{40}Ar) (12%) and calcium-40 (^{40}Ca) (88%). The half-life of potassium-40 is 1.3 billion years; half of the original atoms will decay in 1.3 billion years, and half of the remaining atoms will be gone at the end of the next 1.3 billion years. This decay continues until all radioactive potassium-40 atoms are gone. To measure the age of a rock, one calculates the ratio of remaining potassium-40 atoms to the amount of potassium-40 originally there (the remaining potassium-40 atoms plus the argon-40 and calcium-40 into which they have decayed). Several such isotopes exist for dating purposes, some for determining the age of the earth itself. One of the most useful radioactive clocks depends on decay of uranium into lead. With this method, rocks over 2 billion years old can be dated with a probable error of less than 1%.

The fossil record of macroscopic organisms begins near the start of the Cambrian period of the Paleozoic era, approximately 600 million years BP. Geological time before the Cambrian is called the Precambrian era or Proterozoic eon. Although the Precambrian era occupies 85% of all geological time, it has received much less attention than later eras, partly because oil, which provides a commercial incentive for much geological work, seldom exists in Precambrian formations. The Precambrian era contains well-preserved fossils of bacteria and algae, and casts of jellyfishes, sponge spicules, soft corals, segmented flatworms, and worm trails. Most, but not all, are microscopic fossils.

The more well-known carbon-14 (^{14}C) dating method is of little help in estimating ages of geological formations because its short half-life restricts use of ^{14}C to quite recent events (less than about 40,000 years). It is especially useful, however, for archaeological studies. This method is based on the production of radioactive ^{14}C (half-life of approximately 5570 years) in the upper atmosphere by bombardment of nitrogen-14 (^{14}N) with cosmic radiation. Radioactive ^{14}C enters tissues of living animals and plants, and an equilibrium is established between atmospheric ^{14}C and ^{14}N in living organisms. At death, ^{14}C exchange with the atmosphere stops. In 5570 years, only half of the original ^{14}C remains in a preserved fossil. Its age is found by comparing the ^{14}C content of the fossil with that of living organisms.

Evolutionary Trends

The fossil record allows us to view evolutionary change across the broadest scale of time. Species arise and then become extinct repeatedly throughout the geological history recorded by the fossil record. Animal species typically survive approximately 1 million to 10 million years, although their duration is

figure 1.12

A fossil skeleton from Dinosaur Provincial Park, Alberta, Canada.

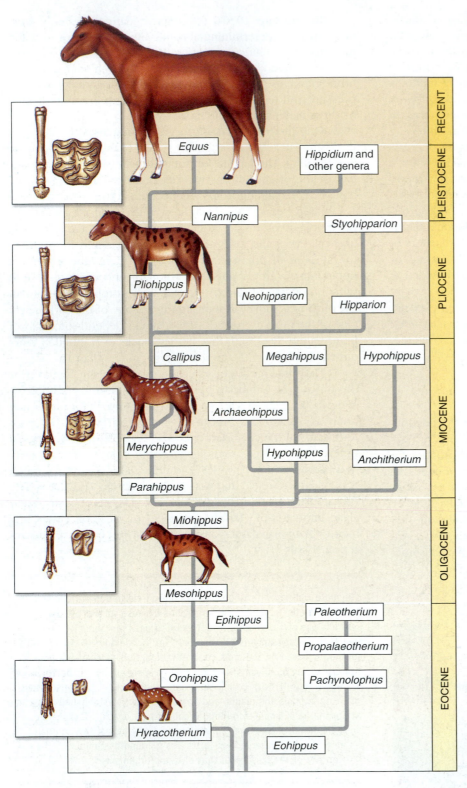

figure 1.13

A reconstruction of genera of horses from the Eocene to present. Evolutionary trends toward increased size, elaboration of molars, and loss of toes are shown together with a hypothetical genealogy of extant and fossil genera.

highly variable. When we study patterns of species or taxon replacement through time, we observe **trends.** Trends are directional changes in characteristic features or patterns of diversity in a group of organisms. Fossil trends clearly demonstrate Darwin's principle of perpetual change.

A well-studied fossil trend is the evolution of horses from the Eocene epoch to present (figure 1.13). Looking back at the Eocene epoch, we see many different genera and species of horses that replaced each other through time (figure 1.13). George Gaylord Simpson (p. 75) showed that this trend is compatible with Darwinian evolutionary theory. Three characteristics that show the clearest trends in horse evolution are body size, foot structure, and tooth structure. Compared to modern horses, those of extinct genera were small, their teeth had a relatively small grinding surface, and their feet had a relatively large number of toes (four). Throughout the subsequent Oligocene, Miocene, Pliocene, and Pleistocene epochs, new genera arose and old ones became extinct. In each case, there was a net increase in body size, expansion of the grinding surface of teeth, and reduction in number of toes. As the number of toes was reduced, the central digit became increasingly more prominent in the foot, and eventually only this central digit remained.

The fossil record shows a net change not only in the characteristics of horses but also variation in numbers of different horse genera (and numbers of species) that exist through time. Many horse genera of past epochs have been lost to extinction, leaving only a single survivor. Evolutionary trends in diversity are observed in fossils of many different groups of animals (figure 1.14).

Our use of the phrase "evolutionary trend" does not imply that more recent forms are superior to older ones or that the changes represent progress in adaptation or organismal complexity. Although Darwin predicted that such trends would show progressive adaptation, many contemporary paleontologists consider progressive adaptation rare among evolutionary trends. Observed trends in the evolution of horses do not imply that contemporary horses are superior in any general sense to their Eocene ancestors.

Trends in fossil diversity through time are produced by different rates of species formation versus extinction through time. Why do some lineages generate large numbers of new species

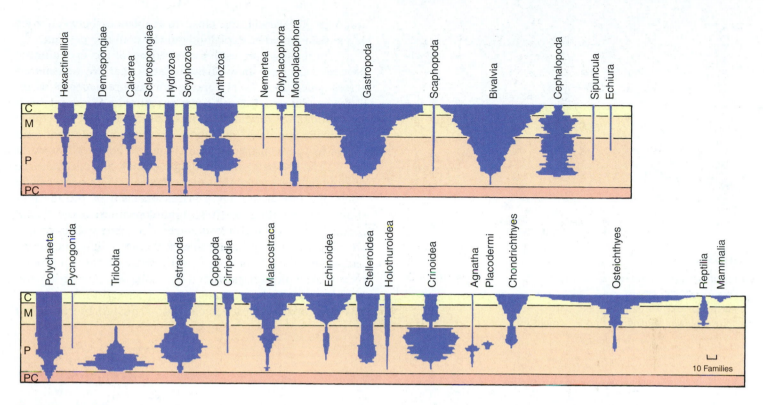

figure 1.14

Diversity profiles of taxonomic families from different animal groups in the fossil record. A scale marks the Precambrian (PC) and Paleozoic (P), Mesozoic (M), and Cenozoic (C) eras. The number of families is indicated by width of each profile.

whereas others generate relatively few? Why do different lineages undergo higher or lower rates of extinction (of species, genera, or families) throughout evolutionary time? To answer these questions, we must turn to Darwin's other four theories of evolution. Regardless of how we answer these questions, however, observed trends in animal diversity clearly illustrate Darwin's principle of perpetual change. Because Darwin's remaining four theories rely on perpetual change, evidence supporting these theories strengthens Darwin's theory of perpetual change.

Common Descent

Darwin proposed that all plants and animals have descended from "some one form into which life was first breathed." Life's history is depicted as a branching tree, called a phylogeny, that gives all of life a unified evolutionary history. Pre-Darwinian evolutionists, including Lamarck, advocated multiple independent origins of life, each of which gave rise to lineages that changed through time without extensive branching. Like all good scientific theories, common descent makes several important predictions that can be tested and potentially used to reject it. According to this theory, we should be able to trace genealogies of all modern species backward until they converge on ancestral lineages shared with other species, both living and extinct. We should be able to continue this process, moving farther backward through evolutionary time, until we

reach a primordial ancestor of all life on earth. All forms of life, including many extinct forms that represent dead branches, connect to this tree somewhere. Although reconstructing a history of life in this manner may seem almost impossible, it has in fact been extraordinarily successful. How has this difficult task been accomplished?

Homology and Reconstruction of Phylogeny

Darwin recognized a major source of evidence for common descent in the concept of **homology.** Darwin's contemporary, Richard Owen (1804–1892), used this term to denote "the same organ in different organisms under every variety of form and function." A classic example of homology is the limb skeleton of vertebrates. Bones of vertebrate limbs maintain characteristic structures and patterns of connection despite diverse modifications for different functions (figure 1.15). According to Darwin's theory of common descent, structures that we call homologies represent characteristics inherited with some modification from a corresponding feature in a common ancestor.

Darwin devoted an entire book, *The Descent of Man and Selection in Relation to Sex,* largely to the idea that humans share common descent with apes and other animals. This idea was repugnant to many Victorians, who responded with outrage (figure 1.16). Darwin built his case mostly on anatomical comparisons revealing homology between humans

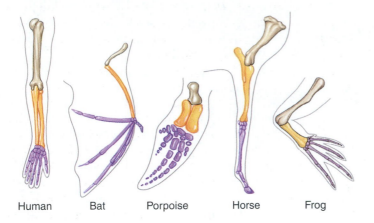

figure 1.15

Forelimbs of five vertebrates show skeletal homologies: green, humerus; yellow, radius and ulna; purple, "hand" (carpals, metacarpals, and phalanges). Homologies of bones and patterns of connection are evident despite evolutionary modification for various uses.

and apes. To Darwin, the close resemblances between apes and humans could be explained only by common descent.

Throughout the history of all forms of life, evolutionary processes generate new characteristics that are transmitted across generations. Every time a new feature becomes established in a lineage destined to be ancestral to others, we see a new homology originate. The pattern formed by these homologies provides evidence for common descent and allows us to reconstruct a branching evolutionary history of life. We can illustrate such evidence using a phylogenetic tree of ground-dwelling flightless birds (figure 1.17). A new skeletal homology (see figure 1.17) arises on each of the lineages shown (descriptions of these homologies are not included because they are highly technical). The different groups of species located at tips of the branches contain different combinations of these homologies, thereby revealing common ancestry. For example, ostriches show homologies 1 through 5 and 8, whereas kiwis show homologies 1, 2, 13, and 15. Branches of this tree combine these species into a **nested hierarchy** of

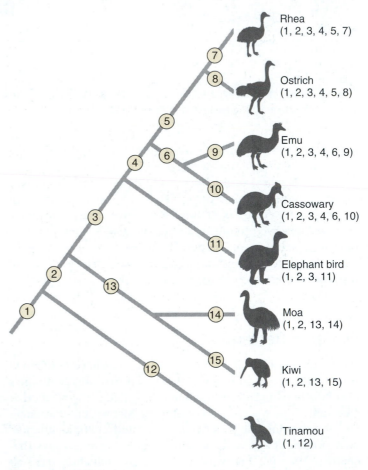

figure 1.17

Phylogenetic pattern specified by 15 homologous structures in skeletons of flightless birds. Homologous features are numbered 1 through 15 and are marked both on the branches on which they arose and on birds that have them. If you erase the tree structure, you should be able to reconstruct it without error from distributions of homologous features shown for birds at the terminal branches.

figure 1.16

This 1873 advertisement for Merchant's Gargling Oil ridicules Darwin's theory of the common descent of humans and apes, which hardly received universal public acceptance during Darwin's lifetime.

groups within groups (see Chapter 4). Smaller groups (species grouped near terminal branches) are contained within larger ones (species grouped by basal branches, including the trunk of the tree). If we erase the tree structure but retain the patterns of homology observed in the terminal groups of species, we will be able to reconstruct the branching structure of the entire tree. Evolutionists test the theory of common descent by observing patterns of homology present in all groups of organisms. The pattern formed by all homologies taken together should specify a single branching tree that represents the evolutionary genealogy of all living organisms.

The nested hierarchical structure of homology is so pervasive in the living world that it forms the basis for our systematic classification of all forms of life (genera grouped into families, families grouped into orders, orders into classes, classes into phyla, and phyla into the animal kingdom). Hierarchical classification even preceded Darwin's theory because this pattern is so evident, but it was not explained scientifically before Darwin. Once common descent was understood, biologists began investigating structural, molecular, and/or chromosomal homologies of animal groups. Taken together, the nested hierarchical patterns uncovered by these studies permit us to reconstruct the evolutionary trees of many groups and to continue investigating others. Use of Darwin's theory of common descent to reconstruct the evolutionary history of life and to classify animals is covered in Chapter 4.

Characters of different organisms that perform similar functions are not necessarily homologous. The wings of bats and birds, although homologous as vertebrate forelimbs, are not homologous as wings. The most recent common ancestor of bats and birds had forelimbs, but the forelimbs were not in the form of wings. Wings of bats and birds evolved independently and have only superficial similarity in their flight structures.

Bat wings are formed by skin stretched over elongated digits, whereas bird wings are formed by feathers attached along the forelimb. Such functionally similar but nonhomologous structures are often termed analogues.

Note that the earlier evolutionary hypothesis that life arose many times, forming unbranched lineages, predicts linear sequences of evolutionary change with no nested hierarchy of homologies among species. Because we do observe nested hierarchies of homologies, that hypothesis is rejected. Note also that a creationist argument, which is not a scientific hypothesis, can make no testable predictions about any pattern of homology.

Ontogeny, Phylogeny, and Recapitulation

Zoologists find in animal development important clues to an animal's evolutionary history. **Ontogeny** is the history of development of an organism through its entire life. Early developmental and embryological features contribute greatly to our knowledge of homology and common descent. Comparative

studies of ontogeny show how evolutionary alteration of developmental timing generates new **phenotypes** (expressed characteristics or appearance of an organism), thereby causing evolutionary divergence among lineages.

Comparisons of gene expression among animals shows that in forms as dissimilar as insects and humans, homologous genes may guide developmental differentiation of anterior versus posterior body segments. Mutations of such genes, termed **homoeotic genes,** in fruit flies may cause awkward developmental changes such as legs appearing in the place of antennae or an extra pair of wings. Such genes provide an evolutionary "tool kit" that can be used to construct new body parts by altering expression of the genes in different parts of a developing embryo. Perhaps the most famous homoeotic genes are those containing a sequence of 180 base pairs, called the **homeobox,** which encodes a protein sequence that binds to other genes, thereby altering their expression.

The German zoologist Ernst Haeckel, a contemporary of Darwin, proposed the influential hypothesis that each successive stage in development of an organism represented an adult form present in its evolutionary history. For example, a human embryo with gill depressions in its neck was considered to signify a fishlike ancestor. On this basis Haeckel gave his generalization: *ontogeny (individual development) recapitulates (repeats) phylogeny (evolutionary descent).* This notion later became known simply as **recapitulation** or the **biogenetic law.** Haeckel based his biogenetic law on a flawed premise that evolutionary change occurs by successively adding stages onto the end of an unaltered ancestral ontogeny, condensing the ancestral ontogeny into earlier developmental stages. This notion was based on Lamarck's concept of inheritance of acquired characteristics (p. 5).

A nineteenth-century embryologist, K. E. von Baer, gave a more satisfactory explanation of the relationship between ontogeny and phylogeny. He argued that early developmental features were simply more widely shared among different animal groups than later ones. For example, figure 1.18 shows early embryological similarities of organisms whose adult forms are very different. Adults of animals with relatively short

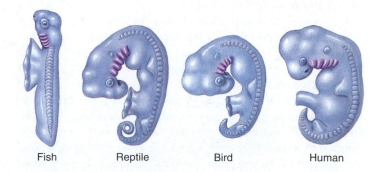

Fish   Reptile   Bird   Human

figure 1.18

Comparison of gill arches of different embryos. All are shown separated from the yolk sac. Note a remarkable similarity of these four embryos at this early stage in development.

and simple ontogenies often resemble pre-adult stages of other animals whose ontogeny is more elaborate, but embryos of descendants do not necessarily resemble adults of their ancestors. Even early development undergoes evolutionary divergence among groups, however, and it is not as stable as von Baer thought.

We now know many parallels between ontogeny and phylogeny, but features of an ancestral ontogeny can be shifted either to earlier or later stages in descendant ontogenies. Evolutionary change in timing of development is called **heterochrony,** a term initially used by Haeckel to denote exceptions to recapitulation. Because the lengthening or shortening of ontogeny can change different parts of an organism independently, we often see a mosaic of different kinds of developmental evolutionary change in a single lineage. Therefore, cases in which an entire ontogeny recapitulates phylogeny are rare.

Despite many changes that have occurred in scientific thinking about relationships between ontogeny and phylogeny, one important fact remains clear. Darwin's theory of common descent is strengthened enormously by the many homologies found among developmental stages of organisms belonging to different species.

Multiplication of Species

Multiplication of species through time is a logical corollary to Darwin's theory of common descent. A branch point on an evolutionary tree means that an ancestral species has split into two different ones. Darwin's theory postulates that variation present within a species, especially variation that occurs between geographically separated populations, provides material from which new species are produced. Because evolution is a branching process, the total number of species produced by evolution increases through time, although most of these species eventually become extinct. A major challenge for evolutionists is to discover processes by which an ancestral species "branches" to form two or more descendant species.

Before we explore multiplication of species, we must decide what we mean by "species." No consensus exists regarding definition of species (Chapter 4). Most biologists would agree, however, that important criteria for recognizing species include (1) descent from a common ancestral population, (2) reproductive compatibility (ability to interbreed) within and reproductive incompatibility between species, and (3) maintenance within species of genotypic and phenotypic cohesion (lack of abrupt differences among populations in allelic frequencies [p. 25]) and organismal appearance). The criterion of reproductive compatibility has received the greatest attention in studies of species formation, also called **speciation.**

Biological factors that prevent different species from interbreeding are called **reproductive barriers.** A primary problem of speciation is to discover how two initially compatible populations evolve reproductive barriers that cause them to become distinct, separately evolving lineages. How do populations diverge from each other in their reproductive properties while maintaining complete reproductive compatibility within each population?

Geographical barriers between populations are not the same thing as reproductive barriers. Geographical barriers refer to spatial separation of two populations. They prevent gene exchange and are usually a precondition for speciation. Reproductive barriers result from evolution and refer to various physical, physiological, ecological, and behavioral factors that prevent interbreeding between different species. Geographical barriers do not guarantee that reproductive barriers will evolve. Reproductive barriers are most likely to evolve under conditions that include small population size, a favorable combination of selective factors, and long periods of geographical isolation. One or both of a pair of geographically isolated populations may become extinct prior to evolution of reproductive barriers between them. Over the vast span of geological time, however, conditions sufficient for speciation have occurred millions of times.

Reproductive barriers between populations usually evolve gradually. Evolution of reproductive barriers requires that diverging populations must be kept physically separate ("isolated") for long periods of time. If the diverging populations were reunited before reproductive barriers were completely formed, interbreeding would occur between the populations and they would merge. Speciation by gradual divergence in animals usually requires perhaps 10,000 to 100,000 years or more. Geographical isolation followed by gradual divergence is the most effective way for reproductive barriers to evolve, and many evolutionists consider geographical separation a prerequisite for branching speciation. Speciation that results from evolution of reproductive barriers between geographically separated populations is called **allopatric speciation,** or geographical speciation.

Evidence for allopatric ("in another land") speciation occurs in many forms, but perhaps most convincing is an occurrence of geographically separated but adjoining, closely related populations that illustrate gradual origin of reproductive barriers. Populations of a salamander, *Ensatina eschscholtzii,* in California are a particularly clear example (figure 1.19). These populations show evolutionary divergence in color pattern and collectively form a geographical ring around California's central valley. Genetic exchange between differentiated, geographically adjoining populations is evident through formation of hybrids and occasionally regions of extensive genetic exchange (called zones of introgression). Two populations at the southern tip of the geographical range (called *E. e. eschscholtzii* and *E. e. klauberi*) make contact but do not interbreed. A gradual accumulation of reproductive differences among contiguous populations around the ring is visible, with the two southernmost populations being separated by strong reproductive barriers.

Additional evidence for allopatric speciation comes from observations of animal diversification on islands. Oceanic islands formed by volcanoes are initially devoid of life. They are colonized gradually by plants and animals from a continent or from other islands in separate invasions. Invaders often

figure 1.19

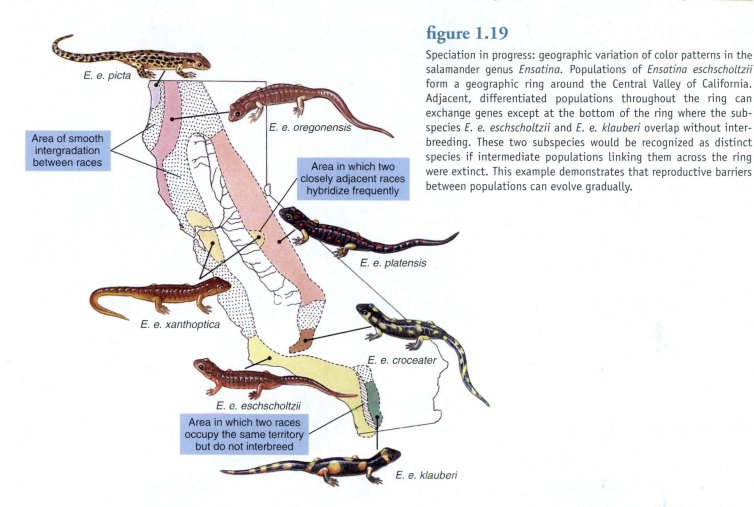

Speciation in progress: geographic variation of color patterns in the salamander genus *Ensatina*. Populations of *Ensatina eschscholtzii* form a geographic ring around the Central Valley of California. Adjacent, differentiated populations throughout the ring can exchange genes except at the bottom of the ring where the subspecies *E. e. eschscholtzii* and *E. e. klauberi* overlap without interbreeding. These two subspecies would be recognized as distinct species if intermediate populations linking them across the ring were extinct. This example demonstrates that reproductive barriers between populations can evolve gradually.

E. e. picta

E. e. oregonensis

Area of smooth intergradation between races

Area in which two closely adjacent races hybridize frequently

E. e. platensis

E. e. xanthoptica

E. e. croceater

E. e. eschscholtzii

Area in which two races occupy the same territory but do not interbreed

E. e. klauberi

encounter situations ideal for evolutionary diversification, because environmental resources that were exploited heavily by other species on the mainland are free for colonization on a sparsely populated island. Because colonization of oceanic islands is rare, populations established on islands are effectively isolated geographically from their parental populations and can undergo divergent evolution, leading to reproductive barriers and speciation. Archipelagoes, such as the Galápagos Islands, greatly increase opportunities for speciation in this manner.

Evolutionists have often wondered whether the isolation of populational gene pools needed for reproductive barriers to evolve may occur sometimes in the absence of geographical isolation. Populations that are reproductively active at different seasons or on different substrates could, in theory, achieve gene-pool isolation without geographic isolation. The term "sympatric speciation" is used to denote species formation not involving geographical isolation. A condition intermediate between allopatric and sympatric speciation is one in which the diverging populations are geographically separate but make contact along a borderline; speciation arising in this manner is called parapatric speciation. Many zoologists doubt that sympatric and parapatric speciation have been important in animal evolution.

Production of many ecologically diverse species from a common ancestral stock is called **adaptive radiation.** Galápagos finches clearly illustrate adaptive radiation on an oceanic archipelago (figures 1.20 and 1.21). Galápagos finches (the name "Darwin's finches" was popularized in the 1940s by the British ornithologist David Lack) are close relatives, but each species differs from others in size and shape of its beak and in feeding habits. Darwin's finches descended from a single ancestral population that arrived from South America and subsequently colonized different islands of the Galápagos archipelago. These finches underwent adaptive radiation, occupying habitats that in South America were denied to them by other species better able to exploit those habitats. Galápagos finches thus acquired characteristics of mainland families as diverse and unfinchlike as warblers and woodpeckers (figure 1.21B). The founding of new island populations by a small number of migrants may have accelerated evolutionary divergence among these island finches (p. 28). A fourteenth species of finch, found on isolated Cocos Island far north of the Galápagos archipelago, represents yet another speciation event in this impressive adaptive radiation.

Gradualism

Darwin's theory of gradualism opposed arguments for a sudden origin of species. Small differences, resembling those that we observe among organisms within populations today, are the raw material from which different major forms of life

figure 1.20

Tentative model for evolution of the 13 species of Darwin's finches on the Galápagos Islands. This model postulates three steps: (1) Immigrant finches from South America reach the Galápagos and colonize an island; (2) after a population becomes established, finches disperse to other islands where they adapt to new conditions and change genetically; and (3) after a period of isolation, secondary contact is established between different populations. Different populations are then recognized as separate species if they cannot interbreed successfully.

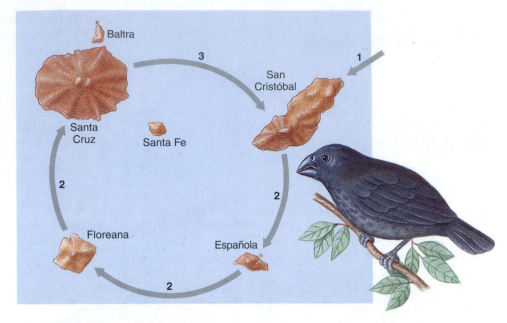

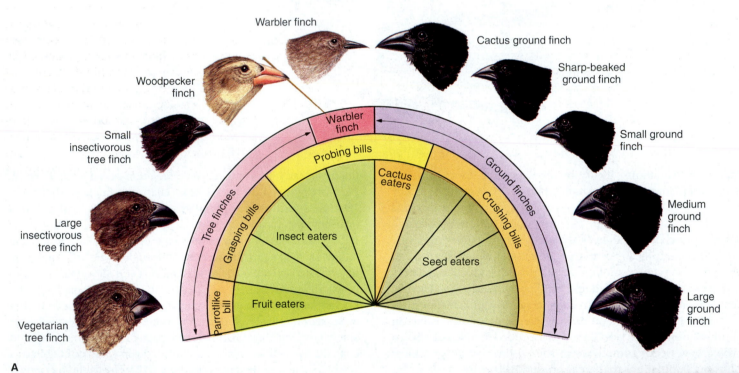

figure 1.21

A, Adaptive radiation in 10 species of Darwin's finches from Santa Cruz, one of the Galápagos Islands. Differences in bills and feeding habits are shown. All apparently descended from a single common ancestral finch from South America. **B,** Woodpecker finch, one of 13 species of Galápagos Islands finches, using a slender twig as a tool for feeding. This finch worked for about 15 minutes before spearing and removing a wood roach from a break in the tree.

evolved. This theory shares with Lyell's uniformitarianism a notion that we must not explain past changes by invoking unusual catastrophic events that are not observable today. If new species originated in single, catastrophic events, we should be able to see these events happening today, and we do not. What we observe instead are small, continuous changes in phenotypes occurring in natural populations. Such continuous changes can produce major differences among species only by accumulating over many thousands to millions of years. A simple statement of Darwin's theory of gradualism is that accumulation of quantitative changes leads to qualitative change.

Phenotypic gradualism was controversial when Darwin first proposed it, and it is still controversial. Not all phenotypic changes are small, incremental ones. Some mutations that appear during artificial breeding, traditionally called "sports," change a phenotype substantially in a single mutational step. Sports that produce dwarfing are observed in many species, including humans, dogs, and sheep, and have been used by animal breeders to achieve desired results; for example, a sport that deforms limbs was used to produce ancon sheep, which cannot jump hedges and are therefore easily contained (figure 1.22). Many colleagues of Darwin who accepted his other theories considered phenotypic gradualism too extreme. If sporting mutations can be used in animal breeding, why must we exclude them from our evolutionary theory? In favor of gradualism, some have replied that sporting mutations always have negative side effects that would prevent them from surviving in natural populations. Indeed, it is questionable whether ancon sheep, despite their attractiveness to farmers, would propagate successfully in the presence of their long-legged relatives without human intervention. Naturalists nonetheless report mutations of large effect that appear adaptive in natural populations; a mutation responsible for a large difference in bill size in the African finch species *Pyrenestes ostrinus* allows large-billed birds to eat hard seeds, whereas small-billed forms eat softer seeds.

When we view Darwinian gradualism on a geological timescale, we may expect to find in the fossil record a long series of intermediate forms bridging phenotypes of ancestral and descendant populations (figure 1.23). This predicted pattern is called **phyletic gradualism.** Darwin recognized that phyletic gradualism is not often revealed by the fossil record. Studies conducted since Darwin's time likewise have failed to produce a continuous series of fossils as predicted by phyletic gradualism. Is the theory of gradualism therefore refuted by the fossil record? Darwin and others claim that it is not, because the fossil record is too imperfect to preserve transitional series. Although evolution is a slow process by our standards, it is rapid relative to the rate at which good fossil deposits accumulate. Others have argued, however, that abrupt origins and extinctions of species in the fossil record force us to conclude that phyletic gradualism is rare.

Niles Eldredge and Stephen Jay Gould proposed **punctuated equilibrium** in 1972 to explain discontinuous evolutionary changes observed throughout geological time. Punctuated equilibrium states that phenotypic evolution is concentrated in relatively brief events of branching speciation, followed by much longer intervals of evolutionary stasis (figure 1.24). Speciation is an episodic event, occurring over a period of approximately 10,000 to 100,000 years. Because species may survive for 5 million to 10 million years, the speciation event is a "geological instant," representing 1% or less of a species' existence. Ten thousand years is plenty of time, however, for Darwinian evolution to accomplish dramatic changes. A small fraction of

figure 1.22

The ancon breed of sheep arose from a "sporting mutation" that caused dwarfing of their legs. Many of his contemporaries criticized Darwin for his claim that such large mutations are not important for evolution by natural selection.

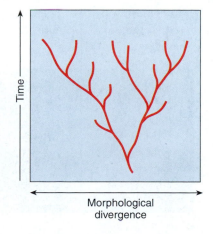

Morphological divergence

figure 1.23

A gradualist model of evolutionary change in morphology, viewed as proceeding more or less steadily through geological time (millions of years). Bifurcations followed by gradual divergence led to speciation.

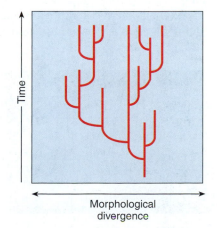

Morphological divergence

figure 1.24

A punctuated equilibrium model shows evolutionary change concentrated in relatively rapid bursts of branching speciation (*lateral lines*) followed by prolonged periods of little change throughout geological time (millions of years).

the total evolutionary history of a group therefore accounts for most of the morphological evolutionary change that occurs.

Evolutionists who lamented the imperfect state of the fossil record were treated in 1981 to the opening of an uncensored page of fossil history in Africa. Peter Williamson, a British paleontologist working in fossil beds 400 m deep near Lake Turkana, documented a remarkably clear record of speciation in freshwater snails. Geological strata of the Lake Turkana basin reveal a history of instability. Earthquakes, volcanic eruptions, and climatic changes caused waters to rise and fall periodically, sometimes by hundreds of feet. Thirteen lineages of snails show long periods of stability interrupted by relatively brief periods of rapid change in shell shape when snail populations were fragmented by receding waters. These populations diverged to produce new species that then remained unchanged through thick deposits before becoming extinct and being replaced by descendant species. The transitions occurred within 5000 to 50,000 years. In the few meters of sediment where speciation occurred, transitional forms were visible. Williamson's study conforms well to a punctuated equilibrium model, which remains an important challenge to gradualism on a geological timescale.

Natural Selection

Many examples show how natural selection alters populations in nature. Sometimes selection can proceed very rapidly as, for example, in evolution of high resistance to insecticides by insects, especially flies and mosquitoes. Doses that at first killed almost all pests later were ineffective in controlling them. As more insects were exposed to insecticides, those most sensitive were killed, leaving more space and less competition for resistant strains to multiply. Thus, as a result of selection, mutations bestowing high resistance, but previously rare in these populations, increased in frequency.

Perhaps the most famous instance of rapid evolution by natural selection is that of industrial melanism (dark pigmentation) in peppered moths of England (figure 1.25). Before 1850, peppered moths were white with black speckling on their wings and body. In 1849, a mutant black form of the species appeared. It became increasingly common, reaching frequencies of 98% in Manchester and other heavily industrialized areas by 1900. Peppered moths, like most moths, are active at night and rest in exposed places during daytime, depending upon cryptic coloration for protection. Experimental studies have shown that, consistent with the hypothesis of natural selection, birds are able to locate and to eat moths that do not match their surroundings, but that birds in the same area frequently fail to find moths that match their surroundings. The mottled pattern of the normal white form blends well with lichen-covered tree trunks. With increasing industrialization, soot from thousands of chimneys darkened bark of trees for miles around centers such as Manchester. Against this dark background, white moths were conspicuous to predatory birds, whereas mutant black forms were camouflaged. When pollution was diminished, frequency of lightly pigmented individuals increased in moth populations (figure 1.25C), consistent with predictions of the hypothesis of natural selection. The hypothesis of natural selection to explain industrial melanism requires further testing, particularly by more detailed studies of the behavioral ecology of peppered moths and their predators.

A recurring criticism of natural selection is that it cannot generate new structures or species but can only modify old ones. Most structures in their early evolutionary stages could not have performed roles that the fully formed structures perform, and therefore it is unclear how natural selection could have favored them. What use is half a wing or the rudiment of a feather for a flying bird? To answer this criticism, evolutionists propose that many structures evolved initially for purposes different from the uses that they have today. Rudimentary feathers could have been useful in thermoregulation, for example. Their role in flying would have evolved later after they incidentally acquired aerodynamic properties that subjected them to

A

B

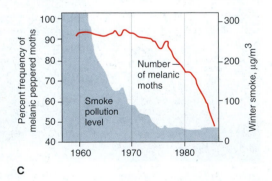

C

figure 1.25

Light and melanic forms of peppered moths, *Biston betularia* on, **A,** an unpolluted lichen-covered tree and, **B,** a soot-covered tree near industrial Birmingham, England. These color variants have a simple genetic basis. **C,** Recent decline in frequency of the melanic form with falling air pollution in industrial areas of England. Frequency of the melanic form still exceeded 90% in 1960, when smoke and sulfur dioxide emissions were still high. Later, as emissions fell and light-colored lichens began to grow again on tree trunks, the melanic form became more conspicuous to predators. By 1986, only 50% of the moths were melanic, the rest having been replaced by the light form.

selection for improvement of flying. Because the anatomical differences observed among organisms from different, closely related species resemble variation observed within species, it is unreasonable to propose that selection will never lead beyond a species boundary.

Revisions of Darwinian Evolutionary Theory

Neo-Darwinism

The most serious weakness in Darwin's argument was his failure to identify correctly a mechanism for inheritance. Darwin saw heredity as a blending phenomenon in which the hereditary factors of parents melded in their offspring. Darwin also invoked the Lamarckian hypothesis that an organism could alter its heredity through use and disuse of body parts and through direct environmental influence. Darwin did not realize that hereditary factors could be discrete and nonblending, and that a new genetic variant therefore could persist unaltered from one generation to the next. The German biologist August Weismann (1834–1914) rejected Lamarckian inheritance by showing experimentally that modifications of an organism during its lifetime do not change its heredity, and he revised Darwinian evolutionary theory accordingly. We now use the term **neo-Darwinism** to denote Darwinian evolutionary theory as revised by Weismann. The genetic basis of neo-Darwinism eventually became what is now called the **chromosomal theory of inheritance,** a synthesis of Mendelian genetics and cytological studies of segregation of chromosomes into gametes.

Gregor Mendel published his theories of inheritance in 1868, 14 years before Darwin died. Darwin presumably never read this work, although it was found in Darwin's extensive library after his death. Had Mendel written Darwin about his results, it is possible that Darwin would have modified his theory accordingly. However, it is also possible that Darwin could not have seen the importance of hereditary mechanisms to the continuous variations and gradual changes that represent the hub of Darwinian evolution. It took other scientists many years to establish the relationship between Mendelian genetics and Darwin's theory of natural selection.

Emergence of Modern Darwinism: A Synthetic Theory

In the 1930s, a new breed of geneticists began to reevaluate Darwinian evolutionary theory from a different perspective. These were population geneticists, scientists who studied variation in natural populations of animals and plants and who had a sound knowledge of statistics and mathematics. Gradually, a new comprehensive theory emerged that brought together population genetics, paleontology, biogeography, embryology, systematics, and animal behavior in a Darwinian framework.

Population geneticists study evolution as change in genetic compositions of populations. With the establishment of population genetics, evolutionary biology became divided into two different subfields. **Microevolution** pertains to evolutionary changes in frequencies of variant forms of genes within populations. **Macroevolution** refers to evolution on a grand scale, encompassing origins of new organismal structures and designs, evolutionary trends, adaptive radiation, phylogenetic relationships of species, and mass extinction. Macroevolutionary research is based in systematics and the comparative method (p. 5). Following this evolutionary synthesis, both macroevolution and microevolution have operated firmly within a tradition of neo-Darwinism, and both have expanded Darwinian theory in important ways.

Microevolution: Genetic Variation and Change within Species

Microevolution is the study of genetic change occurring within natural populations. A population is a reproductively cohesive group of organisms of the same species; populations of sexually reproducing forms show interbreeding among their members. Variant forms of a single gene are called **alleles.** Occurrence of different alleles of a gene in a population is called **polymorphism.** All alleles of all genes possessed by members of a population collectively form its **gene pool.** Polymorphism is potentially enormous in large populations because at observed mutation rates, many different alleles are expected for all genes.

Population geneticists study polymorphism by identifying allelic forms of a gene present in a population and then measuring their relative frequencies in that population. The relative frequency of a particular allele of a gene in a population is known as its **allelic frequency.** For example, in human populations, three different allelic forms occur for the gene encoding ABO blood types: I^A, I^B, and i. Because each individual **genotype** contains two copies of this gene, the total number of copies present in a population is twice the number of individuals. What fraction of this total is represented by each different allelic form? In France, we find the following allelic frequencies: $I^A = 0.46$, $I^B = 0.14$, and $i = 0.40$. In Russia, the corresponding allelic frequencies differ ($I^A = 0.38$, $I^B = 0.28$, and $i = 0.34$), demonstrating microevolutionary divergence between these populations (figure 1.26). Genetically, alleles I^A and I^B are dominant to i, but i is nearly as frequent as I^A and exceeds the frequency of I^B in both populations. Dominance describes the *phenotypic effect* of an allele in **heterozygous** individuals, not its relative abundance in a population of individuals. We will demonstrate that Mendelian inheritance and dominance do not alter allelic frequencies directly or produce evolutionary change in a population.

Genetic Equilibrium

In many human populations, genetically recessive traits, including the O blood type, blond hair, and blue eyes, are very common. Why have genetically dominant alternatives not gradually

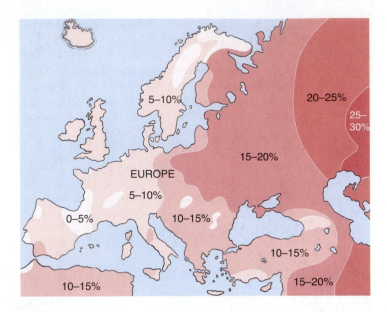

figure 1.26

Frequencies of the blood-type B allele among humans in Europe. This allele is more common in the east and rarer in the west. The allele may have arisen in the east and gradually diffused westward through genetically continuous populations. There is no known selective advantage of this allele, and its changing frequency probably represents random genetic drift.

supplanted these recessive traits? It is a common misconception that a characteristic associated with a dominant allele increases in proportion because of its genetic dominance. This notion is not true because there is a tendency in *large* populations for allelic frequencies to remain in equilibrium generation after generation. This principle is based on the **Hardy-Weinberg equilibrium** (see box, p. 27), which forms a foundation for population genetics. According to this theorem, Mendelian heredity alone does not produce evolutionary change. In large biparental populations, allelic frequencies and genotypic ratios attain an equilibrium in one generation and remain constant thereafter unless disturbed by recurring mutations, natural selection, migration, nonrandom mating, or genetic drift (random sorting). Such disturbances are important sources of microevolutionary change.

A rare allele, according to this principle, does not disappear from a large population merely because it is rare. That is why certain rare traits, such as albinism and cystic fibrosis, persist for endless generations. For example, albinism in humans is caused by a rare recessive allele a. Only one person in 20,000 is an albino, and this individual must be homozygous (a/a) for the recessive allele. Obviously the population contains many carriers, people with normal pigmentation who are heterozygous (A/a) for albinism. What is their frequency? A convenient way to calculate frequencies of genotypes in a population is with a binomial expansion of $(p + q)^2$ (see box p. 27). We will let p represent the allelic frequency of A and q the allelic frequency of a.

Assuming that mating is random (a questionable assumption, but one that we will accept for our example), the distribution of genotypic frequencies is $p^2 = A/A$, $2pq = A/a$, and $q^2 = a/a$. Only the frequency of genotype a/a is known with certainty, 1/20,000; therefore:

$$q^2 = \frac{1}{20,000}$$

$$q = \sqrt{\frac{1}{20,000}} = \frac{1}{141}$$

$$p = 1 - q$$

$$= \frac{141}{141} - \frac{1}{141} = \frac{140}{141}$$

The frequency of carriers is:

$$A/a = 2pq = 2 \times \frac{140}{141} \times \frac{1}{141} = \frac{1}{70}$$

One person in every 70 is a carrier! Although a recessive trait may be rare, it is amazing how common a recessive allele may be in a population. There is a message here for anyone proposing to eliminate a "bad" recessive allele from a population by controlling reproduction. It is practically impossible. Because only homozygous recessive individuals reveal a phenotype against which artificial selection acts (by sterilization, for example), the allele would continue to propagate from heterozygous carriers. For a recessive allele present in 2 of every 100 persons (but homozygous in only 1 in 10,000 persons), 50 generations of complete selection against homozygous recessives are required just to reduce its frequency to 1 in 100 persons.

Eugenics deals with improvement of hereditary qualities in humans by social control of mating and reproduction. While the genetic argument against eugenics cited in the text is compelling in itself, a eugenics program lingered in the United States (and elsewhere) until finally dispatched following Adolph Hitler's attempt to "purify" races in Europe by genocide.

Processes of Evolution: How Genetic Equilibrium Is Upset

Genetic equilibrium is disturbed in natural populations by (1) random genetic drift, (2) nonrandom mating, (3) migration, (4) natural selection, and interactions among these factors. Recurring mutation is the ultimate source of variability in all populations, but it usually requires interaction with one or more other factors to upset genetic equilibrium. We consider these other factors individually.

Hardy-Weinberg Equilibrium: Why Mendelian Heredity Does Not Change Allelic Frequencies

The Hardy-Weinberg law is a logical consequence of Mendel's first law of segregation and expresses a tendency toward equilibrium inherent in Mendelian heredity. Mendel's first law of inheritance states that each organism contains a pair of genetic factors for each variable trait. Different forms of these factors are called alleles. The paired factors in a given organism may be copies of the same allele (homozygous) or different alleles (heterozygous). In either case, formation of gametes involves "segregation" of the paired factors so that each gamete receives only one factor for a given trait. Each gamete from a homozygous organism contains a copy of the same allele. A gamete from a heterozygous organism contains one of its two alleles, and that organism produces its two kinds of gametes in equal frequency.

Let us select for our example a population having a single locus bearing just two alleles, T and t. Phenotypic expression of this gene might be, for example, the ability to taste a chemical compound called phenylthiocarbamide. Individuals in a population will be of three genotypes for this locus, T/T, T/t (both tasters), and t/t (nontasters). In a sample of 100 individuals, suppose we have determined that there are 20 of T/T genotype, 40 of T/t genotype, and 40 of t/t genotype. We could then make a table showing allelic frequencies (remember that every individual's genotype has two copies of a gene):

Genotype	Number of Individuals	Copies of the T Allele	Copies of the t Allele
T/T	20	40	
T/t	40	40	40
t/t	40		80
TOTAL	100	80	120

Of the 200 copies, the proportion of the T allele is $80/200 = 0.4$ (40%); and the proportion of the t allele is $120/200 = 0.6$ (60%). It is customary in presenting this equilibrium to use p and q to represent the two allelic frequencies. The genetically dominant allele is represented by p, and the genetically recessive by q. Thus:

$$p = \text{frequency of } T = 0.4$$
$$q = \text{frequency of } t = 0.6$$
$$\text{Therefore } p + q = 1$$

Having calculated allelic frequencies in this sample, let us determine whether these frequencies will change spontaneously in a new generation of the population. Assuming that mating is random (and this is important; all mating combinations of genotypes must be equally probable), each individual is expected to contribute an equal number of gametes to a common pool from which the next generation is formed. Frequencies of gametes in the pool will be proportional to allelic frequencies in the sample: 40% of the gametes will be T, and 60% will be t (ratio of 0.4:0.6). Both ova and sperm will, of course, show similar frequencies. The next generation is formed as shown here:

		Ova	
Sperm		$T = 0.4$	$t = 0.6$
$T = 0.4$		$T/T = 0.16$	$T/t = 0.24$
$t = 0.6$		$T/t = 0.24$	$t/t = 0.36$

Collecting genotypes, we have:

frequency of $T/T = 0.16$
frequency of $T/t = 0.48$
frequency of $t/t = 0.36$

Next, we determine values of p and q from randomly mated populations. From the table, we see that frequency of T will be the sum of genotypes T/T, which is 0.16, and one-half of the genotype T/t, which is 0.24:

$$T(p) = 0.16 + .5(0.48) = 0.4$$

Similarly, frequency of t will be the sum of genotypes t/t, which is 0.36, and one-half the genotype T/t, which is 0.24:

$$t(p) = 0.36 + .5(0.48) = 0.6$$

The new generation bears exactly the same allelic frequencies as its parent population! Note that no increase has occurred in frequency of the genetically dominant allele T. Thus *in a freely interbreeding, sexually reproducing population, the frequency of each allele would remain constant generation after generation in the absence of natural selection, migration, recurring mutation, and genetic drift* (see text this page). A mathematically minded reader will recognize that the genotype frequencies T/T, T/t, and t/t are actually a binomial expansion of $(p = q)^2$:

$$(p + q)^2 = p^2 + 2pq + q^2 = 1$$

Genetic Drift

Some species, like cheetahs (figure 1.27), contain very little genetic variation, probably because their ancestral lineages sometimes were restricted to very small populations. A small population clearly cannot contain large amounts of genetic variation. Each individual organism has at most two different allelic forms of each gene, and a single breeding pair contains at most four different allelic forms of a gene. Suppose that we have such a breeding pair. We know from Mendelian genetics that chance decides which allelic forms of a gene get passed to offspring. It is therefore possible by chance alone that one or two parental alleles in this example will not be passed to any offspring. It is highly unlikely that the different alleles present in a small ancestral population are all passed to descendants without any change of allelic frequency. This chance fluctuation in allelic frequency from one generation to the next, including loss of alleles from a population, is called **genetic drift.**

Genetic drift occurs to some degree in all populations of finite size. Perfect constancy of allelic frequencies, as predicted

figure 1.27

Cheetahs, whose genetic variability has been depleted to very low levels because of small population size in the past.

by Hardy-Weinberg equilibrium, occurs only in infinitely large populations, and such populations occur only in mathematical models. All populations of animals are finite and therefore experience some effect of genetic drift, which becomes greater, on average, as population size declines. Genetic drift erodes genetic variability of a population. If population size remains small for many generations in a row, genetic variation can be greatly depleted. This loss is harmful to evolutionary success of a species because it restricts potential genetic responses to environmental change. Indeed, biologists are concerned that cheetah populations may have insufficient variation for continued survival.

A large reduction in the size of a population that increases evolutionary change by genetic drift is termed a bottleneck. A bottleneck associated with the founding of a new geographic population is called a founder effect and may be associated with the formation of a new species (p. 21).

During the 1960s, genetic drift was deemphasized as a factor of much importance in evolution because it would be strong enough to oppose natural selection only in small populations. However, evolutionists now realize that most breeding populations of animals are small. A natural barrier, such as a stream or a ravine between two hilltops, may effectively separate a population into two separate and independent evolutionary units.

Nonrandom Mating

If mating is nonrandom, genotypic frequencies deviate from Hardy-Weinberg expectations. For example, if two different alleles of a gene are equally frequent ($p = q = .5$), we expect half of the genotypes to be heterozygous ($2pq = 2 [.5] [.5] = .5$) and one-quarter to be homozygous for each allele ($p^2 = q^2 = [.5]^2 = .25$). For **positive assortative mating,** individuals mate preferentially with others of the same genotype, such as albinos mating with other albinos. Matings among homozygous parents generate offspring that are homozygous like themselves. Matings among heterozygous parents produce on average 50% heterozygous offspring and 50% homozygous offspring (25% of each alternative type) each generation. Positive assortative mating increases the frequency of homozygous genotypes and decreases the frequency of heterozygous genotypes in a population but does not change allelic frequencies.

Preferential mating among close relatives also increases homozygosity and is called **inbreeding.** Whereas positive assortative mating usually affects one or a few traits, inbreeding simultaneously affects all variable traits. Strong inbreeding greatly increases chances that rare recessive alleles will become homozygous and be expressed.

Inbreeding has surfaced as a serious problem in zoos holding small populations of rare mammals. Matings of close relatives tend to bring together genes from a common ancestor and increase the probability that two copies of a deleterious gene will come together in the same organism. The result is "inbreeding depression." Our management solution is to enlarge genetic diversity by bringing together captive animals from different zoos or by introducing new stock from wild populations if possible. Paradoxically, where zoo populations are extremely small and no wild stock can be obtained, deliberate inbreeding is recommended. This procedure selects for genes that tolerate inbreeding; deleterious genes disappear if they kill animals homozygous for them.

Migration

Migration prevents different populations of a species from diverging. If a species is divided into many small populations, genetic drift and selection acting separately in different populations can produce evolutionary divergence between them. A small amount of migration between populations each generation keeps different populations from diverging strongly. For example, French and Russian populations show some genetic divergence (figure 1.26), but continuing migration prevents them from becoming completely distinct.

Natural Selection

Natural selection can change both allelic frequencies and genotypic frequencies in a population. Although effects of selection are often reported for particular polymorphic genes, we must stress that natural selection acts on whole animals, not on isolated traits. The organism that possesses a superior combination of traits is favored. An animal may have some traits that confer no advantage or even a disadvantage, but it is successful overall if its combination of traits is favorable. When we claim that a genotype at a particular gene has a higher relative **fitness** than others, we state that on average that genotype confers an advantage in survival and reproduction in a population. If alternative genotypes have unequal probabilities of survival and reproduction, Hardy-Weinberg equilibrium will be upset.

Some traits and combinations of traits are advantageous for certain aspects of an organism's survival or reproduction and disadvantageous for others. Darwin used the term **sexual selection** to denote selection of traits that are advantageous for obtaining mates but may be harmful for survival. Bright colors and elaborate feathers may enhance a male bird's competitive ability in obtaining mates while simultaneously increasing his vulnerability to predators (figure 1.28). Environmental changes can alter selective values of different traits. Selection on character variation is therefore very complex.

Selection is often studied using quantitative traits, those that show continuous variation with no obvious pattern of Mendelian segregation in their inheritance. Values of a trait in offspring often are intermediate between values of their parents. Such traits are influenced by variation at many genes, each of which follows Mendelian inheritance and contributes a small, incremental amount to a phenotype. Traits that show quantitative variation include tail length in mice, length of a leg segment in grasshoppers, number of gill rakers in sunfishes, number of peas in pods, and height of adult humans. When values are graphed with respect to frequency distribution, they often approximate a normal, bell-shaped curve (figure 1.29A). Most individuals fall near the average, fewer fall well above or below the average, and extreme values form "tails" of the frequency curve with increasing rarity.

Selection can act on quantitative traits to produce three different kinds of evolutionary response (figure 1.29B, C, and D). **Stabilizing selection** favors average values of a trait and disfavors extreme ones (figure 1.29B). **Directional selection** favors an extreme value of a phenotype and causes a population average to shift toward it (figure 1.29C). When we think about natural selection producing evolutionary change, it is usually directional selection that we have in mind, although we must remember that it is not the only possibility. A third alternative is **disruptive selection** in which two different extreme phenotypes are favored simultaneously, but their average is disfavored (figure 1.29D). If inbreeding or positive assortative mating accompanies disruptive selection, this may cause a population to become bimodal, meaning that two very different phenotypes predominate.

Interactions of Selection, Drift, and Migration

Subdivision of a species into small populations that exchange migrants is an optimal one for promoting rapid adaptive evolution of a species. Interactions of genetic drift and selection in

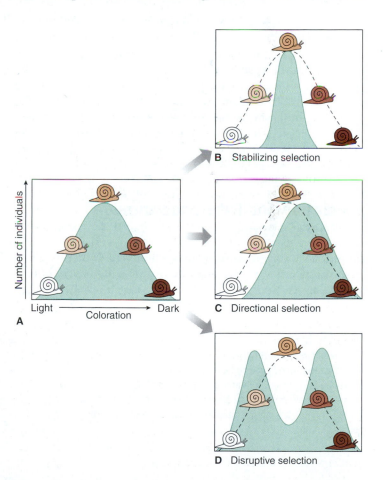

figure 1.29

Responses to selection on a continuous (polygenic) character, coloration in a snail. **A,** A frequency distribution of coloration before selection. **B,** Stabilizing selection culls extreme variants from the population, in this case eliminating unusually light or dark individuals, thereby stabilizing the mean. **C,** Directional selection shifts the population mean, in this case by favoring darkly colored variants. **D,** Disruptive selection favors both extremes but not the mean; the mean is unchanged but the population no longer has a bell-shaped distribution of phenotypes.

figure 1.28

A pair of wood ducks. Brightly colored feathers of male birds probably confer no survival advantage and might even be harmful by alerting predators. Such colors nonetheless confer advantage in attracting mates, which overcomes, on average, any negative consequences of these colors for survival. Darwin used the term "sexual selection" to denote traits that give an individual an advantage in attracting mates, even if these traits are neutral or harmful for survival.

different populations permit many different genetic combinations of many polymorphic genes to be tested against natural selection. Migration among populations permits particularly favorable new genetic combinations to spread throughout the species as a whole. Interactions of selection, genetic drift, and migration in this example produce evolutionary change qualitatively different from what would result if any of these three factors acted alone. Natural selection, genetic drift, mutation, nonrandom mating, and migration interact in natural populations to create an enormous opportunity for evolutionary change; the perpetual stability predicted by Hardy-Weinberg equilibrium almost never lasts across any substantial amount of evolutionary time.

Macroevolution: Major Evolutionary Events

Macroevolution describes large-scale events in organic evolution. Speciation links macroevolution and microevolution. The major trends in the fossil record described earlier (see figures 1.13 and 1.14) fall clearly within the realm of macroevolution. Patterns and processes of macroevolutionary change emerge from those of microevolution, but they acquire some degree of autonomy in doing so. The emergence of new adaptations and species, and the varying rates of speciation and extinction observed in the fossil record go beyond the fluctuations of allelic frequencies within populations.

Speciation and Extinction through Geological Time

Evolutionary change at the macroevolutionary level provides a new perspective on Darwin's theory of natural selection. A species has two possible evolutionary fates: it may give rise to new species or become extinct without leaving descendants. Rates of speciation and extinction vary among lineages, and lineages that have the highest speciation rates and lowest extinction rates produce the greatest number of living forms. The characteristics of a species may make it more or less likely than others to undergo speciation or extinction events. Because many characteristics are passed from ancestral to descendant species (analogous to heredity at the organismal level), lineages whose properties enhance the probability of speciation and confer resistance to extinction should dominate the living world. This species-level process that produces differential rates of speciation and extinction among lineages is analogous in many ways to natural selection. It represents an expansion of Darwin's theory of natural selection.

Species selection is the differential survival and multiplication of species through geological time based on variation among lineages in species-level properties. These species-level properties include mating rituals, social structuring, migration patterns, and geographic distribution. Descendant species usually resemble their ancestors for these properties. For example, a "harem" system of mating in which a single male and several females compose a breeding unit characterizes some mammalian lineages but not others. We expect speciation rates to be enhanced by social systems that promote the founding of new populations by small numbers of individuals. Certain social systems may increase the probability that a species will survive environmental challenges through cooperative action. Such properties would be favored over geological time by species selection.

Mass Extinctions

When we study evolutionary change on an even larger timescale, we observe episodic events in which large numbers of taxa become extinct nearly simultaneously. These events are called **mass extinctions** (figure 1.30). The most cataclysmic of

figure 1.30

Changes in numbers of taxonomic families of marine animals through time from the Cambrian period to the present. Sharp drops represent five major extinctions of skeletonized marine animals. Note that despite these extinctions, the overall number of marine families has increased over the past 600 million years.

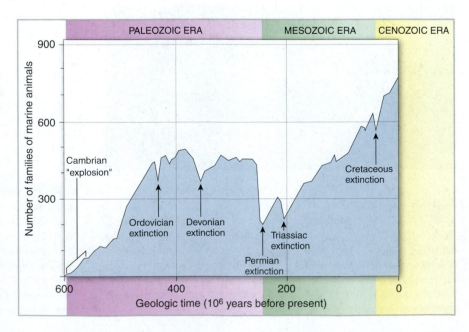

figure 1.31

Twin craters of Clearwater Lakes in Canada show that multiple impacts on the earth are not as unlikely as they might seem. Evidence suggests that at least two impacts within a short time were responsible for the Cretaceous mass extinction.

proposed that the earth was occasionally bombarded by asteroids, causing these mass extinctions (figure 1.31). The drastic effects of such a bombardment of a planet were observed in July 1994 when fragments of Comet Shoemaker-Levy 9 bombarded Jupiter. The first fragment to hit Jupiter was estimated to have the force of 10 million hydrogen bombs. Twenty additional fragments hit Jupiter within the following week, one of which was 25 times more powerful than the first fragment. This bombardment was the most violent event in the recorded history of our solar system. Bombardments by asteroids or comets could change the earth's climate drastically, sending debris into the atmosphere and blocking sunlight. Temperature changes would have challenged ecological tolerances of many species. This hypothesis is being tested in several ways, including a search for impact craters left by asteroids and for altered mineral content of rock strata where mass extinctions occurred. Unusual concentrations of the rare-earth element iridium in some strata imply that this element entered the earth's atmosphere through asteroid bombardment.

these extinction episodes happened about 225 million years ago, when at least half of the families of shallow-water marine invertebrates, and fully 90% of marine invertebrate species disappeared within a few million years. This was the **Permian extinction.** The **Cretaceous extinction,** which occurred about 65 million years ago, marked the end of dinosaurs, as well as numerous marine invertebrates and many small reptilian species.

Causes of mass extinctions and their occurrence at intervals of approximately 26 million years are difficult to explain. Some researchers have proposed biological explanations for these periodic mass extinctions and others consider them artifacts of our statistical and taxonomic analyses. Walter Alvarez

Sometimes lineages favored by species selection are unusually at risk during a mass extinction. Climatic changes produced by hypothesized asteroid bombardments could produce selective challenges very different from those encountered at other times in the earth's history. Selective discrimination of particular biological traits by events of mass extinction is termed **catastrophic species selection.** For example, mammals survived the end-Cretaceous mass extinction that destroyed dinosaurs and other prominent vertebrate and invertebrate groups. Following this event, mammals were able to use environmental resources that previously had been denied them, leading to their adaptive radiation.

Natural selection, species selection, and catastrophic species selection interact to produce macroevolutionary trends that we see in the fossil record. Intensive study of these interacting causal processes has made modern paleontology an active and exciting field.

Summary

Zoology is the scientific study of animals. Science is characterized by a particular approach to acquisition of human knowledge. It is guided by, and is explanatory with reference to, natural law, and it is testable, tentative, and falsifiable. These criteria form the basis for the scientific method, which may be described as hypothetico-deductive in character. On the basis of prior observations, a scientist formulates an explanatory hypothesis. Predictions about future observations based on the hypothesis can support or falsify the hypothesis. A testable hypothesis for which there is a large amount of supporting data, particularly one that explains a very large

number of observations, may be elevated to the status of a theory.

Experimental sciences seek to understand proximate or immediate causes in biological systems and use experimental methodology. Tests of hypotheses are performed by experiments, which must include controls. Evolutionary sciences seek ultimate causes in biological systems by using comparative methodology.

Organic evolution explains diversity of living organisms as the historical outcome of gradual change from previously existing forms. Evolutionary theory is strongly identified with Charles Robert Darwin, who pre-

sented the first credible explanation for evolutionary change. Darwin derived much of the material used to construct his theory from his experiences on a five-year voyage around the world aboard H.M.S. *Beagle.*

Darwin's evolutionary theory has five major components. Its most basic proposition is *perpetual change,* the theory that the world is neither constant nor perpetually cycling but is steadily undergoing irreversible change. The fossil record amply demonstrates perpetual change in continuing fluctuations of animal form and diversity following the Cambrian explosion 600 million years ago. Darwin's theory of *common descent* states that all organisms descend from a common ancestor through a branching of genealogical lineages. This theory explains morphological homologies among organisms as characteristics inherited with modification from a corresponding feature in their common evolutionary ancestor. Patterns of homology formed by common descent with modification permit us to classify organisms according to their evolutionary relationships.

A corollary of common descent is *multiplication of species* through evolutionary time. Allopatric speciation describes evolution of reproductive barriers between geographically separated populations to generate new species. Adaptive radiation denotes proliferation of many adaptively diverse species from a single ancestral lineage. Oceanic archipelagoes, such as the Galápagos Islands, are particularly conducive to adaptive radiation of terrestrial organisms.

Darwin's theory of *gradualism* states that large phenotypic differences between species are produced by accumulation through evolutionary time of many individually small changes. Gradualism is still controversial. Mutations that have large effects on phenotype have been useful in animal breeding, leading some to dispute Darwin's claim that such mutations are not important in evolution. On a macroevolutionary perspective, punctuated equilibrium states that most evolutionary change occurs in relatively brief events of branching speciation, separated by long intervals in which little phenotypic change accumulates.

Darwin's fifth major statement is that *natural selection* is the major guiding force of evolution. This principle is founded on observations that all species overproduce their kind, causing a struggle for limited resources that support existence. Because no two organisms are exactly alike, and because variable traits are at least partially heritable, those whose hereditary endowment enhances their use of resources for survival and reproduction contribute disproportionately to the next generation. Over many generations, the sorting of variation by selection produces new species and new adaptations.

Population geneticists discovered principles by which genetic properties of populations change through time. A particularly important discovery, known as Hardy-Weinberg equilibrium, showed that Mendelian heredity does not change the genetic composition of populations. Important sources of evolutionary change include mutation, genetic drift, nonrandom mating, migration, natural selection, and their interactions. Mutations are the ultimate source of all new variation on which selection acts.

Macroevolution comprises studies of evolutionary change on a geological timescale. Macroevolutionary studies measure rates of speciation, extinction, and changes of diversity through time. These studies have expanded Darwinian evolutionary theory to include higher-level processes that regulate rates of speciation and extinction among lineages, including species selection and catastrophic species selection.

Review Questions

1. What are the essential characteristics of science? Describe how evolutionary studies fit these characteristics. Why are arguments based on supernatural or miraculous explanations, such as "scientific creationism," or "intelligent design," not valid scientific hypotheses?
2. What is the relationship between a hypothesis and a theory?
3. Explain how biologists distinguish between experimental and evolutionary sciences.
4. Briefly summarize Lamarck's hypothesis of evolution. What evidence rejects this hypothesis?
5. What is "uniformitarianism?" How did it influence Darwin's evolutionary theory?
6. Why was the *Beagle's* journey so important to Darwin's thinking?
7. What key idea, contained in Malthus's essay on populations, helped Darwin formulate his theory of natural selection?
8. Explain how each of the following contributes to Darwin's evolutionary theory: fossils, geographic distributions of closely related animals, homology, animal classification.

9. How do modern evolutionists view the relationship between ontogeny and phylogeny?
10. What major evolutionary lesson is illustrated by Darwin's finches on the Galápagos Islands?
11. How does observation of "sporting mutations" in animal breeding challenge Darwin's theory of gradualism? Why did Darwin reject such mutations as having little evolutionary importance?
12. What does the theory of punctuated equilibrium state about occurrence of speciation throughout geological time?
13. Describe the observations and inferences that compose Darwin's theory of natural selection.
14. Identify the random and nonrandom components of Darwin's theory of natural selection.
15. Describe some recurring criticisms of Darwin's theory of natural selection. How can these criticisms be refuted?
16. A common but mistaken notion is that because some alleles are genetically dominant and others are recessive, dominants eventually will replace all recessives in a population. How does the Hardy-Weinberg equilibrium refute this notion?

17. Assume that you are sampling a trait from animal populations; the trait is controlled by a single allelic pair A and a, and you can distinguish all three phenotypes AA, Aa, and aa (intermediate inheritance). Your sample includes:

Population	AA	Aa	aa	TOTAL
I	300	500	200	1000
II	400	400	200	1000

Calculate the distribution of phenotypes in each population as expected under Hardy-Weinberg equilibrium. Is population I in equilibrium? Is population II in equilibrium?
18. If after studying a population for a trait determined by a single pair of alleles you find that this population is not in equilibrium, what possible reasons might explain its departure from equilibrium?
19. Explain why genetic drift is more powerful in small populations.
20. Is it easier for selection to remove a deleterious recessive allele from a randomly mating population or a highly inbred population? Why?
21. Distinguish between microevolution and macroevolution.

Selected References

See also general references on page 415.

Avise, J. C. 2004. Molecular markers, natural history, and evolution, ed. 2. Sunderland, Massachusetts, Sinauer Associates. *An exciting account of using molecular genetics to understand evolution.*

Conner, J. K., and D. L. Hartl. 2004. A primer of ecological genetics. Sunderland, Massachusetts, Sinauer Associates. *An introductory text on population genetics.*

Coyne, J. A., and H. A. Orr. 2004. Speciation. Sunderland, Massachusetts, Sinauer Associates. *A detailed coverage of species formation.*

Darwin, C. 1859. On the origin of species by means of natural selection, or the preservation of favoured races in the struggle for life. London, John Murray. *There were five subsequent editions by the author.*

Desmond, A., and J. Moore. 1991. Darwin. Warner Books, New York. *An interpretive biography of Charles Darwin.*

Freeman, S., and J. C. Herron. 2004. Evolutionary analysis, ed. 3. Upper Saddle River, New Jersey, Pearson/Prentice-Hall. *An introductory textbook on evolutionary biology designed for undergraduate biology majors.*

Futuyma, D. J. 2005. Evolution. Sunderland, Massachusetts, Sinauer Associates. *A challenging introductory textbook on evolution.*

Gould, S. J. 2002. The structure of evolutionary theory. Cambridge, Massachusetts, Belknap Press of Harvard University Press. *A provocative account of what fossils tell us about evolutionary theory.*

Hall, B. K. 1998. Evolutionary developmental biology, ed. 2. New York, Chapman & Hall. *A review of the interaction of genetics and development in evolving lineages, with particular attention to issues of heterochrony, homology, and developmental constraints on evolution.*

Kitcher, P. 1982. Abusing science: the case against creationism. Cambridge, Massachusetts, MIT Press. *A treatise on how knowledge is gained in science and why creationism does not qualify as science. The argument refuted as "scientific creationism" in this book is equivalent to recent arguments termed "intelligent design theory."*

Mayr, E. 2001. What evolution is. New York, Basic Books. *A general survey of evolution by a leading evolutionary biologist.*

Custom Website

The *Animal Diversity* Online Learning Center is a great place to check your understanding of chapter material. Visit www.mhhe.com/hickmanad4e for access to key terms, quizzes, and more! Further enhance your knowledge with web links to chapter-related material.

Links to the Internet

Explore live links for these topics:

Introductory Sites
Writing Papers and Study Tips
Glossaries and Dictionaries
Careers in Science
Utility and Organizational Sites
Scientific Method
Individual Contributions to Environmental Issues
Evolution
Darwinian Evolutionary Theory
Evidences for Evolution
Classification and Phylogeny of Animals
Theories of Evolutionary Patterns

2

Animal Ecology

Talkeetna Mountain Range, Alaska.

Every Species Has Its Niche

The lavish richness of the earth's biomass is organized into a hierarchy of interacting units: an individual organism, a population, a community, and finally an ecosystem, that most bewilderingly complex of all natural systems. Central to ecological study is habitat, the spatial location where an animal lives. What an animal does in its habitat, its profession as it were, is its niche: how it gets its food, how it arranges for its reproductive perpetuity—in short, how it survives and stays adapted in a Darwinian sense.

A niche is a product of evolution and once it is established, no other species in the community can evolve to exploit exactly the same resources. This illustrates the "competitive exclusion principle:" no two species will occupy the same niche. Different species are therefore able to form an ecological community in which each has a different role in their shared environment.

In the mid-nineteenth century, the German zoologist Ernst Haeckel introduced the term **ecology,** defined as the "relation of the animal to its organic as well as inorganic environment." Environment here includes everything external to the animal but most importantly its immediate surroundings. Although we no longer restrict ecology to animals alone, Haeckel's definition is still basically sound. Animal ecology is now a highly synthetic science that incorporates everything we know about behavior, physiology, genetics, and evolution of animals to study interactions between populations of animals and their environments. The major goal of ecological studies is to understand how these diverse interactions determine geographical distributions and abundances of animal populations. Such knowledge is crucial for ensuring continued survival of many populations when their natural environments are altered by human activity.

ecology is studied as a hierarchy of biological systems in interaction with their environments. At the base of the ecological hierarchy is an **organism.** To understand why animals live where they do, ecologists must examine the varied physiological and behavioral mechanisms that animals use to survive, to grow, and to reproduce. A near-perfect physiological balance between production and loss of heat is required for success in certain endothermic species (such as birds and mammals) under extreme temperatures, as found in the Arctic or a desert. Other species succeed in these situations by escaping the most extreme conditions by migration, hibernation, or torpidity. Insects, fishes, and other ectotherms (animals whose body temperature depends on environmental heat) compensate for fluctuating temperatures by behavioral responses (figure 2.1) or by altering biochemical and cellular processes involving enzymes, lipid organization, and the neuroendocrine system. Thus an animal's physiological capacities permit it to live under changing and often adverse environmental conditions. Behavioral responses are important also for obtaining food, finding shelter, escaping enemies and unfavorable environments, finding a mate, courting, and caring for young. Physiological mechanisms and behaviors that improve adaptability to environments assist survival of organisms. Ecologists who focus their studies at the organismal level are called physiological ecologists or behavioral ecologists.

Animals in nature coexist with others of the same species as reproductive units; these groups are called **populations** and have a unique gene pool (p. 25). Populations have properties that cannot be discovered from studying individual animals alone, including genetic variability among individuals (polymorphism), a gene pool, growth in numbers over time, and factors that limit the density of individuals in a given area. Ecological studies at the population level help us to predict the future success of endangered species and to discover controls for pest species.

Just as individuals do not exist alone in nature, populations of different species co-occur in more complex associations called **communities.** The variety of a community is measured as the number of species present, called **species diversity.** The populations of species in a community interact with each other in many ways, the most prevalent of which are **predation, parasitism,** and **competition. Predators** obtain energy and nutrients by killing and eating prey. **Parasites** derive similar benefits from their hosts, but usually do not kill the hosts. Competition occurs when food and/or space are limited and members of different species interfere with each other's use of their shared resources. **Mutualism** occurs when both members of a pair of species benefit from their interactions, usually by avoiding negative interactions with other species. Communities are complex because all of these interactions occur simultaneously, and their individual effects on the community often cannot be isolated.

Most people know that lions, tigers, and wolves are predators, but the world of invertebrates also includes numerous predaceous animals. These predators include unicellular organisms, jellyfish and their relatives, various worms, predaceous insects, sea stars, and many others.

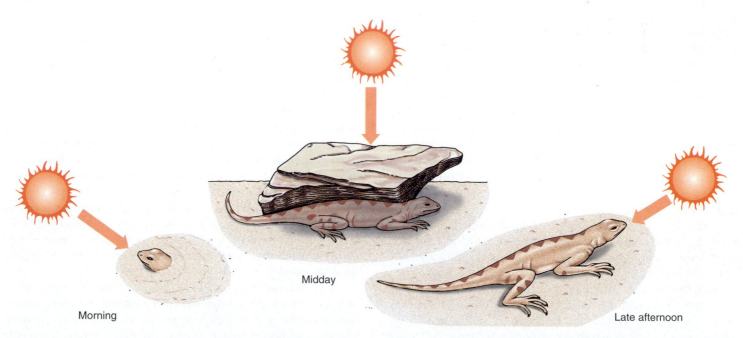

figure 2.1

How a lizard regulates its body temperature behaviorally. In the morning the lizard absorbs the sun's heat through its head while keeping the rest of its body protected from cool morning air. Later it will emerge to bask. At noon, with its body temperature high, it seeks shade from the hot sun. When the air temperature drops in the late afternoon, it emerges and lies parallel to the sun's rays.

Morning   Midday   Late afternoon

Ecological communities are biological components of even larger, more complex entities called **ecosystems.** An ecosystem consists of all populations in a community together with their physical environment. The study of ecosystems reveals two key processes in nature, the flow of energy and the cycling of materials through biological channels. The largest ecosystem is the **biosphere,** the thin veneer of land, water, and atmosphere that envelopes the planet, and that supports all life on earth.

Environment and the Niche

An animal's environment consists of all conditions that directly affect its chances of survival and reproduction. These factors include space, forms of energy such as sunlight, heat, wind and water currents, and also materials such as soil, air, water, and numerous chemicals. The environment also includes other organisms, which can be an animal's food, or its predators, competitors, hosts, or parasites. The environment thus includes both abiotic (nonliving) and biotic (living) factors. Some environmental factors, such as space and food, are utilized directly by an animal, and these are called **resources.**

A resource may be expendable or nonexpendable, depending on how an animal uses it. Food is expendable, because once eaten it is no longer available. Food therefore must be continuously replenished in the environment. Space, whether total living area or a subset such as the number of suitable nesting sites, is not exhausted by being used, and thus is nonexpendable.

The physical space where an animal lives, and that contains its environment, is its **habitat.** Size of habitat is variable. A rotten log is a normal habitat for carpenter ants. Such logs occur in larger habitats called forests where deer also live. However, deer forage in open meadows, so their habitat is larger than the forest. On a larger scale, some migratory birds occupy forests of the north temperate region during summer and move to the tropics during winter. Thus, habitat is defined by an animal's normal activity rather than by arbitrary physical boundaries.

Animals of any species demonstrate environmental limits of temperature, moisture, and food within which they can grow, reproduce, and survive. A suitable environment therefore must meet all requirements for life. A freshwater clam living in a tropical lake could tolerate the temperature of a tropical ocean, but it would be killed by the ocean's salinity. A brittle star living in the Arctic Ocean could tolerate the salinity of a tropical ocean but not its temperature. Thus temperature and salinity are two separate dimensions of an animal's environmental limits. If we add another variable, such as pH, we increase our description to three dimensions (figure 2.2). If we consider all environmental conditions that permit members of a species to survive and to multiply, we define a role for that species in nature as distinct from all others. This unique, multidimensional fingerprint of a species is called its **niche** (p. 34).

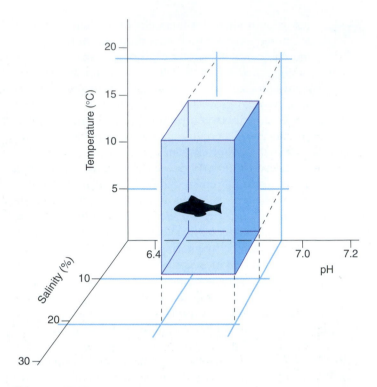

figure 2.2

Three-dimensional niche volume of a hypothetical animal showing three tolerance ranges. This graphic representation is one way to show the multidimensional nature of environmental relations. This representation is incomplete, however, because additional environmental factors also influence growth, reproduction, and survival.

Dimensions of niche vary among members of a species, making the niche subject to evolution by natural selection. The niche of a species undergoes evolutionary changes over successive generations.

Animals may be generalists or specialists with respect to tolerance of environmental conditions. For example, most fish are adapted to either fresh water or seawater, but not both. However, those that live in salt marshes, such as the topminnow *Fundulus heteroclitus,* easily tolerate changes in salinity that occur over tidal cycles in these estuarine habitats as fresh water from land mixes with seawater. Similarly, although most snakes are capable of eating a wide variety of animal prey, others have narrow dietary requirements; for example, the African snake *Dasypeltis scaber* is specialized for eating bird eggs (figure 2.3).

However broad may be the tolerance limits of an animal, it experiences only a single set of conditions at a time. An animal probably will not experience in its lifetime all environmental conditions that it potentially can tolerate. Thus, we must distinguish an animal's **fundamental niche,** which describes its potential role, and its **realized niche,** the subset of potentially suitable environments that an animal actually experiences.

figure 2.3

This African egg-eating snake, *Dasypeltis*, subsists entirely on hard-shelled birds' eggs, which it swallows whole. Its special adaptations are reduced size and number of teeth, enormously expansible jaw provided with elastic ligaments, and teethlike vertebral spurs that puncture the shell. Shortly after the second photograph was taken, the snake punctured and collapsed the egg, swallowed its contents, and regurgitated the crushed shell.

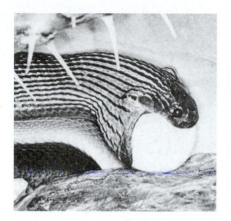

Populations

Animals exist in nature as members of populations. As we saw in Chapter 1, a population is a reproductively interactive group of animals of a single species (p. 20). A species may contain a single, cohesive population or many geographically disjunct populations. Geographically disjunct populations of the same species are called **demes.** Because members of a deme regularly interbreed, they share a common gene pool (p. 25).

Migration of individuals among demes within a species can impart some evolutionary cohesion to the species as a whole. Local environments may change unpredictably, sometimes causing a local deme to become severely depleted or eliminated. Migration is therefore a crucial source of replacement among demes within a region. Extinction of a species may be avoided if risk of extinction is spread among many demes, because simultaneous destruction of the environments of all demes in this manner is unlikely unless a catastrophe is widespread. Interaction among demes in this manner is called **metapopulation dynamics,** with "metapopulation" referring to a population subdivided into multiple genetically interacting demes. In some species, gene flow and recolonization among demes may be nearly symmetrical. If some demes are stable and others more susceptible to extinction, the more stable ones, termed **source demes,** differentially supply migrants to the less stable ones, called **sink demes.**

Each population or deme has a characteristic **age structure, sex ratio,** and **growth rate.** The study of these properties and factors that influence them is called demography. Demographic characteristics (individual size, age, number of offspring) vary according to lifestyles of the species under study. For example, some animals (and most plants) are **modular.** Modular animals, such as sponges, corals, and ectoprocts (p. 265), consist of colonies of genetically identical organisms. Reproduction is by asexual **cloning,** as described for hydrozoans in Chapter 7 (p. 130). Most colonies also have distinct periods of gamete formation and sexual reproduction. Colonies propagate also by fragmentation, as seen on coral reefs during severe storms. Pieces of coral may be scattered by wave action on a reef, forming propagules for new reefs. For these modular animals, age structure and sex ratio are difficult to determine. Changes in colony size can be used to measure growth rate, but counting individuals and measuring their ages and numbers of offspring are more difficult and less meaningful than in **unitary** animals, which are independently living organisms.

Most animals are unitary. However, even some unitary species reproduce by **parthenogenesis.** Parthenogenetic species occur in many animal taxa, including insects, reptiles, and fish. Usually such groups contain only females, which lay unfertilized eggs that hatch into daughters genotypically identical to their mothers. The praying mantid *Bruneria borealis,* common in the southeastern United States, is a parthenogenetic unitary animal.

Parthenogenesis ("virgin origin") is the development of an embryo from an unfertilized egg or one in which the male and female nuclei fail to unite following fertilization. There are many kinds of parthenogenesis, and it is surprisingly widespread in the animal kingdom.

Most metazoans are biparental, and reproduction follows a period of organismal growth and maturation. Each new generation begins with a **cohort** of individuals born at the same time. Of course, individuals of any cohort will not all survive to reproduce. For a population to retain constant size from generation to generation, each adult female must replace herself on average with one daughter that survives to reproduce. If females produce on average more than one viable daughter, the population grows; if fewer than one, the population declines.

Animal species have different characteristic patterns of **survivorship** from birth until death of the last member of a cohort. The three principal types of survivorship are illustrated in figure 2.4. Curve I, in which all individuals die at the same time, probably occurs rarely in nature. Curve II, in which rate of mortality as a proportion of survivors is constant over all ages, is characteristic of some animals that care for their young, as do many birds. Human populations generally fall somewhere between curves I and II, depending on nutrition and medical care.

Survivorship of most invertebrates, and of vertebrates like fish that produce great numbers of offspring, resembles

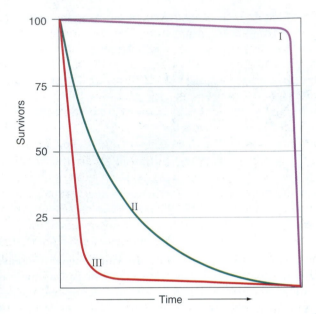

figure 2.4

Three types of theoretical survivorship curves. See text for explanation.

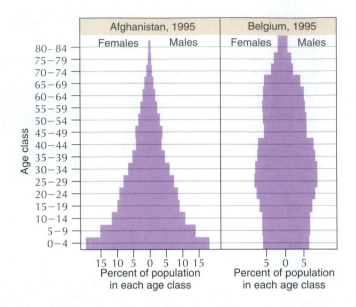

figure 2.5

Age structure profiles of the human populations of Afghanistan and Belgium in 1995 contrast the rapidly growing, youthful population of Afghanistan with the stable population of Belgium, where the fertility rate is below replacement. Countries such as Afghanistan with a large fraction of the population as children are strained to provide adequate child services. With so many children soon to enter their reproductive years, the population will continue to grow rapidly for many years to come.

curve III. For example, a mature female marine prosobranch snail, *Ilyanassa obsoleta*, produces thousands of eggs each reproductive period. Zygotes become free-swimming planktonic veliger larvae, which are scattered far from the mother's habitat by oceanic currents. They form part of the plankton and experience high mortality from numerous animals that consume plankton. Furthermore, larvae require a specific, sandy-bottomed substrate on which to settle and then to metamorphose into an adult snail. The probability of a larva surviving long enough to find a suitable habitat is very low, and most of the cohort dies during the veliger stage. We therefore see a rapid drop in survivorship in the first part of the curve. The few larvae that do survive to become snails have improved odds of surviving further, as shown by the more gentle slope of the curve for older snails. Thus, high reproductive output balances high juvenile mortality.

Many animals survive to reproduce only once before they die, as seen in many insect species of the temperate zone. Here, adults reproduce before the onset of winter and die, leaving only their eggs to overwinter and to repopulate their habitat the following spring. Similarly, Pacific salmon after several years return from the ocean to fresh water to spawn only once, after which all adults of a cohort die. This condition in which an organism reproduces only once during its life history is termed **semelparity.** However, other animals survive long enough to produce multiple cohorts of offspring that may mature and reproduce while their parents are still alive and reproductively active. **Iteroparity** denotes the occurrence of more than one reproductive cycle in an organism's life history. Populations of animals containing multiple cohorts, such as robins, box turtles, and humans, exhibit **age structure.** Analysis of age structure reveals whether a population is actively growing, stable, or declining. Figure 2.5 shows age profiles of

two populations. On a global scale, humans exhibit an age structure similar to curve I in figure 2.4, although age structures vary among regions.

Population Growth and Intrinsic Regulation

Population growth is the difference between rates of birth and death. As Darwin recognized from an essay by Thomas Malthus (p. 9), all populations have an inherent ability to grow exponentially. This ability is called the **intrinsic rate of increase,** denoted by the symbol **r.** The steeply rising curve in figure 2.6 shows this kind of growth. If species continually grew in this fashion, earth's resources soon would be exhausted and mass extinction would follow. A bacterium dividing three times per hour could produce a colony a foot deep over the entire earth after 36 hours, and this mass would be over our heads only one hour later. Animals have much lower potential growth rates than bacteria, but they could achieve the same kind of result over a longer period of time, given unlimited resources. Many insects lay thousands of eggs each year. A single codfish may spawn 6 million eggs in a season, and a field mouse can produce 17 litters of five to seven young each year. Obviously, unrestricted growth is not prevalent in nature.

Even in the most benign environment, a growing population eventually exhausts food or space. Exponential increases such as locust outbreaks or planktonic blooms in lakes must end when food or space is expended. Actually, among all resources that could limit a population, the one in shortest

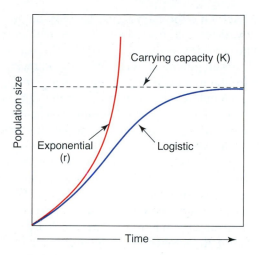

figure 2.6

Population growth, showing exponential growth of a species in an unlimited environment, and logistic growth in a limited environment.

non is called density dependence, and it is the mechanism for intrinsic regulation of populations. We can compare density dependence by negative feedback to the way endothermic animals regulate their body temperatures when environmental temperature exceeds an optimum. If the resource is expendable, as is food, carrying capacity is reached when the rate of resource replenishment equals the rate of depletion by the population; the population is then at K for that limiting resource. According to the logistic model, when population density reaches K, rates of birth and death are equal and growth of the population ceases. If food is being replaced at a rate that supports the current population, but no more, a population of grasshoppers in a green meadow may be at carrying capacity even though we see plenty of unconsumed food.

Although experimental populations of protozoa may fit a logistic growth curve closely, most populations in nature tend to fluctuate above and below carrying capacity. For example, after sheep were introduced to Tasmania around 1800, their numbers changed logistically with small oscillations around an average population size of about 1.7 million; we thereby infer the carrying capacity of the environment to be 1.7 million sheep (figure 2.7A). Ring-necked pheasants introduced on an island in Ontario, Canada exhibited wider oscillations (figure 2.7B).

Why do intrinsically regulated populations oscillate this way? First, the carrying capacity of an environment can change over time, requiring that a population change its density to track a limiting resource. Second, animals always experience a lag between the time that a resource becomes limiting and the time that the population responds by reducing its rate of growth. Third, **extrinsic** factors occasionally may limit a population's growth below carrying capacity. We consider extrinsic factors on p. 40.

supply relative to the needs of the population is depleted before others. This one is termed the **limiting resource.** The largest population that can be supported by the limiting resource in a habitat is called the **carrying capacity** of that environment, symbolized **K.** Ideally, a population would slow its growth rate in response to diminishing resources until it just reaches K, as represented by the sigmoid curve in figure 2.6. The mathematical expressions of exponential and sigmoid (or logistic) growth curves are compared in the box on this page. Sigmoid growth occurs when there is negative feedback between growth rate and population density. This phenome-

Exponential and Logistic Growth

We can describe the sigmoid growth curve (see figure 2.6) by a simple model called the logistic equation. The slope at any point on the growth curve is the growth rate, how rapidly the population size is changing with time. If **N** represents the number of organisms and **t** the time, we can, in the language of calculus, express growth as an instantaneous rate:

$d\text{N}/d\text{t}$ = the rate of change in number of organisms per time at a particular instant in time.

When populations are growing in an environment of unlimited resources (unlimited food and space, and no competition from other organisms), growth is limited only by the inherent capacity of the population to reproduce itself. Under these ideal conditions growth is expressed by the sym-

bol **r,** which is defined as the intrinsic rate of population growth per capita. The index **r** is actually the difference between birth rate and death rate per individual in the population at any instant. The growth rate of the population as a whole is then:

$$d\text{N}/d\text{t} = r\text{N}$$

This expression describes the rapid, **exponential growth** illustrated by the early upward-curving portion of the sigmoid growth curve (see figure 2.6).

Growth rate for populations in the real world slows as the upper limit is approached, and eventually stops altogether. At this point **N** has reached its maximum density because the space being studied has become "saturated" with animals. This limit is called the carrying capacity of the environment and is expressed by the symbol **K.** The

sigmoid population growth curve can now be described by the logistic equation, which is written as follows:

$$d\text{N}/d\text{t} = r\text{N}([\text{K} - \text{N}]/\text{K})$$

This equation states the rate of increase per unit of time ($d\text{N}/d\text{t}$ = rate of growth per capita **(r)** × population size **(N)** × unutilized freedom for growth [**K** − **N**]/**K**). One can see from the equation that when the population approaches carrying capacity, **K** − **N** approaches 0, $d\text{N}/d\text{t}$ also approaches 0, and the curve will flatten.

Populations occasionally overshoot the carrying capacity of the environment so that **N** exceeds **K.** The population then exhausts some resource (usually food or shelter). The rate of growth, $d\text{N}/d\text{t}$, then becomes negative and the population must decline.

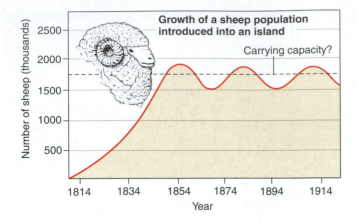

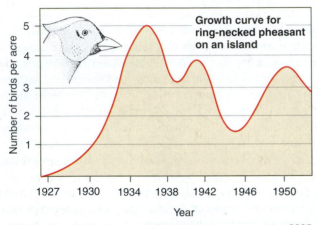

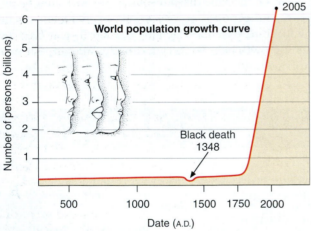

figure 2.7

Growth curves for sheep **A,** ring-necked pheasant **B,** and world human populations **C** throughout history. Note that the sheep population on an island is stable because of human control of the population, but the ring-necked pheasant population oscillates greatly, probably because of large changes in carrying capacity. Where would you place the carrying capacity for the human population?

On a global scale, humans have the longest record of exponential population growth (figure 2.7C). Although famine and war have restrained growth of populations locally, the only dip in global human growth resulted from bubonic plague ("black death"), which decimated much of Europe during the fourteenth

century. What then is the carrying capacity for the human population? The answer is far from simple, and several important factors must be considered when estimating the human K.

With development of agriculture, the carrying capacity of the environment increased, and the human population grew steadily from 5 million around 8000 B.C., when agriculture was introduced, to 16 million around 4000 B.C. Despite the toll of terrible famines, disease, and war, the population reached 500 million by 1650. With the Industrial Revolution in Europe and England in the eighteenth century, followed by a medical revolution, discovery of new lands for colonization, and better agricultural practices, the human carrying capacity increased dramatically. The population doubled to 1 billion around 1850. It doubled again to 2 billion by 1927, to 4 billion in 1974, passed 6 billion in October 1999, and is expected to reach 8.9 billion by 2040. Thus, growth has been exponential and remains high (figure 2.7C).

Recent surveys show that growth of the human population is slackening. Between 1970 and 2004 the annual growth rate decreased from 1.9% to 1.23%. At 1.23%, it will take nearly 57 years for the world population to double rather than 36.5 years at the higher annual growth rate. The decrease is credited to better family planning. Nevertheless, half the global population is under 25 years old and most live in developing countries where reliable contraception is limited or nonexistent. Thus, despite the drop in growth rate, the greatest surge in population lies ahead, with a projected 3 billion people added within the next five decades, the most rapid increase ever in human numbers.

Although rapid advancements in agricultural, industrial, and medical technology have undoubtedly increased the earth's carrying capacity for humans, it also has widened the difference between birth and death rates to increase our rate of exponential growth. Each day we add 208,000 people (net) to the approximately 6.4 billion people currently alive. Assuming that growth remains constant (certainly not a safe assumption, based on the history of human population growth), by the year 2030 more than half a million people will be added each day. In other words, less than 10 days will be required to replace all people who inhabited the world in 8000 B.C.

To estimate carrying capacity for the human species, we must consider not only quantity of resources, but quality of life. Many of the 6.4 billion people alive today are malnourished. At present 99% of our food comes from land, and the tiny fraction that we derive from the sea is decreasing due to overexploitation of fish stocks. Although there is some disagreement on what would constitute the maximum sustainable agricultural output, scientists do not expect food production to keep pace with population growth.

Extrinsic Limits to Growth

We have seen that the intrinsic carrying capacity of a population for an environment prevents unlimited exponential growth of the population. Population growth also can be limited by extrinsic biotic factors, including predation, parasitism (including disease-causing pathogens), and interspecific competition, or by abiotic influences such as floods, fires, and storms. Although abiotic factors certainly reduce populations in nature, they cannot truly regulate population growth because their effect is wholly independent of population size; abiotic limiting factors are **density-independent.** A single hailstorm might kill most of the young of wading bird populations, and a forest fire might eliminate entire populations of many animals, regardless of how many individuals there may be.

In contrast, biotic factors can and do act in a **density-dependent** manner. Predators and parasites respond to changes in density of their prey and host populations, respectively, to maintain those populations at fairly constant sizes. These sizes are below carrying capacity, because populations regulated by predation or parasitism are not limited by their resources. Competition between species for a common limiting resource lowers the effective carrying capacity for each species below that of either one alone.

Community Ecology

Interactions among Populations in Communities

Populations of animals are part of a larger system, known as the **community,** within which populations of different species interact. The number of species that share a habitat is known as **species diversity.** These species interact in a variety of ways that can be detrimental (−), beneficial (+), or neutral (0) to each species, depending on the nature of the interaction. For instance, we can consider a predator's effect on prey as (−), because the survival of the prey animal is reduced. However, the same interaction benefits the predator (+) because the food obtained from prey increases the predator's ability to survive and to reproduce. Thus, the predator-prey interaction is + −. Ecologists use this shorthand notation to characterize interspecific interactions because it shows the direction in which the interaction affects each species.

We see other kinds of + − interactions. One of these is **parasitism,** in which the parasite benefits by using the host as a home and source of nutrition, and the host is harmed. **Herbivory,** in which an animal eats a plant, is another + − relationship. **Commensalism** is an interaction that benefits one species and neither harms nor benefits the other (0 +). Most bacteria that normally inhabit our intestinal tracts do not affect us (0), but the bacteria benefit (+) by having food and a place to live. A classic example of commensalism is the association of pilot fishes and remoras with sharks (figure 2.8). These fishes get the "crumbs" remaining when the host shark makes its kill, but we now know that some remoras also feed on ectoparasites of sharks. Commensalism therefore grades into mutualism.

Organisms engaged in mutualism have a friendlier arrangement than commensalistic species, because the fitness of both is enhanced (+ +). Biologists are finding mutualistic

figure 2.8

Four remoras, *Remora* sp., attached to a shark. Remoras feed on fragments of food left by their shark host, as well as on pelagic invertebrates and small fishes. Although they actually are good swimmers, remoras prefer to be pulled through the water by marine creatures or boats. The shark host may benefit by having embedded copepod skin parasites removed by remoras.

relationships far more common in nature than previously thought (figure 2.9). Some mutualistic relationships are not only beneficial but necessary for survival of one or both species. Such relationships are termed obligatory mutualisms; by contrast, facultative mutualisms are not required for a species' survival. The relationship between a termite and protozoa inhabiting its gut illustrates an obligatory mutualism. The protozoa can digest wood eaten by the termite because the protozoa produce an enzyme, lacking in the termite, that digests cellulose; the termite lives on waste products of protozoan metabolism. In return, the protozoa gain a habitat and food supply. Such absolute interdependence among species can be a liability if one of the participants is lost. *Calvaria* trees native to the island of Mauritius have not reproduced successfully for over 300 years, because their seeds germinate only after being eaten and passed through the gut of a dodo bird, now extinct.

Competition between species reduces fitness of both (− −). Many biologists, including Darwin, considered competition the most common and important interaction in nature. Ecologists have constructed most of their theories of community structure from the premise that competition is the chief organizing factor in species assemblages. Sometimes the effect on one species in a competitive relationship is negligible. This condition is called **amensalism,** or **asymmetric competition** (0 −). For example, two species of barnacles, *Chthamalus stellatus* and *Balanus balanoides,* compete for space on rocky intertidal habitats. A famous experiment by Joseph Connell* demonstrated that *B. balanoides* excluded

*Connell, J. H. 1961. The influence of interspecific competition and other factors on the distribution of the barnacle *Chthamalus stellatus*. Ecology **42**:710–723.

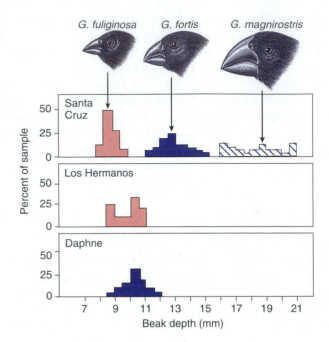

figure 2.10

Displacement of beak sizes in Darwin's finches from the Galápagos Islands. Beak depths are given for the ground finches *Geospiza fuliginosa* and *G. fortis* where they occur together (sympatric) on Santa Cruz Island and where they occur alone on the islands Daphne and Los Hermanos. *G. magnirostris* is another large ground finch that lives on Santa Cruz.

figure 2.9

Among the many examples of mutualism that abound in nature is the whistling thorn acacia of the African savanna and the ants that make their homes in the acacia's swollen galls. The acacia provides both protection for the ants' larvae (*lower photograph of opened gall*) and honeylike secretions used by ants as food. In turn, ants protect the tree from herbivores by swarming out as soon as the tree is touched. Giraffes, however, which love the tender acacia leaves, seem immune to the ants' fiery stings.

C. stellatus from a portion of the habitat, whereas *C. stellatus* had no effect on *B. balanoides*.

We have treated interactions as occurring between pairs of species. However, in natural communities containing populations of many species, a predator may have more than one prey and several animals may compete for the same resource. Thus, ecological communities are quite complex and dynamic, a challenge to ecologists who study this level of natural organization.

Competition and Character Displacement

Competition occurs when two or more species share a **limiting resource.** Simply sharing food or space with another species does not produce competition unless the resource is

in short supply relative to needs of the species that share it. Thus, we cannot assume that competition occurs in nature based solely on sharing of resources. However, we find evidence of competition by investigating different ways that species exploit a resource.

Competing species may reduce conflict by reducing overlap of their niches. **Niche overlap** is the portion of resources shared by the niches of two or more species. For example, if two species of birds eat seeds of exactly the same size, competition eventually will exclude one species from the habitat. This example illustrates the principle of **competitive exclusion:** strongly competing species cannot coexist indefinitely. This principle was discovered in 1932 by the Soviet microbiologist G. F. Gause, who performed experiments on the mechanism of competition between two species of yeast. To coexist in the same habitat, species must specialize by partitioning a shared resource and using different portions of it. Specialization of this kind is called ecological **character displacement.**

Character displacement usually appears as differences in organismal morphology or behavior related to exploitation of a resource. For example, in his classic study of the Galápagos finches (p. 22), English ornithologist David Lack noticed that bill sizes of these birds depended on whether they occurred together on the same island (figure 2.10). On the islands Daphne and Los Hermanos, where *Geospiza fuliginosa* and *G. fortis* occur separately and therefore do not compete with each other, bill sizes are nearly identical; on the island Santa Cruz, where both *G. fuliginosa* and *G. fortis* coexist, their bill sizes do not overlap. These results suggest resource partitioning, because bill size determines the size of seeds selected for food. Recent work by American ornithologist Peter Grant has confirmed what Lack suspected: *G. fuliginosa* with its smaller bill selects smaller seeds than does *G. fortis* with its larger bill. Where the two species coexist, competition between them led to evolutionary displacement of bill sizes to diminish competition. Absence of competition today has been called appropriately "the ghost of competition past."

Character displacement promotes coexistence by reducing niche overlap. When several species share the same general resources by such partioning, they form a **guild.** Just as a guild in medieval times constituted a brotherhood of men sharing a common trade, species in an ecological guild share a common livelihood. The term guild was introduced to ecology by Richard Root in his 1967 paper on niche patterns of the blue-gray gnatcatcher.[†] A classic example of a bird guild is Robert MacArthur's study of a feeding guild consisting of five species of warblers in spruce woods of the northeastern United States.[‡]

[†]Root, R. B. 1967. The niche exploitation pattern of the blue-gray gnatcatcher. Ecological Monographs **37:**317–350.
[‡]MacArthur, R. H. 1958. Population ecology of some warblers of northeastern coniferous forests. Ecology **39:**599–619.

At first glance, we might ask how five birds, very similar in size and appearance, could coexist by feeding on insects in the same tree. However, on close inspection MacArthur found subtle differences among these birds in sites of foraging (figure 2.11). One species searched only on outer branches of spruce crowns; another species used the top 60% of the tree's outer and inner branches away from the trunk; another species concentrated on inner branches closer to the trunk; another species used the midsection from the periphery to the trunk; and still another species foraged in the bottom 20% of the tree. These observations suggest that each warbler's niche within this guild is formed by structural differences in the habitat.

Guilds are not limited to birds. For example, a study done in England on insects associated with Scotch broom plants revealed nine different guilds of insects, including three species of stem miners, two gall-forming species, two that fed on seeds and five that fed on leaves. Another insect guild consists of three species of praying mantids that avoid both competition and predation by differing in sizes of their prey, timing of hatching, and height of vegetation in which they forage.

Predators and Parasites

The ecological warfare waged by predators against their prey causes coevolution: predators get better at catching prey, and prey get better at escaping predators. This is an evolutionary

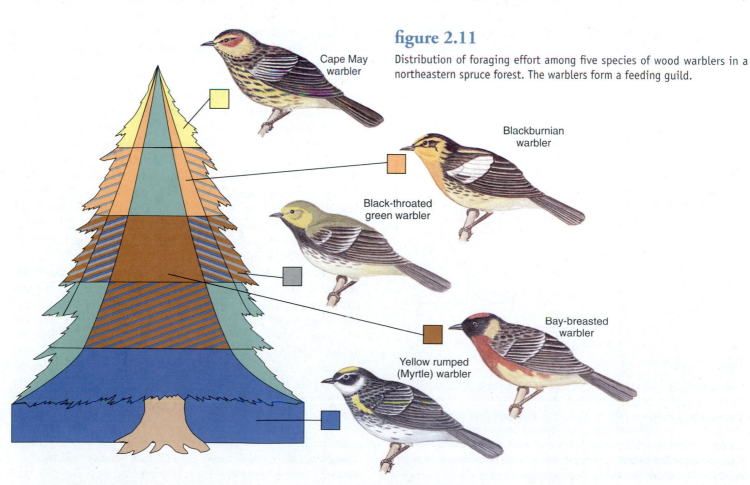

figure 2.11

Distribution of foraging effort among five species of wood warblers in a northeastern spruce forest. The warblers form a feeding guild.

Cape May warbler

Blackburnian warbler

Black-throated green warbler

Bay-breasted warbler

Yellow rumped (Myrtle) warbler

race that a predator cannot afford to win. If a predator became so efficient that it exterminated its prey, the predator species would become extinct. Because most predators feed on more than a single species, specialization on a single prey to the point of extermination is uncommon.

However, when a predator does rely primarily on a single prey species, both populations tend to fluctuate cyclically. First prey density increases, then that of the predator until prey become scarce. At that point, predators must adjust their population size downward by leaving the area, lowering reproduction, or dying. When density of a predator population falls enough to allow reproduction by prey to outpace mortality from predation, the cycle begins again. Thus, populations of both predators and prey show cycles of abundance, but increases and decreases in predator abundance are slightly delayed relative to those of prey because of the time lag in a predator's response to changing prey density. We can illustrate this process in the laboratory with protozoa (figure 2.12). Perhaps the longest documented natural example of a predator-prey cycle is between Canadian populations of snowshoe hares and lynxes (see figure 20.21, p. 405).

The war between predators and prey reaches high art in the evolution of defenses by potential prey. Many animals that are palatable escape detection by matching their background, or by resembling some inedible feature of the environment (such as a bird dropping). Such defenses are called cryptic. In contrast to cryptic defenses, animals that are toxic or distasteful to predators actually advertise their strategy with bright colors and conspicuous behavior. These species are protected because predators learn to recognize and to avoid them after distasteful encounters.

When distasteful prey adopt warning coloration, advantages of deceit arise for palatable prey. Palatable prey can deceive potential predators by mimicking distasteful prey, a phenomenon called Batesian mimicry. Coral snakes and monarch butterflies are both brightly colored, noxious prey. Coral snakes have a venomous bite, and monarch butterflies are poisonous because caterpillars store poison (cardiac glycoside) from milkweed they eat. Both species serve as **models** for other species, called **mimics,** that do not possess toxins of their own but look like the model species that do (figure 2.13A and B). A Batesian mimic essentially parasitizes the model population; a predator that encounters the tasteful mimic first may later harm models, expecting them to be tasteful.

In another form of mimicry, termed Müllerian mimicry, two or more toxic species resemble each other (figure 2.13C). What may an animal that has its own poison should gain by evolving resemblance to another poisonous animal? The answer is that a predator needs only to experience the toxicity of one species to avoid all similar prey. A predator can learn one warning signal more easily than many!

Sometimes the influence of one population on others is so pervasive that its absence drastically changes the entire community. We call such a population a **keystone species.*** On rocky intertidal shores of western North America, the sea star *Pisaster ochraceous* is a keystone species. Sea stars are a major predator of the mussel *Mytilus californianus.* When sea stars were removed experimentally from a patch of Washington State coastline, mussel populations expanded, occupying all space previously used by 25 other invertebrate and algal species (figure 2.14). Keystone predators act by reducing prey populations below a level where resources, such as space, are limiting. The original notion that all keystone species were predators has been broadened to include any species whose removal causes the extinction of others.

By reducing competition, keystone species allow more species to coexist on a resource. Consequently they contribute to maintaining diversity in a community. Keystone species illustrate a more general phenomenon, disturbance. Periodic natural disturbances such as fires and hurricanes also can prevent monopolization of resources and competitive exclusion by a few broadly adapted competitors. Disturbances can permit more species to coexist in such highly diverse communities as coral reefs and rain forests.

Parasites are often considered freeloaders because they appear to benefit from their hosts at no expense. **Ectoparasites** such as ticks and lice infect many different kinds of animals. The host provides nutrition from its body and aids dispersal of the parasite. However, we must consider that the evolutionary pathway to parasitism from free-living forms often has costs as well as benefits. **Endoparasites** such as tapeworms (see Chapter 8), have lost their ability to choose habitats. Also, because they must move among hosts to complete their life cycle, the chance that a single individual will live to

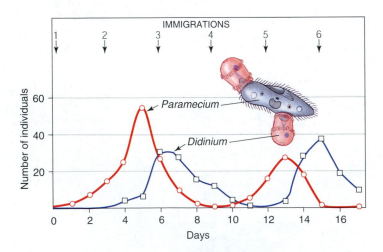

figure 2.12

Classic predator-prey experiment by Russian biologist G. F. Gause in 1934 shows the cyclic interaction between predator (*Didinium*) and prey (*Paramecium*) in laboratory culture. When the *Didinium* find and eat all the *Paramecium,* the *Didinium* themselves starve. Gause could keep the two species coexisting only by occasionally introducing one *Didinium* and one *Paramecium* to the culture (*arrows*); these introductions simulated migration from an outside source.

*Paine, R. T. 1969. A note on trophic complexity and community stability. American Naturalist **103**:91–93.

figure 2.13

Artful guises abound. **A,** A palatable viceroy butterfly (*top*) mimics a distasteful monarch butterfly (*below*). **B,** A harmless clearwing moth (*top photograph*) mimics a yellowjacket wasp, which is armed with a stinger (*lower photograph*). Both **A** and **B** illustrate Batesian mimicry. **C,** Two unpalatable tropical butterflies of different families resemble one another, an example of Müllerian mimicry.

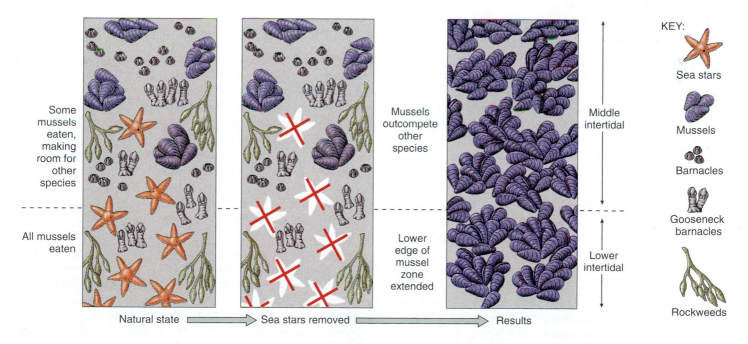

figure 2.14

Experimental removal of a keystone species, the predatory sea star *Pisaster ochraceus,* from an intertidal community completely changes the structure of the community. With their principal predator missing, mussels form dense beds by outcompeting and replacing other intertidal species.

Life without the Sun

For many years, ecologists thought that all animals depended directly or indirectly on primary production from solar energy. However, in 1977 and 1979, dense communities of animals were discovered living on the sea floor adjacent to vents of hot water issuing from rifts (Galápagos Rift and East Pacific Rise) where tectonic plates on the sea floor are slowly spreading apart. Such communities were subsequently found on numerous ocean ridges (Juan de Fuca, East Pacific, Southeast Pacific, Mid Atlantic, and Indian Ocean), back-arc basins (mostly West Pacific), hydrocarbon or hypersaline cold seeps (Gulf of Mexico, West Coast of Africa, Asia), and subduction zones (Mediterranean, west coasts of North and South America). These communities (see photo) include several species of molluscs, some crabs, poly-chaete worms, enteropneusts (acorn worms), and giant pogonophoran worms. The temperature of seawater above and immediately around vents is 7° to 23°C where it is heated by basaltic intrusions, whereas the surrounding normal seawater is 2°C.

The producers in these vent communities are chemoautotrophic bacteria that derive energy from oxidation of large amounts of hydrogen sulfide or methane in vent water and fix carbon dioxide into organic carbon. Some animals in vent communities—for example, bivalve molluscs—are filter feeders that ingest bacteria. Others, such as the giant pogonophoran tubeworms, which lack mouths and digestive tracts, harbor colonies of symbiotic bacteria in their tissues and use the organic carbon that these bacteria synthesize.

A population of giant pogonophoran tubeworms grows in dense profusion near a Galápagos Rift thermal vent, photographed at 2800 m (about 9000 feet) from the deep submersible *Alvin*. Also visible in the photograph are mussels and crabs.

---

reproduce is very low. The more intermediate hosts involved in a parasite's life cycle, the lower the likelihood of success, and the greater reproductive output must be to balance mortality.

Biologists often are puzzled by the complexity of parasite-host relationships. For example, a trematode parasite of the marine gastropod *Ilyanassa obsoleta* actually changes its host's behavior to complete its life cycle. These snails live in sandy-bottomed intertidal habitats in eastern North America. If the snails are exposed to air when the tide recedes, they normally burrow into sand to avoid desiccation. If, however, a snail is infected with the trematode *Gynaecotyla adunca,* it moves shoreward on high tides preceding low night tides to be left on the beach on the receding tide. Then, as in the legend of the Trojan Horse, the snail sheds cercariae into the sand where they can infect the next intermediate host, a beach-living crustacean. The crustacean may then be eaten by a gull or other shorebird, the definitive hosts for this trematode. The life cycle is completed when the bird defecates into water, releasing eggs from which hatch larvae that will infect more snails.

Coevolution between parasite and host is expected to generate an increasingly benign, less virulent relationship. Selection favors a benign relationship, because a parasite's fitness is diminished if its host dies. This traditional view has been challenged in recent years. Virulence is correlated, at least in part, with availability of new hosts. When alternative hosts are common and transmission rates are high, continued colonization of new hosts makes any particular host's life less valuable to the parasite, so that there may be no disadvantage to high virulence.

Ecosystems

Transfer of energy and materials among organisms within ecosystems is the ultimate level of organization in nature. Energy and materials are required to construct and to maintain life, and their incorporation into biological systems is called **productivity.** Productivity is divided into component **trophic levels** based on how organisms obtain energy and materials. Trophic levels are linked together into **food webs** (figure 2.15), which are pathways for transfer of energy and materials among organisms within an ecosystem.

Primary producers are organisms that begin productivity by fixing and storing energy from outside the ecosystem. Primary producers usually are plants or algae that capture solar energy through **photosynthesis** (but see an exception in the box on this page). Powered by solar energy, plants assimilate and organize minerals, water, and carbon dioxide into living tissue. All other organisms survive by consuming this tissue, or by consuming organisms that consumed this tissue. **Consumers** include **herbivores,** which eat plants directly, and **carnivores,** which eat other animals. The most important consumers are **decomposers,** mainly bacteria and fungi that break dead organic matter into its mineral components, returning it to a soluble form available to plants at the base of the nutrient cycle (p. 49). Although important chemicals such as nitrogen and carbon are reused endlessly through biological cycling, all energy ultimately is lost from the ecosystem as heat and not recycled. Thus, no ecosystems are truly closed; all require input of new energy from the sun or hydrothermal vents.

figure 2.15

Midwinter food web in *Salicornia* salt marsh of San Francisco Bay area.

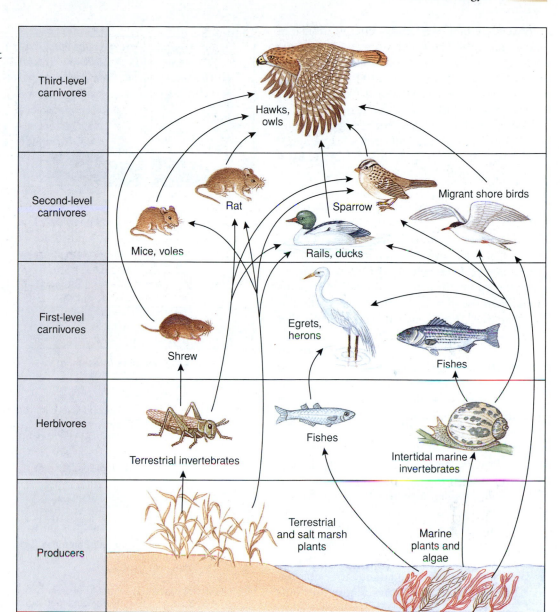

Third-level carnivores	Hawks, owls
Second-level carnivores	Rat, Sparrow, Migrant shore birds, Mice, voles, Rails, ducks
First-level carnivores	Shrew, Egrets, herons, Fishes
Herbivores	Terrestrial invertebrates, Fishes, Intertidal marine invertebrates
Producers	Terrestrial and salt marsh plants, Marine plants and algae

Energy Flow and Productivity

Every organism in nature has an **energy budget.** Just as we each must partition our income for housing, food, utilities, and taxes, each organism must obtain enough energy to meet its metabolic costs, to grow, and to reproduce.

Ecologists divide the budget into three main components: **gross productivity, net productivity,** and **respiration.** Gross productivity is like gross income; it is the total energy assimilated, analogous to your pay before deductions. When an animal eats, food passes through its gut and nutrients are absorbed. Most energy assimilated from these nutrients serves the animal's metabolic demands, which include cellular metabolism and regulation of body heat in endotherms. The energy required for metabolic maintenance is respiration, which is deducted from gross productivity to yield net productivity, an animal's take-home pay. Net productivity is energy stored by an animal in its tissues as **biomass.**

This energy is available for growth, and also for reproduction, which is population growth.

An animal's energy budget is expressed by a simple equation, in which gross and net productivity are represented by P_g and P_n, respectively, and respiration is R:

$$P_n = P_g - R$$

This equation states the first law of thermodynamics in the context of ecology. (The first law of thermodynamics states that energy cannot be created or destroyed. It can change from one form to another, but the total amount of energy in a system remains unchanged.) Its important messages are that the energy budget of every animal is finite and may be limiting, and that energy is available for growth of individuals and populations only after maintenance is satisfied.

The second law of thermodynamics, which states that total disorder or randomness of a system always increases, is important when we study energy transfers between trophic levels in food webs. Energy for maintenance, **R,** usually constitutes more than 90% of the assimilated energy (**P$_g$**) for animal consumers. More than 90% of the energy in an animal's food is lost as heat, and less than 10% is stored as biomass. Each succeeding trophic level therefore contains only 10% of the energy in the next lower trophic level. Most ecosystems are thereby limited to five or fewer trophic levels.

Our ability to feed a growing human population is influenced profoundly by the second law of thermodynamics. Humans, who are at the end of the food chain, may eat grain, fruits, and vegetables of plants that fix the sun's energy in chemical bonds; this very short chain represents an efficient use of potential energy. Humans also may eat beef from animals that eat grass that fixes the sun's energy; the addition of a trophic level decreases available energy by a factor of 10. Ten times as much plant biomass is needed to feed humans as meat eaters as to feed humans as grain eaters. Consider a person who eats a bass that eats a sunfish that eats zooplankton that eats phytoplankton that fixes the sun's energy. The tenfold loss of energy at each trophic level in this five-step chain requires that the pond must produce 5 tons of phytoplankton for a person to gain a pound by eating bass. If the human population depended on bass for survival, we would quickly exhaust this resource.

These figures must be considered as we look to the sea for food. Productivity of oceans is very low and limited largely to estuaries, marshes, reefs, and upwellings where nutrients are available to phytoplankton producers. Such areas constitute a small part of the ocean. The rest is a watery void.

Marine fisheries supply 18% of the world's protein, but much of this protein is used to feed livestock and poultry. If we remember the rule of 10-to-1 loss in energy with each transfer of material between trophic levels, then use of fish as food for livestock rather than humans is poor use of a valuable resource in a protein-deficient world. Fishes that we prefer to eat include flounder, tuna, and halibut, which are three or four levels up the food chain. Every 125 g of tuna requires one metric ton of phytoplankton to produce. If humans are to derive greater benefit from oceans as a food source, we must eat more of the less desirable fishes that are at lower trophic levels.

When we examine the food chain in terms of biomass at each level, we can construct **ecological pyramids** either of numbers or of biomass. A pyramid of numbers (figure 2.16A), also known as **Eltonian pyramid,** depicts numbers of organisms that are transferred between each trophic level. This

"R.F.D.2."

Steve Stinson and Roanoke Times and World-News

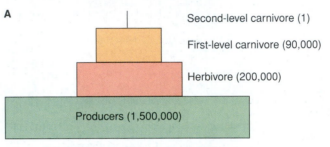

A

Second-level carnivore (1)
First-level carnivore (90,000)
Herbivore (200,000)
Producers (1,500,000)

Pyramid of numbers (grassland)

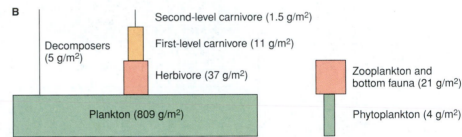

B

Decomposers (5 g/m²)
Second-level carnivore (1.5 g/m²)
First-level carnivore (11 g/m²)
Herbivore (37 g/m²)
Plankton (809 g/m²)

Zooplankton and bottom fauna (21 g/m²)
Phytoplankton (4 g/m²)

Pyramids of biomass (aquatic ecosystems)

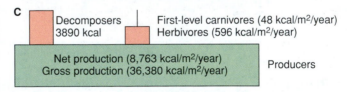

C

Decomposers 3890 kcal
First-level carnivores (48 kcal/m²/year)
Herbivores (596 kcal/m²/year)
Net production (8,763 kcal/m²/year)
Gross production (36,380 kcal/m²/year)
Producers

Pyramid of energy (tropical forest)

figure 2.16

Ecological pyramids of numbers, biomass, and energy. Pyramids are generalized, since the area within each trophic level is not scaled proportionally to quantitative differences in units given.

pyramid provides a vivid impression of the great difference in numbers of organisms involved in each step of the chain, and supports the observation that large predatory animals are rarer than the small animals on which they feed. However, a pyramid of numbers does not indicate actual mass of organisms at each level.

More instructive are pyramids of biomass (figure 2.16B), which depict the total bulk, or "standing crop," of organisms at each trophic level. Such pyramids usually slope upward because mass and energy are lost at each transfer. However, in some aquatic ecosystems in which the producers are algae, which have short life spans and rapid turnover, the pyramid is inverted. Algae tolerate heavy exploitation by zooplankton consumers. Therefore, the base of the pyramid (biomass of phytoplankton) is smaller than the biomass of zooplankton it supports. This inverted pyramid is analogous to a person who weighs far more than the food in a refrigerator, but who can be sustained from the refrigerator because food is constantly replenished.

The concepts of food chains and ecological pyramids were invented and first explained in 1923 by Charles Elton, a young ecologist at Oxford University. Working for a summer on a treeless arctic island, Elton watched arctic foxes as they roamed, noting what they ate and, in turn, what their prey had eaten, until he was able to trace the complex cycling of nitrogen in food throughout the animal community. Elton realized that life in a food chain comes in discrete sizes, because each form had evolved to be much bigger than the thing it eats. He thus explained the common observation that large animals are rare while their smaller prey are common; the ecological pyramids illustrating this phenomenon now bear Elton's name.

A third type of pyramid is a pyramid of energy, which shows rate of energy flow between levels (figure 2.16C). An energy pyramid is never inverted because energy transferred from each level is less than what entered it. A pyramid of energy gives the best overall picture of community structure because it is based on production. Productivity of phytoplankton exceeds that of zooplankton, even though biomass of phytoplankton is less than biomass of zooplankton (because of heavy grazing by zooplankton consumers).

Nutrient Cycles

All elements essential for life are derived from the environment, where they are present in air, soil, rocks, and water. When plants and animals die and their bodies decay, or when organic substances are burned or oxidized, elements and inorganic compounds essential for life processes (nutrients) are released and returned to the environment. Decomposers fulfill an essential role in this process by feeding on the remains of plants and animals and on fecal material. The result is that nutrients flow in a perpetual cycle between biotic and abiotic components of the ecosystem. Nutrient cycles are often called **biogeochemical cycles** because they involve exchanges between living organisms (bio-) and rocks, air, and water of the earths' crust (geo-). Continuous input of energy from the sun keeps nutrients flowing and the ecosystem functioning (figure 2.17).

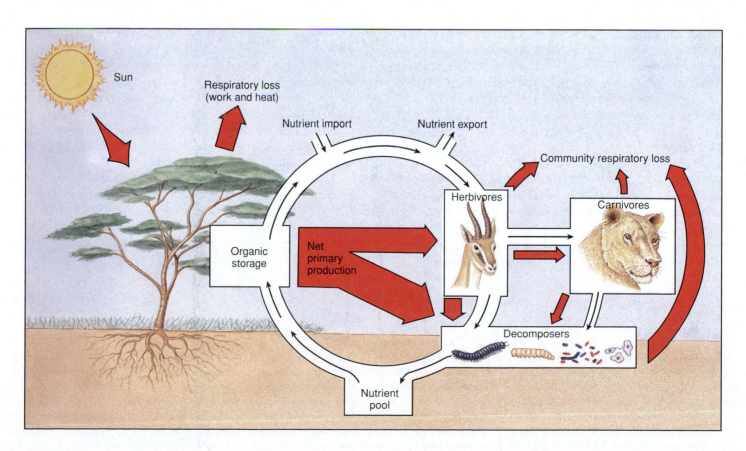

figure 2.17

Nutrient cycles and energy flow in a terrestrial ecosystem. Note that nutrients are recycled, whereas energy flow (*red*) is one way.

We tend to think of biogeochemical cycles in terms of naturally occurring elements, such as water, carbon, phosphorus, and nitrogen. For example, the nutrient pool of the soil shown in figure 2.17 contains a storage of phosphorus, some of which is in compounds that can be extracted by living plants. Phosphorus obtained by plants can be stored in the plants' tissues, returned to the soil, or consumed by herbivorous animals. The herbivores likewise store phosphorus in their bodies, return it to the soil via waste products, or provide it to carnivores that consume them. These carnivores likewise retain some phosphorus, return some to the soil via waste products, and may themselves be the prey of other carnivores. The amount of phosphorus available in the nutrient pool can be a limiting factor in the nutrient cycle of a local community.

Nutrient pools can be altered greatly by fertilizers and other compounds that people synthesize and add to the soil. Our synthetic compounds often challenge nature's nutrient cycling because decomposers have not evolved ways to degrade them. Among the most harmful synthetic compounds are pesticides. Pesticides in natural food webs can be insidious for three reasons. First, many pesticides become concentrated as they travel up succeeding trophic levels. The highest concentrations will occur in top carnivores such as hawks and owls, diminishing their ability to reproduce. Second, many species that are killed by pesticides are not pests, but merely innocent bystanders, called nontarget species. Nontarget effects happen when pesticides move out of the agricultural field to which they were applied, through rainwater runoff, leaching through soil, or dispersal by wind. The third problem is persistence; some chemicals used as pesticides have a long life span in the environment, so that nontarget effects persist long after the pesticides have been applied. Genetic engineering of crop plants aims to improve pest resistance in order to lessen the need for chemical pesticides.

Biodiversity and Extinction

Today, during a period of high economic prosperity and social progress, we face a serious environmental crisis. An unfortunate reality is that as the economy expands, the ecosystem on which the economy depends does not, leading to an increasingly stressed relationship. An important indicator of the planet's health is its biodiversity—the number of species that thrive on earth. An inventory of the earth's biodiversity is currently beyond our grasp. The earth's total number of species is typically estimated to be as high as 10 million, and this number may be too low by an order of magnitude.

Fossil studies of species formation and extinction help us put into evolutionary perspective the consequences of human-mediated ecological changes for biodiversity. Biodiversity exists because rates of speciation on average slightly exceed rates of extinction in earth's evolutionary history. Approximately 99% of all species that have lived are extinct. Extinction events killing at least 5% of existing species have occurred almost continuously throughout geological time. Extinction events killing at least 30% of existing species have occurred episodically approximately every 10 million years, and truly catastrophic events that killed at least 65% of existing species have occurred, on average, every 100 million years. Although the earth's ecosystems have recovered from loss of biodiversity in the past, the time for recovery was enormously long from the perspective of a human life span.

At rates unprecedented during human history, we are losing biodiversity, genetic diversity within species, and habitat diversity. Fragmentation of populations, as seen especially on islands, leads to loss of genetic diversity and makes a species unusually susceptible to extinction. About half of all areas in the world that contain at least two endemic bird species are islands, even though islands constitute less than 10 percent of the earth's terrestrial habitats. Island species are often particularly prone to destruction by introduction of invasive exotic species. For example, diverse land snails of the genus *Partula* on the Tahitian island of Moorea were displaced by introduction of exotic snails. Mainland habitats, such as forests, are fragmented into virtual islands when development clears vast areas of habitat and when introduced species invade these areas. Because tropical regions have high species endemism, human fragmentation of these environments is particularly likely to produce species extinction.

Human activity clearly has induced numerous species extinctions, and we must avoid making the present time one that rivals the great extinction crises of earth's geological history. Only through significant conservation efforts and through laws that limit environmental degradation will we be able to preserve the planet's biodiversity.

Summary

Ecology is the study of relationships between organisms and their environments to explain the distribution and abundance of life on earth. The physical space where an animal lives, and that contains its environment, is its habitat. Within the habitat are physical and biological resources that an animal uses to survive and to reproduce, which constitute its niche.

Animal populations consist of demes of interbreeding members sharing a common gene pool. Cohorts of animals have characteristic patterns of survivorship that represent adaptive trade-offs between parental care and numbers of offspring. Animal populations consisting of overlapping cohorts have age structure that indicates whether they are growing, declining, or at equilibrium.

Every species in nature has an intrinsic rate of increase that gives it the potential for exponential growth. The human population is growing exponentially at about 1.23% each year, and is expected to increase from 6.4 billion to 8.9 billion by the year 2040. Population growth may be regulated intrinsically by the carrying capacity of the environment, extrinsically by competition between species for a limiting resource, or by predators or parasites. Density independent abiotic factors can limit, but not truly regulate, population growth.

Communities consist of populations that interact with one another in any of several ways, including competition, predation, parasitism, commensalism, and mutualism. These relationships are results of coevolution among populations within communities. Guilds of species avoid competitive exclusion by character displacement, the partitioning of limited resources by morphological specialization. Keystone predators are those that control

community structure and reduce competition among prey, which increases species diversity. Parasites and their hosts evolve a benign relationship that ensures their coexistence.

Ecosystems consist of communities and their abiotic environments. Animals occupy the trophic levels of herbivorous and carnivorous consumers within ecosystems. All organisms have an energy budget consisting of gross and net productivity, and respiration. For animals, respiration usually is at least 90% of this budget. Thus, transfer of energy from one trophic level to another is limited to

about 10%, which in turn limits the number of trophic levels in an ecosystem. Ecological pyramids of energy depict how productivity decreases in successively higher trophic levels of food webs.

Ecosystem productivity is a result of energy flow and material cycles within ecosystems. All energy is lost as heat, but nutrients and other materials including pesticides are recycled. No ecosystem, including the global biosphere, is closed because they all depend upon imports and exports of energy and materials from outside.

Review Questions

1. The term "ecology" is derived from a Greek word meaning "house" or "place to live." However, as used by scientists, the terms "ecology" and "environment" do not mean the same thing. What is the distinction between ecology and environment?
2. How would you distinguish between ecosystem, community, and population?
3. What is the distinction between habitat and environment?
4. Define the niche concept. How does the "realized niche" of a population differ from its "fundamental niche?" How does the concept of niche differ from the concept of guild?
5. Populations of independently living (unitary) animals have a characteristic age structure, sex ratio, and growth rate. However, these properties are difficult to determine for modular animals. Why?
6. Explain which of the three survivorship curves in figure 2.4 best fits the following: (a) a population in which mortality as a proportion of survivors is constant; (b) a population in which there is little early death and most individuals live to old age; (c) a population that experiences heavy mortality of the very young but with survivors living to old age. Offer an example from the real world of each survivorship pattern.

7. Contrast exponential and logistic growth of a population. Under what conditions might you expect a population to exhibit exponential growth? Why cannot exponential growth be perpetuated indefinitely?
8. Growth of a population may be controlled by either density-dependent or density-independent mechanisms. Define and contrast these two mechanisms. Offer examples of how growth of the human population might be curbed by either agent.
9. Herbivory is an example of an interspecific interaction beneficial for the animal (+) but harmful to the plant it eats (−). What are some + − interactions among animal populations? What is the difference between commensalism and mutualism?
10. Explain how character displacement can ease competition between coexisting species.
11. Define predation. How does the predator-prey relationship differ from the parasite-host relationship? Why is the evolutionary race between predator and prey one that the predator cannot afford to win?
12. Mimicry of monarch butterflies by viceroys is an example of a palatable species resembling a toxic one. What is the advantage to the viceroy of this

form of mimicry? What is the advantage to a toxic species of mimicking another toxic species?
13. A keystone species has been defined as one whose removal from a community causes extinction of other species. How does this extinction happen?
14. What is a trophic level, and how does it relate to a food chain?
15. Define *productivity* as the word is used in ecology. What is a primary producer? What is the distinction between gross productivity, net productivity, and respiration? What is the relation of net productivity to biomass (or standing crop)?
16. What is a food chain? How does a food chain differ from a food web?
17. How is it possible to have an inverted pyramid of biomass in which the consumers have a greater biomass than the producers? Can you think of an example of an inverted pyramid of numbers of organisms, in which there are, for example, more herbivores than plants on which they feed?
18. The pyramid of energy has been offered as an example of the second law of thermodynamics. Why?
19. Animal communities surrounding deep-sea thermal vents apparently exist in total independence of solar energy. How can this existence be possible?

Selected References

See general references on p. 415.

Chase, J. M., and M. A. Liebold. 2003. Ecological niches: linking classical and contemporary approaches. Chicago, University of Chicago Press. *An insightful coverage of niche concepts in community ecology.*

Krebs, C. J. 2001. Ecology: the experimental analysis of distribution and abundance, ed. 5. New York, Addison Wesley Longman, Inc. *Important treatment of population ecology.*

Molles, M. C., Jr. 2002. Ecology: concepts and applications, ed. 2. New York, McGraw-Hill. *A concise and well-illustrated survey of ecology.*

Pianka, E. R. 2000. Evolutionary ecology, ed. 5. New York, Harper Collins College Publishers. *An introduction to ecology written from an evolutionary perspective.*

Schluter, D. 2000. The ecology of adaptive radiation. Oxford, Oxford University Press. *A synthesis of community ecology and evolution of biodiversity.*

Smith, R. L., and T. M. Smith. 2001. Ecology and field biology, ed. 6. San Francisco, Benjamin Cummings. *Clearly written, well-illustrated general ecology text.*

Zimmer, C. 2000. Parasite Rex. New York, The Free Press (a division of Simon & Schuster). *A fascinating book for nonspecialists by a talented science writer. Many examples of parasites that affect or control their host's behavior and drive host evolution.*

Custom Website

The *Animal Diversity* Online Learning Center is a great place to check your understanding of chapter material. Visit *www.mhhe.com/hickmanad4e* for access to key terms, quizzes, and more! Further enhance your knowledge with web links to chapter-related material.

Explore live links for these topics:

General Ecology Sites
Animal Population Ecology
Life Histories
Population Growth
Human Population Growth
Community Ecology
Competition
Parasitism, Predation, Herbivory
Coevolution
Food Webs
Primary Productivity
Global Ecology

Animal Architecture

New Designs for Living

Zoologists today recognize 32 phyla of multicellular animals, each phylum characterized by a distinctive body plan and array of biological properties that set it apart from all other phyla. Nearly all are the survivors of perhaps 100 phyla that were generated 600 million years ago during the Cambrian explosion, the most important evolutionary event in the history of animal life. Within the space of a few million years virtually all of the major body plans that we see today, together with many other novel plans that we know only from the fossil record, were established. Entering a world sparse in species and mostly free of competition, these new life forms began widespread experimentation, producing new themes in animal architecture. Nothing since then has equaled the Cambrian explosion. Later bursts of speciation that followed major extinction events produced only variations on established themes.

Established themes, in the form of distinctive body plans, are passed down a lineage from an ancestral population to its descendants; molluscs carry a hard shell, bird forelimbs become wings. These ancestral traits limit the morphological scope of descendants no matter what their lifestyle. Although penguin bodies are modified for an aquatic life, the wings and feathers of their bird ancestors might never adapt to aquatic environments as well as fish fins and scales. Despite structural and functional evolution, new forms are constrained by the architecture of their ancestors.

Cnidarian polyps have radial symmetry and cell-tissue grade of organization, (*Dendronephthya* sp.).

The English satirist Samuel Butler proclaimed that the human body was merely "a pair of pincers set over a bellows and a stewpan and the whole thing fixed upon stilts." While human attitudes toward the human body are distinctly ambivalent, most people less cynical than Butler would agree that the body is a triumph of intricate, living architecture. Less obvious, perhaps, is that humans and most other animals share an intrinsic material design and fundamental functional plan despite vast differences in structural complexity. This essential uniformity of biological organization derives from the common ancestry of animals and from their basic cellular construction. In this chapter, we consider the limited number of body plans that underlie the diversity of animal form and examine some of the common architectural themes that animals share.

The Hierarchical Organization of Animal Complexity

Among the different animal groups, we can recognize five major grades of organization (table 3.1). Each grade is more complex than the preceding one and builds upon it in a hierarchical manner.

The unicellular protozoan groups are the simplest animal-like organisms. They are nonetheless complete organisms that perform all basic functions of life as seen in more complex animals. Within confines of their cell, they show remarkable organization and division of labor, possessing distinct supportive structures, locomotor devices, fibrils, and simple sensory structures. The diversity observed among unicellular organisms is achieved by varying the architectural patterns of subcellular structures, organelles, and the cell as a whole (see Chapter 5).

Metazoa, or multicellular animals, evolved greater structural complexity by combining cells into larger units. A metazoan cell is a specialized part of the whole organism and, unlike a protozoan cell, it is not capable of independent existence. Cells of a multicellular organism are specialized for performing the various tasks accomplished by subcellular elements in unicellular forms. The simplest metazoans show the **cellular** grade of organization in which cells demonstrate division of labor but are not strongly associated to perform a specific collective function (table 3.1). In the more complex **tissue** grade, similar cells are grouped together and perform their common functions as a highly coordinated unit. In animals of the tissue-organ grade of organization, tissues are assembled into still larger functional units called **organs.** Usually one type of tissue carries the burden of an organ's chief function, as muscle tissue does in the heart; other tissues—epithelial, connective, and nervous—perform supportive roles. The chief functional cells of an organ are called its **parenchyma** (pa-ren´ka-ma; Gr. *para,* beside, + *enchyma,* infusion). The supportive tissues are its **stroma** (Gr. bedding). For instance, in the vertebrate pancreas the secreting cells are the parenchyma; the capsule and connective tissue framework represent the stroma.

Most metazoa (nemerteans and all more structurally complex phyla) have an additional level of complexity in which different organs operate together as **organ systems.** Eleven different kinds of organ systems are observed in metazoans: skeletal, muscular, integumentary, digestive, respiratory, circulatory, excretory, nervous, endocrine, immune, and reproductive. The great evolutionary diversity of these organ systems is covered in Chapters 8–20.

Animal Body Plans

As described in the prologue to this chapter, ancestral body plans constrain the form of descendant organisms. Animal body plans differ in grade of organization, in number of embryonic germ layers, in form and number of body cavities, and in symmetry of the body. Symmetry can generally be determined from the external appearance of an animal, but other features require more detailed study.

Animal Symmetry

Symmetry refers to balanced proportions, or correspondence in size and shape of parts on opposite sides of a median plane. **Spherical symmetry** means that any plane passing through the center divides the body into equivalent, or mirrored, halves (figure 3.1). This type of symmetry occurs chiefly among some protozoan groups and is rare in metazoans. Spherical forms are best suited for floating and rolling.

Radial symmetry (figure 3.1) applies to forms that can be divided into similar halves by more than two planes passing through the longitudinal axis. These are the tubular, vase, or bowl shapes in which one end of the longitudinal axis is usually the mouth. Examples are the hydras, jellyfishes, sea urchins, and some sponges. A variant form is biradial symmetry, in which, because of some part that is single or paired rather than radial, only one or two planes passing through the longitudinal axis produce mirrored halves. Sea walnuts or comb jellies (phylum Ctenophora), which are roughly globular in form but have a pair of tentacles, are an example. Radial and biradial animals are usually sessile, freely floating, or weakly swimming. The two phyla that are primarily radial, Cnidaria and Ctenophora, are called the **Radiata.** Echinoderms (sea stars and their kin) are primarily bilateral animals (their larvae are bilateral) that have become secondarily radial as adults.

Bilateral symmetry applies to animals that can be divided along a sagittal plane into two mirrored portions, right and left halves (figure 3.1). Some convenient terms used for locating regions of animal bodies (figure 3.2) are **anterior,** used to designate the head end; **posterior,** the opposite or tail end; **dorsal,** the back side; and **ventral,** the front or belly side. **Medial** refers to the midline of the body; **lateral** refers to the sides. **Distal** parts are farther from the middle of the body; **proximal** parts are nearer. A **frontal plane** (also sometimes called coronal plane) divides a bilateral body into dorsal and ventral halves by running through the anteroposterior axis and the right-

Table 3.1   Levels of Organization in Organismal Complexity

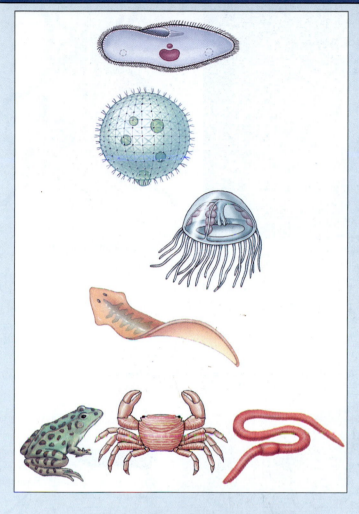

1. *Protoplasmic level of organization.* Protoplasmic organization is found in unicellular organisms. All life functions are confined within the boundaries of a single cell, the fundamental unit of life. Within a cell, living substance is differentiated into organelles that perform specialized functions.

2. *Cellular level of organization.* Cellular organization is an aggregation of cells that are functionally differentiated. A division of labor is evident, so that some cells are concerned with, for example, reproduction, others with nutrition. Such cells have little tendency to become organized into tissues (a tissue is a group of similar cells organized to perform a common function). Some protozoan colonial forms that have distinct somatic and reproductive cells might be placed at the cellular level of organization. Many authorities also place sponges at this level.

3. *Cell-tissue level of organization.* A step beyond the preceding is the aggregation of similar cells into definite patterns or layers, thus becoming a **tissue.** Some authorities assign sponges to this level, although jellyfishes and their relatives (Cnidaria) more clearly demonstrate the tissue plan. Both groups are still largely of the cellular grade of organization because most of the cells are scattered and not organized into tissues. An excellent example of a tissue in cnidarians is the nerve net, in which the nerve cells and their processes form a definite tissue structure, with the function of coordination.

4. *Tissue-organ level of organization.* Aggregation of tissues into organs is a further step in complexity. Organs are usually made up of more than one kind of tissue and have a more specialized function than tissues. This is the organizational level of the flatworms (Platyhelminthes), in which there are well-defined organs such as eyespots, digestive tract, and reproductive organs. In fact, the reproductive organs are well organized into a reproductive system.

5. *Organ-system level of organization.* When organs work together to perform some function we have the most complex level of organization—the organ system. Systems are associated with basic body functions—circulation, respiration, digestion, and others. The simplest animals that show this type of organization are nemertean worms, which have a complete digestive system distinct from the circulatory system. Most animal phyla demonstrate this type of organization.

left axis at right angles to the **sagittal plane,** the plane dividing an animal into right and left halves. A **transverse plane** (also called a cross section) would cut through a dorsoventral and a right-left axis at right angles to both the sagittal and frontal planes and separate anterior and posterior portions.

The appearance of bilateral symmetry in animal evolution was a major innovation because bilateral animals are much better fitted for directional (forward) movement than

are radially symmetrical animals. Bilateral animals are collectively called **Bilateria.** Bilateral symmetry is strongly associated with **cephalization,** the differentiation of a head end. Concentration of nervous tissue and sense organs in a head bestows obvious advantages to an animal moving through its environment head first. This is the most efficient positioning of instruments for sensing the environment and responding to it. Usually the mouth of an animal is located on the head as

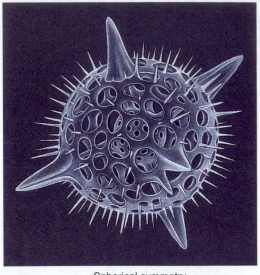

Spherical symmetry

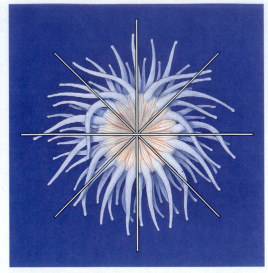

Radial symmetry

Bilateral symmetry

figure 3.1

The planes of symmetry as illustrated by spherically, radially, and bilaterally symmetrical animals.

figure 3.2

Descriptive terms used to identify positions on the body of a bilaterally symmetrical animal.

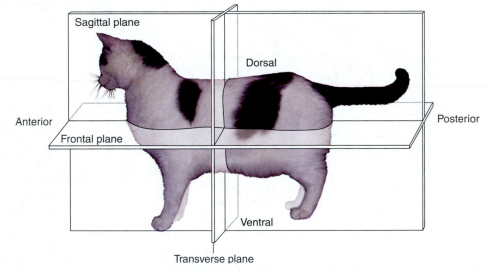

figure 3.3

Radial and spiral cleavage patterns shown at two-, four-, and eight-cell stages. **A,** Radial cleavage, typical of echinoderms, chordates, and hemichordates. **B,** Spiral cleavage, typical of molluscs, annelids, and other protostomes. Arrows indicate clockwise movements of small cells (micromeres) following division of large cells (macromeres).

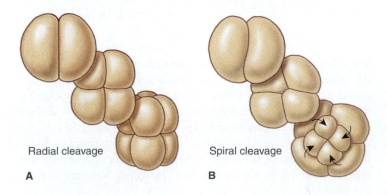

Radial cleavage

A

Spiral cleavage

B

well, since so much of an animal's activity is concerned with procuring food.

Cephalization is always accompanied by differentiation along an anteroposterior axis (**polarity**). Each end of the body forms a pole. The release of substances from each pole may produce gradients of chemical signals along an anteroposterior axis. Such gradients are important during development as regions of the body develop differently in response to the chemical signals they experience.

Development of Animal Body Plans

The body plan of an animal forms as part of an inherited developmental sequence. This sequence begins after fertilization of an egg to form a **zygote.** The zygote is a single large cell until the process of cleavage divides it into a large number of smaller cells called blastomeres.

Cleavage occurs in several different ways: sponges and cnidarians (sea anemones and their kin) lack a distinct cleavage pattern, but bilateral animals typically exhibit either **spiral** or **radial cleavage** (figure 3.3). Although there is much debate about which type of cleavage occurred first, biologists assume that these two types of cleavage each evolved only once. Thus, animals with spiral cleavage are assumed to have shared a recent common ancestor; likewise those with radial cleavage.

In **radial cleavage,** the cleavage planes are symmetrical to the polar axis and produce tiers, or layers, of cells on top of each other in an early embryo. Radial cleavage is also called **regulative cleavage** because each blastomere of the early embryo, if separated from the others, can adjust or "regulate" its development into a complete and well-proportioned (though possibly smaller) embryo (figure 3.4).

Spiral cleavage, found in several phyla, differs from radial in several ways. Rather than an egg dividing parallel or perpendicular to the animal-vegetal axis, it cleaves oblique to this axis and typically produces a quartet of cells that come to lie not on top of each other but in furrows between the cells (see figure 3.3). In addition, spirally cleaving eggs tend to pack their cells tightly together much like a cluster of soap bubbles, rather than just lightly contacting each other as do those in many radially cleaving embryos. Spirally cleaving embryos also differ from radial embryos in having a **mosaic** form of development, in which the organ-forming determinants in the egg cytoplasm become strictly localized in the egg, even before the first cleavage division. The result is that if the early blastomeres are separated, each will continue to develop for a time as though it were still part of the whole. Each forms a defective, partial embryo (figure 3.4).

A curious feature of most spirally cleaving embryos is that at about the 29-cell stage a blastomere called the 4d cell is formed that will give rise to all mesoderm of the embryo. Mesoderm is one of three germ layers usually present in an embryo; the other two germ layers are ectoderm and endoderm. Ultimately, all features of an adult animal are derived from one of these three germ layers, but adult structures do not form until well after cleavage.

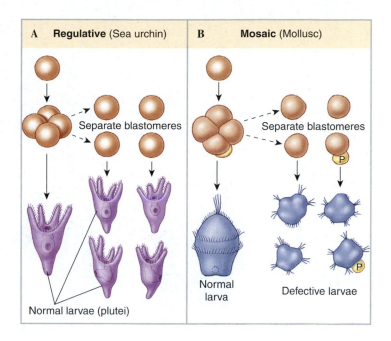

figure 3.4

Regulative and mosaic development. **A,** Regulative development. If the early blastomeres of a sea urchin embryo are separated, each will develop into a complete larva. **B,** Mosaic development. When the blastomeres of a mollusc embryo are separated, each gives rise to a partial, defective larva.

Cleavage proceeds until the zygote is divided into many small cells, typically surrounding a fluid-filled cavity (figure 3.5). At this point the embryo is called a **blastula** and the fluid-filled cavity is a **blastocoel.** In animals other than sponges, the blastula becomes a two-layered stage called a **gastrula** (figure 3.5A). The gastrula has two germ cell layers, ectoderm and endoderm. The outer germ cell layer, ectoderm, surrounds the blastocoel. The endoderm surrounds and defines an inner body cavity called the **gastrocoel.** The gastrocoel will become the gut cavity in an adult animal. Some body cavities, like the gut, persist into adulthood, but other cavities are lost, and still others appear only later in development. Adult sea anemones and their kin are two-layered animals, so they do not develop beyond the gastrula stage, but most animals develop three embryonic germ layers and additional body cavities.

Body Cavities

The most obvious internal space or body cavity is a gut developing from an embryonic gastrocoel (figure 3.5A). The gut always has at least one opening, the blastopore. In adult sea anemones and their kin, this opening takes in food and releases digestive wastes. A gut with only one opening is a two-way, blind, or incomplete gut. Most animals develop a second opening to the gut, creating a tube (figure 3.5B). Tubular guts permit a sequential movement of food from mouth to anus and are called one-way or complete guts.

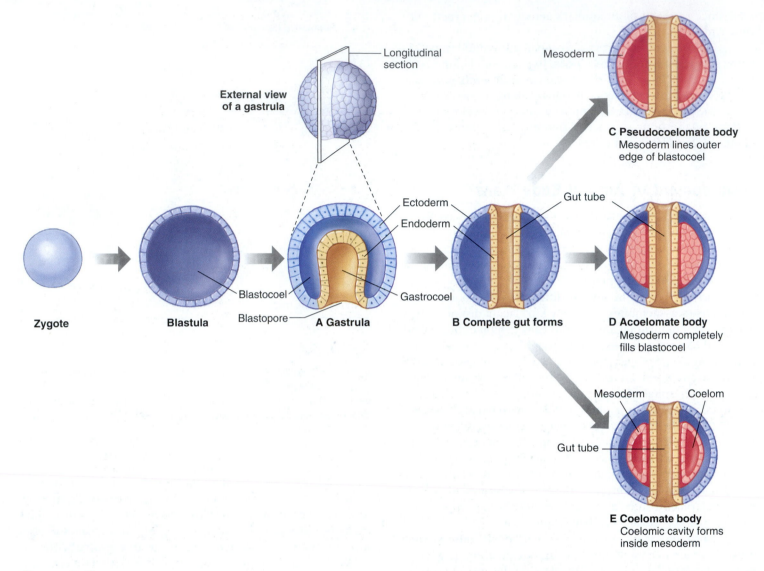

figure 3.5

A generalized developmental sequence beginning with a zygote. All metazoans develop to a blastula stage; arrows between stages A through E illustrate possible later developmental sequences.

In many animals the gut is surrounded by a fluid-filled cavity. There are several ways this cavity can form. For animals with only two germ layers, like cnidarians (sea anemones and their kin), the embryonic blastocoel persists into adulthood as a cavity surrounding the blind gut (figure 3.5B). The blastocoel may also persist in animals with a third germ layer, mesoderm, but the addition of a layer of mesoderm inside the blastocoel causes structural changes.

In some animals, mesoderm lines the outer edge of the blastocoel, lying next to the ectoderm (figure 3.5C); when this occurs, the blastocoel is renamed the **pseudocoelom** in adult animals. "Pseudocoelom" means "false coelom," and the descriptive nature of this name will be clear once a coelom is understood.

In other animals, mesoderm completely fills the blastocoel (figure 3.5D). These animals are **acoelomate,** meaning that the body is without a coelom. The only body cavity is the gut lumen, surrounded by tissues derived from mesoderm.

In the development of most bilaterally symmetrical animals, the blastocoel fills with mesoderm and then a new cavity forms *inside* the mesoderm (figure 3.5E). This new cavity, completely surrounded by mesoderm, is a **coelom.** This is in contrast to a pseudocoelom, which has mesoderm on only the outer edge of the cavity. A coelom may form by one of two methods, outlined on pages 60 and 61, but the end result is the same. The evolution of a coelom, a fluid-filled body cavity between the outer body wall and the gut, was a major development event affecting the body plans of many bilateral animals (figure 3.6).

A coelom provides a tube-within-a-tube arrangement that allows much greater body flexibility than is possible in animals lacking an internal body cavity. A coelom also provides space for visceral organs and permits greater size and complexity by

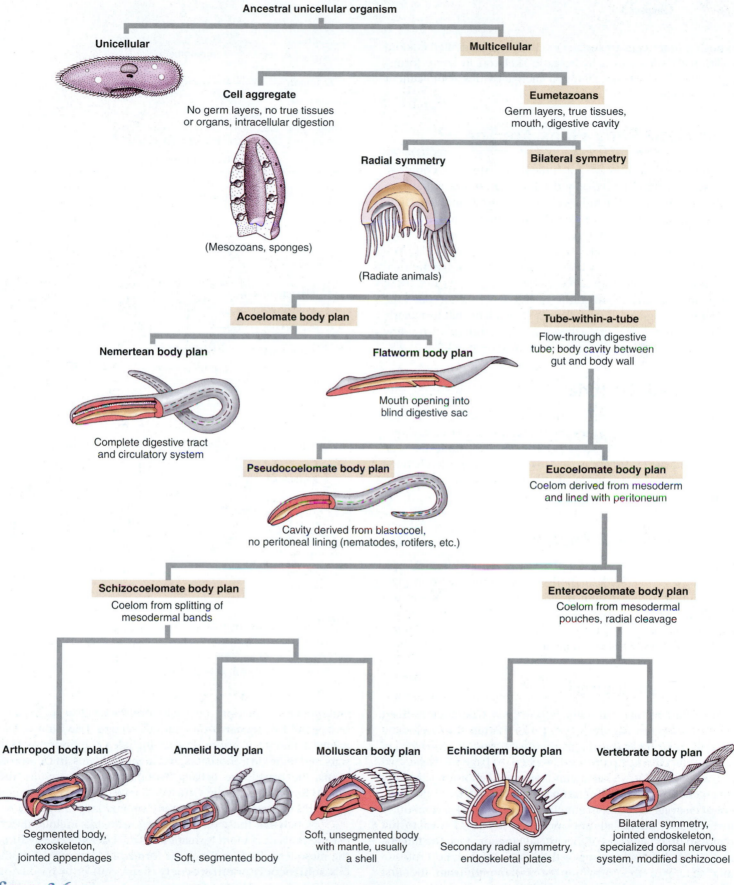

Ancestral unicellular organism

Unicellular

Multicellular

Cell aggregate
No germ layers, no true tissues
or organs, intracellular digestion

(Mesozoans, sponges)

Eumetazoans
Germ layers, true tissues,
mouth, digestive cavity

Radial symmetry

(Radiate animals)

Bilateral symmetry

Acoelomate body plan

Tube-within-a-tube
Flow-through digestive
tube; body cavity between
gut and body wall

Nemertean body plan

Complete digestive tract
and circulatory system

Flatworm body plan

Mouth opening into
blind digestive sac

Pseudocoelomate body plan

Cavity derived from blastocoel,
no peritoneal lining (nematodes, rotifers, etc.)

Eucoelomate body plan
Coelom derived from mesoderm
and lined with peritoneum

Schizocoelomate body plan
Coelom from splitting of
mesodermal bands

Enterocoelomate body plan
Coelom from mesodermal
pouches, radial cleavage

Arthropod body plan

Segmented body,
exoskeleton,
jointed appendages

Annelid body plan

Soft, segmented body

Molluscan body plan

Soft, unsegmented body
with mantle, usually
a shell

Echinoderm body plan

Secondary radial symmetry,
endoskeletal plates

Vertebrate body plan

Bilateral symmetry,
jointed endoskeleton,
specialized dorsal nervous
system, modified schizocoel

figure 3.6

Architectural patterns of animals. These basic body plans have been variously modified during evolutionary descent to fit animals to a great variety of habitats. Ectoderm is shown in gray, mesoderm in red, and endoderm in yellow.

exposing more cells to surface exchange. A fluid-filled coelom additionally serves as a hydrostatic skeleton in some forms, especially many worms, aiding in such activities as movement and burrowing.

How Many Body Plans Are There?

The multicellular animals, collectively the Metazoa, comprise 32 phyla. Of these phyla, only the Porifera (sponges) have no embryonic germ layers and lack tissues in adult animals. A few phyla, including phylum Cnidaria (sea anemones and their kin) and phylum Ctenophora (comb jellies) develop only two embryonic germ layers; these animals typically have radial symmetry. A vast majority of phyla contain bilaterally symmetrical animals developing from three embryonic germ layers; these phyla are collectively called the Bilateria. Within the Bilateria, biologists recognize several distinct body plans that form early in development. We can categorize most phyla into a few groups whose members share a body plan (figure 3.6).

Acoelomate Bilateria

Flatworms and a few others have *no body cavity* surrounding the gut (figure 3.7, *top*); they are "acoelomate" (Gr. *a*, without, + *koiloma*, cavity). The region between the ectodermal epidermis and the endodermal digestive tract is completely filled with mesoderm in the form of a spongy mass of space-filling cells called parenchyma.

Pseudocoelomate Bilateria

Nematodes and several other phyla have a pseudocoelom surrounding the gut, giving the body a tube-within-a-tube arrangement similar to that in animals with a true coelom. However, a pseudocoelom is a persistent blastocoel. It lacks a **peritoneum,** a thin cellular membrane derived from mesoderm that lines the body cavity in animals with a true coelom.

Eucoelomate Bilateria

The remaining bilateral animals possess a true coelom lined with peritoneum (figure 3.7, *bottom*). Within the coelomate animals there are two major evolutionary groupings, marked in part by the cleavage patterns described on page 57. These two cleavage patterns signal a fundamental dichotomy, an early divergence of bilateral metazoans into two separate lineages. Spiral cleavage is found in annelids, molluscs, and several other invertebrate phyla; all are included in the **Protostomia** ("mouth first") division of the animal kingdom (see the illustration inside the front cover of this book). The name Protostomia* refers to the formation of the mouth from the first

---

* Modern molecular analyses suggest that Protostomia should be subdivided into two groups, one containing all animals that molt their exoskeletons and the other containing animals sharing a similar larval form.

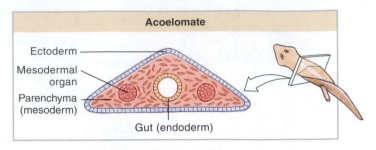

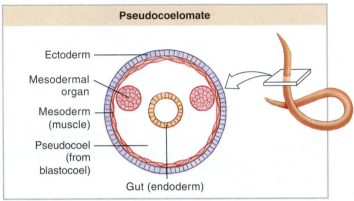

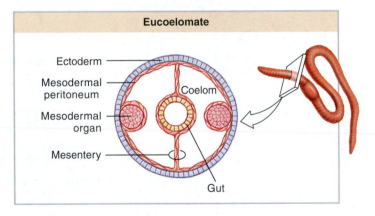

figure 3.7

Acoelomate, pseudocoelomate, and eucoelomate body plans are shown as cross sections of representative animals. Note the relative positions of parenchyma, peritoneum, and body organs.

embryological opening, the blastopore. Radial cleavage characterizes the **Deuterostomia** ("mouth second") division of the animal kingdom, a grouping that includes echinoderms (sea stars and their kin), chordates, and hemichordates. In Deuterostomia, the blastopore usually becomes the anus, while the mouth forms secondarily. Other distinguishing developmental hallmarks of these two divisions are summarized in figure 3.8.

Notice that these distinguishing hallmarks include different methods of coelom formation. A true coelom arises within the mesoderm itself and may be formed by one of two methods, **schizocoely** or **enterocoely** (figure 3.9), or by modifying these methods. The two terms are descriptive, for *schizo* comes from the Greek *schizein,* meaning to split; *entero* is derived from the Greek *enteron,* meaning gut; and *coelous* comes from the Greek *koilos,* meaning hollow or cavity. In

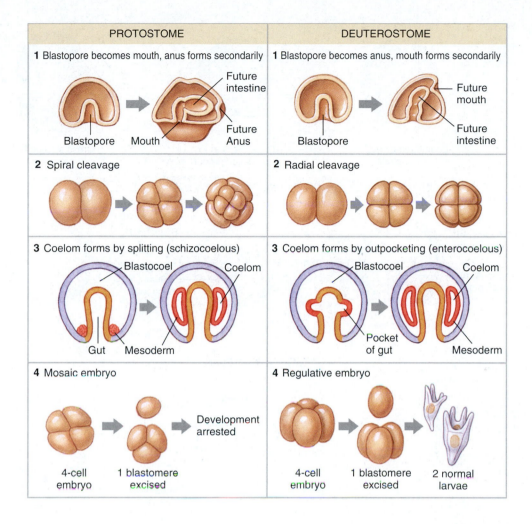

figure 3.8

Developmental tendencies of protostomes and deuterostomes. These tendencies are much modified in some groups, for example, the vertebrates. Cleavage in mammals is rotational rather than radial; in reptiles, birds, and many fishes the cleavage is discoidal. Vertebrates have also evolved a derived form of coelom formation that is basically schizocoelous.

schizocoelous formation the coelom arises, as the word implies, from splitting of mesodermal bands that originate as cells in the blastopore region migrate into the blastocoel. Most protostomes develop a coelom by schizocoely.

In enterocoelous formation, the coelom comes from pouches of the archenteron, or primitive gut, that push outward into the blastocoel (figure 3.9). Most deuterostomes develop a coelom by enterocoely, with the notable exception of the vertebrate chordates. The invertebrate chordates form the coelom by enterocoely as expected for deuterostomes, but the vertebrate chordates use a modified, and independently derived, version of schizocoely.

Once development is complete, the results of schizocoelous and enterocoelous formations are indistinguishable. Both produce a true coelom lined with a mesodermal peritoneum (Gr. *peritonaios*, stretched around) and having mesenteries in which the visceral organs are suspended (figure 3.7, *bottom*).

Metameric or Segmented Body Plans

Metamerism, also called segmentation, is a serial repetition of similar body segments along the longitudinal axis of the body. Each segment is called a metamere, or somite. In forms such as earthworms and other annelids, in which metamerism is most clearly represented, the segmental arrangement includes both external and internal structures of several systems. There is repetition of muscles, blood vessels, nerves, and the setae of locomotion. Some other organs, such as those of sex, may be repeated in only a few segments.

Metameric or segmented body plans occur in both protostomes and deuterostomes, but evolutionary changes have obscured much of the segmentation in many animals, including humans. True metamerism characterizes only three phyla: Annelida, Arthropoda, and Chordata (figure 3.10), although many diverse animals display superficial segmentation of the ectoderm and the body wall.

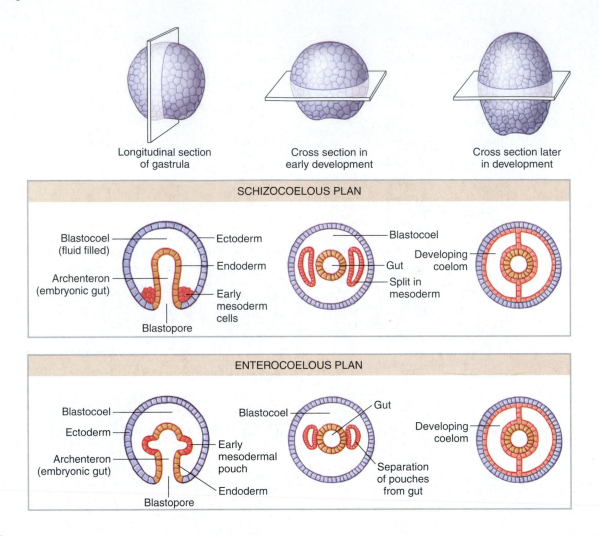

figure 3.9

Types of mesoderm and coelomic formation. In schizocoelous formation, the mesoderm originates from the wall of the embryonic gut near the blastopore and proliferates into a band of tissue which splits to form the coelom. In enterocoelous formation, most mesoderm originates as a series of pouches from the embryonic gut; these pinch off and enlarge to form the coelom. In both formations the coeloms expand to obliterate the blastocoel.

Components of Metazoan Bodies

Metazoan bodies consist of cellular components, the tissues derived from the three embryonic germ layers, as well as extracellular components.

Cellular Components: Tissues

A tissue is a group of cells, together with associated cell products, specialized for performing a common function. During embryonic development, the three germ layers become differentiated into different tissues. All cells in metazoan animals take part in the formation of tissues. Some tissues are composed of several kinds of cells, and some tissues have a great many intercellular materials.

The study of tissues is called **histology** (Gr. *histos,* tissue, + *logos,* discourse). There are four kinds of tissues: epithelial, connective (including vascular), muscular, and nervous tissues (figure 3.11). This is a surprisingly short list of basic tissue types that are able to meet the diverse requirements of animal life.

Epithelial Tissue

An **epithelium** (pl., epithelia) is a sheet of cells that covers an external or internal surface. Outside the body, epithelium forms a protective covering. Inside, an epithelium lines all organs of the body cavity, as well as ducts and passageways through which various materials and secretions move. On many surfaces epithelial cells are often modified into glands that produce lubricating mucus or specialized products such as hormones or enzymes.

Epithelia are classified by cell form and number of cell layers. Simple epithelia (figure 3.12) are found in all metazoan animals, while stratified epithelia (figure 3.13) are mostly restricted to vertebrates. All types of epithelia are supported

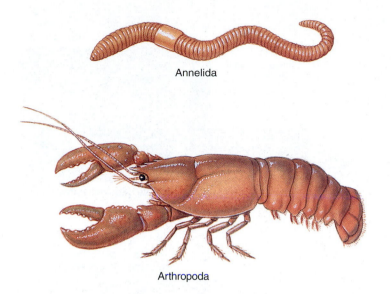

Annelida

Arthropoda

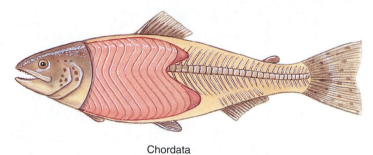

Chordata

figure 3.10

Segmented phyla. These three phyla illustrate an important principle in nature—metamerism, or repetition of structural units. Segmentation brings more varied specialization because segments, especially in arthropods, have become modified for different functions.

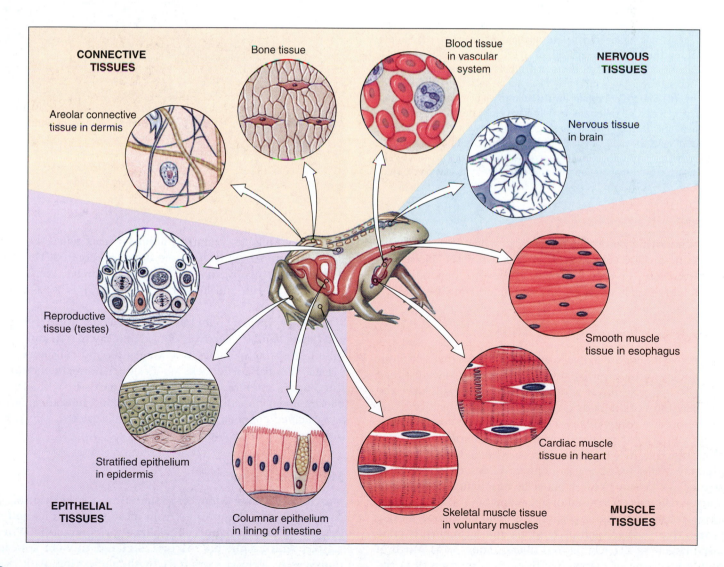

CONNECTIVE TISSUES

Bone tissue

Blood tissue in vascular system

NERVOUS TISSUES

Areolar connective tissue in dermis

Nervous tissue in brain

Reproductive tissue (testes)

Smooth muscle tissue in esophagus

Stratified epithelium in epidermis

Cardiac muscle tissue in heart

EPITHELIAL TISSUES

Columnar epithelium in lining of intestine

Skeletal muscle tissue in voluntary muscles

MUSCLE TISSUES

figure 3.11

Types of tissues in a vertebrate, showing examples of where different tissues are located in a frog.

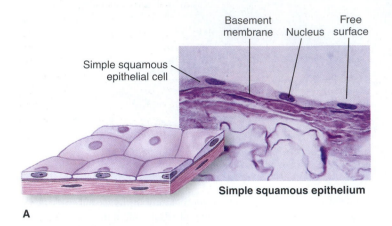

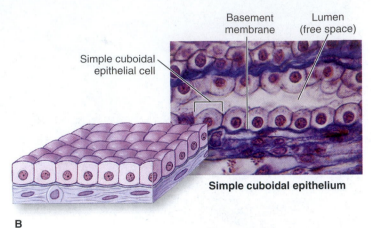

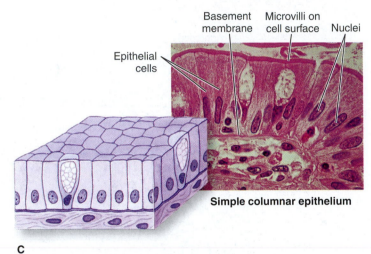

figure 3.12

Types of simple epithelium. **(A) Simple squamous epithelium,** composed of flattened cells that form a continuous delicate lining of blood capillaries, lungs, and other surfaces where it permits the passive diffusion of gases and tissue fluids into and out of cavities. **(B) Simple cuboidal epithelium** is composed of short, boxlike cells. Cuboidal epithelium usually lines small ducts and tubules, such as those of the kidney and salivary glands, and may have active secretory or absorptive functions. **(C) Simple columnar epithelium** resembles cuboidal epithelium, but the cells are taller and usually have elongate nuclei. This type of epithelium is found on highly absorptive surfaces such as the intestinal tract of most animals. The cells often bear minute, fingerlike projections called microvilli that greatly increase the absorptive surface. In some organs, such as female reproductive tract, the cells may be ciliated.

by an underlying basement membrane, which is a condensation of the ground substance of connective tissue. Blood vessels never enter epithelial tissues, so they are dependent on diffusion of oxygen and nutrients from underlying tissues.

Connective Tissue

Connective tissues are a diverse group of tissues that serve various binding and supportive functions. They are so widespread in the body that removal of other tissues would still leave the complete form of the body clearly apparent. Connective tissue is composed of relatively few **cells,** a great many extracellular **fibers,** and a fluid, gel-like, or rigid **ground substance** (also called **matrix**), in which the fibers are embedded. We recognize several different types of connective tissue. Two kinds of **connective tissue proper** occur in vertebrates. **Loose connective tissue** is composed of fibers and both fixed and wandering cells suspended in a syrupy ground substance. **Dense connective tissue,** such as tendons and ligaments, is composed largely of densely packed fibers (figure 3.14). Much of the fibrous tissue of connective tissue is composed of **collagen** (Gr. *kolla,* glue, + *genos,* descent), a protein material of great tensile strength. Collagen is the most abundant protein in

the animal kingdom, found in animal bodies wherever both flexibility and resistance to stretching are required. The connective tissue of invertebrates, as in vertebrates, consists of cells, fibers, and ground substance, but usually it is not as elaborately developed.

Other types of connective tissue include **blood, lymph,** and **tissue fluid** (collectively considered vascular tissue), composed of distinctive cells in a watery ground substance, the plasma. Vascular tissue lacks fibers under normal conditions. **Cartilage** is a semirigid form of connective tissue with closely packed fibers embedded in a gel-like ground substance. **Bone** is a calcified connective tissue containing calcium salts organized around collagen fibers.

Muscular Tissue

Muscle is the most common tissue of most animals. It originates (with few exceptions) from mesoderm, and its unit is the cell or **muscle fiber,** specialized for contraction. When viewed with a light microscope, **striated muscle** appears transversely striped (striated), with alternating dark and light bands (figure 3.15). In vertebrates we recognize two types of striated muscle: **skeletal** and **cardiac muscle.** A

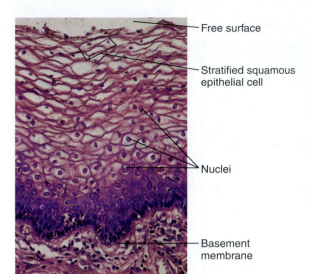

- Free surface
- Stratified squamous epithelial cell
- Nuclei
- Basement membrane

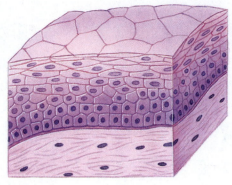

Stratified squamous epithelium

Stratified squamous epithelium consists of two to many layers of cells adapted to withstand mild mechanical abrasion. The basal layer of cells undergoes continuous mitotic divisions, producing cells that are pushed toward the surface where they are sloughed off and replaced by new cells beneath them. This type of epithelium lines the oral cavity, esophagus, and anal canal of many vertebrates, and vagina of mammals.

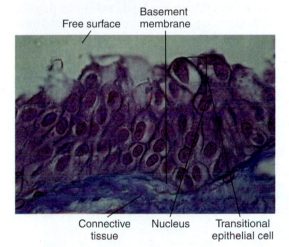

- Free surface
- Basement membrane
- Connective tissue
- Nucleus
- Transitional epithelial cell

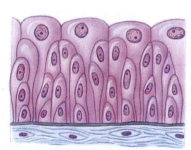

Transitional epithelium—unstretched

Transitional epithelium is a type of stratified epithelium specialized to accommodate great stretching. This type of epithelium is found in the urinary tract and bladder of vertebrates. In the relaxed state it appears to be four or five cell layers thick, but when stretched out it appears to have only two or three layers of extremely flattened cells.

Transitional epithelium—stretched

figure 3.13

Types of stratified epithelium.

third kind of muscle is smooth (or visceral) muscle, which lacks the characteristic alternating bands of the striated type (figure 3.15). Unspecialized cytoplasm of muscles is called **sarcoplasm,** and contractile elements within the fiber are **myofibrils.**

Nervous Tissue

Nervous tissue is specialized for reception of stimuli and conduction of impulses from one region to another. Two basic types of cells in nervous tissue are **neurons** (Gr. nerve), the basic functional unit of the nervous system, and **neuroglia** (nu-rog´ le-a; Gr. nerve, + *glia,* glue), a variety of nonnervous cells that insulate neuron membranes and serve various supportive functions. Figure 3.16 shows the functional anatomy of a typical nerve cell.

Extracellular Components of the Metazoan Body

In addition to the hierarchically arranged cellular structures discussed, metazoan animals contain two important noncellular components: body fluids and extracellular structural elements. In all eumetazoans, body fluids are subdivided into two fluid "compartments:" those that occupy **intracellular space,** within the body's cells, and those that occupy **extracellular space,** outside the cells. In animals with closed vascular systems (such as segmented worms and vertebrates), extracellular fluids are subdivided further into **blood plasma** (the fluid portion of the blood outside the cells; blood cells are really part of the intracellular compartment) and **lymph.** Lymph, also called tissue fluid or interstitial fluid, occupies the space surrounding cells. Many invertebrates have open blood systems,

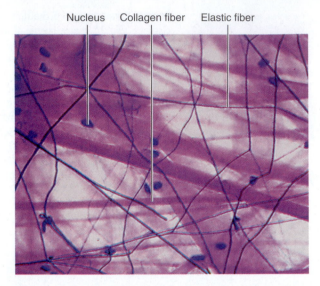

Nucleus   Collagen fiber   Elastic fiber

Loose connective tissue, also called areolar connective tissue, is the "packing material" of the body that anchors blood vessels, nerves, and body organs. It contains fibroblasts that synthesize the fibers and ground substance of connective tissue and wandering macrophages that phagocytize pathogens or damaged cells. The different fiber types include strong collagen fibers (thick and violet in micrograph) and elastic fibers (black and branching in micrograph) formed of the protein elastin. Adipose (fat) tissue is considered a type of loose connective tissue.

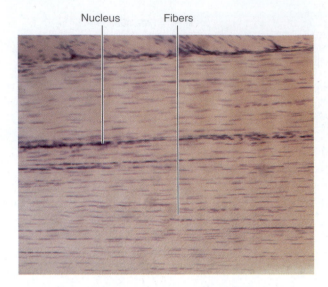

Nucleus   Fibers

Dense connective tissue forms tendon, ligaments, and fasciae (fa'sha), the latter arranged as sheets or bands of tissue surrounding skeletal muscle. In tendon (shown here) the collagenous fibers are extremely long and tightly packed together.

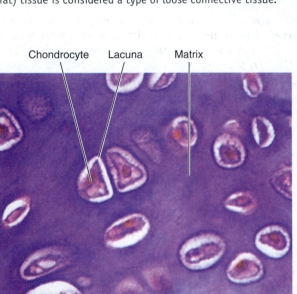

Chondrocyte   Lacuna   Matrix

Cartilage is a vertebrate connective tissue composed of a firm gel ground substance (matrix) containing cells (chondrocytes) living in small pockets called lacunae, and collagen or elastic fibers (depending on the type of cartilage). In hyaline cartilage shown here, both collagen fibers and ground substance are stained uniformly purple, and cannot be distinguished one from the other. Because cartilage lacks a blood supply, all nutrients and waste materials must diffuse through the ground substance from surrounding tissues.

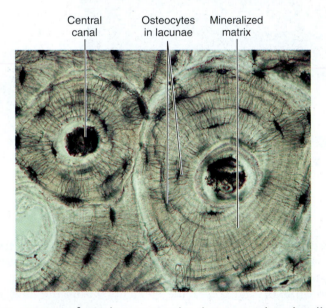

Central canal   Osteocytes in lacunae   Mineralized matrix

Bone, strongest of vertebrate connective tissues, contains mineralized collagen fibers. Small pockets (lacunae) within the matrix contain bone cells, called osteocytes. The osteocytes communicate with blood vessels that penetrate into bone by means of a tiny network of channels called canaliculi. Unlike cartilage, bone undergoes extensive remodeling during an animal's life, and can repair itself following even extensive damage.

figure 3.14

Types of connective tissue

figure 3.15

Types of muscle tissue

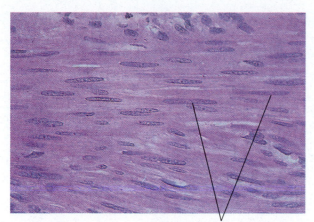

Nuclei of smooth muscle cells

Smooth muscle is nonstriated muscle found in both invertebrates and vertebrates. Smooth muscle cells are long, tapering strands, each containing a single nucleus. Smooth muscle is the most common type of muscle in invertebrates in which it serves as body wall musculature and lines ducts and sphincters. In vertebrates, smooth muscle cells are organized into sheets of muscle circling the walls of the alimentary canal, blood vessels, respiratory passages, and urinary and genital ducts. Smooth muscle is typically slow acting and can maintain prolonged contractions with very little energy expenditure. Its contractions are involuntary and unconscious. The principal functions of smooth muscles are to push the material in a tube, such as the intestine, along its way by active contractions or to regulate the diameter of a tube, such as a blood vessel, by sustained contraction.

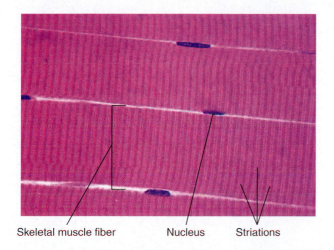

Skeletal muscle fiber   Nucleus   Striations

Skeletal muscle is a type of striated muscle found in both invertebrates and vertebrates. It is composed of extremely long, cylindrical fibers, which are multinucleate cells that may reach from one end of the muscle to the other. Viewed through the light microscope, the cells appear to have a series of stripes, called striations, running across them. Skeletal muscle is called voluntary muscle (in vertebrates) because it contracts when stimulated by nerves under conscious cerebral control.

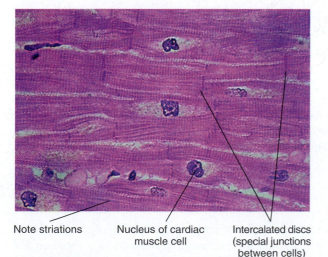

Note striations   Nucleus of cardiac   Intercalated discs
                         muscle cell   (special junctions
                                                     between cells)

Cardiac muscle is another type of striated muscle found only in the vertebrate heart. The cells are much shorter than those of skeletal muscle and have only one nucleus per cell (uninucleate). Cardiac muscle tissue is a branching network of fibers with individual cells interconnected by junctional complexes called intercalated discs. Cardiac muscle is called involuntary muscle because it does not require nerve activity to stimulate contraction. Instead, heart rate is controlled by specialized pacemaker cells located in the heart itself. However, autonomic nerves from the brain may alter pacemaker activity.

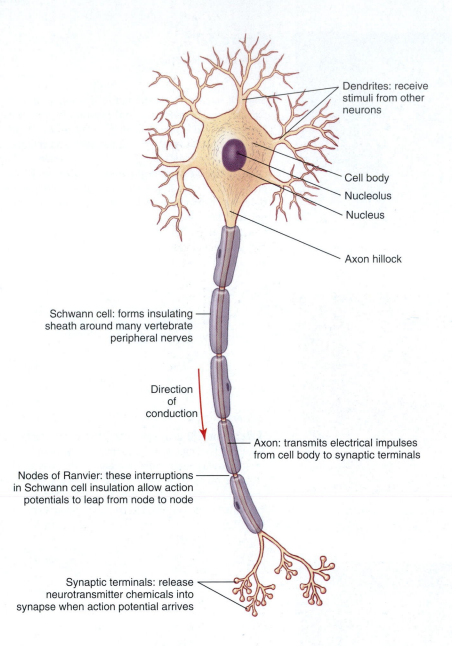

Dendrites: receive
stimuli from other
neurons

Cell body

Nucleolus

Nucleus

Axon hillock

Schwann cell: forms insulating
sheath around many vertebrate
peripheral nerves

Direction
of
conduction

Axon: transmits electrical impulses
from cell body to synaptic terminals

Nodes of Ranvier: these interruptions
in Schwann cell insulation allow action
potentials to leap from node to node

Synaptic terminals: release
neurotransmitter chemicals into
synapse when action potential arrives

figure 3.16

Functional anatomy of a neuron. From the nucleated body, or soma, extend one or more **dendrites** (Gr. *dendron,* tree), which receive electrical impulses from receptors or other nerve cells, and a single **axon** that carries impulses away from the cell body to other nerve cells or to an effector organ. The axon is often called a **nerve fiber.** Nerves are separated from other nerves or from effector organs by specialized junctions called synapses.

however, with no true separation of blood plasma from interstitial fluid.

If we were to remove all specialized cells and body fluids from the interior of the body, we would be left with the third element of the animal body: extracellular structural elements. This is the supportive material of the organism, including loose connective tissue (especially well developed in vertebrates but present in all metazoa), cartilage (molluscs and chordates), bone (vertebrates), and cuticle (arthropods, nematodes, annelids, and others). These elements provide mechanical sta-

bility and protection. In some instances, they act also as a depot of materials for exchange and serve as a medium for extracellular reactions. We will describe the diversity of extracellular skeletal elements characteristic of the different groups of animals in Chapters 10–20.

The term "intercellular," meaning "between cells," should not be confused with the term "intracellular," meaning "within cells."

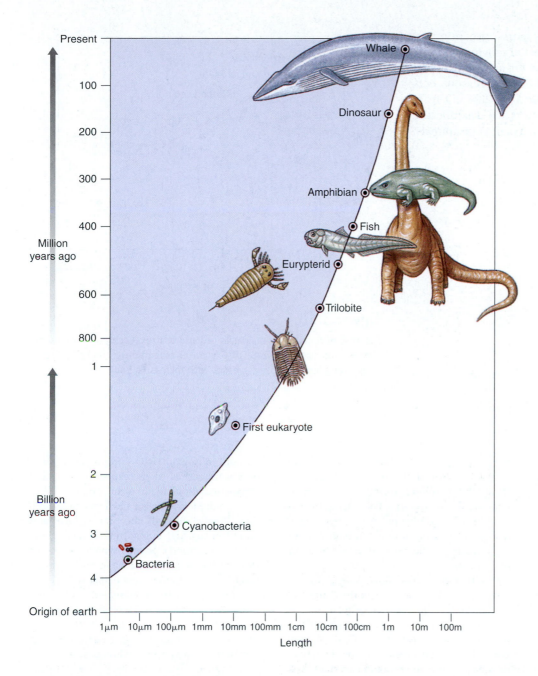

figure 3.17

Graph showing evolution of length increase in the largest organisms present at different periods of life on earth. Note that both scales are logarithmic.

Complexity and Body Size

The most complex grades of metazoan organization permit and to some extent even promote evolution of large body size (figure 3.17). Large size confers several important physical and ecological consequences for an organism. As animals become larger, the body surface increases much more slowly than body volume because surface area increases as the square of body length (length^2), whereas volume (and therefore mass) increases as the cube of body length (length^3). In other words, a large animal will have less surface area relative to its volume than will a small animal of the same shape. The surface area of a large animal may be inadequate for respiration and nutrition by cells located deep within the body. There are two possible solutions to this problem. One solu-

tion is to fold or invaginate the body surface to increase surface area or, as exploited by flatworms, flatten the body into a ribbon or disc so that no internal space is far from the surface. This solution allows the body to become large without internal complexity. However, most large animals adopted a second solution; they developed internal transport systems to shuttle nutrients, gases, and waste products between the cells and the external environment.

Larger size buffers an animal against environmental fluctuations; it provides greater protection against predation and enhances offensive tactics, and it permits a more efficient use of metabolic energy. A large mammal uses more oxygen than a small mammal, but the cost of maintaining its body temperature is less per gram of weight for a large mammal than for a small one. Large animals also can move at less energy cost than can small animals. A large mammal uses more oxygen in running than a small mammal, but energy cost of moving 1 g of its body over a given distance is much less for a large mammal than for a small one (figure 3.18). For all of these reasons, ecological opportunities of larger animals are very different from those of small ones. In subsequent chapters we describe the extensive adaptive radiations observed in taxa of large animals, covered in Chapters 8–20.

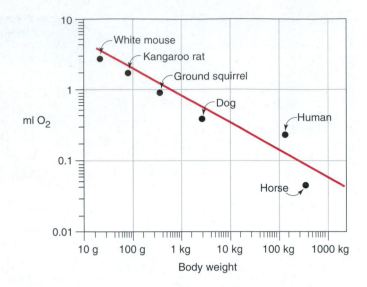

figure 3.18

Net cost of running for mammals of various sizes. Each point represents the cost (measured in rate of oxygen consumption) of moving 1 g of body over 1 km. The cost decreases with increasing body size.

Summary

From the relatively simple organisms that mark the beginnings of life on earth, animal evolution has produced many more intricately organized forms. Adaptive modifications for different lifestyles within a phyletic line of ancestry, however, have been constrained by the ancestral pattern of body architecture.

Every organism has an inherited body plan that may be described in terms of broadly inclusive characteristics, such as symmetry, presence or absence of body cavities, partitioning of body fluids, presence or absence of segmentation, degree of cephalization, and type of nervous system.

Based on several developmental characteristics, bilateral metazoans are divided into two major groups. The Protostomia are characterized by spiral cleavage, mosaic development, and the mouth forming at or near the embryonic blastopore. The Deuterostomia are characterized by radial cleavage, regulative development, and the mouth forming secondarily to the anus and not from the blastopore.

Whereas a unicellular organism performs all life functions within the confines of a single cell, a complex multicellular animal is an organization of subordinate units that are united at successive levels. Cells of metazoans develop into various tissues composed of cells performing common functions. Basic tissue types are nervous, connective, epithe-lial, and muscular. Tissues are organized into larger functional units called organs, and organs are associated to form systems.

In addition to cells, a metazoan body contains fluids, divided into intracellular and extracellular fluid compartments; and extracellular structural elements, which are fibrous or formless elements that serve various structural functions in the extracellular space.

One correlate of increased body complexity in animals is an increase in body size, which offers certain advantages, such as more effective predation, reduced energy cost of locomotion, and improved homeostasis.

Review Questions

1. Name the five levels of organization in animal complexity and explain how each successive level is more complex than the one preceding it.

2. Can you suggest why, during the evolutionary history of animals, there has been a tendency for maximum body size to increase? Do you think it inevitable that complexity should increase along with body size? Why or why not?

3. What is the meaning of the term "parenchyma" as it relates to body organs?

4. Body fluids of eumetazoan animals are separated into fluid "compartments." Name these compartments and explain how compartmentalization may differ in animals with open and closed circulatory systems.

5. What are the four major types of tissues in the body of a metazoan?

6. How would you distinguish simple and stratified epithelium? What characteristic of stratified epithelium might explain why it, rather than simple epithelium, is found lining the oral cavity, esophagus, and vagina?

7. What are the three elements present in all connective tissue? Give some examples of the different types of connective tissue.

8. What are three different kinds of muscle found among animals? Explain how each is specialized for particular functions.
9. Describe the principal structural and functional features of a neuron.
10. Match the animal group with its body plan:

___ Unicellular   a. Nematode
___ Cell aggregate   b. Vertebrate
___ Blind sac, acoelomate   c. Protozoan
     d. Flatworm
___ Tube-within-a-tube, pseudocoelomate   e. Sponge
     f. Arthropod
___ Tube-within-a-tube, eucoelomate   g. Nemertean

11. Distinguish among spherical, radial, biradial, and bilateral symmetry.
12. Use the following terms to identify regions on your body and on the body of a frog: anterior, posterior, dorsal, ventral, lateral, distal, proximal.
13. How would frontal, sagittal, and transverse planes divide your body?

14. What is the difference between radial and spiral cleavage?
15. What are the distinguishing developmental hallmarks of the two major groups of bilateral metazoans, the Protostomia and the Deuterostomia?
16. What is metamerism? Name three phyla showing metamerism.

Selected References

See also general references on p. 415.

Arthur, W. 1997. The origin of animal body plans. Cambridge, United Kingdom, Cambridge University Press. *Explores the genetic, developmental, and population-level processes involved in the evolution of the 35 or so body plans that arose in the geological past.*

Bonner, J. T. 1988. The evolution of complexity by means of natural selection. Princeton, New Jersey, Princeton University Press. *Levels of complexity in organisms and how size affects complexity.*

Grene, M. 1987. Hierarchies in biology. Amer. Sci. **75:**504–510 (Sept.–Oct.). *The term "hierarchy" is used in many different senses in biology. The author points out that current evolutionary theory carries the hierarchical concept beyond the Darwinian restriction to the two levels of gene and organism.*

Kessel, R. G., and R. H. Kardon. 1979. Tissues and organs: a text-atlas of scanning electron microscopy. San Francisco, W. H. Freeman & Company. *Collection of excellent scanning electron micrographs with text.*

McGowan, C. 1999. A practical guide to vertebrate mechanics. New York, Cambridge University Press. *Using many examples from his earlier book,* Diatoms to dinosaurs, 1994, *the author describes principles of biomechanics that underlie functional anatomy. Includes practical experiments and laboratory exercises.*

Radinsky, L. B. 1987. The evolution of vertebrate design. Chicago, University of Chicago Press. *A lucid functional analysis of vertebrate body plans and their evolutionary transformations over time.*

Welsch, U., and V. Storch. 1976. Comparative animal cytology and histology. London, Sidgwick & Jackson. *Comparative histology with good treatment of invertebrates.*

Willmer, P. 1990. Invertebrate relationships: patterns in animal evolution. Cambridge, Cambridge University Press. *Chapter 2 is an excellent discussion of animal symmetry, developmental patterns, origin of body cavities, and segmentation.*

Custom Website

The *Animal Diversity* Online Learning Center is a great place to check your understanding of chapter material. Visit *www.mhhe.com/hickmanad4e* for access to key terms, quizzes, and more! Further enhance your knowledge with web links to chapter-related material.

Explore live links for these topics:

Architectural Pattern and Diversity of Animals
Animal Systems
Cells and Tissues: Histology

Basic Tissue Types
Vertebrate Laboratory Exercises
Principles of Development/Embryology in Vertebrates

Classification and Phylogeny of Animals

Molluscan shells from the collection of Jean Baptiste de Lamarck (1744–1829).

Order in Diversity

Evolution has produced a great diversity of species in the animal kingdom. Zoologists have named more than 1.5 million species of animals, and thousands more are described each year. Some zoologists estimate that species named so far constitute less than 20% of all living animals and less than 1% of all those that have existed.

Despite its magnitude, diversity of animals is not without limits. Many conceivable forms do not exist in nature, as our myths of minotaurs and winged horses show. Animal diversity is not random but has a definite order. Characteristic features of humans and cattle never occur together in a single organism as they do in mythical minotaurs; nor do the characteristic wings of birds and bodies of horses occur together naturally as they do in the mythical horse Pegasus. Humans, cattle, birds, and horses are distinct groups of animals, yet they do share some important features, including vertebrae and homeothermy, that separate them from even more dissimilar forms such as insects and flatworms.

All human cultures classify familiar animals according to patterns in animal diversity. These classifications have many purposes. Some societies classify animals according to their usefulness or destructiveness to human endeavors; others may group animals according to their roles in mythology. Biologists group animals according to their evolutionary relationships as revealed by ordered patterns in their sharing of homologous features. This classification is called a "natural system" because it reflects relationships that exist among animals in nature, outside the context of human activity. A systematic zoologist has three major goals: to discover all species of animals, to reconstruct their evolutionary relationships, and to classify them accordingly.

Darwin's theory of common descent (see Chapter 1) is the underlying principle that guides our search for order in diversity of animal life. Our science of taxonomy ("arrangement law") produces a formal system for naming and classifying species that communicates this order. Animals that have very recent common ancestry share many features in common and are grouped most closely in our taxonomic classification; dissimilar animals that share only very ancient common ancestry are placed in different taxonomic groups except at the "highest" or most inclusive levels of **taxonomy.** Taxonomy is part of a broader science of **systematics,** or comparative biology, in which studies of variation among animal populations are used to understand their evolutionary relationships. Taxonomy predates evolutionary biology, however, and many taxonomic practices are relics of pre-evolutionary world views. Adjusting our taxonomic system to accommodate evolution has produced many problems and controversies. Taxonomy has reached an unusually active and controversial point in its development with several alternative taxonomic systems competing for use. To understand this controversy, we need to review the history of animal taxonomy.

figure 4.1

Carolus Linnaeus (1707–1778). This portrait was made of Linnaeus at age 68, three years before his death.

Linnaeus and Classification

The Greek philosopher and biologist Aristotle was the first to classify organisms based on their structural similarities. Following the Renaissance in Europe, the English naturalist John Ray (1627–1705) introduced a more comprehensive system of classification and a new concept of species. Rapid growth of systematics in the eighteenth century culminated in the work of Carolus Linnaeus (1707–1778; figure 4.1), who produced our current scheme of classification.

Linnaeus was a Swedish botanist at the University of Uppsala. He had a great talent for collecting and classifying organisms, especially flowering plants. Linnaeus produced an extensive system of classification for both plants and animals. This scheme, published in his great work, *Systema Naturae,* used **morphology** (the comparative study of organismal form) for arranging specimens in collections. He divided the animal kingdom into species and gave each one a distinctive name. He grouped species into genera, genera into orders, and orders into classes. Because his knowledge of animals was limited, his lower categories, such as genera, were very broad and included animals that are only distantly related. Much of his classification has been drastically altered, but his basic principles are still followed.

Linnaeus's scheme of arranging organisms into an ascending series of groups of increasing inclusiveness is a **hierarchical system** of classification. The major categories, or **taxa** (sing., **taxon**), into which organisms are grouped are given one of several standard taxonomic ranks to indicate the general inclusiveness of each group. The hierarchy of taxonomic ranks has been expanded considerably since Linnaeus's time (table 4.1). It now includes seven mandatory ranks for the animal kingdom, in descending series: kingdom, phylum, class, order, family, genus, and species. All organisms being classified must be placed into at least seven taxa, one at each of these mandatory ranks. Taxonomists have the option of subdividing these seven ranks even further to recognize more than seven taxa (superclass, subclass, infraclass, superorder, suborder, and others) for any particular group of organisms. More than 30 taxonomic ranks now are recognized. For very large and complex groups, such as fishes and insects, these additional ranks are needed to express different degrees of evolutionary divergence. Unfortunately, they also make taxonomy more complex.

Linnaeus's system for naming species is known as **binomial nomenclature.** Each species has a Latinized name composed of two words (hence binomial) written in italics (underlined if handwritten or typed). The first word is the name of the **genus,** written with a capital initial letter; the second word is the **species epithet,** which is peculiar to the species within the genus and is written with a small initial letter (table 4.1). The name of a genus is always a noun, and the species epithet is usually an adjective that must agree in gender with the genus. For instance, the scientific name of a common robin is *Turdus migratorius* (L. *turdus,* thrush; *migratorius,* of the migratory habit). A species epithet never stands alone; the complete binomial must be used to name a species. Names of genera must refer only to single groups of organisms; a single name cannot be given to two different genera of animals. The same species epithet may be used in different genera, however, to denote different and unrelated species. For example, the scientific name of a white-breasted nuthatch is *Sitta carolinensis.* The species epithet *"carolinensis"* is used in other genera, including *Poecile carolinensis* (Carolina chickadee) and *Anolis carolinensis* (green anole, a lizard) to mean "of Carolina." All ranks above species are designated using uninomial nouns, written with a capital initial letter.

Table 4.1   Examples of Taxonomic Categories to Which Representative Animals Belong

	Human	Gorilla	Southern Leopard Frog	Bush Katydid
Kingdom	Animalia	Animalia	Animalia	Animalia
Phylum	Chordata	Chordata	Chordata	Arthropoda
Subphylum	Vertebrata	Vertebrata	Vertebrata	Uniramia
Class	Mammalia	Mammalia	Amphibia	Insecta
Subclass	Eutheria	Eutheria	—	Pterygota
Order	Primates	Primates	Anura	Orthoptera
Suborder	Anthropoidea	Anthropoidea	—	Ensifera
Family	Hominidae	Hominidae	Ranidae	Tettigoniidae
Subfamily	—	—	Raninae	Phaneropterinae
Genus	*Homo*	*Gorilla*	*Rana*	*Scudderia*
Species	*Homo sapiens*	*Gorilla gorilla*	*Rana sphenocephala*	*Scudderia furcata*
Subspecies	—	—	—	*Scudderia furcata furcata*

Hierarchical classification applied to four species (human, gorilla, Southern leopard frog, and bush katydid). Higher taxa generally are more inclusive than lower-level taxa, although taxa at two different levels may be equivalent in content. Closely related species are united at a lower point in the hierarchy than are distantly related species. For example, humans and gorillas are united at the family (Hominidae) and above; they are united with Southern leopard frogs at the subphylum level (Vertebrata) and with bush katydids at the kingdom (Animalia) level. Mandatory Linnean ranks are shown in bold type.

The person who first describes a type specimen and publishes the name of a species is called the authority. This person's name and date of publication often appear after a species name. Thus, *Didelphis marsupialis* Linnaeus, 1758, tells us that Linnaeus was the first person to publish the species name of opossums. The authority citation is not part of the scientific name but rather is an abbreviated bibliographical reference. Sometimes, generic status of a species is revised following its initial description. In this case, the authority's name is presented in parentheses.

Species

While discussing Darwin's book, *On the Origin of Species,* in 1859, Thomas Henry Huxley (p. 11) asked, "In the first place, what is a species? The question is a simple one, but the right answer to it is hard to find, even if we appeal to those who should know most about it." We have used the term "species" so far as if it had a simple and unambiguous meaning. Actually, Huxley's commentary is as valid today as it was over 140 years ago. Our concepts of species have become more sophisticated, but the diversity of different concepts and disagreements surrounding their use are as evident now as in Darwin's time.

Criteria for Recognition of Species

Despite widespread disagreement about the nature of species, biologists repeatedly have used certain criteria for recognizing species. First, **common descent** is central to nearly all modern concepts of species. Members of a species must trace their ancestry to a common ancestral population although not necessarily to a single pair of parents. Species are thus historical entities. A second criterion is that species must be the *smallest distinct groupings* of organisms sharing patterns of ancestry and descent; otherwise, it would be difficult to separate species from higher taxa whose members also share common descent. Morphological characters traditionally have been important in identifying such groupings, but chromosomal and molecular characters are now extensively used for this purpose. A third important criterion is that of *reproductive community,* which pertains to sexually reproducing organisms; members of a species must form a reproductive community that excludes members of other species. This criterion is very important to many modern concepts of species. For organisms whose reproduction is strictly **asexual,** reproductive community entails occupation of a particular ecological niche in a particular place so that a reproducing population responds as a unit to evolutionary processes such as natural selection and genetic drift (p. 36).

Any species has a distribution through space, called its *geographic range,* and a distribution through time, called its *evolutionary duration.* Species differ greatly from each other in both of these dimensions. When we compare a local population to similar but not identical populations located hundreds of miles away, we often have difficulty deciding whether these populations are parts of a single species or are different species.

Throughout the evolutionary duration of a species, its geographic range might change many times. A geographic range could be either continuous or disjunct, having breaks within it where the species is absent. Suppose that we find two similar but not identical populations living 300 miles apart with no closely related populations between them. Are we observing a single species with a disjunct distribution, or two different but closely related species? Suppose that these populations have been separated historically for 50,000 years. Is this enough time for them to have evolved separate reproductive communities, or do they form parts of a single reproductive community? Species concepts differ in how they attempt to answer these questions.

Concepts of Species

Before Darwin, a species was considered a distinct and immutable entity. The concept that species are defined by fixed, essential features (usually morphological) is called the **typological species concept.** This concept was discarded following establishment of Darwinian evolutionary theory.

The most influential concept of species inspired by Darwinian evolutionary theory is the **biological species concept** formulated by Theodosius Dobzhansky and Ernst Mayr. In 1983, Mayr stated the biological species concept as follows: *"A species is a reproductive community of populations (reproductively isolated from others) that occupies a specific niche in nature."* Note that a species is identified here according to reproductive properties of populations, groups of related organisms inhabiting a particular geographic area, not according to organismal morphology. A species is an *interbreeding* population of individuals having *common descent.* By adding the criterion of **niche,** an ecological concept denoting an organism's role in its ecological community, we recognize that members of a reproductive community constitute an ecological entity in nature. Because reproductive community should maintain genetic cohesiveness, organismal variation should be relatively smooth and continuous within species and discontinuous between them. Although a biological species is based on reproductive properties of populations rather than organismal morphology, morphology nonetheless can help us to diagnose biological species.

The biological species concept has been strongly criticized for several reasons. First, it refers to contemporary populations but ignores the species status of ancestral populations. Second, according to the biological species concept, species do not exist in groups of organisms that reproduce only asexually. It is common taxonomic practice, however, to describe species in all groups of organisms. A third problem is that systematists using the biological species concept often disagree on the amount of reproductive divergence necessary for considering two populations separate species. Reproductive barriers between populations range from strong to weak, and some barriers, such as differences in the preferred habitat or time of mating, may be eliminated by natural selection.

The **evolutionary species concept** was proposed by Simpson (see figure 4.2) in the 1940s to add an evolutionary time dimension to the biological species concept. This concept persists in a modified form today. A current definition of the evolutionary species is *a single lineage of ancestor-descendant populations that maintains its identity from other such lineages and that has its own evolutionary ten-*

figure 4.2

George Gaylord Simpson (1902–1984) formulated the evolutionary species concept and principles of evolutionary taxonomy.

dencies and historical fate. Note that the criterion of common descent is retained here in the need for a species to have a distinct historical identity. Unlike the biological species concept, the evolutionary species concept applies both to sexually and asexually reproducing forms. As long as continuity of diagnostic features is maintained by an evolving lineage, it will be recognized as a single species. Abrupt changes in diagnostic features will mark a boundary between different species in evolutionary time.

The last concept that we present is the **phylogenetic species concept.** The phylogenetic species concept is defined as an *irreducible (basal) grouping of organisms diagnosably distinct from other such groupings and within which there is a parental pattern of ancestry and descent.* This concept also emphasizes common descent, and both asexual and sexual groups are covered. Any population that has become separated from others and has undergone character evolution that distinguishes it will be recognized as a species. The criterion of irreducibility requires that no more than one such population can be placed in a single species. The main difference in practice between the evolutionary and phylogenetic species concepts is that the latter emphasizes recognizing as species the smallest groupings of organisms that have undergone independent evolutionary change. The evolutionary species concept would group into a single species geographically disjunct populations that demonstrate some genetic divergence but are judged similar in their major "evolutionary tendencies," whereas the phylogenetic species concept would treat them as separate species. In general, a larger number of species would be described using the phylogenetic species concept than any other concept. The phylogenetic species concept is intended to encourage us to reconstruct patterns of evolutionary common descent on the finest scale possible.

Current disagreements concerning concepts of species should not be considered trivial or discouraging. Whenever a field of scientific investigation enters a phase of dynamic growth, old concepts are reevaluated and either refined or replaced with newer, more progressive ones. Active debate among systematists shows that this field has acquired unprecedented activity and importance in biology. Just as Thomas Henry Huxley's time was one of enormous advances in biology, so is the present time. Both times are marked by fundamental reconsiderations of the meaning of species. We cannot predict yet which, if any, of these concepts of species will prevail. Understanding conflicting perspectives, rather than learning a single concept of species, is therefore most important for students of zoology.

Some species are divided into subspecies, in which case a trinomial nomenclature is employed (see katydid example, table 4.1 and salamander example, figure 1.19); such species are called polytypic. Generic, specific, and subspecific names are printed in italics (underlined if handwritten or typed). A polytypic species contains one subspecies whose subspecific name is a repetition of the species epithet and one or more additional subspecies whose names differ. Thus, to distinguish geographic variants of *Ensatina eschscholtzii*, one subspecies is named *Ensatina eschscholtzii eschscholtzii*, and different subspecies names are used for each of the six other subspecies (figure 1.19). Both the genus name and the species epithet may be abbreviated as shown in figure 1.19. Formal recognition of subspecies has lost popularity among taxonomists because boundaries between subspecies rarely are distinct. Recognition of subspecies usually is based on one or a few superficial characters that do not always diagnose an evolutionarily distinct unit. Subspecies, therefore, should be viewed as tentative statements indicating that the species status of the populations needs further investigation.

Taxonomic Characters and Reconstruction of Phylogeny

A major goal of systematics is to infer an evolutionary tree or **phylogeny** that relates all extant and extinct species. This tree is constructed by studying organismal features, formally called **characters,** that vary among species. A character is any feature that a taxonomist uses to study variation within or among species. We find potentially useful taxonomic characters in morphological, chromosomal, and molecular features. Taxonomists find characters by observing patterns of similarity among organisms. If two organisms possess similar features, they may have inherited these features from an equivalent one in a common ancestor. Character similarity that results from common ancestry is called **homology** (see p. 17). The character of feathers in birds illustrates homology; all birds have feathers that were inherited with various modifications from those of this group's most recent common ancestor. Similarity does not always reflect common ancestry, however. Independent evolutionary origins of similar features on different lineages produce patterns of similarity among organisms that do not reflect common descent; this occurrence complicates the work of taxonomists. Character similarity that misrepresents common descent is called nonhomologous similarity or **homoplasy.** Convergent evolution (p. 19) of image-forming eyes by cephalopod molluscs and by vertebrates illustrates homoplasy; the most

recent common ancestor of molluscs and vertebrates did not have such eyes.

Using Character Variation to Reconstruct Phylogeny

To reconstruct phylogeny of a group using characters that vary among its members, the first step is to determine which variant form of each character was present in the most recent common ancestor of the entire group. This form is called the **ancestral character state** for the group as a whole. We presume that all other variant forms of the character arose later within the group, and these forms are called evolutionarily **derived character states. Polarity** of a character refers to ancestral/descendant relationships among its different states. For example, if we consider as a character dentition of amniotic vertebrates (reptiles, birds, and mammals), presence versus absence of teeth constitute two different character states. Teeth are absent from birds but present in other **amniotes.** To evaluate polarity of this character, we must determine which character state, presence or absence of teeth, occurred in the most recent common ancestor of amniotes and which state was derived subsequently within amniotes.

The method used to examine polarity of a variable character is called **outgroup comparison.** We begin by selecting an additional group of organisms, called an **outgroup,** that is phylogenetically close but not within the group being studied. Amphibians and different groups of bony fishes constitute appropriate outgroups to amniotes for polarizing variation in dentition of amniotes. Next, we infer that any character state found both within the group being studied and in an outgroup is ancestral for the study group. Teeth are usually present in amphibians and bony fishes; therefore, we infer that presence of teeth is ancestral for amniotes and absence of teeth is derived. Polarity of this character indicates that teeth were lost in an ancestral lineage of all modern birds. Polarity of characters is evaluated most effectively when several different outgroups are used. All character states found in a study group that are absent from appropriate outgroups are considered derived (see figure 4.3 for additional examples).

Organisms or species that share derived character states form subsets within the study group; these subsets are called **clades** (Gr. *klados,* branch). A derived character shared by members of a clade is formally called a **synapomorphy** (Gr. *synapsis,* joining together, + *morphē,* form) of that clade. Taxonomists use synapomorphies as evidence of homology to infer that a particular group of organisms forms a clade. Within amniotes, absence of teeth and presence of feathers are synapomorphies that identify birds as a clade. A clade corresponds to a unit of evolutionary common descent; it includes the most recent common ancestor of a group and all of that ancestor's descendants. The pattern formed by the sharing of derived states of all characters within our study group will take the form of a **nested hierarchy** of clades within clades. The goal is to identify all clades nested within the study group, which would reveal patterns of common descent among all species in the group. Each clade may be formally recognized as a taxon and perhaps given a rank in the Linnean hierarchy.

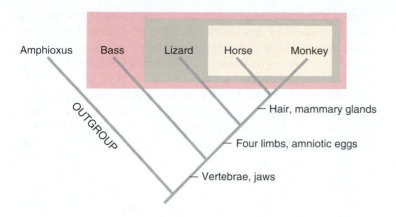

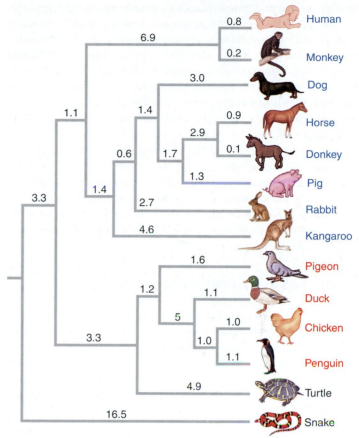

figure 4.3

A cladogram as a nested hierarchy of taxa. *Amphioxus* (p. 300) is the outgroup, and the study group comprises four vertebrates (bass, lizard, horse, and monkey). Four characters that vary among vertebrates are used to generate a simple cladogram: presence versus absence of four legs, amniotic eggs, hair, and mammary glands. For all four characters, absence is considered the ancestral state in vertebrates because this condition occurs in the outgroup, *Amphioxus*; for each character, presence is derived within vertebrates. Because they share presence of four limbs and amniotic eggs as synapomorphies, lizards, horses, and monkeys form a clade relative to bass. This clade is subdivided further by two synapomorphies (presence of hair and mammary glands) that unite horses and monkeys relative to lizards. We know from comparisons involving even more distantly related animals that presence of vertebrae and jaws constitute synapomorphies of vertebrates and that *Amphioxus*, which lacks these features, falls outside the vertebrate clade.

A nested hierarchy of clades is presented as a branching diagram called a **cladogram** (figure 4.3; see also figure 1.17). Taxonomists often make a technical distinction between a cladogram and a **phylogenetic tree.** The branches on a cladogram are only a formal device for indicating a nested hierarchy of clades within clades. The same nested hierarchy is sometimes presented as an indented list of taxon names; for example, the cladogram for major taxa of living gnathostome vertebrates shown in figure 16.2 (p. 313) could be represented:

> Chondrichthyes
>> Holocephali
>> Elasmobranchii
>
> Osteichthyes
>> Actinopterygii
>> Sarcopterygii
>>> Dipnoi (lobe-finned fishes)
>>> Tetrapoda

A cladogram is not strictly equivalent to a phylogenetic tree, whose branches represent real lineages that occurred in the evolutionary past. To obtain a phylogenetic tree, we must add to a cladogram information concerning ancestors, durations of evolutionary lineages, or amounts of evolutionary change that occurred on lineages. Because the branching order of a cladogram matches that of the corresponding phylogenetic tree, however, a cladogram often serves as a first approximation of the structure of the corresponding phylogenetic tree.

figure 4.4

An early phylogenetic tree of representative amniotes based on inferred base substitutions in the gene that encodes the respiratory protein, cytochrome *c*. Numbers on branches are estimated numbers of mutational changes that occurred in this gene along different evolutionary lineages. Publication of this tree by Fitch and Margoliash in 1967 was influential in convincing systematists that molecular sequences contain phylogenetic information. Subsequent work confirms some hypotheses, including monophyly of mammals (blue) and birds (red) while rejecting others; kangaroo, for example, should be outside a branch containing all other mammals sampled.

Sources of Phylogenetic Information

We find characters used to construct cladograms in comparative morphology (including embryology), comparative cytology, and comparative biochemistry. **Comparative morphology** examines varying shapes and sizes of organismal structures, including their developmental origins. As we discuss in later chapters, variable structures of skull bones, limb bones, and integument (scales, hair, feathers) are particularly important for reconstructing the phylogeny of vertebrates. Comparative morphology uses specimens obtained from both living organisms and fossilized remains. **Comparative biochemistry** uses the sequences of amino acids in **proteins** and the sequences of nucleotides in nucleic acids to identify variable characters for constructing a cladogram or phylogenetic tree (figure 4.4).

Recent work has shown that some fossils retain enough DNA for comparative biochemical studies. **Comparative cytology** uses variation in numbers, shapes, and sizes of chromosomes and their parts to obtain variable characters for constructing cladograms. Comparative cytology is used almost exclusively on living rather than fossilized organisms because chromosomal structure is not well preserved in fossils.

To add the evolutionary timescale needed for producing a phylogenetic tree, we must consult the fossil record. We look for the earliest appearance of derived morphological characters in fossils to estimate ages of clades distinguished by those characters. The ages of fossils showing derived characters of a particular clade are determined by radioactive dating (p. 15) to estimate the age of the clade. A lineage representing the most recent common ancestor of all taxa in the clade is then added to the phylogenetic tree.

Theories of Taxonomy

A theory of taxonomy establishes principles that we use to recognize and to rank taxonomic groups. There are two currently popular theories of taxonomy: (1) traditional evolutionary taxonomy and (2) phylogenetic systematics (cladistics). Both are based on evolutionary principles. We will see, however, that these two theories differ on how evolutionary principles are used. These differences have important implications for how we use a taxonomy to study evolutionary processes.

The relationship between a taxonomic group and a phylogenetic tree or cladogram is important for both theories. This relationship can take one of three forms: **monophyly, paraphyly,** or **polyphyly** (figure 4.5). A taxon is monophyletic if it includes the most recent common ancestor of all members of a group and all descendants of that ancestor (figure 4.5A). A taxon is paraphyletic if it includes the most recent

common ancestor of all members of a group and some but not all descendants of that ancestor (figure 4.5B). A taxon is polyphyletic if it does not include the most recent common ancestor of all members of a group; this situation requires the group to have had at least two separate evolutionary origins, usually requiring independent evolutionary acquisition of a diagnostic feature (figure 4.5C). For example, if birds and mammals were grouped in a taxon called Homeothermia, we would have a polyphyletic taxon because birds and mammals descend from two quite separate amniotic lineages that have evolved homeothermy independently. The most recent common ancestor of birds and mammals is not homeothermic and does not occur in the polyphyletic Homeothermia just described. Both evolutionary and cladistic taxonomies accept monophyletic groups and reject polyphyletic groups in their classifications. They differ regarding acceptance of paraphyletic groups.

Traditional Evolutionary Taxonomy

Traditional **evolutionary taxonomy** incorporates two different evolutionary principles for recognizing and ranking higher taxa: (1) common descent and (2) amount of adaptive evolutionary change, as shown on a phylogenetic tree. Evolutionary taxa must have a single evolutionary origin, and must show unique adaptive features.

The mammalian paleontologist George Gaylord Simpson (figure 4.2) was highly influential in developing and formalizing principles of evolutionary taxonomy. According to Simpson, a particular branch on an evolutionary tree is given the status of a higher taxon if it represents a distinct **adaptive zone.** Simpson describes an adaptive zone as "a characteristic reaction and mutual relationship between environment and organism, a way of life and not a place where life is led." By entering a new adaptive zone through a fundamental change

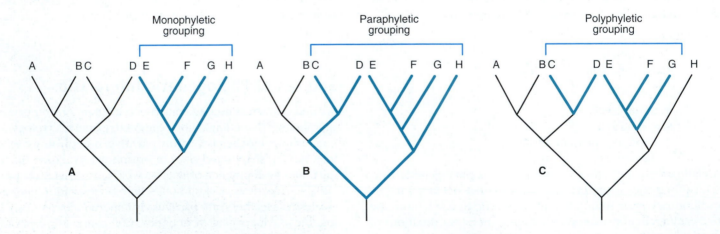

figure 4.5

Relationships between phylogeny and taxonomic groups illustrated for a hypothetical phylogeny of eight species (A through H). **A,** *Monophyly*—a monophyletic group contains the most recent common ancestor of all members of the group and all of its descendants. **B,** *Paraphyly*—a paraphyletic group contains the most recent common ancestor of all members of the group and some but not all of its descendants. **C,** *Polyphyly*—a polyphyletic group does not contain the most recent common ancestor of all members of the group, thereby requiring the group to have at least two separate phylogenetic origins.

in organismal structure and behavior, an evolving population can use environmental resources in a completely new way.

A taxon forming a distinct adaptive zone is termed a **grade.** Simpson gives the example of penguins as a distinct adaptive zone within birds. The lineage immediately ancestral to all penguins underwent fundamental changes in form of the body and wings to permit a switch from aerial to aquatic locomotion (figure 4.6). Aquatic birds that can fly both in air and underwater are somewhat intermediate in habitat, morphology, and behavior between aerial and aquatic adaptive zones. Nonetheless, obvious modifications of a penguin's wings and body for swimming represent a new grade of organization. Penguins are therefore recognized as a distinct taxon within

birds, family Spheniscidae. The Linnean rank of a taxon depends upon breadth of its adaptive zone: the broader the adaptive zone when fully realized by a group of organisms, the higher the rank that the corresponding taxon is given.

Evolutionary taxa may be either monophyletic or paraphyletic. Recognition of paraphyletic taxa requires, however, that taxonomies distort patterns of common descent. An evolutionary taxonomy of **anthropoid** primates provides a good example (figure 4.7). This taxonomy places humans (genus *Homo*) and their immediate fossil ancestors in family Hominidae and places chimpanzees (genus *Pan*), gorillas (genus *Gorilla*), and orangutans (genus *Pongo*) in family Pongidae. However, pongid genera *Pan* and *Gorilla* share more recent common ancestry with

A   **B**

figure 4.6

A, Penguin. **B,** Diving petrel. Penguins (avian family Spheniscidae) were considered by George G. Simpson a distinct adaptive zone within birds because of their adaptations for submarine flight. Simpson believed that the adaptive zone ancestral to penguins resembled that of diving petrels, which display adaptations for combined aerial and aquatic flight. Adaptive zones of penguins and diving petrels are distinct enough to be recognized taxonomically as different families within a common order (Ciconiiformes).

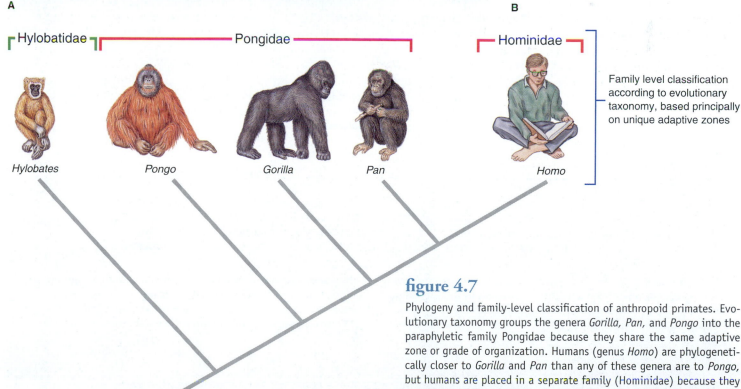

Family level classification according to evolutionary taxonomy, based principally on unique adaptive zones

figure 4.7

Phylogeny and family-level classification of anthropoid primates. Evolutionary taxonomy groups the genera *Gorilla, Pan,* and *Pongo* into the paraphyletic family Pongidae because they share the same adaptive zone or grade of organization. Humans (genus *Homo*) are phylogenetically closer to *Gorilla* and *Pan* than any of these genera are to *Pongo,* but humans are placed in a separate family (Hominidae) because they represent a different grade of organization. Cladistic taxonomy eliminates the paraphyletic family Pongidae and groups *Pongo, Gorilla, Pan,* and *Homo* into a single monophyletic family Hominidae. Gibbons (genus *Hylobates*) form the monophyletic family Hylobatidae, which is compatible with both evolutionary and cladistic classifications.

Hominidae than they do with the remaining pongid genus, *Pongo.* Family Pongidae is therefore paraphyletic because it does not include humans, who also descend from its most recent common ancestor (figure 4.7). Evolutionary taxonomists nonetheless recognize pongid genera as a single, family-level grade of **arboreal,** herbivorous primates having limited mental capacity; in other words, they show a family-level adaptive zone. Humans are terrestrial, omnivorous primates who possess greatly expanded mental and cultural attributes, thereby comprising a distinct adaptive zone at the taxonomic level of a family.

Traditional evolutionary taxonomy has been challenged from two opposite directions. One challenge states that because

phylogenetic trees can be very difficult to obtain, it is impractical to base our taxonomic system on common descent and adaptive evolution. We are told that our taxonomy should represent a more easily measured feature, overall similarity of organisms evaluated without regard to phylogeny. This principle is known as **phenetic taxonomy.** Phenetic taxonomy did not have a strong impact on animal classification, and scientific interest in this approach is in decline.

Despite difficulties of reconstructing phylogeny, zoologists still consider this endeavor a central goal of their systematic work, and they are unwilling to compromise this goal for methodological simplicity.

Phylogenies from DNA Sequences

A simple example illustrates cladistic analysis of DNA sequence data to examine phylogenetic relationships among species. The study group in this example contains three species of chameleons, two from the island of Madagascar (*Brookesia theili* and *B. brygooi*) and one from Equatorial Guinea (*Chamaeleo feae*). The outgroup is a lizard of the genus *Uromastyx,* which is a distant relative of chameleons. Do the molecular data in this example confirm or reject the prior taxonomic hypothesis that the two Madagascan chameleons are more closely related to each other than either one is to the Equatorial Guinean species?

The molecular information in this example comes from a piece of the mitochondrial DNA sequence (57 bases) for each species. Each sequence encodes amino acids 221–239 of a protein called "NADH dehydrogenase subunit 2" in the species from which it was obtained. These DNA base sequences are aligned and numbered as:

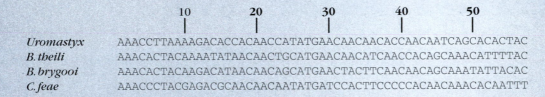

Each column in the aligned sequences constitutes a character that takes one of four states: A, C, G, or T (a fifth possible state, absence of the base, is not observed in this example). Only characters that vary among the three chameleon species potentially contain information on which pair of species is most closely related. Twenty-three of the 57 aligned bases show variation *among chameleons,* as shown here in bold letters:

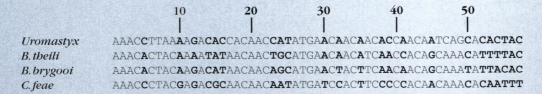

To be useful for constructing a cladogram, a character must demonstrate sharing of derived states (= synapomorphy). Which of these 23 characters demonstrate synapomorphies for chameleons? For each of the 23 variable characters, we must ask whether one of the states observed in chameleons is shared with the outgroup, *Uromastyx.* If so, this state is judged ancestral for chameleons and the alternative state(s) derived. Derived states are identified for 21 of the 23 characters just identified; derived states are shown here in blue:

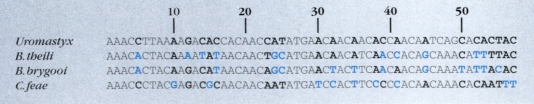

Note that polarity is ambiguous for two variable characters (at positions 23 and 54) whose alternative states in chameleons are not observed in the outgroup.

Of the characters showing derived states, 10 of them show synapomorphies among chameleons. These characters are marked here with numbers 1, 2, or 3 in the appropriate column.

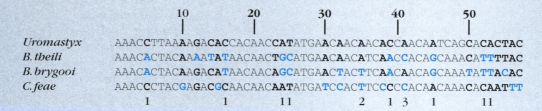

	10	20	30	40	50
Uromastyx	AAACCTTAAA	AAGACACCAC	AACCATATGA	ACAACAACAC	CAACAATCAGCACACTAC
B. theili	AAACACTACA	AAATATAACA	ACTGCATGAA	CAACATCAACC	ACAGCAAACATTTTAC
B. brygooi	AAACACTACA	AGACATAACA	ACAGCATGAA	CTACTTCAAC	AACAGCAAATATTACAC
C. feae	AAACCCTACG	AGACGCAACA	ACAATATGAT	CCACTTCCCCC	ACAACAAACACAATTT
	1	1	11	2 13	1 11

The eight characters marked "1" show synapomorphies grouping the two Madagascan species (*Brookesia theili* and *B. brygooi*) to the exclusion of the Equatorial Guinean species, *Chamaeleo feae*. We can represent these relationships as a cladogram.

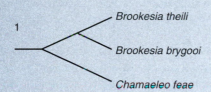

We can explain evolution of all characters favoring this cladogram by placing a single mutational change on the branch ancestral to the two *Brookesia* species. This is the simplest explanation for evolutionary changes of these characters.

Characters marked "2" and "3" disagree with our cladogram and favor alternative relationships:

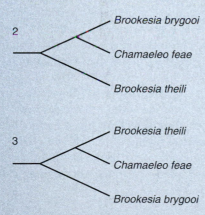

To explain evolutionary changes in characters favoring cladograms 2 or 3 using cladogram 1, we need at least two changes per character. Likewise, if we try to explain evolution of characters favoring cladogram 1 on cladogram 2 or 3, we need at least two changes for each of these characters. These two diagrams show the minimum numbers of changes required for character 5 (which favors cladogram 1) and character 41 (which favors cladogram 3) on cladogram 1; the ancestral state of each character is shown at the root of the tree and the nucleotide states observed in each species at the tips of the branches:

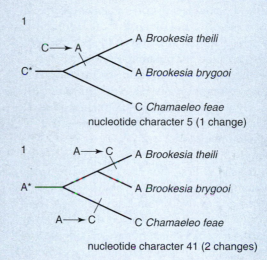

nucleotide character 5 (1 change)

nucleotide character 41 (2 changes)

Systematists often use a principle called **parsimony** to resolve conflicts among taxonomic characters, as seen here. We choose as our best working hypothesis the cladogram that requires the smallest total amount of character change. In our example, cladogram 1 is favored by parsimony. For all 10 phylogenetically informative characters, cladogram 1 requires a total of 12 changes of character state (one for each of the 8 characters favoring it and two for each of the other 2 characters); cladograms 2 and 3 each require at least 19 character-state changes, 7 steps longer than cladogram 1. By choosing cladogram 1, we claim that characters favoring cladograms 2 and 3 show homoplasy in their evolution.

The molecular sequences shown in this example therefore confirm predictions of the prior hypothesis, based on appearance and geography of these chameleons, that the *Brookesia* species shared a common ancestor with each other more recently than either one did with *Chamaeleo feae*.

As a further exercise, you should convince yourself that the 12 characters that vary among chameleons but which do not demonstrate unambiguous sharing of derived states are equally compatible with each of the three possible cladograms shown. For each character, find the minimum total number of changes that must occur to explain its evolution on each cladogram. You will see, if you do this exercise correctly, that the three cladograms do not differ in minimum numbers of changes required for each of these characters. For this reason, the characters are phylogenetically uninformative by the parsimony criterion.

* Ancestral nucleotide state inferred by outgroup comparison.

Data from Townsend, T., and A. Larson. 2002. Molecular phylogenetics and mitochondrial genomic evolution in the Chamaeleonidae (Reptilia, Squamate). Molecular Phylogenetics and Evolution **23**:22–36.

Phylogenetic Systematics/Cladistics

A second and stronger challenge to evolutionary taxonomy is one known as **phylogenetic systematics** or **cladistics.** As the first name implies, this approach emphasizes the criterion of common descent and, as the second name implies, it is based on the cladogram of a group being classified. This approach to taxonomy was first proposed in 1950 by German entomologist Willi Hennig (figure 4.8) and therefore is sometimes called "Hennigian systematics." All taxa recognized by Hennig's cladistic system must be monophyletic. We saw previously how evolutionary taxonomists' recognition of primate families Hominidae and Pongidae distorts genealogical relationships to emphasize adaptive uniqueness of the Hominidae. Because the most recent common ancestor of the paraphyletic family Pongidae is also an ancestor of family Hominidae, recognition of Pongidae is incompatible with cladistic taxonomy. To avoid paraphyly, cladistic taxonomists have discontinued use of the traditional family Pongidae, placing chimpanzees, gorillas, and orangutans with humans in the family Hominidae (as shown for *Gorilla gorilla* in table 4.1).

The disagreement regarding validity of paraphyletic groups may seem trivial at first, but its important consequences become clear when we discuss evolution. For example, claims that amphibians evolved from bony fish, that birds evolved from reptiles, or that humans evolved from apes may be made by an evolutionary taxonomist but are meaningless to a cladist. We imply by these statements that a descendant group (amphibians, birds, or humans) evolved from part of an ancestral group (bony fish, reptiles, and apes, respectively) to which the descendant does not belong. This usage automatically makes the ancestral group paraphyletic, and indeed bony fish, reptiles, and apes as traditionally recognized are paraphyletic groups. How are such paraphyletic groups recognized? Do they share distinguishing features that are not shared by a descendant group?

Paraphyletic groups are usually defined in a negative manner. They are distinguished only by absence of features found in a particular descendant group, because any traits shared from their common ancestry are present also in the excluded descendants (unless secondarily lost). For example, apes are those "higher" primates that are not humans. Likewise, fish are those vertebrates that lack the distinguishing characteristics of tetrapods (amphibians and amniotes). What does it mean then to say that humans evolved from apes? To an evolutionary taxonomist, apes and humans are different adaptive zones or grades of organization; to say that humans evolved from apes states that bipedal, tailless organisms of large brain capacity evolved from arboreal, tailed organisms of smaller brain capacity. To a cladist, however, the statement that humans evolved from apes says essentially that humans evolved from an arbitrary grouping of species that lack the distinctive characteristics of humans, a trivial statement that contains no useful information. An extinct ancestral group is always paraphyletic because it excludes a descendant that shares its most recent common ancestor. Although many such groups have been recognized by evolutionary taxonomists, none are recognized by cladists.

Cladists denote the common descent of different taxa by identifying **sister taxa.** Sister taxa share more recent common ancestry with each other than either one does with any other taxon. The sister group of humans appears to be chimpanzees, with gorillas forming a sister group to humans and chimpanzees combined. Orangutans are the sister group of the clade that contains humans, chimpanzees, and gorillas; gibbons form the sister group of the clade that contains orangutans, chimpanzees, gorillas, and humans (see figure 4.7).

Evolutionary paleontologists traditionally use the family-level adjective **hominid** to denote humans and fossil species closer to humans than to other apes; a cladist's use of "hominid," however, refers to any member of the expanded family Hominidae, which includes humans, chimpanzees, gorillas, orangutans, and fossil forms that are phylogenetically closer to these species than to other living primates.

Current State of Animal Taxonomy

The formal taxonomy of animals that we use today was established using principles of evolutionary systematics and has been revised recently in part using principles of cladistics. Introduction of cladistic principles initially has the effect of replacing paraphyletic groups with monophyletic subgroups while leaving the remaining taxonomy mostly unchanged. A thorough revision of taxonomy along cladistic principles, however, will require profound changes, one of which almost certainly will be abandonment of Linnean ranks. A new taxonomic system called PhyloCode is being developed as an alternative to Linnean taxonomy; this system replaces Linnean ranks with

figure 4.8

Willi Hennig (1913–1976), German entomologist who formulated the principles of phylogenetic systematics/cladistics.

codes that denote the nested hierarchy of monophyletic groups conveyed by a cladogram. In our coverage of animal taxonomy, we emphasize taxa that are monophyletic and therefore consistent with criteria of both evolutionary and cladistic taxonomy. We continue, however, to use Linnean ranks. Some new groups, especially ones comprising single-celled organisms, encompass multiple phyla and are therefore designated as named clades without a Linnean rank. In some cases in which familiar taxa are clearly paraphyletic grades, we note this fact and suggest alternative taxonomic schemes that contain only monophyletic taxa.

When discussing patterns of descent, we avoid statements such as "mammals evolved from reptiles" that imply paraphyly and instead specify appropriate sister-group relationships. We avoid referring to groups of organisms as being primitive, advanced, specialized, or generalized because all groups of animals contain combinations of primitive, advanced, specialized, and generalized features; these terms are best restricted to describing specific characteristics and not an entire group.

Revision of taxonomy according to cladistic principles can cause confusion. In addition to new taxonomic names, we see old ones used in unfamiliar ways. For example, cladistic use of "bony fishes" includes amphibians and amniotes (including reptilian groups, birds, and mammals) in addition to the finned, aquatic animals that evolutionary taxonomists normally group under the term "fish." Cladistic use of "reptiles" includes birds in addition to snakes, lizards, turtles, and crocodilians; however, it excludes some fossil forms, such as synapsids, that evolutionary taxonomists traditionally place in Reptilia (see Chapter 18). Taxonomists must be very careful to specify when using these seemingly familiar terms whether traditional evolutionary taxa or newer cladistic taxa are being discussed.

Major Divisions of Life

From Aristotle's time, people have tried to assign every living organism to one of two kingdoms: plant or animal. Unicellular forms were arbitrarily assigned to one of these kingdoms, whose recognition was based primarily on properties of multicellular organisms. This system has outlived its usefulness. It does not represent common descent among organisms accurately. Under the traditional, two-kingdom system, neither animals nor plants constitute monophyletic groups.

Several alternative systems have been proposed to solve the problem of classifying unicellular forms. In 1866 Haeckel proposed a new kingdom, Protista, to include all single-celled organisms. At first bacteria and cyanobacteria (blue-green algae), forms that lack nuclei bounded by a membrane, were included with nucleated unicellular organisms. Finally, important differences between the anucleate bacteria and cyanobacteria (prokaryotes) and all other organisms that have membrane-bound nuclei (eukaryotes) were recognized. In 1969 R. H. Whittaker proposed a five-kingdom system that incorporated a basic prokaryote-eukaryote distinction. Kingdom Monera contained prokaryotes. Kingdom Protista contained unicellular eukaryotic organisms (protozoa and unicellular eukaryotic algae). Multicellular organisms were split into three kingdoms by mode of nutrition and other fundamental differences in organization. Kingdom Plantae included multicellular photosynthesizing organisms (higher plants and multicellular algae). Kingdom Fungi contained molds, yeasts, and fungi, which obtain their food by absorption. Invertebrates (except the protozoa) and vertebrates form the kingdom Animalia. Most of these forms ingest their food and digest it internally, although some parasitic forms are absorptive.

All of these different systems were proposed without regard to phylogenetic relationships needed to construct evolutionary or cladistic taxonomies. The oldest phylogenetic events in the history of life have been obscure, because different forms of life share very few characters that can be compared among these higher taxa to reconstruct their phylogeny. Recently, however, a cladistic classification of all life forms has been proposed based on phylogenetic information obtained from molecular data (the nucleotide base sequence of ribosomal RNA, figure 4.9). According to this tree, Carl Woese, Otto Kandler, and Mark Wheelis recognized three monophyletic **domains** above the kingdom level: Eucarya (all eukaryotes), Bacteria (the true bacteria), and Archaea (prokaryotes differing from bacteria in structure of membranes and in ribosomal RNA sequences). They did not divide Eucarya into kingdoms, although if we retain Whittaker's kingdoms Plantae, Animalia, and Fungi, Protista is paraphyletic because this group does not contain all descendants of its most recent common ancestor (figure 4.10). To maintain a cladistic classification, Protista must be discontinued by recognizing as separate kingdoms Ciliata, Flagellata, and Microsporidia as shown in figure 4.9, and phylogenetic information must be gathered for additional protistan groups, including amebas. This taxonomic revision has not been made; however, if the phylogenetic tree in figure 4.9 is supported by further evidence, revision of taxonomic kingdoms will be necessary.

Until recently, heterotrophic **prostists** were traditionally studied in zoology courses as the animal phylum Protozoa. Given current knowledge and principles of phylogenetic systematics, this practice commits a taxonomic error because "protozoa" do not form a monophyletic group. Protists, now often arrayed in seven separate phyla, are nonetheless of interest to students of zoology because of their shared eukaryotic ancestry with multicellular animals. We therefore cover them in this book.

Major Subdivisions of the Animal Kingdom

The phylum is the largest category in Linnean classification of the animal kingdom. Animal (= metazoan) phyla are often grouped together to produce additional, informal taxa intermediate between phylum and kingdom. Traditional groupings

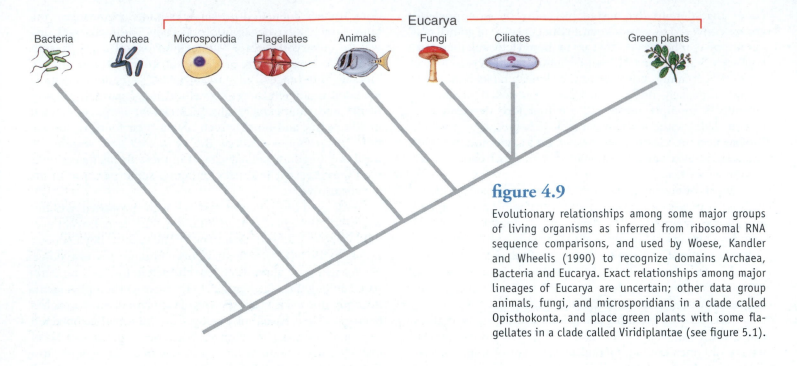

figure 4.9

Evolutionary relationships among some major groups of living organisms as inferred from ribosomal RNA sequence comparisons, and used by Woese, Kandler and Wheelis (1990) to recognize domains Archaea, Bacteria and Eucarya. Exact relationships among major lineages of Eucarya are uncertain; other data group animals, fungi, and microsporidians in a clade called Opisthokonta, and place green plants with some flagellates in a clade called Viridiplantae (see figure 5.1).

figure 4.10

Whittaker's five-kingdom classification superimposed on a phylogenetic tree showing living representatives of these kingdoms. Note that kingdoms Monera and Protista constitute paraphyletic groups (because they do not include all of their descendants) and are therefore unacceptable to cladistic systematics.

based on embryological and anatomical characters that can reveal phylogenetic affinities of different animal phyla are:

Branch A (Mesozoa): phylum Mesozoa, the mesozoa
Branch B (Parazoa): phylum Porifera, the sponges, and phylum Placozoa
Branch C (Eumetazoa): all other phyla
    Grade I (Radiata): phyla Cnidaria, Ctenophora
    Grade II (Bilateria): all other phyla
        Division A (Protostomia): characteristics in figure 4.11
            Acoelomates: phyla Platyhelminthes, Gnathostomulida, Nemertea
            Pseudocoelomates: phyla Rotifera, Gastrotricha, Kinorhyncha, Nematoda, Nematomorpha, Acanthocephala, Entoprocta, Priapulida, Loricifera

Eucoelomates: phyla Mollusca, Annelida, Arthropoda, Echiurida, Sipunculida, Tardigrada, Pentastomida, Onychophora
        Division B (Deuterostomia): characteristics in figure 4.11; phyla Phoronida, Ectoprocta, Chaetognatha, Brachiopoda, Echinodermata, Hemichordata, Chordata

As in the outline, bilateral animals are customarily divided into **Protostomia** and **Deuterostomia** based on their embryological development (figure 4.11). Note, however, that the individual characters listed in figure 4.11 are not completely diagnostic in separating protostomes from deuterostomes. Some phyla are difficult to place into one of these two categories because they possess characteristics of each group.

Recent molecular phylogenetic studies have challenged traditional classification of Bilateria, but results are not yet conclu-

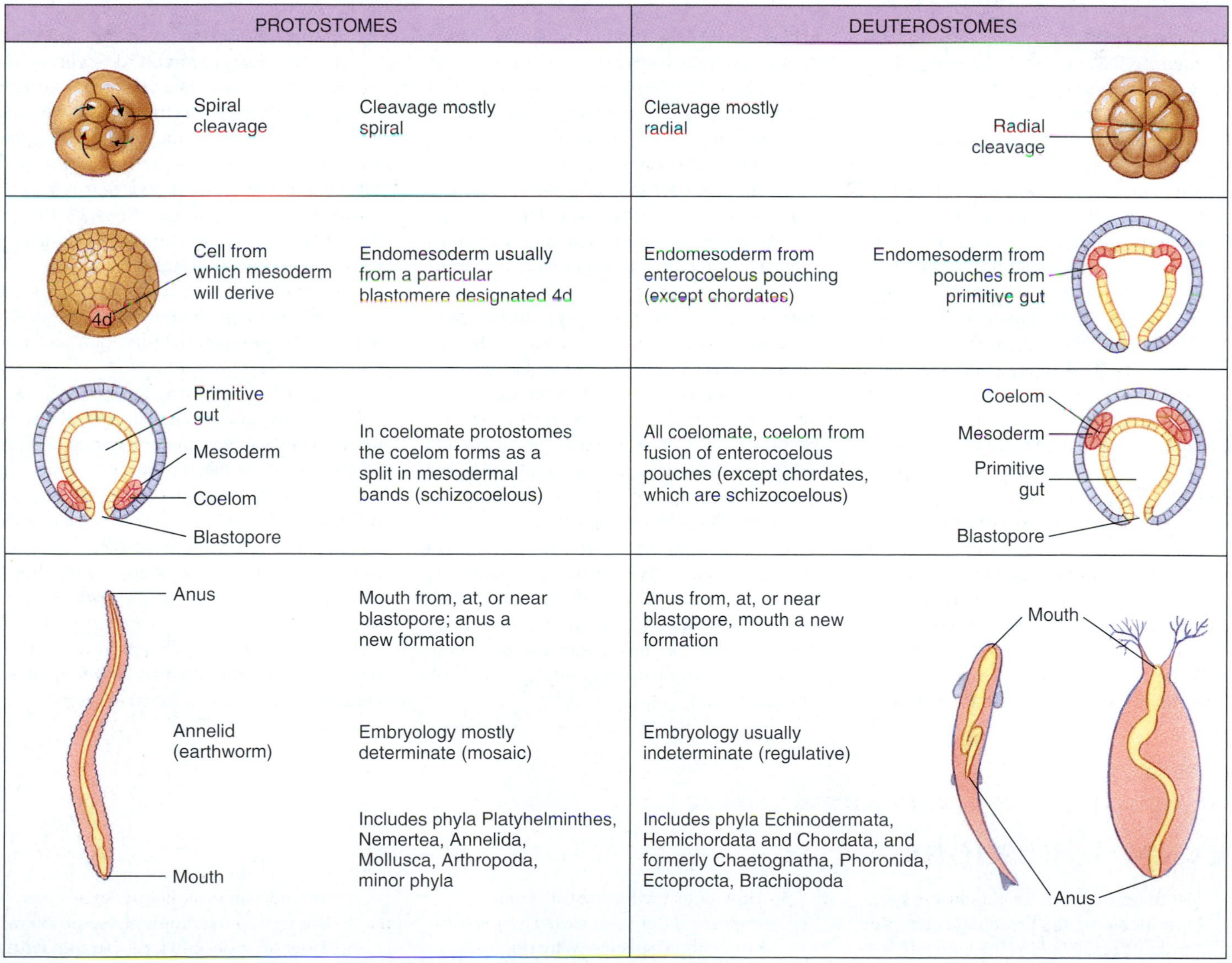

PROTOSTOMES			DEUTEROSTOMES		
Spiral cleavage	Cleavage mostly spiral		Cleavage mostly radial	Radial cleavage	
Cell from which mesoderm will derive (4d)	Endomesoderm usually from a particular blastomere designated 4d		Endomesoderm from enterocoelous pouching (except chordates)	Endomesoderm from pouches from primitive gut	
Primitive gut / Mesoderm / Coelom / Blastopore	In coelomate protostomes the coelom forms as a split in mesodermal bands (schizocoelous)		All coelomate, coelom from fusion of enterocoelous pouches (except chordates, which are schizocoelous)	Coelom / Mesoderm / Primitive gut / Blastopore	
Anus / Annelid (earthworm) / Mouth	Mouth from, at, or near blastopore; anus a new formation Embryology mostly determinate (mosaic) Includes phyla Platyhelminthes, Nemertea, Annelida, Mollusca, Arthropoda, minor phyla		Anus from, at, or near blastopore, mouth a new formation Embryology usually indeterminate (regulative) Includes phyla Echinodermata, Hemichordata and Chordata, and formerly Chaetognatha, Phoronida, Ectoprocta, Brachiopoda	Mouth / Anus	

figure 4.11

Basis for distinctions between divisions of bilateral animals.

sive enough to present a precise hypothesis of phylogenetic relationships among metazoan phyla. Molecular phylogentic results place four phyla traditionally classified as deuterostomes (Brachiopoda, Chaetognatha, Ectoprocta, and Phoronida) in Protostomia. Furthermore, the traditional major groupings of protostome phyla (acoelomates, pseudocoelomates, and eucoelomates) appear not to be monophyletic. Instead, protostomes are divided into two major monophyletic groups called Lophotrochozoa and Ecdysozoa. Reclassification of Bilateria is:

  Grade II: Bilateria
    Division A (Protostomia):
      Lophotrochozoa: phyla Platyhelminthes,
        Nemertea, Rotifera, Gastrotricha,

Acanthocephala, Mollusca, Annelida, Echiurida, Sipunculida, Phoronida, Ectoprocta, Entoprocta, Gnathostomulida, Chaetognatha, Brachiopoda
      Ecdysozoa: phyla Kinorhyncha, Nematoda, Nematomorpha, Priapulida, Arthropoda, Tardigrada, Onychophora, Loricifera, Pentastomida
    Division B (Deuterostomia): phyla Chordata, Hemichordata, Echinodermata

Further study is needed to confirm these new groupings. We organize our survey of animal diversity using the traditional classification, but discuss implications of the newer one.

Summary

Animal systematics has three major goals: (1) to identify all species of animals, (2) to evaluate evolutionary relationships among animal species, and (3) to group animal species hierarchically in taxonomic groups (taxa) that convey evolutionary relationships. Taxa are ranked to denote increasing inclusiveness as follows: species, genus, family, order, class, phylum, and kingdom. All of these ranks can be subdivided to signify taxa that are intermediate between them. The names of species are binomial, with the first name designating the genus to which the species belongs (first letter capitalized) followed by a species epithet (lowercase), both written in italics. Taxa at all other ranks are given single nonitalicized names.

The biological species concept has guided recognition of most animal species. A biological species is defined as a reproductive community of populations (reproductively isolated from others) that occupies a specific niche in nature. A biological species is not immutable through time but changes during its evolution. Because the biological species concept may be difficult to apply in spatial and temporal dimensions, and because it excludes asexually reproducing forms, alternative concepts have been proposed. These alternatives include the evolutionary species concept and the phylogenetic species concept. No single concept of species is universally accepted by all zoologists.

Two major schools of taxonomy are currently active. Traditional evolutionary taxonomy groups species into higher taxa according to joint criteria of common descent and adaptive evolution; such taxa have a single evolutionary origin and occupy a distinctive adaptive zone. A second approach, known as phylogenetic systematics or cladistics, emphasizes common descent exclusively in grouping species into higher taxa. Only monophyletic taxa (those having a single evolutionary origin and containing all descendants of the group's most recent common ancestor) are used in cladistics. In addition to monophyletic taxa, evolutionary taxonomy recognizes some taxa that are paraphyletic (having a single evolutionary origin but excluding some descendants of the most recent common ancestor of the group). Both schools of taxonomy exclude polyphyletic taxa (those having more than one evolutionary origin).

Both evolutionary taxonomy and cladistics require that common descent among species be assessed before higher taxa are recognized. Comparative morphology (including development), cytology, and biochemistry are used to reconstruct the nested hierarchical relationships among taxa that reflect the branching of evolutionary lineages through time. The fossil record provides estimates of ages of evolutionary lineages. Comparative studies and the fossil record jointly permit us to reconstruct a phylogenetic tree representing the evolutionary history of the animal kingdom.

Traditionally, all living forms were placed into two kingdoms (animal and plant) but more recently, a five-kingdom system (animals, plants, fungi, protistans, and monerans) has been followed. Neither system conforms to principles of evolutionary or cladistic taxonomy because both place single-celled organisms into either paraphyletic or polyphyletic groups. Based on our current knowledge of the phylogenetic tree of life, "protozoa" do not form a monophyletic group and they do not belong within the animal kingdom, which comprises multicellular forms (metazoa).

Review Questions

1. List in order, from most inclusive to least inclusive, the principal categories (ranks of taxa) in Carolus Linnaeus's system of classification.
2. Explain why the system for naming species that originated with Linnaeus is "binomial."
3. How does the biological species concept differ from earlier typological concepts of species? Why do evolutionary biologists prefer it to typological species concepts?
4. What problems have been identified with the biological species concept?
   How do other concepts of species attempt to overcome these problems?
5. How do monophyletic, paraphyletic, and polyphyletic taxa differ? How do these differences affect validity of such taxa for both evolutionary and cladistic taxonomies?

6. How are taxonomic characters recognized? How are such characters used to construct a cladogram?

7. What is the difference between a cladogram and a phylogenetic tree? Given a cladogram for a group of species, what additional information is needed to obtain a phylogenetic tree?

8. How would cladists and evolutionary taxonomists differ in their interpretations of the statement that humans evolved from apes, which evolved from monkeys?

9. What are the five kingdoms distinguished by Whittaker? How does their recognition conflict with principles of cladistic taxonomy?

Selected References

See also general references on page 415.

Aguinaldo, A. M. A., J. M. Turbeville, L. S. Linford, M. C. Rivera, J. R. Garey, R. A. Raff, and J. A. Lake. 1997. Evidence for a clade of nematodes, arthropods and other moulting animals. Nature **387**:489–493. *This molecular phylogenetic study challenges traditional classification of the Bilateria.*

Ereshefsky, M. 2001. The poverty of the Linnaean hierarchy. Cambridge, UK, Cambridge University Press. *A philosophical study of biological taxonomy, emphasizing problems with Linnean taxonomy.*

Ereshefsky, M. (ed.). 1992. The units of evolution. Cambridge, Massachusetts, MIT Press. *A thorough coverage of concepts of species, including reprints of important papers on the subject.*

Felsentein, J. 2002. Inferring phylogenies. Sunderland, Massachusetts, Sinauer Associates. *A thorough coverage of phylogenetic methods.*

Hall, B. G. 2001. Phylogenetic trees made easy. Sunderland, Massachusetts, Sinauer Associates, Inc. *A how-to manual for molecular phylogenetics.*

Hall, B. K. 1994. Homology: the hierarchical basis of comparative biology. San Diego, Academic Press. *A collection of papers discussing the many dimensions of homology, the central concept of comparative biology and systematics.*

Hull, D. L. 1988. Science as a process. Chicago, University of Chicago Press. *A study of the working methods and interactions of systematists, containing a thorough review of the principles of evolutionary, phenetic, and cladistic taxonomy.*

Maddison, W. P., and D. R. Maddison. 2003. MacClade 4.06. Sunderland, Massachusetts, Sinauer Associates, Inc. *A program for MacIntosh computers that conducts phylogenetic analyses of systematic characters. The instruction manual stands alone as an excellent introduction to phylogenetic procedures.*

The computer program is user-friendly and excellent for instruction in addition to serving as a tool for analyzing real data.

Mayr, E., and P. D. Ashlock. 1991. Principles of systematic zoology. New York, McGraw-Hill. *A detailed survey of systematic principles as applied to animals.*

Swofford, D. 2002. Phylogenetic analysis using parsimony (and other methods) PAUP* version 4. Sunderland, Massachusetts, Sinauer Associates. *A powerful computer package and instructive manual for constructing phylogenetic trees from data.*

Wagner, G. P. (ed.). 2001. The character concept in evolutionary biology. San Diego, Academic Press. *A thorough coverage of evolutionary character concepts.*

Custom Website

The *Animal Diversity* Online Learning Center is a great place to check your understanding of chapter material. Visit www.mhhe.com/hickmanad4e for access to key terms, quizzes, and more! Further enhance your knowledge with web links to chapter-related material.

Explore live links for these topics:

Classification and Phylogeny of Animals
Methods of Classification
Trees of Life
Speciation

5

Protozoan Groups

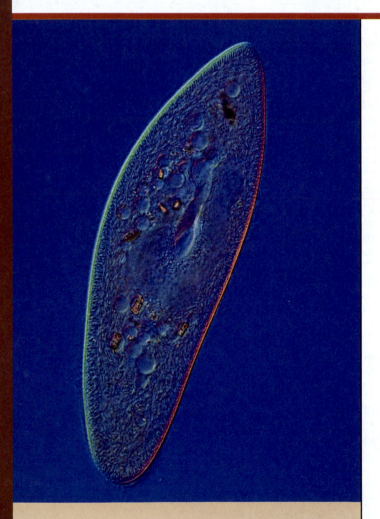

A paramecium.

Emergence of Eukaryotes and a New Life Pattern

The first reasonable evidence for life on earth dates from approximately 3.5 billion years ago. These first cells were prokaryotic, bacteria-like organisms. After an enormous time span of evolutionary diversification at the prokaryotic level, unicellular eukaryotic organisms appeared. Although the origin of single-celled eukaryotes can never be known with certainty, we have good evidence that it occurred through a process of endosymbiosis. Certain aerobic bacteria may have been engulfed by other bacteria unable to cope with the increasing concentrations of oxygen in the atmosphere. The aerobic bacteria had the enzymes necessary for deriving energy in the presence of oxygen, and they would have become the ancestors of mitochondria. Most, but not all, of the genes of the mitochondria would be lost or come to reside in the host-cell nucleus. Almost all present-day eukaryotes have mitochondria and are aerobic.

Some ancestral eukaryotic cells engulfed photosynthetic bacteria, which evolved to become chloroplasts, enabling some eukaryotes to manufacture their own food molecules using energy from sunlight. Green algae and members of the traditional kingdom Plantae represent one such photosynthetic lineage and are now placed together in the clade Viridiplantae. Other groups may have originated by *secondary* endosymbiosis, in which one eukaryotic cell engulfed another eukaryotic cell, and the latter became evolutionarily transformed into an organelle.

Some eukaryotes that did not become residences for chloroplasts, and even some that did, evolved animal-like characteristics and gave rise to a variety of phyla that collectively have been called protozoa. They are distinctly animal-like in several respects: they lack a cell wall, have at least one motile stage in the life cycle, and most ingest their food. Throughout their long history, protozoan groups have radiated to generate a bewildering array of morphological forms within the constraints of a single cell.

A protozoan is a complete organism in which all life activities are carried on within the limits of a single plasma membrane. Protozoans do demonstrate a basic body plan or grade—the single eukaryotic cell—and they amply demonstrate the enormous adaptive potential of that grade. Over 64,000 species have been named, and over half of these are known only from fossils. Some researchers think that there may be 250,000 species of protozoans.

Protozoa are found wherever life exists. They are highly adaptable and easily distributed from place to place. They require moisture, whether they live in marine or freshwater habitats, soil, decaying organic matter, or plants and animals. They may be sessile or free swimming, and they form a large part of the floating plankton. Some species may have spanned geological eras of more than 100 million years.

Protozoa play an enormous role in the economy of nature. Their fantastic numbers are attested by the gigantic oceanic and soil deposits formed by their skeletons. About 10,000 species of protozoa are symbiotic in or on animals or plants, or sometimes even other protozoa. The symbiotic relationship may be **mutualistic** (both partners benefit), **commensalistic** (one partner benefits without affecting the other), or **parasitic** (one partner benefits at the expense of the other) (see Chapter 2). Some of the most important diseases of humans and domestic animals are caused by parasitic protozoa.

For many years, "Protozoa" was considered a single phylum, but phylogenetic studies show that protozoa do not form a monophyletic group. The various phyla of protozoa now recognized are included in this single chapter only for convenience. The traditional phylum Protozoa contained four classes: flagellates, amebas, sporozoans (an important parasitic group including malarial organisms), and ciliates. An enormous amount of information on protozoan structure, life histories, and physiology accumulated, and the Society of Protozoologists published a new classification of protozoa in 1980, recognizing seven separate phyla, the most important of which were

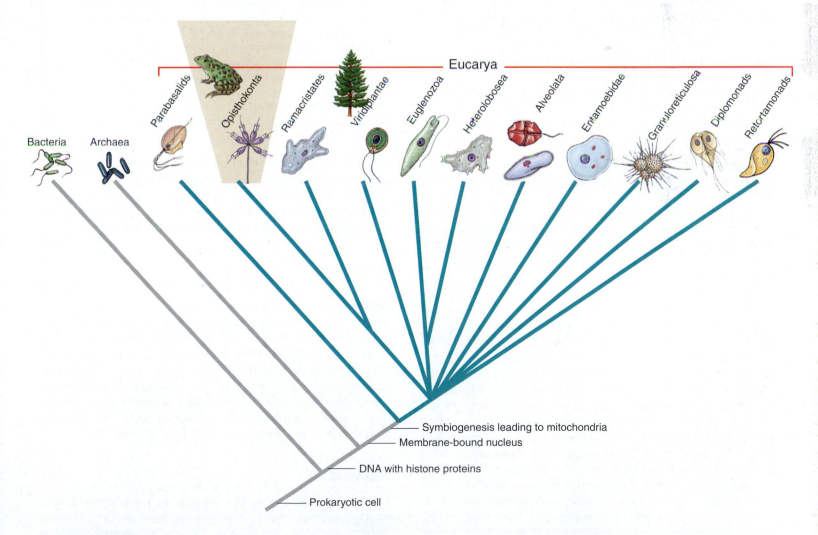

figure 5.1

Cladogram showing two major prokaryotic branches and diversification of eukaryotes. Major eukaryotic clades containing protists are shown, but several clades of amebas and other forms are not shown. The order of branching remains to be determined for most clades. The very large opisthokont clade contains choanoflagellates, microsporidians, fungi, and all multicellular animals.

Position in the Animal Kingdom

A protozoan is a complete organism in which all life activities occur within the limits of a single cell membrane. Because their protoplasmic mass is not subdivided into cells, protozoa sometimes have been termed "acellular," but most people prefer "unicellular" to emphasize the many structural similarities to the cells of multicellular animals.

Evidence from electron microscopy, life-cycle studies, genetics, biochemistry, and molecular biology has shown that the for-mer phylum Protozoa encompassed numerous phyla of varying evolutionary relationships. Combining all animal-like unicellular eukaryotes with the unicellular algae into a kingdom **Protista,** or including the unicells with their multicellular relatives in the Protoctista, simply created another, more massive paraphyletic group. There may be more than 60 clades of eukaryotes, but here we discuss only 11 such groups, all having some unicellular members with animal-like features (see the prologue to this chapter).

Biological Contributions

1. **Intracellular specialization** (division of labor within the cell) involves the organization of functional organelles in the cell.
2. The simplest example of **division of labor between cells** is seen in certain colonial protozoa that have both somatic and reproductive zooids (individuals) in the colony.
3. **Asexual reproduction** by mitotic division appears in unicellular eukaryotes.
4. **True sexual reproduction** with zygote formation is found in some protozoa.

5. The responses (taxes) of protozoa to stimuli represent the **simplest reflexes and instincts** as we know them in metazoans.
6. The simplest animal-like organisms with **exoskeletons** are certain shelled protozoa.
7. **All types of nutrition** are developed in protozoa: autotrophic, saprozoic, and holozoic. **Basic enzyme systems** to accomplish these types of nutrition are developed.
8. Means of **locomotion** in aqueous media are developed.

the Sarcomastigophora (flagellates and amebas), Apicomplexa (sporozoans and related organisms), and Ciliophora (ciliates). However, analyses of sequences of bases in genes, primarily those encoding the small subunit of ribosomal RNA, but also those encoding several proteins, have revolutionized our concepts of phylogenetic affinities and relationships, not only of protozoan groups, but all eukaryotes. The origin of the first eukaryote (see prologue to this chapter) was followed by diversification into the many unicellular and multicellular taxa we see today (figure 5.1). According to some scientists, if we retain classical kingdoms such as animals (Metazoa), fungi, and plants, we must recognize no fewer than *12 additional* kingdoms.[1] In some cases, however, molecular data are available for very few (or only one) species in a group; this seems slender evidence upon which to establish a kingdom of living organisms.

Rather than focus on taxonomic level, we will discuss eleven major eukaryotic clades, including both traditional protozoan phyla and some new groups identified using molecular characters. The clade Opisthokonta contains unicellular groups such as choanoflagellates and microsporidians, as well as fungi and all the multicellular animals (metazoans). Unicellular algae and multicellular plants both belong in clade Viridiplantae (figure 5.1). Some groups traditionally distinguished by their morphology were also identified as monophyletic taxa in molecular phylogenies. Foraminiferans are one such example: these shelled amebas form a monophyletic group now called clade Granuloreticulosa.

With respect to the names of phyla, we will generally follow the system used in comprehensive monographs, such as Hausmann and Hülsmann (1996). Ameboid organisms (former Sarcodina) fall into numerous groups, but for the sake of simplicity, we will discuss these together under an informal heading "Amebas."

Form and Function

Protozoans are unicellular, but they are not simple. They are functionally complete organisms with many complicated microanatomical structures. Their various organelles tend to be more specialized than those of an average cell in a multicellular organism. Particular organelles may perform as skeletons, sensory structures, conducting mechanisms, and feeding structures. Special features of a particular organelle can often be used as defining characters for protozoan clades, as is the case with the shape of the mitochondrial cristae (see p. 98).

Nucleus

As in other eukaryotes, the nucleus is a membrane-bound structure whose interior communicates with the cytoplasm by small pores. Within the nucleus the genetic material (DNA) is borne on chromosomes. Also within the nucleus, one or more **nucleoli** are often present (figures 5.2 and 5.10). Several nuclei, including a large macronucleus and smaller micronuclei, are present in ciliates (see figure 5.17). The persistence of nucleoli during mitosis is one of the defining characters for the Euglenozoa clade.

---

[1] Baldauf, S. L., A. J. Roger, I. Wenk-Siefert, and W. F. Doolittle. 2000. A kingdom-level phylogeny of eukaryotes based on combined protein data. Science **290:** 972–977.

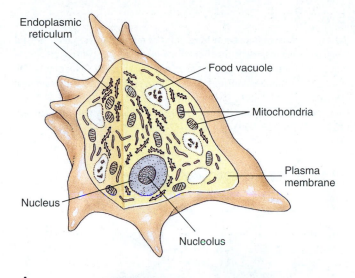

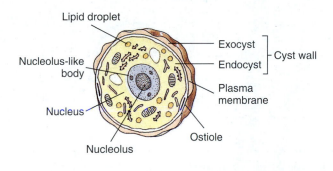

A

B

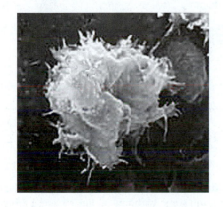

C

figure 5.2

Structure of *Acanthamoeba palestinensis*. **A,** Active, feeding form. **B,** Cyst. **C,** *Acanthamoeba castellanii* kills cells of the human cornea; it is spread by contact lenses that haven't been correctly disinfected.

Mitochondria

A **mitochondrion** is an organelle used in energy acquisition where oxygen serves as the terminal electron acceptor. It contains DNA. The internal membranes of a mitochondrion, called cristae (figure 5.2), vary in form, being flat, tubular, discoid, or branched. Form of the cristae is a homologous character that identifies some protozoan clades, like Ramacristates and Euglenozoa. In cells without mitochondria, **hydrogenosomes** may be present. Hydrogenosomes produce molecular hydrogen when oxygen is absent and are assumed to have evolved from mitochondria. **Kinetoplasts** are also assumed to be mitochondrial derivatives, but they work in association with a kinetosome, an organelle at the base of a flagellum (see p. 92).

Golgi Apparatus

The Golgi apparatus is part of the secretory system of the endoplasmic reticulum. Golgi bodies are also called **dictyosomes** in protozoan literature. **Parabasal bodies** are similar structures with potentially similar functions.

Plastids

Plastids are organelles containing photosynthetic pigments. The original addition of a plastid to eukaryotic cells occurred when a cyanobacterium was engulfed and not digested. **Chloroplasts** (figure 5.13) contain different versions of chlorophylls (*a, b,* or *c*), but other kinds of plastids contain other pigments. For example, red algal plastids contain phycobilins. Particular pigments shared among protozoan lineages may indicate shared ancestry, but plastids could also have been gained by secondary endosymbiosis rather than direct inheritance from a common ancestor (see this chapter's prologue).

Extrusomes

The general term *extrusomes* refers to membrane-bound organelles that protozoans use to extrude something from the cell. The wide variety of structures extruded suggests that extrusomes are not all homologous. The ciliate **trichocyst** (p. 100) is an extrusome.

Locomotor Organelles

Protozoa move chiefly by cilia and flagella and by pseudopodial movement. These mechanisms are extremely important in the biology of higher animals as well.

Many small metazoans use cilia not only for locomotion but also to create water currents for their feeding and respiration. Ciliary movement is vital to many species in such functions as handling food, reproduction, excretion, and osmoregulation (as in flame cells, p. 149).

Cilia and Flagella

No real morphological distinction exists between cilia and flagella (figure 5.3), and some investigators have preferred to call them both undulipodia (L. dim. of *unda*, a wave, + Gr. *podos*, a foot). However, a cilium propels water parallel to the surface to

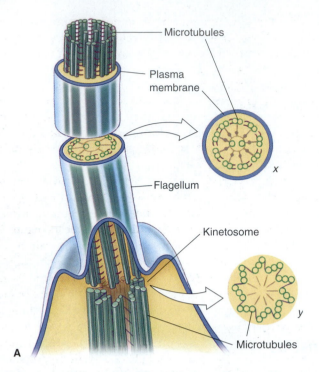

A

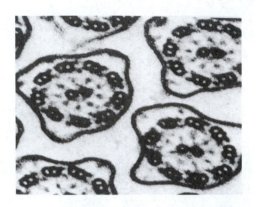

B

figure 5.3

A, The axoneme is composed of nine pairs of microtubules plus a central pair, and it is enclosed within the cell membrane. The central pair ends at about the level of the cell surface in a basal plate (axosome). The peripheral microtubules continue inward for a short distance to compose two of each of the triplets in the kinetosome (at level *y* in **A**). **B,** Electron micrograph of section through several cilia, corresponding to section at *x* in **A**. (× 133,000)

which the cilium is attached, whereas a flagellum propels water parallel to the main axis of the flagellum. Each flagellum or cilium contains nine pairs of longitudinal microtubules arranged in a circle around a central pair (figure 5.3), and this is true for all motile flagella and cilia in the animal kingdom, with a few notable exceptions. This "9 + 2" tube of microtubules in a flagellum or cilium is its **axoneme;** an axoneme is covered by a membrane continuous with the cell membrane covering the rest of the organism. At about the point where an axoneme enters the cell proper, the central pair of microtubules ends at a small plate within the circle of nine pairs. Also at about that point, another microtubule joins each of the nine pairs, so that these form a short tube extending from the base of the flagellum into the cell. The tube consists of nine *triplets* of microtubules and is called the **kinetosome (or basal body).** Kinetosomes are exactly the same in structure as **centrioles** that organize mitotic spindles during cell division. Centrioles of some flagellates may give rise to kinetosomes, or kinetosomes may function as centrioles. All typical flagella and cilia have a kinetosome at their base, whether borne by a protozoan or metazoan cell.

Description of the axoneme as "9 + 2" is traditional, but it is also misleading because there is only a single pair of microtubules in the center. If we were consistent, we would have to describe the axoneme as "9 + 1." A more consistent way to describe the nine sets of doublets and a pair of singlets is "9(2) + 2(1)"; the kinetosome is then 9(3) + 0.

The current explanation for ciliary and flagellar movement is the **sliding microtubule hypothesis.** Movement is powered by the release of chemical bond energy in ATP. Two little arms are visible in electron micrographs on each of the

characteristics
of Protozoan Phyla

1. **Some members of each group are unicellular;** some colonial, and some with multicellular stages in their life cycles
2. **Mostly microscopic,** although some are large enough to be seen with the unaided eye
3. All symmetries represented in the group; shape variable or constant (oval, spherical, or other)
4. **No germ layer present**
5. No organs or tissues, but **specialized organelles** are found; nucleus single or multiple
6. Free living, mutualism, commensalism, parasitism all represented in the groups
7. Locomotion by **pseudopodia, flagella, cilia,** and direct cell movements; some sessile
8. Some provided with a **simple endoskeleton** or **exoskeleton,** but most are naked
9. **Nutrition of all types:** autotrophic (manufacturing own nutrients by photosynthesis), heterotrophic (depending on other plants or animals for food), saprozoic (using nutrients dissolved in the surrounding medium)
10. Aquatic or terrestrial habitat
11. Reproduction **asexually** by fission, budding, and cysts, and **sexually** by conjugation or by syngamy (union of male and female gametes to form a zygote)

pairs of peripheral tubules in the axoneme (level *x* in figure 5.3), and these bear the enzyme adenosine triphosphatase (ATPase), which cleaves the ATP. When bond energy in ATP is released, the arms "walk along" one of the microtubules in the adjacent pair, causing it to slide relative to the other tubule.

Figure 5.4

Ameboid movement leading to phagocytosis. *Top* and *center,* the ameba extends a pseudopodium toward a *Pandorina* colony. *Bottom,* the ameba surrounds the *Pandorina* before engulfing it by phagocytosis.

Shear resistance, causing the axoneme to bend when the microtubules slide past each other, is provided by "spokes" from one of the tubules in each doublet projecting toward the central pair of microtubules. These spokes are also visible in electron micrographs.

Pseudopodia

Although **pseudopodia** are the chief means of locomotion of amebas, they can be formed by a variety of protozoa, as well as by ameboid cells of many invertebrates. In fact, much defense against disease in the human body depends on ameboid white blood cells, and ameboid cells in many other animals, vertebrate and invertebrate, play similar roles.

In protozoa, pseudopodia exist in several forms. The most familiar are lobopodia (figures 5.4 and 5.7), which are rather large, blunt extensions of the cell body containing both central, granular **endoplasm** and **ectoplasm,** a peripheral, nongranular layer. The ectoplasm is in the gel state of a colloid (jellylike semisolid), and the endoplasm is in the sol state (fluid). **Filipodia** are thin extensions, usually branching, and containing only ectoplasm. They are found in amebas such as *Chlamydophrys* (see figure 5.5). **Reticulopodia** (figure 5.5) are distinguished from filipodia in that reticulopodia repeat-

edly rejoin to form a netlike mesh, although some protozoologists consider the distinction between filipodia and reticulopodia artificial. **Axopodia** (figure 5.5) are long, thin pseudopodia supported by axial rods of microtubules (figure 5.6). The microtubules are arranged in a definite spiral or geometrical array, depending on the species, and constitute the axoneme of the axopod. Axopodia can be extended or retracted, apparently by addition or removal of microtubular material. Cytoplasm can flow along the axonemes, toward the body on one side and in the reverse direction on the other.

How pseudopodia work has long attracted the interest of zoologists, but we have only recently gained some insight into the phenomenon. When a typical lobopodium begins to form, an extension of ectoplasm called a **hyaline cap** appears (figure 5.7). As endoplasmic material flows into the hyaline cap, it fountains out to the periphery and changes from the sol (fluid) to the gel (semisolid) state of the colloid; that is, it becomes ectoplasm. Thus the ectoplasm is a tube through which the endoplasm flows as the pseudopodium extends. On the trailing side of the organism, ectoplasm becomes endoplasm. At some point the pseudopodium becomes anchored to the substrate, and the cell is drawn forward. The current hypothesis of pseudopodial movement involves the participation of actin, myosin, and other components. As endoplasm fountains out in the hyaline cap, actin subunits become polymerized into microfilaments, which in turn become cross-linked to each other by actin-binding protein (ABP) to form a gel; the endoplasm becomes ectoplasm. At the "posterior" the ABP releases the microfilaments, which return to the sol state of endoplasm. Before the microfilaments are disassembled into actin subunits, they interact with myosin, contracting and causing the endoplasm to flow in the direction of the pseudopodium by hydrostatic pressure.

Colloidal systems are permanent suspensions of finely divided particles that do not precipitate, such as milk, blood, starch, soap, ink, and gelatin. Colloids in living systems are commonly proteins, lipids, and polysaccharides suspended in the watery fluid of cells (cytoplasm). Such systems may undergo sol-gel transformations, depending on whether the fluid or particulate components become continuous. In the sol state of cytoplasm, solids are suspended in a liquid, and in the semisolid gel state, liquid is suspended in a solid.

Nutrition and Digestion

Protozoa can be categorized broadly into autotrophs, which synthesize their own organic constituents from inorganic substrates, and heterotrophs, which must have organic molecules synthesized by other organisms. Within heterotrophs, those that ingest visible particles of food (**phagotrophs,** or **holozoic** feeders) contrast with those ingesting food in a soluble form (**osmotrophs,** or **saprozoic** feeders).

Holozoic nutrition implies phagocytosis (figures 5.4 and 5.8), in which there is an infolding or invagination of the cell membrane around the food particle. As the invagination extends farther into the cell, it is pinched off at the surface. A

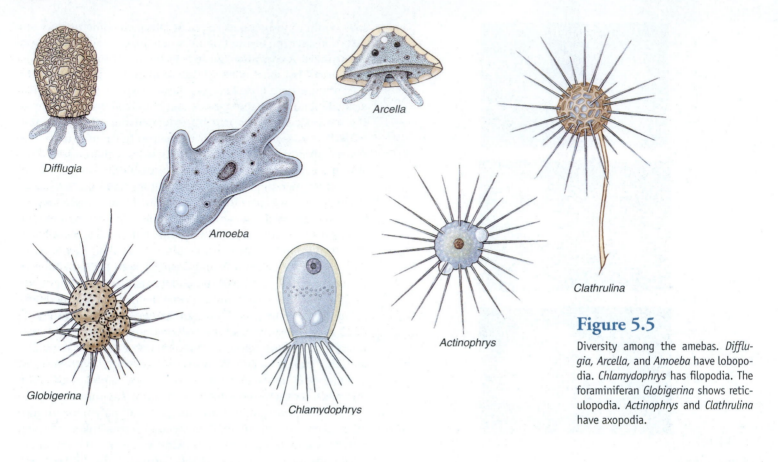

Figure 5.5

Diversity among the amebas. *Difflugia,* *Arcella,* and *Amoeba* have lobopodia. *Chlamydophrys* has filopodia. The foraminiferan *Globigerina* shows reticulopodia. *Actinophrys* and *Clathrulina* have axopodia.

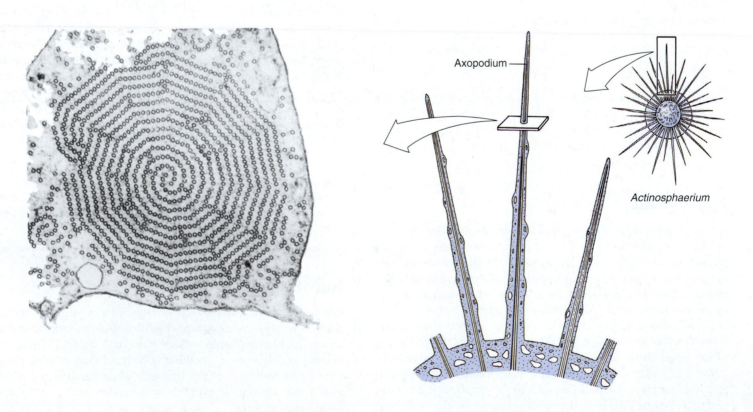

Figure 5.6

Diagram of axopodium *(right)* to show orientation of electron micrograph of axopodium (from *Actinosphaerium nucleofilum*) in cross section *(left)*. Protozoa with axopodia are shown in Figure 5.5 (*Actinophrys* and *Clathrulina*). The axoneme of the axopodium is composed of an array of microtubules, which may vary from three to many in number depending on the species. Some species can extend or retract their axopodia quite rapidly (×99,000).

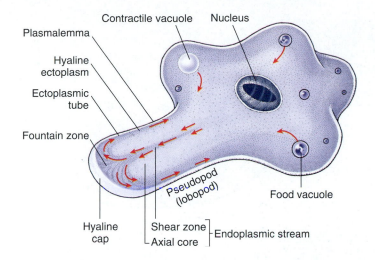

figure 5.7

Ameba in active locomotion. Arrows indicate direction of streaming protoplasm. The first sign of a new pseudopodium is thickening of the ectoplasm to form a clear hyaline cap, into which the fluid endoplasm flows. As endoplasm reaches the forward tip, it fountains out and is converted into ectoplasm, forming a stiff outer tube that lengthens as the forward flow continues. Posteriorly the ectoplasm is converted into fluid endoplasm, replenishing the flow. Substratum is necessary for ameboid movement.

food particle is thus contained in an intracellular, membrane-bound vesicle, the **food vacuole** or **phagosome.** Lysosomes, small vesicles containing digestive enzymes, fuse with the phagosome and pour their contents into it, where digestion begins. As digested products are absorbed across the vacuole membrane, the phagosome becomes smaller. Any undigestible material may be released to the outside by exocytosis, the vacuole again fusing with the cell surface membrane. In most ciliates, many flagellates, and many apicomplexans, the site of phagocytosis is a definite mouth structure, the **cytostome** (figures 5.8 and 5.16). In amebas, phagocytosis can occur at almost any point by envelopment of the particle with pseudopodia. Many ciliates have a characteristic structure for expulsion of waste matter, called a **cytopyge** or **cytoproct** (see figure 5.16), found in a characteristic location.

Figure 5.8

Some feeding methods among protozoa. *Amoeba* surrounds a small flagellate with pseudopodia. *Leidyopsis,* a flagellate living in the intestine of termites, forms pseudopodia and ingests wood chips. *Didinium,* a ciliate, feeds only on *Paramecium,* which it swallows through a temporary cytostome in its anterior end. Sometimes more than one *Didinium* feeds on the same *Paramecium. Podophrya* is a suctorian ciliophoran. Its tentacles attach to its prey and suck prey cytoplasm into its body, where the cytoplasm is pinched off to form food vacuoles. *Codosiga,* a sessile flagellate with a collar of microvilli, feeds on particles suspended in the water drawn through its collar by the beat of its flagellum. The particles are moved to the cell body and ingested (surrounded by small pseudopods). Technically all of these methods are types of phagocytosis.

Excretion and Osmoregulation

Water balance, or osmoregulation, is a function of the one or more **contractile vacuoles** (see figures 5.7, 5.13, and 5.16) possessed by most protozoa, particularly freshwater forms, which live in a hypoosmotic environment. These vacuoles are often absent in marine or parasitic protozoa, which live in a nearly isosmotic medium.

Contractile vacuoles usually are located in the ectoplasm and act as pumps to remove excess water from the cytoplasm. A recent hypothesis suggests that proton pumps on the vacuolar surface and on tubules radiating from it actively transport H^+ and cotransport bicarbonate (HCO_3^-) (figure 5.9), which are osmotically active particles. As these particles accumulate within a vacuole, water would be drawn into the vacuole. Fluid within the vacuole would remain isosmotic to the cytoplasm. Then as the vacuole finally joins its membrane to the surface membrane and empties its contents to the outside, it would expel water, H^+, and HCO_3^-. These ions can be replaced readily by action of the enzyme carbonic anhydrase on CO_2 and H_2O. Carbonic anhydrase is present in the cytoplasm.

Nitrogenous wastes from metabolism apparently diffuse through the cell membrane, but some may also be emptied by way of the contractile vacuoles.

Reproduction

Asexual Reproduction: Fission

The cell multiplication process that results in more genetically identical individuals in protozoa is called **fission.** The most common type of fission is **binary,** in which two essentially identical individuals result (figures 5.10 and 5.17). When a progeny cell is considerably smaller than the parent and then grows to adult size, the process is called **budding.** This process occurs in some ciliates. In **multiple fission,** division of the cytoplasm

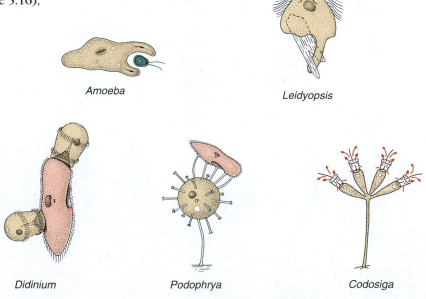

Amoeba

Leidyopsis

Didinium

Podophrya

Codosiga

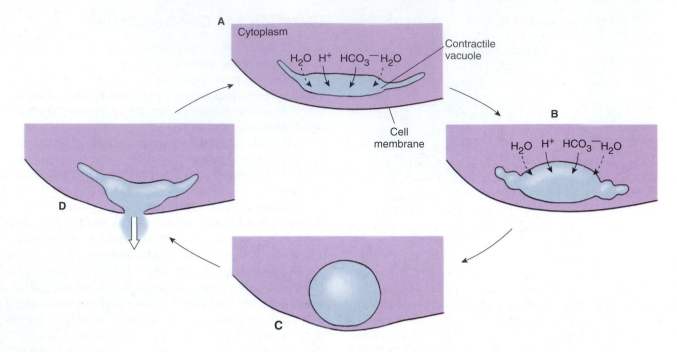

Figure 5.9

Proposed mechanism for operation of contractile vacuoles. **A, B,** Vacuoles are composed of a system of cisternae and tubules. Proton pumps in their membranes transport H^+ and cotransport HCO_3^- into the vacuoles. Water diffuses in passively to maintain an osmotic pressure equal to that in the cytoplasm. When the vacuole fills **C,** its membrane fuses with the cell's surface membrane, expelling water, H^+, and HCO_3^-. **D,** Protons and bicarbonate ions are replaced readily by action of carbonic anhydrase on carbon dioxide and water.

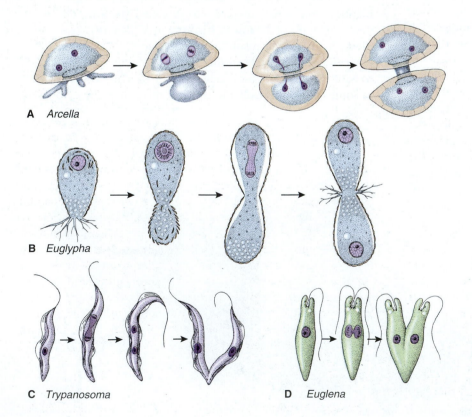

A   *Arcella*

B   *Euglypha*

C   *Trypanosoma*

D   *Euglena*

figure 5.10

Binary fission in some amebas and Euglenozoa. **A,** The two nuclei of *Arcella* divide as some of its cytoplasm is extruded and begins to secrete a new shell for the daughter cell. **B,** The shell of another sarcodine, *Euglypha,* is constructed of secreted platelets. Secretion of the platelets for the daughter cell begins before the cytoplasm begins to move out the aperture. As these form the shell of the daughter cell, the nucleus divides. **C,** *Trypanosoma* has a kinetoplast (part of the mitochondrion) near the kinetosome of its flagellum close to its posterior end in the stage shown. All of these parts must replicate before the cell divides. **D,** Division of *Euglena.*

(cytokinesis) is preceded by several nuclear divisions, so that numerous individuals are produced almost simultaneously (see figure 5.20). Multiple fission, or **schizogony,** is common among the Apicomplexa and some amebas. When multiple fission leads to spore or sporozoite formation, it is called **sporogony.**

Sexual Processes

Although all protozoa reproduce asexually, and some are apparently exclusively asexual, the widespread occurrence of sex among protozoa testifies to the value of genetic recombination. Sexual processes may precede certain phases of asexual reproduction, but embryonic development does not occur; protozoa do not have embryos. The essential features of sexual processes include a reduction division of the chromosome number to half (diploid number to haploid number), development of sex cells (gametes) or at least gamete nuclei, and usually a fusion of the gamete nuclei (p. 102).

The gamete nuclei, or pronuclei, which fuse in fertilization to restore the diploid number of chromosomes, are usually borne in special gamete cells. Gametes may be similar in appearance or unlike.

In animals meiosis usually occurs during or just before gamete formation (called **gametic meiosis**). Such is indeed the case in the Ciliophora and some flagellated and ameboid groups. However, in other flagellated groups and in the Apicomplexa, the first divisions *after* fertilization (zygote formation) are meiotic (**zygotic meiosis**) (see figure 5.20), and all individuals produced asexually (mitotically) in the life cycle up to the next zygote are haploid. In some amebas (foraminiferans) haploid and diploid generations alternate (**intermediary meiosis**), a phenomenon widespread among plants.

Fertilization of an individual gamete by another is **syngamy,** but some sexual phenomena in protozoa do not involve syngamy. Examples are **autogamy,** in which gametic nuclei arise by meiosis and fuse to form a zygote within the same organism that produced them, and **conjugation,** in which an exchange of gametic nuclei occurs between paired organisms (conjugants). Conjugation will be described further in the discussion of *Paramecium.*

Encystment and Excystment

Separated as they are from their external environment only by their delicate external cell membrane, protozoa are surprisingly successful in habitats frequently subjected to extremely harsh conditions. Survival under harsh conditions surely is related to the ability to form cysts: dormant forms marked by the possession of resistant external coverings and a more or less complete shutdown of metabolic machinery. Cyst formation is also important to many parasitic forms that must survive a harsh environment between hosts (see figure 5.2). However, some parasites do not form cysts, apparently depending on direct transfer from one host to another. Reproductive phases such as fission, budding, and syngamy may occur in cysts of some species. Encystment has not been found in *Paramecium,* and it is rare or absent in marine forms.

Cysts of some soil-inhabiting and freshwater protozoa have amazing durability. Cysts of the soil ciliate *Colpoda* can survive 7 days in liquid air and 3 hours at 100°C. Survival of *Colpoda* cysts in dried soil has been shown for up to 38 years, and those of a certain small flagellate (*Podo*) can survive up to 49 years! Not all cysts are so sturdy, however. Those of *Entamoeba histolytica* will tolerate gastric acidity but not desiccation, temperature above 50°C, or sunlight.

The conditions stimulating encystment are incompletely understood, although in some cases cyst formation is cyclic, occurring at a certain stage in the life cycle. In most free-living forms, adverse environmental change favors encystment. Such conditions may include food deficiency, desiccation, increased environmental osmotic pressure, decreased oxygen concentration, or change in pH or temperature.

During encystment a number of organelles, such as cilia or flagella, are resorbed, and the Golgi apparatus secretes cyst wall material, which is carried to the surface in vesicles and extruded.

Although the exact stimulus for excystation (escape from cysts) is usually unknown, a return of favorable conditions initiates excystment in those protozoa in which the cysts are a resistant stage. In parasitic forms an excystment stimulus may be more specific, requiring conditions similar to those found in the host.

Protozoan Taxa

An understanding of protozoan classification can be quite useful if the diagnostic features of each clade are easily recognized. For example, the presence of branching netlike pseudopodia in a shelled ameba identifies the organism as belonging to clade Granuloreticulosa (figure 5.1). The clade Opisthokonta is characterized by flattened mitochondrial cristae and one posterior flagellum on flagellated cells, if such cells are present. This clade contains metazoans and fungi, as well as unicellular microsporidians and choanoflagellates (see *Codosiga* in figure 5.8). The microsporidians are intracellular parasites, and the choanoflagellates are free-living unicells that may form colonies. Choanoflagellate cells resemble certain cells in sponges (phylum Porifera, see p. 105), which may be homologous to them. Although traditionally considered protozoans, these two groups are not discussed further here. Other clades, such as the red algal phylum Rhodophyta, are not discussed here because they are plantlike. Members of Rhodophyta have plastids, are not heterotrophic, and lack flagellated stages (have no motile sperm) in the life cycle. The clades we discuss further contain some animal-like forms, defined as having motile stages in the life cycle, ingesting food, and lacking a cell wall.

Viridiplantae

The clade Viridiplantae contains unicellular and multicellular green algae, bryophytes, and vascular plants. Members of this group have chlorophylls *a* and *b* in chloroplasts, but some move via flagella whereas others do not. As a subset of this clade, the unicellular and multicellular green algae are placed together in phylum Chlorophyta.

Phylum Chlorophyta

This group contains autotropic single-celled algae such as *Chlamydomas* (figure 5.11), as well as colonial forms such as *Gonium* (figure 5.11) and *Volvox* (figure 5.12). *Volvox* is often studied in introductory courses because its mode of development is somewhat similar to the embryonic development of some metazoans. The basic form of *Volvox,* a hollow ball of cells, is reminiscent of the metazoan blastula, leading some to suggest that the first metazoan was a nonphotosynthetic flagellate similar to *Volvox* in body design. *Volvox* is also of interest because certain reproductive zooids develop into eggs or bundles of sperm. Each fertilized egg (zygote) undergoes repeated divisions to produce a small colony, which is released in spring. A number of asexual generations may follow before sexual reproduction occurs again.

Phylum Euglenozoa

The phylum Euglenozoa (figure 5.11) is generally recognized as a monophyletic group, based on the shared persistence of the nucleoli during mitosis, and the presence of discoid mitochondrial cristae. Cristae of this form are found also in a clade of amebas (Heterolobosea), so these groups are shown as sister taxa in figure 5.1. Members of Euglenozoa have a series of longitudinal microtubules just beneath the cell membrane that help to stiffen the membrane into a **pellicle** (figure 5.13). The phylum is divided into two subphyla, the Euglenida and the Kinetoplasta.

Euglenids have chloroplasts with chlorophyll *b.* These chloroplasts are surrounded by a double membrane and are likely to have arisen via secondary endosymbiosis. *Euglena* is a representative of this group. It has a light-sensitive **stigma,** or eyespot, which is a shallow pigment cup that allows light from only one direction to strike a light-sensitive receptor (figure 5.13).

Kinetoplastans are all parasitic in plants and animals. They have a unique organelle known as a **kinetoplast,** which is part of their mitochondrion and is composed of a large disk of DNA. Various species of *Trypanosoma* (Gr. *trypanon,* auger, + *soma,* body) (figure 5.11) cause serious diseases of humans and other animals. Two subspecies of *Trypanosoma brucei* (*T. b. rhodesiense* and *T. b. gambiense*) cause clinically distinct forms of African sleeping sickness in humans. Another subspecies, *T. b. brucei,* along with several other trypanosomes, causes a similar disease in domestic animals. It makes agriculture very difficult in large areas of Africa. *Trypanosoma cruzi* causes American trypanosomiasis, or Chagas' disease, a very serious disease in South and Central America.

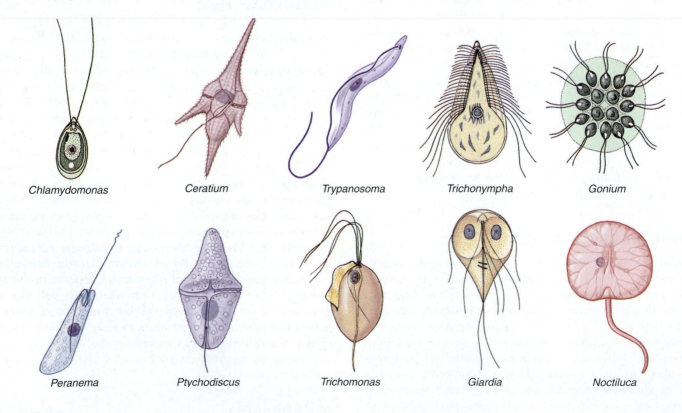

Chlamydomonas   Ceratium   Trypanosoma   Trichonympha   Gonium

Peranema   Ptychodiscus   Trichomonas   Giardia   Noctiluca

figure 5.11

Representatives of several protozoan taxa with flagella: Diplomonads (*Giardia*); Parabasalids (*Trichomonas, Trichonympha*); Chlorophyta (*Chlamydomonas, Gonium*); Euglenozoa (*Peranema, Trypanosoma*); Dinoflagellata (*Ceratium, Ptychodiscus, Noctiluca*)

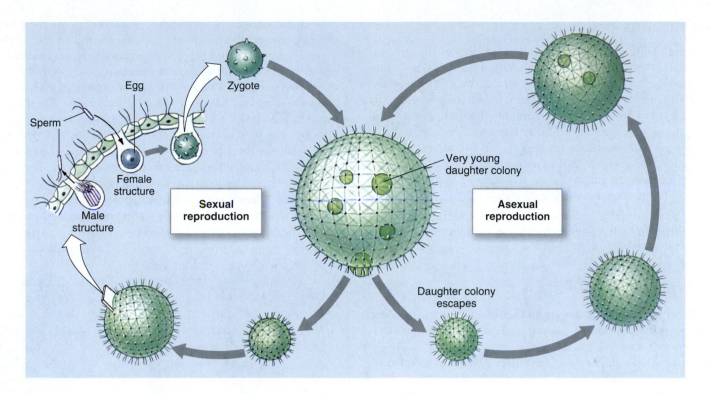

Figure 5.12

Life cycle of *Volvox*. Asexual reproduction occurs in spring and summer when specialized diploid reproductive cells divide to form young colonies that remain in the mother colony until large enough to escape. Sexual reproduction occurs largely in autumn when haploid sex cells develop. The zygote may encyst and so survive the winter, developing into a mature asexual colony in the spring. In some species the colonies have separate sexes; in others both eggs and sperm are produced in the same colony.

figure 5.13

Euglena. Features shown are a combination of those visible in living and stained preparations.

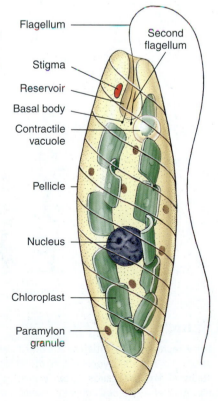

Phylum Retortamonada and the Diplomonads

This former phylum now has been divided into two clades: Retortamonads and Diplomonads. Both groups lack mitochondria, so biologists wondered whether they branched from the main eukaryotic lineage before the mitochondrial symbiosis. However, recent work has shown that there are mitochondrial genes in the cell nucleus* so it much more likely that the absence of mitochondria is due to secondary loss rather than to primary absence.

Retortamonads include commensal and parasitic unicells, such as *Chilomastix* and *Retortamonas. Giardia* (figure 5.11), a representative diplomonad, is a common intestinal parasite of humans and other animals. It is often asymptomatic but may cause a rather discomforting, but not fatal, diarrhea. Cysts are passed in the feces, and new hosts are infected by ingestion of cysts, often in contaminated water.

---

* Roger, A. J. 1999. Reconstructing early events in eukaryotic evolution. Amer. Nat. **154** (supplement): S146–163.

Parabasalids

The Parabasalid clade contains some members of the phylum Axostylata. Members of this phylum have a stiffening rod composed of microtubules, the **axostyle,** that extends along the longitudinal axis of their body. Parabasalids, traditionally part of the class Parabasalea, possess a modified region of the Golgi apparatus called a **parabasal body** (see p. 91) connected to a kinetosome. Mitochondria are absent, but members of order Trichomonadida possess hydrogenosomes (see p. 91).

Some trichomonads (figure 5.11) are of medical or veterinary importance. *Trichomonas vaginalis* infects the urogenital tract of humans and is sexually transmitted. It produces no symptoms in males but is the most common cause of vaginitis in females.

Superphylum Alveolata

Based on molecular and morphological evidence, the next three phyla (Apicomplexa, Ciliophora, Dinoflagellata) are often grouped together in a clade or superphylum, Alveolata. Alveolates possess **alveoli** or related structures, which are membrane-bound sacs that lie beneath the cell membrane and serve structural functions or produce pellicles (ciliates) or thecal plates (dinoflagellates).

Phylum Ciliophora

Ciliates are a large and interesting group, with a great variety of forms living in all types of freshwater and marine habitats. They are the most structurally complex and diversely specialized of all protozoa. The majority are free living, although some are commensal or parasitic. They are usually solitary and motile, but some are colonial and some are sessile. There is a great diversity of shape and size. In general they are larger than most other protozoa, ranging from 10 μm to 3 mm long. At some stage all have cilia that beat in a coordinated rhythmical manner, although the arrangement of the cilia varies. Suctorians are ciliates in which the young possess cilia and are free-swimming, but the adults grow an attachment stalk, become sessile, and lose their cilia.

The **pellicle** of ciliates may consist only of a cell membrane or in some species may form a thickened armor. Cilia are short and usually arranged in longitudinal or diagonal rows. Cilia may cover the surface of the animal or may be restricted to the oral region or to certain bands. In some forms cilia are fused into a sheet called an **undulating membrane** or into smaller **membranelles,** both used to propel food into the **cytopharynx** (gullet). In other forms there may be fused cilia forming stiffened tufts called **cirri,** often used in locomotion by the creeping ciliates (figure 5.14).

The kinetosomes and a structural system of fibrils just beneath the pellicle make up the **infraciliature** (figure 5.15). The infraciliature apparently does not coordinate ciliary beat, as formerly believed. Coordination of ciliary movement seems to be by waves of depolarization of the cell membrane moving down the animal, similar to conduction of a nerve impulse.

Some ciliates have curious small bodies in their ectoplasm between the bases of the cilia. Examples are **trichocysts** (figures 5.15 and 5.16) and **toxicysts.** Upon mechanical or chemical stimulation, these bodies explosively expel a long, threadlike structure. The mechanism of expulsion is unknown. The function of trichocysts is probably defensive. When a paramecium is attacked by a predatory *Didinium*, it expels its trichocysts but to no avail. Toxicysts, however, release a poison that paralyzes the prey of carnivorous ciliates. Toxicysts are structurally quite distinct from trichocysts. Many dinoflagellates have structures very similar to trichocysts.

Most ciliates are holozoic, possessing a cytostome (mouth) that in some forms is a simple opening and in others

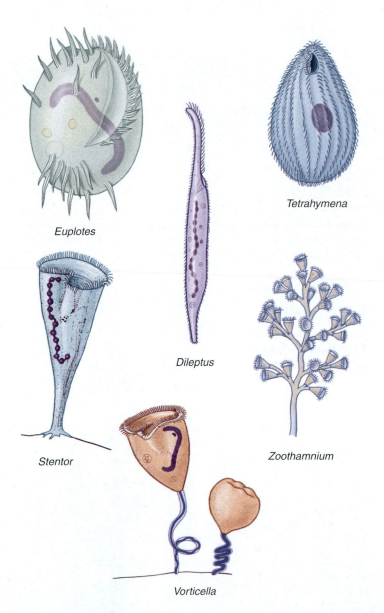

Euplotes

Tetrahymena

Dileptus

Stentor

Zoothamnium

Vorticella

figure 5.14

Some representative ciliates. *Euplotes* have stiff cirri used for crawling about. Contractile myonemes in the ectoplasm of *Stentor* and in the stalks of *Vorticella* allow great expansion and contraction. Note the macronuclei—long and curved in *Euplotes* and *Vorticella,* and shaped like a string of beads in *Stentor.*

is connected to a gullet or ciliated groove. The mouth in some is strengthened with stiff, rodlike structures for swallowing larger prey; in others, such as paramecia, ciliary water currents carry microscopic food particles toward the mouth. *Didinium* has a proboscis for engulfing the paramecia on which it feeds (see figure 5.8). Suctorians paralyze their prey and then ingest their contents through tubelike tentacles by a complex feeding mechanism that apparently combines phagocytosis with a sliding filament action of microtubules in the tentacles (figure 5.8).

Contractile vacuoles are found in all freshwater and some marine ciliates; the number varies from one to many among the different species. In most ciliates the vacuole is fed by one or more collecting canals (figure 5.16). The vacuoles occupy a fixed position, and each discharges through a more-or-less permanent pore.

Ciliates are always multinucleate, possessing at least one **macronucleus** and one **micronucleus,** but varying from one to many of either type. The macronuclei are apparently responsible for metabolic, synthetic, and developmental functions. Macronuclei are varied in shape among the different species (figures 5.14 and 5.16). Micronuclei participate in sexual reproduction and give rise to macronuclei after exchange of micronuclear material between individuals. Micronuclei divide mitotically, and macronuclei divide amitotically.

Reproduction and Life Cycles

Life cycles of ciliates usually involve both **asexual binary fission** (figure 5.17) and a type of sexual reproduction called **conjugation** (figure 5.18). Conjugation is a temporary union of two

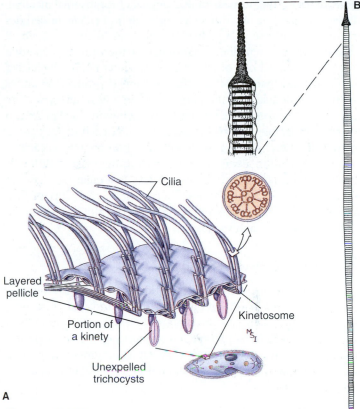

figure 5.15

Infraciliature and associated structures in ciliates. **A,** Structure of the pellicle and its relation to the infraciliature system. **B,** Expelled trichocyst.

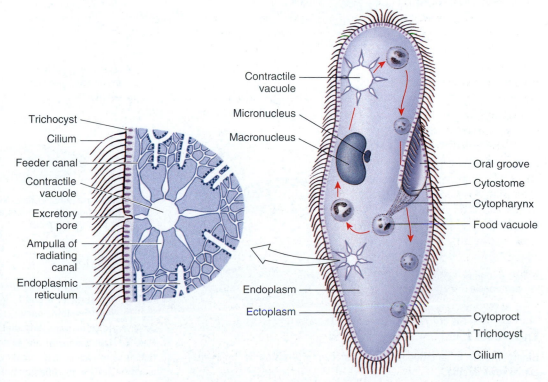

figure 5.16

Left, Enlarged section of a contractile vacuole of *Paramecium.* Water is believed to be collected by the endoplasmic reticulum, emptied into feeder canals and then into the vesicle. The vesicle contracts to empty its contents to the outside, thus serving as an osmoregulatory organelle. *Right, Paramecium,* showing gullet (cytopharynx), food vacuoles, and nuclei.

individuals for the purpose of exchanging chromosomal material. During union the micronucleus of each individual undergoes meiosis, giving rise to four haploid micronuclei, three of which degenerate. The remaining micronucleus then divides into two haploid pronuclei, one of which is exchanged for a pronucleus of the conjugant partner. When the exchanged pronucleus unites with the pronucleus of the partner, the diploid number of chromosomes is restored. The two partners, each with fused pronuclei (now comparable to a zygote), separate, and each divides twice by mitosis, thereby giving rise to four daughter paramecia each.

Conjugation always involves two individuals of different mating types. This prevents inbreeding. Most ciliate species are divided into several varieties, each variety made up of two **mating types.** Mating occurs only between the differing mating types within each variety.

Phylum Dinoflagellata

Dinoflagellates were formerly included by zoologists among plantlike members of Sarcomastigophora, and many are photoautotrophic with chromatophores bearing chlorophylls *a* and *c*. Ecologically, some species are among the most important primary producers in marine environments. They commonly have two flagella, one equatorial and one longitudinal, each

borne at least partially in grooves on the body (see figure 5.11). The body may be naked or covered by cellulose plates or valves. Most dinoflagellates have brown or yellow chromatophores, although some are colorless. Many species can ingest prey through a mouth region between the plates near the posterior area of the body. *Noctiluca* (see figure 5.11), a colorless dinoflagellate, is a voracious predator and has a long, motile tentacle, near the base of which its single, short flagellum emerges. *Noctiluca* is one of many marine organisms that can produce light (bioluminescence).

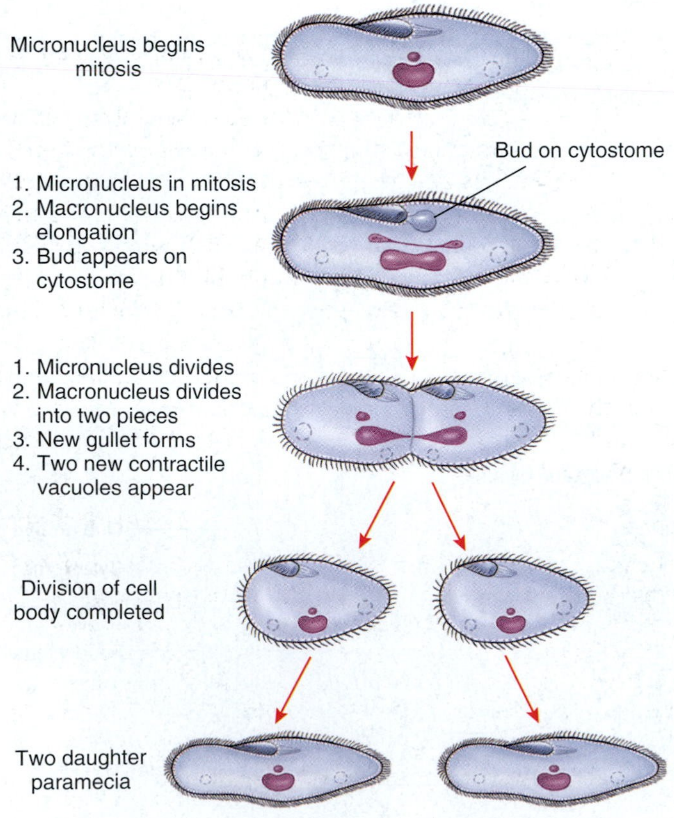

Micronucleus begins mitosis

1. Micronucleus in mitosis
2. Macronucleus begins elongation
3. Bud appears on cytostome

Bud on cytostome

1. Micronucleus divides
2. Macronucleus divides into two pieces
3. New gullet forms
4. Two new contractile vacuoles appear

Division of cell body completed

Two daughter paramecia

figure 5.17

Binary fission in a ciliophoran (*Paramecium*). Division is transverse, across rows of cilia.

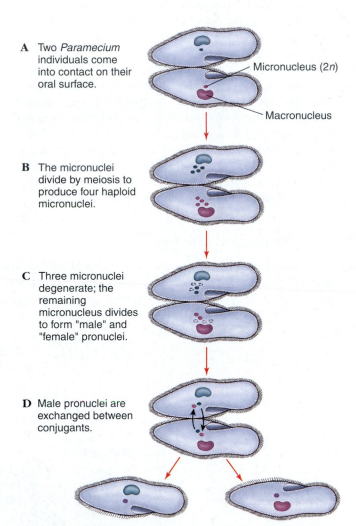

A Two *Paramecium* individuals come into contact on their oral surface.

Micronucleus (2*n*)

Macronucleus

B The micronuclei divide by meiosis to produce four haploid micronuclei.

C Three micronuclei degenerate; the remaining micronucleus divides to form "male" and "female" pronuclei.

D Male pronuclei are exchanged between conjugants.

E Male and female pronuclei fuse, and individuals separate. Subsequently old macronuclei are absorbed and replaced by new macronuclei.

figure 5.18

Scheme of conjugation in *Paramecium*. **A,** Two individuals come in contact on oral surface. **B** and **C,** Micronuclei divide twice (meiosis), resulting in four haploid nuclei in each partner. **D,** Three micronuclei degenerate; remaining one divides to form "male" and "female" pronuclei. **E,** Male pronuclei are exchanged between conjugants. **F,** Male and female pronuclei fuse, and individuals separate. Subsequently old macronuclei are absorbed and replaced by new macronuclei.

Zooxanthellae are dinoflagellates that live in mutualistic association in tissues of certain invertebrates, including other protozoa, sea anemones, horny and stony corals, and clams. The association with stony corals is of ecological and economic importance because only corals with symbiotic zooxanthellae can form coral reefs (see Chapter 7).

Dinoflagellates can damage other organisms, such as when they produce a "red tide." Although this name originally applied to situations in which the organisms reproduced in such profusion (producing a "bloom") that the water turned red from their color, any instance of a bloom producing detectable levels of toxic substances is now called a red tide. The water may be red, brown, yellow, or not remarkably colored at all, and the phenomenon has nothing to do with tides! The toxic substances are apparently not harmful to the organisms that produce them, but they may be highly poisonous to fish, other marine life, and to humans. Red tides have resulted in considerable economic losses to the shellfish industry.

Pfiesteria piscicida is one of several related dinoflagellate species that may affect fish in brackish waters along the Atlantic Coast, south of North Carolina. Much of the time *Pfiesteria* feeds on algae and bacteria, but something in the excreta of large schools of fish causes it to release a powerful, short-lived, toxin. The toxin may stun or kill fish, often creating skin lesions. *Pfiesteria* has flagellated and ameboid forms among its more than 20 body types; some forms feed on fish tissues and blood. Although it does not have chloroplasts, it may sequester chloroplasts from its algal prey and gain energy from them over the short term. This fascinating group of species was discovered in 1988.

Phylum Apicomplexa

All apicomplexans are endoparasites, and their hosts are found in many animal phyla. The life cycle usually includes both asexual and sexual reproduction, and there is sometimes an invertebrate intermediate host. At some point in the life cycle most apicomplexans develop a **spore (oocyst),** which is infective for the next host and is often protected by a resistant coat. The presence of a certain combination of organelles, the **apical complex,** distinguishes this phylum (figure 5.19). The apical complex is usually present only in certain life-cycle stages of the organisms; for example, **merozoites** and **sporozoites** (figure 5.20). Some of the structures, especially the **rhoptries** and **micronemes,** apparently aid in penetrating the host's cells or tissues.

Locomotor organelles are not as obvious in this group as they are in other protozoa. Pseudopodia occur in some intracellular stages, and gametes of some species are flagellated. Tiny contractile fibrils can form waves of contraction across the body surfaces to propel the organism through a liquid medium.

Apicomplexan parasites belong to two groups: **coccidians** (Coccidia [Gr. *kokkos,* kernel or berry]), which infect both vertebrates and invertebrates, and **gregarines,** which live mainly in the digestive tract and body cavity of certain inverte-

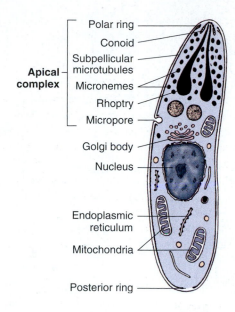

figure 5.19

Diagram of an apicomplexan sporozoite or merozoite at the electron microscopic level, illustrating the apical complex. The polar ring, conoid, micronemes, rhoptries, subpellicular microtubules, and micropore (cytostome) are all considered components of the apical complex.

brates. We discuss two examples of medical importance from class Coccidea: *Toxoplasma* and *Plasmodium.*

Plasmodium is the sporozoan parasite that causes **malaria.** The vectors (carriers) of the parasites are female *Anopheles* mosquitos, and the life cycle is depicted in figure 5.20. Malaria is one of the most important diseases in the world, with almost 500 million cases occurring each year. It causes 1 to 2 million deaths per year worldwide. Although this represents a decline from the situation 45 years ago, recent years have seen a resurgence caused by increasing resistance of mosquitos to insecticides, increasing numbers of *Plasmodium* strains that are resistant to drugs, and socioeconomic conditions and civil strife that interfere with malaria control efforts in tropical countries.

Although malaria has been recognized as a disease and a scourge of humanity since antiquity, the discovery that it was caused by a protozoan was made only 100 years ago by Charles Louis Alphonse Laveran, a French army physician. At that time, the mode of transmission was still mysterious, and "bad air" (hence the name malaria) was a popular candidate. Ronald Ross, an English physician in the Indian Medical Service, determined some years later that the malarial organism was carried by *Anopheles* mosquitos. Because Ross knew nothing about mosquitos or their normal parasites, his efforts were long frustrated by trying to use the wrong kinds of mosquitos and confusion caused by the other parasites he found in them. That he persisted is a tribute to his determination. The momentous discovery earned Ross the Nobel Prize in 1902 and knighthood in 1911.

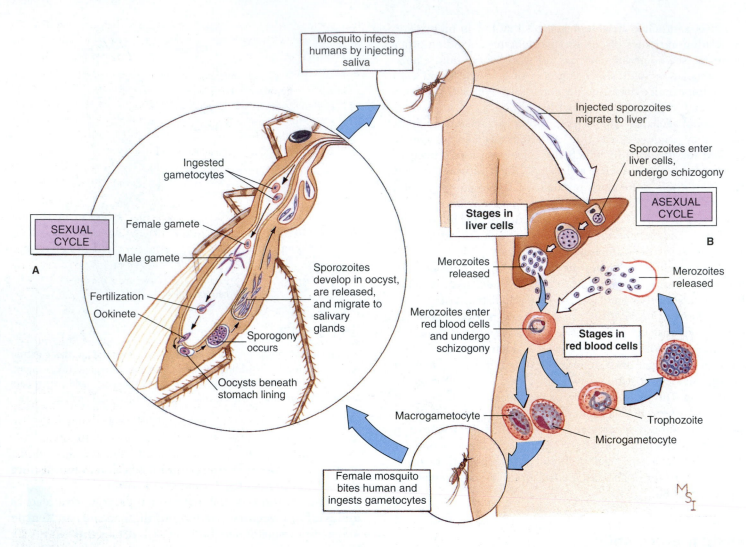

figure 5.20

Life cycle of *Plasmodium vivax*, one of the protozoa (class Coccidea) that causes malaria in humans. **A,** Sexual cycle produces sporozoites in the body of the mosquito. **B,** Sporozoites infect humans and reproduce asexually, first in liver cells, then in red blood cells. Malaria is spread by the *Anopheles* mosquito, which sucks up gametocytes along with human blood, then, when biting another victim, leaves sporozoites in the new wound.

Toxoplasma (Gr. *toxo,* a bow,+ *plasma,* molded) is a common parasite in the intestinal tissues of cats, but this parasite can produce extraintestinal stages as well. The extraintestinal stages can develop in a wide variety of animals other than cats—for example, rodents, cattle, and humans. Gametes and oocysts are not produced by the extraintestinal forms, but they can initiate the intestinal cycle in a cat, if the cat eats infected prey. In humans *Toxoplasma* causes few or no ill effects except in AIDS patients or in a woman infected during pregnancy. Such infection greatly increases the chances of a birth defect in the baby; about 2% of all mental retardation in the United States is a result of congenital toxoplasmosis. Humans can become infected from eating insufficiently cooked beef, pork, or lamb or from accidentally ingesting oocysts from the feces of cats. Pregnant women should not eat raw meat or empty cats' litterboxes.

Some 20% or more of adults in the United States are infected with *Toxoplasma gondii;* we have no symptoms because the parasite is held in check by our immune systems. However, *T. gondii* is one of the most important opportunistic infections in AIDS patients. The latent infection is activated in between 5 and 15% of AIDS patients, often in the brain, with serious consequences. Another coccidian, *Cryptosporidium parvum,* first was reported in humans in 1976. We now recognize it as a major cause of diarrheal disease worldwide, especially in children in tropical countries. Waterborne outbreaks have occurred in the United States, and the diarrhea can be life-threatening in immunocompromised patients (such as those with AIDS). The latest coccidian pathogen to emerge has been *Cyclospora cayetanensis.* About 850 cases of diarrhea due to *Cyclospora* in the United States and Canada were reported in May and June, 1996. We do not yet know how the parasite is transmitted nor even the normal host.

Amebas

Amebas characteristically move and feed by means of pseudopodia (figures 5.5 and 5.7). They are found in both fresh and salt water and in moist soils. Some are planktonic; some prefer a substratum. A few are parasitic.

Nutrition in amebas is holozoic; they ingest and digest liquid or solid foods. Most amebas are omnivorous, living on algae, bacteria, protozoa, rotifers, and other microscopic organisms. An ameba may take in food at any part of its body surface merely by putting out a pseudopodium to enclose the food (**phagocytosis**). The enclosed food particle, along with some environmental water, becomes a food vacuole, which is carried by the streaming movements of endoplasm. As digestion occurs within the vacuole by enzymatic action, water and digested materials pass into the cytoplasm. Undigested particles are eliminated through the cell membrane.

Most amebas reproduce by binary fission. Sporulation and budding occur in some.

Some amebas are covered with a protective **test,** or shell, with openings for the pseudopodia; in others the test is internal. Some tests are constructed of secreted siliceous material reinforced with sand grains (figure 5.5). The **foraminiferans** (L. *foramen,* hole, + *fero,* bear) are mostly marine forms that secrete complex many-chambered tests of calcium carbonate (figure 5.21). They usually add sand grains to the secreted material, using great selectivity in choosing colors. Slender pseudopodia extend through openings in the test and then run together to form a protoplasmic net (reticulum) in which they ensnare their prey. The presence of the protoplasmic net is reflected in the new name of this clade, Granuloreticulosa. They are beautiful little creatures with many pseudopodia radiating out from a central test. **Radiolarians** (L. *radiolus,* small ray) are marine forms, mostly living in plankton, that have intricate and beautiful siliceous skeletons (figure 5.22). A central capsule that separates the inner and outer cytoplasm is perforated to allow cytoplasmic continuity. Unlike foraminiferans, the radiolarians do not form a monophyletic group; animals commonly called radiolarians have been divided into several clades.

Radiolarians are among the oldest known protozoa, and they and foraminiferans have left excellent fossil records. For millions of years tests of dead protozoa have been dropping to the seafloor, forming deep-sea sediments that are estimated to be from 600 to 3600 m deep (approximately 2000 to 12,000 feet) and containing as many as 50,000 foraminiferans per gram of sediment. Many limestone and chalk deposits on land were formed by these small creatures when the land was covered by sea.

How Do We Classify Amebas?

All amebas were once grouped together in Sarcodina, but we now know this was a polyphyletic taxon. The body form of an ameba, as well as the ability to make a test, has evolved independently several times. The shape of pseudopodia formed by each species of ameba was used as a character for classification. In particular, the presence of axopodia, supported by axial rods of microtubules (see p. 94), was used to distinguish the actinopods from the other amebas, the nonactinopods. These descriptive terms are still in use, but several distinct clades contain organisms of each type.

A, **B**

Figure 5.21

A, Air-dried foraminiferan, showing spines extending from the test. **B,** A test of the foraminiferan *Vertebralina striata.* Foraminiferans are ameboid marine protozoans that secrete a calcareous, many-chambered test in which to live, and then extrude protoplasm through pores to form a layer over the outside. The animal begins with one chamber and, as it grows, it secretes a succession of new and larger chambers, continuing this process throughout life. Many foraminiferans are planktonic, and when they die their shells are added to the ooze on the ocean's bottom.

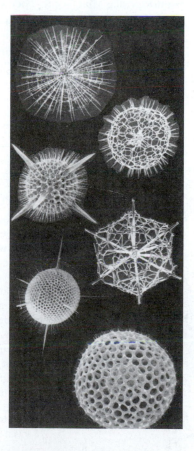

figure 5.22

Types of radiolarian tests. In his study of these beautiful forms collected on the famous *Challenger* expedition of 1872 to 1876, Ernst Haeckel worked out our present concepts of symmetry.

Nonactinopod Amebas Nonactinopod amebas may form lobopodia, filopodia, or rhizopodia (see p. 93). Lobopodia are seen in some members of Ramicristate clade, so named because these animals also possess branched or ramifying mitochondrial cristae. *Acanthamoeba* (figure 5.2) forms lobopodia. Filopodia can be seen in *Chlamydophrys* (figure 5.5). There are many species of rhizopod ameba, including the large *Chaos carolinense*. Rhizopodia are seen also in a clade of endoparasitic amebas (Entamoebae) found in the intestine of humans or other animals. Most are harmless, but *Entamoeba histolytica* (Gr. *entos,* within, + *amoibē,* change; *histos,* tissue, + *lysis,* a loosing), the most common amebic parasite harmful to humans, causes amebic dysentery. In rare cases, certain free-living amebas cause a brain disease that is almost always fatal. They apparently gain access through the nose while a person is swimming in a pond or lake.

Actinopod Amebas This polyphyletic group of amebas has axopod pseudopodia (figures 5.5 and 5.6). The descriptive name *radiolarian* applies to some of these amebas, but taxa formerly placed in each group are now distributed among several clades.

Phylogeny and Adaptive Radiation

Phylogeny

Traditionally, amebas (former Sarcodina) and flagellates (former Mastigophora) were considered separate classes in phylum Protozoa. Observations that some flagellates could form pseudopodia, that some species of amebas had flagellated stages, and that a supposed ameba was really a flagellate without a flagellum, all seemed to support the concept of a phylum Sarcomastigophora. However, analyses of sequences of bases in genes, particularly the gene encoding the small subunit of ribosomal RNA, have provided strong evidence that neither Sarcodina nor Mastigophora are monophyletic groups. Unfor-

tunately, the branching order for most clades containing protozoans is not yet clear.

In fact, molecular evidence has almost completely revised our concepts of protozoan phylogeny. Such evidence has strengthened the hypothesis of a symbiotic origin of eukaryotes and that origins of nuclei, mitochondria, and plastids represent separate endosymbiogenetic events. The common ancestor of green plants and green algae acquired its chlorophyll-bearing plastids by symbiogenesis with a cyanobacterium. These plastids have two membranes, one apparently derived from the ancestral cyanobacterium, and the other from the host vacuole containing it. Some investigators propose that in the Euglenozoa, plastids were added by secondary symbiogenesis. Euglenozoa apparently acquired their plastids after divergence from ancestors of Kinetoplasta. Among alveolates, many dinoflagellates are photoautotrophs, maintaining their chloroplasts, but some have lost them. Apicomplexans have a small circular, plastidlike DNA, localized to a discrete organelle surrounded by four membranes, and evidently a relict plastid inherited from their common ancestor with dinoflagellates. Ancestors of ciliophorans either lost their plastid symbionts or diverged from their common ancestor with dinoflagellates before the secondary symbiogenesis occurred.

Adaptive Radiation

We have described some of the wide range of adaptations of protozoa in the previous pages. Amebas range from bottom-dwelling, naked species to planktonic forms with beautiful, intricate tests such as the foraminiferans and radiolarians. There are many symbiotic species of amebas. Flagellated forms likewise show adaptations for a similarly wide range of habitats, with the added variation of photosynthetic ability in many groups.

Within a single-cell body plan, the division of labor and specialization of organelles are carried furthest by ciliates, the most complex of all protozoa. Specializations for intracellular parasitism have been adopted by Apicomplexa.

classification of Protozoan Phyla

This classification primarily follows Hausmann and Hülsmann (1996) and is abridged from Roberts and Janovy (2005), but some new clades identified by molecular and morphological characters are also included. With few exceptions, we are including only taxa of examples discussed in this chapter.

Much strong evidence indicates that phylum Sarcomastigophora and its subphyla are no longer tenable. Newer monographs consider amebas as belonging to several taxa with various affinities, not all yet determined. Organisms previously assigned to phylum Sarcomastigophora, subphylum Sarcodina, should be placed in at least two phyla, if not more. Nevertheless, amebas form a number of fairly recognizable morphological groups, which we will use for the convenience of readers and not assign such groups to specific taxonomic levels.

Phylum Chlorophyta (klor-off'i-ta) (Gr. *chloros,* green, + *phyton,* plant). Unicellular and multicellular algae; photosynthetic pigments of chlorophyll *a* and *b,* reserve food is starch (characters in common with "higher" plants: bryophytes and vascular plants); all with biflagellated stages; flagella smooth and of equal length; mostly free-living photoautotrophs. Examples: *Chlamydomonas, Volvox.* Members of this phylum are placed in clade Viridiplantae.

Phylum Retortamonada (re-tor'ta-mo'nad-a) (L. *retorqueo,* to twist back, + *monas,* single, unit). Mitochondria and Golgi bodies lacking; three anterior and one recurrent (running toward posterior) flagellum lying in a groove; intestinal parasites or free living in anoxic environments. This phylum is now divided into two clades with two genera in clade Retortamonads.

Class Diplomonadea (di′plo-mon-a′de-a). (Gr. *diploos,* double + L. *monas,* unit). One or two karyomastigonts (group of kinetosomes with a nucleus); individual mastigonts with one to four flagella; mitotic spindle within nucleus; cysts present; free living or parasitic. Nine genera diplomonads constitute the clade Diplomonads.

Order Diplomonadida (di′plo-mon-a′di-da). Two karyomastigonts, each with four flagella, one recurrent; with variety of microtubular bands. Example: *Giardia.*

Phylum Axostylata (ak-so-sty′la-ta) (Gr. *axon,* axle + *stylos,* style, stake). With an axostyle made of microtubules.

Class Parabasalea (par′a-ba-sa′le-a) (Gr. *para,* beside + *basis,* base, refers to parabasal body = Golgi). With very large Golgi bodies associated with karyomastigont; up to thousands of flagella. *Trichomonas* and two other forms constitute clade Parabasalids.

Order Trichomonadida (tri′ko-mon-a′di-da) (Gr. *trichos,* hair + *monas,* unit). Typically at least some kinetosomes associated with rootlet filaments characteristic of trichomonads; parabasal body present; division spindle extranuclear; hydrogenosomes present; no sexual reproduction; true cysts rare; all parasitic. Examples: *Dientamoeba, Trichomonas.*

Phylum Euglenozoa (yu-glen-a-zo′a) (Gr. *eu-,* good, true, + *glēnē,* cavity, socket, + *zōon,* animal). With cortical microtubules; flagella often with paraxial rod (rodlike structure accompanying axoneme in flagellum); mitochondria with discoid cristae; nucleoli persist during mitosis. This phylum is equivalent to the clade Euglenozoa.

Subphylum Euglenida (yu-glen′i-da). With pellicular microtubules that stiffen pellicle.

Class Euglenoidea (yu-glen-oyd′e-a) (Gr. *eu-,* good, true, + *glēnē,* cavity, socket, + *-ōideos,* form of, type of). Two heterokont flagella (flagella with different structures) arising from apical reservoir; some species with light-sensitive stigma and chloroplasts. Example: *Euglena.*

Subphylum Kinetoplasta (ky-neet′o-plas′ta) (Gr. *kinētos,* to move + *plastos,* molded, formed). With a unique mitochondrion containing a large disc of DNA; paraxial rod.

Class Trypanosomatidea (try-pan′o-som-a-tid′e-a) (Gr. *trypanon,* a borer, + *sōma,* the body). One or two flagella arising from pocket; flagella typically with paraxial rod that parallels axoneme; single mitochondrion (nonfunctional in some forms) extending length of body as tube, hoop, or network of branching tubes, usually with single conspicuous DNA-containing kinetoplast located near flagellar kinetosomes; Golgi body typically in region of flagellar pocket, not connected to kinetosomes and flagella; all parasitic. Examples: *Leishmania, Trypanosoma.*

Phylum Apicomplexa (ap′i-com-pleks′a) (L. *apex,* tip or summit, + *complex,* twisted around). Characteristic set of organelles (apical complex) associated with anterior end present in some developmental stages; cilia and flagella absent except for flagellated microgametes in some groups; cysts often present; all parasitic. This phylum is within clade Alveolata.

Class Gregarinea (gre-ga-ryn′e-a) (L. *gregarius,* belonging to a herd or flock). Mature gamonts (individuals that produce gametes) large, extracellular; gametes usually alike in shape and size; zygotes forming oocysts within gametocysts; parasites of digestive tract of body cavity of invertebrates, life cycle usually with one host. Examples: *Monocystis, Gregarina.*

Class Coccidea (kok-sid′e-a) (Gr. *kokkos,* kernel, grain). Mature gamonts small, typically intracellular; life cycle typically with merogony, gametogony, and sporogony; most species in vertebrates. Examples: *Cryptosporidium, Cyclospora, Eimeria, Toxoplasma, Plasmodium, Babesia.*

Phylum Ciliophora (sil-i-of′or-a) (L. *cilium,* eyelash, + Gr. *phora,* bearing). Cilia or ciliary organelles in at least one stage of life cycle; two types of nuclei, with rare exceptions; binary fission across rows of cilia; budding and multiple fission; sexuality involving conjugation and autogamy; nutrition heterotrophic; contractile vacuole typically present; most species free living, but many commensal, some parasitic. (This is a very large group, now divided into three superclasses and numerous classes and orders. Classes are separated on technical characteristics of ciliary patterns, especially around the cytostome, development of the cytostome, and other characteristics.) Examples: *Paramecium, Colpoda, Tetrahymena, Balantidium, Stentor, Blepharisma, Epidinium, Euplotes, Vorticella, Carchesium, Trichodina, Podophrya, Ephelota.* This phylum is within clade Alveolata.

Phylum Dinoflagellata (dy′no-fla-jel-at′a) (Gr. *dinos,* whirling + *flagellum,* little whip). Typically with two flagella, one transverse and one trailing; body usually grooved transversely and longitudinally, each groove containing a flagellum; chromoplasts usually yellow or dark brown, occasionally green or blue-green, bearing chlorophylls *a* and *c*; nucleus unique among eukaryotes in having chromosomes that lack or have low levels of histones; mitosis intranuclear; body form sometimes of spherical unicells, colonies, or simple filaments; sexual reproduction present; members free living, planktonic, parasitic, or mutualistic. Examples: *Zooxanthella, Ceratium, Noctiluca, Ptychodiscus.* This phylum is within clade Alveolata.

Amebas (an informal grouping for former members of Sarcodina to simplify presentation). Amebas move by pseudopodia or locomotive protoplasmic flow without discrete pseudopodia; flagella, when present, usually restricted to development or other temporary stages; body naked or with external or internal test or skeleton; asexual reproduction by fission; sexuality, if present, associated with flagellated or, more rarely, ameboid gametes; most free living.

Rhizopodans Locomotion by lobopodia, filopodia (thin pseudopodia that often branch but do not rejoin), or by protoplasmic flow without production of discrete pseudopodia. Examples: *Amoeba, Endamoeba, Difflugia, Arcella, Chlamydophrys.* These amebas are divided among several clades.

Granuloreticulosans Locomotion by reticulopodia (thin pseudopia that branch and often rejoin [anastomose]); includes foraminiferans. Examples: *Globigerina, Vertebralina.* Clade Granuloreticulosa contains these animals.

Actinopodans Locomotion by axopodia; includes radiolarians. Examples: *Actinophrys, Clathrulina.* These amebas are divided among several clades.

Summary

"Animal-like," single-celled organisms were formerly assigned to the phylum Protozoa. It is now recognized that the "phylum" was composed of numerous phyla of varying phylogenetic relationships. We use the terms *protozoa* and *protozoan* informally to refer to all these highly diverse organisms. They demonstrate the great adaptive potential of the basic body plan: a single eukaryotic cell. They occupy a vast array of niches and habitats. Many species have complex and specialized organelles.

Ciliary movement is important in both protozoa and metazoa. The most widely accepted mechanism to account for ciliary movement is the sliding microtubule hypothesis. Pseudopodial or ameboid movement is a locomotory and food-gathering mechanism in protozoa and plays a vital role as a defense mechanism in metazoa. It is accomplished by microfilaments moving past each other, and it requires expenditure of energy from ATP.

Various protozoa feed by holophytic, holozoic, or saprozoic means. The excess water that enters their bodies is expelled by contractile vacuoles. Respiration and waste elimination are through the body surface. Protozoa reproduce asexually by binary fission, multiple fission, and budding; sexual processes are common.

Members of several phyla have photoautotrophic species, including Chlorophyta, Euglenozoa, and Dinoflagellata. Some of these are very important planktonic organisms. Euglenozoa also includes many nonphotosynthetic species, and some of these cause serious diseases of humans, such as African sleeping sickness and Chagas' disease. Apicomplexa are all parasitic, including *Plasmodium,* which causes malaria. Ciliophora move by means of cilia or ciliary organelles. They are a large and diverse group, and many are complex in structure. Amebas move by pseudopodia and are now assigned to a number of mutually exclusive clades.

Review Questions

1. Explain why a protozoan may be very complex, even though it is composed of only one cell.
2. Distinguish among the following protozoan phyla: Euglenozoa, Apicomplexa, Ciliophora, Dinoflagellata.
3. Explain the transitions of endoplasm and ectoplasm in ameboid movement. What is a current hypothesis regarding the role of actin in ameboid movement?
4. Distinguish lobopodia, filipodia, reticulopodia, and axopodia.
5. Contrast the structure of an axoneme to that of a kinetosome.
6. What is the sliding-microtubule hypothesis?
7. Explain how protozoa eat, digest their food, osmoregulate, and respire.
8. Distinguish the following: sexual and asexual reproduction; binary fission, budding, and multiple fission.
9. What is the survival value of encystment?
10. Contrast and give examples of autotrophic and heterotrophic protozoa.
11. Name three kinds of amebas and tell where they are found (their habitats).
12. Outline the general life cycle of malarial organisms. How do you account for the resurgence of malaria in recent years?
13. What is the public health importance of *Toxoplasma,* and how do humans become infected with it? What is the public health importance of *Cryptosporidium* and *Cyclospora?*
14. Define the following with reference to ciliates: macronucleus, micronucleus, pellicle, undulating membrane, cirri, infraciliature, trichocysts, toxicysts, conjugation.
15. What are indications that apicomplexans descended from a photoautotrophic ancestor?
16. Distinguish primary endosymbiogenesis from secondary endosymbiogensis.

Selected References

See also general references on page 415.

Anderson, O. R. 1988. Comparative protozoology: ecology, physiology, life history. New York, Springer-Verlag. *Good treatment of the aspects mentioned in the subtitle.*

Baldauf, S. L., A. J. Roger, I. Senk-Siefert, and W. F. Doolittle. 2000. A kingdom-level phylogeny of eukaryotes based on combined protein data. Science **290:**972–977. *They contend that combining sequence data for genes encoding several proteins indicate there are 15 kingdoms of organisms.*

Bonner, J. T. 1983. Chemical signals of social amoebae. Sci. Am. **248:**114–120 (April). *Social amebas of two different species secrete different chemical compounds that act as aggregation signals. The evolution of the aggregation signals in these protozoa may provide clues to the origin of the diverse chemical signals (neurotransmitters and hormones) in more complex organisms.*

Cavalier-Smith, T. 1999. Principles of protein and lipid targeting in secondary symbiogenesis: euglenoid, dinoflagellate, and sporozoan plastid origins and the eukaryote family tree. J. Euk. Microbiol. **46:**347–366. *Many organisms are the products of secondary symbiogenesis (a eukaryote is consumed by another eukaryote, both products of primary symbiogenesis, and symbiont becomes an organelle), but tertiary symbiogenesis also has occurred (product of secondary symbiogenesis itself becomes a symbiont . . . and organelle).*

Harrison, G. 1978. Mosquitoes, malaria and man: a history of the hostilities since 1880. New York, E. P. Dutton. *A fascinating story, well told.*

Hausmann, K., and N. Hülsmann. 1996. Protozoology. New York, Thieme Medical Publishers, Inc. *This is the most up-to-date, comprehensive treatment now available.*

Lee, J. 1993. On a piece of chalk—Updated. J. Eukaryotic Microbiol. **40**:395–410. *Excellent summary of current knowledge of foraminiferans.*

Lee, J. J., S. H. Hutner, and E. C. Bovee (eds.). 1985. An illustrated guide to the protozoa. Lawrence, Kansas, Society of Protozoologists. *A comprehensive guide and essential reference for students of the protozoa, but molecular data have rendered its classification obsolete.*

Patterson, D. J. 1999. The diversity of eukaryotes. Amer. Nat. **154** (supplement): S96–S124. *Patterson provides morphological descriptions and synapomorphies for many clades containing protozoans.*

Roberts, L. S., and J. J. Janovy Jr. 2005. Foundations of parasitology, ed. 7. Dubuque, Iowa, McGraw-Hill Higher Education. *Up-to-date and readable information on parasitic protozoa.*

Roger, A. J. 1999. Reconstructing early events in eukaryotic evolution. Amer. Nat. **154** (supplement): S146–S163. *Methods in determining whether absence of mitochondria is primary or due to a secondary loss are discussed here.*

Sleigh, M. A. 1989. Protozoa and other protists. London, Edward Arnold. *Extensively updated version of the author's* The biology of protozoa.

Sogin, M. L., and J. D. Silberman. 1998. Evolution of the protists and protistan parasites from the perspective of molecular systematics. Int. J. Parasitol. **28**:11–20. *Sequence analyses of genes for ribosomal RNAs have revolutionized our concepts of phylogenetic relationships in single-celled eukaryotes.*

Stossel, T. P. 1994. The machinery of cell crawling. Sci. Am. **271**:54–63 (Sept.). *Ameboid movement—how cells crawl—is important throughout the animal kingdom, as well as in Protista. We now understand quite a bit about its mechanism.*

Wainright, P. O., G. Hinkle, M. L. Sogin, and S. K. Stickel. 1993. Monophyletic origins of the Metazoa: an evolutionary link with fungi. Science **260**:340–342. *Their data suggest that green algae and plants are closer to animals and fungi than green algae are to other unicellular groups.*

Custom Website

The *Animal Diversity* Online Learning Center is a great place to check your understanding of chapter material. Visit www.mhhe.com/hickmanad4e for access to key terms, quizzes, and more! Further enhance your knowledge with Web links to chapter-related material.

Explore live links for these topics:

Classification and Phylogeny of Animals
Protistan Groups
Phylum Sarcomastigophora
Phylum Apicomplexa
Phylum Ciliophora
Other Protozoan Phyla
Parasitic Protists
Free-living Protists

Dinoflagellates
Pfisteria
Red Tides
Adaptations for Planktonic Life

6

Sponges
Phylum Porifera

A Caribbean demosponge, *Aplysina fistularis*.

The Advent of Multicellularity

Sponges are the simplest multicellular animals. Because the cell is the elementary unit of life, organisms larger than unicellular protozoa arose as aggregates of such building units. Nature has experimented with producing larger organisms without cellular differentiation—certain large, single-celled marine algae, for example—but such examples are rarities. There are many advantages to multicellularity as opposed to simply increasing the mass of a single cell. Since it is at cell surfaces that metabolic exchange takes place, dividing a mass into smaller units greatly increases the surface area available for metabolic activities. It is impossible to maintain a workable surface-to-mass ratio by simply increasing the size of a single-celled organism. Thus, multicellularity is a highly adaptive path toward increasing body size.

Strangely, while sponges are multicellular, their organization is quite distinct from other metazoans. A sponge body is an assemblage of cells embedded in a gelatinous matrix and supported by a skeleton of minute needlelike spicules and protein. Because sponges neither look nor behave like other animals, it is understandable that they were not completely accepted as animals by zoologists until well into the nineteenth century. Nonetheless, molecular evidence suggests that sponges form a clade with other metazoa.

Sponges belong to phylum Porifera (po-rif′-er-a) (L. *porus,* pore, + *fera,* bearing). Sponges bear myriads of tiny pores and canals that constitute a filter-feeding system adequate for their inactive life habit. They are sessile animals and depend on water currents carried through their unique canal systems to bring them food and oxygen and to carry away their body wastes. Their bodies are little more than masses of cells embedded in a gelatinous matrix and stiffened by a skeleton of minute **spicules** of calcium carbonate or silica and collagen (p. 64). They have no organs or true tissues, and even their cells show a certain degree of independence. As sessile animals with only negligible body movement, they have not evolved a nervous system or sense organs and have only the simplest of contractile elements.

So, although they are multicellular, sponges share few of the characteristics of other metazoan phyla. For this reason they are often called Parazoa (Gr. *para,* beside or alongside of, + *zōon,* animal).

Sponges vary in size from a few millimeters to the great loggerhead sponges, which may reach 2 m or more across. Many sponge species are brightly colored because of pigments in their dermal cells. Red, yellow, orange, green, and purple sponges are not uncommon. However, color fades quickly when sponges are removed from water. Some sponges, including the simplest, are radially symmetrical, but many are quite irregular in shape. Some stand erect, some are branched or lobed, and others are low, even encrusting, in form (figure 6.1). Some bore holes into shells or rocks.

Sponges are an ancient group, with an abundant fossil record extending back to the early Cambrian period and even, according to some claims, the Precambrian. Living poriferans traditionally have been assigned to three classes: Calcarea (with calcareous spicules), Hexactinellida (six-rayed siliceous spicules), and Demospongiae (with a skeleton of siliceous

spicules or spongin [a specialized collagen] or both). A fourth class (Sclerospongiae) was created to contain sponges with a massive calcareous skeleton and siliceous spicules. Some zoologists maintain that known species of sclerosponges can be placed in the traditional classes of sponges (Calcarea and Demospongiae); thus, we do not need a new class.

Ecological Relationships

Most of the 5000 or more sponge species are marine, although some 150 species live in fresh water. Marine sponges are abundant in all seas and at all depths, and a few even exist in brackish water. Although their embryos are free swimming, adults are always attached, usually to rocks, shells, corals, or other submerged objects (figure 6.2). Some bottom-dwelling forms even grow on sand or mud. Their growth patterns often depend on shape of the substratum, direction and speed of water currents, and availability of space, so that the same species may differ markedly in appearance under different environmental conditions. Sponges in calm waters may grow taller and straighter than those in rapidly moving waters.

Many animals (crabs, nudibranchs, mites, bryozoans, and fish) live as commensals or parasites in or on sponges. Larger sponges particularly tend to harbor a large variety of invertebrate commensals. On the other hand, sponges grow on many other living animals, such as molluscs, barnacles, brachiopods, corals, or hydroids. One sponge has been described that preys on shrimp. Some crabs attach pieces of sponge to their carapace for camouflage and for protection, since most predators seem to find sponges distasteful. Certainly one reason for the success of sponges as a group is that they have few enemies. Because of a sponge's elaborate skeletal framework and often

figure 6.1

Some growth habits and forms of sponges.

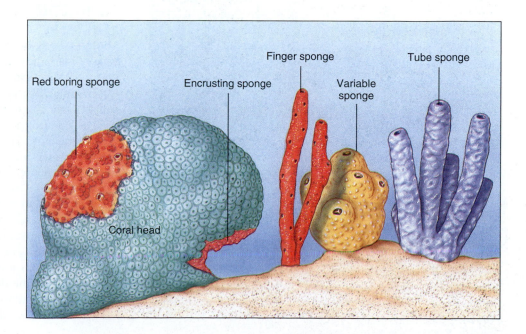

Position in the Animal Kingdom

Multicellular organisms (metazoa) are typically divided into three grades: (1) Mesozoa (a single phylum), (2) Parazoa (phylum Porifera, sponges; and phylum Placozoa), and (3) Eumetazoa (all other phyla). Space limitations preclude covering the very small phyla Mesozoa and Placozoa in this text.

Although Mesozoa and Parazoa are multicellular, their plan of organization is distinct from that in the eumetazoan phyla. Such cellular layers as they possess are not homologous to the germ layers of the Eumetazoa, and neither group has developmental patterns resembling those of other metazoa. The name Parazoa means the "beside-animals."

Biological Contributions

1. Although the simplest in organization of all metazoa, these groups do compose a higher level of morphological and physiological integration than that found in protozoan colonies. Mesozoa and Parazoa may be said to belong to a **cellular level of organization.**
2. Sponges (poriferans) have several types of cells differentiated for various functions, some of which are organized into **incipient tissues** of a low level of integration.
3. Developmental patterns of poriferans are different from those of other phyla, and their embryonic layers are not homologous to the germ layers of Eumetazoa.
4. Sponges have developed a unique system of **water currents** on which they depend for food and oxygen.

figure 6.2

This orange demosponge, *Mycale laevis,* often grows beneath platelike colonies of the stony coral, *Montastrea annularis.* The large oscula of the sponge are seen at the edges of the plates. Unlike some other sponges, *Mycale* does not burrow into the coral skeleton and may actually protect the coral from invasion by more destructive species. Pinkish radioles of a Christmas tree worm, *Spirobranchus giganteus* (phylum Annelida, class Polychaeta), also project from the coral colony. An unidentified reddish sponge can also be seen further to the right of and below the Christmas tree worm.

noxious odor, most potential predators find sampling a sponge about as pleasant as eating a mouthful of glass splinters embedded in evil-smelling gristle. Some reef fishes, however, do graze on shallow-water sponges.

Form and Function

The only body openings of these unusual animals are pores, usually many tiny ones called **ostia** for incoming water, and one to a few large ones called **oscula** (sing., **osculum**) for water outlet. These openings are connected by a system of canals, some of which are lined with peculiar flagellated collar cells called **choanocytes,** whose flagella maintain a current of environmental water through the canals. Water enters the canals through a multitude of tiny incurrent pores (**dermal ostia**) and leaves by way of one or more large oscula. Choanocytes not only keep the water moving but also trap and phagocytize food particles that are carried in the water. Cells lining the passageways are very loosely organized. Collapse of the canals is prevented by the skeleton, which, depending on the species, may be composed of needlelike calcareous or siliceous spicules, a meshwork of organic spongin fibers, or a combination of the two.

Sessile, or almost sessile, animals make few movements; they typically have little in the way of nervous, sensory, or locomotor parts. Sponges apparently have been sessile from their earliest appearance and have never acquired specialized nervous or sensory structures, and they have only the very simplest of contractile systems.

Types of Canal Systems

Most sponges have one of three types of canal systems—asconoid, syconoid, or leuconoid (figure 6.3).

Asconoids—Flagellated Spongocoels

Asconoid sponges have the simplest organization. They are small and tube shaped. Water enters through microscopic dermal pores into a large cavity called a **spongocoel,** which is lined with choanocytes. Choanocyte flagella pull water through the pores and expel it through a single large osculum (figure 6.3). *Leucosolenia* (Gr. *leukos,* white, + *solen,* pipe) is an asconoid type of sponge. Its slender, tubular individuals grow in groups attached by a common **stolon,** or stem, to

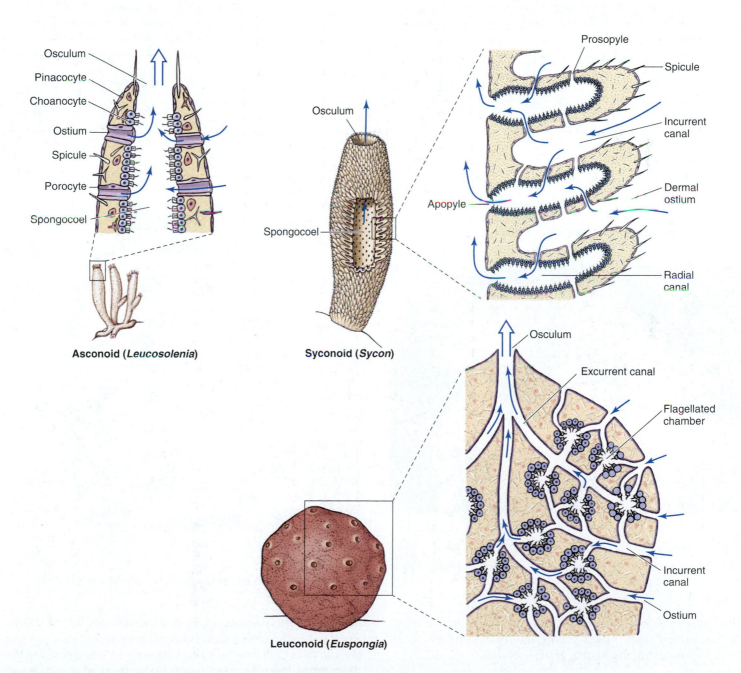

figure 6.3

Three types of canal systems. The degree of complexity from simple asconoid to complex leuconoid type has involved mainly the water canal and skeletal systems, accompanied by outfolding and branching of the collar-cell layer. The leuconoid type is considered the major plan for sponges because it permits greater size and more efficient water circulation.

Characteristics of Phylum Porifera

1. Multicellular; body a loose aggregation of cells
2. Body with **pores (ostia), canals,** and **chambers** that serve for passage of water
3. All aquatic; mostly marine
4. Symmetry radial or none
5. **Epidermis of flat pinacocytes;** most interior surfaces lined with **flagellated collar cells (choanocytes)** that create water currents; a gelatinous protein matrix called mesohyl contains amebocytes, collencytes, and skeletal elements

6. Skeletal structure of **fibrillar collagen** (a protein) and **calcareous** or **siliceous crystalline spicules,** often combined with variously modified **collagen (spongin) fibrils**
7. No organs or true tissues; digestion intracellular; excretion and respiration by diffusion
8. Reactions to stimuli apparently local and independent; nervous system probably absent
9. All adults sessile and attached to substratum
10. Asexual reproduction by buds or gemmules and sexual reproduction by eggs and sperm; free-swimming ciliated larvae

---

figure 6.4

Clathrina canariensis (class Calcarea) is common on Caribbean reefs in caves and under ledges.

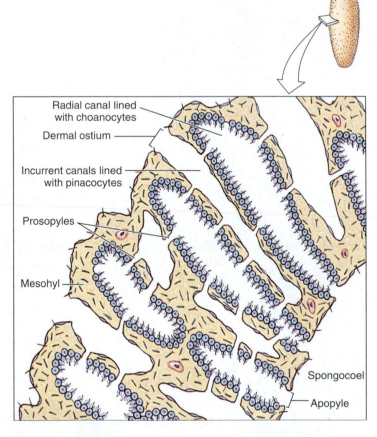

Radial canal lined with choanocytes

Dermal ostium

Incurrent canals lined with pinacocytes

Prosopyles

Mesohyl

Spongocoel

Apopyle

figure 6.5

Cross section through wall of sponge *Sycon,* showing canal system.

---

objects in shallow seawater. *Clathrina* (L. *clathri,* lattice work) is an asconoid with bright yellow, intertwined tubes (figure 6.4). Asconoids are found only in the Calcarea.

Syconoids—Flagellated Canals

Syconoid sponges look somewhat like larger editions of asconoids, from which they were derived. They have a tubular body and a single osculum, but instead of a simple choanocyte

layer lining the spongocoel, as in asconoids, this layer is folded back and forth to make canals in syconoids. The choanocytes line certain folds called **radial canals** (figure 6.3). Water, entering the body through dermal pores, moves first to **incurrent canals** and then into radial canals via small lateral openings called prosopyles (figure 6.5). From the radial canals, filtered water moves through apopyles into the spongocoel, finally exiting by the osculum. The spongocoel in syconoids is lined with epithelial-type cells (figure 6.5) rather than choanocytes as found in asconoids. Syconoids are found in Calcarea. *Sycon* (Gr. *sykon*, a fig) is a commonly studied example of the syconoid type of sponge (figure 6.5).

Leuconoids—Flagellated Chambers

Leuconoid organization is the most complex of the sponge types and permits an increase in sponge size. Most leuconoids form large masses with numerous oscula (see figure 6.2). Clusters of flagellated chambers are filled from incurrent canals and discharge water into excurrent canals that eventually lead to the osculum (see figure 6.3). There is no spongocoel. Most sponges are of the leuconoid type, which occurs in most Calcarea and in all other classes.

These three types of canal systems—asconoid, syconoid, and leuconoid—demonstrate an increase in complexity and efficiency of the water-pumping system, but they do not imply an evolutionary or developmental sequence. The leuconoid grade of construction has evolved independently many times in sponges. Possession of a leuconoid plan is of clear adaptive value; it increases the proportion of flagellated surfaces compared with the volume, thus providing more collar cells to meet food demands.

Types of Cells

Sponge cells are loosely arranged in a gelatinous matrix called **mesohyl** (figure 6.6). The mesohyl is the "connective tissue" of the sponges; in it are found various ameboid cells, fibrils, and skeletal elements. Several types of cells occur in sponges.

Pinacocytes

The nearest approach to a true tissue in sponges is arrangement of the **pinacocyte** (figure 6.6) cells of the external epithelium. These are thin, flat, epithelial-type cells that cover the exterior surface and some interior surfaces. Some are T-shaped, with their cell bodies extending into the mesohyl. Pinacocytes are somewhat contractile and help regulate the surface area of the sponge. Some pinacocytes are modified as contractile **myocytes,** which are usually arranged in circular bands around the oscula or pores, where they help regulate the rate of water flow. Myocytes contain microfilaments similar to those found in muscle cells of other animals.

Porocytes

Tubular cells that pierce the wall of asconoid sponges, through which water flows, are called **porocytes** (see figure 6.3).

Choanocytes

Choanocytes, which line flagellated canals and chambers, are ovoid cells with one end embedded in mesohyl and the other exposed. The exposed end bears a flagellum surrounded by a collar (figures 6.6 and 6.7). Electron microscopy shows that the collar is composed of adjacent

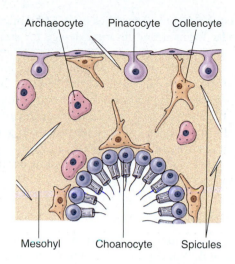

figure 6.6

Small section through sponge wall, showing four types of sponge cells. Pinacocytes are protective and contractile; choanocytes create water currents and engulf food particles; archaeocytes have a variety of functions, including phagocytosis of food particles and differentiation into other cell types; collencytes appear to have a contractile function.

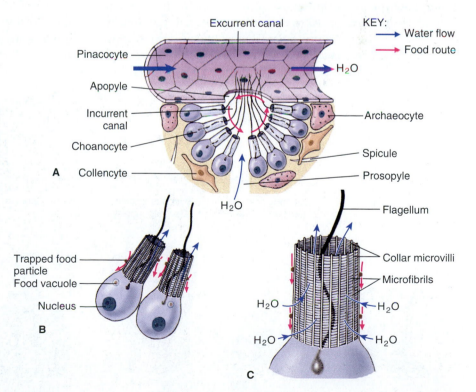

figure 6.7

Food trapping by sponge cells. **A,** Cutaway section of canals showing cellular structure and direction of water flow. **B,** Two choanocytes. **C,** Structure of the collar. Small red arrows indicate movement of food particles.

microvilli, connected to each other by delicate microfibrils, so that the collar forms a fine filtering device for straining food particles from the water (figure 6.7B and C). The beat of a flagellum pulls water through the sievelike collar and forces it out through the open top of the collar. Particles too large to enter the collar become trapped in secreted mucus and slide down the collar to the base where they are phagocytized by the cell body. Larger particles have already been screened out by the small size of the dermal pores and prosopyles. Food engulfed by the cells is passed to a neighboring archaeocyte for digestion.

Archaeocytes

Archaeocytes are ameboid cells that move about in the mesohyl (see figure 6.6) and perform a number of functions. They can phagocytize particles at the external epithelium and receive particles for digestion from choanocytes. Archaeocytes apparently can differentiate into any of the other types of more specialized cells in the sponge. Some, called **sclerocytes,** secrete spicules. Others, called **spongocytes,** secrete the spongin fibers of the skeleton, and **collencytes** secrete fibrillar collagen.

Types of Skeletons

Its skeleton gives support to a sponge, preventing collapse of canals and chambers. The major structural protein in the animal kingdom is collagen, and fibrils of collagen are found throughout the intercellular matrix of all sponges. In addition, various Demospongiae secrete a form of collagen traditionally known as spongin (figure 6.8). Demospongiae also secrete siliceous spicules. Calcareous sponges secrete spicules composed mostly of crystalline calcium carbonate that have one, three, or four rays (figure 6.8). Glass sponges (Hexactinellida) have siliceous spicules with six rays arranged in three planes at right angles to each other. There are many variations in the shape of spicules, and these structural variations are of taxonomic importance.

Sponge Physiology

Sponges feed primarily on particles suspended in water pumped through their canal systems. Detritus particles, planktonic organisms, and bacteria are consumed nonselectively in the size range from 50 μm (average diameter of ostia) to 0.1 μm (width of spaces between the microvilli of the choanocyte collar). Pinacocytes may phagocytize particles at the surface, but most larger particles are consumed in the canals by archaeocytes that move close to the lining of the canals. The smallest particles, accounting for about 80% of the particulate organic carbon, are phagocytized by choanocytes. Digestion is entirely **intracellular** (occurs within cells), a chore performed by the archaeocytes.

Sponges consume a significant portion of their nutrients in the form of organic matter dissolved in water circulating through the system. Such material is apparently taken up by pinocytosis (or potocytosis).

There are no respiratory or excretory organs; these functions are performed by diffusion. Contractile vacuoles have been found in archaeocytes and choanocytes of freshwater sponges.

All life activities of a sponge depend on a current of water flowing through the body. A sponge pumps a remarkable amount of water. Some large sponges can filter 1500 liters of

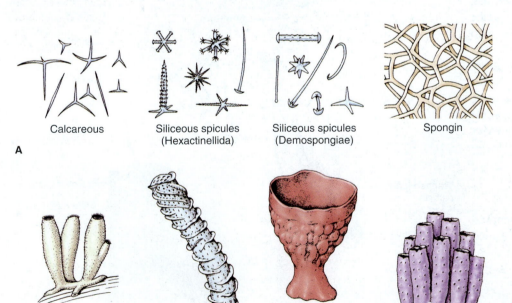

Calcareous

Siliceous spicules
(Hexactinellida)

Siliceous spicules
(Demospongiae)

Spongin

A

Leucosolenia   *Euplectella*   *Poterion*   *Callyspongia*

B

figure 6.8

A, Types of spicules found in sponges. There is amazing diversity, beauty, and complexity of form among the many types of spicules. **B,** Some examples of sponge body forms.

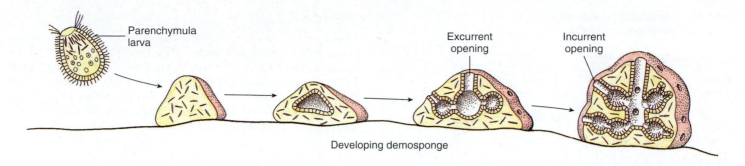

Parenchymula larva

Excurrent opening

Incurrent opening

Developing demosponge

figure 6.9

Development of demosponges.

water a day. At least some sponges can crawl (move laterally over their supporting substratum) at speeds of up to 4 mm per day. This ability may give them an advantage over more sessile encrusting organisms in competition for space.

Reproduction and Development

All sponges are capable of both sexual and asexual reproduction. In **sexual reproduction** most sponges are **monoecious** (have both male and female sex cells in one individual). Sperm arise from transformation of choanocytes. In Calcarea and at least some Demospongiae, oocytes also develop from choanocytes; in other demosponges oocytes apparently are derived from archaeocytes. Sperm are released into the water by one individual and are taken into the canal system of another. There choanocytes phagocytize them, then transform into carrier cells and carry the sperm through the mesohyl to the oocytes.

Other sponges are oviparous and expel both oocytes and sperm into the water. Ova are fertilized by motile sperm (without carrier cells) in the mesohyl. There the zygotes develop into flagellated larvae that break loose and are carried away by water currents. The free-swimming larva of most sponges is a solid-bodied **parenchymula** (figure 6.9). The outwardly directed, flagellated cells on the larval surface migrate to the interior after the larva settles and become choanocytes in the flagellated chambers.

The loose organization of sponges is ideally suited for regeneration of injured and lost parts, and for asexual reproduction. Sponges reproduce asexually by fragmentation and by forming external buds that detach or remain to form colonies. In addition to external buds, which all sponges can form, freshwater sponges and some marine sponges reproduce asexually by the regular formation of internal buds called **gemmules** (figure 6.10). These dormant masses of encapsulated archaeocytes are produced during unfavorable conditions. They can survive periods of drought and freezing and more than three months in the absence of oxygen. Later, with the return of

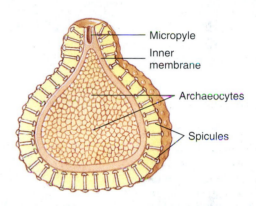

Micropyle

Inner membrane

Archaeocytes

Spicules

figure 6.10

Section through a gemmule of a freshwater sponge (Spongillidae). Gemmules are a mechanism for survival of the harsh conditions of winter. On return of favorable conditions, the archaeocytes exit through the micropyle to form a new sponge. The archaeocytes of the gemmule give rise to all cell types of the new sponge.

favorable conditions for growth, archaeocytes in the gemmules escape and develop into new sponges.

Brief Survey of Sponges

Class Calcarea (Calcispongiae)

Calcarea are calcareous sponges, so called because their spicules are composed of calcium carbonate. Spicules are straight monaxons or have three or four rays (see figure 6.8A). The sponges tend to be small—10 cm or less in height—and tubular or vase shaped. They may be asconoid, syconoid, or leuconoid in structure. Although many are drab, some are bright yellow, red, green, or lavender. *Leucosolenia*, *Clathrina* (see figure 6.4), and *Sycon* are common examples.

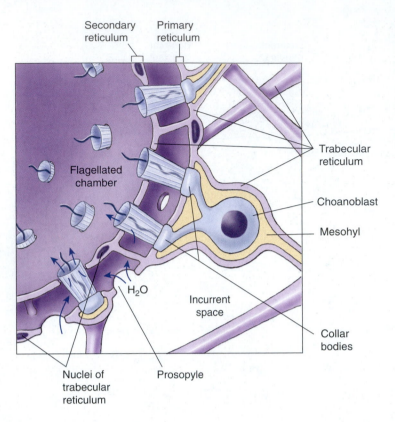

Secondary reticulum
Primary reticulum
Flagellated chamber
Trabecular reticulum
Choanoblast
Mesohyl
H₂O
Incurrent space
Collar bodies
Nuclei of trabecular reticulum
Prosopyle

figure 6.11

Diagram of part of a flagellated chamber of hexactinellids. The primary and secondary reticula are branches of the trabecular reticulum, which is syncytial. Cell bodies of the choanoblasts and their processes are borne by the primary reticulum and are embedded in a thin, collagenous mesohyl. Processes of the choanoblasts end in collar bodies, whose collars extend up through the secondary reticulum. Flagellar action propels water (*arrows*) to be filtered through the mesh of collar microvilli.

Class Hexactinellida (Hyalospongiae)

Glass sponges are nearly all deep-sea forms. Most are radially symmetrical and range from 7 to 10 cm to more than 1 m in length. One distinguishing feature, reflected in the class name, is the skeleton of six-rayed siliceous spicules bound together in an exquisite glasslike latticework (see figure 6.8A).

Their tissue structure differs so dramatically from other sponges that some scientists advocate placing hexactinellids in a subphylum separate from other sponges. The body of hexactinellids consists of a single, continuous syncytial tissue called a **trabecular reticulum.** The trabecular reticulum is the largest, continuous syncytial tissue known in Metazoa. It is bilayered and encloses a thin, collagenous mesohyl between the layers, as well as cellular elements such as archaeocytes, sclerocytes, and **choanoblasts.** Choanoblasts are associated with flagellated chambers, where the layers of the trabecular reticulum separate into a **primary reticulum** (incurrent side) and a **secondary reticulum** (excurrent, or atrial side) (figure 6.11). The spherical choanoblasts are borne by the primary reticulum, and each choanoblast has one or more processes extending to **collar bodies,** the bases of which are also supported by the primary reticulum. Each collar and flagellum extends into the flagellated chamber through an opening in the secondary reticulum. Water is drawn into the space between primary and secondary reticula through prosopyles in the primary reticulum, then through the collars into the lumen of the flagellated chamber. Collar bod-

ies do not participate in phagocytosis; that process is accomplished by the primary and secondary reticula.

The latticelike network of spicules found in many glass sponges is of exquisite beauty, such as that of *Euplectella* (NL. from Gr. *euplektos,* well-plaited), a classic example of Hexactinellida (figure 6.8B).

Class Demospongiae

Demospongiae comprise approximately 80% of all sponge species, including most larger sponges. Their skeletons may be of siliceous spicules, spongin fibers, or both (figure 6.8A). All members of the class are leuconoid, and all are marine except one family, the freshwater Spongillidae. Freshwater sponges are widely distributed in well-oxygenated ponds and streams, where they are found encrusting plant stems and old pieces of submerged wood. They resemble a bit of wrinkled scum, pitted with pores, and are brownish or greenish in color. Freshwater sponges die and disintegrate in late autumn, leaving gemmules to survive the winter.

Marine Demospongiae are varied in both color and shape. Some are encrusting; some are tall and fingerlike; and some are shaped like fans, vases, cushions, or balls (figure 6.12). Some sponges bore into and excavate molluscan shells and coral skeletons. Loggerhead sponges may grow several meters in diameter. So-called bath sponges belong to the group called horny sponges, which have only spongin skeletons. They can be cultured by cutting out pieces of the individual sponges,

A   B   C

figure 6.12

Marine Demospongiae on Caribbean coral reefs. **A,** *Pseudoceratina crassa* is a colorful sponge growing at moderate depths. **B,** *Ectyoplasia ferox* is irregular in shape and its oscula form small, volcano-like cones. It is toxic and may cause skin irritation if touched. **C,** *Monanchora unguifera* with commensal brittle star, *Ophiothrix suensoni* (phylum Echinodermata, class Ophiuroidea).

fastening them to a weight, and dropping them into the proper water conditions. It takes many years for them to grow to market size. Most commercial "sponges" now on the market are synthetic, but the harvest and use of bath sponges persist.

Phylogeny and Adaptive Radiation

Phylogeny

Sponges originated before the Cambrian period. Two groups of calcareous spongelike organisms occupied early Paleozoic reefs. The Devonian period saw rapid development of many glass sponges. The possibility that sponges arose from choanoflagellates (protozoa that bear collars and flagella) earned support for a time. However, many zoologists object to that hypothesis because sponges do not acquire collars until later in their embryological development. The outer cells of the larvae are flagellated but not collared, and they do not become collar cells until they become internal. Also, collar cells are found in certain corals and echinoderms, so they are not unique to the sponges.

However, these objections are countered by evidence based on the sequences of ribosomal RNA. This evidence supports the hypothesis of a common ancestor for choanoflagellates and metazoans (see figure 5.1). It suggests also that sponges and Eumetazoa are sister groups, with Porifera having split off before the origin of the radiates.

Adaptive Radiation

Porifera are a highly successful group that includes several thousand species and a variety of marine and freshwater habitats. Their diversification centers largely on their unique water-current system and its various degrees of complexity. Proliferation of flagellated chambers in leuconoid sponges was more favorable to an increase in body size than that of asconoid and syconoid sponges because facilities for feeding and gaseous exchange were greatly enlarged.

One very novel way of feeding has evolved within a family of sponges inhabiting nutrient-poor deep-water caves. Each sponge has a fine coating of tiny hooklike spicules over its highly branched body. The spicule layer entangles the appendages of tiny crustaceans swimming near the sponge surface. Later, filaments of the sponge body grow over the prey, enveloping and digesting them. These animals are carnivores, not suspension feeders; they lack choanocytes and internal canals, but have silicious spicules like typical members of class Demospongiae. In addition to capturing prey, some augment their diets with nutrients obtained from symbiotic methanotrophic bacteria. To colonize such a nutrient-poor habitat initially, the ancestors of this group must have had at least one alternate feeding system, either carnivory or chemoautotrophy, already in place. Presumably, after the alternative method of food capture was in use, the choanocytes and internal canals were no longer formed.

classification of Phylum Porifera

Class Calcarea (cal-ca're-a) (L. *calcis,* lime, + Gr. *spongos,* sponge) **(Calcispongiae).** Have spicules of calcium carbonate that often form a fringe around the osculum; spicules needle-shaped or three- or four-rayed; all three types of canal systems (asconoid, syconoid, leuconoid) represented; all marine. Examples: *Sycon, Leucosolenia, Clathrina.*

Class Hexactinellida (hex-ak-tin-el'i-da) (Gr. *hex,* six, + *aktis,* ray) **(Hyalospongiae).** Have six-rayed, siliceous spicules extending at right angles from a central point; spicules often united to form network; body often cylindrical or funnel shaped.

Flagellated chambers in simple syconoid or leuconoid arrangement. Habitat mostly deep water; all marine. Examples: Venus' flower basket *(Euplectella), Hyalonema.*

Class Demospongiae (de-mo-spun'je-e) (tolerated misspelling of Gr. *desmos,* chain, tie, bond, + *spongos,* sponge). Have skeleton of siliceous spicules that are not six-rayed, or spongin, or both. Leuconoid-type canal systems. One family found in fresh water; all others marine. Examples: *Thenea, Cliona, Spongilla, Myenia,* and all bath sponges.

Summary

Sponges (phylum Porifera) are an abundant marine group with some freshwater representatives. They have various specialized cells, but these cells are not organized into tissues or organs. They depend on the flagellar beat of their choanocytes to circulate water through their bodies for gathering food and exchange of respiratory gases. They are supported by secreted skeletons of fibrillar collagen, collagen in the form of large fibers or filaments (spongin), calcareous or siliceous spicules, or a combination of spicules and spongin in most species.

Most sponges are monoecious but produce sperm and oocytes at different times. Embryogenesis is unusual with a migration of flagellated cells from the embryo surface to the interior of the larva. The larva is typically a free-swimming parenchymula form. Sponges have great regenerative capabilities and may reproduce asexually by budding, fragmentation, and gemmules (internal buds).

Sponges are an ancient group, remote phylogenetically from other metazoa, but molecular evidence suggests that they are the sister group to Eumetazoa. Their adaptive radiation is centered on elaboration of the water circulation and filter-feeding system, with the exception of a unique group of cave-dwelling carnivorous sponges that do not filter-feed.

Review Questions

1. Give six characteristics of sponges.
2. Briefly describe asconoid, syconoid, and leuconoid body types in sponges.
3. What sponge body type is most efficient and makes possible the largest body size?
4. Define the following: ostia, osculum, spongocel, mesohyl.
5. Define the following: pinacocytes, choanocytes, archaeocytes, sclerocytes, collencytes.
6. What material is found in the skeleton of all sponges?
7. Describe the skeletons of each of the classes of sponges.
8. Describe how sponges feed, respire, and excrete.
9. What is a gemmule?
10. Describe how gametes are produced and the process of fertilization in most sponges.
11. What is the largest class of sponges, and what is its body type?
12. What are possible ancestors to sponges? Justify your answer.
13. It has been suggested that despite being large, multicellular animals, sponges function more like protozoa. What aspects of sponge biology support this statement and how? Consider, for example, nutrition, reproduction, gas exchange, and cellular organization.

Selected References

See also general references on page 415.

Bergquist, P. R. 1978. Sponges. Berkeley, California, University of California Press. *Excellent monograph on sponge structure, classification, evolution, and general biology.*

Bond, C. 1997. Keeping up with the sponges. Nat. Hist. **106**:22–25. *Sponges are not fixed in permanent position; they can crawl on their substrate.* Haliclona loosanoffi *can move over 4 mm/day.*

Gould, S. J. 1995. Reversing established orders. Nat. Hist. **104**(9):12–16. *Describes several anomalous animal relationships, including the sponge that preys on shrimp.*

Leys, S. P. 1999. The choanosome of hexactinellid sponges. Invert. Biol. **118**: 221–235. *Choanosomes are flagellated chambers and associated tissues. This author supports the position that Hexactinellida should constitute a separate subphylum. She includes an excellent description of trabecular reticulum.*

Reiswig, H. M., and T. L. Miller. 1998. Freshwater sponge gemmules survive months of anoxia. Invert. Biol. **117**:1–8. *Hatchability of gemmules kept in the absence of oxygen was equal to controls, but they would not hatch unless oxygen was present.*

Vacelet, J., and N. Boury-Esnault. 1995. Carnivorous sponges. Nature **373**: 333–335. *A fascinating article on a novel feeding method. Later work demonstrates that symbiotic methanotrophic bacteria produce a second source of nutrition for the sponges.*

Wood, R. 1990. Reef-building sponges. Am. Sci. **78**:224–235. *The author presents evidence that the known sclerosponges belong to either the Calcarea or the Demospongiae and that a separate class Sclerospongiae is not needed.*

Wyeth, R. C. 1999. Video and electron microscopy of particle feeding in sandwich cultures of the hexactinellid sponge,* Rhabdocalyptus dawsoni. *Invert. Biol. **118**:236–242. *Phagocytosis is not by choanoblasts but by trabecular reticulum, especially primary reticulum. He places Hexactinellida in subphylum Symplasma and the rest of Porifera in subphylum Cellularia.*

Custom Website

The *Animal Diversity* Online Learning Center is a great place to check your understanding of chapter material. Visit www.mhhe.com/hickmanad4e for access to key terms, quizzes, and more! Further enhance your knowledge with Web links to chapter-related material.

Explore live links for these topics:

Classification and Phylogeny of Animals
Phylum Porifera

7

Radiate Animals
Cnidarians and Ctenophores

Tentacles of a Caribbean sea anemone, *Condylactis gigantea*.

A Fearsome Tiny Weapon

Although members of phylum Cnidaria are more highly organized than sponges, they are still relatively simple animals. Most are sessile; those that are unattached, such as jellyfish, can swim only feebly. None can chase their prey. Indeed, we might easily get the false impression the cnidarians were placed on earth to provide easy meals for other animals. The truth is, however, many cnidarians are very effective predators that are able to kill and eat prey that are much more highly organized, swift, and intelligent. They manage these feats because they possess tentacles that bristle with tiny, remarkably sophisticated weapons called nematocysts.

As it is secreted within the cell that contains it, a nematocyst is endowed with potential energy to power its discharge. It is as though a factory manufactured a gun, cocked and ready with a bullet in its chamber, as it rolls off the assembly line. Like the cocked gun, the completed nematocyst requires only a small stimulus to make it fire. Rather than a bullet, a tiny thread bursts from the nematocyst. Achieving a velocity of 2 m/sec and an acceleration of 40,000 × gravity, it instantly penetrates its prey and injects a paralyzing toxin. A small animal unlucky enough to brush against one of the tentacles is suddenly speared with hundreds or even thousands of nematocysts and quickly immobilized. Some nematocyst threads can penetrate human skin, resulting in sensations ranging from minor irritation to great pain, even death, depending on the species. A fearsome, but wondrous, tiny weapon.

Phylum Cnidaria

Phylum Cnidaria (ny-dar′e-a) (Gr. *knidē,* nettle, + L. *aria* [pl. suffix]; like or connected with) is an interesting group of more than 9000 species. It takes its name from cells called **cnidocytes,** which contain the stinging organelles (cnidae) characteristic of the phylum. Cnidae come in several types, including the common **nematocysts.** Nematocysts are *formed and used only by cnidarians.* Another name for the phylum, Coelenterata (se-len′te-ra′ta) (Gr. *koilos,* hollow, + *enteron,* gut, + L. *ata* [pl. suffix], characterized by), is used less commonly than formerly, and it sometimes now refers to both radiate phyla, (Cnidaria and Ctenophora) because its meaning is equally applicable to both.

Cnidarians are generally regarded as originating close to the basal stock of the metazoan line. They are an ancient group with the longest fossil history of any metazoan, reaching back more than 700 million years. Although their organization has a structural and functional simplicity not found in other metazoans, they form a significant proportion of the biomass in some locations. They are widespread in marine habitats, and there are a few in fresh water. Although they are mostly sessile or, at best, fairly slow moving or slow swimming, they are quite efficient predators of organisms that are much swifter and more complex.

The phylum includes some of nature's strangest and loveliest creatures: branching, plantlike hydroids; flowerlike sea anemones; jellyfishes; and those architects of the ocean floor, gorgonian corals (sea whips, sea fans, and others), and stony corals, whose thousands of years of calcareous housebuilding have produced great reefs and coral islands. Algae frequently live as mutuals in tissues of cnidarians, notably in some freshwater hydras and in reef-building corals. The presence of algae in reef-building corals limits the occurrence of coral reefs to relatively shallow, clear water where sunlight is sufficient for photosynthetic requirements of the algae. These corals are an essential component of coral reefs, and reefs are extremely important habitats in tropical waters. Coral reefs are discussed further later in the chapter (see p. 139).

We recognize four classes of Cnidaria: Hydrozoa (the most variable class, including hydroids, fire corals, Portuguese man-of-war, and others), Scyphozoa ("true" jellyfishes), Cubozoa (cube jellyfishes), and Anthozoa (the largest class, including sea anemones, stony corals, soft corals, and others).

Position in the Animal Kingdom

The two phyla Cnidaria and Ctenophora contain the radiate animals, which are characterized by **primary radial** or **biradial symmetry.** Radial symmetry, in which body parts are arranged concentrically around an oral-aboral axis, is particularly suitable for sessile or sedentary animals and for free-floating animals because they approach their environment (or it approaches them) from all sides equally. Biradial symmetry is basically a type of radial symmetry in which only two planes through the oral-aboral axis divide the animal into mirror images because of the presence of some part that is paired. All other eumetazoans have a primary bilateral symmetry; they are bilateral or were derived from an ancestor that was bilateral.

Neither phylum has advanced generally beyond the **tissue level of organization,** although a few organs occur. In general, ctenophores are structurally more complex than cnidarians.

Biological Contributions

1. Both phyla have developed two well-defined **germ layers,** ectoderm and endoderm; a third, middle layer, which is derived embryologically from ectoderm, is present in some. The body plan is saclike, and the body wall is composed of two distinct layers, epidermis and gastrodermis, derived from ectoderm and endoderm, respectively. The gelatinous matrix, mesoglea, between these layers may be structureless, may contain a few cells and fibers, or may be composed largely of connective tissue and muscle fibers.

2. An internal body cavity, the **gastrovascular cavity,** is lined by gastrodermis and has a single opening, the mouth, which also serves as the anus.

3. **Extracellular digestion** occurs in the gastrovascular cavity, and intracellular digestion takes place in gastrodermal cells. Extracellular digestion allows ingestion of larger food particles.

4. Most radiates have **tentacles,** or extensible projections around the oral end, that aid in capturing food.

5. Radiates are the simplest animals to possess true **nerve cells** (protoneurons), but nerves are arranged as a nerve net, with no central nervous system.

6. Radiates are the simplest animals to possess sense organs, which include well-developed statocysts (organs of equilibrium) and ocelli (photosensitive organs).

7. Locomotion in free-moving forms is achieved by either **muscular contractions** (cnidarians) or **ciliary comb plates** (ctenophores). However, both groups are still better adapted to floating or being carried by currents than to strong swimming.

8. Some unique features are found in these phyla, such as **nematocysts** (stinging organelles) in cnidarians and **colloblasts** (adhesive organelles) and **ciliary comb plates** in ctenophores.

characteristics
of Phylum Cnidaria

1. Entirely aquatic, some in fresh water but mostly marine
2. **Radial symmetry** or biradial symmetry around a longitudinal axis with **oral** and **aboral** ends; no definite head
3. Two basic types of individuals: **polyps** and **medusae**
4. **Polymorphism**[1] in cnidarians has widened their ecological possibilities. In many species the presence of both a polyp (sessile and attached) stage and a medusa (free-swimming) stage permits occupation of a benthic (bottom) and a pelagic (open-water) habitat by the same species. Polymorphism among polyp types also widens the possibilities of structural complexity.
5. Exoskeleton or endoskeleton of chitinous, calcareous, or protein components in some
6. Body with two layers, epidermis and gastrodermis, with mesoglea **(diploblastic);** mesoglea with cells and connective tissue (ectomesoderm) in some
7. **Gastrovascular cavity** (often branched or divided with septa) with a single opening that serves as both mouth and anus; extensible tentacles usually encircling the mouth or oral region
8. Special stinging-cell organelles called cnidae, in particular nematocysts, in either epidermis or gastrodermis or in both; nematocysts abundant on tentacles, where they may form batteries or rings
9. **Nerve net** with symmetrical and asymmetrical synapses; with some sensory organs; diffuse conduction
10. Muscular system (epitheliomuscular type) of an outer layer of longitudinal fibers at base of epidermis and an inner one of circular fibers at base of gastrodermis; modifications of this plan in some cnidarians, such as separate bundles of independent fibers in mesoglea
11. Asexual reproduction by budding (in polyps) or sexual reproduction by gametes (in all medusae and some polyps); sexual forms monoecious or dioecious; **planula larva;** holoblastic indeterminate cleavage
12. No excretory or respiratory system
13. No coelomic cavity

---

[1]Note that polymorphism here refers to more than one structural form of individual within a species, as contrasted with the use of the word in genetics, in which it refers to different allelic forms of a gene in a population.

Ecological Relationships

Cnidarians are found most abundantly in shallow marine habitats, especially in warm temperatures and tropical regions. There are some freshwater species, but no terrestrial species. Colonial hydroids are usually found attached to mollusc shells, rocks, wharves, and other animals in shallow coastal water, but some species are found at great depths. Floating and free-swimming medusae are found in open seas and lakes, often far from shore. Floating colonies such as the Portuguese man-of-war and *Velella* (L. *velum,* veil, + *ellus,* dim. suffix) have floats or sails by which the wind carries them.

Some ctenophores, molluscs, and flatworms eat hydroids bearing nematocysts and use these stinging structures for their own defense. Some other animals, such as some molluscs and fishes, feed on cnidarians, but cnidarians rarely serve as food for humans.

Cnidarians sometimes live symbiotically with other animals, often as commensals on the shell or other surface of their host. Certain hydroids (figure 7.1) and sea anemones commonly live on snail shells inhabited by hermit crabs, providing the crabs some protection from predators.

Although many cnidarians have little economic importance, reef-building corals are an important exception. Fish and other animals associated with reefs provide substantial amounts of food for humans, and reefs are of economic value as tourist attractions. Precious coral is used for jewelry and ornaments, and coral rock serves for building purposes.

Planktonic medusae may be of some importance as food for fish that are of commercial value; the reverse is also true—young fish fall prey to cnidarians.

Form and Function

Dimorphism and Polymorphism in Cnidarians

One of the most interesting—and sometimes puzzling—aspects of this phylum is the dimorphism and often polymorphism displayed by many of its members. All cnidarian forms fit into one of two morphological types (dimorphism): a **polyp,** or hydroid form, which is adapted to a sedentary or sessile life, and a **medusa,** or jellyfish form, which is adapted for a floating or free-swimming existence (figure 7.2).

Most polyps have tubular bodies. A mouth surrounded by tentacles defines the oral end of the body. The mouth leads into a blind gut or gastrovascular cavity (figure 7.2). The aboral end of the polyp is usually attached to a substratum by a pedal disc or other device.

Polyps may reproduce asexually by budding, fission, or pedal laceration. In **budding,** a knob of tissue forms on the side of an existing polyp and develops a functional mouth and tentacles (see figure 7.7). A bud that detaches from the polyp that made it is a clone. Clones can be formed also by **fission,** where one-half of a polyp pulls away from the other, or by **pedal laceration,** where tissue torn from the pedal disc develops into new, tiny polyps (p. 137). Polyps that do not bud are solitary; others form clones or colonies. The distinction between clones and colonies may be blurred when a colony fragments.

When buds stay attached to the polyp that made them, a colony forms and food may be shared through a common gastrovascular cavity (figure 7.1). A shared gastrovascular cav-

A

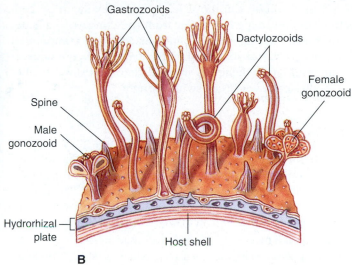

B

figure 7.1

A, A hermit crab with its cnidarian mutuals. The shell is blanketed with polyps of the hydrozoan *Hydractinia milleri*. The crab gets some protection from predation by the cnidarians, and the cnidarians get a free ride and bits of food from their host's meals. **B,** Portion of a colony of *Hydractinia,* showing the types of zooids and the stolon (hydrorhiza) from which they grow.

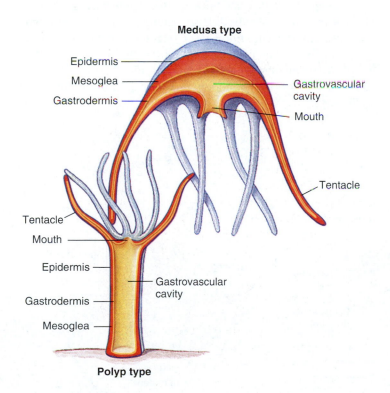

figure 7.2

Comparison between the polyp and medusa types of individuals.

ity permits some polyps (often called zooids) to specialize for particular functions (figure 7.1) such as feeding (gastrozooids), defense (dactylozooids), or making sexually reproducing stages (gonozooids). Gastrozooids and dactylozooids do not collect food, but are fed through the shared gastrovas-

cular cavity. A colony that contains several body forms is polymorphic; such colonies are common in class Hydrozoa.

The name "medusa" was suggested by a fancied resemblance to the Gorgon Medusa, a mythological creature with snaky tresses that turned to stone any who gazed upon her.

Medusae are usually free swimming and have bell-shaped or umbrella-shaped bodies and tetramerous symmetry (body parts arranged in fours). The mouth is usually centered on the concave side, and tentacles extend from the rim of the umbrella.

Superficially, polyps and medusae seem very different. But actually each has retained the saclike body plan basic to the phylum (figure 7.2). A medusa is essentially an unattached polyp with the tubular portion widened and flattened into the bell shape.

Both polyp and medusa possess two tissue layers with an acellular mesoglea between them, but the jellylike layer of mesoglea is much thicker in a medusa, constituting the bulk of the animal and making it more buoyant. Because of this mass of mesoglea ("jelly"), medusae are commonly called jellyfishes.

Locomotion

Colonial polyps are permanently attached, but hydras can move about freely by gliding on their basal disc, aided by mucus secretions. Sea anemones can move similarly on their basal discs. Hydras can also use a "measuring worm" movement, looping along a surface by bending over and attaching their tentacles to the substratum. They may even turn handsprings or detach and, by forming a gas bubble on the basal disc, float to the surface.

Most medusae can move freely, and they swim by contracting the bell, expelling water from the concave, oral side. The muscular contractions are antagonized by the compressed mesoglea and elastic fibers within it. Usually, they contract several times and move generally upward, then sink slowly. Cubozoan medusae, however, can swim strongly.

Life Cycles

In a cnidarian life cycle, polyps and medusae play different roles. The exact life cycle varies among the classes, but in general a zygote develops into a free-swimming **planula** larva. The planula settles and metamorphoses into a polyp. A polyp may make other polyps asexually, but polyps in classes Hydrozoa and Scyphozoa eventually make medusae. Although medusae are made asexually from the polyp body, they will reproduce sexually. Medusae develop into either male or female individuals (dioecious) and produce gametes. Fertilization typically occurs in open water, although some species brood their zygotes until the planula stage.

Sea anemones and corals (class Anthozoa) are all polyps: hence, both sexual and asexual reproduction occurs in the polyp phase. The true jellyfishes (class Scyphozoa) have a conspicuous medusoid form, but many have a polypoid larval stage. Colonial hydroids of class Hydrozoa, however, sometimes have life histories that feature both a polyp stage and a free-swimming medusa stage—rather like a Jekyll-and-Hyde existence. A species that has both an attached polyp and a floating medusa within its life history can take advantage of the feeding and distribution possibilities of both pelagic (open-water) and benthic (bottom) environments.

Feeding and Digestion

The mouth opens into the **gastrovascular cavity** (coelenteron), which communicates with cavities in the tentacles. The mouth may be surrounded by an elevated **manubrium** or by elongated **oral lobes.** Cnidarians prey on a variety of organisms of appropriate size; larger species are usually capable of killing and eating larger prey. Normally prey organisms are drawn into the gastrovascular cavity into which gland cells discharge enzymes. Digestion is started in the gastrovascular cavity **(extracellular digestion),** but nutritive-muscular cells phagocytize many food particles for **intracellular digestion.** Ameboid cells may carry undigested particles to the gastrovascular cavity, where they are eventually expelled with other indigestible matter.

Body Wall

The body wall surrounding the gastrovascular cavity consists of an outer **epidermis** (ectodermal) and an inner **gastrodermis** (endodermal) with **mesoglea** between them (see figure 7.3). Each layer and the cells it contains are discussed briefly.

Mesoglea   Mesoglea lies between the epidermis and gastrodermis and adheres to both layers (figure 7.2). It is gelatinous, or jellylike, and has no fibers or cellular elements in hydrozoan polyps. It is thicker in medusae and has elastic fibers, and in scyphozoan medusae it has ameboid cells. The mesoglea of anthozoans contains ameboid cells and epitheliomuscular cells (see p. 127).

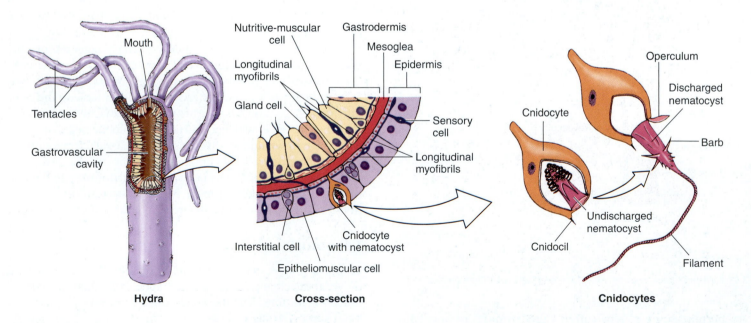

figure 7.3

At right, structure of a stinging cell. Center, portion of the body wall of a hydra. Cnidocytes, which contain the nematocysts, arise in the epidermis from interstitial cells.

Gastrodermis   The gastrodermis, a layer of cells lining the gastrovascular cavity, is made up chiefly of large, ciliated, columnar epithelial cells with irregular flat bases (figure 7.3). Cells of the gastrodermis include nutritive-muscular, interstitial, and gland cells and, in classes other than Hydrozoa, cnidocytes Gonads are gastrodermal in most cnidarians.

Nutritive-muscular cells are usually tall columnar cells that have laterally extended bases containing myofibrils. In hydrozoans the myofibrils run at right angles to the body or tentacle axis and so form a circular muscle layer. However, this muscle layer is very weak, and longitudinal extension of the body and tentacles is achieved mostly by increasing the volume of water in the gastrovascular cavity. Water is brought in through the mouth by the beating of cilia on the nutritive-muscular cells in hydrozoans or by ciliated cells in the pharynx of anthozoans. Thus, water in the gastrovascular cavity serves as a **hydrostatic skeleton.** The two cilia on the free end of each cell also serve to circulate food and fluids in the digestive cavity. The cells often contain large numbers of food vacuoles. Gastrodermal cells of green hydras (*Chlorohydra* [Gr. *chlōros,* green, + *hydra,* a mythical nine-headed monster slain by Hercules]), bear green algae **(zoochlorellae),** but in marine cnidarians they are a type of dinoflagellate (p. 102) **(zooxanthellae).** Both are cases of mutualism, with the algae furnishing organic compounds they have synthesized to their cnidarian hosts.

Interstitial cells scattered among the bases of the nutritive cells can transform into other cell types. **Gland cells** are tall cells that secrete digestive enzymes.

Epidermis   The epidermal layer contains epitheliomuscular, interstitial, gland, cnidocyte, and sensory and nerve cells. Gonads are epidermal in Hydrozoa.

Epitheliomuscular cells   (figure 7.4) form most of the epidermis and serve both for covering and for muscular contraction. The bases of most of these cells are extended parallel to the tentacle or body axis and contain myofibrils, thus forming a layer of longitudinal muscle next to the mesoglea. Contraction of these fibrils shortens the body or tentacles.

Interstitial cells are undifferentiated stem cells found among the bases of the epitheliomuscular cells. Differentiation of interstitial cells gives rise to cnidoblasts, sex cells, buds, nerve cells, and others, but generally not to epitheliomuscular cells (which reproduce themselves).

Gland cells are particularly abundant around the mouth and in the pedal disc of hydra. They secrete mucus or adhesive material.

Cnidocytes containing cnidae are found throughout the epidermis (figure 7.3). They may be between the epitheliomuscular cells or housed in invaginations of these cells, and they are most abundant on the tentacles. Over 20 different types of cnidae have been described in cnidarians so far; they are important in taxonomic determinations. The cnida is a tiny capsule composed of material similar to chitin and containing a coiled tubular "thread" or filament, which is a continuation of the narrowed end of the capsule. This end of the capsule is covered by a little lid, or **operculum.** The inside of the undischarged thread may bear tiny barbs, or spines, as in the most common cnida, the **nematocyst.** There are three functional types of cnidae in hydras: those that penetrate prey and inject poison (penetrants, see figure 7.3); those that recoil and entangle prey (volvents, see figure 7.5); and those that secrete an adhesive substance used in locomotion and attachment (glutinants).

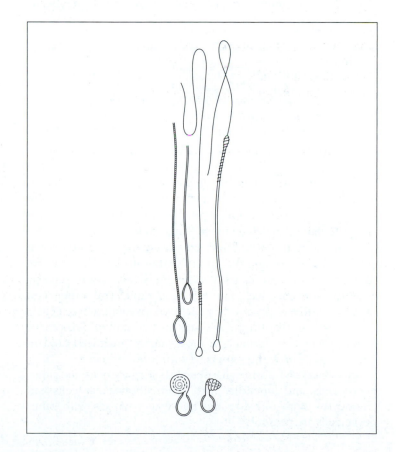

figure 7.5

Several types of cnidae shown after discharge. At bottom are two views of a type that does not impale the prey; it recoils like a spring, catching any small part of the prey in the path of the recoiling thread.

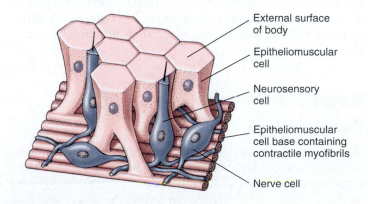

External surface of body

Epitheliomuscular cell

Neurosensory cell

Epitheliomuscular cell base containing contractile myofibrils

Nerve cell

figure 7.4

Epitheliomuscular and nerve cells in hydra.

A cnida is enclosed in the cell that has produced it (during its development, a cnidocyte is properly called a **cnidoblast**). Except in Anthozoa, cnidocytes are equipped with a triggerlike **cnidocil,** which is a modified cilium. Anthozoan cnidocytes have a somewhat different ciliary mechanoreceptor. In some sea anemones, and perhaps other cnidarians, small organic molecules from a prey "tune" the mechanoreceptors, sensitizing them to the frequency of vibration caused by the prey swimming. Tactile stimulation causes a nematocyst to discharge. Cnidocytes are borne in invaginations of ectodermal cells and, in some forms, in gastrodermal cells, and they are especially abundant on the tentacles. When a nematocyst has discharged, its cnidocyte is absorbed and a new one replaces it.

The mechanism of nematocyst discharge is remarkable. Present evidence indicates that discharge is due to a combination of tensional forces generated during nematocyst formation and also to an astonishingly high osmotic pressure within the nematocyst: 140 atmospheres. When stimulated to discharge, permeability of the nematocyst changes, and the high internal osmotic pressure causes water to rush into the capsule. The operculum opens, and the rapidly increasing *hydrostatic pressure* within the capsule forces the thread out with great force, the thread turning inside out as it goes. At the everting end of the thread, the barbs flick to the outside like tiny switchblades. This minute but awesome weapon then injects poison when it penetrates prey.

Note the distinction between osmotic and hydrostatic pressure. A nematocyst is never required actually to contain 140 atmospheres of hydrostatic pressure within itself; such a hydrostatic pressure would doubtless cause it to explode. As water rushes in during discharge, osmotic pressure falls rapidly, while hydrostatic pressure rapidly increases.

Nematocysts of most cnidarians are not harmful to humans and are a nuisance at worst. However, stings of the Portuguese man-of-war (see figure 7.12) and certain jellyfish are quite painful and sometimes dangerous.

Sensory cells are scattered among the other epidermal cells, especially around the mouth and tentacles. The free end of each sensory cell bears a flagellum, which is the sensory receptor for chemical and tactile stimuli. The other end branches into fine processes, which synapse with nerve cells.

Nerve cells of the epidermis are often multipolar (have many processes), although in more highly organized cnidarians the cells may be bipolar (with two processes). Their processes (axons) form synapses with sensory cells and other nerve cells, and junctions with epitheliomuscular cells and cnidocytes. Both one-way and two-way synapses with other nerve cells are present.

Nerve Net

The nerve net of the cnidarians is one of the best examples of a diffuse nervous system in the animal kingdom. This plexus of nerve cells is found both at the base of the epidermis and at the base of the gastrodermis, forming two interconnected nerve nets. Nerve processes (axons) end on other nerve cells at synapses or at junctions with sensory cells or effector organs (nematocysts or epitheliomuscular cells). Nerve impulses are transmitted from one cell to another by release of a neurotransmitter from small vesicles on one side of the synapse or junction. One-way transmission between nerve cells in higher animals is ensured because the vesicles are located on only one side of the synapse. However, cnidarian nerve nets are peculiar in that many of the synapses have vesicles of neurotransmitters on both sides, allowing transmission across the synapse in either direction. Another peculiarity of cnidarian nerves is the absence of any sheathing material (myelin) on the axons.

There is no concentrated grouping of nerve cells to suggest a "central nervous system." Nerves are grouped, however, in the "ring nerves" of hydrozoan medusae and in the marginal sense organs of scyphozoan medusae. In some cnidarians the nerve nets form two or more systems: in Scyphozoa there is a fast conducting system to coordinate swimming movements and a slower one to coordinate movements of tentacles.

Note that there is little adaptive value for a radially symmetrical animal to have a central nervous system with a brain. The environment approaches from all sides equally, and there is no control over the direction of approach to a prey organism.

Nerve cells of the net have synapses with slender sensory cells that receive external stimuli, and the nerve cells have junctions with epitheliomuscular cells and nematocysts. Together with the contractile fibers of epitheliomuscular cells, the sensory cell and nerve net combination is often considered a **neuromuscular system,** an important landmark in the evolution of nervous systems. The nerve net arose early in metazoan evolution, and it has never been completely lost phylogenetically. Annelids have it in their digestive systems. In the human digestive system it is represented by nerve plexuses in the musculature. The rhythmical peristaltic movements of the stomach and intestine are coordinated by this counterpart of the cnidarian nerve net.

Survey of Cnidarians

Class Hydrozoa

Most Hydrozoa are marine and colonial in form, and the typical life cycle includes both an asexual polyp and a sexual medusa stage. Some, however, such as the freshwater hydras, have no medusa stage. Some marine hydroids do not have free medusae (figure 7.6), whereas some hydrozoans occur only as medusae and have no polyp.

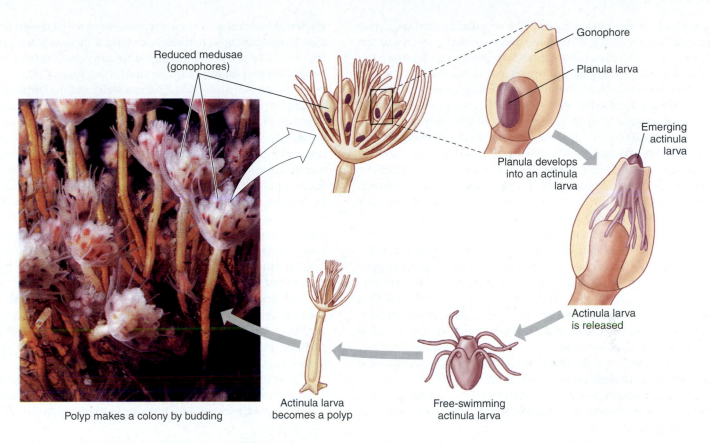

figure 7.6

In some hydroids, such as this *Tubularia crocea*, medusae are reduced to gonadal tissue and do not detach. These reduced medusae are known as gonophores.

figure 7.7

Hydra with developing bud on the right and ovary on the left.

Hydras, although not typical hydrozoans, have become favorites as an introduction to Cnidaria because of their size and ready availability. Combining study of a hydra with that of a representative colonial marine hydroid such as *Obelia* (Gr. *obelias*, round cake) gives an excellent idea of class Hydrozoa.

Hydra: A Freshwater Hydrozoan   The common freshwater hydra (figure 7.7) is a solitary polyp and one of the few cnidarians found in fresh water. Its normal habitat is the underside of aquatic leaves and lily pads in cool, clean fresh water of pools and streams. The hydra family is found throughout the world, with 16 species occurring in North America.

Over 230 years ago, Abraham Trembley was astonished to discover that isolated sections of the stalk of hydra could regenerate and each become a complete animal. Since then, over 2000 investigations of hydra have been published, and the organism has become a classic model for the study of morphological differentiation. The mechanisms governing morphogenesis have great practical importance, and the simplicity of hydra lends itself to these investigations. Substances controlling development (morphogens), such as those determining which end of a cut stalk will develop a mouth and tentacles, have been discovered, and they may be present in the cells in extremely low concentrations (10^{-10}M).

The body of a hydra can extend to a length of 25 to 30 mm or can contract to a tiny, jellylike mass. It is a cylindrical tube with the lower (aboral) end drawn out into a slender stalk, ending in a **basal** (or pedal) **disc** for attachment. This basal disc has gland cells that enable a hydra to adhere to a substratum and also to secrete a gas bubble for floating. In the center of the disc there may be an excretory pore. The **mouth,** located on a coni-cal elevation called the **hypostome,** is encircled by six to ten hollow tentacles that, like the body, can greatly extend when the animal is hungry. The mouth opens into the gastrovascular cav-ity, which communicates with the cavities in the tentacles.

Hydras feed on a variety of small crustaceans, insect lar-vae, and annelid worms. The hydra awaits its prey with tenta-cles extended (figure 7.8). A food organism that brushes against its tentacles may find itself harpooned by scores of nematocysts that render it helpless, even though it may be larger than the hydra. The tentacles move the prey toward the mouth, which slowly widens. Well moistened with mucous secretions, the mouth glides over and around the prey, totally engulfing it.

The activator that actually causes the mouth to open is the reduced form of **glutathione,** which is found to some extent in all living cells. Glutathione is released from the prey through the wounds made by the nematocysts, but only ani-mals releasing enough of the chemical to activate a feeding response are eaten by a hydra. This explains how a hydra dis-tinguishes between *Daphnia,* which it relishes, and some other forms that it refuses. If we place glutathione in water containing hydras, each hydra will go through the motions of feeding, even though no prey is present.

In asexual reproduction, buds appear as outpocketings of the body wall and develop into young hydras that eventually detach from the parent. In sexual reproduction, temporary gonads (see figure 7.7) usually appear in autumn, stimulated by lower temperatures and perhaps also by reduced aeration of stagnant waters. Testes or ovaries, when present, appear as rounded projections on the surface of the body (see figure 7.7). Eggs in the ovary usually mature one at a time and are fer-tilized by sperm shed into the water. A cyst forms around the embryo before it breaks loose from the parent, enabling it to survive the winter. Young hydras hatch in spring when the weather is favorable.

figure 7.8

Hydra catches an unwary water flea with the nematocysts of its tenta-cles. This hydra already contains one water flea eaten previously.

Hydroid Colonies   Far more representative of class Hydrozoa than hydras are those hydroids that have a medusa stage in their life cycle. *Obelia* is often used in laboratory exercises for begin-ning students to illustrate the hydroid type (figure 7.9).

A typical hydroid has a base, a stalk, and one or more ter-minal polyps (zooids). The base by which colonial hydroids are attached to the substratum is a rootlike stolon, which gives rise to one or more stalks. The living cellular part of the stalks secretes a nonliving chitinous sheath. Attached to the ends of the branches of the stalks are individual zooids. Most zooids are feeding polyps called **hydranths,** or **gastrozooids.** They may be tubular, bottle shaped, or vaselike, but all have a terminal mouth and a circlet of tentacles. In some forms, such as *Obelia,* the chitinous sheath continues as a protective cup around the polyp into which the polyp can withdraw for protection (figure 7.9). In others the polyp is naked. **Dactylozooids** are polyps specialized for defense in some species (see figure 7.1).

Hydranths, much like hydras, capture and ingest prey, such as tiny crustaceans, worms, and larvae, thus providing nutrition for the entire colony. After partial digestion in a hydranth, the digestive broth passes into the common gas-trovascular cavity where intracellular digestion occurs.

Circulation within the gastrovascular cavity is a function of the ciliated gastrodermis, but rhythmical contractions and pulsations of the body, which occur in many hydroids, also aid circulation.

In contrast to hydras, new individuals that bud do not detach from the parent; thus the size of the colony increases. New polyps may be hydranths or reproductive polyps known as **gonangia.** Medusae are produced by budding within the gonangia. Young medusae leave the colony as free-swimming individuals that mature and produce gametes (eggs and sperm) (figure 7.9). In some species medusae remain attached to the colony and shed their gametes there. In other species medusae never develop, gametes being shed by male and female gonophores. Development of a zygote produces a ciliated plan-ula larva that swims for a time. Then it settles onto a substra-tum, becoming a minute polyp that gives rise, by asexual budding, to the hydroid colony, thus completing the life cycle.

Hydroid medusae are usually smaller than their scypho-zoan counterparts, ranging from 2 or 3 mm to several centime-ters in diameter (figure 7.10). The margin of the bell projects inward as a shelflike **velum,** which partly closes the open side of the bell and is used in swimming (figure 7.11). Muscular pulsa-tions that alternately fill and empty the bell propel the animal for-ward, aboral side first, with a sort of "jet propulsion." Tentacles attached to the bell margin are richly supplied with nematocysts.

The mouth opening at the end of a suspended **manubrium** leads to a stomach and four radial canals that connect with a ring canal around the margin. This in turn con-nects with the hollow tentacles. Thus the coelenteron is con-tinuous from mouth to tentacles, and the entire system is lined with gastrodermis. Nutrition is similar to that of hydranths.

The nerve net is usually concentrated into two nerve rings at the base of the velum. The bell margin is liberally supplied with sensory cells. It usually also bears two kinds of specialized

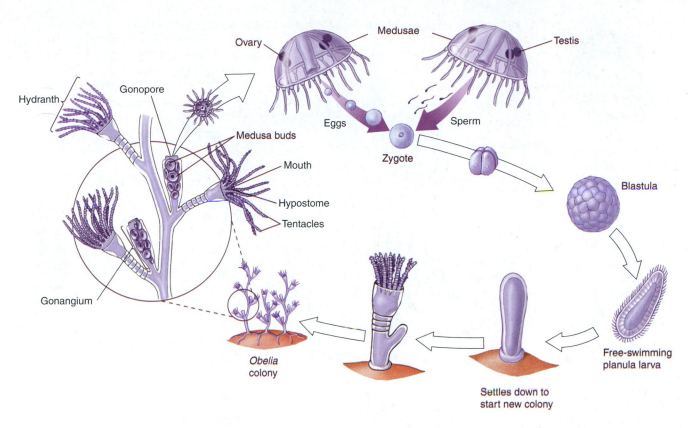

figure 7.9

Life cycle of *Obelia,* showing alternation of polyp (asexual) and medusa (sexual) stages. *Obelia* is a calyptoblastic hydroid; that is, its polyps as well as its stems are protected by continuations of the nonliving covering.

figure 7.10

Bell medusa, *Polyorchis penicillatus,* medusa stage of an unknown attached polyp.

sense organs: **statocysts,** which are small organs of equilibrium (figure 7.11), and **ocelli,** which are light-sensitive organs.

Other Hydrozoans Some hydrozoans known as siphonophores, form floating colonies, such as *Physalia* (Gr. *physallis,* bladder), the Portuguese man-of-war (figure 7.12). These colonies include several types of modified medusae and polyps. *Physalia* has a rainbow-hued float, probably a modified polyp, which carries it along at the mercy of winds and currents. It contains an air sac filled with secreted gas and acts as a carrier for the generations of individuals that bud from it and hang suspended in the water. There are several types of individuals, including feeding polyps, reproductive polyps, long stinging tentacles, and so-called jelly polyps. Many swimmers have experienced the uncomfortable sting that these colonial floaters can inflict. The pain, along with the panic of the swimmer, can increase the danger of drowning.

Other hydrozoans secrete massive calcareous skeletons that resemble true corals (figure 7.13). They are sometimes called **hydrocorals.**

Class Scyphozoa

Class Scyphozoa (si-fo-zo'a) (Gr. *skyphos,* cup) includes most of the larger jellyfishes, or "cup animals." A few, such as *Cyanea*

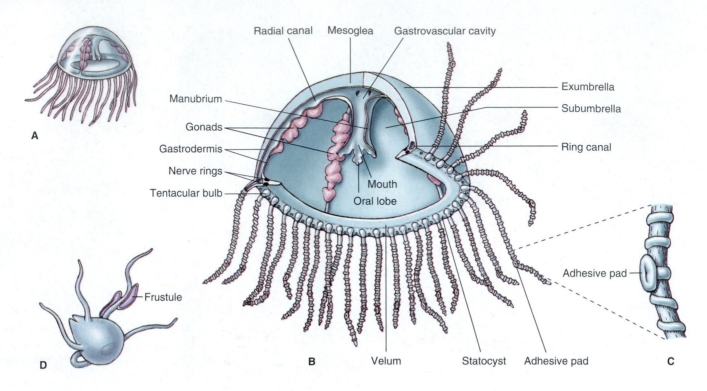

figure 7.11

Structure of *Gonionemus*. **A,** Medusa with typical tetramerous arrangement. **B,** Cutaway view showing morphology. **C,** Portion of a tentacle with its adhesive pad and ridges of nematocysts. **D,** Tiny polyp, or hydroid stage, that develops from the planula larva. It can produce more polyps by budding (frustules) or produce medusa buds.

figure 7.12

A Portuguese man-of-war colony, *Physalia physalis* (order Siphonophora, class Hydrozoa). Colonies often drift onto southern ocean beaches, where they are a hazard to bathers. Each colony of medusa and polyp types is integrated to act as one individual. As many as a thousand zooids may be found in one colony. The nematocysts secrete a powerful neurotoxin.

(Gr. *kyanos,* dark-blue substance), may attain a bell diameter exceeding 2 m and tentacles 60 to 70 m long (figure 7.14). Most scyphozoans, however, range from 2 to 40 cm in diameter. Most are found floating in the open sea, some even at depths of 3000 m, but one unusual order is sessile and attaches by a stalk to seaweeds and other objects on the sea bottom (figure 7.15). Their coloring may range from colorless to striking orange and pink hues.

Scyphomedusae, unlike hydromedusae, have no velum. Bells of different species vary in depth from a shallow saucer shape to a deep helmet or goblet shape, and in many the margin is scalloped, each notch bearing a sense organ called a **rhopalium** and a pair of lobelike projections called lappets. *Aurelia* (L. *aurum,* gold) has eight such notches (figures 7.16 and 7.17); others may have four or sixteen. Each rhopalium bears a statocyst for balance, two sensory pits containing concentrations of sensory cells, and sometimes an ocellus (simple eye) for photoreception. The mesoglea is thick and contains cells as well as fibers. The stomach is usually divided into pouches containing small tentacles with nematocysts.

The mouth is centered on the subumbrellar side. The manubrium is usually drawn out into four frilly **oral lobes** used in food capture and ingestion. Marginal tentacles may be

A **B**

figure 7.13

These hydrozoans form calcareous skeletons that resemble true coral. **A,** *Stylaster roseus* (order Stylasterina) occurs commonly in caves and crevices in coral reefs. These fragile colonies branch in only a single plane and may be white, pink, purple, red, or red with white tips. **B,** Species of *Millepora* (order Milleporina) form branching or platelike colonies and often grow over the horny skeleton of gorgonians. They have a generous supply of powerful nematocysts that produce a burning sensation on human skin, justly earning the common name fire coral.

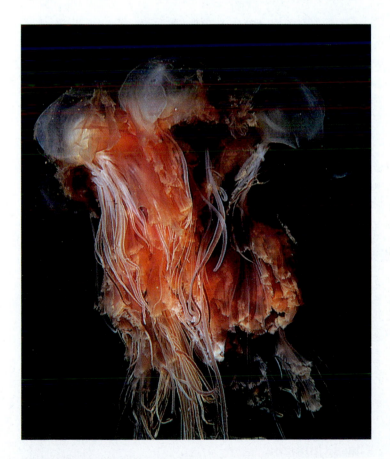

many or few and may be short, as in *Aurelia,* or long, as in *Cyanea.* Tentacles, manubrium, and often the entire body surface of scyphozoans are well supplied with nematocysts. Scyphozoans feed on all sorts of small organisms, from protozoa to fishes. Capture of prey involves stinging and manipulation with tentacles and oral arms, but methods vary. *Aurelia* feeds on small planktonic animals. These are caught in mucus of the umbrella surface, carried to "food pockets" on the umbrella margin by cilia, and picked up from the pockets by the oral lobes whose cilia carry the food to the gastrovascular cavity.

figure 7.14

Giant jellyfish, *Cyanea capillata* (order Semeaeostomeae, class Scyphozoa). A North Atlantic species of *Cyanea* reaches a bell diameter exceeding 2 m. It is known as "sea blubber" by fishermen.

figure 7.15

Thaumatoscyphus hexaradiatus (order Stauromedusae, class Scyphozoa). Members of this order are unusual scyphozoans in that the medusae are sessile and attached to seaweed or other objects.

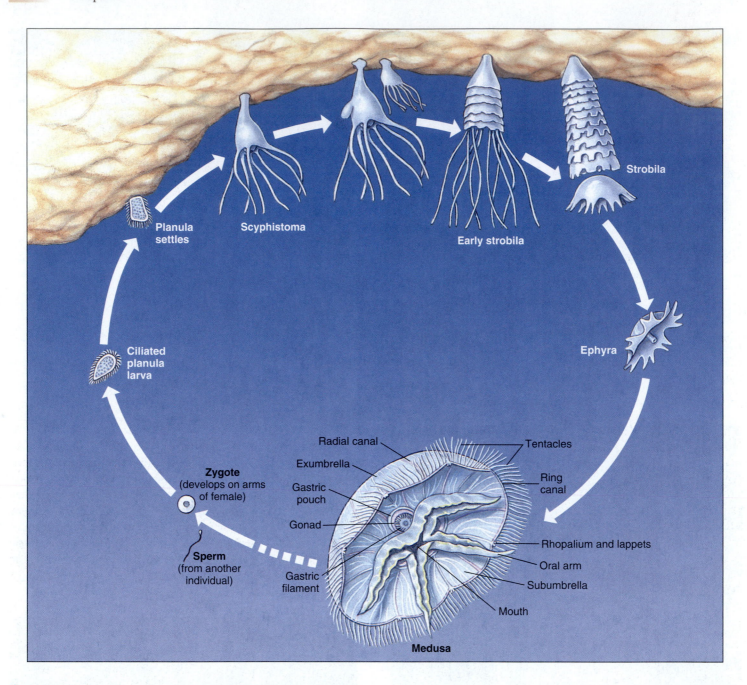

figure 7.16

Life cycle of *Aurelia*, a marine scyphozoan medusa.

Cilia on the gastrodermis keep a current of water moving to bring food and oxygen into the stomach and carry out wastes.

Internally four **gastric pouches** containing nematocysts connect with the stomach in scyphozoans, and a complex system of **radial canals** that branch from the pouches to the **ring canal** (see figure 7.16) completes the gastrovascular cavity, through which nutrients circulate.

Sexes are separate, with gonads located in the gastric pouches. Fertilization is internal, with sperm being carried by ciliary currents into the gastric pouch of the female. Zygotes may develop in seawater or may be brooded in folds of the oral arms. The ciliated planula larva becomes attached and devel-

ops into a **scyphistoma,** a hydralike form (figure 7.16) that may bud to make other polyps. By a process of **strobilation** the scyphistoma of *Aurelia* forms a series of saucerlike buds, **ephyrae,** and is now called a **strobila** (figure 7.16). When the ephyrae break loose, they grow into mature jellyfish.

Class Cubozoa

Cubozoa formerly were considered an order (Cubomedusae) of Scyphozoa. The medusa is the predominant form (figure 7.18); the polyp is inconspicuous and in most cases unknown. In transverse section the bells are almost square. A tentacle or

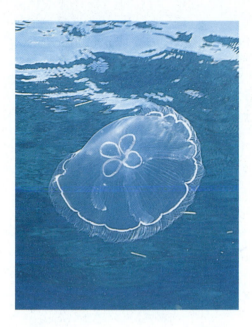

figure 7.17

Moon jellyfish *Aurelia aurita* (class Scyphozoa) is cosmopolitan in distribution. It feeds on planktonic organisms caught in mucus on its umbrella.

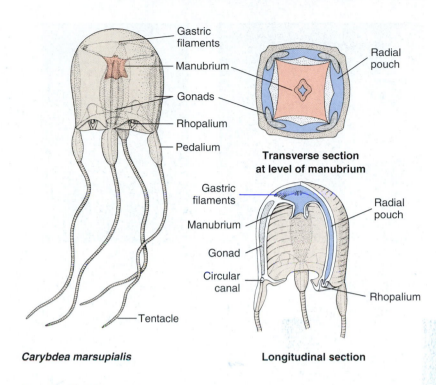

Carybdea marsupialis

figure 7.18

Carybdea, a cubozoan medusa.

group of tentacles is found at each corner of the square at the umbrella margin. The base of each tentacle is differentiated into a flattened, tough blade called a **pedalium** (figure 7.18). Rhopalia are present. The umbrella margin is not scalloped, and the subumbrella edge turns inward to form a **velarium.** The velarium functions as a velum does in hydrozoan medusae, increasing swimming efficiency, but it differs structurally. Cubomedusae are strong swimmers and voracious predators, feeding mostly on fish.

Class Anthozoa

Anthozoans, or "flower animals," are polyps with a flowerlike appearance (figure 7.19). There is no medusa stage. Anthozoa are all marine and are found in both deep and shallow water and in polar seas as well as tropical seas. They vary greatly in size and may be solitary or colonial. Many are supported by skeletons.

Chironex fleckeri (Gr. *cheir*, hand, + *nexis*, swimming) is a large cubomedusa known as the sea wasp. Its stings are quite dangerous and sometimes fatal. Most fatal stings have been reported from tropical Australian waters, usually following quite massive stings. Witnesses have described victims as being covered with "yards and yards of sticky wet string." Stings are very painful, and death, if it is to occur, ensues within a matter of minutes. If death does not occur within 20 minutes after stinging, complete recovery is likely.

The class has three subclasses: **Zoantharia** (or **Hexacorallia**), composed of the sea anemones, hard corals, and others;

figure 7.19

Sea anemones (order Actiniaria, subclass Zoantharia) are the familiar and colorful "flower animals" of tide pools, rocks, and pilings of the intertidal zone. Most, however, are subtidal, their beauty seldom revealed to human eyes. These are rose anemones, *Tealia piscivora*.

Ceriantipatharia, which includes only tube anemones and thorny corals; and **Octocorallia** (or **Alcyonaria**), containing soft and horny corals, such as sea fans, sea pens, sea pansies, and others. Zoantharians and ceriantipatharians have a **hexamerous** plan (of six or multiples of six) or polymerous symmetry

A

B

figure 7.20

A, Orange sea pen *Ptilosarcus gurneyi* (order Pennatulacea, subclass Octocorallia, class Anthozoa). Sea pens are colonial forms that inhabit soft bottoms. The base of the fleshy body of the primary polyp is buried in sediment. It gives rise to numerous secondary, branching polyps. **B,** Close-up of a gorgonian. The pinnate tentacles characteristic of the subclass Octocorallia are apparent.

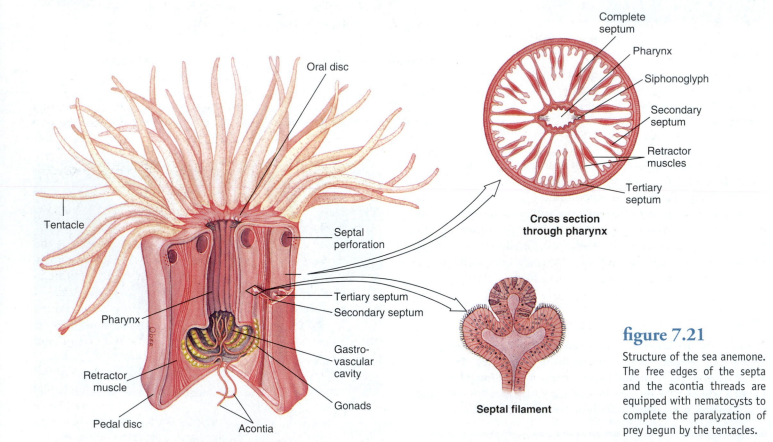

Cross section through pharynx

Septal filament

figure 7.21

Structure of the sea anemone. The free edges of the septa and the acontia threads are equipped with nematocysts to complete the paralyzation of prey begun by the tentacles.

and have simple tubular tentacles arranged in one or more circlets on the oral disc. Octocorallians are **octomerous** (built on a plan of eight) and always have eight pinnate (featherlike) tentacles arranged around the margin of the oral disc (figure 7.20).

The gastrovascular cavity is large and partitioned by septa, or mesenteries, that are inward extensions of the body wall. Where one septum extends into the gastrovascular cavity from the body wall, another extends from the diametrically opposite side; thus, they are said to be **coupled.** In Zoantharia, septa are not only coupled, they are also **paired** (figure 7.21). The muscular arrangement varies among different groups, but there are usually circular muscles in the body wall and longitudinal and transverse muscles in the septa.

Subclass Ceriantipatharia has been created from the Ceriantharia and Antipatharia, formerly considered orders of Zoantharia. Ceriantharians are tube anemones and live in soft bottom sediments, buried to the level of the oral disc. Antipatharians are thorny or black corals. They are colonial and have a hard axial skeleton. Both of these groups are small in numbers of species and are limited to warmer waters of the sea.

There is a general tendency towards biradial symmetry in the septal arrangement and in the shape of the mouth and pharynx. There are no special organs for respiration or excretion.

Sea Anemones   Sea anemone polyps (subclass, Zoantharia, order Actiniaria) are larger and heavier than hydrozoan polyps (figures 7.19 and 7.21). Most range from 5 mm or less to 100 mm in diameter, and from 5 mm to 200 mm long, but some grow much larger. Some are quite colorful. Anemones are found in coastal areas all over the world, especially in warmer waters, and they attach by means of their pedal discs to shells, rocks, timber, or whatever submerged substrata they can find. Some burrow in mud or sand.

Sea anemones are cylindrical with a crown of tentacles arranged in one or more circles around the mouth on the flat **oral disc** (figure 7.21). The slit-shaped mouth leads into a **pharynx**. At one or both ends of the mouth is a ciliated groove called a **siphonoglyph**, which extends into the pharynx. Siphonoglyphs create water currents directed into the pharynx. Cilia elsewhere on the pharynx direct water outward. Currents thus created carry in oxygen and remove wastes. They also help to maintain an internal fluid pressure providing a hydrostatic skeleton that serves as a support for opposing muscles.

The pharynx leads into a large gastrovascular cavity divided into radial chambers by pairs of septa that extend vertically from the body wall toward the pharynx (figure 7.21). These chambers communicate with each other and are open below the pharynx. In many anemones the lower ends of the septal edges are prolonged into **acontia threads,** also provided with nematocysts and gland cells, that can be protruded through the mouth or through pores in the body wall to overcome prey or to provide defense. The pores also aid in the rapid discharge of water from the body when the animal is endangered and contracts to a small size.

Anemones form some interesting mutualistic relationships with other organisms. Many anemones house unicellular algae in their tissues (as do reef-building corals), from which they undoubtedly derive some nutrients. Some hermit crabs place anemones on the snail shells in which the crabs live, gaining some protection from predators by the presence of the anemone, while the anemone dines on particles of food dropped by the crab. Anemone fishes (figure 7.22) of the tropical Indo-Pacific form associations with large anemones. An unknown property of the skin mucus of the fish causes the anemone's nematocysts not to discharge, but if some other fish is so unfortunate as to brush the anemone's tentacles, it is likely to become a meal.

Sea anemones are carnivorous, feeding on fish or almost any live animals of suitable size. Some species live on minute forms caught by ciliary currents.

Sexes are separate in some sea anemones (dioecious), and some anemones are monoecious (hermaphroditic). Gonads are

figure 7.22

Orangefin anemone fish *(Amphiprion chrysopterus)* nestles in the tentacles of its sea anemone host. Anemone fishes do not elicit stings from their hosts but may lure unsuspecting other fish to become meals for the anemone.

arranged on the margins of the septa. Fertilization is external in some species, whereas in others the sperm enter the gastrovascular cavity to fertilize eggs. The zygote develops into a ciliated larva that will settle to become a polyp. Asexual reproduction commonly occurs by pedal laceration or by transverse fission.

Zoantharian Corals   Zoantharian corals belong to order Scleractinia of subclass Zoantharia, sometimes known as the true or stony corals. Stony corals might be described as miniature sea anemones that live in calcareous cups they themselves have secreted (figures 7.23 and 7.24). Like that of anemones, a coral polyp's gastrovascular cavity is subdivided by septa arranged in multiples of six (hexamerous) and its hollow tentacles surround the mouth, but there is no siphonoglyph.

Instead of a pedal disc, the epidermis at the base of the column secretes a limy skeletal cup, including sclerosepta, which project up into the polyp between its true septa (figure 7.24). Living polyps can retract into the safety of their cup when not feeding. Since the skeleton is secreted below the living tissue rather than within it, the calcareous material is an exoskeleton. In many colonial corals, the skeleton may become massive, accumulating over many years, with the living coral forming a sheet of tissue over the surface (figure 7.25). The gastrovascular cavities of the polyps are all connected through this sheet of tissue.

Octocorallian Corals   Octocorals include soft corals, sea pansies, sea pens, and sea fans and other gorgonian corals (horny corals). They have strict octomerous symmetry, with eight pinnate tentacles and eight unpaired, complete septa

A

B

C

figure 7.23

A, Cup coral *Tubastrea* sp. The polyps form clumps resembling groups of sea anemones. Although often found on coral reefs, *Tubastrea* is not a reef-building coral (ahermatypic) and has no symbiotic zooxanthellae in its tissues. **B,** The polyps of *Montastrea cavernosa* are tightly withdrawn in the daytime but open to feed at night, as in **C** (order Scleractinia, subclass Zoantharia).

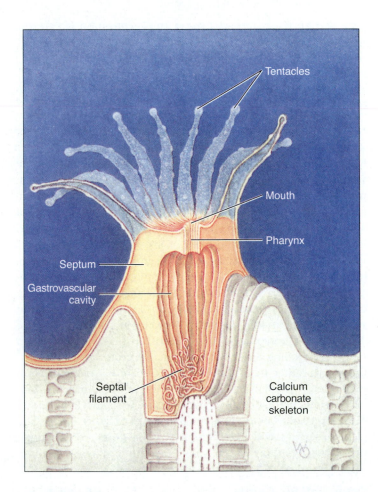

Tentacles

Mouth

Pharynx

Septum

Gastrovascular cavity

Septal filament

Calcium carbonate skeleton

figure 7.24

Polyp of a zoantharian coral (order Scleractinia) showing calcareous cup (exoskeleton), gastrovascular cavity, septa, and septal filaments.

figure 7.25

Boulder star coral, *Montastrea annularis* (subclass Zoantharia, class Anthozoa). Colonies can grow up to 10 feet (3 m) high.

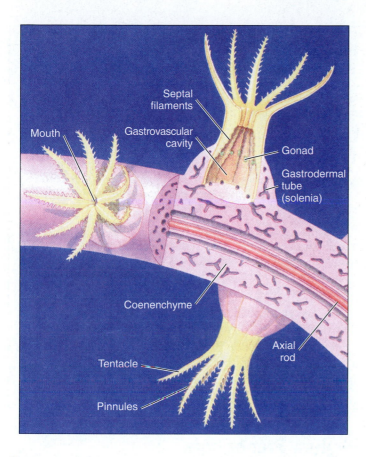

figure 7.26

Polyps of an octocorallian coral. Note the eight pinnate tentacles, coenenchyme, and solenia. They have an endoskeleton of limy spicules often with a horny protein, which may be in the form of an axial rod.

figure 7.27

A soft coral, *Dendronephthya* sp. (order Alcyonacea, subclass Octocorallia, class Anthozoa), on a Pacific coral reef. The showy hues of this soft coral vary from pink and yellow to bright red and contribute much color to Indo-Pacific reefs.

(figure 7.26). They are almost all colonial, and the gastrovascular cavities of polyps in colonies communicate through a system of gastrodermal tubes called **solenia.** The tubes run through an extensive mesoglea in most octocorals, and the surface of the colony is covered by epidermis.

The graceful beauty of octocorals—in hues of yellow, red, orange, and purple—helps create the "submarine gardens" of coral reefs (figure 7.27). Octocoral colonies vary in shape. They may be thick encrusting mats or ribbons, tall forms resembling an antique quill pen, or whiplike clusters of long vertical branches. Soft corals have fleshy bodies with calcareous spicules in the mesoglea, but skeletons of other octocorals typically combine fused or separate spicules with a stiffened, yet flexible, protein called gorgonin. Gorgonin has chemical similarities to keratin and to collagen; it is responsible for structural support in taxa designated as horny corals, including gorgonians. Because skeletal structures are secreted within the mesoglea, most octocorals have an endoskeleton.

Coral Reefs   Coral reefs are among the most productive of all ecosystems, and their diversity of life forms is rivaled only by tropical rain forests. Coral reefs are large formations of calcium carbonate (limestone) in shallow tropical seas laid down by living organisms over thousands of years; living plants and animals are confined to the top layer of reefs where they add more calcium carbonate to that deposited by their predecessors. The most important organisms that take dissolved calcium and carbonate ions from seawater and precipitate it as limestone to form reefs are **reef-building corals** and **coralline algae.** Reef-building corals have mutualistic algae (zooxanthellae) living in their tissues. Coralline algae are several types of red algae, and they may be encrusting or form upright, branching growths. Not only do they contribute to the total mass of calcium carbonate, but their deposits help to hold the reef together. Some octocorals and hydrozoans (especially *Millepora* [L. *mille,* a thousand, + *porus,* pore] spp., the "fire coral," see figure 7.13) contribute in some measure to the calcareous material, and an enormous variety of other organisms contributes small amounts. However, reef-building corals seem essential to the formation of large reefs, since such reefs do not occur where these corals cannot live.

Because zooxanthellae are vital to reef-building corals, and water absorbs light, reef-building corals rarely live below a depth of 30 m (100 feet). Interestingly, some deposits of coral-reef limestone, particularly around Pacific islands and atolls, reach great thickness—even thousands of feet. Clearly the corals and other organisms could not have grown from the bottom in the abyssal blackness of the deep sea and reached shallow water where light could penetrate. Charles Darwin was the first to realize that such reefs began their growth in *shallow* water around volcanic islands; then, as the islands slowly sank beneath the sea, growth of the reefs kept up with the rate of sinking, thus explaining the depth of the deposits.

classification of Phylum Cnidaria

Class Hydrozoa (hi-dro-zo′a) (Gr. *hydra*, water serpent, *zōon*, animal). Solitary or colonial; asexual polyps and sexual medusae, although one type may be suppressed; hydranths with no mesenteries; medusae (when present) with a velum; both fresh water and marine. Examples: *Hydra, Obelia, Physalia, Tubularia*.

Class Scyphozoa (si-fo-zo′a) (Gr. *skyphos*, cup, + *zōon*, animal). Solitary; polyp stage sometimes reduced or absent; bell-shaped medusae without velum; gelatinous mesoglea much enlarged; margin of bell or umbrella typically with eight notches that are provided with sense organs; all marine. Examples: *Aurelia, Cassiopeia, Cyanea, Rhizostoma*.

Class Cubozoa (ku′bo-zo′a) (Gr. *kybos*, a cube, + *zōon*, animal). Solitary; polyp stage reduced; bell-shaped medusae square in cross section, with tentacle or group of tentacles hanging from a bladelike pedalium at each corner of the umbrella; margin of umbrella entire, without velum but with velarium; all marine. Examples: *Tripedalia, Carybdea, Chironex, Chiropsalmus*.

Class Anthozoa (an-tho-zo′a) (Gr. *anthos*, flower, + *z⁻oon*, animal). All polyps; no medusae; solitary or colonial; enteron subdivided by mesenteries or septa bearing nematocysts; gonads endodermal; all marine.

Subclass Zoantharia (zo′an-tha′re-a) (N.L. from Gr. *zōon*, animal, + *anthos*, flower, + L. *aria*, like or connected with) **(Hexacorallia)**. With simple unbranched tentacles; mesenteries in pairs, in multiples of six; sea anemones, hard corals, and others. Examples: *Metridium, Anthopleura, Tealia, Astrangia, Acropora, Montastrea, Tubastrea*.

Subclass Ceriantipatharia (se′re-ant-ip′a-tha′re-a) (N.L. combination of Ceriantharia and Antipatharia). With simple unbranched tentacles; mesenteries unpaired, initially six; tube anemones and black or thorny corals. Examples: *Cerianthus, Antipathes, Stichopathes*.

Subclass Octocorallia (ok′to-ko-ral′e-a) (L. *octo*, + Gr. *korallion*, coral) **(Alcyonaria)**. With eight pinnate tentacles; eight complete, unpaired mesenteries; soft and horny corals. Examples: *Tubipora, Alcyonium, Gorgonia, Plexaura, Renilla, Ptilosarcus*.

The distribution of coral reefs in the world is limited to locations that offer optimal conditions for their zooxanthellae. They require warmth, light, and the salinity of undiluted seawater, thus limiting coral reefs to shallow waters between 30° N and 30° S latitude and excluding them from areas with upwelling of cold water or areas near major river outflows with attendant low salinity and high turbidity. Photosynthesis and fixation of carbon dioxide by the zooxanthellae furnish food molecules for their hosts. Zooxanthellae recycle phosphorus and nitrogenous waste compounds that otherwise would be lost, and they enhance the ability of the coral to deposit calcium carbonate.

Despite their great intrinsic and economic value, coral reefs in many areas are threatened by a variety of factors, mostly of human origin. These include overenrichment with nutrients (from sewage and runoff of agricultural fertilizer from nearby land) and overfishing of herbivorous fishes, both of which contribute to overgrowth of multicellular algae. Agricultural pesticides, sediment from tilled fields and dredging, and oil spills contribute to reef degradation. When such environmental stresses do not kill corals directly, they may make the organisms more susceptible to the numerous coral diseases observed in recent years. Coral reefs are apparently suffering from effects of global warming. When their surrounding water becomes too warm, corals expel their zooxanthellae (coral "bleaching") for reasons that are not yet clear. Instances of coral bleaching are becoming increasingly common around the world. Furthermore, higher atmospheric concentrations of carbon dioxide (from burning hydrocarbon fuels) tend to acidify ocean water, which makes precipitation of $CaCO_3$ by corals more difficult metabolically.

Phylum Ctenophora

Ctenophora (te-nof′o-ra) (Gr. *kteis, ktenos*, comb, + *phora*, pl. of bearing) contains fewer than 100 species. All are marine forms occurring in all seas but especially in warm waters. They take their name from the eight rows of comblike plates they bear for

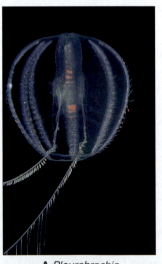

A *Pleurobrachia*   **B** *Mnemiopsis*

figure 7.28

A, Comb jelly *Pleurobrachia* sp. (order Cydippida, class Tentaculata). Its fragile beauty is especially evident at night when it luminesces from its comb rows. **B,** *Mnemiopsis* sp. (order Lobata, class Tentaculata).

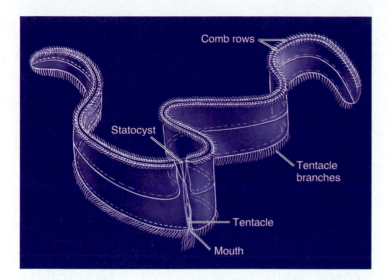

figure 7.29

Venus' girdle (*Cestum* sp.), a highly modified ctenophore. It may reach a length of 5 feet but is usually much smaller.

locomotion. Common names for ctenophores are "sea walnuts" and "comb jellies." Ctenophores, along with cnidarians, represent the only two phyla having primary radial symmetry, in contrast to other metazoans, which have primary bilateral symmetry. In both phyla, the oral/aboral axis is used as a reference point because there is no concentration of sensory or nerve cells to form a head. In common with cnidarians, ctenophores have not advanced beyond the tissue grade of organization. There are no definite organ systems in the strict meaning of the term.

Ctenophores may make adhesive cells called *colloblasts,* but they do not make nematocysts. However, one ctenophore species (*Haeckelia rubra,* after Ernst Haeckel, nineteenth-century German zoologist) carries nematocysts on certain regions of its tentacles and lacks colloblasts. These nematocysts apparently are appropriated from cnidarian medusae on which it feeds.

Except for a few creeping and sessile forms, ctenophores are free-swimming (see figure 7.28). Although they are feeble swimmers and are more common in surface waters, ctenophores are sometimes found at considerable depths. Highly modified forms such as *Cestum* (L. *cestus,* girdle) use sinuous body movements as well as their comb plates in locomotion (figure 7.29).

The fragile, transparent bodies of ctenophores are easily seen at night when they emit light (luminesce).

Form and Function

Pleurobrachia (Gr. *pleuron,* side, + L. *brachia,* arms), is a representative ctenophore (figure 7.29). Its surface bears eight longitudinal rows of transverse plates bearing long fused cilia and called **comb plates.** The beating of the cilia in each row starts at the aboral end and proceeds along the rows to the oral end, thus propelling the animal forward. All rows beat in unison. A reversal of the wave direction drives the animal backward. Ctenophores may be the largest animals that swim exclusively by cilia.

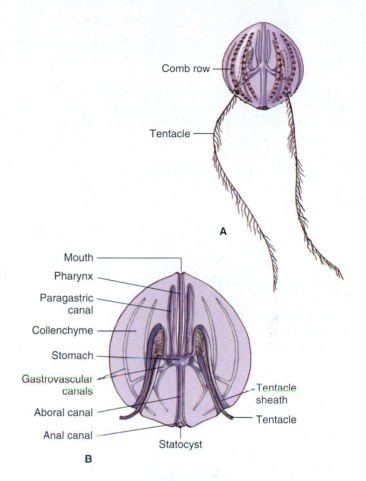

figure 7.30

The comb jelly *Pleurobrachia,* a ctenophore. **A,** Hemisection. **B,** External view.

Two long tentacles are carried in a pair of tentacle sheaths (figure 7.30) from which they can stretch to a length of perhaps 15 cm. The surface of the tentacles bears specialized glue cells called **colloblasts,** which secrete a sticky substance that facilitates catching small prey. When covered with food, the tentacles contract and food is wiped onto the mouth. The gastrovascular cavity consists of a pharynx, stomach, and a system of gastrovascular canals. Rapid digestion occurs in the pharynx; then partly digested food circulates through the rest of the system where digestion is completed intracellularly. Residues are regurgitated or expelled through small pores in the aboral end.

A nerve net system similar to that of the cnidarians includes a subepidermal plexus concentrated under each comb plate.

Since the 1980s population explosions of *Mnemiopsis leidyi* in the Black and Azov Seas have led to catastrophic declines in fisheries there. Inadvertently introduced from the coast of the Americas with ballast water of ships, the ctenophores feed on zooplankton, including small crustaceans and eggs and larvae of fish. The normally inoffensive *M. leidyi* is kept in check in the Atlantic by certain specialized predators, but introduction of such predators into the Black Sea carries its own dangers.

The sense organ at the aboral pole is a **statocyst,** or organ of equilibrium, and is also concerned with beating of the comb rows but does not trigger their beat. Other sensory cells are abundant in the epidermis.

All ctenophores are monoecious, bearing both an ovary and a testis. Gametes are shed into the water, except in a few species that brood their eggs, and there is a free-swimming larva.

Phylogeny and Adaptive Radiation

Phylogeny

Although the origin of the cnidarians and ctenophores is obscure, one hypothesis suggests that the radiate phyla arose from a radially symmetrical, planula-like ancestor with a sessile or free-floating habitat where radial symmetry is a selective advantage. A planula larva in which an invagination formed to become the gastrovascular cavity would correspond roughly to a cnidarian with an ectoderm and an endoderm.

One of the central questions in cnidarian evolution is: Which came first, the polyp or the medusa? Some researchers believe that the life cycles of a specialized group of hydrozoans suggest that the medusa came first. In trachyline hydrozoans, the planula develops to a second larva called an actinula and then directly into a medusa without an intervening polyp stage. If the ancestral cnidarian had this life cycle, polyps would have been added to the life cycle later, and anthozoans would have lost the medusa stage.

The alternative is that the anthozoan life cycle was ancestral and the medusoid form was added to the life cycle later. This would suggest that hydrozoans, scyphozoans, and cubozoans shared a common ancestor that had added the medusa to the life cycle. The medusa would have been lost in the lineage leading to *Hydra.* Molecular evidence currently supports this hypothesis, and some researchers group the three classes with medusae into the Medusozoa (figure 7.31).

In the past it was assumed that ctenophores arose from a medusoid cnidarian, but this assumption has been questioned recently. Similarities between the groups are mostly of a general nature and do not seem to indicate a close relationship. Some molecular evidence suggests that ctenophores branched off the metazoan line after sponges but before cnidarians. Ctenophores have a more derived and stereotypical cleavage

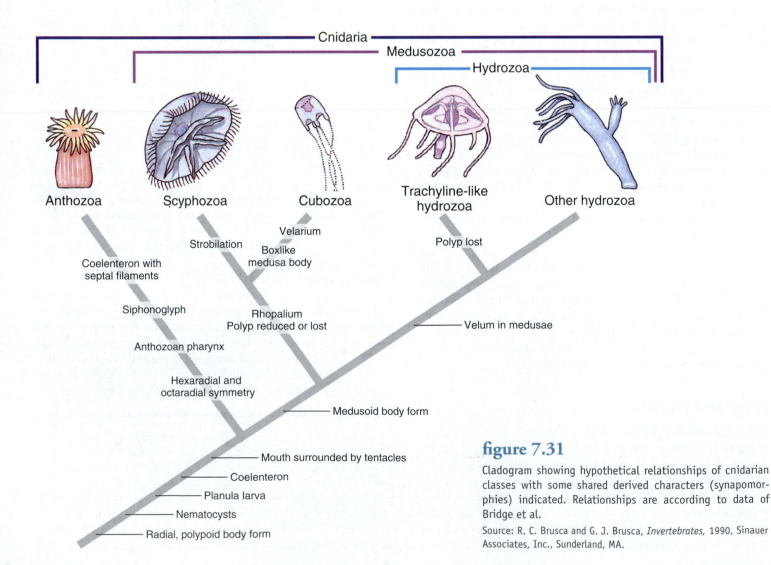

figure 7.31

Cladogram showing hypothetical relationships of cnidarian classes with some shared derived characters (synapomorphies) indicated. Relationships are according to data of Bridge et al.

Source: R. C. Brusca and G. J. Brusca, *Invertebrates,* 1990, Sinauer Associates, Inc., Sunderland, MA.

pattern than cnidarians. Other analyses indicate a nearest common ancestor of ctenophores and cnidarians, which together would form the sister taxon to bilateral metazoans.

Adaptive Radiation

In their evolution neither phylum has deviated far from its basic structure. In Cnidaria, both polyp and medusa are constructed on the same scheme. Nonetheless, cnidarians have achieved large numbers of individuals and species, demonstrating a surprising degree of diversity considering the simplicity of their basic body plan. They are efficient predators, many feeding on prey quite large in relation to themselves. Some are adapted for feeding on small particles. The colonial form of life is well explored, with some colonies growing to great size among corals, and others, such as siphonophores, showing astonishing polymorphism and specialization of individuals within the colony.

Ctenophores have adhered to the arrangement of the comb plates and their biradial symmetry, but some, like Venus' girdle, have a modified body plan.

Summary

Phyla Cnidaria and Ctenophora have a primary radial symmetry; radial symmetry is an advantage for sessile or free-floating organisms because environmental stimuli come from all directions equally. Cnidaria are surprisingly efficient predators because they possess stinging organelles called cnidae. Both phyla are essentially diploblastic, with a body wall composed of epidermis and gastrodermis and a mesoglea between these cell layers. The digestive-respiratory (gastrovascular) cavity has a mouth and no anus. Cnidarians are at the tissue level of organization. Many species have two basic body types (polyp and medusa), and in many hydrozoans and scyphozoans the life cycle involves both an asexually reproducing polyp and a sexually reproducing medusa.

That unique organelle, the cnida, is produced by a cnidoblast (which becomes the cnidocyte) and is coiled within a capsule. When discharged, some types of cnidae, called nematocysts, penetrate prey and inject poison. Discharge is effected by a change in permeability of the capsule and an increase in internal hydrostatic pressure because of high osmotic pressure within the capsule.

Most hydrozoans are colonial and marine, but the solitary freshwater hydras are commonly demonstrated in class laboratories. They have a typical polypoid form, are not colonial, and have no medusoid stage. Most marine hydrozoans are in the form of a branching colony of many polyps (hydranths). Most colonies are sessile, but a few drift or swim in the sea. Hydrozoan medusae may be free-swimming or remain attached to the colony.

Scyphozoans are typical jellyfishes, in which the medusa is the dominant body form, and many have an inconspicuous polypoid stage. Cubozoans are predominantly medusoid. They include the dangerous sea wasps.

Anthozoans are all marine and are polypoid; there is no medusoid stage. The most important subclasses are Zoantharia (with hexamerous or polymerous symmetry) and Octocorallia (with octomerous symmetry). The largest zoantharian orders contain sea anemones, which are solitary and do not have a skeleton, and stony corals, which are mostly colonial and secrete a calcareous exoskeleton. Stony corals are the critical component in coral reefs, which are habitats of great beauty, productivity, and ecological and economic value. Octocorallia contain the soft and horny corals, many of which are important and beautiful components of coral reefs.

Ctenophora are biradial and swim by means of eight comb rows of fused cilia. Colloblasts, with which they capture small prey, are characteristic of the phylum.

Cnidaria and Ctenophora are probably derived from an ancestor that resembled the planula larva of the cnidarians. Despite their relatively simple level of organization, cnidarians are an important phylum economically, environmentally, and biologically.

Review Questions

1. Explain the selective advantage of radial symmetry for sessile and free-floating animals.
2. What characteristics of phylum Cnidaria do you think are most important in distinguishing it from other phyla?
3. Name and distinguish the classes in phylum Cnidaria.
4. Distinguish between polyp and medusa forms.
5. Explain the mechanism of nematocyst discharge. How can a hydrostatic pressure of one atmosphere be maintained within a nematocyst until it receives an expulsion stimulus?
6. What is an unusual feature of the nervous system of cnidarians?
7. Diagram a hydra and label the main body parts.
8. Name and give the functions of the main cell types in the epidermis and in the gastrodermis of hydra.
9. What stimulates feeding behavior in hydras?
10. Give an example of a highly polymorphic, floating, colonial hydrozoan.
11. Distinguish the following from each other: statocyst and rhopalium; scyphomedusae and hydromedusae; scyphistoma, strobila, and ephyrae; velum, velarium, and pedalium; Zoantharia and Octocorallia.
12. Define the following with regard to sea anemones: siphonoglyph; primary septa or mesenteries; incomplete septa; septal filaments; acontia threads; pedal laceration.
13. Describe three specific symbiotic interactions of anemones with other organisms.
14. Contrast the skeletons of zoantharian and octocorallian corals.
15. Coral reefs generally are limited in geographic distribution to shallow marine waters. How do you account for this?

16. Specifically, what kinds of organisms are most important in deposition of calcium carbonate on coral reefs?

17. How do zooxanthellae contribute to the welfare of reef-building corals?

18. What characteristics of Ctenophora do you think are most important in distinguishing it from other phyla?

19. How do ctenophores swim, and how do they obtain food?

20. What are the arguments in support of the ideas that the fist cnidarian was either a medusa or a polyp?

Selected References

See also general references on page 415.

Bridge, D., C. W. Cunningham, R. DeSalle, and L. W. Buss. 1995. Class-level relationships in the phylum Cnidaria: Molecular and morphological evidence. Mol. Biol. Evol. **12:** 679–689. *Analyses of ribosomal DNA sequences support the basal position of anthozoans within Cnidaria.*

Brown, B. E., and J. C. Ogden. 1993. Coral bleaching. Sci. Am. **268:**64–70 (Jan.). *Abnormally warm water is apparently the cause of reef corals losing their zooxanthellae.*

Buddemeier, R. W., and S. V. Smith. 1999. Coral adaptation and acclimatization: a most ingenious paradox. Amer. Zool. **39:**1–9. *First of a series of papers in this issue dealing with effects of climatic and temperature changes on coral reefs.*

Crossland, C. J., B. G. Hatcher, and S.V. Smith. 1991. Role of coral reefs in global ocean production. Coral Reefs **10:**55–64. *Because of extensive recycling of nutrients within reefs, their net energy production for export is relatively minor. However, they play a major role in inorganic carbon precipitation by biologically-mediated processes.*

Humann, P. 1992. Reef creature identification. Florida, Caribbean, Bahamas. Jacksonville, Florida, New World Publications, Inc. *This is the best field guide available for identification of "non-coral" cnidarians of the Caribbean.*

Humann, P. 1993. Reef coral identification. Florida, Caribbean, Bahamas. Jacksonville, Florida, New World Publications, Inc. *Superb color photographs and accurate identifications make this by far the best field guide to corals now available; includes sea grasses and some algae.*

Kenchington, R., and G. Kelleher. 1992. Crown-of-thorns starfish management conundrums. Coral Reefs **11:**53–56. *The first article of an entire issue on the starfish:* Acanthaster planci, *a predator of corals. Another entire issue was devoted to this predator in 1990.*

Lesser, M. P. 1997. Oxidative stress causes coral bleaching during exposure to elevated temperatures. Coral Reefs **16:**187–192. *Evidence that reactive types of oxygen molecules, perhaps produced by the zooxanthellae in response to increased temperatures, cause cell damage and expulsion of zooxanthellae.*

Pennisi, E. 1998. New threat seen from carbon dioxide. Science **279:**989. *Increase in atmospheric CO_2 is acidifying ocean water, making it more difficult for corals to deposit $CaCO_3$. If CO_2 doubles in the next 70 years, as expected, reef formation will decline by 40%, and by 75% if CO_2 doubles again.*

Rosenberg, W., and Y. Loya. 1999. *Vibrio shiloi* is the etiological (causative) agent of *Oculina patagonica* bleaching: general implications. Reef encounter **25:**8–10. *These investigators believe all coral bleaching is due to bacteria, not just that of* O. patagonica. *The bacterium that they report (*Vibrio shiloi*) needs high temperatures to thrive.*

Winnepenninckx, B. M. H., Y. Van de Peer, and T. Backeljau. 1998. Metazoan relationships on the basis of 18S rRNA sequences: a few years later . . . Amer. Zool. **38:**888–906. *Some analyses suggest that sponges and ctenophores form a clade that branched from their common ancestor with cnidarians, all of them forming a sister group to bilateral metazoans.*

Custom Website

The *Animal Diversity* Online Learning Center is a great place to check your understanding of chapter material. Visit www.mhhe.com/hickmanad4e for access to key terms, quizzes, and more! Further enhance your knowledge with Web links to chapter-related material.

Explore live links for these topics:

Classification and Phylogeny of Animals
Radiate Animals
Phylum Cnidaria
Class Hydrozoa
Class Scyphozoa

Class Anthozoa
Coral Reefs
Coral Bleaching
Class Cubozoa
Phylum Ctenophora

Acoelomate Bilateral Animals
Flatworms, Ribbon Worms, and Jaw Worms

Getting Ahead

For animals that spend their lives sitting and waiting, as do most members of the two radiate phyla we considered in the preceding chapter, radial symmetry is ideal. One side of the animal is just as important as any other for snaring prey coming from any direction. But if an animal is active in seeking food, shelter, home sites, and reproductive mates, it requires a different set of strategies and a new body organization. Active, directed movement requires an elongated body form with head (anterior) and tail (posterior) ends. In addition, one side of the body faces up (dorsal) and the other side, specialized for locomotion, faces down (ventral). What results is a bilaterally symmetrical animal in which the body could be divided along only one plane of symmetry to yield two halves that are mirror images of each other. Furthermore, because it is better to determine where one is going than where one has been, sense organs and centers for nervous control have come to be located at the anterior end. This process is called cephalization. Thus cephalization and primary bilateral symmetry occur together in almost all complex animals.

Thysanozoon nigropapillosum, a marine turbellarian (order Polycladida).

The term "worm" has been applied loosely to elongated, bilateral invertebrate animals without appendages. At one time zoologists considered worms (Vermes) a taxon that included a highly diverse assortment of forms. This unnatural assemblage has been reclassified into various phyla. By tradition, however, zoologists still refer to the various groups of these animals as flatworms, ribbon worms, roundworms, segmented worms, and the like. In this chapter we consider Platyhelminthes (Gr. *platys,* flat, + *helmins,* worm), or flatworms, Nemertea (Gr. *Nemertes,* one of the nereids, unerring one), or ribbon worms, and Gnathostomulida (Gr. *gnathos,* jaw + *stoma,* mouth + L. *ulus,* diminutive), or jaw worms. Platyhelminthes is by far the most widely prevalent and abundant of the three phyla, and some flatworms are very important pathogens of humans and domestic animals.

Phylum Platyhelminthes

Platyhelminthes were derived from an ancestor that probably had many cnidarian-like characteristics, including a gelatinous mesoglea. Nonetheless, replacement of gelatinous mesoglea with a cellular, mesodermal **parenchyma** laid the basis for a more complex organization. Parenchyma is a form of tissue containing more cells and fibers than mesoglea of cnidarians.

Flatworms range in size from a millimeter or less to some tapeworms that are many meters in length. Their typically flattened bodies may be slender, broadly leaflike, or long and ribbonlike. There are four classes within Platyhelminthes. Class Turbellaria contains nonparasitic forms, whereas members of classes Monogenea, Trematoda, and Cestoda are parasitic.

Ecological Relationships

Free-living flatworms are found exclusively in class Turbellaria. A few turbellarians are symbiotic (commensals or parasites), but the majority are adapted as bottom dwellers in marine or fresh water or live in moist places on land. Many, especially larger species, are found on undersides of stones and other hard objects in freshwater streams or in littoral zones of the ocean.

Relatively few members of class Turbellaria live in fresh water, but freshwater planarians, such as *Dugesia* (order Tricladida) are used extensively in introductory laboratory courses. Planarians and some others frequent streams and spring pools; others prefer flowing water of mountain streams. Some species occur in moderately hot springs. Terrestrial turbellarians live in fairly moist places under stones and logs (figure 8.2).

All members of classes Monogenea and Trematoda (flukes) and class Cestoda (tapeworms) are parasitic. Most Monogenea are ectoparasites, but all the trematodes and cestodes

Position in the Animal Kingdom

1. Platyhelminthes, or flatworms, Nemertea, or ribbon worms, and Gnathostomulida, or jaw worms, are the simplest animals to have **primary bilateral symmetry.**
2. Members of these phyla are termed **triploblastic** because they have three well-defined germ layers.
3. These phyla have only one internal space, a digestive cavity, with the region between the ectoderm and endoderm filled with mesoderm in the form of muscle fibers and mesenchyme (parenchyma). Since they lack a coelom or a pseudocoelom, they are termed **acoelomate animals** (figure 8.1). Ribbon worms do have a true coelomic cavity, the rhynchocoel, surrounding the proboscis. However, there is

debate as to whether this cavity is homologous to the coelom of most eucoelomates. Some zoologists still consider ribbon worms acoelomate, whereas others place these animals with coelomate protostomes.

4. Acoelomates show more specialization and division of labor among their organs than do radiate animals because having mesoderm makes more elaborate organs possible. Thus acoelomates are said to have reached the **organ-system level of organization.**
5. They belong to the protostome division of the Bilateria and have spiral cleavage, and at least platyhelminths and nemerteans have mosaic (determinate) cleavage.

Biological Contributions

1. Acoelomates show the basic **bilateral** plan of organization that has been widely exploited in the animal kingdom.
2. **Mesoderm** develops into a well-defined embryonic germ layer **(triploblastic),** making available a great source of tissues, organs, and systems.
3. Along with bilateral symmetry, **cephalization** is established. There is some centralization of the nervous system evident in the **ladder type of system** found in flatworms.
4. Along with the subepidermal musculature, there is also a mesenchymal system of muscle fibers.
5. They are the simplest animals with an **excretory system.**
6. Nemerteans are the simplest animals to have a **circulatory system** with blood and a **one-way alimentary canal.**
7. Unique and specialized structures occur in all three phyla. The parasitic habit of many flatworms has led to many specialized adaptations, such as organs of adhesion.

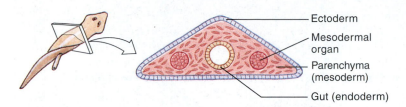

- Ectoderm
- Mesodermal organ
- Parenchyma (mesoderm)
- Gut (endoderm)

figure 8.1

Acoelomate body plan.

figure 8.2

Terrestrial turbellarian (order Tricladida) from the Amazon Basin, Peru.

are endoparasitic. Many species have indirect life cycles with more than one host; the first host is often an invertebrate, and the final host is usually a vertebrate. Humans serve as hosts for a number of species. Certain larval stages may be free-living.

Form and Function

Tegument, Muscles

Most turbellarians have a ciliated cellular epidermis resting on a basement membrane. The epidermis typically contains rod-shaped **rhabdites** that swell to form a protective mucous sheath around the body when discharged with water; single-celled mucous glands may also be present (figure 8.3). Many orders of turbellarians also possess an interesting locomotory system that allows them to quickly attach and detach from surfaces. In such cases, the epidermis contains **dual-gland adhesive organs** consisting of three cell types: viscid and releasing gland cells and anchor cells (figure 8.4). Secretions of the viscid gland cells apparently fasten microvilli of the anchor cells to the substrate, and secretions of the releasing gland cells provide a quick, chemical detaching mechanism.

In contrast to the ciliated cellular epidermis of most turbellarians, all members of the parasitic classes possess a nonciliated body covering called a **syncytial tegument.** The term **syncytial** means that many nuclei are present without any intervening cell membranes. It might seem that a completely new body covering appeared in parasites, but two atypical forms of the epidermis similar to tegument are found within turbellarians. A few turbellarian species have a syncytial

Characteristics
of Phylum Platyhelminthes

1. Three germ layers **(triploblastic)**
2. **Bilateral symmetry;** definite polarity of anterior and posterior ends
3. **Body flattened dorsoventrally** in most; oral and genital apertures mostly on ventral surface
4. Epidermis may be cellular or syncytial (ciliated in some); **rhabdites** in epidermis of most Turbellaria; epidermis a syncytial **tegument** in Monogenea, Trematoda, Cestoda, and some Turbellaria
5. Muscular system of mesodermal origin, in the form of a sheath of circular, longitudinal, and oblique layers beneath the epidermis or tegument
6. No internal body space (acoelomate) other than digestive tube; spaces between organs filled with parenchyma
7. Digestive system incomplete (gastrovascular type); absent in some
8. Nervous system consisting of a **pair of anterior ganglia** with **longitudinal nerve cords** connected by transverse nerves and located in the parenchyma in most forms; similar to cnidarians in some forms
9. Simple sense organs; eyespots in some
10. Excretory system of two lateral canals with branches bearing **flame cells (protonephridia);** lacking in some forms
11. Respiratory, circulatory, and skeletal systems lacking; lymph channels with free cells in some trematodes
12. Most forms monoecious; reproductive system complex, usually with well-developed gonads, ducts, and accessory organs; internal fertilization; life cycle simple in free-swimming forms and those with single hosts; complicated life cycle often involving several hosts in many internal parasites
13. Class Turbellaria mostly free-living; classes Monogenea, Trematoda, and Cestoda entirely parasitic

epidermis, and at least one species has a syncytial "insunk" epidermis where cell bodies (containing the nuclei) are located beneath the basement membrane of the epidermis and communicate with the distal (surface) cytoplasm through cytoplasmic channels. The term *insunk* is a misnomer because the cell bodies below the basement membrane send extensions upward to the epidermis, rather than the other way around. The extensions then fuse to form the syncytial covering, much as they do in the parasitic classes.

Adults of all members of Trematoda, Monogenea, and Cestoda share a syncytial covering that entirely lacks cilia and is designated a **tegument** (figure 8.5). This distinctive tegumental plan is the basis for uniting trematodes, monogeneans, and cestodes in a taxon known as **Neodermata.** It is a peculiar epidermal arrangement and may be related to adaptations for parasitism in ways that are still unclear.

In the body wall below the basement membrane of flatworms are layers of **muscle fibers** that run circularly, longitudinally, and diagonally. A meshwork of **parenchyma** cells, developed from mesoderm, fills the spaces between muscles

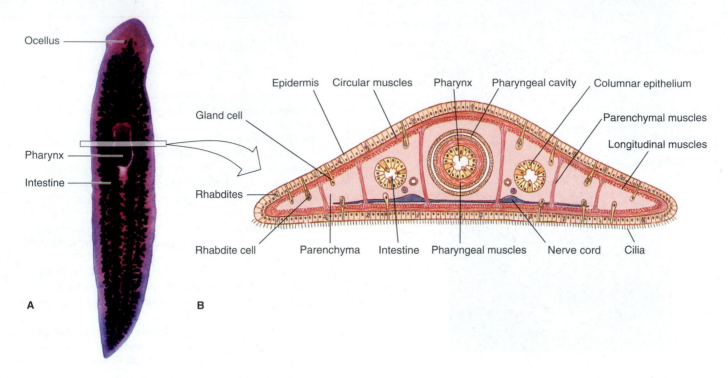

figure 8.3

Cross section through pharyngeal region of a planarian showing relationships of body structures.

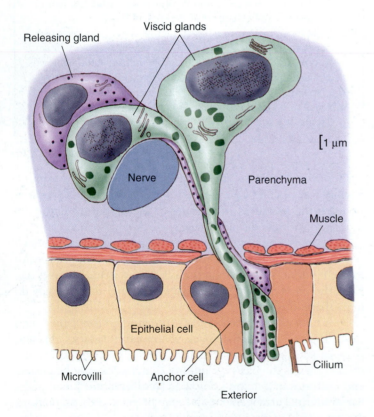

figure 8.4

Reconstruction of dual-gland adhesive organ of a turbellarian *Haplopharynx* sp. There are two viscid glands and one releasing gland, which lie beneath the body wall. The anchor cell lies within the epidermis, and one of the viscid glands and the releasing gland are in contact with a nerve.

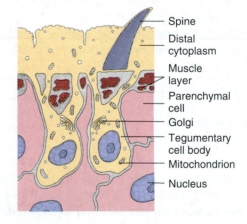

figure 8.5

Tegument of a trematode *Fasciola hepatica*.

and visceral organs. Parenchyma cells in some, perhaps all, flatworms are not a separate cell type but are noncontractile portions of muscle cells.

Nutrition and Digestion

The digestive system includes a mouth, pharynx, and intestine. In turbellarians the muscular **pharynx** opens posteriorly just inside the mouth, through which it can extend (figure 8.6). The mouth is usually at the anterior end in flukes, and the pharynx is not protrusible. The intestine may be simple or branched.

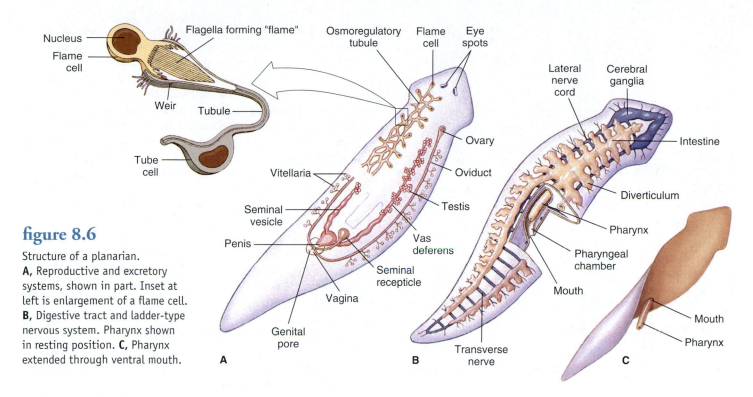

figure 8.6

Structure of a planarian.
A, Reproductive and excretory systems, shown in part. Inset at left is enlargement of a flame cell. **B,** Digestive tract and ladder-type nervous system. Pharynx shown in resting position. **C,** Pharynx extended through ventral mouth.

Intestinal secretions contain proteolytic enzymes for some **extracellular digestion.** Food is sucked into the intestine, where cells of the gastrodermis often phagocytize it and complete digestion **(intracellular).** Undigested food is egested through the pharynx. The entire digestive system is lacking in tapeworms. They must absorb all of their nutrients as small molecules (predigested by the host) directly through their tegument.

Excretion and Osmoregulation

Except in some turbellarians, the osmoregulatory system consists of canals with tubules that end in **flame cells (protonephridia)** (figure 8.6A). Each flame cell surrounds a small space into which a tuft of flagella projects. In some turbellarians and in the other classes of flatworms, the protonephridia form a **weir** (Old English *wer,* a fence placed in a stream to catch fish). In a weir the rim of the cup formed by the flame cell bears fingerlike projections that interdigitate with similar projections of a tubule cell. The beat of the flagella (resembling a flickering flame) provides a negative pressure to draw fluid through the weir into the space (lumen) enclosed by the tubule cell. The lumen continues into collecting ducts that finally open to the outside by pores. The wall of the duct beyond the flame cell commonly bears folds or microvilli that probably function in reabsorption of certain ions or molecules. It is likely that this system is osmoregulatory in most forms; it is reduced or absent in marine turbellarians, which do not have to expel excess water.

Metabolic wastes are largely removed by diffusion through the body wall.

Nervous System and Sense Organs

Flatworms are cephalized but vary in complexity of the nervous system. The simplest system, found in some turbellarians, is a **subepidermal nerve plexus** resembling the nerve net of the cnidarians. Other flatworms have, in addition to a nerve plexus, one to five pairs of **longitudinal nerve cords** lying under the muscle layer (figure 8.6A). Connecting nerves form a "ladder-type" pattern. Their brain is a mass of ganglion cells arising anteriorly from the nerve cords. Except in simpler turbellarians, which have a diffuse system, the neurons are organized into sensory, motor, and association types—an important advance in evolution of nervous systems.

Tactile cells and chemoreceptive cells are abundant over the body, and in planarians they form definitive organs on the **auricles** (the earlike lobes on the sides of the head). Some also have **statocysts** for equilibrium and **rheoreceptors** for sensing water current direction. **Ocelli,** or light-sensitive eyespots, are common in turbellarians (figure 8.6A), monogeneans, and larval trematodes.

Reproduction

Many flatworms reproduce both asexually and sexually. Many freshwater turbellarians can reproduce asexually by fission, merely constricting behind the pharynx and separating into two animals, each of which regenerates the missing parts. In some forms such as *Stenostomum* and *Microstomum,* individuals do not separate at once but remain attached, forming chains of zooids (figure 8.7B and C). Flukes reproduce asexually in their snail intermediate host (p. 152), and some tapeworms,

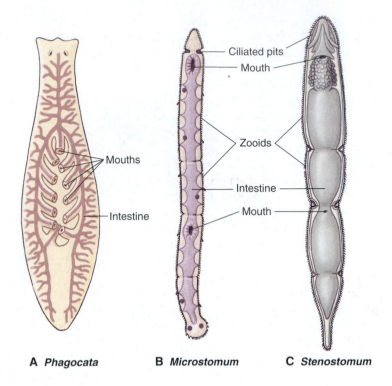

A *Phagocata*   **B** *Microstomum*   **C** *Stenostomum*

figure 8.7

Some small freshwater turbellarians. **A,** *Phagocata* has numerous mouths and pharynges. **B** and **C,** Incomplete fission results for a time in a series of attached zooids.

such as *Echinococcus,* can bud off thousands of juveniles in their intermediate host.

Most flatworms are monoecious (hermaphroditic) and practice cross-fertilization. In some turbellarians yolk for nutrition of the developing embryo is contained within the egg cell itself **(endolecithal),** just as it is normally in other phyla of animals. Possession of endolecithal eggs is considered ancestral for flatworms. Other turbellarians plus all trematodes, monogeneans, and cestodes share a derived condition in which female gametes contain little or no yolk, and yolk is contributed by cells released from separate organs **(yolk glands).** Usually a number of yolk cells surrounds the zygote within an eggshell **(ectolecithal).**

Development following formation of the zygote may be direct or indirect. In some flatworms, an embryo becomes a larva, which may be ciliated or not, according to the group.

Class Turbellaria

Turbellarians are mostly *free-living worms that range in length from 5 mm to nearly 50 cm. Their mouth is on the ventral side and leads into a gut in all groups except acoels, which lack a gut entirely. The orders within class Turbellaria are distinguished by the form of the gut (present or absent; simple or branched; pattern of branching) and pharynx (simple; folded;

bulbous). Except for order Polycladida (Gr. *poly,* many, + *klados,* branch), turbellarians with endolecithal eggs have a simple gut or no gut and a simple pharynx. In a few turbellarians there is no recognizable pharynx. Polyclads have a folded pharynx and a gut with many branches. Polyclads include many marine forms of moderate to large size (3 to more than 40 mm) (p. 145), and a highly branched intestine is correlated with larger size in turbellarians. Members of order Tricladida (Gr. *treis,* three, + *klados,* branch), which are ectolecithal and include freshwater planaria, have a three-branched intestine.

Turbellarians are typically creeping forms that combine muscular with ciliary movements to achieve locomotion. Very small planaria swim by means of their cilia. Others move by gliding, head slightly raised, over a slime track secreted by the marginal adhesive glands. Beating of epidermal cilia in the slime track moves the animal forward, while rhythmical muscular waves can be seen passing backward from the head. Large polyclads and terrestrial turbellarians crawl by muscular undulations, much in the manner of a snail.

Some turbellarians have simple life cycles without a distinct larval stage. In some freshwater planarians egg capsules are attached by little stalks to undersides of stones or plants, and embryos emerge as juveniles that resemble miniature adults. Marine turbellarians have a ciliated larva that is very similar to the trochophore (see p. 182) seen in other phyla.

As traditionally recognized, turbellarians form a paraphyletic group. Several synapomorphies, such as "insunk" epidermis and ectolecithal development, show that some turbellarians are phylogenetically closer to Trematoda, Monogenea, and Cestoda than they are to other turbellarians. Ectolecithal turbellarians therefore appear to form a clade with trematodes, monogeneans, and cestodes to the exclusion of endolecithal turbellarians (see figure 8.21). Endolecithal turbellarians also are paraphyletic; presence of a dual-gland adhesive system in some endolecithal turbellarians indicates a clade with ectolecithal flatworms to the exclusion of other endolecithal turbellarian lineages. Turbellaria therefore describes an artificial group, and the term is used here only because it is prevalent in zoological literature.

Members of order Acoela (Gr. *a,* without, + *koilos,* hollow) were long regarded as having changed least from the ancestral form. Their bodies are small and have a mouth but no gastrovascular cavity or excretory system. Food is merely passed through the mouth into temporary spaces that are surrounded by mesenchyme, where gastrodermal phagocytic cells digest food intracellularly. Molecular evidence suggests that acoels should not be placed in phylum Platyhelminthes and that they represent the sister taxon to all other Bilateria (p. 55). Acoels and a second group of very simple turbellarian-like flatworms collectively belong to a group now called acoelomorphs. Platyhelminthes is not monophyletic when acoelomorphs are retained within it. Without acoelomorphs, class Turbellaria is paraphyletic because the ancestor of the three parasitic classes arose within it, as described in the preceding paragraph.

Class Trematoda

Trematodes are all parasitic flukes, and as adults they are almost all found as endoparasites of vertebrates. They are chiefly leaflike in form and are structurally similar in many respects to more complex Turbellaria. A major difference is found in the tegument (described earlier, figure 8.5).

Other structural adaptations for parasitism are apparent: various penetration glands or glands to produce cyst material; organs for adhesion such as suckers and hooks; and increased reproductive capacity. Otherwise, trematodes share several characteristics with turbellarians, such as a well-developed alimentary canal (but with the mouth at the anterior, or cephalic, end) and similar reproductive, excretory, and nervous systems, as well as a musculature and parenchyma that differ only slightly from those of turbellarians. Sense organs are poorly developed.

Of the three subclasses of Trematoda, two are small and poorly known groups, but Digenea (Gr. *dis,* double, + *genos,* descent) is a large group with many species of medical and economic importance.

Subclass Digenea

With rare exceptions, digenetic trematodes have a complex life cycle, the first **(intermediate)** host being a mollusc and the final **(definitive)** host being a vertebrate. A definitive host is one in which the parasite reproduces sexually. In some species a second, and sometimes even a third, intermediate host intervenes. The group has many species, and they can inhabit diverse sites in their hosts: all parts of the digestive tract, respiratory tract, circulatory system, urinary tract, and reproductive tract.

Among the world's most amazing phenomena are digenean life cycles. Although cycles of different species vary widely in detail, a typical example would include an adult, egg (shelled embryo), miracidium, sporocyst, redia, cercaria, and metacercaria stages (figure 8.8). The shelled embryo or larva usually passes from the definitive host in excreta and must reach water to develop further. There, it hatches to a free-swimming, ciliated larva, the **miracidium.** The miracidium penetrates the tissues of a snail, where it transforms into a **sporocyst.** Sporocysts reproduce asexually to yield either more sporocysts or a number of **rediae.** Rediae, in turn, reproduce asexually to produce more rediae or to produce **cercariae.** In this way a single zygote can give rise to an enormous number of progeny. Cercariae emerge from the snail and penetrate a second intermediate host or encyst on vegetation or other objects to become **metacercariae,** which are juvenile flukes. Adults grow from the metacercaria when that stage is eaten by a definitive host.

Some of the most serious parasites of humans and domestic animals belong to the Digenea (table 8.1).

Clonorchis (Gr. *clon,* branch, + *orchis,* testis) (figure 8.8) is the most important liver fluke of humans and is common in many regions of the Orient, especially in China, southern Asia, and Japan. Cats, dogs, and pigs are also often infected. Adult flukes live in the bile passages, and shelled miracidia are passed in feces. If ingested by certain freshwater snails, sporocyst and redia stages develop, and free-swimming cercariae emerge. Cercariae that manage to find a suitable fish encyst in the skin or muscles as metacercariae. When the fish is eaten raw, the juveniles migrate up the bile duct to mature and may survive there for 15 to 30 years. The effect of the flukes on humans depends mainly on extent of infection. A heavy infection may cause a pronounced cirrhosis of the liver and death.

Schistosomiasis, an infection with blood flukes of genus *Schistosoma* (Gr. *schistos,* divided, + *soma,* body) (figure 8.9), ranks as one of the major infectious diseases in the world, with 200 million people infected. The disease is widely prevalent over much of Africa and parts of South America, West Indies, Middle East, and Far East. It is spread when shelled miracidia shed in human feces and urine get into water containing host snails (figure 8.9B). Cercariae that contact human skin penetrate through the skin to enter blood vessels, which they follow to certain favorite regions depending on the type of fluke. *Schistosoma*

Table 8.1 Examples of Flukes Infecting Humans

Common and scientific names	Means of infection; distribution and prevalence in humans
Blood flukes (*Schistosoma* spp.); three widely prevalent species, others reported	Cercariae in water penetrate skin; 200 million people infected with one or more species
S. mansoni	Africa, South and Central America, caribbean
S. haematobium	Africa
S. japonicum	Eastern Asia
Chinese liver flukes (*Clonorchis sinensis*)	Eating metacercariae in raw fish; about 30 million cases in Eastern Asia
Lung flukes (*Paragonimus* spp.), seven species, most prevalent is *P. westermani*	Eating metacercariae in raw freshwater crabs, crayfish; Asia and Oceania, sub-Saharan Africa, South and Central America; several million cases in Asia
Intestinal fluke (*Fasciolopsis buski*)	Eating metacercariae on aquatic vegetation; 10 million cases in Eastern Asia
Sheep liver fluke (*Fasciola hepatica*)	Eating metacercariae on aquatic vegetation; widely prevalent in sheep and cattle, occasional in humans

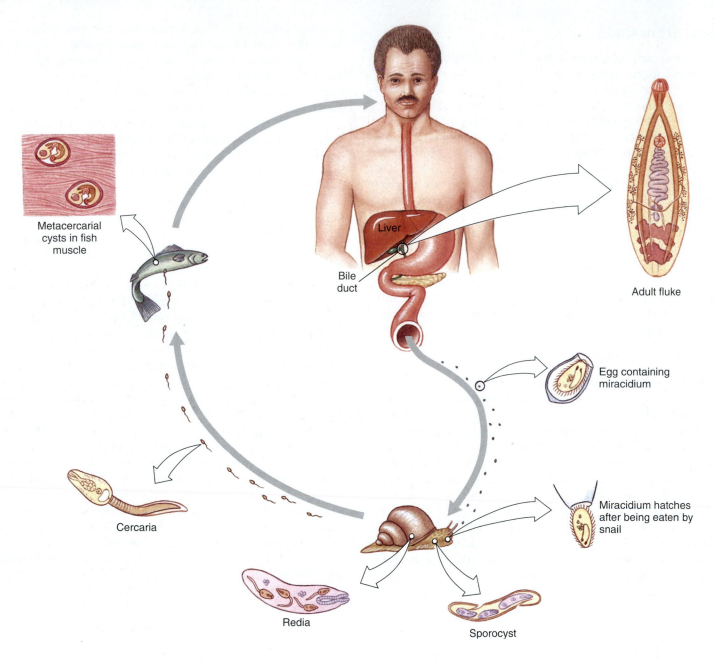

Metacercarial cysts in fish muscle

Liver

Bile duct

Adult fluke

Egg containing miracidium

Miracidium hatches after being eaten by snail

Cercaria

Redia

Sporocyst

figure 8.8

Life cycle of human liver fluke, *Clonorchis sinensis.*

mansoni lives in venules draining the large intestine; *S. japonicum* localizes more in venules draining the small intestine; and *S. baematobium* lives in venules draining the urinary bladder. In each case many eggs released by female worms do not find their way out of the body but lodge in the liver or other organs. There they are sources of chronic inflammation. The inch-long adults may live for years in a human host, and their eggs cause such disturbances as severe dysentery, anemia, liver enlargement, bladder inflammation, and brain damage.

Cercariae of several genera whose normal hosts are birds often enter the skin of human bathers in their search for a suitable bird host, causing a skin irritation known as "swimmer's itch" (figure 8.10). In this case humans are a dead end in the fluke's life cycle because the fluke cannot develop further in humans.

Class Monogenea

Monogenetic flukes traditionally have been placed as an order of Trematoda, but they are sufficiently different to deserve a separate class. Cladistic analysis places them closer to Cestoda. Monogeneans are mostly external parasites that clamp onto the gills and external surfaces of fish using a hooked attachment organ called an **opisthaptor** (figure 8.11). A few are found in urinary bladders of frogs and turtles, and one has been reported from the eye of a hippopotamus. Although widespread and common, monogeneans seem to cause little damage to their hosts under natural conditions. However, like numerous other fish pathogens, they become a serious threat when their hosts are crowded together, as in fish farming.

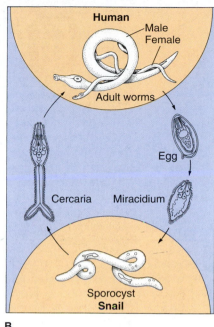

figure 8.9

A, Adult male and female *Schistosoma japonicum* in copulation. Blood flukes differ from most other flukes in being dioecious. Males are broader and heavier and have a large, ventral gynecophoric canal, posterior to the ventral sucker. The gynecophoric canal embraces the long, slender female (darkly stained individual) during insemination and oviposition. **B,** Life cycle of *Schistosoma* spp.

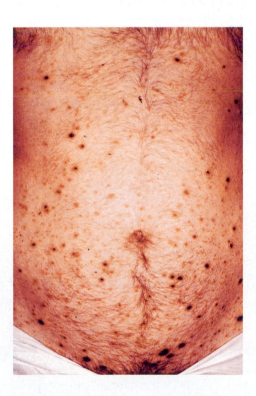

figure 8.10

Human abdomen, showing schistosome dermatitis caused by penetration of schistosome cercariae that are unable to complete development in humans. Sensitization to allergenic substances released by cercariae results in rash and itching.

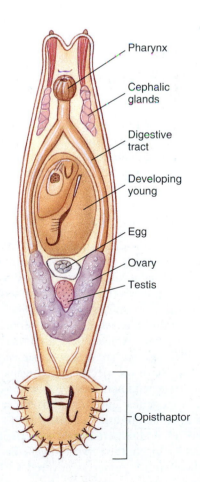

figure 8.11

Gyrodactylus sp. (class Monogenea), ventral view.

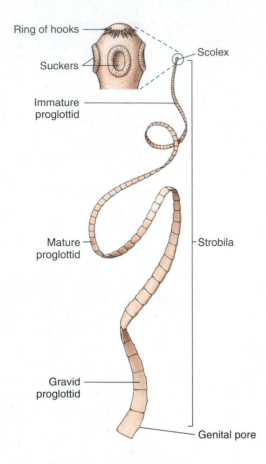

figure 8.12

A tapeworm, showing strobila and scolex. The scolex is the organ of attachment.

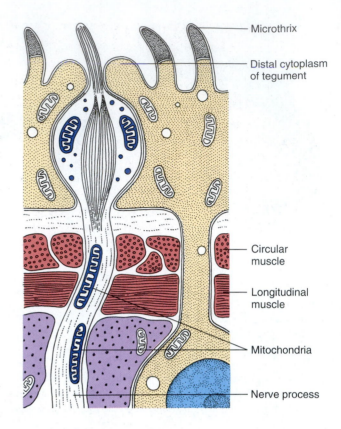

figure 8.13

Schematic drawing of a longitudinal section through a sensory ending in the tegument of *Echinococcus granulosus*.

Life cycles of monogeneans are simple, with a single host, as suggested by the name of the group, which means "single descent." The egg hatches a ciliated larva that attaches to a host or swims around awhile before attachment.

Class Cestoda

Cestoda, or tapeworms, differ in many respects from the preceding classes: they usually have long flat bodies composed of a **scolex,** for attachment to the host, followed by many reproductive units or **proglottids** (figure 8.12). The scolex, or holdfast, is usually provided with suckers or suckerlike organs and often with hooks or spiny tentacles. Tapeworms entirely lack a digestive system, but they have well-developed muscles, and their excretory and nervous systems are somewhat similar to those of other flatworms. They have no special sense organs, but do have sensory endings in the tegument that are modified cilia (figure 8.13).

As in Monogenea and Trematoda, there are no external, motile cilia in adults, and the tegument is of a distal cytoplasm with sunken cell bodies beneath the superficial layer of muscle (see figure 8.5). In contrast to monogeneans and trematodes, however, their entire surface is covered with minute projections that are similar in certain respects to microvilli of the vertebrate small intestine (figure 8.13). These **microtriches** (sing. **microthrix**) greatly amplify the surface area of the tegument,

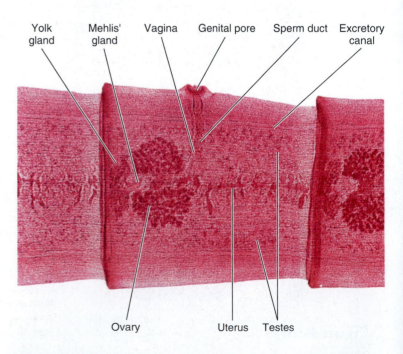

figure 8.14

Mature proglottid of *Taenia pisiformis,* a dog tapeworm. Portions of two other proglottids also shown.

Table 8.2 Common Cestodes of Humans

Common and scientific names	Means of infection; prevalence in humans
Beef tapeworm (*Taenia saginata*)	Eating rare beef; most common of large tapeworms in humans
Pork tapeworm (*Taenia solium*)	Eating rare pork; less common than *T. saginata*
Fish tapeworm (*Diphyllobothrium latum*)	Eating rare or poorly cooked fish; fairly common in Great Lakes region of United States, and other areas of world where raw fish is eaten
Dog tapeworm (*Dipylidium caninum*)	Unhygienic habits of children (juveniles in flea and louse); moderate frequency
Dwarf tapeworm (*Hymenolepis nana*)	Juveniles in flour beetles; common
Unilocular hydatid (*Echinococcus granulosus*)	Cysts of juveniles in humans; infection by contact with dogs; common wherever humans are in close relationship with dogs and ruminants
Multilocular hydatid (*Echinococcus multilocularis*)	Cysts of juveniles in humans; infection by contact with foxes; less common than unilocular hydatid

which is a vital adaptation for a tapeworm, since it must absorb all its nutrients across the tegument.

Tapeworms are nearly all monoecious. The main body of a cestode is a chain of proglottids called a **strobila** (figure 8.12). Typically, there is a **germinative zone** just behind the scolex where new proglottids form. As new proglottids differentiate in front of it, each individual unit moves posteriorly in the strobila, and its gonads mature (figure 8.14). A proglottid usually is fertilized by another proglottid in the same or a different strobila. Shelled embryos form in the uterus of the proglottid, and they are either expelled through a uterine pore, or the entire proglottid detaches from the worm as it reaches the posterior end.

About 4000 species of tapeworms are known to parasitologists. With rare exceptions, all cestodes require at least two hosts, and adults are parasitic in the digestive tract of vertebrates. Often one of the intermediate hosts is an invertebrate. Almost all vertebrate species are infected. Normally,

adult tapeworms do little harm to their hosts. Table 8.2 lists the most common tapeworms in humans.

In the beef tapeworm, *Taenia saginata* (Gr. *tainia*, band, ribbon), shelled larvae shed from the human host are ingested by cattle (figure 8.15). The six-hooked larvae (oncospheres) hatch, burrow into blood or lymph vessels, and migrate to skeletal muscle where they encyst to become "bladder worms" (**cysticerci**). Each of these juveniles develops an invaginated scolex and remains quiescent until the uncooked muscle is eaten by humans. In the new host the scolex evaginates, attaches to the intestine, and matures in 2 or 3 weeks; then ripe proglottids may be expelled daily for many years. Humans become infected by eating raw or rare infested ("measly") beef. Adult worms may attain a length of 7 m or more, folded back and forth in the host intestine.

The pork tapeworm, *Taenia solium*, uses humans as definitive hosts and pigs as intermediate hosts. Humans can

classification of Phylum Platyhelminthes

Class Turbellaria (tur'bel-lar'e-a) (L. *turbellae* [pl.], stir, bustle, + *-aria*, like or connected with). **Turbellarians.** Usually free-living forms with soft flattened bodies; covered with ciliated epidermis containing secreting cells and rodlike bodies (rhabdites); mouth usually on ventral surface sometimes near center of body; no body cavity except intercellular spaces in parenchyma; mostly hermaphroditic; some have asexual fission. A paraphyletic taxon. Examples: *Dugesia* (planaria), *Microstomum*, *Planocera*.

Class Trematoda (trem'a-to'da) (Gr. *trēma*, a hole, + *eidos*, form). **Digenetic flukes.** Body covered with a syncytial tegument without cilia; leaflike or cylindrical in shape; usually with oral and ventral suckers, no hooks; alimentary canal usually with two main branches; mostly monoecious; life cycle complex, with first host a mollusc, final host usually a vertebrate; parasitic in all classes of vertebrates. Examples: *Fasciola, Clonorchis, Schistosoma.*

Class Monogenea (mon'o-gen'e-a) (Gr. *mono*, single, + *genos*, descent). **Monogenetic flukes.** Body covered with a syncytial tegument without cilia; body usually leaflike to cylindrical in shape; posterior attachment organ with hooks, suckers, or clamps, usually in combination; monoecious; life cycle simple, with single host and usually with free-swimming, ciliated larva; all parasitic, mostly on skin or gills of fish. Examples: *Dactylogyrus, Polystoma, Gyrodactylus.*

Class Cestoda (ses-to'da) (Gr. *kestos*, girdle, + *eidos*, form). **Tapeworms.** Body covered with nonciliated, syncytial tegument; general form of body tapelike; scolex with suckers or hooks, sometimes both, for attachment; body usually divided into series of proglottids; no digestive organs; usually monoecious; parasitic in digestive tract of all classes of vertebrates; life cycle complex, with two or more hosts; first host may be vertebrate or invertebrate. Examples: *Diphyllobothrium, Hymenolepis, Taenia.*

figure 8.15

Life cycle of beef tapeworm, *Taenia saginata*. Ripe proglottids break off in the human intestine, leave the body in feces, crawl out of the feces onto grass, and are ingested by cattle. Eggs hatch in the cow's intestine, freeing oncospheres, which penetrate into muscles and encyst, developing into "bladder worms." A human eats infected rare beef, and cysticercus is freed in intestine where it attaches to the intestinal wall, forms a strobila, and matures.

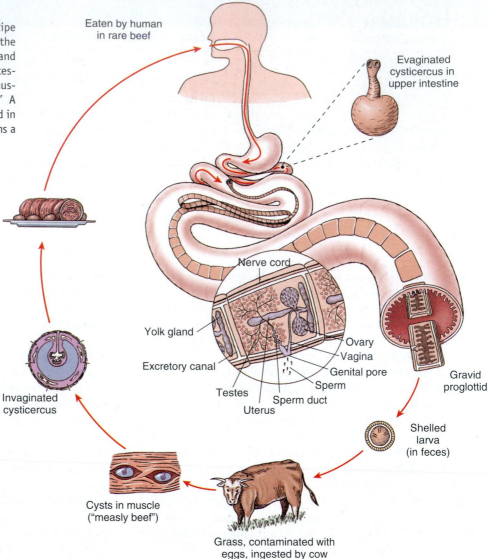

Eaten by human in rare beef

Evaginated cysticercus in upper intestine

Nerve cord

Yolk gland

Excretory canal

Ovary

Vagina

Genital pore

Sperm

Testes

Sperm duct

Uterus

Gravid proglottid

Shelled larva (in feces)

Invaginated cysticercus

Cysts in muscle ("measly beef")

Grass, contaminated with eggs, ingested by cow

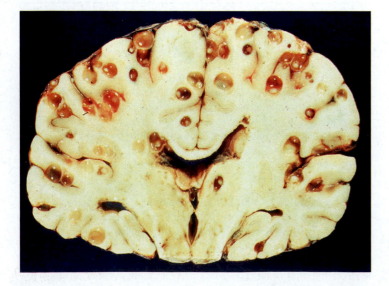

figure 8.16

Section through the brain of a person who died of cerebral cysticercosis, an infection with cysticerci of *Taenia solium*.

also serve as intermediate hosts by ingesting shelled larvae from contaminated hands or food or, in persons harboring an adult worm, by regurgitating segments into the stomach. Cysticerci may encyst in the central nervous system, where great damage may result (figure 8.16).

Phylum Nemertea (Rhynchocoela)

Nemerteans are often called ribbon worms. Their name (Gr. *Nemertes*, one of the nereids, unerring one) refers to the unerring aim of the proboscis, a long muscular tube (figures 8.17 and 8.18) that can be thrust out swiftly to grasp the prey. The phylum is also called Rhynchocoela (Gr. *rhynchos*, beak, + *koilos*, hollow), which also refers to the proboscis. They are thread-shaped or ribbon-shaped predatory worms. There are about 1000 species in the group; nearly all are marine.

Nemertean worms are usually less than 20 cm long, although a few are several meters in length (figure 8.19). Some are brightly colored, although most are dull or pallid. Some live in secreted gelatinous tubes.

figure 8.17

Ribbon worm *Amphiporus bimaculatus* (phylum Nemertea) is 6 to 10 cm long, but other species range up to several meters. The proboscis of this specimen is partially extended at the top; the head is marked by two brown spots.

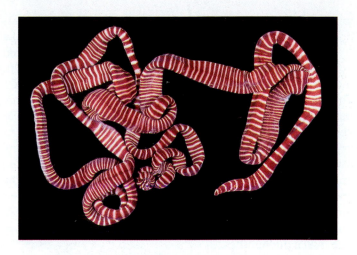

figure 8.19

Baseodiscus is a genus of nemerteans whose members typically measure several meters in length. This *B. mexicanus* from the Galápagos Islands was over 10 m long.

With few exceptions, the general body plan of nemerteans is similar to that of turbellarians. Like the latter, their epidermis is ciliated and has many gland cells. Another striking similarity is the presence of flame cells in the excretory system. Rhabdites occur in several nemerteans, but some work indicates that they are not homologous to rhabdites in flatworms. In marine forms there is a ciliated larva that has some resemblance to trochophore larvae found in annelids and molluscs. All present evidence places nemerteans in Lophotrochozoa with Platyhelminthes. Nevertheless, they differ from flatworms in several important aspects.

Form and Function

Many nemerteans are difficult to examine because they are so long and fragile. *Amphiporus* (Gr. *amphi,* on both sides, + *poros,* pore), one of the smaller genera that ranges from 2 to 10 cm in length, is fairly typical of nemertean structure (see figure 8.18). Its body wall consists of ciliated epidermis and layers of circular and longitudinal muscles. Locomotion consists largely of gliding over a slime track, although larger species move by muscular contractions.

The mouth is anterior and ventral, and the **digestive tract** is **complete,** extending the full length of the body and ending at an anus. Presence of an anus marks a significant advancement over the gastrovascular systems of flatworms and radiates, because ingestion and egestion can occur simultaneously. Cilia move food through the intestine. Digestion is largely extracellular.

Nemerteans are carnivorous, feeding primarily on annelids and other small invertebrates. They seize their prey with a **proboscis** that lies in an interior cavity of its own, the **rhynchocoel,** above the digestive tract (but not connected with it). The proboscis itself is a long, blind muscular tube that opens at the anterior end at a proboscis pore above the mouth. (In a few nemerteans the esophagus opens through the proboscis pore rather than through a separate mouth.) Muscular pressure on fluid in the rhynchocoel causes the long

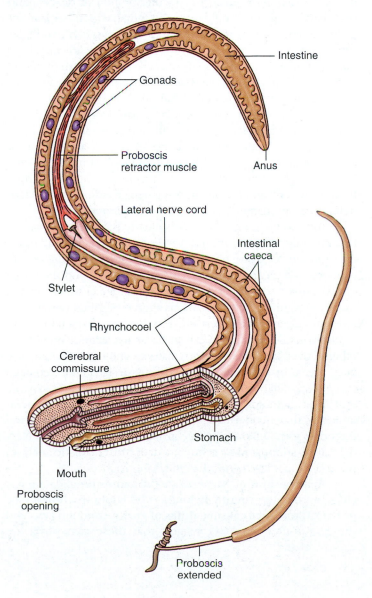

Intestine

Gonads

Proboscis retractor muscle

Anus

Lateral nerve cord

Intestinal caeca

Stylet

Rhynchocoel

Cerebral commissure

Stomach

Mouth

Proboscis opening

Proboscis extended

figure 8.18

A, Internal structure of female ribbon worm (diagrammatic). Dorsal view to show proboscis. **B,** *Amphiporus* with proboscis extended to catch prey.

tubular proboscis to be everted rapidly through the proboscis pore. Eversion of the proboscis exposes a sharp barb, called a stylet (absent in some nemerteans). The sticky, slime-covered proboscis coils around the prey and stabs it repeatedly with the stylet, while pouring potent neurotoxins on the prey. Then, retracting its proboscis, a nemertean draws the prey near its mouth, through which the esophagus is thrust to engulf the prey.

Unlike other acoelomates, nemerteans have a true circulatory system, and an irregular flow is maintained by the contractile walls of the vessels. Many flame-bulb protonephridia are closely associated with the circulatory system, so that their function appears to be truly excretory (for disposal of metabolic wastes), in contrast to their apparently osmoregulatory role in Platyhelminthes.

Nemerteans have a pair of nerve ganglia, and one or more pairs of longitudinal nerve cords are connected by transverse nerves.

Some species reproduce asexually by fragmentation and regeneration. In contrast to flatworms, most nemerteans are dioecious.

Phylum Gnathostomulida

The first species of Gnathostomulida (nath′o-sto-myu′lid-a) (Gr. *gnatho,* jaw, + *stoma,* mouth, + L.-*ulus,* diminutive) was observed in 1928 in the Baltic, but its description was not published until 1956. Since then these animals have been found in many parts of the world, including the Atlantic coast of the United States, and over 80 species in 18 genera have been described.

Gnathostomulids are delicate wormlike animals 0.5 to 1 mm long (figure 8.20). They live in interstitial spaces of very fine sandy coastal sediments and silt and can endure conditions of very low oxygen.

Lacking a pseudocoel, a circulatory system, and an anus, gnathostomulids show some similarities to turbellarians and were at first included in that group. Although the epidermis is ciliated, each epidermal cell has only one cilium, which is a condition not found in other bilateral animals except in some gastrotrichs (p. 166). The gnathostomulid parenchyma is poorly developed, and the pharynx is reminiscent of the rotifer mastax (p. 163). The gnathostomulid pharynx is armed with a pair of lateral jaws used to scrape fungi and bacteria off the substratum. The gnathostomulid pharynx and jaws are so similar to those of rotifers and micrognathozoans (see p. 165) that some researchers have proposed a new clade, Gnathifera, to unite Rotifera, Acanthocephala, Micrognathozoa, and Gnathostomulida.

Phylogeny and Adaptive Radiation

Phylogeny

Bilaterally symmetrical flatworms may have derived from a radial ancestor, perhaps one very similar to the planula larva of cnidarians. Some investigators propose that this **planuloid**

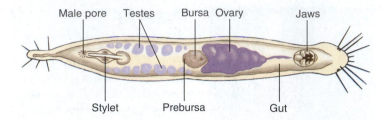

figure 8.20

Gnathostomula jenneri (phylum Gnathostomulida) is a tiny member of the interstitial fauna between grains of sand or mud. It belongs to a family of commonly encountered jaw worms, living in shallow water and down to depths of several hundred meters.

ancestor may have given rise to one branch of descendants that were sessile or free floating and radial, which became the Cnidaria, and another that acquired a creeping habit and bilateral symmetry. Bilateral symmetry is a selective advantage for creeping or swimming animals because sensory structures are concentrated anteriorly (cephalization), which is the end that first encounters environmental stimuli.

A recent report[1] cites sequence data from small-subunit rDNA, embryonic cleavage patterns and mesodermal origins, and nervous system structure as evidence that acoelomorphs are not, in fact, members of the phylum Platyhelminthes. These investigators conclude that acoelomorphs are the sister group to all other Bilateria. If this arrangement is sound, then Platyhelminthes as currently constituted is not monophyletic.

Although class Turbellaria is clearly paraphyletic, we are retaining the taxon for the present because presentation based on thorough cladistic analysis would require introduction of many more taxa and characteristics that are beyond the scope of this book and not yet common in zoological literature. For example, ectolecithal turbellarians should be allied with trematodes, monogeneans, and cestodes in a sister group to endolecithal turbellarians. Some ectolecithal turbellarians share a number of other derived characters with trematodes and cestodes and have been placed by Brooks (1989) in a group designated Cercomeria (Gr. *kerkos,* tail, + *meros,* part) (figure 8.21). Several synapomorphies, including the unique architecture of the tegument, indicates that neodermatans (trematodes, monogeneans, and cestodes) form a monophyletic group, and monophyly of Neodermata is supported by sequence data. Most cladistic analyses place phylum Platyhelminthea in Lophotrochozoa.

Relationships of Nemertea and Gnathostomulida are not clear, but they apparently belong in the Lophotrochozoa (see p. 163). Ultrastructural similarities of gnathostomulid jaws and those of rotifers (p. 163) suggest that these two phyla are closely related.

---

[1]Ruiz-Trillo et al. 2004. Molecular Phylogenetics and Evolution **33:** 321–332.

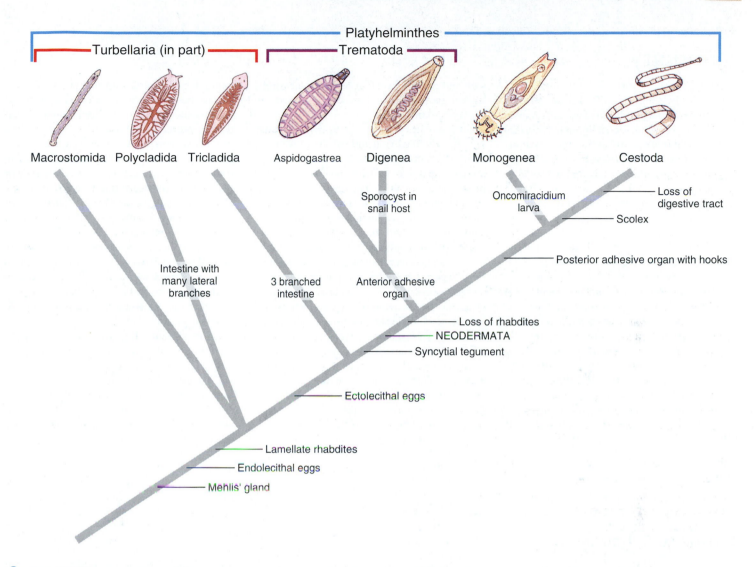

figure 8.21

Hypothetical relationships among parasitic Platyhelminthes. The traditionally accepted class Turbellaria is paraphyletic. Some turbellarians have ectolecithal development and, together with the Trematoda, Monogenea, and Cestoda, form a clade and a sister group of the endolecithal turbellarians. For the sake of simplicity, the synapomorphies of those turbellarians and of the Aspidogastrea, as well as many others given by Brooks (1989) are omitted. All of these organisms form a clade (called Cercomeria) with a posterior adhesive organ.

Source: Modified from D. R. Brooks. The phylogeny of the Cercomeria (Platyhelminthes: Rhabdocola) and general evolutionary principles. *Journal of Parasitology* **75:**606–616, 1989.

Adaptive Radiation

The flatworm body plan, with its creeping adaptation, placed a selective advantage on bilateral symmetry and further development of cephalization, ventral and dorsal regions, and caudal differentiation. Because of their body shape and metabolic requirements, early flatworms must have been well predisposed toward parasitism and gave rise to symbiotic descendants in Neodermata. These descendants radiated abundantly as parasites, and many flatworms became highly specialized for that mode of existence. Specializations include loss of the head, reduced sensory organs, and loss of the gut in tapeworms.

Ribbon worms have stressed the proboscis apparatus in their evolutionary diversity. Its use in capturing prey may have been secondarily evolved from its original function as a highly sensitive organ for exploring the environment. Although ribbon worms have evolved beyond flatworms in their complexity of organization, they have been dramatically less abundant as a group.

Likewise, jaw worms have not radiated nor become nearly as abundant or diverse as flatworms. However, they have exploited marine interstitial environments, particularly zones of very low oxygen concentration.

Summary

Platyhelminthes, Nemertea, and Gnathostomulida are the simplest phyla that are bilaterally symmetrical, a condition of adaptive value for actively crawling or swimming animals. They are triploblastic and at the organ-system level of organization. They have neither a coelom nor a pseudocoel and are thus acoelomate, although argument remains regarding the status of the nemertean rhynchocoel.

The body surface of turbellarians is a cellular epithelium, at least in part ciliated, containing mucous cells and rod-shaped rhabdites that function together in locomotion. Members of all other classes of flatworms are covered by a nonciliated, syncytial tegument with a vesicular distal cytoplasm and cell bodies beneath superficial muscle layers. Digestion is extracellular and intracellular in most; cestodes must absorb predigested nutrients across their tegument because they have no digestive tract. Osmoregulation is by flame-cell protonephridia, and removal of metabolic wastes and respiration occur across the body wall. Except for some turbellarians, flatworms have a ladder-type nervous system with motor, sensory, and association neurons. Most flatworms are hermaphroditic, and asexual reproduction occurs in some groups.

Class Turbellaria is a paraphyletic group with mostly free-living and carnivorous members. Digenetic trematodes have mollusc intermediate hosts and almost always a vertebrate definitive host. The great amount of asexual reproduction that occurs in their intermediate host helps to increase the chance that some offspring will reach a definitive host. Aside from the tegument, digeneans share many basic structural characteristics with the Turbellaria. Digenea includes a number of important parasites of humans and domestic animals. Digenea contrast with Monogenea, which are important ectoparasites of fishes and have a direct life cycle (without intermediate hosts).

Cestodes (tapeworms) generally have a scolex at their anterior end, followed by a long chain of proglottids, each of which contains a complete set of reproductive organs of both sexes. Cestodes live as adults in the digestive tract of vertebrates. They have microvillus-like microtriches on their tegument, which increase its surface area for absorption. The shelled larvae are passed in the feces of their host, and the juveniles develop in a vertebrate or invertebrate intermediate host.

Flatworms and cnidarians may have evolved from a common planuloid ancestor, some of whose descendants became sessile or free floating and radial (cnidarians), while others became creeping and bilateral (flatworms).

Sequence analysis of rDNA, as well as some developmental and morphological criteria, suggest that acoelomorphs, previously considered turbellarians, diverged from a common ancestor shared with other Bilateria and may be the sister group of all other bilateral phyla.

Members of Nemertea have a complete digestive system with an anus and a true circulatory system. They are free-living, mostly marine, and they ensnare prey with their long, eversible proboscis.

Gnathostomulida are a curious phylum of little wormlike marine animals living among sand grains and silt. They have no anus, and they share certain characteristics such as jaw structure with rotifers.

Review Questions

1. Why is bilateral symmetry of adaptive value for actively motile animals?
2. Match the terms in the right column with the classes in the left column:
   ___ Turbellaria   a. Endoparasitic
   ___ Monogenea   b. Free-living and
   ___ Trematoda   commensal
   ___ Cestoda   c. Ectoparasitic
3. Examine the text to discover that there is no unique feature to distinguish Platyhelminthes.
4. Distinguish two mechanisms by which flatworms supply yolk for their embryos. Which system is evolutionarily ancestral for flatworms and which one is derived?
5. How do flatworms digest their food?
6. Briefly describe the osmoregulatory system and the nervous system and sense organs of platyhelminths.
7. Contrast asexual reproduction in Turbellaria and Trematoda.
8. Contrast a typical life cycle of a monogenean with that of a digenetic trematode.
9. Describe and contrast the tegument of turbellarians and the other classes of platyhelminths. Could this be evidence that the trematodes, monogeneans, and cestodes form a clade within the Platyhelminthes? Why?
10. Answer the following questions with respect to both *Clonorchis* and *Schistosoma:* (a) How do humans become infected? (b) What is the general geographical distribution? (c) What are the main disease conditions produced?
11. Define each of the following with reference to cestodes: scolex, microvilli, proglottids, strobila.
12. Why is *Taenia solium* a more dangerous infection than *Taenia saginata?*
13. What are some of the adaptations possessed by flukes and tapeworms for a parasitic lifestyle?
14. Give three differences between nemerteans and platyhelminths.
15. Where do gnathostomulids live?
16. Explain how a planuloid ancestor could have given rise to both Cnidaria and Bilateria.
17. What is an important character of Nemertea that might suggest that the phylum is closer to other bilateral protostomes than to Platyhelminthes? What are some characters of nemerteans suggesting that they be grouped with platyhelminths?
18. Recent evidence suggests that acoelomorphs are not members of Platyhelminthes but constitute a sister group for all other Bilateria. If so, what is a consequence to phylogenetic integrity of Platyhelminthes if acoelomorphs remain within that phylum? What evidence indicates that the traditional class Turbellaria is paraphyletic?

Selected References

See also general references on page 415.

Arme, C., and P. W. Pappas (eds.). 1983. Biology of the Eucestoda, 2 vols. New York, Academic Press, Inc. *The most up-to-date reference available on the biology of tapeworms. Advanced; not an identification manual.*

Brooks, D. R. 1989. The phylogeny of the Cercomeria (Platyhelminthes: Rhabdocoela) and general evolutionary principles. J. Parasitol. **75:**606–616. *Cladistic analysis of parasitic flatworms.*

Dalton, J. P., P. Skelly, and D. W. Halton. 2004. Role of the tegument and gut in nutrient uptake by parasitic platyhelminthes. Can. J. Zool. **82:**211–232. *The function of the tegument and the evolution of an endoparasitic lifestyle are discussed here.*

Desowitz, R. S. 1981. New Guinea tapeworms and Jewish grandmothers. New York, W. W. Norton & Company. *Accounts of parasites and parasitic diseases of humans. Entertaining and instructive. Recommended for all students.*

Eernisse, D. J., and K. J. Peterson. 2004. The history of animals, pp. 197–208. In J. Cracraft and M. J. Donoghue, eds., Assembling the tree of life. Oxford University Press. *A general overview of current metazoan phylogeny with discussion of the major clades and "problem" taxa including acoelomorph flatworms.*

Rieger, R. M., and S. Tyler. 1995. Sister-group relationship of Gnathostomulida and Rotifera-Acanthocephala. Invert. Biol. **114:**186–188. *Evidence that gnathostomulids are the sister group of a clade containing rotifers and acanthocephalans.*

Ruiz-Trullo, M., M. Riutort, H. M. Fourcade, J. Baguna, and J. L. Boore. 2004. Mitochondrial genome data support the basal position of Acoelomorpha and the polyphyly of the Platyhelminthes. Mol. Phylogenet. Evol. **33:**321–332. *This paper adds to the growing body of evidence that acoelomorphs are the sister taxon to all other bilateral metazoans and not members of Platyhelminthes.*

Schell, S. C. 1985. Handbook of trematodes of North America north of Mexico. Moscow, Idaho, University Press of Idaho. *Good for trematode identification.*

Schmidt, G. D. 1985. Handbook of tapeworm identification. Boca Raton, Florida, CRC Press. *Good for cestode identification.*

Strickland, G. T. 2000. Hunter's tropical medicine and emerging infectious diseases, ed. 8. Philadelphia, W. B. Saunders Company. *A valuable source of information on parasites of medical importance.*

Turbeville, J. M. 2002. Progress in nemertean biology: development and phylogeny. Integ. and Comp. Biol. **42:**692–703. *A discussion of nemertean features that suggest shared ancestry with flatworms and other lophotrochozoan phyla.*

Custom Website

The *Animal Diversity* Online Learning Center is a great place to check your understanding of chapter material. Visit www.mhhe.com/hickmanad4e for access to key terms, quizzes, and more! Further enhance your knowledge with Web links to chapter-related material.

Explore live links for these topics:

Classification and Phylogeny of Animals
Phylum Platyhelminthes

Class Turbellaria
Class Trematoda
Class Cestoda
Class Monogenea
Phylum Nemertea

9

Pseudocoelomate Animals

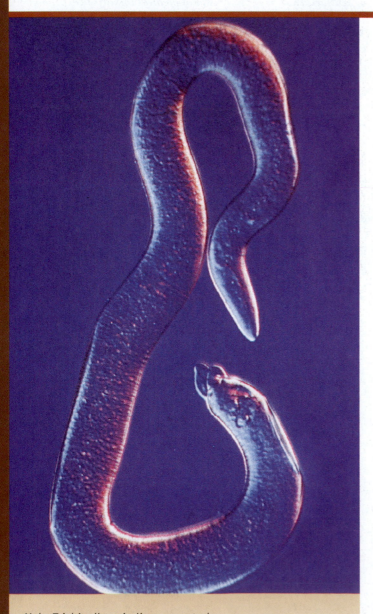

Male *Trichinella spiralis,* a nematode.

A World of Nematodes

Without any doubt, nematodes are the most important pseudocoelomate animals in numbers and their impact on humans. Nematodes are abundant over most of the world, yet most people are only occasionally aware of them as parasites of humans or of their pets. We are not aware of the billions of these worms in the soil, in ocean and freshwater habitats, in plants, and in all kinds of other animals. Their dramatic abundance moved N. A. Cobb to write in 1914:

> If all the matter in the universe except the nematodes were swept away, our world would still be dimly recognizable, and if, as disembodied spirits, we could then investigate it, we should find its mountains, hills, vales, rivers, lakes and oceans represented by a thin film of nematodes. The location of towns would be decipherable, since for every massing of human beings there would be a corresponding massing of certain nematodes. Trees would still stand in ghostly rows representing our streets and highways. The location of the various plants and animals would still be decipherable, and, had we sufficient knowledge, in many cases even their species could be determined by an examination of their erstwhile nematode parasites.[1]

---

[1]N. A. Cobb. 1914. *Yearbook of the United States Department of Agriculture,* 1914.

This chapter covers nine animal phyla with pseudo-coelomate bodies and two related animal groups, Micrognathozoa and Cycliophora, which may be accepted as new phyla eventually. These taxa are considered protostomes, but recent phylogenetic analyses using molecular characters strongly suggest that there are at least two large clades or superphyla within the protostomes. These two clades are Ecdysozoa, whose members typically go through a series of molts during development, and Lophotrochozoa, whose members possess either a lophophore feeding structure (see p. 263) or a trochophore-like larva (see p. 182). Some pseudocoelomates, like Rotifera, Acanthocephala, and Entoprocta, apparently belong in Lophotrochozoa. However, recent studies indicate that Rotifera and Acanthocephala are closely related to the acoelomate Gnathostomulida (see p. 158), along with newly discovered animals called micrognathozoans. These four taxa are united as clade Gnathifera within Lophotrochozoa. Ecdysozoan pseudocoelomates include Nematoda, Nematomorpha, Kinorhyncha, and Priapulida. Some evidence suggests that Kinorhyncha, Priapulida, and Loricifera form a clade within Ecdysozoa. The placement of Gastrotricha and Cycliophora is yet undetermined.

We consider this heterogeneous assemblage of pseudocoelomates in one chapter as a convenience. Most are small; some are microscopic; some are fairly large. Some, such as nematodes, are found in freshwater, marine, terrestrial, and parasitic habitats; others, such as Acanthocephala, are strictly parasitic. Some have unique characteristics such as the lacunar system of acanthocephalans and the ciliary corona of rotifers.

Even in such a diversified grouping, however, a few characteristics are shared. All have a body wall of epidermis (often syncytial) and muscles surrounding a pseudocoel. The digestive tract is complete (except in Acanthocephala and Micrognathozoa), and it, along with gonads and excretory organs, is within the pseudocoel and bathed in perivisceral fluid. The epidermis in many secretes a nonliving cuticle with some specializations such as bristles and spines.

A constant number of cells or nuclei in individuals of a species, a condition known as **eutely,** is present in several groups, and in most of them there is an emphasis on the longitudinal muscle layer.

Phylum Rotifera

Rotifera (ro-tif′e-ra) (L. *rota,* wheel, + *fero,* to bear) derive their name from their characteristic ciliated crown, or *corona,* which, when beating, often gives the impression of rotating wheels. Rotifers range in size from 40 μm to 3 mm in length, but most are between 100 and 500 μm long. Some have beautiful colors, although most are transparent, and some have odd and bizarre shapes. Their shapes are often correlated with their mode of life. Floaters are usually globular and saclike; creepers and swimmers are somewhat elongated and wormlike; and sessile types are commonly vaselike, with a cuticular envelope. Some are colonial.

Rotifers are a cosmopolitan group of about 2000 species, some of which are found throughout the world. Most species are freshwater inhabitants, a few are marine, some are terrestrial, and some are epizoic (live on the surface of other animals) or parasitic. Most often they are benthic, occurring on the bottom or in vegetation of ponds or along the shores of large freshwater lakes where they swim or creep on vegetation.

The body is usually composed of a **head,** a **trunk,** and a **foot.** The head region bears the corona. The corona may form a ciliated funnel with its upper edges folded into lobes bearing bristles, or the corona may be made up of a pair of ciliated discs (figure 9.1). The cilia create currents of water toward the mouth that draw in small planktonic forms for food. The corona may be retractile. The mouth, surrounded by some part of the corona, opens into a modified muscular pharynx called a **mastax,** which is unique to rotifers. The mastax has a set of intricate jaws used for grasping and chewing. The trunk contains visceral organs, and a terminal foot, when present, is segmented and, in some, is ringed with joints that can telescope to shorten. The one to four toes secrete a sticky substance from the **pedal glands** for attachment.

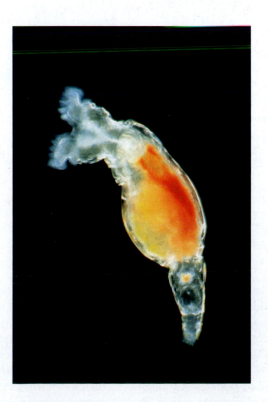

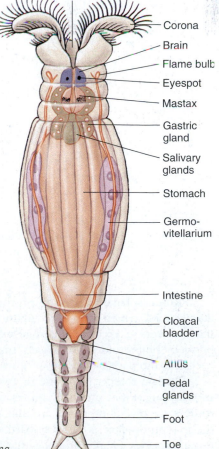

Mouth
Corona
Brain
Flame bulb
Eyespot
Mastax
Gastric gland
Salivary glands
Stomach
Germo-vitellarium
Intestine
Cloacal bladder
Anus
Pedal glands
Foot
Toe

figure 9.1

A live *Philodina,* a common rotifer; structure of *Philodina.*

Position in the Animal Kingdom

In the nine phyla covered in this chapter, the original blastocoel of the embryo persists as a space, or body cavity, between the enteron and body wall. Because this cavity lacks the peritoneal lining found in the true coelomates, it is called a **pseudocoel,** and the animals possessing it are called **pseudocoelomates** (figure 9.2). Pseudocoelomates belong to the Protostomia division of the bilateral animals, with some pseudocoelomate phyla assigned to Lophotrochozoa and some to Ecdysozoa.

Biological Contributions

1. The pseudocoel is a distinct gradation in body plan compared with the solid body structure of acoelomates. The pseudocoel may be filled with fluid or may contain a gelatinous substance with some mesenchyme cells. In common with a true coelom, it presents certain adaptive potentials, although these are by no means realized in all members: (1) greater freedom of movement; (2) space for development and differentiation of digestive, excretory, and reproductive systems; (3) a simple means of circulation or distribution of materials throughout the body; (4) a storage place for waste products to be discharged to the outside by excretory ducts; and (5) a hydrostatic organ. Since most pseudocoelomates are quite small, the most important functions of the pseudocoel are probably in circulation and as a means to maintain a high internal hydrostatic pressure.
2. A complete, mouth-to-anus digestive tract is found in most of these phyla and in all more complex phyla.

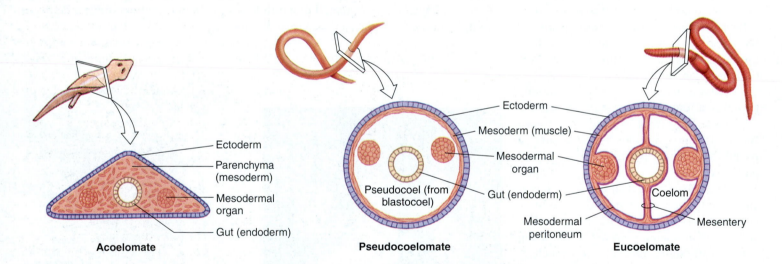

Acoelomate

- Ectoderm
- Parenchyma (mesoderm)
- Mesodermal organ
- Gut (endoderm)

Pseudocoelomate

- Ectoderm
- Mesoderm (muscle)
- Mesodermal organ
- Pseudocoel (from blastocoel)
- Gut (endoderm)

Eucoelomate

- Coelom
- Mesentery
- Mesodermal peritoneum

figure 9.2

Acoelomate, pseudocoelomate, and eucoelomate body plans. Notice that peritoneum and mesenteries are present only in eucoelomate animals.

Rotifers have a pair of protonephridial tubules with flame bulbs, which eventually drain into a **cloacal bladder** that collects excretory and digestive wastes. They have a bilobed "brain" and sense organs that include eyespots, sensory pits, and papillae.

Female rotifers have one or two syncytial ovaries (**germovitellaria**) that produce yolk as well as oocytes. Although rotifers are dioecious, males are unknown in many species; in these reproduction is entirely parthenogenetic. In parthenogenetic species females produce only diploid eggs that have not undergone reduction division, cannot be fertilized, and develop only into females. Such eggs are called **amictic eggs.** Other species can produce two kinds of eggs—amictic eggs, which develop parthenogenetically into females, and **mictic eggs,** which have undergone meiosis and are haploid. Mictic eggs, if unfertilized, develop quickly and parthenogenetically into males; if fertilized, they secrete a thick shell and become dormant for several months before developing into females. Such dormant eggs can withstand desiccation and other adverse conditions and permit rotifers to live in temporary ponds.

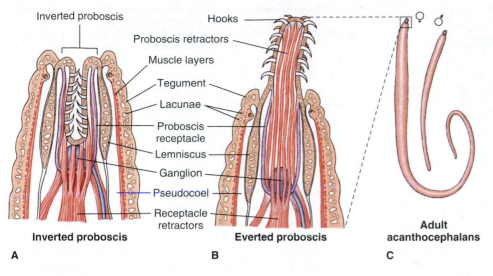

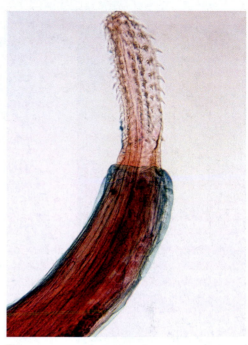

figure 9.3

Structure of a spiny-headed worm (phylum Acanthocephala). **A** and **B,** Eversible spiny proboscis by which the parasite attaches to the intestine of its host, often doing great damage. Because they lack a digestive tract, food is absorbed through their tegument. **C,** Male is typically smaller than female. **D,** A live acanthocephalan (*Leptorhynchoides thecatus*) with proboscis everted.

Mictic (Gr., *miktos,* mixed, blended) refers to the capacity of haploid eggs to be fertilized (that is, "mixed") with the male's sperm nucleus to form a diploid embryo. Amictic ("without mixing") eggs are already diploid and can develop only parthenogenetically.

Phylum Acanthocephala

Members of phylum Acanthocephala (a-kan′tho-sef′a-la) (Gr. *akantha,* spine or thorn, + *kephalē,* head) are commonly known as "spiny-headed worms." The phylum derives its name from one of its most distinctive features, a cylindrical invaginable **proboscis** (figure 9.3) bearing rows of recurved spines, by which the worms attach themselves to the intestine of their hosts. All acanthocephalans are endoparasitic, living as adults in intestines of vertebrates. Molecular evidence suggests that acanthocephalans are highly modified parasitic rotifers. However, taking into account the vast differences in morphology, habitat, and life cycle of the two groups, we retain the traditional phyla. Acanthocephalans and rotifers share a syncytial epidermis, so they are sometimes united in clade Syndermata.

Over 1100 species of acanthocephalans are known, most of which parasitize fishes, birds, and mammals, and the phylum is worldwide in distribution. However, no species is normally a parasite of humans, although rarely humans are infected with species that usually occur in other hosts.

Various species range in size from less than 2 mm to over 1 m in length, with females of a species usually being larger than males. In life, the body is usually bilaterally flattened, with numerous transverse wrinkles. They are usually cream color but may be yellowish or brown from absorption of pigments from intestinal contents.

The body wall is syncytial, and its surface is punctuated by minute crypts 4 to 6 μm deep, which greatly increase the surface area of the tegument. About 80% of the thickness of the tegument is the radial fiber zone, which contains a **lacunar system** of ramifying fluid-filled canals (figure 9.3). The function of the lacunar system is unclear, but it may serve in distribution of nutrients to the peculiar, tubelike muscles in the body wall of these organisms.

Excretion is across the body wall in most species. In one family there is a pair of **protonephridia** with flame cells that unite to form a common tube that opens into the sperm duct or uterus.

Because acanthocephalans have no digestive tract, they must absorb all nutrients through their tegument. They can absorb various molecules by specific membrane transport mechanisms, and their tegument can perform pinocytosis.

Sexes are separate. Males have a protrusible penis, and at copulation sperm travel up the genital duct and escape into the pseudocoel of females. Zygotes develop into shelled acanthor larvae. Shelled larvae escape from the vertebrate host in feces, and, if eaten by a suitable arthropod, they hatch and work their way into the hemocoel where they grow to juvenile acanthocephalans. Either development ceases until the arthropod is eaten by a suitable host, or they may pass through several transport hosts in which they encyst until eaten.

Clade Micrognathozoa

The first and only micrognathozoan species, *Limnognathia maerski,* was collected from Greenland in 1994 but not formally described until 2000. Micrognathozoans are tiny animals that are interstitial (living between sand grains) and about 142 μm long. The body consists of a two-part head, a thorax, and an abdomen

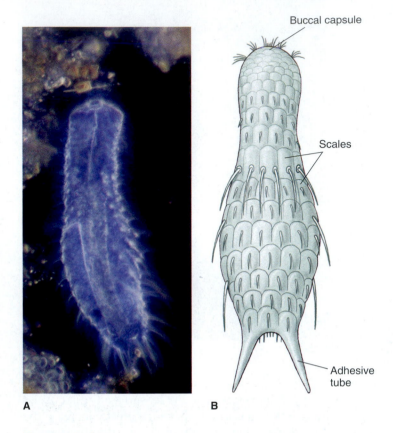

A, A live *Chaetonotus simrothi,* a common gastrotrich. **B,** External structure of *chaetonotus.*

figure 9.4

with a short tail. The cellular epidermis has dorsal plates but lacks plates ventrally. These animals move using cilia and also possess a unique ventral ciliary adhesive pad that produces glue. They have a very complicated jaw system of plates and associated teeth; the mouth leads into a relatively simple gut. An anus opens to the outside only periodically.

Phylum Gastrotricha

The phylum Gastrotricha (gas-trot′re-ka) (Gr. *gaster,* belly, + *thrix,* hair) is a small group (about 460 species) of microscopic animals, approximately 65 to 500 μm long (figure 9.4). They are usually bristly or scaly in appearance, are flattened on the ventral side, and many move by gliding on ventral cilia. Others move in a leechlike fashion by briefly attaching the posterior end by means of adhesive glands. There are marine and freshwater species, and they are common in lakes, ponds, and seashore sands. They feed on bacteria, diatoms, and small protozoa. Gastrotrichs are hermaphroditic. However, the male system of many freshwater species is nonfunctional, and the female system produces offspring parthenogenetically.

Phylum Entoprocta

Entoprocta (en′to-prok′ta) (Gr. *entos,* within, + *proktos,* anus) is a small phylum of less than 100 species of tiny, sessile animals that, superficially, look much like hydroid cnidarians, but

In December, 1995, P. Funch and R. M. Kristensen reported that they had found some very strange little creatures clinging to the mouthparts of Norway lobsters (*Nephrops norvegicus*), so strange that they did not fit into any known phylum (Nature **378:**711–714). Funch and Kristensen concluded that the organisms, only 0.35 mm long, represented a new phylum, which was named **Cycliophora.** The name refers to a crown of compound cilia, reminiscent of rotifers, with which the organisms feed. They were described as "acoelomate," although whether they might have a pseudocoel is unclear, and they do have a cuticle. Their life cycle seems bizarre. The sessile feeding stages on the lobster's mouthparts undergo internal budding to produce motile stages: (1) larvae containing new feeding stages; (2) dwarf males, which become attached to feeding stages that contain developing females; and (3) females, which also attach to the lobster's mouthparts, then produce dispersive larvae and degenerate.

Whether the proposed new phylum will withstand scrutiny of further research is unknown, and its possible relationships to other phyla are quite unclear. Funch and Kristensen think that the organisms are protostomes and see affinities with Entoprocta and Ectoprocta. Little short of astonishing, however, is their abundance on mouthparts of a host as well known as Norway lobsters. How could biologists have failed to notice them before? At a time when habitat destruction drives many species to extinction every year, we wonder if there are phyla suffering the same fate. S. Conway Morris pondered the possibility of further undiscovered phyla (Nature **378:**661–662), suggesting you may need a couple of zoology textbooks and a decent microscope when you next dine at your favorite seafood restaurant: "Who knows what might be found lurking under the lettuce?"

SEM image kindly supplied by Dr. Peter Funch, University of Aarhus

their tentacles are ciliated and tend to roll inward (figure 9.5). Most entoprocts are microscopic, and none is more than 5 mm long. They are all stalked and sessile forms; some are colonial, and some are solitary. All are ciliary feeders.

With the exception of one genus, all entoprocts are marine forms that have a wide distribution from polar regions to tropics. Most marine species are restricted to coastal and brackish waters and often grow on shells and algae. Some are commensals on marine annelid worms. Freshwater entoprocts (figure 9.5) occur on undersides of rocks in running water.

The body of entoprocts is cup shaped, bears a crown, or circle, of ciliated tentacles, and attaches to a substratum by a stalk in solitary species. In colonial species, the stalks of individuals connect to the stolon of the colony. Both tentacles and stalk are continuations of the body wall.

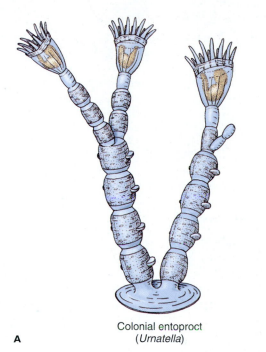

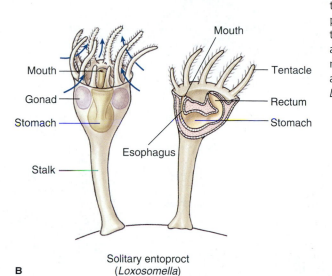

figure 9.5

A, *Urnatella,* a freshwater ento-proct, forms small colonies of two or three stalks from a basal plate. **B,** *Loxosomella,* a solitary entoproct. Both solitary and colonial entoprocts can reproduce asexually by budding, as well as sexually. **C,** A live *Loxosomella.*

Mouth

Mouth

Tentacle

Gonad

Rectum

Stomach

Stomach

Esophagus

Stalk

Colonial entoproct
(*Urnatella*)

A

B

Solitary entoproct
(*Loxosomella*)

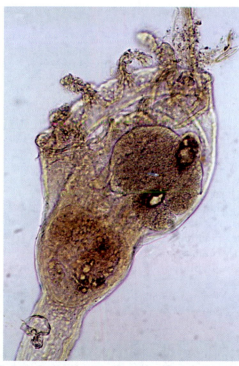

C

The gut is U-shaped and ciliated, and both the mouth and anus open within the circle of tentacles. They capture food particles in the current created by tentacular cilia and then pass particles along the tentacles to the mouth. Entoprocts have a pair of protonephridia but no circulatory or respiratory organs.

Some species are monoecious, some are dioecious, and some seem to be protandrous hermaphrodites in which the gonad at first produces sperm and then eggs. Cleavage is modified spiral and mosaic, and a trochophore-like larva is produced, similar to that of some mollusks and annelids (p. 182).

Entoprocts were once included with phylum Ectoprocta in a phylum called Bryozoa, but ectoprocts are true coelomate

Libbie Henrietta Hyman (1888–1969), author of the monumental multi-volume treatise, *The Invertebrates,* was a distinguished American zoologist. She earned all her academic degrees from the University of Chicago, receiving her PhD in 1915 and remaining there doing research until 1931. She published a laboratory manual for elementary zoology in 1919 and one for comparative vertebrate anatomy in 1922. The latter remained popular for the next 50 years and allowed her to live modestly on proceeds from her books while pursuing her first love, the study of invertebrates. She became a research associate at the American Museum of Natural History and published 135 research papers, chiefly on turbellarians, and the treatise for which she is best remembered, *The Invertebrates.* She completed six volumes of this treatise, covering the protozoan groups through mollusks (to gastropods). Thousands of invertebrate zoologists owe her a debt of gratitude.

animals, so most zoologists place them in a separate group. Ectoprocts are still often called bryozoans. Sequence analysis places entoprocts among lophotrochozoan phyla.

Phylum Nematoda: Roundworms

Nematodes, (Gr., *nematos,* thread) occur in nearly every conceivable kind of ecological niche. Approximately 12,000 species have been named, but Libbie Hyman estimated that, if all species were known, the number would be nearer 500,000.[2] They live in the sea, in fresh water, and in soil, from polar regions to tropics, and from mountaintops to the depths of seas. Good topsoil may contain billions of nematodes per acre. Nematodes also parasitize virtually every type of animal and many plants. The effects of nematode infestation on crops,

[2]Hyman, L. H. 1951. The invertebrates, vol. III. Acanthocephala, Aschelminthes, and Entoprocta, the pseudocoelomate Bilateria. New York, McGraw-Hill Book Company.

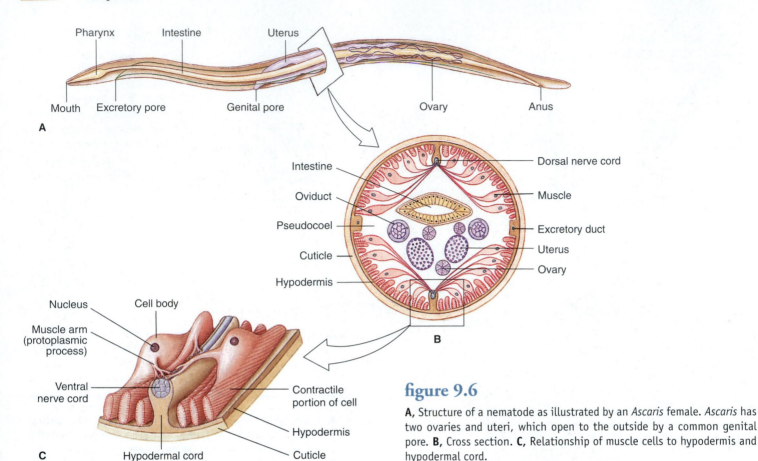

figure 9.6

A, Structure of a nematode as illustrated by an *Ascaris* female. *Ascaris* has two ovaries and uteri, which open to the outside by a common genital pore. **B,** Cross section. **C,** Relationship of muscle cells to hypodermis and hypodermal cord.

domestic animals, and humans make this phylum one of the most important of all parasitic animal groups.

Form and Function

Most nematodes are under 5 cm long, and many are microscopic, but some parasitic nematodes are over a meter in length.

Use of the pseudocoel as a hydrostatic skeleton is a conspicuous feature of nematodes, and we can best understand much of their functional morphology in the context of the high **hydrostatic pressure** within the pseudocoel.

The outer body covering is a relatively thick, noncellular **cuticle,** secreted by the underlying epidermis (**hypodermis**). A cuticle is of great functional importance, serving to contain the high hydrostatic pressure exerted by fluid in the pseudocoel. Layers of the cuticle are primarily of **collagen,** a structural protein present throughout the Metazoa, and abundant in vertebrate connective tissue. Crisscrossing fibers make up three of the layers, conferring some longitudinal elasticity on the worm but severely limiting its capacity for lateral expansion.

Beneath the hypodermis is a layer of **longitudinal muscles.** There are no circular muscles in the body wall. The muscles are arranged in four bands, or quadrants, separated by four epidermal cords that project inward to the pseudocoel (figure 9.6). Another unusual feature of body-wall muscles in nematodes is that the muscles extend processes to synapse with nerve cords, rather than nerves extending an axon to synapse with muscle.

characteristics
of Phylum Nematoda

1. Body bilaterally symmetrical, cylindrical in shape
2. Body covered with a secreted, flexible, nonliving cuticle
3. Motile **cilia and flagella completely lacking;** some sensory endings derived from cilia present
4. **Muscles in body wall running in longitudinal direction only**
5. Excretory system of either one or more **gland cells** opening by an excretory pore, a **canal system** without gland cells, or both gland cells and canals together; **flame cell protonephridia lacking**
6. **Pharynx** usually muscular and **triradiate** in cross section
7. Male reproductive tract opening into rectum to form a cloaca; female reproductive tract opening a separate gonopore
8. Fluid in pseudocoel enclosed by cuticle forming a **hydrostatic skeleton**
9. Juvenile stages separated by molting of cuticle.

The fluid-filled pseudocoel, in which internal organs lie, constitutes a hydrostatic skeleton. Hydrostatic skeletons are found in many invertebrates; they lend support by transmitting the force of muscle contraction to the enclosed, noncompressible fluid. Normally, muscles are arranged antagonistically, so that as movement is effected in one direction by contraction of one group of muscles, movement back in the opposite direction is effected by contraction of an antagonistic set of muscles. However, nematodes do not have circular body-wall muscles to antagonize the longitudinal muscles; therefore, the cuticle must serve that function. As muscles on one side of the body contract, they compress the cuticle on that side, and the force of the contraction is transmitted (by the fluid in the pseudocoel) to the other side of the nematode, stretching the cuticle on that side. This compression and stretching of the cuticle serve to antagonize the muscle and are the forces that return the body to resting position when the muscles relax; this action produces the characteristic thrashing motion seen in nematode movement. An increase in efficiency of this system can be achieved only by an increase in hydrostatic pressure. Consequently, the hydrostatic pressure in a nematode's pseudocoel is much higher than is usually found in other kinds of animals with hydrostatic skeletons but that also have antagonistic muscle groups.

The alimentary canal of a nematode consists of a **mouth** (figure 9.6), a muscular **pharynx,** a long nonmuscular **intestine,** a short **rectum,** and a terminal **anus.** The cylindrical pharynx has radial muscles that insert on the cuticular lining of its lumen and on a basement membrane at its periphery. At rest the lumen is closed. When muscles in the anterior of the pharynx contract, they open its lumen and suck food material inside. Relaxation of muscles anterior to the food mass closes the pharyngeal lumen and forces the food posteriorly toward the intestine. The intestine is only one cell-layer thick and has no muscles. Food matter moves posteriorly in the intestine by body movements and by additional food being passed into the intestine from the pharynx. Defecation is accomplished by muscles that simply pull the anus open, and pseudocoelomic pressure surrounding the rectum provides expulsive force.

Adults of many parasitic nematodes have an anaerobic energy metabolism; a Krebs cycle and cytochrome system characteristic of aerobic metabolism are absent. They derive energy through glycolysis and some additional electron-transport sequences. Interestingly, some free-living nematodes and free-living stages of parasitic nematodes are obligate aerobes (require oxygen) and have a Krebs cycle and cytochrome system.

A **ring of nerve tissue** and ganglia around the pharynx gives rise to small nerves to the anterior end and to two **nerve cords,** one dorsal and one ventral. Some sensory organs around the lips and around the posterior end are rather elaborate.

In 1963 Sydney Brenner started studying a free-living nematode, *Caenorhabditis elegans,* which was the beginning of some extremely fruitful research. Now this small worm has become one of the most important experimental models in biology. The origin and lineage of all cells in its body (959) have been traced from zygote to adult, and the complete "wiring diagram" of its nervous system is known—all neurons and all connections between them. Its genome has been completely mapped, and scientists have sequenced its entire genome of 3 million bases comprising 19,820 genes. Many basic discoveries of gene function, such as genes encoding proteins essential for programmed cell death, have been and will be made using *C. elegans.*

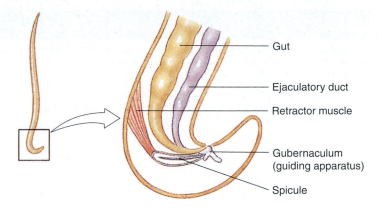

figure 9.7

Posterior end of a male nematode.

Copulatory spicules of male nematodes are not true intromittent organs, since they do not conduct sperm, but are another adaptation to cope with the high internal hydrostatic pressure. Spicules must hold the vulva of a female open while ejaculatory muscles in the male reproductive tract overcome the hydrostatic pressure in the female and rapidly inject sperm into her reproductive tract. Furthermore, nematode spermatozoa are unique among those studied in the animal kingdom in that they lack a flagellum and acrosome. Within the female reproductive tract, sperm become ameboid and move by pseudopodial movement.

Most nematodes are dioecious. Males are smaller than females, and their posterior end usually bears a pair of copulatory spicules (figure 9.7). Fertilization is internal, and shelled zygotes or embryos are stored in the uterus until deposition. After embryonic development, a juvenile worm hatches from an egg. There are four juvenile stages, each separated by a molt, or shedding, of the cuticle. Many parasitic nematodes have free-living juvenile stages, and others require an intermediate host to complete their life cycles.

Some Nematode Parasites

Nearly all vertebrates and many invertebrates are parasitized by nematodes. A number of these are very important pathogens of humans and domestic animals. A few nematodes are common in humans in North America (table 9.1), but they and many others usually abound in tropical countries. We have space to mention only a few.

Ascaris lumbricoides: The Large Roundworm of Humans

The large human roundworm (Gr. *askaris,* intestinal worm) is one of the most common worm parasites of humans (figure 9.8). Recent surveys have shown a prevalence up to 25% in some areas of the southeastern United States, and it is estimated that 1.27 billion people in the world (20% of the world's population)

Table 9.1   Common Parasitic Nematodes of Humans in North America

Common and scientific names	Means of infection; prevalence in humans
Hookworm (*Ancylostoma duodenale* and *Necator americanus*)	Contact with juveniles in soil that burrow into skin; common in southern states
Pinworm (*Enterobius vermicularis*)	Inhalation of dust with ova and by contamination with fingers; most common worm parasite in United States
Intestinal roundworm (*Ascaris lumbricoides*)	Ingestion of embryonated ova in contaminated food; common in rural areas of Appalachia and southeastern states
Trichina worm (*Trichinella spiralis*)	Ingestion of infected muscle; occasional in humans throughout North America
Whipworm (*Trichuris trichiura*)	Ingestion of contaminated food or by unhygienic habits; usually common wherever Ascaris is found

have this worm. ***Ascaris suum*** (figure 9.8B), a parasite of pigs, is morphologically similar and was long considered the same species. Females of both are up to 30 cm in length and can produce 200,000 eggs a day. Adults live in their host's small intestine, and eggs leave the host's body in feces. They are extremely resistant to adverse conditions other than direct sunlight and high temperatures and can survive for months or years in soil.

When shelled juveniles are eaten with uncooked vegetables or when children put soiled fingers or toys in their mouths, the tiny juveniles are consumed and hatch in the host's intestine. They penetrate the intestinal wall and travel through the heart in blood to the lungs, where they break out into alveoli. They may cause a serious pneumonia at this stage. From alveoli, juveniles make their way up the bronchi, trachea, and pharynx, to be swallowed and finally reach the intestine again and grow to maturity. In the intestine, worms cause abdominal symptoms and allergic reactions, and in large numbers they may produce intestinal blockage.

Hookworms

Hookworms are so named because their anterior end curves dorsally, suggesting a hook. The most common species is *Necator americanus* (L. *necator,* killer), whose females are up to 11 mm long. Males can reach 9 mm in length. Large plates in their mouth (figure 9.9) cut into the intestinal mucosa of the host where they suck blood and pump it through their intestine, partially digesting it and absorbing the nutrients. They suck much more blood than they need for food, and heavy infections cause anemia in the patient. Hookworm disease in children may retard mental and physical growth and cause a general loss of energy.

Shelled embryos leave the host's body in feces, and juveniles hatch in the soil, where they live on bacteria. When human skin contacts infested soil, the juveniles burrow through the skin to the blood, and reach the lungs and finally the intestine in a manner similar to that described for *Ascaris.*

Trichina Worm

Trichinella spiralis (Gr. *trichinos,* of hair, + *-ella,* diminutive) is one of the species of tiny nematodes responsible for the

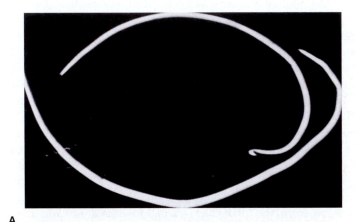

A

B

figure 9.8

A, Intestinal roundworm *Ascaris lumbricoides,* male and female. Male *(top)* is smaller and has characteristic sharp kink in the end of the tail. The females of this large nematode may be over 30 cm long. **B,** Intestine of a pig, nearly completely blocked by *Ascaris suum.* Such heavy infections are also fairly common with *A. lumbricoides* in humans.

potentially lethal disease trichinosis. Adult worms burrow in the mucosa of the small intestine where females produce living young. Juveniles penetrate blood vessels and are carried throughout the body, where they may be found in almost any tissue or body space. Eventually, they penetrate skeletal muscle

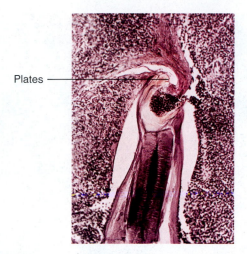

Plates

figure 9.9

Section through anterior end of hookworm attached to dog intestine. Note cutting plates of mouth pinching off bit of mucosa from which the thick muscular pharynx sucks blood. Esophageal glands secrete an anti-coagulant to prevent blood clotting.

figure 9.10

Section of muscle infected with trichina worm *Trichinella spiralis*, human case. Juveniles lie within muscle cells that the worms have induced to transform into nurse cells (commonly called cysts). An inflammatory reaction is evident around the nurse cells. Juveniles may live 10 to 20 years, and nurse cells eventually may calcify.

cells, becoming one of the largest known intracellular parasites. Juveniles cause astonishing redirection of gene expression in their host cell, which loses its striations and becomes a **nurse cell** that nourishes the worm (figure 9.10). When meat containing live juveniles is swallowed, the worms are liberated into the intestine where they mature.

Trichinella spp. can infect a wide variety of mammals in addition to humans, including hogs, rats, cats, and dogs. Hogs become infected by eating garbage containing pork scraps with juveniles or by eating infected rats. In addition to *T. spiralis,* we now know there are four other sibling species in the genus. They differ in geographic distribution, infectivity to different host species, and freezing resistance. Heavy infections may cause death, but lighter infections are much more common— about 25 cases per year are reported in the United States.

Other ascarids are common in wild and domestic animals. Species of *Toxocara,* for example, are found in dogs and cats. Their life cycle is generally similar to that of *Ascaris,* but juveniles often do not complete their tissue migration in adult dogs, remaining in the host's body in a stage of arrested development. Pregnancy in the female dog, however, stimulates juveniles to wander, and they infect the embryos in the uterus. Puppies are then born with worms. These ascarids also survive in humans but do not complete their development, leading to an occasionally serious condition in children known as **visceral larva migrans.** This condition is a good reason for pet owners to practice hygienic disposal of canine wastes!

Pinworms

Pinworms, *Enterobius vermicularis* (Gr. *enteron*, intestine, + *bios*, life), cause relatively little disease, but they are the most

common helminth parasite in the United States, estimated at 30% in children and 16% in adults. Adult parasites (figure 9.11) live in the large intestine and cecum. Females reach about 12 mm in length and migrate to the anal region at night to lay their eggs (figure 9.11). Scratching the resultant itch effectively contaminates hands and bedclothes. Eggs develop rapidly and become infective within 6 hours at body temperature. After they are swallowed, they hatch in the duodenum, and the worms mature in the large intestine.

Members of this order of nematodes have **haplodiploidy,** a characteristic shared with a few other animal groups, notably many hymenopteran insects (p. 249). Males are haploid and are produced parthenogenetically; females are diploid and arise from fertilized eggs.

Diagnosis of most intestinal roundworms is usually by examination of a small bit of feces under a microscope and finding characteristic shelled embryos or juveniles. However, pinworm eggs are not often found in feces because females deposit them on the skin around the anus. The "Scotch tape method" is more effective. The sticky side of cellulose tape is applied around the anus to collect the shelled embryos, then the tape is placed on a glass slide and examined under a microscope. Several drugs are effective against this parasite, but all members of a family should be treated at the same time, since the worm easily spreads through a household.

Filarial Worms

At least eight species of filarial nematodes infect humans, and some cause major diseases. Some 250 million people in tropical

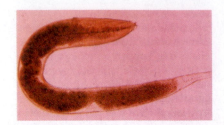

A

B

figure 9.11

Pinworms, *Enterobius vermicularis*. **A,** Female worm from human large intestine (slightly flattened in preparation), magnified about 10 times. **B,** Group of shelled juveniles of pinworms, which are usually discharged at night around the anus of the host, who, by scratching during sleep, may get fingernails and clothing contaminated. This may be the most common and widespread of all human helminth parasites.

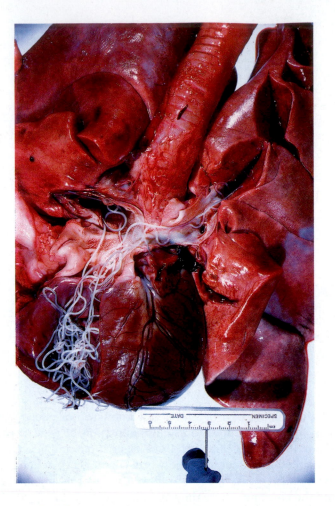

figure 9.13

Dirofilaria immitis in a dog's heart. This nematode is a major menace to the health of dogs in North America. Adults live in the heart, and juveniles circulate in blood where they are picked up and transmitted by mosquitos.

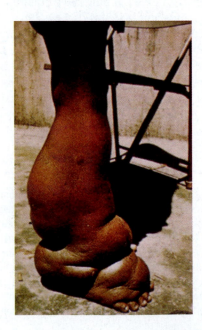

figure 9.12

Elephantiasis of leg caused by adult filarial worms of *Wuchereria bancrofti,* which live in lymph passages and block the flow of lymph. Tiny juveniles, called microfilariae, are picked up in a blood meal of a mosquito, where they develop to the infective stage and are transmitted to a new host.

countries are infected with *Wuchereria bancrofti* (named for Otto Wucherer) or *Brugia malayi* (named for S. L. Brug), which places these species among the scourges of humanity. The worms live in the lymphatic system, and females are as long as 100 mm. Disease symptoms are associated with inflammation and obstruction of the lymphatic system. Females release live young, tiny microfilariae, into the blood and lymph. Microfilariae are ingested by mosquitos as the insects feed, and they develop in mosquitos to the infective stage. They escape from the mosquito when it is feeding again on a human and penetrate the wound made by the mosquito bite.

Dramatic manifestations of elephantiasis are occasionally produced after long and repeated exposure to the worms. The condition is marked by an excessive growth of connective tissue and enormous swelling of affected parts, such as the scrotum, legs, arms, and more rarely, the vulva and breasts (figure 9.12).

Another filarial worm causes river blindness (onchocerciasis) and is carried by blackflies. It infects more than 30 million people in parts of Africa, Arabia, Central America, and South America.

The most common filarial worm in the United States is probably the dog heartworm, *Dirofilaria immitis* (figure 9.13). Carried by mosquitos, it also can infect other canids, cats, ferrets, sea lions, and occasionally humans. Along the Atlantic and Gulf Coast states and northward along the Mississippi River throughout the midwestern states, prevalence in dogs reaches 45%. It occurs in other states at a lower prevalence. This worm causes a very serious disease among dogs, and no responsible owner should fail to provide "heartworm pills" for a dog during mosquito season.

Phylum Nematomorpha

The popular name for the Nematomorpha (nem′a-to-mor′fa) (Gr. *nema, nematos,* thread, + *morphē,* form) is "horsehair worms," based on an old superstition that the worms arise from horsehairs that happen to fall into water; and indeed they resemble hairs from a horse's tail. They were long included with nematodes, with which they share the structure of the cuticle, presence of epidermal cords, longitudinal muscles only, and pattern of nervous system. Several recent studies indicate that nematomorphs are the sister taxon to nematodes; the two groups are united in clade Nematoidea.

About 250 species of horsehair worms have been named. Worldwide in distribution, horsehair worms are free-living as adults and parasitic in arthropods as juveniles. Early larval forms of some species have a striking resemblance to priapulids. A single molt has been reported during development of some species. Juveniles do not emerge from an arthropod host unless water is nearby, and adults are often seen wriggling slowly about in ponds or streams. Juveniles of freshwater forms use various terrestrial insects as hosts, while marine forms use certain crabs. Adults have a vestigial digestive tract and do not feed, but they can live almost anywhere in wet or moist surroundings if oxygen is adequate.

Phylum Kinorhyncha

There are only 179 described species of Kinorhyncha (Gr. *kineo,* to move, + *rhynchos,* beak or snout). They are tiny marine worms, usually less than 1 mm long, that prefer mud bottoms. They have no external cilia, but their cuticle is divided into 13 segments (figure 9.14). The cuticle is molted during growth. Kinorhynchs burrow into mud by extending the head, anchoring it by its recurved spines **(scalids),** and drawing their body forward until their head is retracted. They feed on organic sediment in mud, and some feed on diatoms. Kinorhynchs are dioecious.

Phylum Priapulida

Priapulida (pri′a-pyu′li-da) (Gr. *priapos,* phallus, + *-ida,* pl. suffix) is a small group (only nine species) of marine worms found chiefly in colder water of both hemispheres. Their cylindrical bodies are rarely more than 12 to 15 cm long. Most are burrowing predaceous animals that usually orient themselves upright in mud with their mouth at the surface.

Long considered pseudocoelomate, they were judged coelomate when nuclei were found in membranes lining the

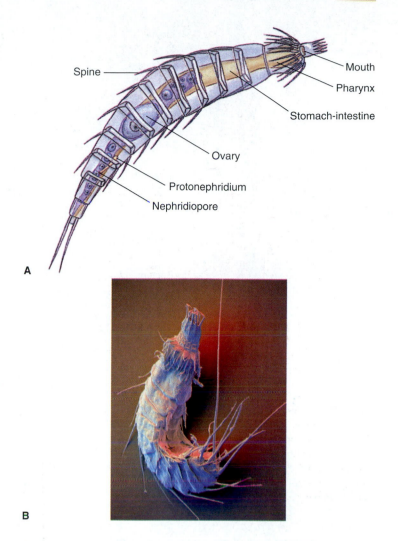

A

B

figure 9.14

A, *Echinoderes,* a kinorhynch, is a minute marine worm. Segmentation is superficial. The head with its circle of spines is retractile. **B,** Colored scanning electron micrograph (SEM) of kinorhynch *Antigomonas sp.*

body cavity, the membranes thus representing a peritoneum. However, electron microscopy showed that nuclei of their muscle cells were peripheral, and that muscles secreted an extracellular membrane. The muscle nuclei and extracellular membrane gave the appearance of an epithelial lining.

The body includes a proboscis, a trunk, and usually one or two caudal appendages (figure 9.15). The eversible proboscis usually ends with rows of curved spines around the mouth; it is used to sample the surroundings as well as to capture small soft-bodied prey. Priapulids are not metameric.

There is no circulatory system, but coelomocytes in body fluids contain the respiratory pigment hemerythrin. There is a nerve ring and ventral cord with nerves and a protonephridial tubule that serves also as a gonoduct. The anus and urogenital pores open at the end of the trunk.

Sexes are separate, and embryogenesis is only poorly known. Their relationship to other coelomates is obscure. Some authorities consider them remnants of groups that were once more successful and widely distributed.

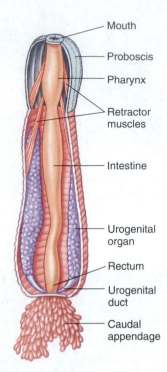

figure 9.15

Major internal structures of *Priapulus*.

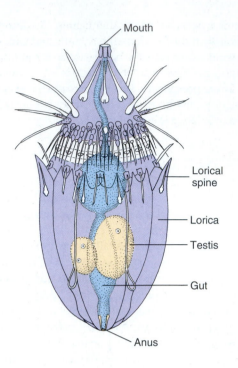

figure 9.16

Dorsal view of an adult loriciferan, *Nanoloricus mysticus*.

Phylum Loricifera

Loricifera (L. *lorica,* corselet, + *fero,* to bear) are a very recently described phylum of animals (1983). They live in spaces between grains of marine gravel, to which they cling tightly. Although they were described from specimens collected off the coast of France, they are apparently widely distributed in the world. About 100 species have been found. They have oral styles and scalids rather similar to those of the kinorhynchs, and the entire forepart of the body can be retracted into the circular lorica (figure 9.16). Loriciferans are dioecious, and life cycles contain multiple stages. Larvae molt as they grow.

Phylogeny and Adaptive Radiation

Phylogeny

Evidence from a sequence analysis of the small-subunit ribosomal gene suggests that some time after ancestral deuterostomes diverged from ancestral protostomes in the Precambrian, protostomes split again into two large groups (or superphyla): Ecdysozoa and Lophotrochozoa. Within the Lophotrochozoa are several clades containing pseudocoelomate animals. Gene sequence data place Acanthocephala and Rotifera together as clade Syndermata, sharing a eutelic syncytial epidermis. Syndermata is placed with Micrognathozoa and Gnathostomulida (acoelomates; see p. 158) in clade Gnathifera, whose members share cuticular jaws having a homologous microstructure. Some work suggests that Cycliophora, Gastrotricha, and Platyhelminthes are closely related to the Gnathifera, but this idea is controversial. Members of Entoprocta have typical protostome cleavage patterns (modified spiral mosaic cleavage) and a trochophore-like larva, so this phylum belongs within Lophotrochozoa, but outside clade Gnathifera.

Within Ecdysozoa, pseudocoelomate animals belong to at least two groups: Nematoda and Nematomorpha are considered sister taxa within clade Nematoidea, and Kinorhyncha and Priapulida are closely related to each other and likely form another clade. Placement of Loricifera is uncertain; it may belong with the latter two phyla, but loriciferans also have similarities to larval nematomorphs. Loriciferans molt during development, so they are likely to remain within Ecdysozoa.

Adaptive Radiation

Certainly the most impressive adaptive radiation in this group of phyla is by nematodes. They are by far the most numerous in terms of both individuals and species, and they have adapted to almost every habitat available to animal life. Their basic pseudocoelomate body plan, with the cuticle, hydrostatic skeleton, and longitudinal muscles, has proved generalized and plastic enough to adapt to an enormous variety of physical conditions. Free-living lines gave rise to parasitic forms on at least several occasions, and virtually all potential hosts have been exploited. All types of life cycle occur: from simple and direct to complex, with intermediate hosts; from normal dioecious reproduction to parthenogenesis, hermaphroditism, and alternation of free-living and parasitic generations. A major factor contributing to evolutionary opportunism of the nematodes is their extraordinary capacity to survive suboptimal conditions, for example, developmental arrests in many free-living and animal parasitic species and ability to undergo cryptobiosis (survival in harsh conditions by assuming a very low metabolic rate) in many free-living and plant parasitic species.

Summary

Phyla covered in this chapter possess a body cavity called a pseudocoel, which is derived from the embryonic blastocoel, rather than a secondary cavity in the mesoderm (coelom). Several of the groups exhibit eutely, a constant number of cells or nuclei in adult individuals of a given species. This chapter also includes two other taxa, Micrognathozoa and Cycliophora, whose members are likely related to the phyla discussed.

Phylum Rotifera comprises small, mostly freshwater organisms with a ciliated corona, which creates currents of water to draw planktonic food toward the mouth. The mouth opens into a muscular pharynx, or mastax, that is equipped with jaws.

Acanthocephalans are all parasitic in the intestine of vertebrates as adults, and their juvenile stages develop in arthropods. They have an anterior, invaginable proboscis armed with spines, which they embed in the intestinal wall of their host. They do not have a digestive tract and so must absorb all nutrients across their tegument.

Micrognathozoa consists of a single species of tiny animals living between sand grains. These animals have complex jaws similar to those of rotifers and acoelomate gnathostomulids.

Gastrotricha, Kinorhyncha, and Loricifera are small phyla of tiny, aquatic pseudocoelomates. Gastrotrichs move by cilia or adhesive glands, and kinorhynchs anchor and then pull themselves by the spines on their head. Loriciferans can withdraw their bodies into the lorica. Priapulids are marine burrowing worms.

Entoprocta are small, sessile, aquatic animals with a crown of ciliated tentacles encircling both the mouth and anus.

By far the largest and most important of these phyla are nematodes, of which there may be as many as 500,000 species in the world. They are more or less cylindrical, tapering at the ends, and covered with a tough, secreted cuticle. Their body-wall muscles are longitudinal only, and to function well in locomotion, such an arrangement must enclose a volume of fluid in the pseudocoel at high hydrostatic pressure. This fact of nematode life has a profound effect on most of their other physiological functions, for example, ingestion of food, egestion of feces, excretion, copulation, and others. Most nematodes are dioecious, and there are four juvenile stages, each separated by a molt of the cuticle. Nematodes have undergone much greater adaptive radiation than any other pseudocoelomate phylum.

Almost all invertebrate and vertebrate animals and many plants have nematode parasites, and many other nematodes are free-living in soil and aquatic habitats. Some parasitic nematodes have part of their life cycle free-living, some undergo a tissue migration in their host, and some have an intermediate host in their life cycle. Some parasitic nematodes cause severe diseases in humans and other animals.

Nematomorpha or horsehair worms are related to the nematodes and have parasitic juvenile stages in arthropods, followed by a free-living, aquatic, nonfeeding adult stage.

The pseudocoelomate taxa are protostomes. Phylogenetic relationships within the protostomes are still under revision, but it appears that rotifers, acanthocephalans, micrognathozoans, gastrotrichs, cycliophorans, and entoprocts are lophotrochozoan protostomes, whereas nematodes, nematomorphs, kinorhynchs, priapulids, and loriciferans are ecdysozoan protostomes. Within Lophotrochozoa, sequence analysis suggests that acanthocephalans are highly derived rotifers, and these two groups belong with micrognathozoans and gnathostomulids in clade Gnathifera.

Review Questions

1. Explain the difference between a true coelom and a pseudocoel.
2. What is the normal size of a rotifer; where is it found; and what are its major body features?
3. Explain the difference between mictic and amictic eggs of rotifers, and tell the adaptive value of each.
4. What is eutely?
5. What are the approximate lengths of loriciferans, priapulids, gastrotrichs, and kinorhynchs? Where are they found?
6. A skeleton is a supportive structure. Explain how a hydrostatic skeleton supports an animal.
7. What feature of body-wall muscles in nematodes requires a high hydrostatic pressure in the pseudocoelomic fluid for efficient function?
8. Explain how high pseudocoelomic pressure affects feeding and defecation in nematodes. How could ameboid sperm be an adaptation to high hydrostatic pressure in the pseudocoel?
9. Explain the interaction of cuticle, body-wall muscles, and pseudocoelomic fluid in the locomotion of nematodes.
10. Outline the life cycle of each of the following: *Ascaris lumbricoides,* hookworm, *Enterobius vermicularis, Trichinella spiralis, Wuchereria bancrofti.*
11. Where in a human body are adults of each species in question 10 found?
12. Where are juveniles and adults of nematomorphs found?
13. The evolutionary ancestry of acanthocephalans is particularly obscure. Describe some characters of acanthocephalans that make it surprising that they could be highly derived rotifers.
14. How do acanthocephalans get food?
15. What characteristics distinguish entoprocts among pseudocoelomates?
16. Other than base sequences in genes, how are Ecdysozoa and Lophotrochozoa distinguished?

Selected References

See also general references on page 415.

Aguinaldo, A. M. A., J. M. Turbeville, L. S. Linford, M. C. Rivera, J. J. F. R. Garey, R. A. Raff, and J. A. Lake. 1997. Evidence for a clade of nematodes, arthropods and other moulting animals. Nature **387**:489–493. *Sequence analysis to support a superphylum Ecdysozoa.*

Balavoine, G., and A. Adoutte. 1998. One or three Cambrian radiations? Science. **280**:397–398. *Discusses radiation of superphyla Ecdysozoa, Lophotrochozoa, and Deuterostomia.*

Bird, A. F., and J. Bird. 1991. The structure of nematodes, ed. 2. New York, Academic Press. *The most authoritative reference available on nematode morphology. Highly recommended.*

Chan, M. S. 1997. The global burden of intestinal nematode infections—fifty years on. Parasitol. Today. **13**:438–443. *According to this author, most recent estimates are 1.273 billion infections (24% prevalence) with* Ascaris, *0.902 billion (17% prevalence) with* Trichuris, *and 1.277 billion (24% prevalence with hookworms. Worldwide prevalence of these nematodes has remained essentially unchanged in 50 years!*

Despommier, D. D. 1990. *Trichinella spiralis:* the worm that would be virus. Parasitol.

Today **6**:193–196. *Juveniles of* Trichinella *are among the largest of all intracellular parasites.*

Duke, B. O. L. 1990. Onchocerciasis (river blindness)—can it be eradicated? Parasitol. Today **6**:82–84. *Despite the introduction of a very effective drug, the author predicts that this parasite will not be eradicated in the foreseeable future.*

Giribet, G., M. V. Sorenson, P. Funch, R. M. Kristensen, and W. Sterrer. 2004. Investigations into the phylogenetic position of Micrognathozoa using four molecular loci. Cladistics **20**:1–13. *Research supports the placement of micrognathozoans outside any known phylum.*

Kristensen, R. M. 2002. An introduction to Loricifera, Cycliophora, and Micrognathozoa. Integ. and Comp. Biol. **42**:641–651. *A clear and informative description of these little-known animal groups.*

Neuhaus, B., and R. P. Higgins. 2002. Ultrastructure, biology, and phylogenetic relationships of Kinorhyncha. Integ. and Comp. Biol. **42**:619–632. *A detailed summary of the biology and morphology of kinorhynchs.*

Poinar, G. O., Jr. 1983. The natural history of nematodes. Englewood Cliffs, New Jersey, Prentice-Hall, Inc. *Contains a great deal*

of information about these fascinating creatures, including free-living and plant and animal parasites.

Roberts, L. 1990. The worm project. Science **248**:1310–1313. *A free-living nematode,* Caenorhabditis elegans, *has been of great value in studies of development and genetics.*

Taylor, M. J., and A. Hoerauf. 1999. *Wolbachia* bacteria of filarial nematodes. Parasitol. Today **15**:437–442. *All filarial parasites of humans have endosymbiotic* Wolbachia, *and most filarial nematodes of all kinds are infected. Nematodes can be "cured" by treatment with the antibiotic tetracycline. If cured, they cannot reproduce. Bacteria are passed vertically from females to offspring.*

Wallace, R. L. 2002. Rotifers: exquisite metazoans. Integ. and Comp. Biol. **42**:660–667. *This paper summarizes recent work on rotifers, but assumes basic knowledge of the group.*

Welch, M. D. B. 2000. Evidence from a protein-coding gene that acanthocephalans are rotifers. Invert. Biol. **119**:17–26. *Sequence analysis of a gene coding for a heat-shock protein supports a position of acanthocephalans within Rotifera. Other molecular and morphological evidence is cited that supports this position.*

Custom Website

The *Animal Diversity* Online Learning Center is a great place to check your understanding of chapter material. Visit www.mhhe.com/hickmanad4e for access to key terms, quizzes, and more! Further enhance your knowledge with Web links to chapter-related material.

Explore live links for these topics:

Classification and Phylogeny of Animals
Pseudocoelomate Animals
Phylum Rotifera
Phylum Kinorhyncha
Phylum Loricifera

Phylum Priapulida
Phylum Nematoda
Human Diseases Caused by Nematodes
Caenorhabditis elegans
Phylum Nematomorpha
Phylum Acanthocephala
Phylum Entoprocta

Molluscs

A Significant Space

Long ago in the Precambrian era, the most complex animals populating the seas were acoelomate. They must have been inefficient burrowers, and they were unable to exploit the rich subsurface ooze. Any that developed fluid-filled spaces within the body would have had a substantial advantage because these spaces could serve as a hydrostatic skeleton and improve burrowing efficiency.

The simplest, and probably first, mode of achieving a fluid-filled space within the body was retention of the embryonic blastocoel, as in pseudocoelomates. This evolutionary solution was not ideal because organs lay loose in the body cavity. Improved efficiency of a pseudocoel as a hydrostatic skeleton depended on increasingly high hydrostatic pressure, a condition that severely limited the potential for adaptive radiation.

Some descendants of Precambrian acoelomate organisms evolved a more elegant arrangement: a fluid-filled space *within* the mesoderm, the *coelom*. This space was lined with mesoderm, and organs were suspended by mesodermal membranes, the *mesenteries*. Not only could the coelom serve as an efficient hydrostatic skeleton, with circular and longitudinal body-wall muscles acting as antagonists, but a more stable arrangement of organs with less crowding resulted. Mesenteries provided an ideal location for networks of blood vessels, and the alimentary canal could become more muscular, more highly specialized, and more diversified without interfering with other organs.

Development of a coelom was a major step in the evolution of larger and more complex forms. All major groups in chapters to follow are coelomates, although some, like the molluscs, have only a small coelom.

Fluted giant clam, *Tridacna squamosa*.

Mollusca (mol-lus′ka) (L. *molluscus,* soft) is one of the largest animal phyla after Arthropoda. There are nearly 75,000 named living species and some 35,000 fossil species. Many more molluscs await formal description. The name Mollusca indicates one of their distinctive characteristics, a soft body.

This very diverse group includes organisms as different as chitons, snails, clams, and octopuses (figure 10.1). The group ranges from fairly simple organisms to some of the most complex invertebrates, and in size from almost microscopic to giant squid *Architeuthis harveyi* (Gr. *archi,* primitive, + *teuthis,* squid). The body of this huge species may grow up to 18 m long with tentacles extended and may weigh up to 454 kg (1000 pounds). The shells of some giant clams *Tridacna gigas* (Gr. *tridaknos,* eaten at three bites) (see figure 10.21), which inhabit the Indo-Pacific coral reefs, reach 1.5 m in length and weigh over 225 kg. These are extremes, however, since probably 80% of all molluscs are less than 5 cm in maximum shell size.

The enormous variety, great beauty, and availability of the shells of molluscs have made shell collecting a popular pastime. However, many amateur shell collectors, although able to name hundreds of the shells that grace our beaches, know very little about the living animals that created those shells and once lived in them. The largest classes of molluscs are Gastropoda (snails and their relatives), Bivalvia (clams, oysters, and others), Polyplacophora (chitons), and Cephalopoda (squids, octopuses, nautiluses). Monoplacophora, Scaphopoda (tusk shells), Caudofoveata, and Solenogastres are much smaller classes.

Ecological Relationships

Molluscs occupy a great range of habitats, from tropics to polar seas, at altitudes exceeding 7000 m, in ponds, lakes, and streams, on mudflats, in pounding surf, and in open ocean from the surface to abyssal depths. Most live in the sea, and they represent a variety of lifestyles, including bottom feeders, burrowers, borers, and pelagic forms. The phylum includes some of the most sluggish and some of the swiftest and most active invertebrates. It includes herbivorous grazers, predaceous carnivores, and ciliary filter feeders.

According to fossil evidence, molluscs originated in the sea, and most have remained there. Much of their evolution occurred along shores, where food was abundant and habitats were varied. Only bivalves and gastropods moved on to brackish and freshwater habitats. As filter feeders, bivalves are unable to leave aquatic surroundings; however, snails (gastropods) actually invaded land and may have been the first animals to do so. Terrestrial snails are limited in range by their need for humidity, shelter, and calcium in the soil.

Economic Importance

A group as large as molluscs would naturally affect humans in some way. A wide variety of molluscs are used as food. Pearls, both natural and cultured, are produced in the shells of clams and oysters, most of them in a marine oyster, found in eastern Asia (see figure 10.4B).

Position in the Animal Kingdom

1. Molluscs are a major group of true **coelomate** animals.
2. They belong to the **protostome** branch, or schizocoelous coelomates, and have spiral cleavage and mosaic development.
3. Many molluscs have a **trochophore larva** similar to the trochophore larva of marine annelids and other marine protostomes. Developmental evidence thus suggests that

molluscs and annelids share a common ancestor within the lophotrochozoan branch of protostomes.
4. Molluscs show no evidence of the segmentation (metamerism) so clearly present in annelids (see p. 203); they may have diverged from their common ancestor with annelids before the advent of segmentation.
5. All **organ systems** are present and well developed.

Biological Contributions

1. In molluscs gaseous exchange occurs not only through the body surface as in phyla discussed previously, but also in specialized **respiratory organs** in the form of **gills** or a **lung.**
2. Most classes have an **open circulatory system** with pumping **heart,** vessels, and blood sinuses. In most cephalopods the circulatory system is closed.
3. Efficiency of the respiratory and circulatory systems in cephalopods has made greater body size possible. Invertebrates reach their largest size in some cephalopods.
4. They have a fleshy **mantle** that in most cases secretes a shell and is variously modified for a number of functions. Other features unique to the phylum are the rasping, tongue-like **radula** and the muscular **foot.**
5. The highly developed **eye** of cephalopods is similar to the eye of vertebrates but arises as a skin derivative in contrast to the brain eye of vertebrates.

figure 10.1

Molluscs: a diversity of life forms. The basic body plan of this ancient group has become variously adapted for different habitats. **A,** A chiton *(Tonicella lineata),* class Polyplacophora. **B,** A marine snail *(Calliostoma annulata),* class Gastropoda. **C,** A nudibranch *(Chromodoris kuniei),* class Gastropoda. **D,** Pacific giant clam *(Panope abrupta),* with siphons to the left, class Bivalvia. **E,** An octopus *(Octopus briareus),* class Cephalopoda, forages at night on a Caribbean coral reef.

Some molluscs are destructive. Burrowing shipworms (see figure 10.24), which are bivalves of several species, do great damage to wooden ships and wharves. To prevent the ravages of shipworms, wharves must be either creosoted or built of concrete. Snails and slugs often damage garden and other vegetation. In addition, many snails serve as intermediate hosts for serious parasites such as trematodes causing schistosomiasis (see p. 151). A certain boring snail, the oyster drill, rivals sea stars in destroying oysters.

Form and Function

Body Plan

Reduced to its simplest dimensions, a mollusc body consists of a **head-foot** portion and a **visceral mass** portion (figure 10.2). The head-foot region contains feeding, cephalic sensory, and locomotor organs. It depends primarily on muscular action for its function. The visceral mass is the portion containing digestive, circulatory, respiratory, and reproductive organs, and it depends primarily on ciliary tracts for its functioning. Two folds of skin, outgrowths of the dorsal body wall, make up a protective **mantle,** which encloses a space between the mantle and body wall called a **mantle cavity.** The mantle cavity houses **gills** or a lung, and in many molluscs the mantle secretes a protective **shell** over the visceral mass and head-foot. Modifications of structures that form the head-foot and the visceral mass produce the great profusion of different patterns seen in this major group of animals.

Head-Foot

Most molluscs have a well-developed head, which bears a mouth and some specialized sensory organs. Photosensory receptors range from fairly simple to the complex eyes of cephalopods. Tentacles are often present. Within the mouth is a structure unique to molluscs, the radula, and usually posterior to the mouth is the chief locomotor organ, or foot.

Radula The **radula** is a rasping, protrusible, tonguelike organ found in all molluscs except bivalves and some gastropods and solenogasters. It is a ribbonlike membrane on which are mounted rows of tiny teeth that point backward (figure 10.3). Complex muscles move the radula and its supporting cartilages (**odontophore**) in and out while the membrane is partly rotated over the tips of the cartilages. There may be a few or as many as 250,000 teeth, which, when protruded, can scrape, pierce, tear, or cut particles of food material, and the radula may serve as a rasping file for carrying particles in a continuous stream toward the digestive tract.

Foot The molluscan foot may be variously adapted for locomotion, for attachment to a substratum, or for a combination of functions. It is usually a ventral, solelike structure in which waves of muscular contraction effect a creeping locomotion. However, there are many modifications, such as the attachment disc of limpets, the laterally compressed "hatchet foot" of bivalves, or the funnel for jet propulsion in squids and octopuses. Secreted mucus is often used as an aid to adhesion or as a slime track by small molluscs that glide on cilia.

figure 10.2

Generalized mollusc. Although this construct is often presented as a "hypothetical ancestral mollusc" (HAM), most experts now agree that it never actually existed. It is an abstraction used to facilitate description of the general body plan of molluscs.

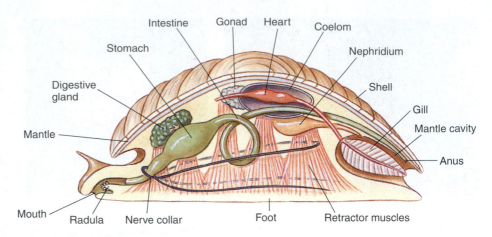

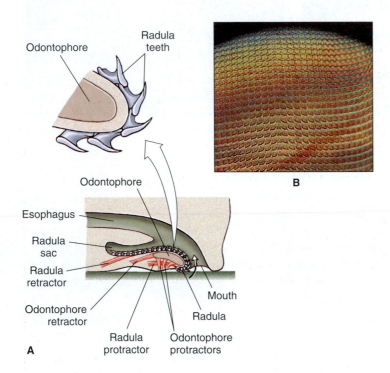

figure 10.3

A, Diagrammatic longitudinal section of gastropod head showing the radula and radula sac. The radula moves back and forth over the odontophore cartilage. As the animal grazes, the mouth opens, the odontophore is thrust forward, the radula gives a strong scrape backward bringing food into the pharynx, and the mouth closes. The sequence is repeated rhythmically. As the radula ribbon wears out anteriorly, it is continually replaced posteriorly. **B,** Radula of a snail prepared for microscopic examination.

Visceral Mass

Mantle and Mantle Cavity   The mantle is a sheath of skin extending from the visceral hump that hangs down on each side of the body, protecting the soft parts and creating between itself and the visceral mass a space called the mantle cavity. The outer surface of the mantle secretes the shell.

The mantle cavity plays an enormous role in the life of a mollusc. It usually houses respiratory organs (gills or a lung), which develop from the mantle, and the mantle's own exposed surface serves also for gaseous exchange. Products from the digestive, excretory, and reproductive systems empty into the mantle cavity. In aquatic molluscs a continuous current of water, kept moving by surface cilia or by muscular pumping, brings in oxygen, and in some forms, food; flushes out wastes; and carries reproductive products out to the environment. In aquatic forms the mantle is usually equipped with sensory receptors for sampling the environmental water. In cephalopods (squids and octopuses) the muscular mantle and its cavity create jet propulsion used in locomotion.

Shell   The shell of a mollusc, when present, is secreted by the mantle and is lined by it. Typically there are three layers (figure 10.4). The **periostracum** is the outer organic layer, composed of a resistant protein called conchiolin. It helps protect the underlying calcareous layers from erosion by boring organisms. It is secreted by a fold of the mantle edge, and growth occurs only at the margin of the shell. On the older parts of the shell the periostracum often becomes worn away. The middle **prismatic layer** is composed of densely packed prisms of calcium carbonate laid down in a protein matrix. It is secreted by the glandular margin of the mantle, and increase in shell size occurs at the shell margin as the animal grows. The inner **nacreous layer** of the shell is composed of calcium carbonate sheets laid down over a thin protein matrix. This layer is secreted continuously by the mantle surface, so that it becomes thicker during the life of the animal.

Freshwater molluscs usually have a thick periostracum that gives some protection against acids produced in water by decay of leaf litter. In some marine molluscs the periostracum is thick, but in others it is relatively thin or absent. There is a great range in variation in shell structure. Calcium for the shell comes from environmental water or soil or from food. The first shell appears during the larval period and grows continuously throughout life.

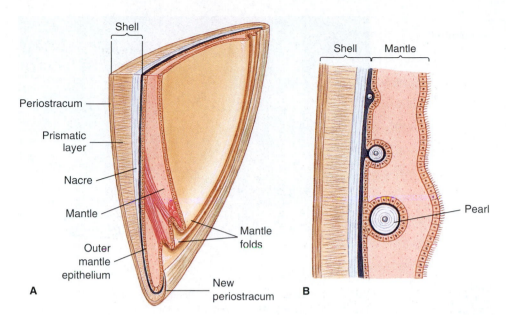

figure 10.4

A, Diagrammatic vertical section of shell and mantle of a bivalve. The outer mantle epithelium secretes the shell; the inner epithelium is usually ciliated. **B,** Formation of pearl between mantle and shell as a parasite or bit of sand under the mantle becomes covered with nacre.

Pearl production is a by-product of a protective device used by a mollusc when a foreign object, such as a grain of sand or a parasite, becomes lodged between the shell and mantle. The mantle secretes many layers of nacre around the irritating object (see figure 10.4). Pearls are cultured by inserting small spheres, usually made from pieces of the shells of freshwater clams, in the mantle of a certain species of oyster and by maintaining the oysters in enclosures. The oyster deposits its own nacre around the "seed" in a much shorter time than would be required to form a pearl normally.

Internal Structure and Function

Gaseous exchange occurs through the body surface, particularly the mantle, and in specialized respiratory organs such as gills or lungs. There is an **open circulatory system** with a pumping **heart,** blood vessels, and blood sinuses. Most cephalopods have a closed blood system with heart, vessels, and capillaries. The digestive tract is complex and highly specialized according to feeding habits of the various molluscs. Most molluscs have a pair of **kidneys (metanephridia),** a type of nephridium in which the inner end opens into the coelom; ducts of the kidneys in many forms serve also for discharge of eggs and sperm. The **nervous system** consists of several pairs of ganglia with connecting nerve cords. There are various types of highly specialized sense organs.

Most molluscs are dioecious, although some gastropods are hermaphroditic. Many aquatic molluscs pass through free-swimming **trochophore** (figure 10.5) and **veliger** (figure 10.6) larval stages. A veliger is the free-swimming larva of most marine snails, tusk shells, and bivalves. It develops from a trochophore and has the beginning of a foot, shell, and mantle.

---

characteristics
of Phylum Mollusca

1. Body bilaterally symmetrical (bilateral asymmetry in some); unsegmented; usually with definite head
2. Ventral body wall specialized as a **muscular foot,** variously modified but used chiefly for locomotion
3. Dorsal body wall forms the **mantle,** which encloses the **mantle cavity,** is modified into **gills** or a **lung,** and secretes the **shell** (shell absent in some)
4. Surface epithelium usually ciliated and bearing mucous glands and sensory nerve endings
5. Coelom mainly limited to area around heart (pericardial cavity)
6. Complex digestive system; rasping organ (**radula**) usually present (see figure 10.3); anus usually emptying into mantle cavity
7. **Open circulatory system** of heart (usually three chambered, two in most gastropods), blood vessels, and sinuses in all classes other than Cephalopoda; respiratory pigments in blood
8. Gaseous exchange by gills, lung, mantle, or body surface
9. Usually one or two kidneys (**metanephridia**) opening into the pericardial cavity and usually emptying into the mantle cavity
10. Nervous system of paired cerebral, pleural, pedal, and visceral ganglia, with nerve cords and subepidermal plexus; ganglia centralized in nerve ring in polyplacophorans, gastropods, and cephalopods
11. Sensory organs of touch, smell, taste, equilibrium, and vision (in some); eyes highly developed in cephalopods

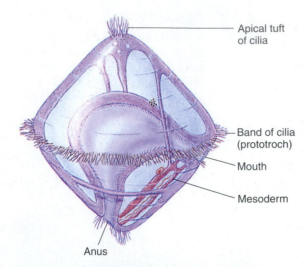

A

Apical tuft
of cilia

Band of cilia
(prototroch)

Mouth

Mesoderm

Anus

figure 10.6

Veliger of a snail, *Pedicularia,* swimming. The adults are parasitic on corals. The ciliated process (velum) develops from the prototroch of the trochophore (figure 10.5).

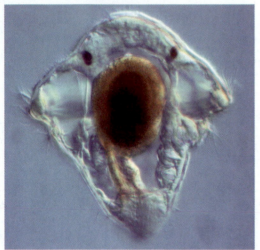

B

figure 10.5

A, Generalized trochophore larva. Molluscs and annelids with primitive embryonic development have trochophore larvae, as do several other phyla. **B,** Trochophore of a Christmas tree worm, *Spirobranchus spinosus* (Annelida).

Trochophore larvae (figure 10.5) are minute, translucent, more or less top-shaped, and have a prominent circlet of cilia (prototroch) and sometimes one or two accessory circlets. They are found in molluscs and annelids with primitive embryonic development and considered one of the evidences for common phylogenetic origin of the two phyla. Some form of trochophore-like larva also occurs in marine turbellarians, nemertines, brachiopods, phoronids, sipunculids, and echiurids. Possession of a trochophore or trochophore-like larva supports assignment of these phyla to superphylum Lophotrochozoa.

Classes Caudofoveata and Solenogastres

Caudofoveates and the solenogasters (see figure 10.35, p. 198) are often united in class Aplacophora because they are both wormlike and shell-less, with calcareous scales or spicules in their integument. Members of both groups have a reduced head and lack nephridia. Caudofoveates have one pair of gills and are dioecious; their body plan may have more features in common with those of ancestral molluscs than any other living group. They burrow in marine sediments, feeding on microorganisms and detritus. Class Caudofoveata is sometimes known as Chaetodermomorpha. In contrast to caudofoveates, solenogasters usually have no true gills, and they are hermaphroditic. Solenogasters live freely on the ocean bottom and often feed on cnidarians. Class Solenogastres is sometimes known as Neomeniomorpha.

Class Monoplacophora

Until 1952 Monoplacophora (mon-o-pla-kof′o-ra) were known only from Paleozoic shells. However, in that year living specimens of *Neopilina* (Gr. *neo,* new, + *pilos,* felt cap) were dredged up from the ocean bottom near the west coast of Costa Rica. These molluscs are small and have a low, rounded shell and a creeping foot (figure 10.7). They have a superficial resemblance to limpets (see figure 10.18A), but unlike in most other molluscs, a number of organs, such as metanephridia, gonads, and gills, are serially repeated. Serial repetition occurs to a more limited extent in chitons. Why should there be repeated sets of body structures in these animals? Body structures repeat in each segment of an annelid worm (see p. 203)—are these repeated structures an indication that molluscs had a segmented (metameric) ancestor? Some authors have considered monoplacophorans truly segmented, but most others argue that *Neopilina* shows only pseudometamerism and that molluscs did not have a metameric ancestor.

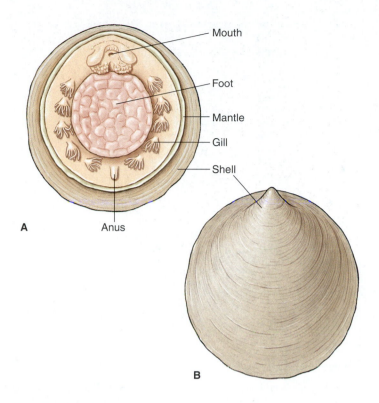

figure **10.7**

Neopilina, class Monoplacophora. Living specimens range from 3 mm to about 3 cm in length. **A,** Ventral view. **B,** Dorsal view.

figure **10.8**

Mossy chiton, *Mopalia muscosa.* The upper surface of the mantle, or "girdle," is covered with hairs and bristles, an adaptation for defense.

Class Polyplacophora: Chitons

Chitons are somewhat flattened and have a convex dorsal surface that bears eight (rarely seven) articulating calcareous **plates,** or **valves,** which give them their name (figures 10.8 and 10.9). The term Polyplacophora means "bearing many plates," in contrast to Monoplacophora, which bear one shell (*mono,* single). The plates overlap posteriorly and are usually dull colored like the rocks to which chitons cling.

Most chitons are small (2 to 5 cm); the largest rarely exceeds 30 cm. They commonly occur on rocky surfaces in intertidal regions, although some live at great depths. Chitons are stay-at-home organisms, straying only very short distances for feeding. In feeding, a sensory subradular organ protrudes from their mouth to explore for algae or colonial organisms. When some are found, the radula projects to scrape algae off the rocks. A chiton clings tenaciously to its rock with the broad flat foot. If detached, it can roll up like an armadillo for protection.

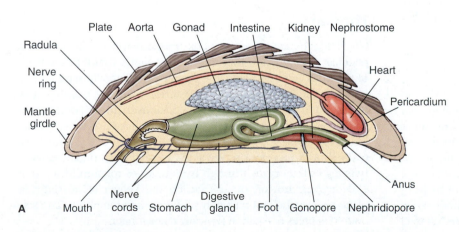

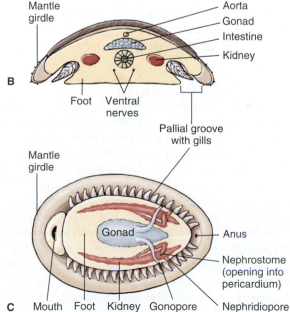

figure **10.9**

Anatomy of a chiton (class Polyplacophora). **A,** Longitudinal section. **B,** Transverse section. **C,** External ventral view.

The mantle forms a **girdle** around the margin of the plates, and in some species mantle folds cover part or all of the plates. On each side of the broad ventral foot and lying between the foot and the mantle is a row of gills suspended from the roof of the mantle cavity. With the foot and the mantle margin adhering tightly to the substrate, these grooves become closed chambers, open only at the ends. Water enters the grooves anteriorly, flows across the gills, and leaves posteriorly, thus bringing a continuous supply of oxygen to the gills.

Blood pumped by the three-chambered heart reaches the gills by way of an aorta and sinuses. Two kidneys carry waste from the pericardial cavity to the exterior. Two pairs of longitudinal nerve cords are connected in the buccal region. Sense organs include shell eyes on the surface of the shell (in some) and a pair of **osphradia** (chemosensory organs for sampling water).

Sexes are separate in chitons. Eggs are released singly or in strings or masses of jelly; they may be shed into the female mantle cavity or into the sea. Sperm shed by males into the excurrent water may enter the gill grooves of females by incurrent openings. Trochophore larvae metamorphose directly into juveniles, without a second larval stage.

Class Scaphopoda

Scaphopoda (ska-fop'o-da), commonly called tusk shells or tooth shells, are sedentary marine molluscs that have a slender body covered with a mantle and a tubular shell open at both ends. Here the molluscan body plan has taken a new direction, with the mantle wrapped around the viscera and fused to form a tube. Most scaphopods are 2.5 to 5 cm long, although they range from 4 mm to 25 cm long.

The foot, which protrudes through the larger end of the shell, functions in burrowing into mud or sand, always leaving the small end of the shell exposed to water above (figure 10.10). Respiratory water circulates through the mantle cavity both by movements of the foot and by ciliary action. Gaseous exchange occurs in the mantle. Most food is detritus and protozoa from the substrate. Cilia of the foot or on the mucus-covered, ciliated knobs of long tentacles catch the food.

Class Gastropoda

Among molluscs class Gastropoda (gas-trop'o-da) (Gr. *gastēr*, stomach, + *pous, podos,* foot) is by far the largest and most diverse, containing about 40,000 living and 15,000 fossil species. Its members differ so widely that there is no single general term in our language that can apply to them as a group. They include snails, limpets, slugs, whelks, conchs, periwinkles, sea slugs, sea hares, sea butterflies, and others. They range from some marine molluscs with many primitive characters to highly evolved, air-breathing snails and slugs.

Gastropods are often sluggish, sedentary animals because most of them have heavy shells and slow locomotor organs. When present, the shell is almost always of one piece (univalve) and may be coiled or uncoiled. Some snails have an

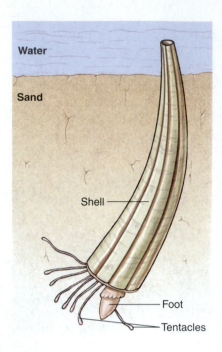

figure 10.10

The tusk shell, *Dentalium,* a scaphopod. It burrows into soft mud or sand and feeds by means of its prehensile tentacles. Respiratory currents of water are drawn in by ciliary action through the small open end of the shell, then expelled through the same opening by muscular action.

operculum, a hard proteinaceous plate that covers the shell aperture when the body withdraws into the shell. It protects the body and prevents water loss. These animals are basically bilaterally symmetrical, but because of **torsion,** a twisting process that occurs during development, the visceral mass has become asymmetrical. Gastropods develop through a trochophore and a veliger larval stage.

Form and Function

Torsion

Of all molluscs, only gastropods undergo torsion. Torsion is a peculiar phenomenon that moves the mantle cavity to the front of the body. Before torsion, the embryo is bilaterally symmetrical with an anterior mouth and a posterior anus and mantle cavity (figure 10.11). During torsion, there is uneven growth of the right and left muscles that attach the shell to the headfoot, as well as differences in muscle contraction on the two sides of the body. Relatively shorter muscles on one side of the visceral organs cause a twisting of the organs through two stages of rotation from 90 to 180 degrees. Torsion occurs during the veliger larval stage; in some species the first part may take only a few minutes. The second 90 degrees of rotation typically takes longer.

After torsion, the anus and mantle cavity become anterior and open above the mouth and head. The left gill, kidney, and heart auricle are now on the right side, whereas the original right gill, kidney, and heart auricle (lost in most modern gastropods) are now on the left, and the nerve cords have

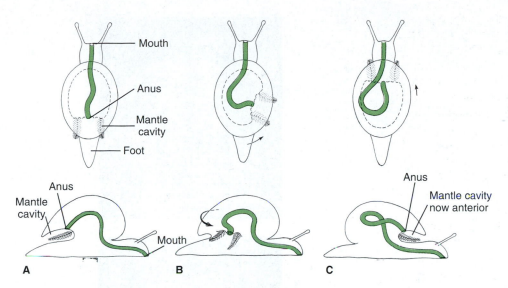

figure 10.11

Torsion in gastropods. **A,** Ancestral condition before torsion. **B,** Intermediate condition. **C,** Early gastropod, torsion complete; direction of crawling now tends to carry waste products back into the mantle cavity, resulting in fouling.

been twisted into a figure eight. Because of the space available in the mantle cavity, the animal's sensitive head end can now be withdrawn into the protection of the shell, with the tougher foot forming a barrier to the outside.

The curious arrangement that results from torsion poses a serious sanitation problem by creating the possibility of wastes being washed back over the gills (**fouling**) and causes us to wonder what evolutionary factors favored such a strange realignment of the body. Several explanations have been proposed, none entirely satisfying. For example, sense organs of the mantle cavity (osphradia) would better sample water when turned in the direction of travel, and as mentioned already, the head could be withdrawn into the shell. Certainly the consequences of torsion and the resulting need to avoid fouling have been very important in the subsequent evolution of gastropods. We cannot explore these consequences, however, until we describe another unusual feature of gastropods—coiling of the shell and visceral mass.

Coiling

Coiling, or spiral winding, of the shell and visceral hump is not the same as torsion. Coiling may occur in the larval stage at the same time as torsion, but the fossil record shows that coiling was a separate evolutionary event and originated in gastropods earlier than torsion did. Nevertheless, all living gastropods have descended from coiled, torted ancestors, whether or not they now show these characteristics.

Early gastropods had a bilaterally symmetrical shell with all whorls lying in a single plane (figure 10.12A). Such a shell was not very compact, since each whorl had to lie completely outside the preceding one. Curiously, a few modern species have secondarily returned to that form. The compactness problem of the planospiral shell was solved by a shape in which each succeeding whorl was at the side of the preceding one (figure 10.12B). However, this shape clearly was unbalanced, hanging as it did with much weight over to one side. They achieved better weight distribution by shifting the shell upward and posteriorly, with the shell axis oblique to the longitudinal axis of the foot (figure 10.12C and D). The weight

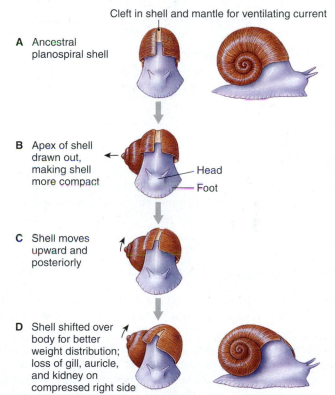

Cleft in shell and mantle for ventilating current

A Ancestral planospiral shell

B Apex of shell drawn out, making shell more compact

Head

Foot

C Shell moves upward and posteriorly

D Shell shifted over body for better weight distribution; loss of gill, auricle, and kidney on compressed right side

figure 10.12

Evolution of shell in gastropods. **A,** Earliest coiled shells were planospiral, each whorl lying completely outside the preceding whorl. Interestingly, the shell has become planospiral secondarily in some living forms. **B,** Better compactness was achieved by snails in which each whorl lay partially to the side of the preceding whorl. **C** and **D,** Better weight distribution resulted when shell was moved upward and posteriorly.

and bulk of the main body whorl, the largest whorl of the shell, pressed on the right side of the mantle cavity, however, and apparently interfered with the organs on that side. Accordingly, the gill, auricle, and kidney of the right side have been lost in all except a few living gastropods, leading to a condition of **bilateral asymmetry.**

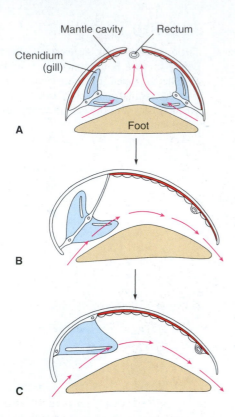

figure 10.13

Evolution of gills in gastropods. **A,** Primitive condition in prosobranchs with two gills and excurrent water leaving the mantle cavity by a dorsal slit or hole. **B,** Condition after one gill had been lost. **C,** Derived condition in prosobranchs, in which filaments on one side of remaining gill are lost, and axis is attached to mantle wall.

Adaptations to Avoid Fouling

Although loss of the right gill was probably an adaptation to the mechanics of carrying a coiled shell, that condition made it possible to reduce the effects of fouling (see p. 185). In most modern gastropods, there is a one-way flow of water; it is brought into the left side of the mantle cavity and out the right side, carrying with it the wastes from the anus and nephridiopore, which lie near the right side (figure 10.13). Some gastropods with primitive characteristics (those with two gills, such as abalone) (figure 10.14A) avoid fouling by venting excurrent water through a dorsal slit or hole in the shell above the anus (see figure 10.12). Opisthobranchs (nudibranchs and others) have evolved an even more curious "twist;" after undergoing torsion as larvae, they develop various degrees of *detorsion* as adults.

Feeding Habits

Feeding habits of gastropods are as varied as their shapes and habitats, but all include the use of some adaptation of the radula. Many gastropods are herbivorous, rasping off particles of algae from a substrate. Some herbivores are grazers, some are browsers, some are planktonic feeders. Abalones (figure 10.14)

figure 10.14

A, Red abalone, *Haliotus rufescens*. This huge, limpetlike snail is prized as food and extensively marketed. Abalones are strict vegetarians, feeding especially on sea lettuce and kelp. **B,** Moon snail, *Polinices lewisii*. A common inhabitant of West Coast sand flats, the moon snail is a predator of clams and mussels. It uses its radula to drill neat holes through its victim's shell, through which the proboscis is then extended to eat the bivalve's fleshy body.

hold seaweed with the foot and break off pieces with their radula. Some snails are scavengers, living on dead and decayed flesh; others are carnivorous, tearing prey apart with their radular teeth. Some, such as oyster borers and moon snails (figure 10.14B), have an extensible proboscis for drilling holes in the shells of bivalves whose soft parts they find delectable. Some even have a spine for opening shells. Most pulmonates (air-breathing snails) (see figure 10.20, p. 189) are herbivorous, but some feed on earthworms and other snails. After maceration by the radula or by some grinding device, such as the so-called gizzard in sea hares (figure 10.15) and in others, digestion is usually extracellular in the lumen of the stomach or digestive glands.

Some sessile gastropods, such as slipper shells, are ciliary feeders that use the gill cilia to draw in particulate matter, which they roll into a mucous ball and carry to their mouth. Some sea butterflies secrete a mucous net to catch small planktonic forms and then draw the web into their mouth. In ciliary feeders the stomachs are sorting regions and most digestion is intracellular in the digestive gland.

Rhinophore   Oral tentacle

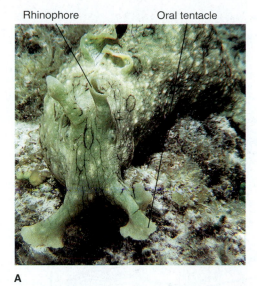

A

B

figure 10.15

A, Sea hare, *Aplysia dactylomela,* crawls and swims across a coral reef, assisted by large, winglike parapodia, here curled above the body. **B,** When attacked, sea hares squirt a copious protective secretion from their "purple gland" in the mantle cavity.

Among the most interesting predators are venomous cone shells (figure 10.16), which feed on vertebrates or other invertebrates, depending on the species. When *Conus* senses its prey, a single radular tooth slides into position at the tip of the proboscis. When the proboscis strikes prey, it expels the tooth like a harpoon, and the poison tranquilizes or kills the prey at once. Some species can deliver very painful stings, and the sting of several species is lethal to humans. The venom consists of a series of toxic peptides, and each *Conus* species carries peptides **(conotoxins)** specific for the neuroreceptors of its preferred prey.

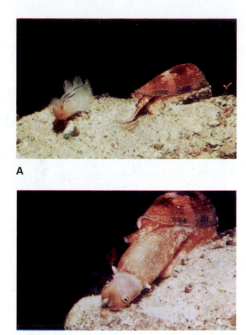

A

B

figure 10.16

A, *Conus* extends its long, wormlike proboscis. When the fish attempts to consume this tasty morsel, the *Conus* stings it in the mouth and kills it. **B,** The snail engulfs the fish with its distensible stomach, then regurgitates the scales and bones some hours later.

Internal Form and Function

Respiration in most gastropods is carried out by a gill (two gills in a few), although some aquatic forms lack gills and depend on the skin. Pulmonates (most freshwater and terrestrial snails) have lost their gill altogether, and the vascularized mantle wall has become a lung. The anus and nephridiopore open near the opening of the lung to the outside (**pneumostome;** see figure 10.20B), and waste is expelled forcibly with air or water from the lung. Freshwater pulmonates must surface to expel a bubble of gas from the lung and to curl the edge of the mantle around the pneumostome to form a siphon for air intake.

Most gastropods have a single nephridium (kidney). The circulatory and nervous systems are well developed (figure 10.17). The nervous system includes three pairs of ganglia connected by nerves. Sense organs include eyes, statocysts, tactile organs, and chemoreceptors.

There are both dioecious and hermaphroditic gastropods. Many forms perform courtship ceremonies. During copulation in hermaphroditic species there is sometimes an exchange of **spermatophores** (bundles of sperm), so that self-fertilization is avoided. Most land snails lay their eggs in holes in the ground or under logs. Some aquatic gastropods lay their eggs in gelatinous masses; others enclose them in gelatinous capsules or in parchment egg cases. Most marine gastropods go through a free-swimming veliger larval stage during which torsion and coiling occur. Others develop directly into a juvenile within an egg capsule.

figure 10.17

Anatomy of a pulmonate snail.

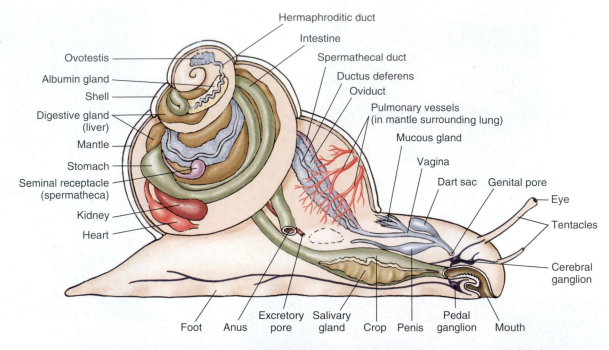

Labels: Ovotestis, Albumin gland, Shell, Digestive gland (liver), Mantle, Stomach, Seminal receptacle (spermatheca), Kidney, Heart, Hermaphroditic duct, Intestine, Spermathecal duct, Ductus deferens, Oviduct, Pulmonary vessels (in mantle surrounding lung), Mucous gland, Vagina, Dart sac, Genital pore, Eye, Tentacles, Cerebral ganglion, Foot, Anus, Excretory pore, Salivary gland, Crop, Penis, Pedal ganglion, Mouth

figure 10.18

A, Keyhole limpet, *Diodora aspera,* a prosobranch gastropod with a hole in the apex through which water leaves the mantle cavity. **B,** Flamingo tongues, *Cyphoma gibbosum,* are showy inhabitants of Caribbean coral reefs, where they are associated with gorgonians. These snails have a smooth creamy orange to pink shell that is normally covered by the brightly marked mantle. Here the flexible foot is visible as the snail crawls along a gorgonian coral.

A

B

Major Groups of Gastropods

Traditional classification of class Gastropoda recognized three subclasses: Prosobranchia, Opisthobranchia, and Pulmonata. Prosobranchia is by far the largest subclass, and its members are almost all marine. Familiar examples of marine prosobranchs are periwinkles, limpets (figure 10.18A), whelks, conchs, abalones (see figure 10.14A), slipper shells, oyster borers, rock shells, and cowries.

Opisthobranchia is an assemblage of marine gastropods including sea slugs, sea hares, nudibranchs, and canoe shells. At present 8 to 12 groups of opisthobranchs are recognized. Some have a gill and a shell, although the latter may be vestigial, and some have no shell or true gill. Large sea hares *Aplysia* (see figure 10.15) have large earlike anterior tentacles and a vestigial shell. Nudibranchs have no shell as adults and rank among the most beautiful and colorful of molluscs (figure 10.19). Having lost the gill, the body surface of some nudibranchs is often increased for gaseous exchange by small projections (**cerata**), or a ruffling of the mantle edge.

figure 10.19

Phyllidia ocellata, a nudibranch. Like other *Phyllidia* spp., it has a hard body with dense calcareous spicules and bears its gills along the sides, between its mantle and foot.

A

B

figure 10.20

A, Pulmonate land snail. Note two pairs of tentacles; the second larger pair bears the eyes. **B,** Banana slug, *Ariolimax columbianus.*

The third major group, Pulmonata, contains most land and freshwater snails and slugs. Usually lacking gills, their mantle cavity has become a lung, which fills with air by contraction of the mantle floor. Aquatic and a few terrestrial species have one pair of nonretractile tentacles, at the base of which are eyes; land forms usually have two pairs of tentacles, with the posterior pair bearing eyes (figures 10.17 and 10.20). The few nonpulmonate species of gastropods that live in fresh water usually can be distinguished from pulmonates because they have an operculum, which is lacking in pulmonates.

Currently, gastropod taxonomy is in a state of flux, and some workers regard any attempt to present a classification of the class as premature. A phylogeny of the gastropods based on morphological characters divided the class into two groups differing in form of the radula and other features. However, a preliminary molecular analysis did not support this grouping. The molecular data generally support a clade (Euthyneura) combining opisthobranchs and pulmonates, but Opisthobranchia appears to be paraphyletic.

Class Bivalvia (Pelecypoda)

Bivalvia (bi-val′ve-a) are also known as Pelecypoda (pel-e-sip′o-da) (Gr. *pelekus,* hatchet, + *pous, podus,* foot). They are bivalved (two-shelled) molluscs that include mussels, clams, scallops, oysters, and shipworms and range in size from tiny seed shells 1 to 2 mm in length to the giant, South Pacific clams *Tridacna,* mentioned previously (figure 10.21). Most bivalves are sedentary **suspension feeders** that depend on ciliary currents produced by the gills to bring in food materials. Unlike gastropods, they have no head, no radula, and very little cephalization (figure 10.22).

Most bivalves are marine, but many live in brackish water and in streams, ponds, and lakes.

Form and Function

Shell

Bivalves are laterally compressed, and their two shells **(valves)** are held together dorsally by a **hinge ligament** that causes the valves to gape ventrally. **Adductor muscles** work in opposition

figure 10.21

Clam (*Tridacna gigas*) lies buried in coral rock with greatly enlarged siphonal area visible. These tissues are richly colored and bear enormous numbers of symbiotic single-celled algae (zooxanthellae) that provide much of the clam's nutriment.

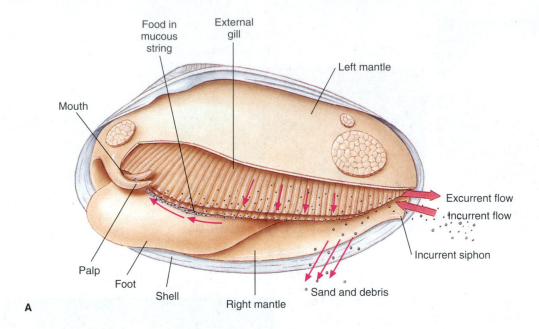

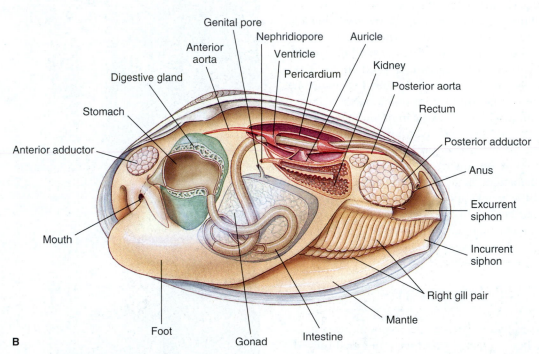

figure 10.22

A, Feeding mechanism of freshwater clam. Left valve and mantle are removed. Water enters the mantle cavity posteriorly and is drawn forward by ciliary action to the gills and palps. As water enters the tiny openings of the gills, food particles are sieved out and caught up in strings of mucus that are carried by cilia to the palps and directed to the mouth. Sand and debris drop into the mantle cavity and are removed by cilia. **B,** Clam anatomy.

to the hinge ligament and draw the valves together (figure 10.23C and D). Projecting above the hinge ligament on each valve is the **umbo,** which is the oldest part of the shell. The valves function largely for protection, but those of shipworms (figure 10.24) have microscopic teeth for rasping wood, and rock borers use spiny valves for boring into rock. A few bivalves such as scallops (figure 10.25) use their shells for locomotion by clapping the valves together so that they move in spurts.

Body and Mantle

The **visceral mass** is suspended from the dorsal midline, and the muscular foot is attached to the visceral mass anteroventrally. The gills hang down on each side, each covered by a fold of the mantle. The posterior edges of the mantle folds form dorsal excurrent and ventral incurrent openings (figures 10.23A and 10.26). In some marine bivalves part of the mantle is drawn

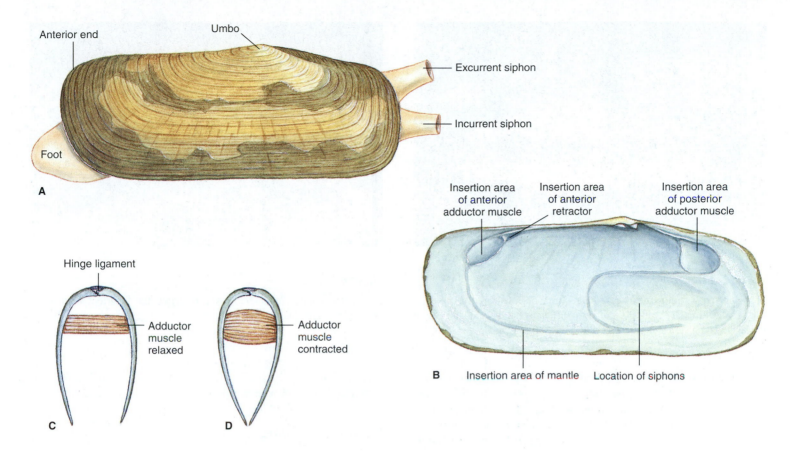

figure 10.23

Tagelus plebius, the stubby razor clam (class Bivalvia). **A,** External view of left valve. **B,** Inside of right shell showing scars where muscles were attached. The mantle was attached to part of the shell leaving a pallial line visible at the edge of the mantle insertion area. **C** and **D,** Sections showing function of adductor muscles and hinge ligament. In **C** the adductor muscle is relaxed, allowing the hinge ligament to pull the valves apart. In **D** the adductor muscle is contracted, pulling the valves together.

figure 10.24

A, Shipworms (*Teredo, Bankia,* and others) are bivalves that burrow in wood, causing great damage to unprotected wooden hulls and piers. **B,** The two small, anterior valves, seen at left, are used as rasping organs to extend the burrow.

figure 10.25

Representing a group that has evolved from burrowing ancestors, the surface-dwelling bay scallop *Aequipecten irradians* has developed sensory tentacles and a series of blue eyes along its mantle edges.

figure 10.26

In northwest ugly clams, *Entodesma navicula,* the incurrent and excurrent siphons are clearly visible.

out into long muscular siphons to allow the clam to burrow into the mud or sand and extend the siphons to the water above. Cilia on the gills and inner surface of the mantle direct the flow of water over the gills bringing in food and oxygen.

Bivalves have a three-chambered heart that pumps blood through the gills and mantle for oxygenation and to the kidneys for waste elimination (figure 10.27). They have three pairs of widely separated ganglia and poorly developed sense organs. A few bivalves have ocelli. The steely blue eyes of some

scallops (see figure 10.25), located around the mantle edge, are remarkably complex, equipped with cornea, lens, and retina.

Feeding and Digestion

Most bivalves are suspension feeders. Their respiratory currents bring both oxygen and organic materials to their gills where ciliary tracts direct them to the tiny pores of the gills. Gland cells on the gills and labial palps secrete copious

figure 10.27

Section through heart region of a freshwater clam to show relation of circulatory and respiratory systems. Respiratory water currents: water is drawn in by cilia, enters gill pores, and then passes up water tubes to suprabranchial chambers and out excurrent aperture. Blood in gills exchanges carbon dioxide and oxygen. Blood circulation: ventricle pumps blood forward to sinuses of foot and viscera, and posteriorly to mantle sinuses. Blood returns from mantle to auricles; it returns from viscera to the kidney, and then goes to the gills, and finally to the auricles.

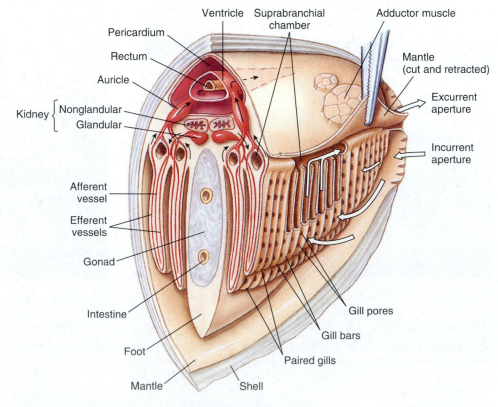

figure 10.28

Mussels, *Mytilus edulis,* occur in northern oceans around the world; they form dense beds in the intertidal zone. A host of marine creatures live protected beneath attached mussels.

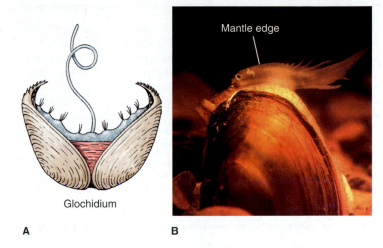

Glochidium

A   B

figure 10.29

A, Glochidium, or larval form, for some freshwater clams. When larvae are released from brood pouch of mother, they may become attached to a fish's gill by clamping their valves closed. They remain as parasites on the fish for several weeks. Their size is approximately 0.3 mm. **B,** Some clams have adaptations that help their glochidia find a host. The mantle edge of this female pocketbook mussel (*Lampsilis ovata*) mimics a small minnow, complete with eye. When a smallmouth bass comes to dine, it gets doused with glochidia.

amounts of mucus, which entangles food particles suspended in the water going through gill pores. Ciliary tracts move the particle-laden mucus to the mouth (see figure 10.22).

In the stomach the mucus and food particles are kept whirling by a rotating gelatinous rod, called a **crystalline style.** Solution of layers of the rotating style frees digestive enzymes for extracellular digestion. Ciliated ridges of the stomach sort food particles and direct suitable particles to the **digestive gland** for intracellular digestion.

Shipworms (see figure 10.24) feed on the particles they excavate as they burrow in wood. Symbiotic bacteria live in a special organ in these bivalves and produce cellulase to digest wood. Other bivalves such as giant clams gain much of their nutrition from the photosynthetic products of symbiotic algae (zooxanthellae, p. 103) living in their mantle tissue (see figure 10.21).

Locomotion

Most bivalves move by extending their slender muscular foot between the valves (see figure 10.23A). They pump blood into the foot, causing it to swell and to act as an anchor in mud or sand, then longitudinal muscles contract to shorten the foot and pull the animal forward. In most bivalves the foot is used for burrowing, but a few creep. Some bivalves are sessile: oysters attach their shells to a surface by secreting cement, and mussels (figure 10.28) attach themselves by secreting a number of slender **byssal threads.** The foot is often reduced in sessile bivalves.

Reproduction

Sexes are separate, and fertilization is usually external. Marine embryos typically go through three free-swimming larval stages—**trochophore, veliger larva,** and young **spat**—before reaching adulthood. In freshwater clams fertilization is internal, and some gill tubes become temporary brood chambers. There larvae develop into specialized veligers called **glochidia,** which are discharged with the excurrent flow (figure 10.29). If glochidia come in contact with a passing fish, they hitchhike a ride as parasites in the fish's gills for the next 20 to 70 days before sinking to the bottom to become sedentary adults.

Freshwater clams were once abundant and diverse in streams throughout the eastern United States, but they are now easily the most jeopardized group of animals in the country. Of the more than 300 species once present, 16 are extinct, 70 are listed as threatened or endangered, and many more may be listed soon. A combination of causes is responsible, of which a decline in water quality is among the most important. Pollution and sedimentation from mining, industry, and agriculture are among the culprits. Habitat destruction by altering natural water courses and damming is an important factor. Poaching to supply the Japanese cultured pearl industry is partially to blame (see note on p. 181). In addition to everything else, the prolific zebra mussels (see next note) attach in great numbers to native clams, exhausting food supplies (phytoplankton) in the surrounding water.

Zebra mussels, *Dreissena polymorpha*, are a recent and potentially disastrous biological introduction into North America. They were apparently picked up as veligers with ballast water by one or more ships in freshwater ports in northern Europe and then expelled between Lake Huron and Lake Erie in 1986. This 4 cm bivalve spread throughout the Great Lakes by 1990, and by 1994 it was as far south on the Mississippi as New Orleans, as far north as Duluth, Minnesota, and as far east as the Hudson River in New York. It attaches to any firm surface and filter feeds on phytoplankton. Large numbers accumulate rapidly. They foul water intake pipes of municipal and industrial plants, impede intake of water for municipal supplies, and have far-reaching effects on the ecosystem (see preceding note). Zebra mussels will cost billions of dollars to control.

Class Cephalopoda

Cephalopoda (sef-a-lop'o-da) are the most complex of the molluscs—in fact, in some respects they are the most complex of all invertebrates. They include squids, octopuses, nautiluses, and cuttlefishes. All are marine, and all are active predators.

Cephalopods (Gr. *kephalē*, head, + *pous, podos,* foot) have an odd body plan (figure 10.30) that develops as the embryonic head and foot become indistinguishable. The ring around the mouth bearing the arms and tentacles is derived from the anterior margin of the head, whereas the circle of arms or tentacles itself is derived from the anterior margin of the foot. The foot also takes the form of a **funnel** for expelling water from the mantle cavity.

Cephalopods range in size from 2 to 3 cm up to the giant squid, *Architeuthis,* which is the largest invertebrate known. The squid *Loligo* (L., cuttlefish) is about 30 cm long (figure 10.30A).

Cephalopods are predaceous, feeding chiefly on small fishes, molluscs, crustaceans, and worms. Their arms, which are used in food capture and handling, have a complex musculature and are capable of delicately controlled movements. They are highly mobile and swiftly seize prey and bring it to the mouth. Octopods and cuttlefishes have salivary glands that secrete a venom for immobilizing prey. Strong, beaklike **jaws** grasp prey, and the **radula** tears off pieces of flesh (figure 10.30B). Digestion is extracellular and occurs in the stomach and cecum.

The enormous giant squid, *Architeuthis,* is very poorly known because no one has ever been able to study a living specimen. The anatomy has been studied from stranded animals, from those captured in nets of fishermen, and from specimens found in stomachs of sperm whales. The mantle length is 5 to 6 m, and the head is up to one meter long. They have the largest eyes in the animal kingdom: up to 25 cm (10 inches) in diameter. They apparently eat fish and other squids, and they are an important food item for sperm whales. They are thought to live on or near the ocean bottom at a depth of 1000 m, but some have been seen swimming at the surface.

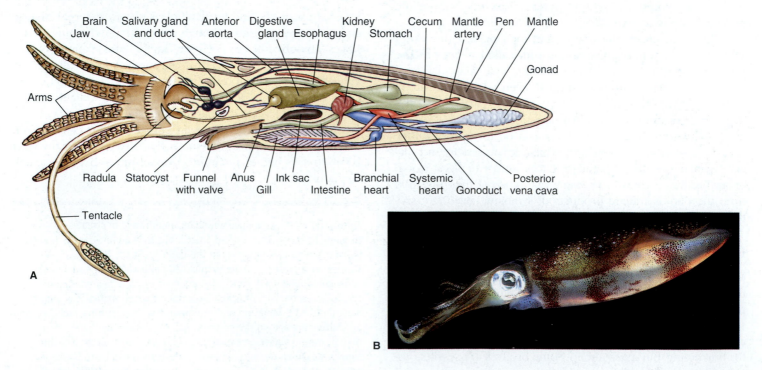

figure 10.30

A, Lateral view of squid anatomy, with the left half of the mantle removed. **B,** Squid *Sepioteuthis lessoniana.*

Form and Function

Shell

Fossil records of cephalopods go back to Cambrian times. The earliest shells were straight cones. Cephalopods without shells or with internal shells (such as octopuses and squids) probably evolved from a straight-shelled ancestor. Other fossil cephalopods had curved or coiled shells as in ammonoids and nautiloids. The only remaining members of the once-flourishing nautiloids are the modern *Nautilus* species (Gr. *nautilos*, sailor) (figure 10.31). Ammonoids were widely prevalent in the Mesozoic era but became extinct by the end of the Cretaceous period. They had chambered shells analogous to nautiloids, but the septa were more complex. Reasons for their extinction remain a mystery. Present evidence suggests that they were gone before the asteroid bombardment at the end of the Cretaceous period (p. 31).

Although early nautiloid and ammonoid shells were heavy, they were made buoyant by a series of **gas chambers,** as is that of *Nautilus* (figure 10.31B), enabling the animal to swim while carrying its shell. The shell of *Nautilus,* although coiled, is quite different from that of a gastropod. Transverse septa divide the shell into internal chambers (figure 10.31B). The living animal inhabits only the last chamber. As it grows, it moves forward, secreting behind it a new septum. The chambers are connected by a cord of living tissue called a **siphuncle,** which extends from the visceral mass. Cuttlefishes also have a small coiled or curved shell, but it is entirely enclosed by the mantle (figure 10.32). In squids most of the shell has disappeared, leaving only a thin, flexible strip called a **pen,** which the mantle encloses. In *Octopus* (Gr. *oktos,* eight, + *pous, podos,* foot) the shell is absent.

After *Nautilus* secretes a new septum, the new chamber is filled with fluid similar in ionic composition to that of the *Nautilus'* blood (and of seawater). Fluid removal involves the active secretion of ions into tiny intercellular spaces in the siphuncular epithelium, so that a very high local osmotic pressure is produced, and the water is drawn from the chamber by osmosis. The gas in the chamber is only the respiratory gas from the siphuncle tissue that diffuses into the chamber as the fluid is removed. Thus the gas pressure in the chamber is 1 atmosphere or less because it is in equilibrium with the gases dissolved in the seawater surrounding the *Nautilus*. These dissolved gases are in turn in equilibrium with air at the surface of the sea, despite the fact that the *Nautilus* may be swimming at 400 m beneath the surface. That the shell can withstand implosion by the surrounding 41 atmospheres (about 600 pounds per square inch), and that the siphuncle can remove water against this pressure are marvelous feats of natural engineering!

Body and Mantle

In *Nautilus* the head with its 60 to 90 or more tentacles can be extruded from the opening of the body compartment of the shell (see figure 10.31). Its tentacles have no suckers but adhere to prey by secretions. The tentacles search for, sense, and grasp food. Beneath the head is the funnel. The shell shelters the mantle, mantle cavity, and visceral mass. Two pairs of gills are located in the mantle cavity.

Cephalopods other than nautiloids have only one pair of gills. Octopods have 8 arms with suckers; squids and cuttlefishes (decapods) have 10 arms: 8 arms with suckers and a pair of long retractile tentacles. The thick mantle covering the trunk fits loosely at the neck region allowing intake of water into the mantle cavity. When the mantle edges contract closely

A

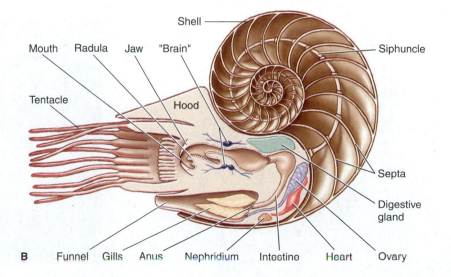

B

figure 10.31

Nautilus, a cephalopod. **A,** Live *Nautilus,* feeding on a fish. **B,** Longitudinal section, showing gas-filled chambers of shell, and diagram of body structure.

about the neck, water is expelled through the funnel. The water current thus created provides oxygenation for the gills in the mantle cavity, jet power for locomotion, and a means of carrying wastes and sexual products away from the body.

The active habits of cephalopods are reflected in their internal anatomy, particularly their respiratory and circulatory systems. Ciliary propulsion would not circulate enough water over the gills for an active animal, so cephalopods ventilate their gills by muscular action of the mantle wall. They have a closed circulatory system with a network of vessels, and blood flows through the gills via capillaries. The plan of circulation of the ancestral mollusc places the entire systemic circulation before blood reaches the gills, which means that the blood pressure at the gills is too low for rapid gaseous exchange. This functional problem has been solved by the evolution of **accessory** or **branchial (gill) hearts** (see figure 10.30A).

Cephalopods have well-developed nervous systems. They have the most complex brain among invertebrates (see figure 10.30A). Except for *Nautilus,* which has relatively simple eyes, cephalopods have elaborate eyes with cornea, lens, chambers, and retina (figure 10.32)—similar to the camera-type eye of vertebrates.

Color Changes

There are special pigment cells called **chromatophores** in the skin of most cephalopods, which by expanding and con-

tracting produce color changes. They are controlled by the nervous system and perhaps by hormones. Some color changes are protective to match background hues; most are behavioral and are associated with alarm or courtship. Many deep-sea squids are bioluminescent.

Ink Production

Most cephalopods other than nautiloids have an ink sac that empties into the rectum. The sac contains an ink gland that secretes a dark fluid containing the pigment melanin. When the animal is alarmed, it releases a cloud of ink through the anus to form a "smokescreen" to confuse an enemy.

Locomotion

Most cephalopods swim by forcefully expelling water from the mantle cavity through a ventral **funnel**—a sort of jet-propulsion method. The funnel is mobile and can be pointed forward or backward to control direction; the force of water expulsion determines speed.

Squids and cuttlefishes are excellent swimmers. The squid body is streamlined and built for speed (figure 10.30). Cuttlefishes swim more slowly (figure 10.33). Both squids and cuttlefishes have lateral fins that can serve as stabilizers, but they are held close to the body for rapid swimming. The gas-filled chambers of *Nautilus* keep the shell upright. Although not as fast as squids, they move surprisingly well.

Octopus has a rather globular body and no fins (see figure 10.1E). Octopuses can swim backwards by spurting jets of water from their funnel, but they are better adapted to crawling over rocks and coral, using the suction discs on their arms to pull or to anchor themselves. Some deep-water octopods have fins and arms webbed like an umbrella; they swim in a medusa-like fashion.

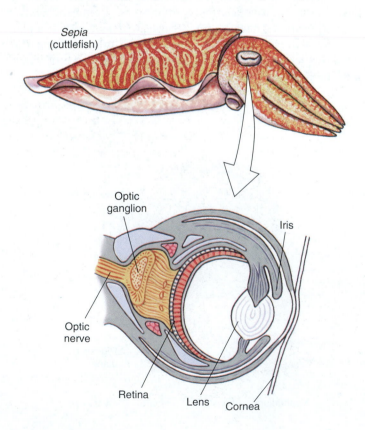

Sepia (cuttlefish)

Optic ganglion
Iris
Optic nerve
Retina
Lens
Cornea

figure 10.32

Eye of a cuttlefish (*Sepia*). The structure of cephalopod eyes shows a high degree of convergent evolution with eyes of vertebrates.

figure 10.33

Cuttlefish, *Sepia latimanus,* has an internal shell familiar to keepers of caged birds as "cuttlebone."

Reproduction

Sexes are separate in cephalopods. Before copulation males often undergo color displays, apparently directed against rival males and for courtship of females. In the male seminal vesicle spermatozoa are encased in spermatophores and stored in a sac that opens into the mantle cavity. During copulation one arm of an adult male plucks a spermatophore from his own mantle cavity and inserts it into the mantle cavity of a female near the oviduct opening (figure 10.34). Eggs are fertilized as they leave the oviduct and are usually attached to stones or other objects to develop. Some octopods tend their eggs.

Phylogeny and Adaptive Radiation

The first molluscs probably arose during Precambrian times because fossils attributed to Mollusca have been found in geological strata as old as the early Cambrian period. A "hypothetical ancestral mollusc" (see figure 10.2) was long viewed as representing the original mollusc ancestor, but neither a solid shell nor a broad, crawling foot are now considered universal characters for Mollusca. The primitive ancestral mollusc was probably a wormlike organism with a ventral gliding surface and a dorsal mantle with a chitinous cuticle and calcareous scales (figure 10.35). It had a posterior mantle cavity with two gills, a radula, a ladderlike nervous system, and an open circulatory system with a heart. Among living molluscs the primitive condition is most nearly approached by caudofoveates, although the foot is reduced to an oral shield in members of this class. Assuming an ancestral form similar to modern caudofoveates, solenogasters have lost the gills, and the foot is represented by the ventral groove. Both aplacophoran classes probably branched from primitive ancestors before the development of a solid shell, a distinct head with sensory organs, and a ventral muscularized foot. Polyplacophorans probably also branched early from the main lines of molluscan evolution before the veliger was established as a second larval stage.

Some workers consider the shells of polyplacophorans not homologous to shells of other molluscs because they differ structurally and developmentally. Polyplacophora and the remaining classes, collectively called Conchifera, are sister groups (figure 10.36).

Cladistic analysis suggests that Gastropoda and Cephalopoda form the sister group to Monoplacophora (see figure 10.36). Both gastropods and cephalopods have a greatly expanded visceral mass. The mantle cavity was brought toward the head by torsion in gastropods, but in cephalopods the mantle cavity was extended ventrally. Evolution of a chambered shell in cephalopods was a very important contribution to their freedom from the substratum and their ability to swim. Elaboration of their respiratory, circulatory, and nervous systems is correlated with their predatory and swimming habits.

Scaphopods and bivalves have an expanded mantle cavity that essentially envelops the body. Adaptations for burrowing characterize this clade: spatulate foot and reduction of the head and sense organs.

Most diversity among molluscs is related to their adaptation to different habitats and modes of life and to a wide variety of feeding methods, ranging from sedentary filter feeding to active predation. There are many adaptations for food gathering within the phylum and an enormous variety in radular structure and function, particularly among gastropods.

The versatile glandular mantle has probably shown more plastic adaptive capacity than any other molluscan structure. Besides secreting the shell and forming the mantle cavity, it is variously modified into gills, lungs, siphons, and apertures, and it sometimes functions in locomotion, in feeding processes, or in a sensory capacity. The shell, too, has undergone a variety of evolutionary adaptations.

To which other phyla are the molluscs closely related? On the basis of such shared features as spiral cleavage, mesoderm from the 4d blastomere, and trochophore larva, most zoologists have accepted Mollusca as protostomes, allied with annelids and arthropods. Opinions differ, however, as to whether

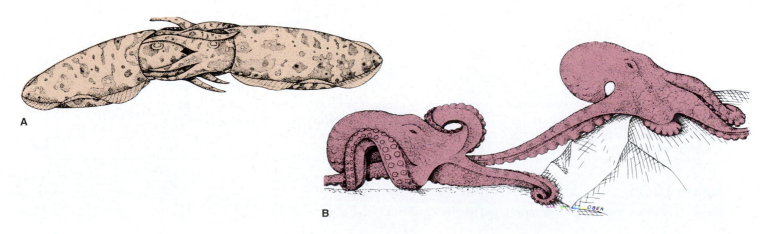

figure 10.34

Copulation in cephalopods. **A,** Mating cuttlefishes. **B,** Male octopus uses modified arm to deposit spermatophores in female mantle cavity to fertilize her eggs. Octopuses often tend their eggs during development.

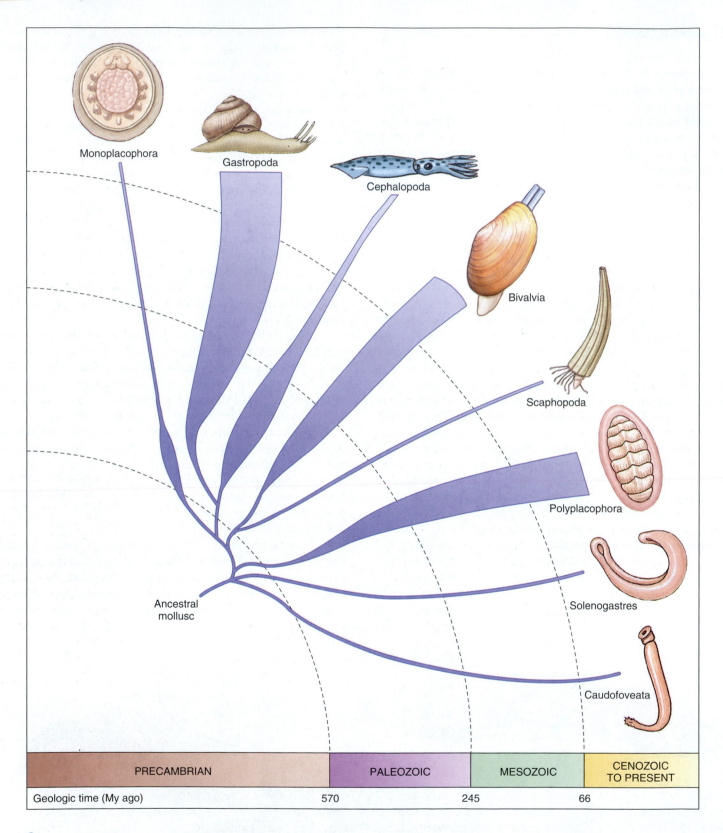

figure 10.35

Classes of Mollusca, showing their derivations and relative abundance.

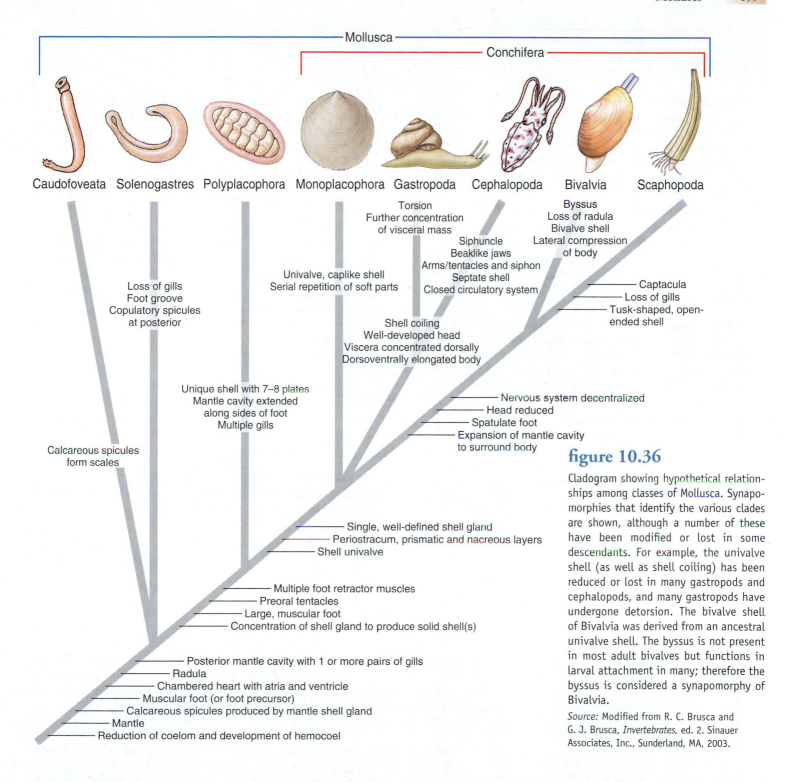

figure 10.36

Cladogram showing hypothetical relationships among classes of Mollusca. Synapomorphies that identify the various clades are shown, although a number of these have been modified or lost in some descendants. For example, the univalve shell (as well as shell coiling) has been reduced or lost in many gastropods and cephalopods, and many gastropods have undergone detorsion. The bivalve shell of Bivalvia was derived from an ancestral univalve shell. The byssus is not present in most adult bivalves but functions in larval attachment in many; therefore the byssus is considered a synapomorphy of Bivalvia.

Source: Modified from R. C. Brusca and G. J. Brusca, *Invertebrates,* ed. 2. Sinauer Associates, Inc., Sunderland, MA, 2003.

molluscs were derived from a flatwormlike ancestor independent of annelids, share an ancestor with annelids after the advent of the coelom, or share a segmented common ancestor with annelids. This last hypothesis is strengthened if the repeated body parts present in *Neopilina* (class Monoplacophora) and some chitons can be considered evidence of metamerism. However, recent morphological and developmental studies strongly suggest that replication of body parts is pseudometamerism and was not inherited from a metameric ancestor. A reasonable hypothesis is that molluscs branched

from the annelid line after the coelom arose but before the advent of metamerism. Some analyses suggest that molluscs and annelids are more closely related to each other than either is to arthropods. This contention is strengthened by molecular evidence that places annelids and molluscs in Lophotrochozoa and arthropods in Ecdysozoa (p. 163). However, this placement means that metamerism arose at least twice independently within protostomes, or that genes for segmentation were present in basal bilaterians and have been suppressed several times (see discussion on p. 216).

classification of Phylum Mollusca

Class Caudofoveata (kaw′do-fo-ve-at′a) (L. *cauda,* tail, + *fovea,* small pit): **caudofoveates.** Wormlike; shell, head, and excretory organs absent; radula usually present; mantle with chitinous cuticle and calcareous scales; oral pedal shield near anterior mouth; mantle cavity at posterior end with pair of gills; sexes separate; often united with solenogasters in class Aplacophora. Examples: *Chaetoderma, Limifossor.*

Class Solenogastres (so-len′o-gas′trez) (Gr. *solēn,* pipe, + *gast ̄er,* stomach): **solenogasters.** Wormlike; shell, head, and excretory organs absent; radula usually present; mantle usually covered with scales or spicules; mantle cavity posterior, without true gills, but sometimes with secondary respiratory structures; foot represented by long, narrow, ventral pedal groove; hermaphroditic. Example: *Neomenia.*

Class Monoplacophora (mon′o-pla-kof′o-ra) (Gr. *monos,* one, + *plax,* plate, + *phora,* bearing): **monoplacophorans.** Body bilaterally symmetrical with a broad flat foot; a single limpetlike shell; mantle cavity with five or six pairs of gills; large coelomic cavities; radula present; six pairs of nephridia, two of which are gonoducts; separate sexes. Example: *Neopilina* (see figure 10.7).

Class Polyplacophora (pol′y-pla-kof′o-ra) (Gr. *polys,* many, several, + *plax,* plate, + *phora,* bearing): **chitons.** Elongated, dorsoventrally flattened body with reduced head; bilaterally symmetrical; radula present; shell of seven or eight dorsal plates; foot broad and flat; gills multiple, along sides of body between foot and mantle edge; sexes usually separate, with a trochophore but no veliger larva. Examples: *Mopalia* (see figure 10.8), *Chaetopleura.*

Class Scaphopoda (ska-fop′o-da) (Gr. *skaphē,* trough, boat, + *pous, podos,* foot): **tusk shells.** Body enclosed in a one-piece tubular shell open at both ends; conical foot; mouth with radula and tentacles; head absent; mantle for respiration; sexes separate; trochophore larva. Example: *Dentalium* (see figure 10.10).

Class Gastropoda (gas-trop′o-da) (Gr. *gastēr,* belly, + *pous, podos,* foot): **snails and relatives.** Body asymmetrical; usually in a coiled shell (shell uncoiled or absent in some); head well developed, with radula; foot large and flat; dioecious or monoecious, some with trochophore, typically with veliger, some without larva. Examples: *Busycon, Polinices* (see figure 10.14B), *Physa, Helix, Aplysia* (see figure 10.15).

Class Bivalvia (bi-val′ve-a) (L. *bi,* two, + *valva,* folding door, valve) **(Pelecypoda): bivalves.** Body enclosed in a two-lobed mantle; shell of two lateral valves of variable size and form, with dorsal hinge; head greatly reduced but mouth with labial palps; no radula; no cephalic eyes; gills platelike; foot usually wedge shaped; sexes usually separate, typically with trochophore and veliger larvae. Examples: *Mytilus* (see figure 10.27), *Venus, Bankia* (see figure 10.24).

Class Cephalopoda (sef′a-lop′o-da) (Gr. *kephalē,* head, + *pous, podos,* foot): **squids and octopuses.** Shell often reduced or absent; head well developed with eyes and a radula; head with arms or tentacles; foot modified into a funnel; nervous system of well-developed ganglia, centralized to form a brain; sexes separate, with direct development. Examples: *Loligo, Sepioteuthis* (see figure 10.30), *Octopus* (see figure 10.1E), *Sepia* (see figure 10.32), *Nautilus* (see figure 10.31).

Summary

Mollusca is one of the largest and most diverse phyla, its members ranging in size from very small organisms to the largest of invertebrates. Their basic body divisions are the head-foot and the visceral mass, usually covered by a shell. The majority are marine, but some are freshwater, and a few are terrestrial. They occupy a variety of niches; a number are economically important, and a few are medically important as hosts of parasites.

Molluscs are coelomate (have a coelom), although their coelom is limited to the area around the heart. The evolutionary development of a coelom was important because it enabled better organization of visceral organs and, in many of the animals that have it, an efficient hydrostatic skeleton. Reduction of the coelom in molluscs may have been correlated with the evolution of a protective shell. The coelom could not function as a hydrostatic skeleton once a hard shell covered the body.

The mantle and mantle cavity are important characteristics of molluscs. The mantle secretes the shell and overlies a part of the visceral mass to form a cavity housing the gills. The mantle cavity has been modified into a lung in some molluscs. The foot is usually a ventral, solelike, locomotory organ, but it may be variously modified, as in the cephalopods, where it has become a funnel and arms. Most molluscs except bivalves and some solenogasters and gastropods have a radula, which is a protrusible, tonguelike organ with teeth used in feeding. The circulatory system of molluscs is open, with a heart and blood sinuses, except in cephalopods, which have a closed circulatory system. Molluscs usually have a pair of nephridia connecting with the coelom and a complex nervous system with a variety of sense organs. The primitive larva of molluscs is the trochophore, and most marine molluscs have a more derived larva, the veliger.

Classes Caudofoveata and Solenogastres are small groups of wormlike molluscs with calcareous spines or scales, but no shell. Scaphopoda is a slightly larger class with a tubular shell, open at both ends, and the mantle wrapped around the body.

Class Monoplacophora is a tiny, univalve marine group showing pseudometamerism or vestiges of true metamerism. Polyplacophora are more common and diverse marine organisms with shells in the form of a series of seven or eight plates. They are rather sedentary animals with a row of gills along each side of their foot.

Gastropoda is the largest class of molluscs. Gastropods are the only molluscs to exploit terrestrial environments. Their interesting evolutionary history includes torsion, or the twisting of the posterior end to the anterior so that the anus and head are at the same end, and coiling, an elongation and spiraling of the visceral mass. Torsion led to the survival problem of fouling, which is the release of excreta over the head and in front of the gills. This problem was solved in various ways among different gastropods. Among the solutions to fouling were bringing water into one side of the mantle cavity and out the other (many gastropods), some degree of detorsion (opisthobranchs and pulmonates), and conversion of the mantle cavity into a lung (pulmonates).

Class Bivalvia is marine and freshwater. Bivalve shells are divided into two valves joined by a dorsal ligament and held together by an adductor muscle. Most bivalves are suspension feeders, drawing water through their gills by ciliary action.

Members of class Cephalopoda are the most complex molluscs; they are all predators and many can swim rapidly. Their tentacles and arms capture prey by adhesive secretions or by suckers. They swim by forcefully expelling water from their mantle cavity through a funnel, which was derived from the foot.

There is strong embryological and molecular evidence that molluscs share a common ancestor with annelids more recently than either of these phyla do with arthropods or deuterostome phyla, although molluscs are not metameric. Molluscs are usually classified as lophotrochozoan protostomes, as are annelids.

Review Questions

1. How does a coelom develop embryologically? Why was the evolutionary development of a coelom important?
2. Members of phylum Mollusca are extremely diverse, yet the phylum clearly constitutes a monophyletic group. What evidence can you cite in support of this statement?
3. How are molluscs important to humans?
4. Distinguish among the following classes of molluscs: Polyplacophora, Gastropoda, Bivalvia, Cephalopoda.
5. Define the following: radula, odontophore, periostracum, prismatic layer, nacreous layer, trochophore, veliger.
6. Briefly describe the habitat and habits of a typical chiton.
7. Define the following with respect to gastropods: operculum, torsion, fouling, bilateral asymmetry.
8. Torsion in gastropods created a selective disadvantage: fouling. Suggest one or more potential selective advantages that could have offset the disadvantage. How have gastropods evolved to avoid fouling?
9. Distinguish between opisthobranchs and pulmonates.
10. Briefly describe how a typical bivalve feeds and how it burrows.
11. What is the function of the siphuncle of *Nautilus*?
12. Describe how cephalopods swim and eat.
13. Cephalopods are actively swimming predators, but they evolved from a slow-moving, probably grazing ancestor. Describe evolutionary modifications of the ancestral plan that make the cephalopod lifestyle possible.
14. To what other major invertebrate groups are molluscs related, and what is the nature of the evidence for the relationship?
15. Briefly describe the ancestral condition of the shell, radula, foot, mantle cavity and gills, circulatory system and head based on characteristics of modern caudofoveates, and explain how typical members of the other classes differ from this plan.

Selected References

See also general references on page 415.

Abbott, R. T., and P. A. Morris (R. T. Peterson [editor]). 2001. A field guide to shells: Atlantic and Gulf Coasts and the West Indies. Boston, Houghton Mifflin Company. *A popular handbook for shell collectors.*

Barinaga, M. 1990. Science digests the secrets of voracious killer snails. Science **249:**250–251. *Describes research on the toxins produced by cone snails.*

Colgan, D. J., W. F. Ponder, and P. E. Eggler. 2000. Gastropod evolutionary rates and phylogenetic relationships assessed using partial 28S rDNA and histone H3 sequences. Zool. Sci. **29:**29–63. *Phylogenies constructed using molecular data do not agree well with hypotheses about gastropod evolution developed from morphological studies.*

Fleischman, J. 1997. Mass extinctions come to Ohio. Discover **18**(5):84–90. *Of the 300 species of freshwater bivalves in the Mississippi River Basin, 161 are extinct or endangered.*

Gosline, J. M., and M. D. DeMont. 1985. Jet-propelled swimming in squids. Sci. Amer. **252:**96–103 (Jan.). *Mechanics of swimming in squid are analyzed; elasticity of collagen in mantle increases efficiency.*

Hanlon, R. T., and J. B. Messenger. 1996. Cephalopod behaviour. Cambridge University Press, 232 pp. *Intended for nonspecialists and specialists.*

Holloway, M. 2000. Cuttlefish say it with skin. Nat. Hist. **109**(3):70–76. *Cuttlefish and other cephalopods can change texture and color of their skin with astonishing speed. Fifty-four components of cuttlefish "vocabulary" have been described, including color display, skin texture, and a variety of arm and fin signals.*

Ponder, W. F., and D. R. Lindberg. 1997. Towards a phylogeny of gastropod molluscs: an analysis using morphological characters. Zool. J. Linn. Soc. **119:** 83–265. *Analysis of 117 morphological characters from 40 taxa suggested several new gastropod subgroups.*

Roper, C. R. E., and K. J. Boss. 1982. The giant squid. Sci. Am. **246:**96–105 (April). *Many mysteries remain about the deep-sea squid, Architeuthis, because it has never* been studied alive. It can reach a weight of 1000 pounds and a length of 18 m, and its eyes are as large as automobile headlights.

Ross, J. 1994. An aquatic invader is running amok in U.S. waterways. Smithsonian 24(11):40–50 (Feb.). *A small bivalve, the zebra mussel, apparently introduced into the Great Lakes with ballast water from ships, is clogging intake pipes and municipal water supplies. It will take billions of dollars to control.*

Ward, P. D. 1998. Coils of time. Discover 19(3):100–106. *Present Nautilus has apparently existed essentially unchanged for 100 million years, and all other species known were derived from it, including king nautilus (Allonautilus), a recent derivation.*

Ward, P., L. Greenwald, and O. E. Greenwald, 1980. The buoyancy of the chambered nautilus. Sci Am. 243:190–203 (Oct.). *Reviews discoveries on how the nautilus removes the* water from a chamber after secreting a new septum.

Yonge, C. M. 1975. Giant clams. Sci. Am. 232:96–105 (Apr.). *Details the fascinating morphological adaptations of this bivalve for life with its mutualistic zooxanthellae.*

Zorpette, G. 1996. Mussel mayhem, continued. Sci. Am. 275:22–23 (Aug.). *Some benefits, though dubious, of the zebra mussel invasion have been described, but these are outweighed by the problems created.*

Custom Website

The *Animal Diversity* Online Learning Center is a great place to check your understanding of chapter material. Visit www.mhhe.com/hickmanad4e for access to key terms, quizzes, and more! Further enhance your knowledge with Web links to chapter-related material.

Explore live links for these topics:

Classification and Phylogeny of Animals
Phylum Mollusca

Primitive Classes of Molluscs
Class Polyplacophora
Class Gastropoda
Class Bivalvia
Class Cephalopoda

Segmented Worms
Annelids, Including Pogonophorans

Dividing the Body

Although a spacious, fluid-filled coelom provided an efficient hydrostatic skeleton for burrowing, precise control of body movements was not possible in the earliest coelomates. The force of muscle contraction in one area was carried throughout the body by fluid in an undivided coelom. In contrast, there were distinct coelomic compartments in each body segment of ancestral annelids. Each fluid-filled compartment could respond individually to local muscle contraction—one segment being long and thin, and another being short and round. The annelid body illustrates *metamerism,* it is composed of serially repeated units called segments or metameres. Each segment is separated from its neighbors by partitions called septa, and each contains components of most organ systems, such as circulatory, nervous, and excretory systems.

The evolutionary advent of metamerism was highly significant because it made possible development of much greater complexity in structure and function. Metamerism not only increased efficiency of burrowing, it made possible independent and separate movements by separate segments. Fine control of movements led, in turn, to evolution of a more sophisticated nervous system. Moreover, repetition of body parts gave the organisms a built-in redundancy, as in some human-made systems. This redundancy provided a safety factor: if one segment should fail, others could still function. Thus an injury to one part would not necessarily be fatal.

The evolutionary potential of a metameric body plan is amply demonstrated by the large and diverse phylum Arthropoda, where it may have arisen independently. Metamerism also arose independently in the deuterostome line, which includes the numerous and adaptively diverse vertebrates.

Cloeia sp., a polychaete.

Annelida (an-nel'i-da) (L. *annellus,* little ring, + *-ida,* suffix) consists of the segmented worms. It is a large phylum, numbering approximately 15,000 species, the most familiar of which are earthworms and freshwater worms (class Oligochaeta) and leeches (class Hirudinea). However, the less familiar marine worms (class Polychaeta) form approximately two-thirds of the phylum. Some polychaetes are strange, even grotesque, whereas others are graceful and beautiful. They include clam worms, plumed worms, parchment worms, scaleworms, lugworms, and many others. Beardworms, or pogonophorans, have recently been added to this class.

Annelids are true coelomates and belong to the protostome branch, with spiral and mosaic cleavage. They are a highly developed group in which the nervous system is more centralized and the circulatory system more complex than those of phyla we have studied so far.

Annelids are sometimes called "bristle worms" because, except for leeches, most annelids bear tiny chitinous bristles called **setae** (L. *seta,* hair or bristle). Short needlelike setae help to anchor segments during locomotion to prevent backward slipping; long, hairlike setae aid aquatic forms in swimming. Because many annelids are either burrowers or live in secreted tubes, stiff setae also aid in preventing a worm from being pulled or washed out of its home. Robins know from experience how effective an earthworm's setae are.

Ecological Relationships

Annelids are worldwide in distribution, occurring in the sea, fresh water, and terrestrial soil. Some marine annelids live quietly in tubes or burrow into bottom mud or sand. Some feed on organic matter in the mud through which they burrow; others feed on suspended particles with elaborate ciliary or mucous devices for trapping food. Many are predators, either pelagic or hiding in crevices of coral or rock except when hunting. Freshwater annelids burrow in mud or sand, live among vegetation, or swim freely. The most familiar annelids are terrestrial earthworms, which move through moist soil. Some leeches are bloodsuckers, and others are carnivores; most of them live in fresh water.

Economic Importance

Much economic importance of annelids is indirect, deriving from their ecological roles. Many are members of grazing food chains or detritus food chains, serving as prey for other organisms of more direct interest to humans, such as fishes. Consequently, there is a thriving market for some polychaetes and oligochaetes as fish bait. Burrows of earthworms increase drainage and aeration of soils, and migrations of worms help mix the soil and distribute organic matter to deeper layers

Position in the Animal Kingdom

1. Annelids belong to the **protostome** branch of the animal kingdom and have **spiral cleavage** and **mosaic development,** characters shared with and indicating relationship to molluscs. Marine annelids and many molluscs also share a trochophore larval form.
2. Annelids as a group show a metamerism with comparatively few differences between different segments.
3. Characters shared with arthropods include an outer secreted cuticle and similar nervous system.

Biological Contributions

1. **Metamerism** represents the greatest innovation in this phylum.
2. A true coelomic cavity reaches a high stage of development in this group.
3. Specialization of the head region into differentiated organs, such as tentacles, palps, and eyespots of polychaetes, is carried further in some annelids than in other invertebrates so far considered.
4. There are modifications of the **nervous system,** with cerebral ganglia (brain), two closely fused ventral nerve cords with giant fibers running the length of the body, and various ganglia with their lateral branches.
5. The circulatory system is much more complex than any we have so far considered. It is a closed system with muscular blood vessels and aortic arches ("hearts") for propelling blood.
6. The appearance of fleshy **parapodia,** with their respiratory and locomotor functions, parallels the paired appendages and specialized gills found in arthropods.
7. Well-developed **nephridia** in most segments remove waste from the blood and coelom.
8. Annelids are the most highly organized animals capable of complete regeneration. However, this ability varies greatly within the group.

characteristics
of Phylum Annelida

1. Body **metameric or segmented;** symmetry bilateral
2. Body wall with outer circular and inner longitudinal muscle layers; outer transparent moist cuticle secreted by epithelium
3. **Epidermal chitinous setae,** often present on fleshy appendages called **parapodia;** setae absent in leeches
4. **Coelom (schizocoel) well developed** and divided by septa, except in leeches; coelomic fluid supplies turgidity and functions as a hydrostatic skeleton
5. Blood system closed and **segmentally arranged;** respiratory pigments (hemoglobin, hemerythrin, or chlorocruorin) often present; amebocytes in blood plasma
6. Digestive system complete and not metamerically arranged
7. Respiratory gas exchange through skin, gills, or parapodia
8. Excretory system typically a pair of nephridia for each segment
9. Nervous system with a double ventral nerve cord and a pair of ganglia with lateral nerves in each metamere; brain, a pair of dorsal cerebral ganglia with connectives to cord
10. Sensory system of tactile organs, taste buds, statocysts (in some), photoreceptor cells, and eyes with lenses (in some)
11. Hermaphroditic or separate sexes; larvae, if present, are trochophore type; asexual reproduction by budding in some; spiral and mosaic cleavage; mesoderm from 4d blastomere (see figure 4.11)

(p. 212). Some marine annelids that burrow serve an analogous role; lugworms (*Arenicola,* see figure 11.10) are sometimes called "earthworms of the sea." New medical uses for leeches (p. 215) have revived the market in blood-sucking leeches and established "leech farms" where these organisms are raised in captivity.

Body Plan

An annelid body typically has a two-part head, composed of **prostomium** and a **peristomium,** a series of segments (sometimes called metameres), and a terminal **pygidium** bearing an anus (figure 11.1). Neither the head nor the pygidium is

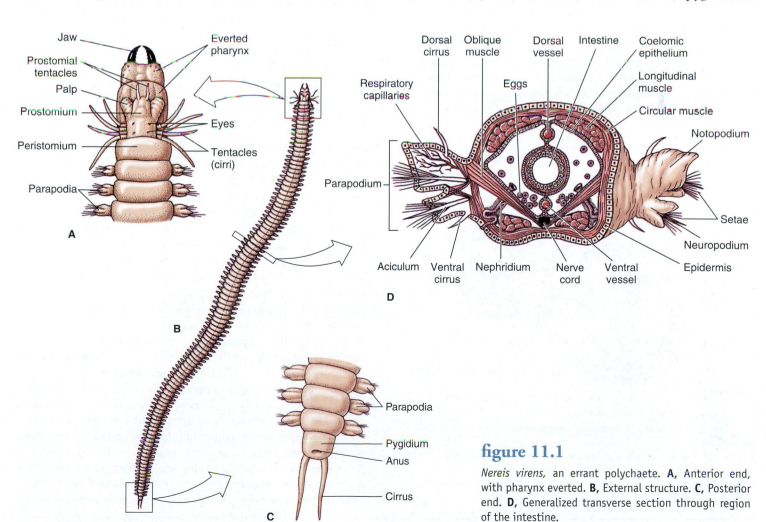

figure 11.1

Nereis virens, an errant polychaete. **A,** Anterior end, with pharynx everted. **B,** External structure. **C,** Posterior end. **D,** Generalized transverse section through region of the intestine.

considered a true segment. During growth, new segments form just in front of the pygidium; thus the oldest segments are at the anterior end and the youngest segments are at the posterior end. Each segment typically contains circulatory, respiratory, nervous, and excretory structures, as well as a coelom.

In most annelids the coelom develops embryonically as a split in the mesoderm on each side of the gut (schizocoel), forming a pair of coelomic compartments in each segment. Each compartment is surrounded by peritoneum (a layer of mesodermal epithelium), which lines the body wall, forms dorsal and ventral mesenteries (double-membrane partitions that support the gut), and covers all organs (figure 11.2). Where peritonea of adjacent segments meet, septa are formed. The body wall surrounding the peritoneum and coelom contains strong circular and longitudinal muscles adapted for swimming, crawling, and burrowing.

Except in leeches, the coelom is filled with fluid and serves as a hydrostatic skeleton. Because the volume of fluid in a coelomic compartment is essentially constant, contraction of longitudinal body-wall muscles causes a segment to shorten and to become larger in diameter, whereas contraction of circular muscles causes it to lengthen and to become thinner. The presence of septa means that widening or elongation occurs in restricted areas; crawling motions are effected by *alternating* waves of contraction by longitudinal and circular muscles passing down the body (peristaltic contractions). Some segments in which longitudinal muscles contract, widen and anchor themselves against burrow walls or other substrata. Other segments, in which circular muscles contract, elongate and stretch forward. Forces powerful enough for burrowing as well as locomotion can thus be generated. Swimming forms use undulatory rather than peristaltic movements in locomotion.

An annelid body has a thin, outer layer of nonchitinous cuticle surrounding the epidermis (figure 11.2). Paired epidermal setae (figure 11.2) are considered ancestral for annelids, although they have been reduced or lost in some. The annelid digestive system is not segmented: the gut runs the length of body perforating each septum (figure 11.2). Longitudinal dorsal and ventral blood vessels follow the same path, as does a ventral nerve cord.

Annelids are divided among three classes: Polychaeta, Oligochaeta, and Hirudinea. In this edition we consider pogonophorans members of class Polychaeta in light of recent morphological and molecular analyses. Polychaeta is a paraphyletic class because ancestors of oligochaetes and hirudineans (leeches) arose from within the polychaetes. Oligochaetes and leeches together form a monophyletic group called Clitellata, characterized by the presence of a reproductive structure called the clitellum. The clitellum is visible in earthworms as a distinctive fat band present about one-third of the way down the body. The clitellum in leeches is visible only in the reproductive season.

Class Polychaeta

Polychaetes (Gr. *polys,* many, + *chaitē,* long hair) are the largest class of annelids, with more than 10,000 described species, mostly marine. Although the majority are from 5 to 10 cm long, some are less than a millimeter, and others may be as long as 3 m. Some are brightly colored in reds and greens; others are dull or iridescent. Some are picturesque, such as "featherduster" worms (figure 11.3).

Polychaetes live under rocks, in coral crevices, or in abandoned shells, or they burrow into mud or sand; some build their own tubes on submerged objects or in bottom material; some adopt tubes or homes of other animals; some are pelagic, forming part of the planktonic population. They are extremely abundant in some areas; for example, a square meter of mudflat may contain thousands of polychaetes. They play a significant part in marine food chains because they are eaten by fish, crustaceans, hydroids, and many others.

Polychaetes differ from other annelids in having a well-differentiated head with specialized sense organs; paired, paddlelike appendages (parapodia) on most segments; and no clitellum (see p. 205, figure 11.1). As their name implies, they have many setae, usually arranged in bundles on the parapodia. They show a pronounced differentiation of some body segments and a specialization of sensory organs practically unknown among clitellates (oligochaetes and leeches).

A polychaete typically has a prostomium, which may or may not be retractile and which often bears eyes, antennae, and sensory palps (figure 11.1). The peristomium surrounds the mouth and may bear setae, palps, or, in predatory forms, chitinous jaws. Suspension feeders may bear a tentacular crown that may be opened like a fan or withdrawn into the tube (see figure 11.3).

The trunk is segmented and most segments bear parapodia, which may have lobes, cirri, setae, and other structures on them (figure 11.1D). Parapodia are composed of two main parts—a dorsal notopodium and ventral neuropodium—either of which may be prominent or reduced in a given species.

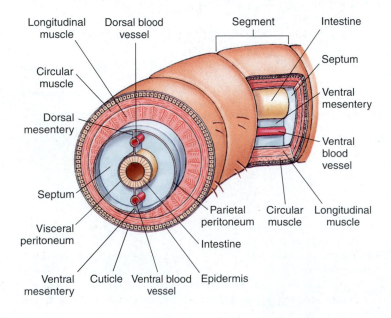

Longitudinal muscle — Dorsal blood vessel — Segment — Intestine — Septum — Ventral mesentery — Circular muscle — Dorsal mesentery — Ventral blood vessel — Septum — Parietal peritoneum — Circular muscle — Longitudinal muscle — Visceral peritoneum — Intestine — Ventral mesentery — Cuticle — Ventral blood vessel — Epidermis

figure 11.2

Annelid body plan.

A

B

figure 11.3

Tube-dwelling sedentary polychaetes. **A,** Christmas-tree worm, *Spirobranchus giganteus,* lives in a calcareous tube. On its head are two whorls of modified tentacles (radioles) used to collect suspended food particles from the surrounding water. Notice the finely branched filters visible on the edge of one radiole. **B,** Sabellid polychaetes, *Bispira brunnea,* live in leathery tubes.

Parapodia are used in crawling, swimming, or anchoring in tubes. They usually serve as the chief respiratory organs, although some polychaetes also may have gills. *Amphitrite,* for example, has three pairs of branched gills and long extensible tentacles (figure 11.4). *Arenicola,* the lugworm (see figure 11.10), which burrows through sand leaving characteristic castings at the entrance to its burrow, has paired gills on certain segments.

Sense organs are more highly developed in polychaetes than in oligochaetes and include eyes and statocysts. Eyes, when present, may range from simple eyespots to well-developed organs. Usually eyes consist of retinal cups, with rodlike photoreceptor cells lining the cup wall and directed toward the lumen of the cup.

In contrast to clitellates, polychaetes have no permanent sex organs, possess no permanent ducts for their sex cells, and usually have separate sexes. Gonads appear as temporary swellings of the peritoneum and shed their gametes into the coelom. Gametes are carried outside through gonoducts, through nephridia, or by rupture of the body wall. Fertilization is external, and development is indirect with a trochophore larva preceding the adult.

Some polychaetes are free-moving pelagic forms, some are active burrowers and crawlers, and some are sedentary, living in tubes or burrows that they rarely (or never) leave. An example of an active, predatory worm is *Nereis* (Greek mythology, a Nereid, or daughter of Nereus, ancient sea god), the clam worm (see figure 11.1). *Nereis* has a muscular, eversible pharynx equipped with jaws that can be thrust out with surprising speed and dexterity for capturing prey. Scale worms (figure 11.5) often live as commensals with other invertebrates, and fireworms (figure 11.6) feed on gorgonians and stony corals.

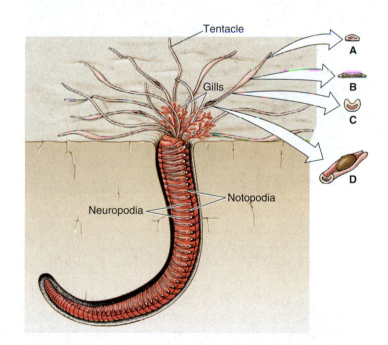

figure 11.4

Amphitrite, which builds its tubes in mud or sand, extends long grooved tentacles out over the mud to pick up bits of organic matter. The smallest particles are moved along food grooves by cilia and larger particles by peristaltic movement. Its plumelike gills are blood red. **A,** Section through exploratory end of a tentacle. **B,** Section through a tentacle in an area adhering to substratum. **C,** Section showing ciliary groove. **D,** Particle being carried toward mouth.

figure 11.5

The scale worm *Hesperonoe adventor* normally lives as a commensal in the tubes of *Urechis* (phylum Echiura, p. 262).

figure 11.6

A fireworm *Hermodice carunculata* feeds on gorgonians and stony corals. Its setae are like tiny glass fibers and serve to repel predators.

Most sedentary tube and burrow dwellers are particle feeders, using ciliary or mucoid methods of obtaining food. The principal food source is plankton and detritus. Some, like *Amphitrite* (Greek mythology, sea goddess) (see figure 11.4), with head peeping out of the mud, send out long extensible tentacles over the surface. Cilia and mucus on the tentacles entrap particles found on the sea bottom and move them toward the mouth (deposit feeding).

Fanworms, or "featherduster" worms, are beautiful tube-worms, fascinating to watch as they emerge from their secreted tubes and unfurl their lovely tentacular crowns to feed. A slight disturbance, sometimes even a passing shadow, causes them to duck quickly into the safety of the homes they have built. Food attracted to the feathery arms, or **radioles,** by ciliary action is trapped in mucus and carried down ciliated

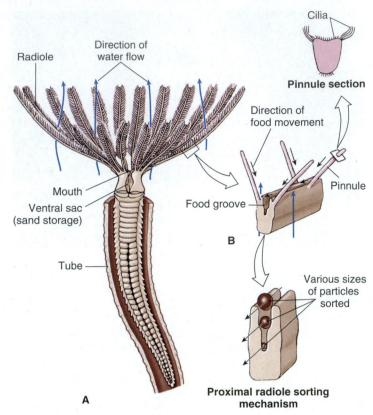

figure 11.7

Sabella, a polychaete suspension feeder. **A,** Anterior view of the crown. Cilia direct small food particles along grooved radioles to the mouth and discard larger particles. Sand grains are directed to storage sacs and later used in tube building. **B,** Distal portion of radiole showing ciliary tracts of pinnules and food grooves.

food grooves to the mouth (figure 11.7). Particles too large for the food grooves are carried along the margins and discarded. Further sorting may occur near the mouth where only small particles of food enter the mouth, and sand grains are stored in a sac to be used later in enlarging the tube.

Some polychaetes live most of the year as sexually unripe animals called *atokes,* but during the breeding season a portion of the body develops into a sexually ripe worm called an *epitoke,* which is swollen with gametes (figure 11.8). For example, palolo worms live in burrows among coral reefs of the South Seas. During the reproductive cycle, their posterior segments become swollen with gametes. During the swarming period, which occurs at the beginning of the last quarter of the October-November moon, these epitokes break off and swim to the surface. Just before sunrise, the sea is literally covered with them, and at sunrise they burst, freeing eggs and sperm for fertilization. The anterior portions of the worms regenerate new posterior sections. A related form swarms in the Atlantic in the third quarter of the June-July moon. Swarming is of great adaptive value because synchronous maturation of all epitokes ensures the maximum number of fertilized eggs. However, it is very hazardous; many types of predators have a feast. In the meantime, the atoke remains safe in its burrow to produce another epitoke at the next cycle!

Some worms, such as *Chaetopterus* (Gr. *chaitē*, hair or mane, + *pteron*, wing), secrete mucous filters through which they pump water to collect edible particles (figure 11.9). Lugworms *Arenicola* (L. *arena*, sand, + *colere*, to inhabit) live in L-shaped burrows in which, by peristaltic movements, they cause water to flow. Sand at the front of their burrows collects filtered food particles. The worms ingest the food-laden sand (figure 11.10).

Tube dwellers secrete many types of tubes. Some are parchmentlike; some are firm, calcareous tubes attached to rocks or other surfaces (see figure 11.3A); and some are simply grains of sand or bits of shell or seaweed cemented together with mucous secretions. Many burrowers in sand and mud flats simply line their burrows with mucus (figure 11.10).

Clade Siboglinidae (Pogonophorans)

Members of former phylum Pogonophora (po-go-nof'e-ra) (Gr. *pōgōn*, beard, + *pherō*, to bear), or beardworms, were entirely unknown before the twentieth century. The first specimens to be described were collected from deep-sea dredgings in 1900 off the coast of Indonesia. They have since been discovered in several seas, including the western Atlantic off the eastern coast of the United States. Some 150 species have been described so far. The taxonomic status of pogonophorans has been controversial for some time, but recent molecular and morphological analyses indicate that these worms are derived

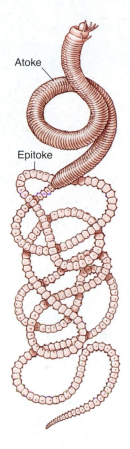

figure 11.8

Eunice viridis, the Samoan palolo worm. Posterior segments make up the epitokal region, consisting of segments packed with gametes. Each segment has one eyespot on the ventral side. Once a year the worms swarm, and the epitokes detach, rise to the surface, and discharge their ripe gametes, leaving the water milky. By the next breeding season, epitokes are regenerated.

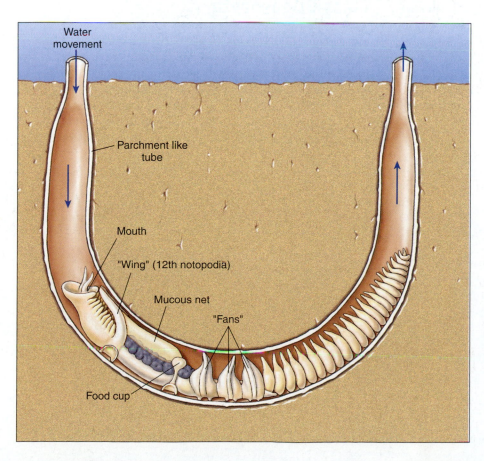

figure 11.9

Chaetopterus, a sedentary polychaete, lives in a U-shaped tube in the sea bottom. It pumps water through a parchmentlike tube (of which one-half has been cut away here) with its three pistonlike fans. The fans beat 60 times per minute to keep water currents moving. Winglike notopodia of the twelfth segment continuously secrete a mucous net that strains out food particles. As the net fills with food, the food cup rolls it into a ball, and when the ball is large enough (about 3 mm), the food cup bends forward and deposits the ball in a ciliated groove to be carried to the mouth and swallowed.

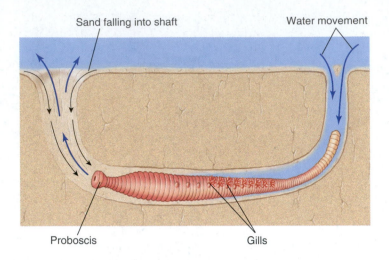

figure 11.10

Arenicola, the lugworm, lives in an L-shaped burrow in intertidal mud-flats. It burrows by successive eversions and retractions of its proboscis. By peristaltic movements it keeps water coming in the open end of the tube and filtering through the sand at the head end. The worm then ingests the food-laden sand.

polychaetes. Segmentation is present only in one body region, but the cuticle and setae of pogonophores are homologous to those of polychaetes. As a clade within Polychaeta, this group is now called Siboglinidae.

Most siboglinids live in bottom ooze on the ocean floor, usually at depths of more than 200 m. Their usual length varies from 5 to 85 cm, with a diameter usually of less than a millimeter. They are sessile and secrete very long chitinous tubes in which they live, probably extending the anterior end only for feeding.

The body has a short forepart, a long, very slender trunk, and a small, segmented opisthosoma (figure 11.11). It is covered with a cuticle and bears setae on the trunk and opisthosoma. A series of coelomic compartments divides the body. The forepart bears from one to many tentacles.

Siboglinids are remarkable in having no mouth or digestive tract, making their mode of nutrition a puzzling matter. They absorb some nutrients dissolved in seawater, such as glucose, amino acids, and fatty acids, through pinnules and microvilli of their tentacles. They apparently derive most of their energy, however, from a mutualistic association with chemoautotrophic bacteria. These bacteria oxidize hydrogen sulfide to provide energy to produce organic compounds from carbon dioxide. An expanded region of the midgut fills with endoderm to form an organ called a **trophosome** that bears the bacteria. All traces of a foregut and hindgut are absent in adults.

Among the most amazing animals found in the deep-water, Pacific rift communities are giant pogonophorans, *Riftia pachyptila*. Much larger than any pogonophorans reported before, they measure up to 1.5 m in length and 4 cm in diameter. (Some authors considered them a separate phylum, **Vestimentifera** but they are now placed within clade Siboglinidae in class Polychaeta). Trophosomes of other siboglinids are confined to the posterior part of the trunk, which is buried in sulfide-rich sediments, but the trophosome of *Riftia* occupies most of its large trunk. It has a much larger supply of hydrogen sulfide, enough to nourish its large body, in the effluent of hydrothermal vents.

There is a well-developed, closed, blood vascular system. Photoreceptor cells are very similar to those of other annelids (oligochaetes and leeches). Sexes are separate.

Class Oligochaeta

More than 3000 species of oligochaetes (Gr. *oligos*, few, + *chait*, long hair) occur in a great variety of habitats. They include the familiar earthworms and many species that live in fresh water. Most are terrestrial or freshwater forms, but some are parasitic, and a few live in marine or brackish water.

Oligochaetes, with few exceptions, bear setae, which may be long or short, straight or curved, blunt or needlelike, or

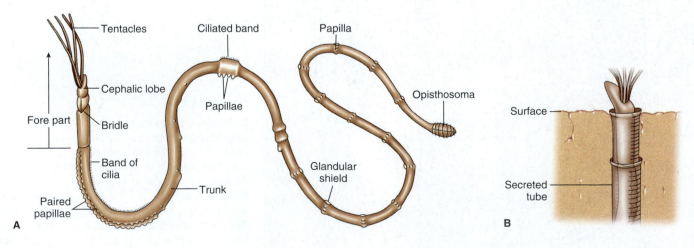

figure 11.11

Diagram of a typical siboglinid. **A,** External features. The body, in life, is much more elongated than shown in this diagram. **B,** Position in its tube.

arranged singly or in bundles. Whatever the type, they are less numerous in oligochaetes than in polychaetes. Aquatic forms usually have longer setae than do earthworms.

Earthworms

The most familiar of oligochaetes are earthworms ("night crawlers"), which burrow in moist, rich soil, emerging at night to feed on surface detritus and vegetation and to breed. In damp, rainy weather they stay near the surface, often with mouth or anus protruding from their burrow. In very dry weather they may burrow several feet underground, coil in a slime chamber, and become dormant. *Lumbricus terrestris* (L. *lumbricum,* earthworm), the form commonly studied in school laboratories, is approximately 12 to 30 cm long (figure 11.12). Giant tropical earthworms may have from 150 to 250 or more segments and may grow to as much as 3 to 4 m in length. They usually live in branched and interconnected tunnels.

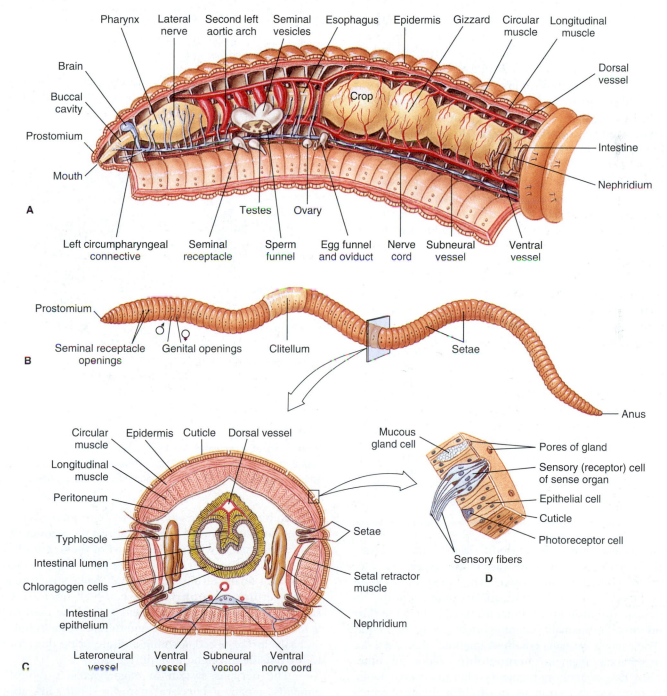

figure 11.12

Earthworm anatomy. **A,** Internal structure of anterior portion of worm. **B,** External features, lateral view. **C,** Generalized transverse section through region posterior to clitellum. **D,** Portion of epidermis showing sensory, glandular, and epithelial cells.

Form and Function

In earthworms a fleshy prostomium overhangs the mouth at the anterior end, with the anus on the terminal end (figure 11.12B). In most earthworms, each segment bears four pairs of chitinous setae (figure 11.12C), although in some oligochaetes each segment may have up to 100 or more setae. Each seta is a bristlelike rod set in a sac within the body wall and moved by tiny muscles (figure 11.13). The setae project through small pores in the cuticle to the outside. In locomotion and burrowing, setae anchor parts of the body to prevent slipping. Earthworms move by peristaltic movement. Contraction of circular muscles in the anterior end lengthens the body, thus pushing the anterior end forward where it is anchored by setae; contractions of longitudinal muscles then shorten the body, pulling the posterior end forward. As these waves of contraction pass along the entire body, it gradually is moved forward.

The food of earthworms is mainly decayed organic matter and bits of vegetation drawn in by the muscular **pharynx** (see figure 11.12A). As in other annelids, the digestive tract is unsegmented and extends the length of the worm. The intestine has a U-shaped fold, the **typhlosole,** that increases surface area for nutrient absorption. **Chloragogen tissue** is found in the typhlosole and around the intestine. Chloragogen cells synthesize glycogen and fat and can break free to distribute these nutrients through the coelom. Chloragogen cells also serve an excretory function.

Aristotle called earthworms the "intestines of the soil." Some 22 centuries later Charles Darwin published his observations in his classic *The Formation of Vegetable Mould Through the Action of Worms.* He showed how worms enrich soil by bringing subsoil to the surface and mixing it with topsoil. An earthworm can ingest its own weight in soil every 24 hours, and Darwin estimated that from 10 to 18 tons of dry earth per acre pass through their intestines annually, thus bringing up potassium and phosphorus from the subsoil and also adding to the soil nitrogenous products from their own metabolism. They expose the mold to the air and sift it into small particles. They also drag leaves, twigs, and organic substances into their burrows, closer to the roots of plants. Their activities are important in aerating the soil. Darwin's views were at odds with his contemporaries, who thought earthworms were harmful to plants. Later research has amply confirmed Darwin's findings, and earthworm management is now practiced in many countries.

Annelids have a double transport system—the coelomic fluid and circulatory system. Food, wastes, and respiratory gases are carried by both in varying degrees. The blood is in a closed system of blood vessels, including capillary systems in the tissues. There are five main longitudinal trunks, of which the **dorsal blood vessel** is the main pumping organ (see figure 11.12A and C). Their blood contains colorless ameboid cells and a dissolved respiratory pigment, **hemoglobin.** Blood of other annelids may have respiratory pigments other than hemoglobin.

The organs of excretion are **nephridia,** a pair of which occurs in each segment except the first three and the last one. Each one occupies parts of two successive segments (figure 11.14). A ciliated funnel, known as a **nephrostome,** lies

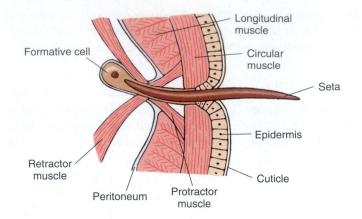

figure 11.13

Seta with its muscle attachments showing relation to adjacent structures. Setae lost by wear and tear are replaced by new ones, which develop from formative cells.

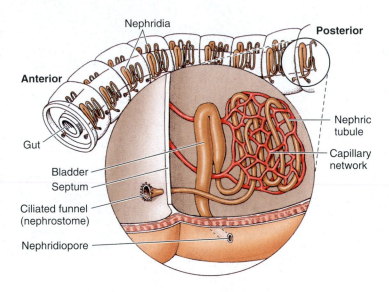

figure 11.14

Nephridium of earthworm. Wastes are drawn into the ciliated nephrostome in one segment, then passed through loops of the nephridium, and expelled through the nephridiopore of the next segment.

just anterior to an intersegmental septum and leads by a small ciliated tubule through the septum into the segment behind, where it connects with the main part of the nephridium. This part of the nephridium is made of several loops of increasing size, which finally terminate in a bladderlike structure leading to an aperture, or **nephridiopore,** which opens to the outside near the ventral row of setae. Cilia draw fluid from the coelom into the nephrostome and tubule. In the tubule, water and salts are resorbed, forming a dilute urine that discharges to the outside through the nephridiopore.

The nervous system in oligochaetes is typical of all annelids. A pair of **cerebral ganglia** (brain) lies above the pharynx joined to a ventral nerve cord by a pair of connectives around the pharynx (see figure 11.12A). The ventral nerve cord bears a pair of ganglia in each segment, giving off segmental nerves containing both sensory and motor fibers. For rapid escape movements most annelids have one or more very large axons commonly called **giant axons,** or giant fibers, in the ventral nerve cord. Speed of conduction in these giant nerve fibers is much greater than that in small axons.

Earthworms are hermaphroditic and exchange sperm during copulation, which usually occurs at night. When mating, worms extend their anterior ends from their burrows and bring their ventral surfaces together (figure 11.15). They are held together by mucus secreted by each worm's **clitellum** and by special ventral setae, which penetrate each other's bodies in the regions of contact. Sperm are discharged and travel to the seminal receptacles of the other worm in its seminal grooves. After sperm exchange, the worms separate. Each worm secretes around its clitellum, first a mucous tube and then a tough, chitinlike band that forms a **cocoon** (figure 11.15). The cocoon slides forward along the body. As it moves, eggs from the oviducts, albumin from

In the dorsal median giant fiber of *Lumbricus,* which is 90 to 160 μm in diameter, speed of conduction has been estimated at 20 to 45 m/second, several times faster than in ordinary neurons of this species. This is also much faster than in polychaete giant fibers, probably because in earthworms the giant fibers are enclosed in myelinated sheaths. Speed of conduction may be altered by changes in temperature.

skin glands, and sperm from the mate (stored in the seminal receptacles) are poured into it. Fertilization of eggs now takes place within the cocoon. When the cocoon slides off the head end of the worm, its ends close, producing a lemon-shaped body. Embryonic development occurs within the cocoon, and the form that hatches from the egg is a young worm similar to the adult. It does not develop a clitellum until it is sexually mature.

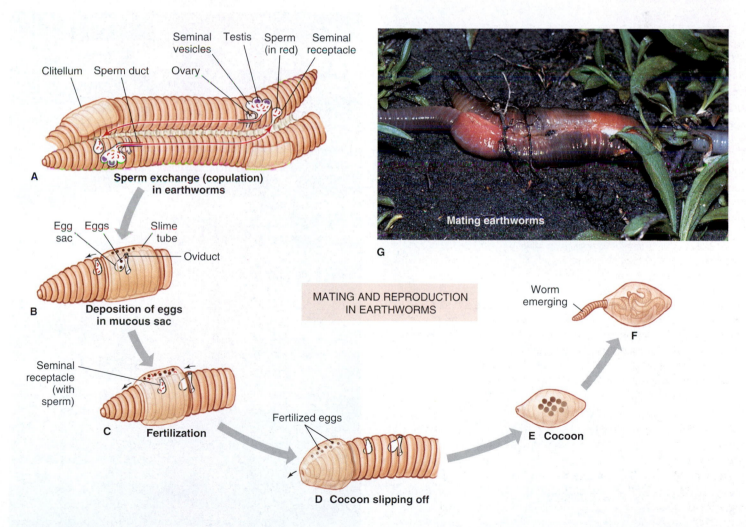

figure 11.15

Earthworm copulation and formation of egg cocoons. **A,** Mutual insemination; sperm from genital pore (segment 15) pass along seminal grooves to seminal receptacles (segments 9 and 10) of each mate. **B** and **C,** After worms separate, a slime tube formed over the clitellum passes forward to receive eggs from oviducts and sperm from seminal receptacles. **D,** As cocoon slips off over anterior end, its ends close and seal. **E,** Cocoon is deposited near burrow entrance. **F,** Young worms emerge in two to three weeks. **G,** Two earthworms in copulation. Their anterior ends point in opposite directions as their ventral surfaces are held together by mucous bands secreted by the clitella.

Freshwater Oligochaetes

Freshwater oligochaetes usually are smaller and have more conspicuous setae than do earthworms. They are more mobile than earthworms and tend to have better developed sense organs. They are generally benthic forms that creep across a substrate or burrow into soft mud. Aquatic oligochaetes provide an important food source for fishes. A few are ectoparasitic.

Some aquatic forms have **gills.** In some, gills are long, slender projections from the body surface. Others have ciliated posterior gills (figure 11.16D), which they extend from their tubes and use to keep water moving. Most forms respire through their skin as do earthworms.

The chief foods are algae and detritus, which worms may gather by extending a mucus-coated pharynx. Burrowers swallow mud and digest the organic material. Some, such as *Aeolosoma,* are ciliary feeders that use currents produced by cilia at the anterior end of the body to sweep food particles into their mouth (figure 11.16B).

Class Hirudinea

Leeches, numbering over 500 species, are found predominantly in freshwater habitats, but a few are marine, and some have even adapted to terrestrial life in moist, warm areas. Most leeches are between 2 and 6 cm in length, but some are smaller, and some reach 20 cm or more (figure 11.17). They are found in a variety of patterns and colors—black, brown, red, or olive green. They are usually flattened dorsoventrally (figure 11.18).

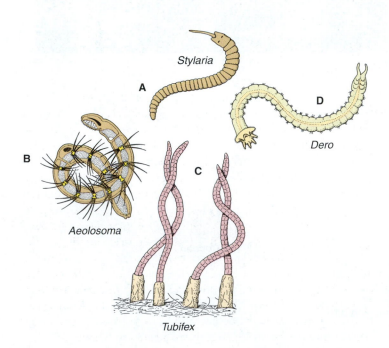

figure 11.16

Some freshwater oligochaetes. **A,** *Stylaria* has its prostomium drawn out into a long snout. **B,** *Aeolosoma* uses cilia around its mouth to sweep in food particles, and it buds off new individuals asexually. **C,** *Tubifex* lives head down in long tubes. **D,** *Dero* has ciliated anal gills.

Form and Function

Leeches have a fixed number of segments, usually 34, and typically have both an anterior and a posterior sucker. They have no parapodia and, except in one genus (*Acanthobdella*), they have no setae. Another primitive character of *Acanthobdella* is its five anterior coelomic compartments separated by septa; septa have disappeared in all other leeches. Their coelom has become filled with connective tissue and muscle, substantially reducing its effectiveness as a hydrostatic skeleton.

Many leeches live as carnivores on small invertebrates; some are temporary parasites, sucking blood from vertebrates; and some are permanent parasites, never leaving their host. Most leeches have a muscular, protrusible proboscis or a muscular pharynx with three jaws armed with teeth. They feed on body juices of their prey, penetrating its surface with their proboscis or jaws and sucking fluids with their powerful, muscular pharynx. Bloodsucking leeches (figure 11.19) secrete an

figure 11.17

The world's largest leech, *Haementeria ghilianii,* on the arm of Dr. Roy K. Sawyer, who found it in French Guiana, South America.

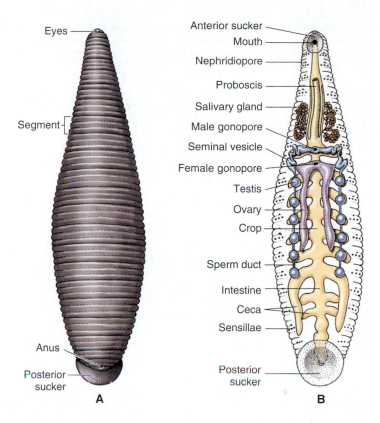

figure 11.18

Structure of a leech, *Placobdella*. **A,** External appearance, dorsal view. **B,** Internal structure, ventral view.

Labels for figure A: Eyes, Segment, Anus, Posterior sucker

Labels for figure B: Anterior sucker, Mouth, Nephridiopore, Proboscis, Salivary gland, Male gonopore, Seminal vesicle, Female gonopore, Testis, Ovary, Crop, Sperm duct, Intestine, Ceca, Sensillae, Posterior sucker

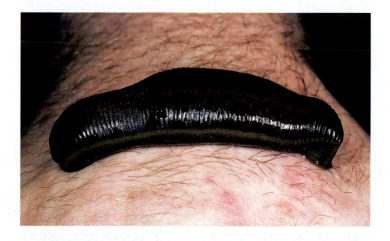

figure 11.19

Hirudo medicinalis feeding on blood from a human arm.

anticoagulant in their saliva. Predatory leeches feed frequently, but those that feed on blood of vertebrates consume large meals (up to several times their body weight) and digest their food slowly. The slow digestion of their meals results from an absence of amylases, lipase, or endopeptidases in gut secretions. In fact, they apparently depend mostly on bacteria in their gut for digestion of a blood meal.

Leeches are hermaphroditic but practice cross-fertilization during copulation. Sperm are transferred by a penis or by hypo-dermic impregnation. Leeches have a clitellum, but it is evident only during the breeding season. After copulation, the clitellum secretes a cocoon that receives eggs and sperm. Cocoons are buried in bottom mud, attached to submerged objects or, in terrestrial species, placed in damp soil. Development is similar to that of oligochaetes.

For centuries "medicinal leeches" *(Hirudo medicinalis)* were used for blood letting because of the mistaken idea that bodily disorders and fevers were caused by an excess of blood. A 10 to 12 cm long leech can extend to a much greater length when distended with blood, and the amount of blood it can suck is considerable. Leech collecting and leech culture in ponds were practiced in Europe on a commercial scale during the nineteenth century. Wordsworth's poem "The Leech-Gatherer" was based on this use of leeches.

Leeches are once again being used medicinally. When fingers or toes are severed, microsurgeons often can reconnect arteries but not the more delicate veins. Leeches are used to relieve congestion until veins can grow back into a healing digit.

Leeches are highly sensitive to stimuli associated with the presence of a prey or host. They are attracted by and will attempt to attach to an object smeared with appropriate host substances, such as fish scales, oil secretions, or sweat. Those that feed on the blood of mammals are attracted by warmth; terrestrial haemadipsid leeches of the tropics will converge on a person standing in one place.

Phylogeny and Adaptive Radiation

Phylogeny

Annelida was a well-accepted monophyletic group until relatively recently, but molecular and morphological analyses suggest that the phylum is paraphyletic without the addition of pogonophorans and, perhaps, echiurans. In the present edition, pogonophorans have been added as clade Siboglinidae within class Polychaeta. Siboglinids include both pogonophorans and vestimentiferans. Unlike siboglinids, echiurans are not segmented (see p. 262), so their inclusion within polychaetes would require a loss of segmentation for this branch of the annelid tree. We have not included echiurans within Polychaeta in this edition.

Class Polychaeta is a paraphyletic taxon, even with the addition of siboglinids, because we now have evidence that ancestral oligochaetes arose from within the polychaete clade. Polychaeta will have to be redefined to exclude these ancestral oligochaetes. Recent work has also shown Oligochaeta to be paraphyletic because leeches (hirudineans) arose from within the oligochaete clade; this class must also be redefined. However, Oligochaeta and Hirudinea together form the clade Clitellata, distinguished by the presence of a clitellum. Proposed relationships among annelid taxa are indicated by a cladogram in figure 11.20.

Which phylum is likely to be the sister taxon for Annelida? One possible answer is phylum Mollusca, and another is

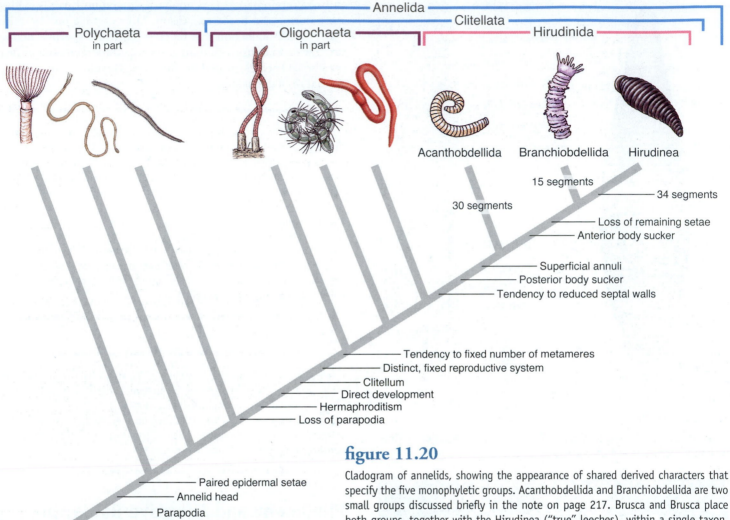

Annelida
Clitellata
Hirudinida
Polychaeta *in part*
Oligochaeta *in part*

Acanthobdellida   Branchiobdellida   Hirudinea

15 segments
34 segments
30 segments
Loss of remaining setae
Anterior body sucker

Superficial annuli
Posterior body sucker
Tendency to reduced septal walls

Tendency to fixed number of metameres
Distinct, fixed reproductive system
Clitellum
Direct development
Hermaphroditism
Loss of parapodia

Paired epidermal setae
Annelid head
Parapodia

Common ancestor of annelid-arthropod clade
(wormlike, metameric, with mesoderm from 4d embryonic cell)

figure 11.20

Cladogram of annelids, showing the appearance of shared derived characters that specify the five monophyletic groups. Acanthobdellida and Branchiobdellida are two small groups discussed briefly in the note on page 217. Brusca and Brusca place both groups, together with the Hirudinea ("true" leeches), within a single taxon, Hirudinida. This clade has several synapomorphies: tendency toward reduction of septal walls, appearance of a posterior sucker, and subdivision of body segments by superficial annuli. Note also that, Oligochaeta has no defining synapomorphies; that is, it is defined solely by retained primitive characters, and thus is paraphyletic. Polychaeta is also a paraphyletic taxon. Parapodia are only present in polychaetes; here they are considered an ancestral trait, lost in Clitellata.
Modified from: R. C. Brusca and G. J. Brusca, Invertebrates, *1990, Sinauer Associates, Inc., Sunderland, MA.*

phylum Arthropoda. Annelids share with arthropods a similar nervous sytem, but the most important resemblance is a shared metameric body plan. What evidence would indicate that segmentation in both phyla was inherited from a common ancestor? Homology would be supported if the same mechanisms of genetic control and chemical signaling were used in both phyla, so research is under way. Very preliminary comparisons of annelids with arthropods suggest that annelids and arthropods do not share the mechanisms of segmentation.[1] Annelids share with molluscs similar features of early embryogenesis, and a very similar trochophore larva form. Such shared characters mesh well with the evolutionary relationships suggested from analysis of base sequences in the gene encoding small-subunit ribosomal

RNA (p. 00); both annelids and molluscs are placed in clade Lophotrochozoa, whereas arthropods are in clade Ecdysozoa.

Adaptive Radiation

Annelids are an ancient group that has undergone extensive adaptive radiation. A basic adaptive feature in evolution of annelids is their septal arrangement, resulting in fluid-filled coelomic compartments. Fluid pressure in these compartments is used as a hydrostatic skeleton in precise movements such as burrowing and swimming. Powerful circular and longitudinal muscles can flex, shorten, and lengthen the body. The basic body structure, particularly of polychaetes, lends itself to great modification. As marine worms, polychaetes have a wide range of habitats in an environment that is not physically or physiologically demanding. Unlike

[1]Seaver, E. C. 2003. Int. J. Dev. Biol. **47**:583–595.

earthworms, whose environment imposes strict physical and physiological demands, polychaetes have been free to experiment and thus have achieved a wide range of adaptive features. In polychaetes the parapodia have been adapted in many ways and for a variety of functions, chiefly locomotion and respiration.

Feeding adaptations show great variation, from the sucking pharynx of oligochaetes and the chitinous jaws of carnivorous polychaetes to the specialized tentacles and radioles of particle feeders. The evolution of a trophosome to house the chemoautotrophic bacteria that provide nutrients to sibiglinids is an adaptation to deep-sea life.

In leeches many adaptations, such as suckers, cutting jaws, pumping pharynx, distensible gut, and the secretion of anticoagulants, are related to their predatory and blood-sucking habits.

Branchiobdellida, a group of small annelids that are parasitic or commensal on crayfish and show similarities to both oligochaetes and leeches, are here placed with oligochaetes, but they are considered a separate class by some authorities. They have 14 or 15 segments and bear a head sucker.

One genus of leech, *Acanthobdella*, has some characteristics of leeches and some of oligochaetes; it is sometimes separated from other leeches into a special class, Acanthobdellida, that characteristically has 27 segments, setae on the first five segments, and no anterior sucker.

---

classification of Phylum Annelida

Class Polychaeta (pol'e-ke'ta) (Gr. *polys,* many, + *chaitē,* long hair). Mostly marine; head distinct and bearing eyes and tentacles; most segments with parapodia (lateral appendages) bearing tufts of many setae; clitellum absent; sexes usually separate; gonads transitory; asexual budding in some; trochophore larva usually; mostly marine. Examples: *Nereis* (see figure 11.1), *Aphrodita, Glycera, Arenicola* (see figure 11.10), *Chaetopterus* (see figure 11.9), *Amphitrite* (see figure 11.4).

Class Oligochaeta (ol'i-go-ke'ta) (Gr. *oligos,* few, + *chaitē,* long hair). Body with conspicuous segmentation; number of segments variable; setae few per segment; no parapodia; head absent; coelom spacious and usually divided by intersegmental septa; hermaphroditic; development direct, no larva; chiefly terrestrial and freshwater. Examples: *Lumbricus* (see figure 11.12), *Stylaria* (see figure 11.16A), *Aeolosoma* (see figure 11.16B), *Tubifex* (see figure 11.16C).

Class Hirudinea (hir'u-din'e-a) (L. *hirudo,* leech, + *-ea,* characterized by): **leeches.** Body with fixed number of segments (usually 34) with many annuli; body usually with anterior and posterior suckers; clitellum present; no parapodia; setae absent (except *Acanthobdella*); coelom closely packed with connective tissue and muscle; septa incomplete in most of the body; development direct; hermaphroditic; terrestrial, freshwater, and marine. Examples: *Hirudo, Placobdella* (see figure 11.18), *Macrobdella.*

Summary

Phylum Annelida is a large, cosmopolitan group containing marine polychaetes, earthworms and freshwater oligochaetes, and leeches. Certainly the most important structural innovation underlying diversification of this group is metamerism, a division of the body into a series of similar segments, each of which contains a repeated arrangement of many organs and systems. The coelom is also highly developed in annelids, and this, together with the septal arrangement of fluid-filled compartments and a well-developed body-wall musculature, is an effective hydrostatic skeleton for precise burrowing and swimming movements. Further metameric specialization occurs in arthropods, to be considered in Chapter 12.

Polychaetes are the largest class of annelids and are mostly marine. On each segment they have many setae, which are borne on paired parapodia. Parapodia show a wide variety of adaptations among polychaetes, including specialization for swimming, respiration, crawling, maintaining position in a burrow, pumping water through a burrow, and accessory feeding. Some polychaetes are mostly predaceous and have an eversible pharynx with jaws. Other polychaetes rarely leave the burrows or tubes in which they live. Several styles of deposit and filter feeding are shown among members of this group. Sibiglinids (pogonophorans) were recently added to this class. Polychaetes are dioecious, have a primitive reproductive system, no clitellum, external fertilization, and a trochophore larva.

Class Oligochaeta contains earthworms and many freshwater forms; they have a small number of setae per segment (compared to Polychaeta) and no parapodia. They have a closed circulatory system, and a dorsal blood vessel is the main pumping organ. Paired nephridia occur in most segments. Earthworms contain the typical annelid nervous system: dorsal cerebral ganglia connected to a double, ventral nerve cord with segmental ganglia running the length of the worm.

Oligochaetes are hermaphroditic and practice cross-fertilization. The clitellum plays an important role in reproduction, including secretion of mucus to surround the worms during copulation and secretion of a cocoon to receive eggs and sperm and in which embryonation occurs. A small, juvenile worm hatches from the cocoon.

Leeches (class Hirudinea) are mostly freshwater, although a few are marine and a few are terrestrial. They feed mostly on fluids; many are predators, some are temporary parasites, and a few are permanent parasites. The hermaphroditic leeches reproduce in a fashion similar to oligochaetes, with cross-fertilization and cocoon formation by the clitellum.

Embryological evidence places annelids with molluscs and arthropods in Protostomia. Recent molecular evidence suggests that annelids and molluscs are more closely related to each other (in superphylum Lophotrochozoa) than either phylum is to arthropods (in superphylum Ecdysozoa).

Review Questions

1. What characteristics of phylum Annelida distinguish it from other phyla?
2. Distinguish among the classes of the phylum Annelida.
3. Describe the annelid body plan, including body wall, segments, coelom and its compartments, and coelomic lining.
4. Explain how the hydrostatic skeleton of annelids helps them to burrow. How is the efficiency for burrowing increased by metamerism?
5. Describe three ways that various polychaetes obtain food.
6. Define each of the following: prostomium, peristomium, radioles, parapodium, setae.
7. Explain functions of each of the following in earthworms: pharynx, typhlosole, chloragogen tissue.
8. Compare the main features of each of the following in each class of annelids: circulatory system, nervous system, excretory system.
9. Describe functions of the clitellum and cocoon.
10. How are freshwater oligochaetes generally different from earthworms?
11. Describe how leeches obtain food.
12. What are the main differences in reproduction and development among the three classes of annelids?
13. What was the evolutionary significance of metamerism and the coelom to its earliest possessors?
14. What are the phylogenetic relationships between molluscs, annelids, and arthropods? What evidence supports these relationships?
15. What is the largest siboglinid known? Where is it found and how is it nourished?

Selected References

See also general references on page 415.

Childress, J. J., H. Felbeck, and G. N. Somero. 1987. Symbiosis in the deep sea. Sci. Am. **256:**114–120 (May). *The amazing story of how the animals around deep-sea vents, including* Riftia pachyptila, *manage to absorb hydrogen sulfide and transport it to their mutualistic bacteria. For most animals, hydrogen sulfide is highly toxic.*

Conniff, R. 1987. The little suckers have made a comeback. Discover **8:**84–94 (Aug.). *Describes medical uses for leeches in microsurgery.*

Fischer, A., and U. Fischer. 1995. On the life-style and life-cycle of the luminescent polychaete *Odontosyllis enopla* (Annelida: Polychaeta). Invert. Biol. **114:**236–247. *If epitokes of this species survive their spawning swarm, they can return to a benthic existence.*

Halanych, K. M., T. D. Dahlgren, and D. McHugh. 2002. Unsegmented annelids? Possible origins of four lophotrochozoan worm taxa. Integ. and Comp. Biol. **42:**678–684. *A nice summary of current morphological and molecular studies on classification of pogonophorans, echiurids, myzostomids, and sipunculans.*

Lent, C. M., and M. H. Dickinson. 1988. The neurobiology of feeding in leeches. Sci. Am. **258:**98–103 (June). *Feeding behavior in leeches is controlled by a single neurotransmitter (serotonin).*

McHugh, D. 2000. Molecular phylogeny of Annelida. Can. J. Zool. **78:**1873–1884. *Descriptions of monophyletic groups within Annelida supported by molecular data.*

Mirsky, S. 2000. When good hippos go bad. Sci. Am. **282:**28 (Jan.). Placobdelloides jaegerskioeldi *is a parasitic leech that breeds only in the rectum of hippopotamuses.*

Pernet, B. 2000. A scaleworm's setal snorkel. Invert. Biol. **119:**147–151. Sthenelais berkeleyi *is an apparently rare but large (20 cm) polychaete that buries its body in sediment and communicates with water above just by its anterior end. Ciliary movement on parapodia pumps water into the burrow for ventilation. The worm remains immobile for long periods, except when prey comes near; it then rapidly everts its pharynx to capture prey.*

Rouse, G. W. 2001. A cladistic analysis of Siboglinidae Caullery, 1914 (Polychaeta: Annelida): Formerly the phyla Pogonophora and Vestimentifera. Zool. J. Linn. Soc. **132:**55–80. *Diagnostic features of Siboglinidae and its subgroups are provided.*

Rouse, G. W., and K. Fauchald. 1998. Recent views on the status, delineation, and classification of the Annelida. Amer. Zool. **38:**953–964. *A discussion of analyses that make Polychaeta paraphyletic, as well as other groups that should/could go into phylum Annelida. These authors conclude that sequence analysis is "clearly no panacea."*

Seaver, E. C. 2003. Segmentation: mono- or polyphyletic. Int. J. Dev. Biol. **47:**583–595. *Preliminary comparisons of the segmentation process in annelids, arthropods, and chordates suggest that annelids and arthropods do not share mechanisms of segmentation, but vertebrates and arthropods may share some mechanisms.*

Winnepenninckx, B. M. H., Y. Van de Peer, and T. Backeljau. 1998. Metazoan relationships on the basis of 18S rRNA sequences: a few years later . . . Amer. Zool. **38:**888–906. *Their calculations and analysis support monophyly of Clitellata but cast doubt on monophyly of Polychaeta.*

Custom Website

The *Animal Diversity* Online Learning Center is a great place to check your understanding of chapter material. Visit www.mhhe.com/hickmanad4e for access to key terms, quizzes, and more! Further enhance your knowledge with Web links to chapter-related material.

Explore live links for these topics:

Classification and Phylogeny of Animals
Phylum Annelida

Class Polychaeta
Class Oligochaeta
Class Hirudinea
Phylum Pogonophora

Arthropods

12

A Winning Combination

Tunis, Algeria—Treating it as an invading army, Tunisia, Algeria, and Morocco have mobilized to fight the most serious infestation of locusts in over 30 years. Billions of the insects have already caused extensive damage to crops and are threatening to inflict great harm to the delicate economies of North Africa.
Source: *New York Times,* April 20, 1988

Humans suffer staggering economic losses due to insects, of which outbreaks of billions of locusts in Africa are only one example. In the western United States and Canada, an outbreak of mountain pine beetles in the 1980s and 1990s killed pines on huge acreages, and the 1973 to 1985 outbreak of spruce budworm in fir/spruce forests killed millions of conifer trees. These examples serve to remind us of our ceaseless struggle with the dominant group of animals on earth today: insects. Insects far outnumber all other species of animals in the world combined, and numbers of individuals are equally enormous. Some scientists have estimated that there are 200 million insects for every human alive today! Insects have an unmatched ability to adapt to all land environments and to virtually all climates. Having originally evolved as land animals, insects developed wings and invaded the air 150 million years before flying reptiles, birds, or mammals. Many have exploited freshwater and saltwater (shoreline) habitats, where they are now widely prevalent; only in the seas are insects almost nonexistent, but there are vast numbers of crustaceans in marine habitats.

How can we account for the enormous success of these creatures? Arthropods have a combination of valuable structural and physiological adaptations, including a versatile exoskeleton, metamerism, an efficient respiratory system, and highly developed sensory organs. In addition, many have a waterproofed cuticle and have extraordinary abilities to survive adverse environmental conditions. We describe these adaptations and others in this chapter.

Spotted spiny lobster, *Panulirus guttatus.*

Phylum Arthropoda (ar-throp′o-da) (Gr. *arthron*, joint, + *pous, podos,* foot) embraces the largest assemblage of living animals on earth. It includes spiders, scorpions, ticks, mites, crustaceans, millipedes, centipedes, insects, and some smaller groups. In addition there is a rich fossil record extending back to the mid-Cambrian period (figure 12.1).

Arthropods are eucoelomate protostomes with well-developed organ systems, and their cuticular exoskeleton containing chitin and tanned protein is a prominent characteristic. Like annelids, they are conspicuously segmented; their primitive body pattern is a linear series of similar segments, each with a pair of jointed appendages. However, unlike annelids, arthropods have embellished the segmentation theme: variation occurs in the pattern of segments and appendages in the phylum. Often segments are combined or fused into functional groups, called **tagmata,** for specialized purposes. The head and thorax are two such tagmata. Appendages, too, are frequently differentiated and specialized for walking, swimming, flying, or eating. Arthropods also differ from annelids in having a very reduced coelom.

Few arthropods exceed 60 cm in length, and most are far below this size. The largest is a Japanese crab (*Macrocheira kaempferi*), which has approximately a 3.7 m span; the smallest is a parasitic mite, which is less than 0.1 mm long.

Arthropods are usually active, energetic animals. However we judge them, whether by their great diversity or their wide ecological distribution or their vast numbers of species, the answer is the same: they are the most abundant and diverse of all animals.

Although arthropods compete with us for food supplies and spread serious diseases, they are essential in pollination of many food plants, and they also serve as food, yield drugs and dyes, and create such products as silk, honey, and beeswax.

Ecological Relationships

Arthropods occur in all types of environment from low ocean depths to very high altitudes and from the tropics far into both north and south polar regions. Some species are adapted for life on land or in fresh, brackish, and marine waters; others live in or on plants and other animals. Most species use flight to varying degrees to move among their favored habitats. Some live in places where no other animal could survive.

Although all feeding types—carnivorous, omnivorous, and herbivorous—occur in this vast group, the majority are herbivorous. Most aquatic arthropods depend on algae for their nourishment, and most land forms live chiefly on plants. There are many parasites. In diversity of ecological distribution arthropods have no rivals.

Why Have Arthropods Achieved Such Great Diversity and Abundance?

Arthropods have achieved great diversity by any measure: number of species, wide distribution, variety of habitats and feeding habits, and power of adaptation to changing conditions. These are some of the structural and physiological patterns that have been helpful to them:

A

B

figure 12.1

Fossils of early arthropods. **A,** Trilobite fossils, dorsal view. These animals were abundant in the mid-Cambrian period. **B,** Eurypterid fossil; eurypterids flourished in Europe and North America from Ordovician to Permian periods.

1. **A versatile exoskeleton.** Arthropods possess an exoskeleton that is highly protective without sacrificing mobility. The skeleton is the cuticle, an outer covering secreted by underlying epidermis.

   The cuticle consists of an inner and thicker **procuticle** and an outer, relatively thin **epicuticle.** Both the procuticle and the epicuticle are composed of several layers each. The outer epicuticle is composed of protein and often lipids. The protein is stabilized and hardened by a chemical process called tanning, adding further

position in the Animal Kingdom

1. Arthropods are coelomate protostomes with a unique cleavage pattern. Growth is accompanied by a series of molts, so arthropods are placed within Ecdysozoa.
2. Evolution of a hard cuticular exoskeleton was followed or accompanied by arthropodization, which included loss of intersegmental septa; development of a hemocoel and loss of

closed circulatory system; jointed appendages; conversion of body-wall muscles to insert on the cuticle.
3. Like annelids, arthropods have conspicuous segmentation, but their segments have greater variety and more grouping for specialized purposes. Specialization of appendages, with pronounced division of labor, results in greater variety of action.

Biological Contributions

1. Cephalization is pronounced, with centralization of fused ganglia and sensory organs in the head.
2. **Segments** are **specialized** for a variety of purposes, forming functional groups **(tagmata).**
3. The presence of paired **jointed appendages** diversified for numerous uses results in greater adaptability.
4. Locomotion is by extrinsic limb muscles. Striated muscles confer rapidity of movement.
5. Although **chitin** is found in a few groups other than arthropods, its use is extensive in arthropods. The **cuticular exoskeleton,** containing chitin, is a great innovation, making possible a wide range of adaptations.

6. The **tracheae** (found only in arthropods) represent a breathing mechanism more efficient than that of most invertebrates.
7. The alimentary canal is highly specialized, having chitinous teeth, compartments, and gastric ossides in various arthropods.
8. Behavioral patterns are much more complex than those of most invertebrates, with a wider occurrence of **social organization.**
9. Many arthropods have well-developed protective coloration and protective resemblances.

---

protection. The procuticle is divided into **exocuticle,** which is secreted before a molt, and **endocuticle,** which is secreted after molting. Both layers of the procuticle contain **chitin** bound with protein. Chitin is a tough, resistant, nitrogenous polysaccharide that is insoluble in water, alkalis, and weak acids. Thus the procuticle not only is flexible and lightweight but also affords protection, particularly against dehydration. In most crustaceans, some areas of the procuticle are also impregnated with **calcium salts,** which reduce flexibility. In the hard shells of lobsters and crabs, for instance, this calcification is extreme.

The cuticle may be soft and permeable or may form a veritable coat of armor. Between body segments and between segments of appendages it is thin and flexible, permitting free movement of joints. In crustaceans and insects the cuticle forms ingrowths for muscle attachment. It may also line the foregut and hindgut, line and support the trachea, and be adapted for a variety of purposes.

The nonexpansible cuticular exoskeleton does, however, impose important conditions on growth. To grow, an arthropod must shed its outer covering at intervals and grow a larger one—a process called **ecdysis,** or **molting.** Arthropods molt from four to seven times before reaching adulthood, and some continue to molt after that. Much of an arthropod's physiology centers on molting, particularly in young animals—preparation, molting itself, and then all processes that must be completed in the postmolt period. More details of the molting process are given for crustaceans (p. 230) and for insects (p. 245).

An exoskeleton is also relatively heavy and becomes proportionately heavier with increasing size. Weight of the exoskeleton tends to limit ultimate body size.

2. **Segmentation and appendages for more efficient locomotion.** Typically each segment has a pair of jointed appendages, but this arrangement is often modified, with both segments and appendages specialized for adaptive functions. Limb segments are essentially hollow levers that are moved by muscles, most of which are striated for rapid action. The jointed appendages are equipped with sensory hairs and are variously modified for sensory functions, food handling, and swift and efficient walking or swimming.

3. **Air piped directly to cells.** Most land arthropods have a highly efficient tracheal system of air tubes, which delivers oxygen directly to tissues and cells and makes a high metabolic rate possible. Aquatic arthropods breathe mainly by some form of gill.

4. **Highly developed sensory organs.** Sensory organs are found in great variety, from compound (mosaic) eyes to senses of touch, smell, hearing, balancing, and chemical reception. Arthropods are keenly alert to environmental stimuli.

5. **Complex behavior patterns.** Arthropods exceed most other invertebrates in complexity and organization of their activities. Innate (unlearned) behavior unquestionably controls much of what they do, but learning also plays an important part in the lives of many arthropods.

characteristics
of Phylum Arthropoda

1. Bilateral symmetry; metameric body, **tagmata** of head and trunk; head, thorax, and abdomen; or cephalothorax and abdomen
2. **Appendages jointed;** primitively, one pair to each segment (metamere), but number often reduced; appendages often modified for specialized functions
3. **Exoskeleton of cuticle** containing protein, lipid, chitin, and often calcium carbonate secreted by underlying epidermis and shed (molted) at intervals, so **growth occurs with ecdysis.**
4. Muscular system complex, with exoskeleton for attachment; striated muscles for rapid action; smooth muscles for visceral organs; **no cilia**
5. Coelom reduced; most of body cavity consisting of **hemocoel** (sinuses, or spaces, in the tissues) filled with blood
6. Complete digestive system; mouthparts modified from appendages and adapted for different methods of feeding
7. **Circulatory system open,** with dorsal contractile heart, arteries, and hemocoel
8. Respiration by body surface, gills, tracheae (air tubes), or book lungs
9. Paired excretory glands called coxal, antennal, or maxillary glands present in some; some with other excretory organs, called Malpighian tubules
10. Nervous system similar to annelid plan, with dorsal brain connected by a ring around the pharynx to a double nerve chain of ventral ganglia; fusion of ganglia in some species; well-developed sensory organs
11. Sexes usually separate, with paired reproductive organs and ducts; usually internal fertilization; oviparous or ovoviviparous; often with metamorphosis; parthenogenesis in a few forms.

6. **Reduced competition through metamorphosis.** Many arthropods pass through metamorphic changes, including a larval form quite different from adults in structure. Larval forms are often adapted for eating a different kind of food from that of adults and occupy a different space, resulting in less competition within a species.

Arthropoda is an astonishingly diverse group, so diverse that zoologists have argued over whether all its members belong within a single phylum. Currently, most biologists accept Arthropoda as a monophyletic group, but there is much discussion of which arthropod subphyla are valid. Previous editions of this text divided the arthropods among four subphyla, the extinct Trilobita, and three extant taxa: Chelicerata, Crustacea, and Uniramia. The subphylum Uniramia is diagnosed by the shared presence of appendages with a single branch (uniramous) and includes insects and myriapods (centipedes, millipedes, and a few less well-known forms). Recent work has called the validity of this group into question, so here we divide uniramians into two subphyla: Myriapoda and Hexapoda. Hexapoda contains class Insecta and one other very small class.

Arthropod subphyla are relatively easy to distinguish based on the number of tagmata and types of appendages present, as outlined next.

Subphylum Trilobita

Trilobites (figure 12.1A) probably had their beginnings a million or more years before the Cambrian period in which they flourished. They have been extinct some 200 million years, but were abundant during the Cambrian and Ordovician periods. Their name refers to the trilobed shape of the body, caused by a pair of longitudinal grooves. They were bottom dwellers, probably scavengers. Most of them could roll up like pill bugs.

Subphylum Chelicerata

Chelicerate arthropods are a very ancient group that includes eurypterids (extinct), horseshoe crabs, spiders, ticks and mites, scorpions, sea spiders, and others. They are characterized by presence of two tagmata and six pairs of appendages that include a pair of **chelicerae,** a pair of **pedipalps,** and **four pairs of walking legs** (a pair of chelicerae and five pairs of walking legs in horseshoe crabs). They have **no mandibles** and **no antennae.** Most chelicerates suck liquid food from their prey.

Class Merostomata

Subclass Eurypterida

Eurypterids, or giant water scorpions (figure 12.1B), lived 200 to 500 million years ago and some were perhaps the largest of all arthropods, reaching a length of 3 m. They had some resemblances to marine horseshoe crabs (figure 12.2) and to scorpions, their terrestrial counterparts.

Subclass Xiphosurida: Horseshoe Crabs

Xiphosurids are an ancient marine group that dates from the Cambrian period. There are only three genera (five species) living today. *Limulus* (L. *limus,* sidelong, askew) (figure 12.2), which lives in shallow water along the North American Atlantic coast, appears in the fossil record practically unchanged back to the Triassic period. Horseshoe crabs have an unsegmented, horseshoe-shaped **carapace** (hard dorsal shield) and a broad abdomen, which has a long spinelike **telson,** or tailpiece. On some abdominal appendages **book gills** (flat leaflike gills) are exposed. Horseshoe crabs can swim awkwardly by means of their abdominal plates and can walk on their walking legs. They feed at night on worms and small molluscs and are harmless to humans.

Class Pycnogonida: Sea Spiders

Pycnogonids are curious little marine animals that are much more common than most of us realize. They move on four pairs of long, thin walking legs, sucking juices from hydroids and soft-bodied animals with their large suctorial proboscis

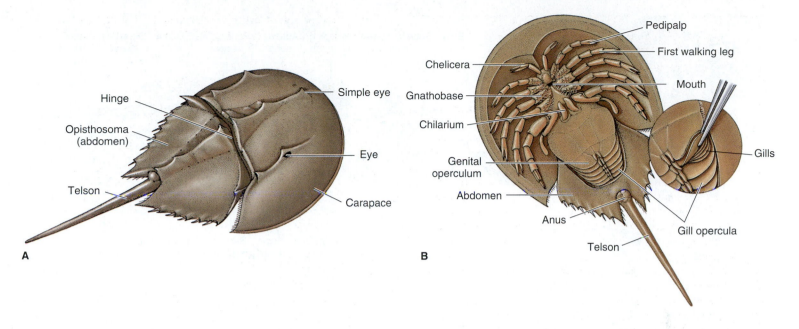

figure 12.2

A, Dorsal view of horseshoe crab *Limulus* (class Merostomata). They grow to 0.5 m in length. **B,** Ventral view.

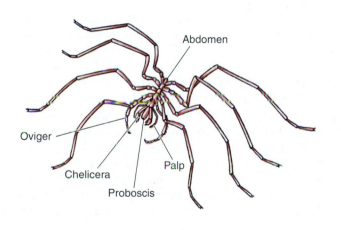

figure 12.3

Pycnogonid, *Nymphon* sp. In this genus all anterior appendages (chelicerae, palps, and ovigers) are present in both sexes, although ovigers are often not present in females of other genera.

(figure 12.3). They often have a pair of ovigerous legs (**ovigers**) with which males carry the egg masses. Their odd appearance is enhanced by the much reduced abdomen attached to an elongated cephalothorax. Most are only a few millimeters long, although some are much larger. They are common in all oceans.

Class Arachnida

Arachnids (Gr. *arachnē*, spider) are a numerous and diverse group, with over 50,000 species described so far. They include spiders, scorpions, pseudoscorpions, whip scorpions, ticks, mites, harvestmen (daddy longlegs), and others. The arachnid tagmata are a cephalothorax and an abdomen.

Order Araneae: Spiders

Spiders are a large group of 35,000 recognized species, distributed all over the world. The cephalothorax and abdomen show no external segmentation, and the tagmata are joined by a narrow, waistlike **pedicel** (figure 12.4).

All spiders are predaceous and feed largely on insects (figure 12.5). Their chelicerae function as fangs and bear ducts from their venom glands, with which they effectively dispatch their prey. Some spiders chase their prey, others ambush them, and many trap them in a net of silk. After a spider seizes its prey with its chelicerae and injects venom, it liquefies tissues with a digestive fluid and sucks the resulting broth into the stomach. Spiders with teeth at the bases of their chelicerae crush or chew prey, aiding digestion by enzymes from their mouth. Many spiders provision their young with previously captured prey.

Spiders breathe by **book lungs** or **tracheae** or both. Book lungs, which are unique to spiders, consist of many parallel air pockets extending into a blood-filled chamber (see figure 12.4). Air enters the chamber by a slit in the body wall. Tracheae are a system of air tubes that carry air directly to tissues from openings called **spiracles.** Tracheae are similar to those of insects (p. 243), but are much less extensive.

Spiders and insects have a unique excretory system of **Malpighian tubules** (see figure 12.4), which work in conjunction with specialized rectal glands. Potassium and other solutes and waste materials are secreted into the tubules, which drain the fluid, or "urine," into the intestine. Rectal glands reabsorb most of the potassium and water, leaving behind such wastes as uric acid. By this cycling of water and potassium, species living in dry environments conserve body fluids, producing a nearly dry mixture of urine and feces. Many spiders also have **coxal glands,** which are modified nephridia that open at the coxa, or base, of the first and third walking legs.

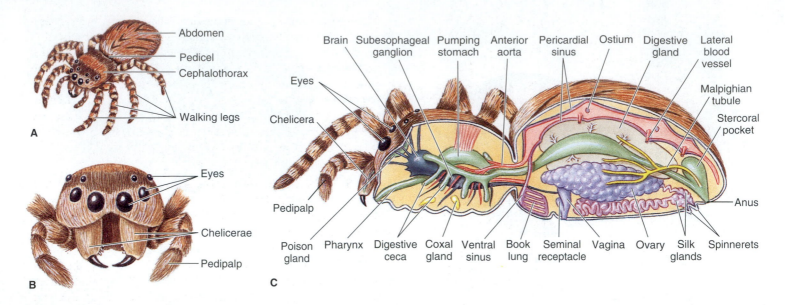

figure 12.4

A, External anatomy of a jumping spider. **B,** Anterior view of head. **C,** Internal anatomy of a spider.

Spiders usually have eight **simple eyes,** each provided with a lens, optic rods, and a retina (see figure 12.4B). Chiefly the eyes perceive moving objects, but some, such as those of the hunting and jumping spiders, may form images. Because vision is usually poor, a spider's awareness of its environment depends especially on its hairlike **sensory setae.** Every seta on its surface is useful in communicating some information about its surroundings, air currents, or changing tensions in the spider's web. By sensing vibrations of its web, a spider can judge the size and activity of its entangled prey or can receive a message tapped on a silk thread by a prospective mate.

Web-Spinning Habits   The ability to spin silk is an important factor in the lives of spiders, as it is in some other arachnids. Two or three pairs of spinnerets containing hundreds of microscopic tubes connect to special abdominal **silk glands** (see fig-

ure 12.4C). A protein secretion emitted as a liquid hardens on contact with air to form a silk thread. Spiders' silk threads are stronger than steel threads of the same diameter and are said to be second in tensional strength only to fused quartz fibers. The threads will stretch one-fifth of their length before breaking.

The spider web used for trapping insects is familiar to most people. Webs of some species consist merely of a few strands of silk radiating out from a spider's burrow or place of retreat. Other species spin beautiful, geometric orb webs. However, spiders use silk threads for many purposes besides web making. They use silk threads to line their nests; form sperm webs or egg sacs; build draglines; make bridge lines, warning threads, molting threads, attachment discs, or nursery webs; or to wrap prey securely (figure 12.6). Not all spiders spin webs for traps. Some, such as the wolf spiders, jumping spiders (see figure 12.5B), and fisher spiders (figure 12.7), simply chase and catch their prey.

figure 12.5

A, A camouflaged crab spider, *Misumenoides* sp., awaits its insect prey. Its coloration matches the petals among which it lies, thus deceiving insects that visit the flowers in search of pollen or nectar. **B,** A jumping spider, *Eris aurantius.* This species has excellent vision and stalks an insect until it is close enough to leap with unerring precision, fixing its chelicerae into its prey.

A   B

figure 12.6

Grasshopper, snared and helpless in the web of a golden garden spider (*Argiope aurantia*), is wrapped in silk while still alive. If the spider is not hungry, its prize will be saved for a later meal.

figure 12.7

A fisher spider, *Dolomedes triton,* feeds on a minnow. This handsome spider feeds mostly on aquatic and terrestrial insects but occasionally captures small fishes and tadpoles. It pulls its paralyzed victim from the water, pumps in digestive enzymes, then sucks out the predigested contents.

Reproduction A courtship ritual usually precedes mating, and mating itself is indirect in that a male spider uses the second pair of appendages (pedipalps) to transfer sperm to the female. Before mating, a male spins a small web, deposits a drop of sperm on it, and then picks the package up and stores

it in special cavities of his pedipalps. When he mates, he inserts a pedipalp into a female's genital opening. Sperm are stored in a seminal receptacle, sometimes for weeks or months, until eggs are ready. A female lays her fertilized eggs in a silken cocoon, which she may carry or may attach to a web or plant. A cocoon may contain hundreds of eggs, which hatch in approximately two weeks. Young usually remain in their egg sac for a few weeks and molt once before leaving it. Several molts occur before adulthood.

Are Spiders Really Dangerous? It is truly amazing that such small and helpless creatures as spiders have generated so much unreasoned fear in humans. Spiders are timid creatures, which, rather than being dangerous enemies to humans, are actually allies in our continuing conflict with insects. The venom produced to kill prey is usually harmless to humans. Even the most venomous spiders bite only when threatened or when defending their eggs or young. American tarantulas (figure 12.8), despite their fearsome appearance, are not dangerous. They rarely bite, and their bite is not considered serious.

Two genera in the United States can give severe or even fatal bites: *Latrodectus* (L. *latro*, robber, + *dektes*, biter), and *Loxosceles* (Gr. *loxos*, crooked, + *skelos*, leg). The most important species are *Latrodectus mactans*, or **black widows**, and *Loxosceles reclusa*, or **brown recluse.** Black widows are moderate to small in size and shiny black, with a bright orange or red "hourglass" on the underside of their abdomen (figure 12.9). Their venom is neurotoxic; that is, it acts on the nervous system. About four or five of each 1000 bites reported are fatal.

Brown recluse spiders, which are smaller than black widows, are brown, and bear a violin-shaped dorsal stripe on their cephalothorax (figure 12.9). Their venom is hemolytic rather than neurotoxic, destroying tissues and skin surrounding a bite. Their bite can be mild to serious and occasionally fatal.

Some spiders in other parts of the world are dangerous, for example, funnel-web spiders *Atrax robustus* in Australia. Most dangerous of all are certain ctenid spiders in South America, for example, *Phoneutria fera.* In contrast to most spiders, these are quite aggressive.

figure 12.8

A tarantula, *Brachypelma vagans.*

A

B

figure 12.9

A, A black widow spider, *Latrodectus mactans,* suspended on her web. Note the orange "hourglass" on the ventral side of her abdomen. **B,** The brown recluse spider, *Loxosceles reclusa,* is a small venomous spider. Note the small violin-shaped marking on its cephalothorax. The venom is hemolytic and dangerous.

Order Scorpionida: Scorpions

Although scorpions are more common in tropical and subtropical regions, some occur in temperate zones. Scorpions are generally secretive, hiding in burrows or under objects by day and feeding at night. They feed largely on insects and spiders, which they seize with clawlike pedipalps and rip with jawlike chelicerae.

A scorpion's body consists of a rather short cephalothorax, which bears appendages and from one to six pairs of eyes, and a clearly segmented abdomen without appendages. The abdomen is divided into a broader **preabdomen** and tail-like **postabdomen,** which ends in a stinging apparatus used to inject venom (figure 12.10). Venom of most species is not harmful to humans, although that of certain species of *Androctonus* in Africa and *Centruroides* in Mexico, Arizona, and New Mexico can be fatal unless antivenom is available.

Scorpions bear living young, which their mother carries on her back until after their first molt.

Order Opiliones: Harvestmen

Harvestmen, often known as "daddy longlegs," are common in the United States and other parts of the world (figure 12.10). These curious creatures are easily distinguished from spiders by a broad joining of their abdomen and cephalothorax without constriction of a pedicel, and by presence of external segmentation of their abdomen. They have four pairs of long, spindly legs, and without apparent ill effect, they can cast off one or more legs if they are grasped by a predator (or human hand). The ends of their chelicerae are pincerlike, and they feed much more as scavengers than do most arachnids.

Order Acari: Ticks and Mites

Acarines differ from all other arachnids in having their cephalothorax and abdomen completely fused, with no sign of external division or segmentation (figure 12.11). Their mouthparts are carried on a little anterior projection, or **capitulum.** They occur almost everywhere—in both fresh and salt water, on vegetation, on the ground, and parasitic on vertebrates and invertebrates. Over 25,000 species have been described, many of which are important to humans, but this is probably only a fraction of the species that exist.

figure 12.10

A, An emperor scorpion (order Scorpionida), *Paninus imperator,* with young, which stay with their mother until their first molt. **B,** A harvestman (order Opiliones). Harvestmen run rapidly on their stiltlike legs. They are especially noticeable during the harvesting season, hence the common name.

A

B

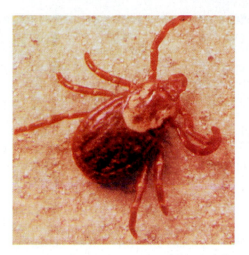

figure 12.11

A wood tick, *Dermacentor variabilis* (order Acarina).

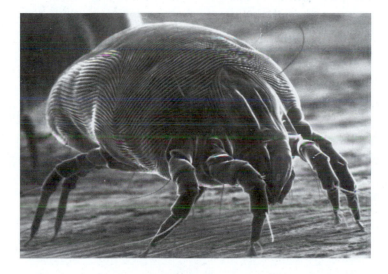

figure 12.12

Scanning electron micrograph of house dust mite, *Dermatophagoides farinae*.

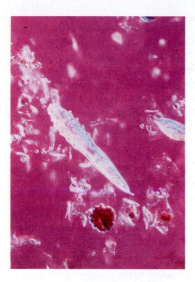

figure 12.13

Demodex follicuorum, human follicle mite. This tiny mite (100 to 400 μm) lives in follicles, particularly around the nose and eyes. Its prevalence ranges from about 20% in persons 20 years of age or younger to nearly 100% in the aged.

are apparently nonpathogenic in humans; they infect most of us although we are unaware of them. Other species of *Demodex* and other genera of mites cause mange in domestic animals.

The inflamed welt and intense itching that follows a chigger bite is not the result of a chigger burrowing into the skin, as is popularly believed. Rather the chigger bites through skin with its chelicerae and injects a salivary secretion containing powerful enzymes that liquefy skin cells. Human skin responds defensively by forming a hardened tube that the larva uses as a sort of drinking straw and through which it gorges itself with host cells and fluid. Scratching usually removes the chigger but leaves the tube, which is a source of irritation for several days.

Many species of mites are entirely free living. *Dermatophagoides farinae* (Gr. *dermatos*, skin, + *phago*, to eat, + *eidos*, likeness of form) (figure 12.12) and related species are denizens of house dust all over the world, sometimes causing allergies and dermatoses. Some mites are marine, but most aquatic species live in fresh water. They have long, hairlike setae on their legs for swimming, and their larvae may be parasitic on aquatic invertebrates. Such abundant organisms have to be important ecologically, but many acarines have more direct effects on our food supply and health. Spider mites (family Tetranychidae) are serious agricultural pests on fruit trees, cotton, clover, and many other plants. Larvae of genus *Trombicula* are called chiggers or redbugs. They feed on dermal tissues of terrestrial vertebrates, including humans, and cause an irritating dermatitis; some species of chiggers transmit a disease called Asiatic scrub typhus. Hair-follicle mites, *Demodex* (figure 12.13),

Ticks are usually larger than mites. They pierce the skin of vertebrates and suck blood until enormously distended; then they drop off and digest their meal. After molting, they are ready for another meal. In addition to disease conditions that they themselves cause, ticks are among the world's premier disease vectors, ranking second only to mosquitos. They carry a greater variety of infectious agents than any other arthropods; such agents include protozoan, rickettsial, viral, bacterial, and fungal organisms. Species of *Ixodes* carry the most common arthropod-borne infection in the United States, Lyme disease (see accompanying note). Species of *Dermacentor* (see figure 12.11) and other ticks transmit Rocky Mountain spotted fever, a poorly named disease because most cases occur in the eastern United States. *Dermacentor* also transmits tularemia and agents of several other diseases. Texas cattle fever, also called red-water fever, is caused by a protozoan parasite transmitted by the cattle tick *Boophilus annulatus.* Many more examples could be cited.

In the 1970s people in the town of Lyme, Connecticut, experienced an epidemic of arthritis. Subsequently known as Lyme disease, it is caused by a bacterium and carried by ticks of the genus *Ixodes*. Now thousands of cases are reported each year in Europe and North America, and other cases have been reported from Japan, Australia, and South Africa. Many people bitten by infected ticks recover spontaneously or do not suffer any ill effects. Others, if not treated at an early stage, develop a chronic, disabling disease. Lyme disease is now the leading arthropod-borne disease in the United States.

Subphylum Crustacea

Crustaceans traditionally were included as a class in subphylum Mandibulata, along with insects and myriapods. Members of all of these groups have, at least, a pair of antennae, mandibles, and maxillae on the head. Whether Mandibulata constitutes a monophyletic grouping has been debated, and we discuss this question further on page 255.

The 67,000 or more species of Crustacea (L. *crusta,* shell) include lobsters, crayfishes, shrimp, crabs, water fleas, copepods, and barnacles. It is the only arthropod class that is primarily aquatic; they are mainly marine, but many freshwater and a few terrestrial species are known. The majority are free living, but many are sessile, commensal, or parasitic. Crustaceans are often very important components of aquatic ecosystems, and several have considerable economic importance.

Crustaceans are the only arthropods with **two pairs of antennae** (figure 12.14). In addition to antennae and **mandibles,**

they have **two pairs of maxillae** on the head, followed by a pair of appendages on each body segment (although appendages on some segments are absent in some groups). All appendages, except perhaps the first antennae (antennules), are primitively **biramous** (two main branches), and at least some appendages of all present-day adults show that condition. Organs specialized for respiration, if present, are in the form of gills; Malpighian tubules are absent.

Crustaceans primitively have 60 segments or more, but most tend to have between 16 and 20 segments and increased tagmatization. The major tagmata are **head, thorax,** and **abdomen,** but these are not homologous throughout the subphylum (or even within some classes) because of varying degrees of fusion of segments, for example, as in the cephalothorax.

In many crustaceans, the dorsal cuticle of the head extends posteriorly and around the sides of the animal to cover or fuse with some or all thoracic and abdominal segments. This covering is called a **carapace.** In some groups the carapace forms clamshell-like valves that cover most or all of the body. In decapods (including lobsters, shrimp, crabs, and others) the carapace covers the entire cephalothorax but not the abdomen.

Form and Function

Appendages

Some modifications of crustacean appendages may be illustrated by those of crayfishes and lobsters (class Malacostraca, order Decapoda, p. 229). **Swimmerets,** or abdominal appendages, retain the primitive biramous condition. Such an appendage consists of inner and outer branches, called the **endopod** and **exopod,** which are attached to one or more basal segments collectively called a **protopod** (figure 12.15).

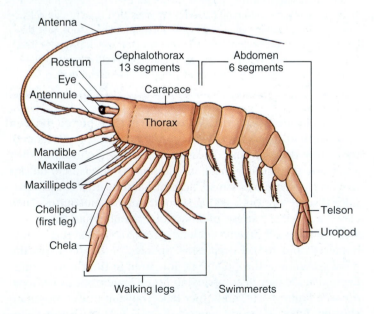

figure 12.14

Archetypical plan of Malacostraca. Note that maxillae and maxillipeds have been separated diagrammatically to illustrate general plan. Typically in living animals only the third maxilliped is visible externally. In order Decapoda the carapace covers the cephalothorax, as shown here.

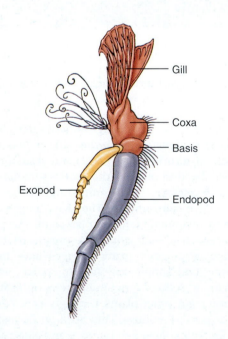

figure 12.15

Parts of a biramous crustacean appendage (third maxilliped of a crayfish).

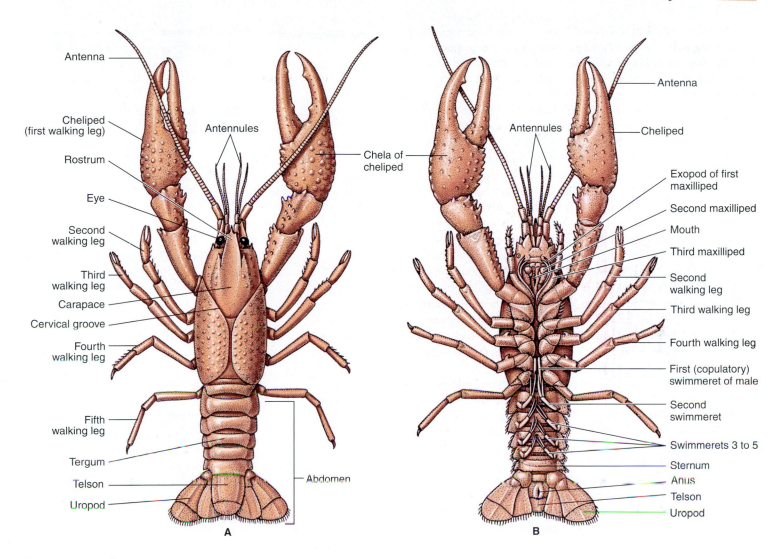

figure 12.16

External structure of crayfishes. **A,** Dorsal view. **B,** Ventral view.

There are many modifications of this plan. In the primitive character state for crustaceans, all trunk appendages are rather similar in structure and adapted for swimming. The evolutionary trend, shown in crayfishes, has been toward reduction in number of appendages and toward a variety of modifications that fit them for many functions. Some are foliaceous (flat and leaflike), as are the maxillae; some are biramous, as are the swimmerets, maxillipeds, uropods, and antennae; some have lost one branch and are **uniramous,** as are the walking legs.

In crayfishes we find the first three pairs of thoracic appendages, called **maxillipeds,** serving along with the two pairs of maxillae as food handlers; the other five pairs of appendages are lengthened and strengthened for walking and defense (figure 12.16). The first pair of walking legs, called **chelipeds,** are enlarged with a strong claw, or chela, for defense. Abdominal swimmerets serve not only for locomotion, but in males the first pair, called gonopods, are modified for copulation, and in females they all serve as a nursery for attached eggs and young. The last pair of appendages, or

uropods, are wide and serve as paddles for swift backward movements, and, with the telson, they form a protective device for eggs or young on the swimmerets.

Terminology applied by various workers to crustacean appendages has not been blessed with uniformity. At least two systems are in wide use. Alternative terms to those we use, for example, are protopodite, exopodite, endopodite, basipodite, coxopodite, and epipodite. The first and second pairs of antennae may be called antennules and antennae, and the first and second maxillae are often called maxillules and maxillae. A rose by any other name . . .

Ecdysis

The problem of growth despite a restrictive exoskeleton is solved in crustaceans, as in other arthropods, by ecdysis (Gr. *ekdysis,* to strip off), a periodic shedding of old cuticle and formation of a larger new one. Molting occurs most frequently during larval stages and less often as an animal reaches adulthood. Although actual shedding of the cuticle is periodic, the molting process and preparations for it, involving a storage of reserves and changes in the integument, are a continuous process lasting most of an animal's life.

During each **premolt** period the old cuticle becomes thinner as inorganic salts are withdrawn from it and stored in tissues. Other reserves, both organic and inorganic, also accumulate and are stored. The underlying epidermis begins to grow by cell division; it secretes first a new inner layer of epicuticle and then enzymes that digest away the inner layers of old endocuticle (figure 12.17). Gradually a new cuticle forms inside the degenerating old one. Finally actual ecdysis occurs as the old cuticle ruptures, usually along the middorsal line, and the animal backs out (figure 12.18). By taking in air or water the animal swells to stretch the new larger cuticle to its full size. During the **postmolt** period the cuticle thickens, its outer layer hardens by tanning, and its inner layer is strengthened as salvaged inorganic salts and other constituents are redeposited. Usually an animal is very secretive during its postmolt period when its defenseless condition makes it particularly vulnerable to predation.

That ecdysis is under hormonal control has been demonstrated in both crustaceans and insects, but the process is often initiated by a stimulus perceived by the central nervous system. The action of the stimulus in decapods is to decrease production of a **molt-inhibiting hormone** from neurosecretory cells in the **X-organ** of the eyestalk. The sinus gland, also in the eyestalk, releases the hormone. When the level of molt-inhibiting hormone drops, **Y-organs** near the mandibles produce **molting hormone.** This hormone initiates processes leading to premolt. Y-organs are homologous to prothoracic glands of insects, which produce ecdysone.

Neurosecretory cells are modified nerve cells that secrete hormones. They are widespread in invertebrates and also occur in vertebrates. Cells in the vertebrate hypothalamus and in the posterior pituitary are good examples.

Other Endocrine Functions

Body color of crustaceans is largely a result of pigments in special branched cells (**chromatophores**) in the epidermis. Chromatophores change color by concentrating pigment granules in the center of each cell, which causes a lightening effect, or by dispersing pigment throughout each cell, which causes darkening. Neurosecretory cells in the eyestalk control pigment behavior. Neurosecretory hormones also control pigment in the eyes for light and dark adaptation, and other neurosecretory hormones control rate and amplitude of heartbeat.

Androgenic glands, which are not neurosecretory, occur in male malacostracans, and their secretion stimulates expression of male sexual characteristics. If androgenic glands are artificially implanted in a female, her ovaries transform to testes and begin to produce sperm, and her appendages begin to acquire male characteristics at the next molt.

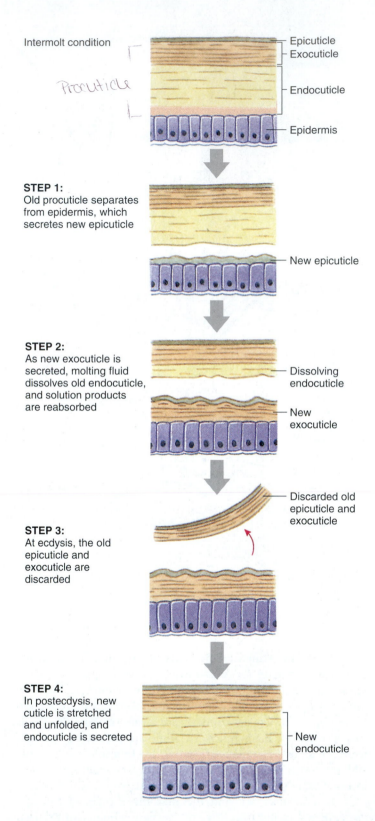

Intermolt condition — Epicuticle / Exocuticle / Endocuticle / Epidermis

Procuticle

STEP 1: Old procuticle separates from epidermis, which secretes new epicuticle — New epicuticle

STEP 2: As new exocuticle is secreted, molting fluid dissolves old endocuticle, and solution products are reabsorbed — Dissolving endocuticle / New exocuticle

STEP 3: At ecdysis, the old epicuticle and exocuticle are discarded — Discarded old epicuticle and exocuticle

STEP 4: In postecdysis, new cuticle is stretched and unfolded, and endocuticle is secreted — New endocuticle

figure 12.17

Cuticle secretion and reabsorption in ecdysis.

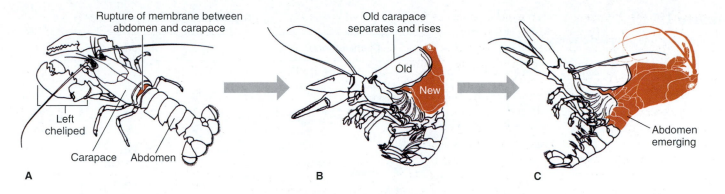

figure 12.18

Molting sequence in a lobster, *Homarus americanus*. **A,** Membrane between carapace and abdomen ruptures, and carapace begins a slow elevation. This step may take up to two hours. **B** and **C,** Head, thorax, and finally abdomen are withdrawn. This process usually takes no more than 15 minutes. Immediately after ecdysis, chelipeds are desiccated and the body is very soft. The lobster now begins rapid absorption of water so that within 12 hours its body increases about 20% in length and 50% in weight. Tissue water will be replaced by protein in succeeding weeks.

Feeding Habits

Feeding habits and adaptations for feeding vary greatly among crustaceans. Many forms can shift from one type of feeding to another depending on environment and food availability, but fundamentally the same set of mouthparts is used by all. Mandibles and maxillae are involved in actual ingestion; maxillipeds hold and crush food. In predators the walking legs, particularly chelipeds, serve in food capture.

Many crustaceans, both large and small, are predatory, and some have interesting adaptations for killing prey. One shrimplike form, *Lygiosquilla,* has on one of its walking legs a specialized digit that can be drawn into a groove and released suddenly to pierce passing prey. Pistol shrimp, *Alpheus,* have one enormously enlarged chela that can be cocked like the hammer of a gun and snapped shut at great speed, forming a cavitation bubble that implodes with a snap sufficient to stun its prey.

Food of crustaceans ranges from plankton, detritus, and bacteria, used by **suspension feeders,** to larvae, worms, crustaceans, snails, and fishes, used by predators, and dead animal and plant matter, used by **scavengers.** Suspension feeders, such as fairy shrimps, water fleas, and barnacles, use their legs, which bear a thick fringe of setae, to create water currents that sweep food particles through the setae. Mud shrimps, *Upogebia,* use long setae on their first two pairs of thoracic appendages to strain food material from water circulated through their burrow by movements of their swimmerets.

Crayfishes have a two-part stomach. The first contains a **gastric mill** in which food, already shredded by the mandibles, can be ground further by three calcareous teeth into particles fine enough to pass through a filter of setae in the second part of the stomach; food particles then pass into the intestine for chemical digestion.

Respiration, Excretion, and Circulation

Gills of crustaceans vary in shape—treelike, leaflike, or filamentous—all provided with blood vessels or sinuses. They usually are attached to appendages and kept ventilated by movement of appendages through water. The overlapping carapace usually protects the **branchial chambers.** Some smaller crustaceans breathe through their general body surface.

Excretory and **osmoregulatory** organs in crustaceans are paired glands located in their head, with excretory pores opening at the base of either antennae or maxillae, thus called **antennal glands** or **maxillary glands,** respectively (figure 12.19). Antennal glands of decapods are also called **green glands.** They resemble coxal glands of chelicerates. Waste products are mostly ammonia with some urea and uric acid. Some wastes diffuse through the gills as well as through the excretory glands.

Circulation, as in other arthropods, is an **open system** consisting of a heart, either compact or tubular, and arteries, which transport blood to different areas of the hemocoel. Some smaller crustaceans lack a heart. An open circulatory system depends less on heartbeats for circulation because movement of organs and limbs circulates blood more effectively in open sinuses than in capillaries. Blood may contain as respiratory pigments either hemocyanin or hemoglobin (hemocyanin in decapods), and it has the property of clotting to prevent loss of blood in minor injuries.

Nervous and Sensory Systems

A cerebral ganglion above the esophagus sends nerves to the anterior sense organs and connects to a subesophageal ganglion by a pair of connectives around the esophagus. A double ventral nerve cord has a ganglion in each segment that sends nerves to viscera, appendages, and muscles (figure 12.19). Giant fiber systems are common among crustaceans.

Sensory organs are well developed. There are two types of eyes—a median, or nauplius, eye and compound eyes. A **median eye** consists usually of a group of three pigment cups containing retinal cells, and it may or may not have a lens. Median eyes are found in nauplius larvae and in some adult forms, and they may be an adult's only eye, as in copepods.

figure 12.19

Internal structure of a male crayfish.

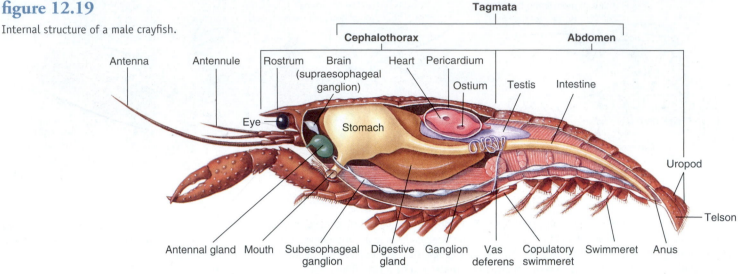

Most crustaceans have **compound eyes** similar to insect eyes. In crabs and crayfishes they are on the ends of movable eyestalks (figure 12.19). Compound eyes are precise instruments, different from vertebrate eyes, yet especially adept at detecting motion; they can analyze polarized light. The convex corneal surface gives a wide visual field, particularly in stalked eyes where the surface may cover an arc of 200 degrees or more.

Compound eyes are composed of many tapering units called **ommatidia** set close together (figure 12.20). Facets, or corneal surfaces, of ommatidia give the surface of the eye an appearance of a fine mosaic. Most crustacean eyes are adapted either to bright or to dim light, depending on their diurnal or nocturnal habits, but some are able, by means of screening pigments, to adapt, to some extent at least, to both bright and dim light. The number of ommatidia varies from a dozen or two in some small crustaceans to 15,000 or more in a large lobster. Some insects have approximately 30,000.

Other sensory organs include statocysts, tactile setae on the cuticle of most of the body, and chemosensitive setae, especially on antennae, antennules, and mouthparts.

Reproduction and Life Cycles

Most crustaceans have separate sexes, and numerous specializations for copulation occur among different groups. Barnacles are monoecious but generally practice cross-fertilization. In some ostracods males are scarce, and reproduction is usually parthenogenetic. Most crustaceans brood their eggs in some manner—branchiopods and barnacles have special brood chambers, copepods have egg sacs attached to the sides of their abdomen (see figure 12.24), and malacostracans usually carry eggs and young attached to their appendages.

A hatchling of a crayfish is a tiny juvenile similar in form to the adult and has a complete set of appendages and segments. However, most crustaceans produce larvae that must go through a series of changes, either gradual or abrupt over a series of molts, to assume adult form (metamorphosis). The ancestral larva of crustaceans is a **nauplius** (figure 12.21). It has an unsegmented body, a frontal eye, and three pairs of appendages, representing the two pairs of antennae and the mandibles. Developmental stages and postlarvae of different groups of Crustacea are varied and have special names.

figure 12.20

Compound eye of an insect. A single ommatidium is shown enlarged to the right.

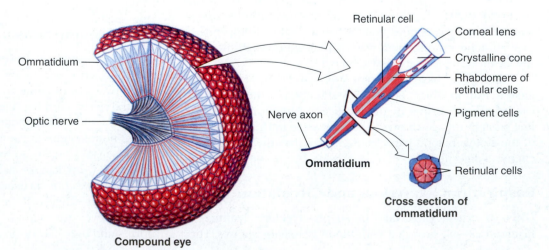

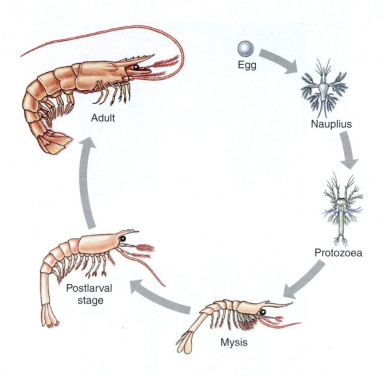

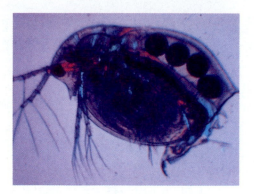

figure 12.22

A water flea, *Daphnia* (order Cladocera), photographed with polarized light. These tiny forms occur in great numbers in northern lakes and are an important component of the food chain leading to fishes.

figure 12.21

Life cycle of a Gulf shrimp *Penaeus*. Penaeids spawn at depths of 40 to 90 m. Young larval forms are planktonic and move inshore to water of lower salinity to develop as juveniles. Older shrimp return to deeper water offshore.

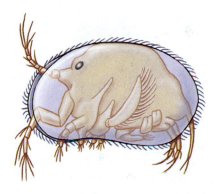

figure 12.23

An ostracod (subclass Ostracoda, class Maxillopoda).

Class Branchiopoda

Members of class Branchiopoda (bran'kec-op'o-da) (Gr. *branchia,* gills, + *pous, podos,* foot) have several primitive characteristics. Four orders are recognized: **Anostraca** (fairy shrimp and brine shrimp), which lack a carapace; **Notostraca** (tadpole shrimp such as *Triops*), whose carapace forms a large dorsal shield covering most trunk segments; **Conchostraca** (clam shrimp such as *Lynceus*), whose carapace is bivalved and usually encloses the entire body; and **Cladocera** (water fleas such as *Daphnia,* figure 12.22), with a carapace typically covering the entire body but not the head. Branchiopods have reduced first antennae and second maxillae. Their legs are flattened and leaflike **(phyllopodia)** and are the chief respiratory organs (hence, the name branchiopods). Legs also are used in suspension feeding in most branchiopods, and in groups other than cladocerans, they are used for locomotion as well. The most important and diverse order is Cladocera, which often forms a large segment of freshwater zooplankton.

Class Maxillopoda

Class Maxillopoda includes a number of crustacean groups traditionally considered classes themselves. Specialists have recognized evidence that these groups form a clade within Crustacea. They basically have five cephalic, six thoracic, and usually four abdominal segments plus a telson, but reductions are common. No typical appendages occur on the abdomen.

Eyes of nauplii (when present) have a unique structure termed a **maxillopodan eye.**

Members of subclass **Ostracoda** (os-trak'o-da) (Gr. *ostrakodes,* testaceous, that is, having a shell) are, like conchostracans, enclosed in a bivalved carapace and resemble tiny clams, 0.25 to 8 mm long (figure 12.23). Ostracods show considerable fusion of trunk segments, and numbers of thoracic appendages are reduced to two or none. Most ostracods crawl or burrow in marine or freshwater sediments. They scavenge food, feed on detritus, or collect suspended particles from the water.

Subclass **Copepoda** (ko-pep'o-da) (Gr. *kōpē,* oar, + *pous, podos,* foot) is an important group of Crustacea, second only to Malacostraca in number of species. Copepods are small (usually a few millimeters or less in length), rather elongate, tapering toward the posterior end, lacking a carapace, and retaining a simple, median, nauplius eye in adults (figure 12.24). They have four pairs of rather flattened, biramous, thoracic swimming appendages, and a fifth, reduced pair. Their abdomen bears no legs. Free-living copepods occur in planktonic and benthic habitats. Planktonic species are essential members of marine and freshwater food webs, often dominating the primary consumer level (herbivore) in aquatic communities.

figure 12.24

A copepod with attached ovisacs (subclass Copepoda, class Maxillopoda).

figure 12.25

Fish louse (subclass Branchiura, class Maxillopoda).

Many symbiotic species are known, and parasitic forms may be so highly modified as adults (and may depart so far from the description just given) that they can hardly be recognized as arthropods.

Subclass **Branchiura** (bran-kee-u'ra) (Gr. *branchia,* gills, + *ura,* tail) is a small group of primarily fish parasites, which, despite its name, has no gills (figure 12.25). Members of this group are usually between 5 and 10 mm long and parasitize marine or freshwater fish. They typically have a broad, shieldlike carapace, compound eyes, four biramous thoracic appendages for swimming, and a short, unsegmented abdomen. The second maxillae have become modified as suction cups.

Subclass **Cirripedia** (sir-i-ped'i-a) (L. *cirrus,* curl of hair, + *pes, pedis,* foot) includes barnacles, individuals of which are usually enclosed in a shell of calcareous plates, as well as three smaller orders of burrowing or parasitic forms. Barnacles are sessile as adults and may be attached to their substrate by a stalk (gooseneck barnacles) (figure 12.26B) or directly (acorn barnacles) (figure 12.26A). Typically, a carapace (mantle) surrounds their body and secretes a shell of calcareous plates. Their head is reduced, their abdomen absent, and their thoracic legs are long, many-jointed cirri with hairlike setae. Cirri are extended through an opening between the calcareous plates to filter from water small particles on which the animal feeds (figure 12.26B).

A

B

figure 12.26

A, Acorn barnacles, *Semibalanus cariosus* (subclass Cirripedia) are found on rocks along the Pacific Coast of North America. **B,** Common gooseneck barnacles, *Lepas anatifera.* Note the feeding legs, or cirri, on *Lepas.* Barnacles attach themselves to a variety of firm substrates, including rocks, pilings, and boat bottoms.

Barnacles frequently foul ship bottoms by settling and growing there. So great may be their number that the speed of a ship may be reduced 30% to 40%, requiring expensive drydocking of the ship to remove them.

Class Malacostraca

Class Malacostraca (mal'a-kos'tra-ka) (Gr. *malakos,* soft, + *ostrakon,* shell) is the largest class of Crustacea and shows great diversity. We mention only 4 of its 12 to 13 orders. The trunk of malacostracans usually has eight thoracic and six abdominal segments, each with a pair of appendages. There are many marine and freshwater species.

Isopoda (i-sop'o-da) (Gr. *isos,* equal, + *pous, podos,* foot) are commonly dorsoventrally flattened, lack a carapace, and have sessile compound eyes. Their abdominal appendages bear gills. Common land forms are sow bugs or pill bugs (*Porcellio* and *Armadillidium,* figure 12.27A), which live under stones and in other damp places. *Asellus* is common in fresh water, and *Ligia* is abundant on sea beaches and rocky shores. Some isopods are parasites of other crustaceans or of fish (figure 12.28).

figure 12.27

A, Four pill bugs, *Armadillidium vulgare* (order Isopoda), common terrestrial forms. **B,** Freshwater sow bug, *Caecidotea* sp., an aquatic isopod.

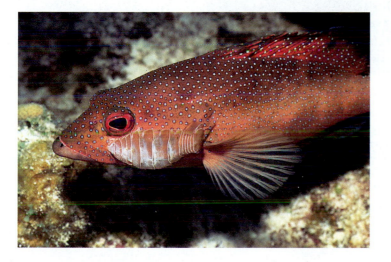

figure 12.28

An isopod parasite (*Anilocra* sp.) on a coney (*Cephalopholis fulvus*) inhabiting a Caribbean coral reef (order Isopoda, class Malacostraca).

Amphipoda (am-fip'o-da) (Gr. *amphis*, on both sides, + *pous, podos,* foot) resemble isopods in that members have no carapace and have sessile compound eyes. However, they are usually compressed laterally, and their gills are in the thoracic position, as in other malacostracans. There are many marine amphipods (figure 12.29), such as beach fleas, *Orchestia*, and numerous freshwater species.

Euphausiacea (yu-faws'i-a'se-a) (Gr. *eu*, well, + *phausi*, shining bright, + *acea*, L. suffix, pertaining to) is a group of only about 90 species, important as oceanic plankton known as "krill." They are about 3 to 6 cm long (figure 12.30) and commonly occur in great oceanic swarms, where they are eaten by baleen whales and many fishes.

Decapoda (de-cap'o-da) (Gr. *deka*, ten, + *pous, podos,* foot) have five pairs of walking legs of which the first is often

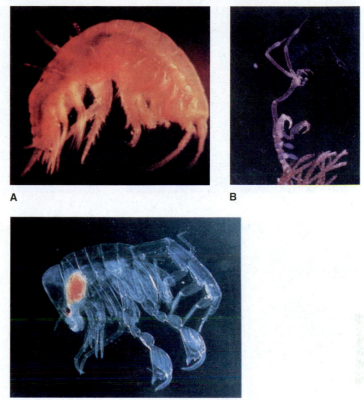

figure 12.29

Marine amphipods. **A,** Free-swimming amphipod, *Anisogammarus* sp. **B,** Skeleton shrimp, *Caprella* sp., shown on a bryozoan colony, resemble praying mantids. **C,** *Phronima*, a marine pelagic amphipod, takes over the tunic of a salp (subphylum Urochordata, Chapter 15). Swimming by means of its abdominal swimmerets, which protrude from the opening of the barrel-shaped tunic, the amphipod maneuvers to catch its prey. The tunic is not seen (order Amphipoda, class Malacostraca).

figure 12.30

Meganyctiphanes, order Euphausiacea, "northern krill."

modified to form pincers **(chelae)** (see figures 12.14 and 12.16). These are lobsters, crayfishes (see figure 12.14), shrimps (see figure 12.21), and crabs, the largest of the crustaceans (figure 12.31). True crabs differ from others in having a broader carapace and a much reduced abdomen (figure 12.31A and C). Familiar examples are fiddler crabs, *Uca*, which burrow in sand just below high-tide

figure 12.31

Decapod crustaceans. **A,** A bright orange tropical rock crab, *Grapsus grapsus,* is a conspicuous exception to the rule that most crabs exhibit cryptic coloration. **B,** A hermit crab, *Elassochirus gilli,* which has a soft abdominal exoskeleton, lives in a snail shell that it carries about and into which it can withdraw for protection. **C,** A male fiddler crab, *Uca* sp., uses its enlarged cheliped to wave territorial displays and in threat and combat. **D,** A red night shrimp, *Rhynchocinetes rigens,* prowls caves and overhangs of coral reefs, but only at night. **E,** Spiny lobster *Panulirus argus* (order Decapoda, class Malacostraca).

level (figure 12.31C), decorator crabs, which cover their carapaces with sponges and sea anemones for camouflage, and spider crabs, such as *Libinia.* Hermit crabs (figure 12.31B) have become adapted to live in snail shells; their abdomen, which lacks a hard exoskeleton, is protected by the snail shell.

Subphylum Myriapoda

The term **myriapod** (Gr. *myrias,* a myriad, + *podos,* foot) refers to several classes that have evolved a pattern of two tagmata—head and trunk—with paired appendages on most or all trunk segments. Myriapods include Chilopoda (centipedes), Diplopoda (millipedes), Pauropoda (pauropods), and Symphyla (symphylans).

The head of myriapods resembles the crustacean head but has only **one pair of antennae,** instead of two. It also has **mandibles** and two pairs of **maxillae** (one pair of maxillae in millipedes). The legs are all **uniramous.**

Respiratory exchange is by body surface and tracheal systems, although juveniles, if aquatic, may have gills.

Class Chilopoda: Centipedes

Centipedes are active predators with a preference for moist places such as under logs or stones, where they feed largely on earthworms and insects. Their bodies are somewhat flattened dorsoventrally, and they may contain from a few to 177 segments (figure 12.32). Each segment, except the one behind the head and the last two, bears one pair of appendages. Those of the first body segment are modified to form venom claws, which they use to kill their prey. Most species are harmless to humans.

Their head bears a pair of eyes, each consisting of a group of ocelli (simple eyes). Respiration is by tracheal tubes with a pair of spiracles in each segment. Sexes are separate, and all species are oviparous. Young are similar to adults. Common house centipedes *Scutigera,* with 15 pairs of legs, and *Scolopendra* (figure 12.32), with 21 pairs of legs, are familiar genera.

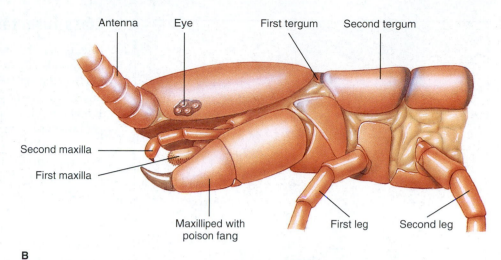

figure 12.32

A, A centipede, *Scolopendra* (class Chilopoda) from the Amazon Basin, Peru. Most segments have one pair of appendages each. First segment bears a pair of poison claws, which in some species can inflict serious wounds. Centipedes are carnivorous. **B,** Head of centipede.

Class Diplopoda: Millipedes

Diplopods, or "double-footed" arthropods, are commonly called millipedes, which literally means "thousand feet" (figure 12.33). Although they do not have a thousand legs, they do have a great many. Their cylindrical bodies contain 25 to 100 segments. The four thoracic segments bear only one pair of legs each, but abdominal segments each have two pairs, a condition that may have evolved from fusion of segments. Two pairs of spiracles occur on each abdominal segment, each opening into an air chamber that gives rise to tracheal tubes.

Millipedes are less active than centipedes and are generally herbivorous, living on decayed plant and animal matter and sometimes living plants. They prefer dark moist places under stones and logs. Females lay eggs in a nest and guard them carefully. Larval forms have only one pair of legs per segment.

Subphylum Hexapoda

This subphylum is named for the presence of six legs in members of the group. All legs are uniramous. Hexapods have three tagmata—head, thorax, and abdomen—with appendages on the head and thorax. Abdominal appendages are greatly reduced or absent. The subphylum has two classes: Entognatha, a small group whose members have the bases of mouthparts enclosed within the head capsule, and Insecta, an enormous group whose members have the bases of mouthparts visible outside the head capsule (hence, ectognathous mouthparts).

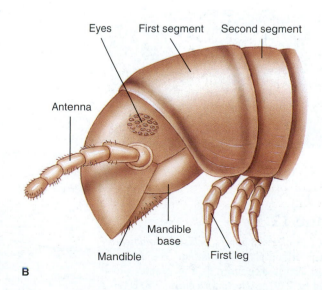

figure 12.33

A, A tropical millipede with warning coloration. Note the typical doubling of appendages on most segments, hence diplosegments. **B,** Head of millipede.

figure 12.34

Pie diagram indicating relative numbers of species of insects to the rest of the animal kingdom and protozoan groups.

Class Insecta: Insects

Insects are the most numerous and diverse of all groups of arthropods (figure 12.34). There are more species of insects than species in all the other classes of animals combined. The number of insect species named has been estimated at close to 1 million, with thousands, perhaps millions, of other species yet to be discovered and classified.

It is difficult to appreciate fully the significance of this extensive group and its role in the biological pattern of animal life. The study of insects (**entomology**) occupies the time and resources of thousands of skilled men and women all over the world. The struggle between humans and insect pests seems to be endless, yet paradoxically, insects are so interwoven into the economy of nature in so many roles that we would have a difficult time without them.

Insects differ from other arthropods in having **three pairs of legs** and usually **two pairs of wings** on the thoracic region of the body (figure 12.35), although some have one pair of wings or none. In size insects range from less than 1 mm to 20 cm in length, the majority being less than 2.5 cm long.

Distribution and Adaptability

Insects have spread into practically all habitats that can support life, but only a relatively few are marine. They are common in brackish water, in salt marshes, and on sandy beaches. They are abundant in fresh water, soils, forests, and plants, and they occur even in deserts and wastelands, on mountaintops, and as parasites in and on the bodies of plants and animals, including other insects.

Their wide distribution is made possible by their powers of flight and their highly adaptable nature. In many cases they

A

figure 12.35

A, A pair of grasshoppers, *Schistocerca obscura* (order Orthoptera), copulating. The African desert locust mentioned in the chapter prologue (p. 219) is *Schistocerca gregaria*. **B,** External features of a female grasshopper. The terminal segment of a male with external genitalia is shown in inset.

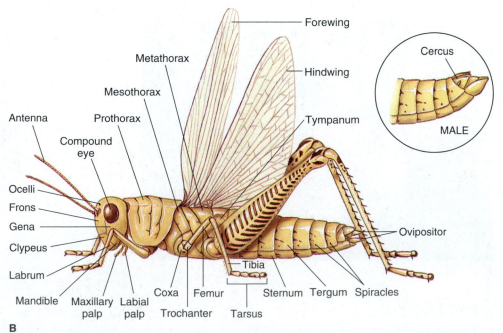

B

can easily surmount barriers that are impassable to many other animals. Their small size and well-protected eggs allow them to be carried great distances by wind, water, and other animals.

The amazing adaptability of insects is evidenced by their wide distribution and enormous diversity of species. Such diversity enables this vigorous group to take advantage of all available resources of food and shelter.

Much success of insects is due to adaptive qualities of their cuticular exoskeleton, as is the case in other arthropods. However, the great exploitation of terrestrial environments by insects has been made possible by an array of adaptations they possess to withstand its rigors. For example, their epicuticle has a waxy and a varnish layer and they can close their spiracles; both characteristics minimize evaporative water loss. They extract maximal fluid from food and fecal material, and many can retain water produced in oxidative metabolism. Many can enter a resting stage (diapause) and lie dormant during inhospitable conditions.

External Features

Insect tagmata are **head, thorax,** and **abdomen.** The cuticle of each body segment is typically composed of four plates (**sclerites**), a dorsal notum (**tergum**), a ventral **sternum,** and a pair of lateral **pleura.** Pleura of abdominal segments are membranous rather than sclerotized (hardened).

The head usually bears a pair of relatively large compound eyes, a pair of antennae, and usually three ocelli. Mouthparts typically consist of a **labrum,** a pair each of **mandibles** and **maxillae,** a **labium,** and a tonguelike **hypopharynx.** The type of mouthparts an insect possesses determines how it feeds. We discuss some of these modifications in a later section.

The thorax is composed of three segments: **prothorax, mesothorax,** and **metathorax,** each bearing a pair of legs (figure 12.35). In most insects the mesothorax and metathorax each bear a pair of wings. Wings consist of a double membrane that contains veins of thicker cuticle, which serve to strengthen the wing. Although these veins vary in their patterns among different species, they are constant within a species and serve as one means of classification and identification.

Legs of insects are often modified for special purposes. Terrestrial forms have walking legs with terminal pads and claws as in beetles. These pads may be sticky for walking upside down, as in house flies. Hindlegs of grasshoppers and crickets are adapted for jumping (figure 12.35B). Mole crickets have the first pair of legs modified for burrowing in the ground. Water bugs and many beetles have paddle-shaped appendages for swimming. For grasping prey, the forelegs of a praying mantis are long and strong (figure 12.36).

There are more than 30 orders of insects. Some of the most common ones, such as Diptera and Hymenoptera, are briefly outlined beginning on page 252.

Wings and the Flight Mechanism

Insects share the power of flight with birds and flying mammals. However, wings evolved independently in birds, bats, and insects. Insect wings are formed by outgrowth from the body wall of the mesothoracic and metathoracic segments and are composed of cuticle.

Most insects have two pairs of wings, but Diptera (true flies) have only one pair (figure 12.37), the hindwings being represented by a pair of small **halteres** (balancers) that vibrate and are responsible for equilibrium during flight. Males in order Strepsiptera have only a hind pair of wings and an anterior pair of halteres. Males of scale insects also have one pair of wings but no halteres. Some insects are wingless. Ants and termites, for example, have wings only on males, and on females during certain periods; workers (females) are always wingless. Lice and fleas are always wingless.

Wings may be thin and membranous, as in flies and many others (figure 12.37); thick and stiff, as in forewings of beetles (see figure 12.51); parchmentlike, as in forewings of grasshoppers; covered with fine scales, as in butterflies and moths; or with hairs, as in caddisflies.

Wing movements are controlled by a complex of thoracic muscles. **Direct flight muscles** are attached to a part of the wing itself. **Indirect flight muscles** are not attached to the wing and cause wing movement by altering the shape of the thorax. The wing is hinged at the thoracic tergum and also slightly laterally on a pleural process, which acts as a fulcrum (figure 12.38). In all insects, the upstroke of a wing is effected by contracting indirect muscles that pull the tergum down toward the sternum

A

B

figure 12.36

A, Praying mantis (order Orthoptera), feeding on an insect. **B,** Praying mantis laying eggs.

figure 12.37

House fly *Musca domestica* (order Diptera). House flies can become contaminated with over 100 human pathogens, and there is strong circumstantial evidence for mechanical transmission of many of them.

(figure 12.38A). Dragonflies and cockroaches accomplish the downstroke by contracting direct muscles attached to the wings lateral to the pleural fulcrum. In Hymenoptera and Diptera all flight muscles are indirect. The downstroke occurs when sternotergal muscles relax and longitudinal muscles of the thorax arch the tergum (figure 12.38B), pulling the tergal articulations upward relative to the pleura. The downstroke in beetles and grasshoppers involves both direct and indirect muscles.

Contraction of flight muscles has two basic types of neural control: **synchronous** and **asynchronous.** Larger insects such as dragonflies and butterflies have synchronous muscles, in which a single volley of nerve impulses stimulates a muscle contraction and thus one wing stroke. Asynchronous muscles are found in more specialized insects. Their mechanism of action is complex and depends on storage of potential energy in resilient parts of the thoracic cuticle. As one set of muscles contracts (moving the wing in one direction), they stretch the antagonistic set of muscles, causing them to contract (and move the wing in the other direction). Because the muscle contractions are not phase-related to nervous stimulation, only occasional nerve impulses are necessary to keep the muscles responsive to alternating stretch activation. Thus extremely rapid wing beats are possible. For example, butterflies (with synchronous muscles) may beat as few as four times per second. Insects with asynchronous muscles, such as flies and bees, may vibrate at 100 beats per second or more. Fruit flies, *Drosophila* (Gr. *drosos,* dew, + *philos,* loving), can fly at 300 beats per second, and midges have been clocked at more than 1000 beats per second!

Obviously flying entails more than a simple flapping of wings; a forward thrust is necessary. As the indirect flight muscles alternate rhythmically to raise and lower the wings, the direct flight muscles alter the angle of the wings so that they act as lifting airfoils during both upstroke and downstroke, twisting the leading edge of the wings downward during downstroke and upward during upstroke. This modulation produces a figure-eight movement (figure 12.38C) that aids in spilling air from the trailing edges of the wings. The quality of the forward thrust depends, of course, on several factors, such as variations in wing venation, how much the wings are tilted, and how they are tapered.

Flight speeds vary. The fastest flyers usually have narrow, fast-moving wings with a strong tilt and a strong figure-eight component. Sphinx moths and horse flies are said to achieve approximately 48 km (30 miles) per hour and dragonflies approximately 40 km (25 miles) per hour. Some insects are capable of long continuous flights. Migrating monarch butter-

figure 12.38

A, Flight muscles of insects such as cockroaches, in which upstroke is by indirect muscles and downstroke is by direct muscles. **B,** In insects such as flies and bees, both upstroke and downstroke are by indirect muscles. **C,** The figure-eight path followed by the wing of a flying insect during the upstroke and downstroke.

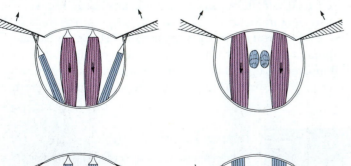

Direct flight muscles
of locusts and dragonflies

Indirect flight muscles
of flies and midges

A   B   C

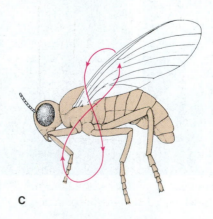

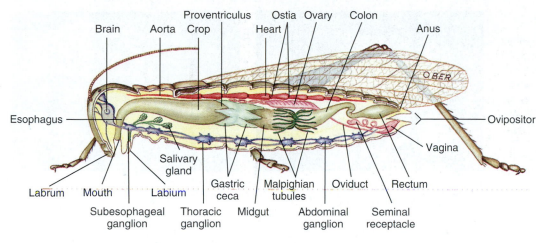

figure 12.39

Internal structure of female grasshopper.

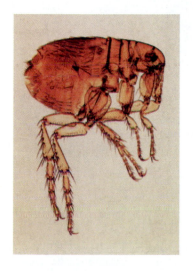

figure 12.40

Female human flea, *Pulex irritans* (order Siphonaptera).

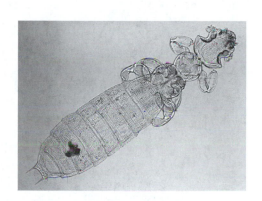

figure 12.41

Gliricola porcelli (order Mallophaga), a chewing louse of guinea pigs. Antennae are normally held in deep grooves on the sides of the head.

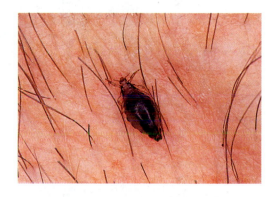

figure 12.42

The head and body louse of humans *Pediculus humanus* (order Anoplura) feeding.

flies, *Danaus plexippus* (Gr. after Danaus, mythical king of Arabia) (see figure 12.46) travel south for hundreds of miles in the fall, flying at a speed of approximately 10 km (6 miles) per hour.

Internal Form and Function

Nutrition   The digestive system (figure 12.39) consists of a **foregut** (mouth with salivary glands, **esophagus, crop** for storage, and **proventriculus** for grinding), **midgut** (stomach and gastric ceca), and **hindgut** (intestine, rectum, and anus). The foregut and hindgut are lined with cuticle, so absorption of food is confined largely to the midgut, although some absorption may occur in all sections. Most insects feed on plant juices and plant tissues. Such a food habit is called **phytophagous.** Some insects feed on specific plants; others, such as grasshoppers, can eat almost any plant. Caterpillars of many moths and butterflies eat foliage of only certain plants. Certain species of ants and termites cultivate fungus gardens as a source of food.

Many beetles and larvae of many insects live on dead animals **(saprophagous).** A number of insects are **predaceous,** catching and eating other insects as well as other types of animals.

Many insects are parasitic as adults, as larvae, or, in some cases, both juveniles and adults are parasites. For example, fleas (figure 12.40) live on blood of mammals as adults, but their larvae are free-living scavengers. Lice (figures 12.41 and 12.42) are parasitic throughout their life cycle. Many parasitic insects are themselves parasitized by other insects, a condition known as **hyperparasitism.** Larvae of many varieties of wasps live inside the bodies of spiders or other insects (figure 12.43B), consuming their hosts and eventually killing them. Because they always destroy their hosts, they are known as **parasitoids** (a particular type of parasite); typical parasites normally do not kill their hosts. Parasitoid insects are enormously important in controlling populations of other insects.

figure 12.43

A, Hornworm, larval stage of a sphinx moth (order Lepidoptera). The more than 100 species of North American sphinx moths are strong fliers and mostly nocturnal feeders. Their larvae, called hornworms because of the large fleshy posterior spine, are often pests of tomatoes, tobacco, and other plants. **B,** Hornworm parasitized by a tiny wasp, *Apanteles*, which laid its eggs inside the caterpillar. The wasp larvae have emerged, and their pupae are on the caterpillar's skin. Young wasps emerge in 5 to 10 days, and the caterpillar usually dies.

A

B

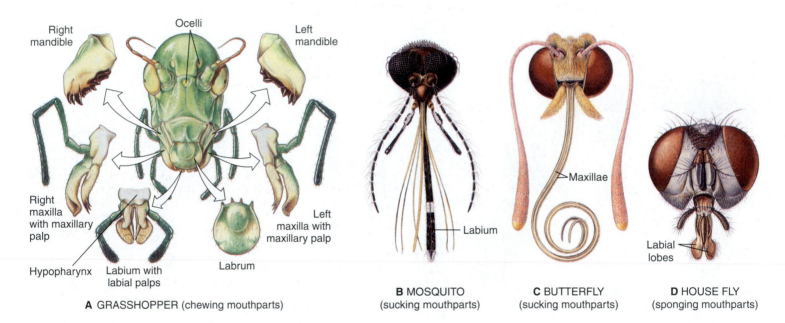

A GRASSHOPPER (chewing mouthparts)   **B** MOSQUITO (sucking mouthparts)   **C** BUTTERFLY (sucking mouthparts)   **D** HOUSE FLY (sponging mouthparts)

figure 12.44

Four types of insect mouthparts.

Feeding habits of insects are determined to some extent by their mouthparts, which are highly specialized for each type of feeding.

Biting and **chewing mouthparts,** such as those of grasshoppers and many herbivorous insects, are adapted for seizing and crushing food (figure 12.44). Mandibles of chewing insects are strong, toothed plates whose edges can bite or tear while the maxillae hold the food and pass it toward the mouth. Enzymes secreted by the salivary glands add chemical action to the chewing process.

Sucking mouthparts are greatly varied. House flies and fruit flies have no mandibles; their labium is modified into two soft lobes containing many small tubules that soak up liquids with a capillary action much as the holes of a commercial sponge do (figure 12.44). Horse flies, however, are fitted not only to soak up surface liquids but to bite into skin with slender, tapering mandibles and then collect blood. Mosquitos combine **piercing** by means of needlelike stylets and sucking through a food channel (figure 12.44). In honey bees the labium forms a flexible and contractile "tongue" covered with many hairs. When a bee plunges its proboscis into nectar, the tip of the tongue bends upward and moves back and forth rapidly. Liquid enters the tube by capillarity and is drawn up the tube continuously by a pumping pharynx. In butterflies and moths mandibles are usually absent, and maxillae are modified into a long sucking proboscis (figure 12.44) for drawing nectar from flowers. At rest, the proboscis is coiled into a flat spiral. In feeding it extends, and pharyngeal muscles pump fluid.

Circulation A tubular heart in the pericardial cavity (see figure 12.39) moves **hemolymph** (blood) forward through the only blood vessel, a dorsal aorta. The heartbeat is a peristaltic wave. Accessory pulsatory organs help to move hemolymph into wings and legs, and flow is also facilitated by various body movements. Hemolymph consists of plasma and amebocytes and apparently has little to do with oxygen transport. Hemolymph functions to distribute substances such as molting hormones and nutrients throughout the body. Oxygen transport is done by the **tracheal system,** an unusual system of tubes that pipe air directly to each cell.

Gas Exchange Terrestrial animals obtain oxygen from air as it dissolves across a wet membrane. Maintaining a wet membrane in a terrestrial environment is difficult, so gas exchange typically occurs in an internal cavity. The twin goals of an efficient respiratory system are to permit rapid oxygen–carbon dioxide exchange and to restrict water loss. In insects this is the function of the tracheal system, an extensive network of thin-walled tubes that branch into every part of the body (figure 12.45). Tracheal trunks open to the outside by paired **spiracles,** usually two on the thorax and seven or eight on the abdomen. A spiracle may be merely a hole in the integument, as in primitively wingless insects, but it is usually provided with a valve or other closing mechanism that decreases water loss. Evolution of such a device must have been very important in enabling insects to move into drier habitats.

Tracheae are composed of a single layer of cells and are lined with cuticle that is shed, along with the outer cuticle, during molts. Spiral thickenings of the cuticle, called **taenidia,** support the tracheae and prevent their collapse. Tracheae branch out into smaller tubes, ending in very fine, fluid-filled tubules called **tracheoles** (not lined with cuticle), which branch into a fine network over the cells. Scarcely any living cell is located more than a few micrometers away from a tracheole. In fact, the ends of some tracheoles actually indent the membranes of cells they supply, so that they terminate close to mitochondria. The tracheal system affords an efficient system of transport without use of oxygen-carrying pigments in hemolymph.

In some very small insects gas transport occurs entirely by diffusion along a concentration gradient. As oxygen is used, a partial vacuum develops in the tracheae, and air is sucked in through the spiracles. Larger or more active insects employ some ventilation device for moving air in and out of the tubes. Usually muscular movements in the abdomen perform the pumping action that draws air in or expels it.

The tracheal system is primarily adapted for breathing air, but many insects (nymphs, larvae, and adults) live in water. In small, soft-bodied aquatic nymphs, gaseous exchange may occur by diffusion through the body wall, usually into and out of a tracheal network just under the integument. Aquatic

Although diving beetles, *Dytiscus* (G. *dytikos,* able to swim), can fly, they spend most of their life in water as excellent swimmers. They use an "artificial gill" in the form of a bubble of air held under the first pair of wings. The bubble is kept stable by a layer of hairs on top of the abdomen and is in contact with the spiracles on the abdomen. Oxygen from the bubble diffuses into the tracheae and is replaced by diffusion of oxygen from the surrounding water. However, nitrogen from the bubble diffuses into the water, slowly decreasing the size of the bubble; therefore, diving beetles must surface every few hours to replace the air. Mosquito larvae are not good swimmers but live just below the surface, putting out short breathing tubes like snorkels to the surface for air. Spreading oil on the water, a favorite method of mosquito control, clogs the tracheae with oil and suffocates the larvae. "Rattailed maggots" of syrphid flies have an extensible tail that can stretch as much as 15 cm to the water surface.

figure 12.45

A, Relationship of spiracle, tracheae, taenidia (chitinous bands that strengthen the tracheae), and tracheoles (diagrammatic). **B,** Generalized arrangement of insect tracheal system (diagrammatic). Air sacs and tracheoles not shown. **C,** Spiracles and tracheae of a caterpillar visible through its transparent cuticle.

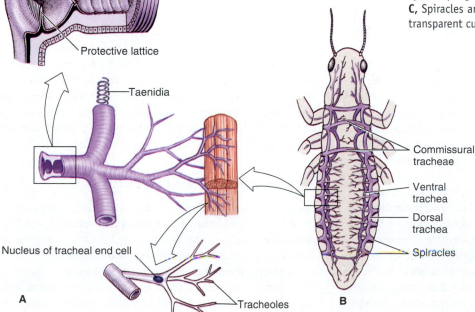

nymphs of stoneflies and mayflies are equipped with **tracheal gills,** which are thin extensions of the body wall containing a rich tracheal supply. Gills of dragonfly nymphs are ridges in the rectum (rectal gills) where gas exchange occurs as water enters and leaves.

Excretion and Water Balance   Malpighian tubules (see figure 12.39) are typical of most insects. As in spiders (p. 223), Malpighian tubules are very efficient, both as excretory organs and as a means of conserving body fluids—an important factor in the success of terrestrial animals.

Because water requirements vary among different types of insects, this ability to cycle water and salts is very important. Insects living in dry environments may resorb nearly all water from the rectum, producing a nearly dry mixture of urine and feces. Leaf-feeding insects take in and excrete large quantities of fluid. Freshwater larvae need to excrete water and conserve salts. Insects that feed on dry grains need to conserve water and excrete salt.

Nervous System   The nervous system in general resembles that of larger crustaceans, with a similar tendency toward fusion of ganglia (see figure 12.39). Some insects have a giant fiber system. There is also a visceral nervous system that corresponds in function with the autonomic nervous system of vertebrates. Neurosecretory cells located in various parts of the brain have an endocrine function, but, except for their role in molting and metamorphosis, little is known of their activity.

Sense Organs   Sensory perceptions of insects are usually keen. Organs receptive to mechanical, auditory, chemical, visual, and other stimuli are well developed. They are scattered over the body but are especially numerous on appendages.

Photoreceptors include both ocelli and compound eyes. Compound eyes are large and constructed of ommatidia, like those of crustaceans (p. 232). Apparently, visual acuity in insect eyes is much lower than that of human eyes, but most flying insects rate much higher than humans in flicker-fusion tests. Flickers of light become fused in human eyes at a frequency of 45 to 55 per second, but in bees and blow flies they do not fuse until 200 to 300 per second. This would be an advantage in analyzing a fast-changing landscape.

Most insects have three ocelli on their head, and they also have dermal light receptors on their body surface, but not much is known about them.

Sounds may be detected by sensitive hairlike **sensilla** or by tympanic organs sensitive to sonic or ultrasonic sound. Sensilla are modifications in the cuticular surface for reception of sensory stimuli other than light and are supplied with one or more neurons. Tympanic organs, found in grasshoppers (figure 12.35B), crickets, cicadas, butterflies, and moths, involve a number of sensory cells extending to a thin tympanic membrane that encloses an air space in which vibrations can be detected.

Chemoreceptive sensilla, which are peglike or setae, are especially abundant on the antennae, mouthparts, or legs. Mechanical stimuli, such as contact pressure, vibrations, and

tension changes in the cuticle, are detected by sensilla or by sensory cells in the epidermis. Insects also sense temperature, humidity, body position (proprioception), gravity, etc.

Reproduction   Sexes are separate in insects, and fertilization is usually internal. Insects have various means of attracting mates. Female moths emit a chemical (pheromone) that can be detected for a great distance by males—several miles from females. Fireflies use flashes of light; some insects find each other by means of sounds or color signals and by various kinds of courtship behavior.

Sperm are usually deposited in the vagina of females at the time of copulation (see figure 12.35A). In some orders sperm are encased in spermatophores that may be transferred at copulation or deposited on the substratum to be collected by a female. A male silverfish deposits a spermatophore on the ground, then spins signal threads to guide a female to it. During the evolutionary transition from aquatic to terrestrial life, spermatophores were widely used, with copulation evolving much later.

Usually sperm are stored in the seminal receptacle of a female in numbers sufficient to fertilize more than one batch of eggs. Many insects mate only once during their lifetime, and none mates more than a few times. Sperm storage allows fertilization to occur much later.

Insects usually lay a great many eggs. A queen honey bee, for example, may lay more than 1 million eggs during her lifetime. On the other hand, some flies are ovoviviparous and bring forth only a single offspring at a time. Forms that make no provision for care of young usually lay many more eggs than those that provide for young or those that have a very short life cycle.

Most species lay their eggs in a particular habitat to which they are guided by visual, chemical, or other clues. Butterflies and moths lay their eggs on the specific kind of plant on which their caterpillars must feed. A tiger moth may look for a pigweed, a sphinx moth for a tomato or tobacco plant, and a monarch butterfly for a milkweed plant (figure 12.46). Insects whose immature stages are aquatic lay their eggs in water. A tiny braconid wasp lays her eggs on a caterpillar of the sphinx moth where they will feed and pupate in tiny white cocoons (see figure 12.43B). An ichneumon wasp, with unerring accuracy, seeks out a certain kind of larva in which her young will live as internal parasites. Her long ovipositors may have to penetrate 1 to 2 cm of wood to find and deposit her eggs in the larva of a wood wasp or a wood-boring beetle (figure 12.47).

Metamorphosis and Growth

Although many animals undergo a metamorphosis, insects illustrate it more dramatically than any other group. The transformation of a caterpillar into a beautiful moth or butterfly is indeed an astonishing morphological change.

Early development occurs within the eggshell, and hatching young escape from the capsule in various ways. During postembryonic development most insects change in form; they undergo **metamorphosis.** A number of molts are neces-

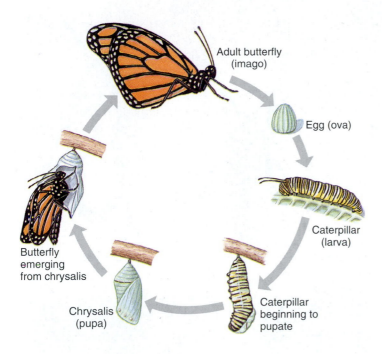

figure 12.46

Holometabolous (complete) metamorphosis in a butterfly, *Danaus plexippus* (order Lepidoptera). Eggs hatch to produce first of several larval instars. Last larval instar molts to become a pupa. Adult emerges at pupal molt.

figure 12.47

An ichneumon wasp with the end of the abdomen raised to thrust her long ovipositor into wood to find a tunnel made by the larva of a wood wasp or wood-boring beetle. She can bore 13 mm or more into wood to lay her eggs in the larva of the wood-boring beetle, which will become host for the ichneumon larvae. Other ichneumon species attack spiders, moths, flies, crickets, caterpillars, and other insects.

sary during the growth period, and each stage between molts is called an **instar.**

Approximately 88% of insects go through **holometabolous (complete) metamorphosis** (Gr. *holo,* complete, + *metabolē,* change) (see figure 12.46), which separates physiological processes of growth **(larva)** from those of differentiation **(pupa)** and reproduction **(adult).** Each stage functions efficiently without competition with other stages, because the larvae often live in entirely different surroundings and eat different foods from adults. The wormlike larvae, which usually have chewing mouthparts, are known by various common names, such as caterpillars, maggots, bagworms, fuzzy worms, and grubs. After a series of instars a larva forms a case or cocoon about itself and becomes a pupa, or chrysalis, a nonfeeding stage in which many insects pass the winter. When the final molt occurs a full-grown adult emerges (see figure 12.46), pale and with wings wrinkled. In a short time the wings expand and harden, and the insect is on its way. The stages are **egg, larva** (several instars), **pupa,** and **adult.** Adults undergo no further molting.

Some insects undergo **hemimetabolous (gradual, or incomplete) metamorphosis** (Gr. *hemi,* half, + *metabolē* change) (figure 12.48). These include insects such as bugs, scale insects, lice, and grasshoppers, which have terrestrial young, and mayflies, stoneflies (figure 12.49A), and dragonflies (figure 12.49B), which lay their eggs in water. The young are called **nymphs** or simply juveniles (figure 12.49C), and their wings develop externally as budlike outgrowths in the early instars and increase in size as the animal grows by successive

molts and becomes a winged adult (figure 12.50). Aquatic nymphs have tracheal gills or other modifications for aquatic life. The stages are **egg, nymph** (several instars), and **adult.**

A few insects, such as silverfish (see figure 12.60) and springtails, undergo direct development. The young, or juveniles, are similar to the adults except in size and sexual maturation. The stages are egg, juvenile, and adult. Such insects include the primitively wingless insects.

The *biological* meaning of the word "bug" is much more restrictive than in common English usage. People often refer to all insects as "bugs," even extending the word to include such nonanimals as bacteria, viruses, and glitches in computer programs. Strictly speaking, however, a bug is a member of order Hemiptera and nothing else.

Hormones control and regulate metamorphosis in insects. Three major endocrine organs guide development through juvenile instars and eventually emergence of adults. These organs and the hormones they produce are the **brain (ecdysiotropin), ecdysial (prothoracic) glands (ecdysone),** and **corpora allata (juvenile hormone).** Hormonal control of molting and metamorphosis is the same in holometabolous and hemimetabolous insects.

Diapause   Many animals can enter a state of dormancy during adverse conditions, and there are periods in the life cycle of many insects when a particular stage can remain dormant

figure 12.48

Life history of a
hemimetabolous insect.

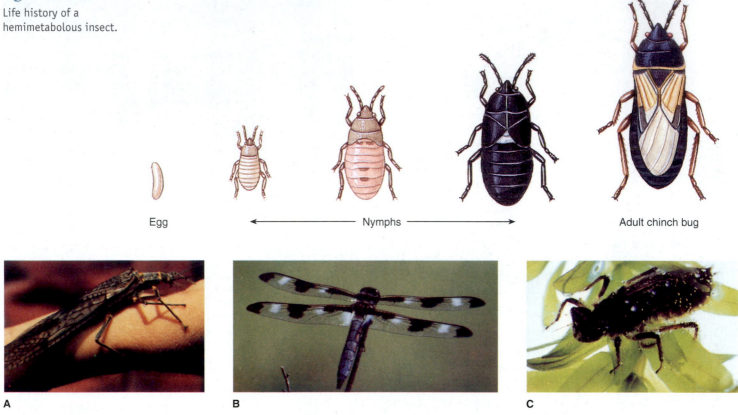

Egg   ←——————— Nymphs ———————→   Adult chinch bug

A   B   C

figure 12.49

A, A stonefly, *Perla* sp. (order Plecoptera). **B,** A ten-spot dragonfly, *Libellula pulchella* (order Odonata). **C,** Nymph (juvenile) of a dragonfly. Both stone-flies and dragonflies have aquatic larvae that undergo gradual metamorphosis.

figure 12.50

A, Ecdysis in a cicada, *Tibicen davisi* (order Homoptera). The old cuticle splits along a dorsal midline as a result of increased blood pressure and of air forced into the thorax by muscle contraction. The emerging insect is pale, and its new cuticle is soft. The wings will be expanded by hemolymph pumped into veins, and the insect enlarges by taking in air. **B,** An adult *Tibicen davisi*.

A   B

for a long time because external climatic conditions are too harsh for normal activity. Most insects enter such a stage facultatively when some environmental factor, such as temperature, becomes unfavorable, and the state continues until conditions again become favorable.

However, some species have a prolonged arrest of growth that is internally programmed and is usually seasonal. This type of dormancy is called **diapause** (di′a-poz) (Gr. *dia,* through, dividing into two parts, + *pausis,* a stopping), and it is an important adaptation to survive adverse environmental conditions. Diapause usually is triggered by some external sig-

nal, such as shortening day length. Diapause always occurs at the end of an active growth stage of the molting cycle so that, when the diapause period is over, the insect is ready for another molt.

Behavior and Communication

The keen sensory perceptions of insects make them extremely responsive to many stimuli. Stimuli may be internal (physiological) or external (environmental), and responses are governed by both the physiological state of the animal and the pattern of

figure 12.51

Dung beetles *Canthon pilularis* (order Coleoptera), chew off a bit of dung, roll it into a ball, and then roll it to where they will bury it in soil. One beetle pushes while the other pulls. Eggs are laid in the ball, and larvae feed on dung. Dung beetles are black, an inch or less in length, and common in pastures.

nerve pathways involved. Many responses are simple, such as orientation toward or away from a stimulus, for example, attraction of a moth to light, avoidance of light by a cockroach, or attraction of carrion flies to the odor of dead flesh.

Much behavior of insects, however, is not a simple matter of orientation but involves a complex series of responses. A pair of dung beetles chew off a bit of dung, roll it into a ball, and roll the ball laboriously to where they intend to bury it, after laying their eggs in it (figure 12.51). A female cicada slits the bark of a twig and then lays an egg in each of the slits. A female potter wasp *Eumenes* scoops up clay into pellets, carries them one by one to her building site, and fashions them into dainty little narrow-necked clay pots, into each of which she lays an egg. Then she hunts and paralyzes a number of caterpillars, pokes them into the opening of a pot, and closes the opening with clay. Each egg, in its own protective pot, hatches to find a well-stocked larder of food awaiting it.

Some insects can memorize and perform in sequence tasks involving multiple signals in various sensory areas. Worker honeybees have been trained to walk through mazes that involved five turns in sequence, using such clues as color of a marker, distance between two spots, or angle of a turn. The same is true of ants. Workers of one species of *Formica* learned a six-point maze at a rate only two or three times slower than that of laboratory rats. Foraging trips of ants and bees often wind and loop in a circuitous route, but once the forager has found food, the return trip is relatively direct. One investigator suggested that the continuous series of calculations necessary to figure the angles, directions, distance, and speed of the trip and to convert it into a direct return could involve a stopwatch, a compass, and integral vector calculus. How an insect does it is unknown.

Much of such behavior is "innate," meaning that entire sequences of actions apparently have been programmed. How-

ever, much more learning is involved than we once thought. A potter wasp, for example, must learn where she has left her pots if she is to return to fill them with caterpillars one at a time. Social insects, which have been studied extensively, are capable of most of the basic forms of learning used by mammals. An exception is insight learning. Apparently insects, when faced with a new problem, cannot reorganize their memories to construct a new response.

Insects communicate with other members of their species by chemical, visual, auditory, and tactile signals. **Chemical signals** take the form of **pheromones,** which are substances secreted by one individual that affect behavior or physiological processes of another individual. Examples of pheromones include sex attractants, releasers of certain behavior patterns, trail markers, alarm signals, and territorial markers. Like hormones, pheromones are effective in minute quantities. Social insects, such as bees, ants, wasps, and termites, can recognize a nestmate—or an alien in the nest—by means of identification pheromones. Pheromones determine caste in termites, and to some extent in ants and bees. In fact, pheromones are probably a primary integrating force in populations of social insects. Many insect pheromones have been extracted and chemically identified.

Sound production and **reception** (phonoproduction and phonoreception) in insects have been studied extensively, and although a sense of hearing is not present in all insects, this means of communication is meaningful to insects that use it. Sounds serve as warning devices, advertisement of territorial claims, or courtship songs. Sounds of crickets and grasshoppers function in courtship and aggression. Male crickets scrape the modified edges of their forewings together to produce their characteristic chirping. The long, drawn-out sound of male cicadas, a call to attract females, is produced by vibrating membranes in a pair of organs located on the ventral side of the basal abdominal segment.

There are many forms of **tactile communication,** such as tapping, stroking, grasping, and antennae touching, which evoke responses varying from recognition to recruitment and alarm. Certain kinds of flies, springtails, and beetles manufacture their own **visual signals** in the form of **bioluminescence.** The best known of luminescent beetles are fireflies, or lightning bugs (which are neither flies nor bugs, but beetles), in which a flash of light helps to locate a prospective mate. Each species has its own characteristic flashing rhythm produced on the ventral side of the last abdominal segments. Females flash an answer to the species-specific pattern to attract males. This interesting "love call" has been adopted by species of *Photuris*, which prey on male fireflies of other species they attract (figure 12.52).

Social Behavior   Insects rank very high in the animal kingdom in their organization of social groups, and cooperation within more complex groups depends heavily on chemical and tactile communication. Social communities are not all complex, however. Some community groups are temporary and uncoordinated, as are hibernating associations of carpenter bees or feeding gatherings of aphids (figure 12.53). Some are coordinated

figure 12.52

Firefly femme fatale, *Photuris versicolor*, eating a male *Photinus tany-toxus*, which she has attracted with false mating signals.

figure 12.53

An ant (order Hymenoptera) tending a group of aphids (order Homoptera). The aphids feed copiously on plant juices and excrete the excess as a clear liquid rich in carbohydrates ("honey-dew"), which is cherished as a food by ants.

for only brief periods, such as the tent caterpillars *Malacosoma,* that join in building a home web and a feeding net. However, all these are still open communities with social behavior.

In true societies of some orders, such as Hymenoptera (honeybees and ants) and Isoptera (termites), a complex social

figure 12.54

Queen bee surrounded by her court. The queen is the only egg layer in the colony. The attendants, attracted by her pheromones, constantly lick her body. As food is transferred from these bees to others, the queen's presence is communicated throughout the colony.

life occurs. Such societies are closed. They involve all stages of the life cycle, communities are usually permanent, all activities are collective, and there is reciprocal communication. There is a high degree of efficiency in division of labor. Such a society is essentially a family group in which the mother or perhaps both parents remain with young, sharing duties of the group in a cooperative manner. The society usually demonstrates polymorphism, or **caste** differentiation.

Honey bees have one of the most complex social organizations in the insect world. Instead of lasting one season, their organization continues for a more or less indefinite period. As many as 60,000 to 70,000 honey bees may live in a single hive. Of these, there are three castes—a single sexually mature female, or **queen,** a few hundred **drones,** which are sexually mature males, and thousands of **workers,** which are sexually inactive genetic females (figure 12.54).

Workers take care of young, secrete wax with which they build the six-sided cells of the honeycomb, gather nectar from flowers, manufacture honey, collect pollen, and ventilate and guard the hive. One drone, sometimes more, fertilizes the queen during the mating flight, at which time enough sperm are stored in her seminal receptacle to last her lifetime.

Castes are determined partly by fertilization and partly by what is fed to the larvae. Drones develop parthenogeneti-

A   B

figure 12.55

A, Termite workers, *Reticulitermes flavipes* (order Isoptera), eating yellow pine. Workers are wingless sterile adults that tend the nest and care for the young. **B,** Termite queen becomes a distended egg-laying machine. The queen and several workers and soldiers are shown here.

cally from unfertilized eggs (and consequently are haploid); queens and workers develop from fertilized eggs (and thus are diploid; see haplodiploidy, p. 171). Female larvae that will become queens are fed **royal jelly,** a secretion from the salivary glands of nurse workers. Royal jelly differs from the "worker jelly" fed to ordinary larvae, but components in it that are essential for queen determination have not yet been identified. Honey and pollen are added to worker diet about the third day of larval life. Pheromones in "queen substance," which is produced by the queen's mandibular glands, prevent female workers from maturing sexually. Workers produce royal jelly only when the level of "queen substance" pheromone in the colony drops. This change occurs when the queen becomes too old, dies, or is removed. Then workers start enlarging a larval cell and feeding a larva royal jelly that produces a new queen.

Honey bees have evolved an efficient system of communication by which, through certain body movements, their scouts inform workers of the location and quantity of food sources.

Termite colonies contain several castes, consisting of fertile individuals, both males and females, and sterile individuals (figure 12.55). Some fertile individuals may have wings and may leave the colony, mate, lose their wings, and as **king** and **queen** start a new colony. Wingless fertile individuals may under certain conditions substitute for the king or queen. Sterile members are wingless and become **workers** and **soldiers.** Soldiers have large heads and mandibles and serve for defense of the colony. As in bees and ants, extrinsic factors cause caste differentiation. Reproductive individuals and soldiers secrete inhibiting pheromones that pass throughout the colony to nymphs through a mutual feeding process, called **trophallaxis,** so that they become sterile workers. Workers also produce pheromones, and if the level of "worker substance" or "soldier substance" falls, as might happen after an attack by marauding predators, for example, the next generation produces compensating proportions of the appropriate caste.

Ants also have highly organized societies. Superficially, they resemble termites, but they are quite different (belong to a different order) and can be distinguished easily. In contrast to termites, ants are usually dark in color, are hard bodied, and have a constriction posterior to their first abdominal segment.

Entomologists have chosen to describe insect societies by borrowing terms commonly used to describe human societies: queen, king, royal, soldier, worker, and caste. This usage can be misleading by implying a correspondence between human and insect societies that does not exist. For example, "queen" suggests a position of political power in human societies, which has no correspondence to the reproductive female designated a "queen" of a bee colony. Confusion of these terms led a famous population geneticist, Ronald Fisher, to argue that human societies could achieve greater stability by emulating insect societies, specifically by concentrating reproduction among members of upper classes. This argument is now considered an embarrassment of his otherwise highly regarded and influential book, *The Genetical Theory of Natural Selection* (1930).

In ant colonies males die soon after mating and the queen either starts her own new colony or joins an established colony and does the egg laying. Sterile females are wingless workers and soldiers that do the work of the colony—gather food, care for young, and protect the colony. In many larger colonies there may be two or three types of individuals within each caste.

Ants have evolved some striking patterns of "economic" behavior, such as making slaves, farming fungi, herding "ant cows" (aphids or other homopterans, see figure 12.53), sewing their nests together with silk (figure 12.56), and using tools.

Insects and Human Welfare

Beneficial Insects   Although most of us think of insects primarily as pests, humanity would have great difficulty surviving if all insects were suddenly to disappear. Insects are necessary for cross-fertilization (pollination) of many crops. Bees pollinate over $10 billion worth of food crops per year in the United States alone, and this value does not include pollination of forage crops for livestock or pollination by other insects. In addition, some insects produce useful materials: honey and beeswax from bees, silk from silkworms, and shellac from a wax secreted by lac insects.

Very early in their evolution insects and flowering plants formed a relationship of mutual adaptations that have been to each other's advantage. Insects exploit flowers for food, and

figure 12.56

A weaver ant nest in Australia.

flowers exploit insects for pollination. Each floral development of petal and sepal arrangement is correlated with the sensory adjustment of certain pollinating insects. Among these mutual adaptations are amazing devices of allurements, traps, specialized structure, and precise timing.

Many predaceous insects, such as tiger beetles, aphid lions, ant lions, praying mantids, and ladybird beetles, destroy harmful insects (figure 12.57A, B). Some insects control harmful ones by parasitizing them or by laying their eggs where their young, when hatched, may devour the host (figure 12.57C).

Dead animals are quickly consumed by maggots hatched from eggs laid on carcasses.

Insects and their larvae serve as an important source of food for many birds, fish, and other animals.

Harmful Insects   Harmful insects such as grasshoppers, chinch bugs, corn borers, boll weevils, grain weevils, San Jose scale, and scores of others (figure 12.58) include those that eat and destroy plants and fruits. Practically every cultivated crop has several insect pests. Humans expend enormous resources in all agricultural activities, in forestry, and in the food industry to counter insects and the damage they engender. Outbreaks of bark beetles or defoliators such as spruce budworms and gypsy moths have generated tremendous economic losses and have become a major element in determining the composition of forests in the United States. Gypsy moths, introduced into the United States in 1869 in an ill-advised attempt to breed a better silkworm, have spread throughout the Northeast as far south as Virginia. They defoliate oak forests in years when there are outbreaks. In 1981, they defoliated 13 million acres in 17 northeastern states.

Lice, blood-sucking flies, warble flies, bot flies, and many others attack humans or domestic animals or both. Malaria, carried by the *Anopheles* mosquito (figure 12.59), is still one of the world's major diseases; mosquitos also transmit yellow fever and lymphatic filariasis. Fleas carry plague, which at times in history has eradicated significant portions of human populations. House flies are vectors of typhoid, as are lice for typhus fever; tsetse flies carry African sleeping sickness; and certain blood-sucking bugs, *Rhodnius* and related genera, transmit Chagas' disease.

There is tremendous destruction of food, clothing, and property by weevils, cockroaches, ants, clothes moths, termites, and carpet beetles. Not the least of the insect pests are bed bugs, *Cimex,* blood-sucking hemipterous insects that humans contracted from bats that shared their caves early in human evolution.

A

B

C

figure 12.57

Some beneficial insects. **A,** A predaceous stink bug (order Hemiptera) feeds on a caterpillar. Note the sucking proboscis of the bug. **B,** A ladybird beetle ("ladybug," order Coleoptera). Adults (and larvae of most species) feed voraciously on plant pests such as mites, aphids, scale insects, and thrips. **C,** A parasitic wasp (*Larra bicolor*) attacking a mole cricket. The wasp drives the cricket from its burrow, then stings and paralyzes it. After the wasp deposits her eggs, the mole cricket recovers and resumes an active life—until it is killed by developing wasp larvae.

A   B   C

figure 12.58

Insect pests. **A,** Japanese beetles, *Popillia japonica* (order Coleoptera) are serious pests of fruit trees and ornamental shrubs. They were introduced into the United States from Japan in 1917. **B,** Longtailed mealybug, *Pseudococcus longispinus* (order Homoptera). Many mealybugs are pests of commercially valuable plants. **C,** Corn ear worms, *Heliothis zea* (order Lepidoptera). An even more serious pest of corn is the infamous corn borer, an import from Europe in 1908 or 1909.

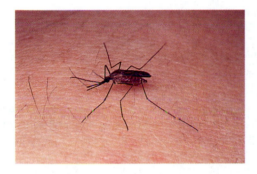

figure 12.59

A mosquito, *Anopheles quadrimaculatus* (order Diptera). *Anopheles* spp. are vectors of malaria.

West Nile virus is a disease agent spread by mosquitoes that affects mammals and birds throughout the world. First identified in Uganda in 1937, it spread to North America in 1999. Birds serve as a reservoir for the virus: a bird bitten by infected mosquito plays host to the virus for one to four days, during which time the virus can be picked up by other mosquitoes and spread to new hosts. Human response to infection varies, with about 80% showing no symptoms, close to 20% exhibiting flu-like symptoms of fever and aches, and less than 1% developing potentially fatal encephalitis or other possibly permanent neurological effects. Preventing mosquito bites is the best way to avoid infection, so mosquito behavior is of interest. Researchers wonder whether mosquitoes that initially feed on infected birds tend to choose another bird for the next bite, or simply feed opportunistically on any available animal. Understanding virus transmission makes it possible to develop mathematical models to predict how and when the disease will spread.

Control of Insects Because all insects are an integral part of the ecological communities to which they belong, their total destruction would probably do more harm than good. Food chains would be disturbed, some of our favorite birds would disappear, the biological cycles by which dead animal and plant matter disintegrates and returns to enrich the soil would be seriously impeded. The beneficial roles of insects in our environment are often overlooked, and in our zeal to control the pests we spray the landscape indiscriminately with extremely effective "broad-spectrum" insecticides that eradicate good, as well as harmful, insects. We have also found, to our dismay, that many chemical insecticides persist in the environment and accumulate as residues in the bodies of animals higher in the food chain. Furthermore, many insects have developed a resistance to insecticides in common use.

In recent years, methods of control other than chemical insecticides have been under intense investigation, experimentation, and development. Economics, concern for the environment, and consumer demand are causing thousands of farmers across the United States to use alternatives to strict dependence on chemicals.

Several types of biological controls have been developed and are under investigation. All of these areas present problems but also show great possibilities. One is the use of bacterial, viral, and fungal pathogens. A bacterium, *Bacillus thuringiensis,* is quite effective in control of lepidopteran pests (cabbage looper, imported cabbage worm, tomato worm, gypsy moth). Other strains of *B. thuringiensis* attack insects in other orders, and the species diversity of target insects is being widened by techniques of genetic engineering. Genes coding for the toxin produced by *B. thuringiensis* (Bt) also have been introduced into DNA of the plants themselves, which makes the plants resistant to insect attack, and Bt is harmless to humans. Genes for Bt and for herbicide resistance have been incorporated into

classification of Subphylum Hexapoda

Hexapods are divided into orders chiefly on the basis of morphology and developmental features. Entomologists do not all agree on the names of orders or on the limits of each order. Some tend to combine and others to divide groups. However, the following synopsis of major orders is one that is rather widely accepted.

Class Entognatha

Order Protura (pro-tu′ra) (Gr. *protos,* first, + *oura,* tail*)*. Minute (1 to 1.5 mm); no eyes or antennae; appendages on abdomen as well as thorax; live in soil and dark, humid places; slight, gradual metamorphosis.

Order Diplura (dip-lu′ra) (Gr. *diploos,* double, + *oura,* tail): **japygids.** Usually less than 10 mm; pale, eyeless; a pair of long terminal filaments or pair of caudal forceps; live in damp humus or rotting logs; development direct.

Order Collembola (col-lem′bo-la) (Gr. *kolla,* glue, + *embolon,* peg, wedge): **springtails** and **snowfleas.** Small (5 mm or less); no eyes; respiration by trachea or body surface; a springing organ folded under the abdomen for leaping; abundant in soil; sometimes swarm on pond surface film or on snowbanks in spring; development direct.

Class Insecta

Order Thysanura (thy-sa-nu′ra) (Gr. *thysanos,* tassel, + *oura,* tail): **silverfish** (figure 12.60) and **bristletails.** Small to medium size; large eyes; long antennae; three long terminal cerci; live under stones and leaves and around human habitations; development direct.

Order Ephemeroptera (e-fem-er-op′ter-a) (Gr. *ephēmeros,* lasting but a day, + *pteron,* wing): **mayflies** (figure 12.61). Wings membranous; forewings larger than hindwings; adult mouthparts vestigial; nymphs aquatic, with lateral tracheal gills, hemimetabolous development.

Order Odonata (o-do-na′ta) (Gr. *odontos,* tooth, + *ata,* characterized by): **dragonflies** (see figure 12.49B), **damselflies.** Large; membranous wings are long, narrow, net veined, and similar in size; long and slender body; aquatic nymphs with aquatic gills and prehensile labium for capture of prey; hemimetabolous development.

Order Orthoptera (or-thop′ter-a) (Gr. *orthos,* straight, + *pteron,* wing): **grasshoppers, locusts, crickets, cockroaches, walkingsticks, praying mantids** (see figures 12.35 and 12.36). Wings when present, with forewings thickened and hindwings folded like a fan under forewings; chewing mouthparts; hemimetabolous development.

Order Isoptera (i-sop′ter-a) (Gr. *isos,* equal, + *pteron,* wing): **termites** (see figure 12.55). Small; membranous, narrow wings similar in size with few veins; wings shed at maturity; erroneously called "white ants"; distinguishable from true ants by broad union of thorax and abdomen; complex social organization; hemimetabolous development.

Order Mallophaga (mal-lof′a-ga) (Gr. *mallos,* wool, + *phagein,* to eat): **biting lice** (see figure 12.41). As large as 6 mm; wingless; chewing mouthparts; legs adapted for clinging to host; live on birds and mammals; hemimetabolous development.

figure 12.60

Silverfish *Lepisma* (order Thysanura) is often found in homes.

A   **B**

figure 12.61

Mayfly (order Ephemeroptera). **A,** Nymph. **B,** Adult.

Order Anoplura (an-o-plu′ra) (Gr. *anoplos,* unarmed, + *oura,* tail): **sucking lice** (see figure 12.42). Depressed body; as large as 6 mm; wingless; mouthparts for piercing and sucking; adapted for clinging to warm-blooded host; includes the head louse, body louse, crab louse, others; hemimetabolous development.

Order Hemiptera (he-mip′ter-a) (Gr. *hemi,* half + *pteron,* wing) **(Heteroptera): true bugs** (see figure 12.57A). Size 2 to 100 mm; wings present or absent; forewings with basal portion leathery, apical portion membranous; hindwings membranous; at rest, wings held flat over abdomen; piercing-sucking mouthparts; many with odorous scent glands; include water scorpions, water striders, bed bugs, squash bugs, assassin bugs, chinch bugs, stink bugs, plant bugs, lace bugs, others; hemimetabolous.

Order Homoptera (ho-mop′ter-a) (Gr. *homos,* same, + *pteron,* wing): **cicadas** (figure 12.50), **aphids** (see figure 12.53), **scale insects, mealybugs** (see figure 12.58B), **leafhoppers, treehoppers** (figure 12.62). (Often included as suborder under Hemiptera.) If winged, either membranous or thickened forewings and membranous hindwings; wings held rooflike over body; piercing-sucking mouthparts; all plant eaters; some destructive; a few serving as source of shellac, dyes, etc.; some with complex life histories; hemimetabolous.

Order Neuroptera (neu-rop′ter-a) (Gr. *neuron,* nerve, + *pteron,* wing): **dobsonflies, ant lions** (figure 12.63), **lacewings.** Medium to large size; similar, membranous wings with many cross veins; chewing mouthparts; dobsonflies with greatly enlarged mandibles in males, and with aquatic larvae; ant lion larvae (doodlebugs) make craters in sand to trap ants; holometabolous.

Order Coleoptera (ko-le-op′ter-a) (Gr. *koleos,* sheath, + *pteron,* wing): **beetles** (see figures 12.57B; 12.58A), **fireflies** (see figure 12.52), **weevils.** The largest order of animals; forewings (elytra) thick, hard, opaque; membranous hindwings folded under forewings at rest; mouthparts for biting and chewing; includes ground beetles, carrion beetles, whirligig beetles, darkling beetles, stag beetles, dung beetles (see figure 12.51), diving beetles, boll weevils, fireflies, ladybird beetles (ladybugs), others; holometabolous.

Order Lepidoptera (lep-i-dop′ter-a) (Gr. *lepidos,* scale, + *pteron,* wing): **butterflies** and **moths** (see figures 12.46 and 12.58C). Membranous wings covered with overlapping scales, wings coupled at base; mouthparts a sucking tube, coiled when not in use; larvae (caterpillars) with chewing mandibles for plant eating, stubby prolegs on the abdomen, and silk glands for spinning cocoons; antennae knobbed in butterflies and usually plumed in moths; holometabolous.

Order Diptera (dip′ter-a) (Gr. *dis,* two, + *pteron,* wing): **true flies.** Single pair of wings, membranous and narrow; hindwings reduced to inconspicuous balancers (halteres); sucking mouthparts or adapted for sponging or lapping or piercing; legless larvae called maggots or, when aquatic, wigglers; include crane flies, mosquitos (see figure 12.59), moth flies, midges, fruit flies, flesh flies, house flies, horse flies (see figure 12.37), bot flies, blow flies, gnats, and many others; holometabolous.

Order Trichoptera (tri-kop′ter-a) (Gr. *trichos,* hair, + *pteron,* wing): **caddisflies.** Small, soft bodied; wings well-veined and hairy, folded rooflike over hairy body; chewing mouthparts; aquatic larvae construct cases of leaves, sand, gravel, bits of shell, or plant matter, bound together with secreted silk or cement; some make silk feeding nets attached to rocks in streams; holometabolous.

Order Siphonaptera (si-fon-ap′ter-a) (Gr. *siphon,* a siphon, + *apteros,* wingless): **fleas** (see figure 12.40). Small; wingless; bodies laterally compressed; legs adapted for leaping; no eyes; ectoparasitic on birds and mammals; larvae legless and scavengers; holometabolous.

Order Hymenoptera (hi-men-op′ter-a) (Gr. *hymen,* membrane, + *pteron,* wing): **ants, bees** (figure 12.54), **wasps.** Very small to large; membranous, narrow wings coupled distally; subordinate hindwings; mouthparts for biting and lapping up liquids; ovipositor sometimes modified into stinger, piercer, or saw; both social and solitary species; most larvae legless, blind, and maggotlike; holometabolous.

figure 12.62

Oak treehoppers *Platycotis vittata* (order Homoptera).

figure 12.63

Adult ant lion (order Neuroptera).

classification Phylum Arthropoda

Subphylum Trilobita (tri′lo-bi′ta) (Gr. *tri-,* three, + *lobos,* lobe): **trilobites.** All extinct forms; Cambrian to Permian; body divided by two longitudinal furrows into three lobes; distinct head, thorax, and abdomen; biramous appendages.

Subphylum Chelicerata (ke-lis′e-ra′ta) (Gr. *chēle,* claw, + *keratos,* a horn): **eurypterids, horseshoe crabs, spiders, ticks.** First pair of appendages modified to form chelicerae; pair of pedipalps and four pairs of legs; no antennae, no mandibles; cephalothorax and abdomen often with segments fused.

Class Merostomata (mer′o-sto′ma-ta) (Gr. *meros,* thigh, + *stomatos,* mouth): **aquatic chelicerates.** Cephalothorax and abdomen; compound lateral eyes; appendages with gills; sharp telson; **subclasses Eurypterida** (all extinct) and **Xiphosurida,** the horseshoe crabs.

Class Pycnogonida (pik′no-gon′i-da) (Gr. *pyknos,* compact, + *gonia,* knee, angle): **sea spiders.** Small (3 to 4 mm), but some reach 500 mm; body chiefly cephalothorax; tiny abdomen; usually four pairs of long walking legs (some with five or six pairs); one pair of subsidiary legs (ovigers) for egg bearing; mouth on long proboscis; four simple eyes; no respiratory or excretory system. Example: *Pycnogonum.*

Class Arachnida (ar-ack′ni-da) (Gr. *arachnē,* spider): **scorpions, spiders, mites, ticks, harvestmen.** Four pairs of legs; segmented or unsegmented abdomen with or without appendages and generally distinct from cephalothorax; respiration by gills, tracheae, or book lungs; excretion by Malpighian tubules or coxal glands; dorsal bilobed brain connected to ventral ganglionic mass with nerves; simple eyes; sexes separate; chiefly oviparous; no true metamorphosis. Examples: *Argiope, Centruroides.*

Subphylum Crustacea (crus-ta′she-a) (L. *crusta,* shell, + *acea,* group suffix): **crustaceans.** Mostly aquatic, with gills; cephalothorax usually with dorsal carapace; biramous appendages, modified for various functions; head appendages consisting of two pairs of antennae, one pair of mandibles, and two pairs of maxillae; sexes usually separate; development primitively with nauplius stage.

Class Branchiopoda (bran′kee-op′o-da) (Gr. *branchia,* gills, + *pous, podos,* foot): **branchiopods.** Flattened, leaflike swimming appendages (phyllopodia) with respiratory function. Examples: *Triops, Lynceus, Daphnia.*

Class Maxillopoda (maks′i-lop′o-da) (L. *maxilla,* jawbone, + Gr. *pous, podos,* foot): **ostracods, copepods, branchiurans, barnacles.** Five cephalic, six thoracic, and usually four abdominal segments; no typical appendages on abdomen; unique maxillopodan eye. Examples: *Cypris, Cyclops, Ergasilus, Argulus, Balanus.*

Class Malacostraca (mal′a-kos′tra-ka) (Gr. *malakos,* soft, + *ostrakon,* shell): **shrimps, crayfishes, lobsters, crabs.** Usually with eight thoracic and six abdominal segments, each with a pair of appendages. Examples: *Armadillidium, Gammarus, Megacytiphanes, Grapsus, Homarus, Panulirus.*

Subphylum Myriapoda (mir-ee-ap′o-da) (Gr. *myrias,* a myriad, + *pous, podus,* foot): **myriapods.** All appendages uniramous; head appendages consisting of one pair of antennae, one pair of mandibles, and one or two pairs of maxillae.

Class Diplopoda (di-plop′o-da) (Gr. *diploos,* double, + *pous, podos,* foot): **millipedes.** Subcylindrical body; head with short antennae and simple eyes; body with variable number of segments; short legs, usually two pairs of legs to a segment; separate sexes. Examples: *Julus, Spirobolus.*

Class Chilopoda (ki-lop′o-da) (Gr. *cheilos,* lip, + *pous, podos,* foot): **centipedes.** Dorsoventrally flattened body; variable number of segments, each with one pair of legs; one pair of long antennae; separate sexes. Examples: *Cermatia, Lithobius, Geophilus.*

Class Pauropoda (pau-rop′o-da) (Gr. *pauros,* small, + *pous, podos,* foot): **pauropods.** Minute (1 to 1.5 mm), cylindrical body consisting of double segments and bearing nine or ten pairs of legs; no eyes. Example: *Pauropus.*

Class Symphyla (sim′fi-la) (Gr. *syn,* together, + *phylon,* tribe): **garden centipedes.** Slender (1 to 8 mm) with long, threadlike antennae; body consisting of 15 to 23 segments with 10 to 12 pairs of legs; no eyes. Example: *Scutigerella.*

Subphylum Hexapoda (hek-′sap′oda) (Gr. *hex,* six + *pous, podus,* foot): **hexapods.** Body with distinct head, thorax, and abdomen; pair of antennae; mouthparts modified for different food habits; head of six fused segments; thorax of three segments; abdomen with variable number, usually 11 somites; thorax with two pairs of wings (sometimes one pair or none) and three pairs of jointed legs; separate sexes; usually oviparous; gradual or abrupt metamorphosis. (Insect orders on pp. 252–253). Bases of mouthparts exposed and exiting head capsule; mandibles generally have two regions of articulation.

Class Entognatha (en′tog-natha) (Gr. *entos,* within, inside + *gnathos,* jaw): **entognaths.** Base of mouth parts lies within head capsule; mandibles have one articulation. Example: *Entomobrya.*

Class Insecta (in-sek′ta) (L. *insectus,* cut into): **insects.** Bases of mouthparts exposed and exiting head capsule; mandibles generally have two regions of articulation. Examples: *Drosophila, Bombus, Anopheles* (insect orders listed on pp. 252–253).

much of soybeans, corn, cotton, and canola produced in the United States, thus reducing need for hazardous chemical sprays. Concerns about health risks of consuming genetically modified crops have arisen, especially in Europe, but such fears are supported by little or no scientific evidence.

Introduction of natural predators or parasites of the insect pests has had some success. In the United States, vedalia beetles from Australia help to control the cottony-cushion scale on citrus plants, and numerous instances of control by use of insect parasites have been recorded. Introduction of exotic species for control of insect pests may have unexpected negative consequences, however, and should be done with caution.

Another approach to biological control is to interfere with reproduction or behavior of insect pests with sterile males or with naturally occurring organic compounds that act as hormones or pheromones. Such research, although very promising, is slow because of our limited understanding of insect behavior and the problems of isolating and identifying complex compounds that are produced in such minute amounts. Nevertheless, pheromones may play an important role in biological pest control in the future.

A systems approach called **integrated pest management** is practiced with many crops. This approach involves integrated utilization of all possible, practical techniques to contain pest infestations at a tolerable level—for example, cultural techniques (resistant plant varieties, crop rotation, tillage techniques, timing of sowing, planting, or harvesting, and others), use of biological controls, and sparing use of insecticides.

Phylogeny and Adaptive Radiation

Phylogeny

Whether phylum Arthropoda is monophyletic has long been controversial. Some scientists have contended that Arthropoda is polyphyletic and that some or all current subphyla are derived from different annelid-like ancestors that have undergone "arthropodization," the crucial hardening of the cuticle to form a stiffened exoskeleton. However, most other zoologists argue that derived similarities of the arthropod subphyla strongly support monophyly of the phylum. Debate continues regarding relationships among the four extant subphyla.

Insects and myriapods were united in subphylum Uniramia, but molecular data indicate that these taxa are not as closely related as previously assumed, so Hexapoda and Myriapoda are considered subphyla now. Crustaceans were traditionally allied with insects and myriapods in a group called Mandibulata because they all have mandibles, as contrasted with chelicerae. Critics of this traditional grouping have argued that the mandibles in each group were so different that they could not have been inherited from a common ancestor. However, advocates of the "mandibulate hypothesis" maintain that these differences are not so fundamental that they could not have been derived from a common ancestral mandible during the 550-million-year history of mandibulate taxa.

Another explanation for the numerous similarities between crustaceans and insects, such as basic structure of ommatidia and head primitively composed of five segments each with a pair of appendages, is that these two taxa form a clade within Arthropoda. Evidence of a close relationship between crustaceans and insects emerged from several studies using molecular data and prompted a reevaluation of morphological data. We depict crustaceans and hexapods as sister taxa (figure 12.64), but some evidence suggests that hexapods arose from within the crustacean group, making Crustacea paraphyletic unless it includes hexapods. The taxon Pancrustacea includes crustaceans and hexapods.

Phylogenetic placement of subphylum Myriapoda is highly controversial; according to the mandibulate hypothesis, the sister taxon to Pancrustacea would be Myriapoda, but other work suggests that Myriapoda and Chelicerata are sister taxa, together forming the sister taxon to Pancrustacea. Awaiting further research, we do not depict a branch order for these taxa, but instead show a polytomy (multiple branches coming from a single point) in figure 12.64.

Evolution within hexapods involved specialization of the first three postcephalic somites to become locomotor segments (thorax) and a loss or reduction of appendages on the rest of the body (abdomen). The wingless orders traditionally have been regarded as having the most primitive characteristics. Three wingless orders (Diplura, Collembola, Protura) have their mandibles and first maxillae located deeply in pouches in the head, a condition known as **entognathy.** All other insects are **ectognathous,** including the wingless order Thysanura. Ectognathous insects do not have their mandibles and maxillae in pouches, and they share other synapomorphies. Entognathous and ectognathous insects form sister groups, and Thysanura diverged from a common ancestor of ectognathous insects before the advent of flight, which unites the remaining ectognathous orders (figure 12.65).

An ancestral winged insect gave rise to three lines, which differed in their ability to flex their wings (figure 12.65). Two of these (Odonata and Ephemeroptera) have outspread wings. The other line has wings that can fold back over the abdomen. It branched into three groups: one with hemimetabolous metamorphosis and chewing mouthparts (Orthoptera, Dermaptera, Isoptera, Embioptera); one group with hemimetabolous metamorphosis and usually sucking mouthparts (Thysanoptera, Hemiptera, Homoptera, Mallophaga, Anoplura); and a group with holometabolous metamorphosis. Insects in the last group have the most elaborate life history and apparently form a clade.

How did the first arthropod evolve, and to which other phyla are arthropods closely related? Do all segmented phyla share a segmented common ancestor? A long-held hypothesis states that both annelids and arthropods originated from a common line of coelomate segmented protostomes from which two or more lines then diverged; a protoannelid line with laterally located parapodia and one or more protoarthropod lines with more ventrally located appendages. However, molecular evidence now supports a dramatically contrasting hypothesis: placement of annelids in superphylum Lophotrochozoa and

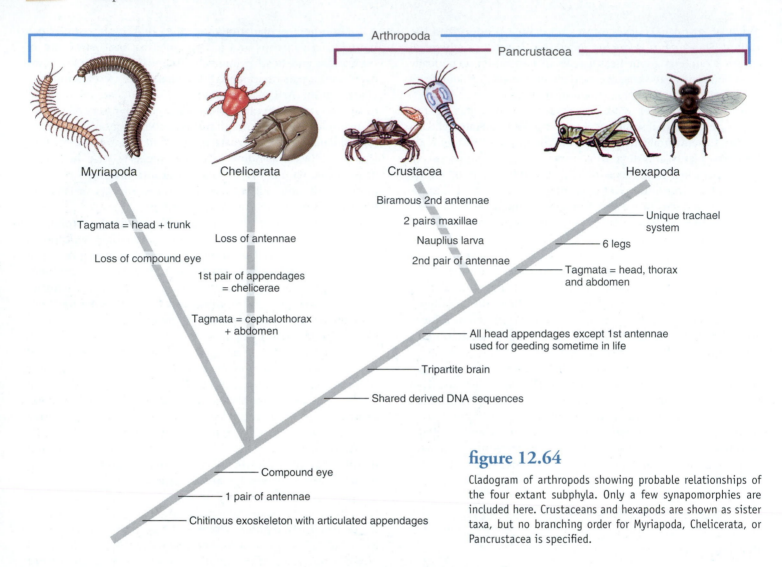

figure 12.64

Cladogram of arthropods showing probable relationships of the four extant subphyla. Only a few synapomorphies are included here. Crustaceans and hexapods are shown as sister taxa, but no branching order for Myriapoda, Chelicerata, or Pancrustacea is specified.

arthropods in superphylum Ecdysozoa. Alignment in separate superphyla not only requires that we abandon our view that annelids and arthropods are closely related, it implies that metamerism arose independently in the two groups and is a convergent character. Were metameric bodies convergent, the genetic controls and chemical signals used during development of a segmented body should differ among phyla. As discussed on p. 216, very preliminary comparisons suggest that annelids and arthropods do not share mechanisms of segmentation.

Assuming that arthropods belong within Ecdysozoa and are not closely related to annelids, several other phyla do seem to be allied with them. Phylum Tardigrada may be the sister taxon to arthropods, with phylum Onychophora being the sister taxon to the combined Arthropoda and Tardigrada (see Chapter 13). A cladogram depicting possible relationships is presented in Chapter 13 (p. 270).

The taxon Panarthropoda includes tardigrades and onychophorans with traditional arthropod subphyla. Placement of the peculiar wormlike pentastomids (p. 267) near arthropods has been controversial, but a recent study of gene arrangements indicates that phylum Pentastomida actually belongs *within* Arthropoda (see p. 269). It appears that these vertebrate parasites are highly derived crustaceans.

Adaptive Radiation

The adaptive trend in arthropods has been toward tagmatization of the body by differentiation or fusion of segments, giving rise in more derived groups to such tagmata as head and trunk; head, thorax, and abdomen; or cephalothorax (fused head and thorax) and abdomen. A series of similar appendages, one pair on each trunk segments, is the primitive character state, still retained by some crustaceans and by myriapods. More derived forms have appendages specialized for specific functions, and some appendages are lost entirely.

Much of the amazing diversity in arthropods seems to have developed because of modification and specialization of their cuticular exoskeleton and their jointed appendages, thus producing a wide variety of locomotor and feeding adaptations. Whether it be in the area of habitat, feeding adaptations, means of locomotion, reproduction, or general mode of living, adaptive achievements of arthropods are truly remarkable.

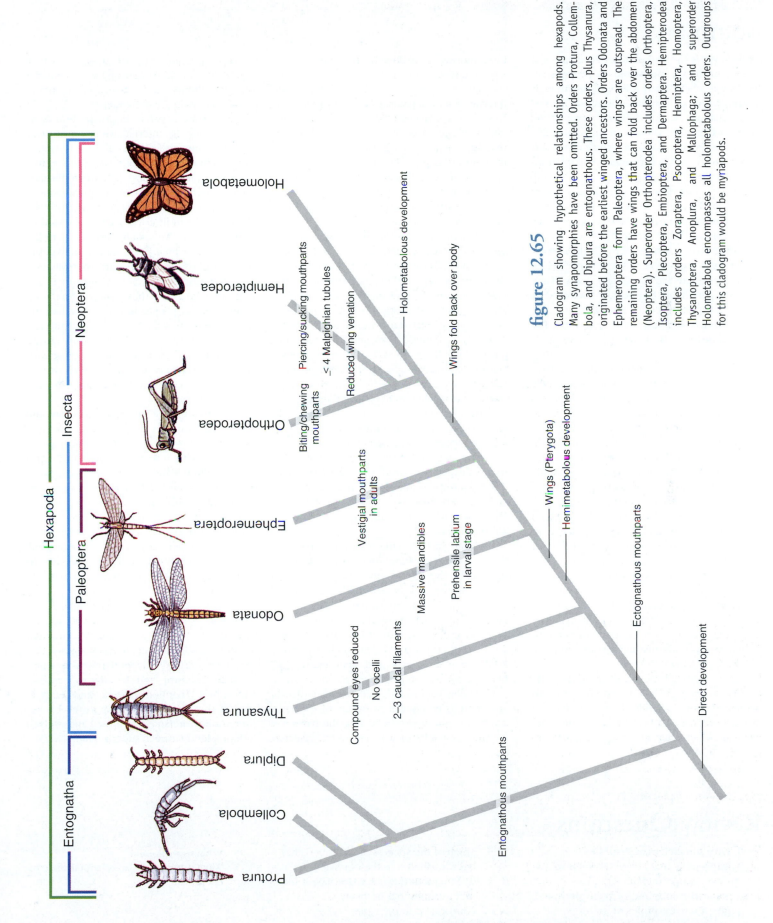

figure 12.65

Cladogram showing hypothetical relationships among hexapods. Many synapomorphies have been omitted. Orders Protura, Collembola, and Diplura are entognathous. These orders, plus Thysanura, originated before the earliest winged ancestors. Orders Odonata and Ephemeroptera form Paleoptera, where wings are outspread. The remaining orders have wings that can fold back over the abdomen (Neoptera). Superorder Orthopterodea includes orders Orthoptera, Isoptera, Plecoptera, Embioptera, and Dermaptera. Hemipterodea includes orders Zoraptera, Psocoptera, Hemiptera, Homoptera, Thysanoptera, Anoplura, and Mallophaga; and superorder Holometabola encompasses all holometabolous orders. Outgroups for this cladogram would be myriapods.

Summary

Arthropoda is the largest, most abundant and diverse phylum in the world. Arthropods are metameric, coelomate protostomes with well-developed organ systems. Most show marked tagmatization. They are extremely diverse and occur in all habitats capable of supporting life. Perhaps more than any other single factor, success of arthropods is explained by adaptations made possible by their cuticular exoskeleton. Other important elements in their success are jointed appendages, tracheal respiration, efficient sensory organs, complex behavior, metamorphosis, and the ability to fly.

All arthropods must periodically cast off their cuticle (ecdysis) and grow larger before the newly secreted cuticle hardens. Premolt and postmolt periods are hormonally controlled, as are several other structures and functions.

Adaptive radiation of the arthropods has been enormous, and they are extremely abundant.

Members of subphylum Chelicerata have no antennae, and their main feeding appendages are chelicerae. In addition, they have a pair of pedipalps (which may be similar to walking legs) and four pairs of walking legs. The great majority of living chelicerates are in class Arachnida: spiders (order Araneae), scorpions (order Scorpionida), harvestmen (order Opiliones), and ticks and mites (order Acarina).

Tagmata of spiders (cephalothorax and abdomen) show no external segmentation and join by a waistlike pedicel. Chelicerae of spiders have poison glands for paralyzing or killing their prey. Spiders can spin silk, which they use for a variety of purposes.

The cephalothorax and abdomen of ticks and mites are completely fused, and the anterior capitulum bears the mouthparts. Ticks and mites are the most numerous arachnids; some are important disease carriers, and others are serious plant pests.

Crustacea is a large, primarily aquatic subphylum of arthropods. Crustaceans bear two pairs of antennae, mandibles, and two pairs of maxillae on the head. Their appendages are primitively biramous, and major tagmata are head, thorax, and abdomen. Many have a carapace and respire with gills.

There are many predators, scavengers, filter feeders, and parasites among Crustacea. Respiration is through the body surface or by gills, and excretory organs take the form of maxillary or antennal glands. Circulation, as in other arthropods, is through an open system of sinuses (hemocoel), and a dorsal, tubular heart is the chief pumping organ. Most crustaceans have compound eyes composed of units called ommatidia.

Members of class Maxillopoda, subclass Copepoda, lack a carapace and abdominal appendages. They are abundant and among the most important of primary consumers in many freshwater and marine ecosystems. Malacostraca is the largest crustacean class, and the most important orders are Isopoda, Amphipoda, Euphausiacea, and Decapoda. All have both abdominal and thoracic appendages. Decapods include crabs, shrimp, lobster, crayfish, and others; they have five pairs of walking legs (including chelipeds) on their thorax.

Former subphylum Uniramia grouped arthropods having uniramous appendages, one pair of antennae, a pair of mandibles, and two pairs of maxillae (one pair of maxillae in millipedes) on the head. Uniramia has been divided into two subphyla: Myriapoda, where animals have head and trunk tagmata, and Hexapoda, where tagmata are head, thorax, and abdomen.

Hexapoda is the largest subphylum of the world's largest phylum. Insects are easily recognized by the combination of their tagmata and possession of three pairs of thoracic legs.

Radiation and abundance of insects are largely explained by several features allowing them to exploit terrestrial habitats, such as waterproof cuticle and other mechanisms to minimize water loss and the ability to become dormant during adverse conditions.

Feeding habits vary greatly among insects, and there is an enormous variety of specialization of mouthparts reflecting the particular feeding habits of a given insect. Insects breathe by means of a tracheal system, which is a system of tubes that opens by spiracles on the thorax and abdomen. Excretory organs are Malpighian tubules.

Sexes are separate in insects, and fertilization is usually internal. Almost all insects undergo metamorphosis during development. In hemimetabolous (gradual) metamorphosis, larval instars are called nymphs, and adults emerge at the last nymphal molt. In holometabolous (complete) metamorphosis, the last larval molt gives rise to a nonfeeding stage (pupa). An adult, usually winged, emerges at the final, pupal, molt. Both types of metamorphosis are hormonally controlled.

Insects are important to human welfare, particularly because they pollinate food crop plants, control populations of other, harmful insects by predation and parasitism, and serve as food for other animals. Many insects are harmful to human interests because they feed on crop plants, and many are carriers of important diseases affecting humans and domestic animals.

Preponderance of morphological and molecular evidence supports monophyly of phylum Arthropoda. Traditional alliance of Crustacea with insects and myriapods in a group called Mandibulata is supported by some morphological evidence, but some molecular data support a clade composed of crustaceans and insects with myriapods as the sister taxon to this clade, or as the sister taxon to chelicerates. Wings, hemimetabolous metamorphosis, and holometabolous metamorphosis evolved among ectognathous insects.

Until recently zoologists believed that Annelida and Arthropoda were closely related and shared a metameric common ancestor. Molecular evidence suggests that they were derived independently from a protostome ancestor and belong to separate superphyla (Lophotrochozoa and Ecdysozoa). If this hypothesis is supported by further studies, metamerism must have evolved independently in each superphylum.

Review Questions

1. Give some characteristics of arthropods that clearly distinguish them from Annelida.
2. Name the subphyla of arthropods, and give a few examples of each.
3. Much of the success of arthropods has been attributed to their cuticle. Why do you think this is so? Describe some other factors that probably contributed to their success.
4. What is a trilobite?
5. What appendages are characteristic of chelicerates?

6. Briefly describe the appearance of each of the following: eurypterids, horseshoe crabs, pycnogonids.
7. Tell the mechanism of each of the following with respect to spiders: feeding, excretion, sensory reception, webspinning, reproduction.
8. Distinguish each of the following orders from each other: Araneae, Scorpionida, Opiliones, Acarina.
9. People fear spiders and scorpions, but ticks and mites are far more important medically and economically. Why? Give examples.
10. What are tagmata and appendages on the head of crustaceans? What are some other important characteristics of Crustacea?
11. Among classes of Crustacea, Branchiopoda, Maxillopoda, and Malacostraca are the most important. Distinguish them from each other.
12. Distinguish among subclasses Ostracoda, Copepoda, Branchiura, and Cirripedia of class Maxillopoda.
13. Copepods sometimes have been called "insects of the sea" because marine planktonic copepods probably are the most abundant animals in the world. What is their ecological importance?

14. Define each of the following: swimmeret, maxilliped, cheliped, nauplius.
15. Describe molting in Crustacea, including action of hormones.
16. Explain the mechanism of each of the following with respect to crustaceans: feeding, respiration, excretion, circulation, sensory reception, reproduction.
17. Distinguish the following from each other: Diplopoda, Chilopoda, Insecta.
18. Define each of the following with respect to insects: sclerite, notum, tergum, sternum, pleura, labrum, labium, hypopharynx, haltere, instar, diapause.
19. Explain why wings powered by indirect flight muscles can beat much more rapidly than those powered by direct flight muscles.
20. What different modes of feeding are found in insects, and how are these reflected in their mouthparts?
21. Describe each of the following with respect to insects: respiration, excretion and water balance, sensory reception, reproduction.
22. Explain the difference between holometabolous and hemimetabolous

metamorphosis in insects, including stages in each.
23. Describe and give an example of each of four ways insects communicate with each other.
24. What are castes found in honey bees and in termites, and what is the function of each? What is trophallaxis?
25. Name several ways in which insects are beneficial to humans and several ways they are detrimental.
26. For the past 50 or more years, people have relied on toxic insecticides for control of harmful insects. What problems have arisen resulting from such reliance on insecticides? What are the alternatives? What is integrated pest management?
27. What evidence suggests that metamerism evolved independently in Annelida and Arthropoda?
28. We hypothesize that the earliest insects were wingless, making lack of wings the primitive condition. Does winglessness characterize a hexapod class? In what sense is it useful for classification?

Selected References

See also general references on page 415.

Arnett, R. H., Jr., and M. C. Thomas, eds. 2000. American Beetles, vol. 1. Boca Raton, Florida, CRC Press.

Arnett, R. H., Jr., M. C. Thomas, P. E. Skelley, and J. H. Franks, eds. 2002. American Beetles, vol. 2. Polyphaga: Scarabaeoidea through Curculionoidea. Boca Raton, Florida, CRC Press. *These two volumes present recent detailed keys to families of American beetles; the second edition of volume 1 (2001) includes additions and corrections.*

Berenbaum, M. R. 1995. Bugs in the system. Reading, Massachusetts, Addison-Wesley Publishing Company. *How insects impact human affairs. Well written for a wide audience, highly recommended.*

Boore, J. L., D. V. Lavrov, and W. M. Brown. 1998. Gene translocation links insects and crustaceans. Nature **392**:667–668. *A single mitochondrial gene translocation, indicative of a recent common ancestor, is shared by insects and crustaceans, but not present in chelicerates or myriapods.*

Brown, K. 2001. Seeds of concern. Sci. Am. **284**:52–57 (Apr.). *The first of four articles that examine the pros and cons of genetically modified crops, including those with genes for insecticidal toxin.*

Brusca, R. C. 2001. Unraveling the history of arthropod biodiversification. Ann. Missouri Bot. Garden **87**:13–25. *An easy-to-read discussion of recent changes in our understanding of arthropod evolution.*

Downs, A. M. R., K. A. Stafford, and G. C. Coles. 1999. Head lice: prevalence in schoolchildren and insecticide resistance. Parasitol. Today **15**:1–4. *This report is mainly concerned with England, but head lice are one of the most common parasites of American schoolchildren.*

Giribet, G., G. D. Edgecombe, and W. C. Wheeler. 2001. Arthropod phylogeny based eight molecular loci and morphology. Nature **413**:157–161. *This phylogenetic analysis indicates a close relationship between crustaceans and hexapods, and supports a mandibulate clade.*

Gullan, P. J., and P. S. Cranston. 2005. The insects: an outline of entomology, ed. 3. Malden, Massachusetts, Blackwell Publishing. *An easy-to-use text for general entomology with good figures, modern phylogenies, and data on biogeography.*

Hubbell, S. 1997. Trouble with honeybees. Nat. Hist. **106**:32–43. *Parasitic mites (Varroa jacobsoni on bee larvae and Acarapis woodi in the trachea of adults) cause serious losses among honeybees.*

Johnson, N. F., and C. Triplehorn. 2005. Borror and DeLong's introduction to the study of insects, ed. 7. Belmont, California, Brooks/Cole Publishing Company. *An up-to-date reference text for the study of insects; keys are included, but some require specialized knowledge of morphological characters. See Arnett and Thomas for keys to the American beetle families.*

Lane, R. P., and R. W. Crosskey (eds). 1993. Medical insects and arachnids. London, Chapman & Hall. *This is the best book currently available on medical entomology.*

Luoma, J. R. 2001. The removable feast. Audubon **103(3)**:48-54. *During May and June large numbers of horseshoe crabs ascend the shores of U.S. Atlantic states to breed and lay eggs. Since the 1980s they have been heavily harvested to be chopped up and used for bait. This practice has led to serious declines in* Limulus *populations, with accompanying declines in populations of migrating shore birds that feed on* Limulus *eggs.*

Mallatt, J., J. R. Garey, and J. W. Shultz. 2004. Ecdysozoan phylogeny and Bayesian inference: first use of nearly complete 28s and 18s rRNA gene sequences to classify arthropods and their kin. Mol. Phylogenet. Evol. **31**:178-191. *Results indicate the Crustacea is paraphyletic without hexapods, but Pancrustacea is a monophyletic group, that chelicerates and myriapods are sister taxa, and that Panarthropoda is a monophyletic group. There was no support for a mandibulate clade.*

O'Brochta, D. A., and P. W. Atkinson. 1998. Building the better bug. Sci. Am. **279**:90-95 (Dec.). *Inserting new genes into certain insect species could render them incapable of being vectors for disease, help agriculture, and have other applications.*

O'Neill, S. L., A. A. Hoffmann, and J. H. Werren. 1997. Influential passengers. Oxford, Oxford University Press. *Symbiotic bacteria* Wolbachia *can determine reproduction and evolution of their insect, crustacean, or nematode hosts.*

Ostfeld, R. S. 1997. The ecology of Lyme-disease risk. Amer. Sci **85**:338-346. *Lyme disease, caused by a bacterium transmitted by ticks, has been reported in 48 of the 50 United States and seems to be increasing in frequency and geographic range.*

Suter, R. B. 1999. Walking on water. Amer. Sci. **87**:154-159. *Fishing spiders* (Dolomedes) *depend on surface tension to walk on water.*

Versluis, M., B. Schmitz, A. von der Heydt, and D. Lohse. 2000. How snapping shrimp snap: through cavitating bubbles. Science **289**:2114-2117. *When* Alpheus *snaps the claw on its chela shut, it generates a jet of water with a velocity so high that the pressure decreases below the vapor pressure of water, producing a cavitation bubble. As the water pressure rises again, the bubble collapses with a force sufficient to stun or kill prey.*

Whiting, R. 2004. Phylogenetic relationships and evolution of insects, pp. 330-344. In J. Cracraft and M. J. Donoghue, eds., Assembling the tree of life. New York, Oxford University Press. *A detailed discussion of current hypotheses for insect evolution, including the suggestion that entognathous hexapods do not form a monophyletic group, unlike ectognathous forms.*

Custom Website

The *Animal Diversity* Online Learning Center is a great place to check your understanding of chapter material. Visit www.mhhe.com/hickmanad4e for access to key terms, quizzes, and more! Further enhance your knowledge with Web links to chapter-related material.

Explore live links for these topics:

Classification and Phylogeny of Animals
Phylum Arthropoda
Subphylum Chelicerata
Subphylum Crustacea
Subphylum Myriapoda

Subphylum Hexapoda
Class Insecta

Lesser Protostomes

Some Evolutionary Experiments

The early Cambrian period, about 570 million years ago, was the most fertile time in evolutionary history. For 3 billion years before this period, evolution had forged little more than bacteria and blue-green algae. Then, within the space of a few million years, all major phyla, and probably all smaller phyla, became established. This was the Cambrian explosion, the greatest evolutionary "bang" the world has known. In fact, the fossil record suggests that more phyla existed in the Paleozoic era than exist now, but some disappeared during major extinction events that punctuated evolution of life on earth. The greatest of these disruptions was the Permian extinction about 230 million years ago. Thus evolution has led to many "experimental models." Some of these models failed because they were unable to survive in changing conditions. Others gave rise to abundant and dominant species and individuals that inhabit the world today. Some persisted with small numbers of species, whereas others were formerly more abundant but are now in decline.

The great evolutionary flow that began with the appearance of the coelom and led to the three huge phyla of molluscs, annelids, and arthropods produced other lines as well. Most of those that survived are small and lack great economic and ecological importance; they are sometimes called "lesser protostomes." Under this rather arbitrary category, we consider lophotrochozoan phyla Sipuncula, Echiura, Brachiopoda, Ectoprocta, and Phoronida; and ecdysozoan phyla Pentastomida, Onychophora, Tardigrada, and Chaetognatha.

Three phyla—Phoronida, Ectoprocta, and Brachiopoda—possess a crown of ciliated tentacles, called a lophophore, used in food capture and respiration. Brachiopods were abundant in the Paleozoic but began to decline thereafter. The exception to the common theme of this chapter is phylum Ectoprocta. It arose in the Cambrian, became widespread in the Paleozoic, and remains a prevalent group today.

Ectoprocts and other animals fouling a boat bottom.

This chapter includes brief discussions of nine coelomate phyla whose position in the phylogeny of the animal kingdom has long been problematic. Sipuncula and Echiura have some annelid-like characters, and molecular evidence now supports their placement with annelids in superphylum Lophotrochozoa. Ectoprocta, Phoronida, and Brachiopoda, grouped together by virtue of having a structure called a lophophore (p. 263), likewise are positioned in Lophotrochozoa, despite having been considered deuterostomes based on developmental and morphological criteria. Pentastomida, Onychophora, and Tardigrada show some arthropod-like characters; they often have been grouped together because they have unjointed limbs with claws (at some stage) and a cuticle that undergoes molting. Molecular and other evidence now supports placement of these phyla near Arthropoda in Ecdysozoa. Chaetognatha is another phylum formerly positioned in Deuterostomia, but sequence analysis indicates placement of chaetognaths near nematodes within Ecdysozoa.

The existence of phyla whose members seem to possess a patchwork of the morphological and molecular characters found in other taxa forces biologists to confront difficult problems. How many times did a metameric body evolve? Could segmentation be lost or suppressed in some lineages? What features were present in bilaterally symmetrical animals before the split between protostomes and deuterostomes? Tantalizing clues from the biology of the nine phyla discussed here may help us answer these questions.

Phylum Sipuncula

Phylum Sipuncula (sy-pun′kyu-la) (L. *sipunculus,* little siphon) consists of about 250 species of benthic marine worms, predominantly littoral or sublittoral. Sometimes called "peanut worms," they live sedentary lives in burrows in mud or sand (figure 13.1), occupy borrowed snail shells, or live in coral crevices or among vegetation. Some species construct their own rock burrows by chemical and perhaps mechanical means. Most species are restricted to tropical zones. Some are tiny, slender worms, but the majority range from 15 to 30 cm in length.

Sipunculans are not metameric, nor do they possess setae. Their head forms an **introvert,** which is crowned by ciliated tentacles surrounding the mouth (figure 13.1). They are largely deposit feeders, extending their introvert and tentacles from their burrow to explore and to feed. They have a cerebral ganglion, nerve cord, and pair of nephridia. They have a U-shaped gut lying within a spacious coelom. The coelomic fluid contains red blood cells bearing a respiratory pigment called hemerythrin.

Sipunculan larvae are trochophores (see figure 10.5, p. 182), and their early embryological development indicates affinities to Annelida, Mollusca, and Echiura.

Phylum Echiura

Phylum Echiura (ek-ee-yur′a) (Gr. *echis,* viper, + *oura,* tail) consists of marine worms that burrow into mud or sand or live in empty snail shells, sand dollar tests, or rocky crevices. They live

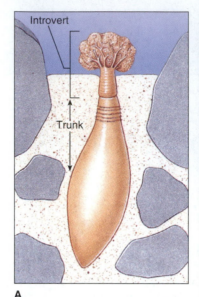
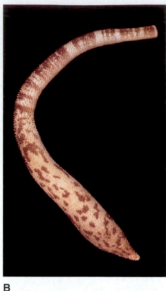

figure 13.1

A, *Themiste,* a sipunculan, has extended its introvert. **B,** *Phascolosoma,* another sipunculan, has the introvert tucked into the trunk.

in all oceans, most commonly in littoral zones of warm waters. They vary in length from a few millimeters to 40 to 50 cm.

Although there are only about 160 species, echiurans are more morphologically diverse than sipunculans. Their bodies are cylindrical. Anterior to the mouth is a flattened, extensible proboscis, which, unlike the introvert of sipunculans, cannot be retracted into the trunk. Echiurans are often called "spoonworms" because of the shape of the contracted proboscis in some of them. The proboscis has a ciliated groove leading to the mouth. There is a complete gut with a posterior anus. While an animal lies buried, its proboscis can extend out over the mud for exploration and deposit feeding (figure 13.2). A different feeding method is used by *Urechis* (Gr. *oura,* tail, + *echis,* viper); it secretes a mucous net in a U-shaped burrow through which it pumps water and strains food particles. *Urechis* is sometimes called the "fat innkeeper" because it has characteristic species of commensals living with it in its burrow, including a crab, fish, clam, and polychaete annelid.

Echiurans, with the exception of *Urechis,* have a **closed circulatory system** with a contractile vessel; most have one to three pairs of nephridia (some have many pairs), and all have a nerve ring and ventral nerve cord. Many species have paired epidermal setae. A pair of anal sacs arises from the rectum and opens into the coelom; they are probably respiratory in function and possibly accessory nephridial organs.

Early cleavage and trochophore stages are very similar to those of annelids and sipunculans. Molecular data suggest a strong relationship to annelids, but lack of segmentation poses a problem. There are some serially repeated structures in echiuran larvae, such as nerve cord ganglia and mucus glands, and a few in adult species (nephridia); these could be remnant structures from a metameric ancestor (see discussion on p. 269). Some biologists accept the relationship indicated by

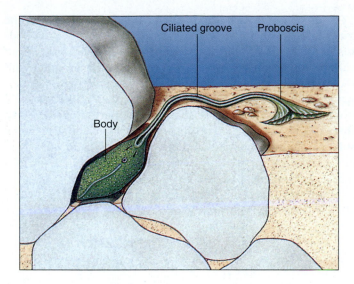

figure 13.2

Bonellia (phylum Echiura) is a detritus feeder. Lying in its burrow, it explores the surface with its long proboscis, which picks up organic particles and carries them along a ciliated groove to its mouth.

molecular characters and now consider echiurans derived polychaetes that lost segmentation.

In some species of echiurans sexual dimorphism is pronounced, with females being much larger than males. *Bonellia* has an extreme sexual dimorphism, and sex is determined in a very interesting way. At first free-swimming larvae are sexually undifferentiated. Those that come into contact with the proboscis of a female become tiny males (1 to 3 mm long) that migrate to the female uterus. About 20 males are usually found in a single female. Larvae that do not contact a female proboscis metamorphose into females. The stimulus for development into males is apparently a pheromone produced by the female proboscis.

Lophophorates

Three apparently disparate phyla are called lophophorates. **Phoronids** (phylum Phoronida) are wormlike marine forms that live in secreted tubes in sand or mud or attached to rocks or shells. **Ectoprocts** (phylum Ectoprocta) are minute forms, mostly colonial, whose protective cases often form encrusting masses on rocks, shells, or plants. **Brachiopods** (phylum Brachiopoda) are bottom-dwelling marine forms that superficially resemble molluscs because of their bivalved shells. All have a free-swimming larval stage but are sessile as adults.

One may wonder why these three apparently widely different types of animals are considered together. They are all coelomate; all have some deuterostome and some protostome characteristics; all are sessile; and none has a distinct head. But these characteristics are also shared by other phyla. What really sets them apart from other phyla is a common possession of a ciliary feeding device called a **lophophore** (Gr. *lophas*, crest or tuft, + *phorein*, to bear).

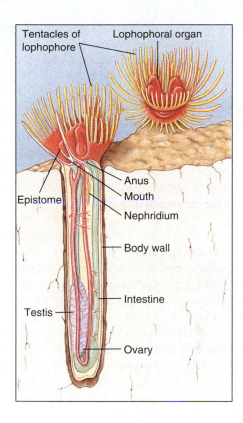

figure 13.3

Internal structure of *Phoronis* (phylum Phoronida), in diagrammatic vertical section.

A lophophore is a unique arrangement of ciliated tentacles borne on a ridge (a fold of the body wall), which surrounds the mouth but not the anus (figures 13.3 and 13.5). A lophophore with its crown of tentacles contains within it an extension of the coelom, and the thin, ciliated walls of the tentacles not only constitute an efficient feeding device but also serve as a respiratory surface for exchange of gases between environmental water and coelomic fluid. A lophophore can usually be extended for feeding or withdrawn for protection.

In addition, all three phyla have a U-shaped alimentary canal, with their anus placed near their mouth but outside the lophophore. Their coelom is divided primitively into three compartments, **protocoel, mesocoel** and **metacoel,** and the mesocoel extends into the hollow tentacles of the lophophore. Their protocoel, where present, forms a cavity in a flap over their mouth, called an **epistome.** The portion of the body that contains their mesocoel is called a **mesosome,** and that containing their metacoel is a **metasome.**

Phylum Phoronida

Phylum Phoronida (fo-ron'i-da) (Gr. *phoros*, bearing, + L. *nidus*, nest) comprises approximately 10 species of small wormlike animals that live on the bottom of shallow coastal waters, especially in temperate seas. The phylum name refers to their tentacled lophophore. Phoronids range from a few millimeters to 30 cm in length. Each worm secretes a leathery or chitinous tube in which it lies free, but which it never leaves (figure 13.3). Their tubes may be anchored singly or in a tangled mass on rocks, shells, or pilings, or buried in the sand. Tentacles on the lophophore are thrust out for feeding, but if an animal is disturbed it can withdraw completely into its tube.

Their lophophore has two parallel ridges curved in a horseshoe shape, the bend located ventrally and the mouth lying between the two ridges. Cilia on the tentacles direct a water current toward a groove between the two ridges, which leads toward their mouth. Plankton and detritus caught in this current become entangled in mucus and are carried by cilia to their mouth.

Mesenteric partitions divide their coelomic cavity into proto-, meso-, and metacoel, similar to compartments of deuterostomes. Phoronids have a closed system of contractile blood vessels but no heart; their red blood contains hemoglobin. There is a pair of metanephridia. A nerve ring sends nerves to tentacles and the body wall.

There are both monoecious (the majority) and dioecious species of Phoronida, and at least one species reproduces asexually. Cleavage combines features of both spiral and radial types.

Phylum Brachiopoda

Brachiopoda (brak'i-op'o-da) (Gr. *brachīon*, arm, + *pous, podos,* foot), or lamp shells, is an ancient group. Compared with the fewer than 300 species now living, some 30,000 fossil species, which flourished in the Paleozoic and Mesozoic seas, have been described. Brachiopods were once very abundant, but they are now apparently in decline. Some modern forms have changed little from early ones. Genus *Lingula* (L., little tongue) (figure 13.4A) has existed virtually unchanged for over 400 million years. Most modern brachiopod shells range between 5 and 80 mm, but some fossil forms reached 30 cm in length.

Brachiopods are all attached, bottom-dwelling, marine forms that mostly prefer shallow water. Externally brachiopods resemble bivalve molluscs in having two calcareous shell valves secreted by a mantle. They were, in fact, grouped with molluscs until the middle of the nineteenth century, and their name refers to arms of their lophophore, which were thought homologous to the mollusc foot. Brachiopods, however, have **dorsal** and **ventral valves** instead of right and left lateral valves as do bivalve molluscs and, unlike bivalves, most of them are attached to a substrate either directly or by means of a fleshy stalk called a **pedicel** (or pedicle).

In most brachiopods the ventral (pedicel) valve is slightly larger than the dorsal (brachial) valve, and one end pro-

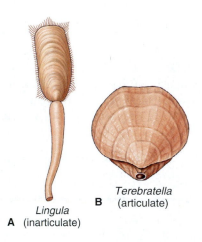

Lingula
A (inarticulate)

Terebratella
B (articulate)

figure 13.4

Brachiopods. **A,** *Lingula,* an inarticulate brachiopod that normally occupies a burrow. Its contractile pedicel can withdraw the body into the burrow. **B,** An articulate brachiopod, *Terebratella*. Its valves have a tooth-and-socket articulation and a short pedicel projects through one valve to attach to the substratum (pedicel shown in figure 13.5).

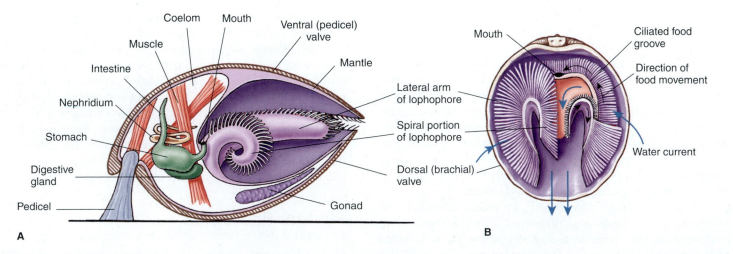

figure 13.5

Brachiopod anatomy. **A,** An articulate brachiopod (longitudinal section). **B,** Feeding and respiratory currents. Large arrows show water flow over its lophophore; small arrows indicate food movement toward its mouth in the ciliated food groove.

jects in the form of a short pointed beak that is perforated where the fleshy stalk passes through (figure 13.4B). In many the shape of the pedicel valve is that of a classical oil lamp of Greek and Roman times, so that brachiopods came to be known as "lamp shells."

There are two classes of brachiopods based on shell structure. Shell valves of Articulata are connected by a hinge with an interlocking tooth-and-socket arrangement (articular process); those of Inarticulata lack the hinge and are held together by muscles only.

The body occupies only the posterior part of the space between the valves (figure 13.5A), and extensions of the body wall form mantle lobes that line and secrete the shell. The large horseshoe-shaped lophophore in the anterior mantle cavity bears long ciliated tentacles used in respiration and feeding. Ciliary water currents carry food particles between the gaping valves and over the lophophore. Food is caught in mucus on the tentacles and carried in a ciliated food groove along an arm of the lophophore to the mouth (figure 13.5).

There is no cavity in the epistome of articulates, but in inarticulates there is a protocoel in the epistome that opens into a mesocoel. As in other lophophorates, the posterior metacoel bears the viscera. One or two pairs of nephridia open into the coelom and empty into the mantle cavity. There is an open circulatory system with a contractile heart. There is a nerve ring with a small dorsal and a larger ventral ganglion.

Sexes are separate and paired gonads discharge gametes through the nephridia. Development of brachiopods is similar in some ways to that of deuterostomes, with radial, mostly equal, holoblastic cleavage and coelom forming enterocoelically in articulates. Free-swimming larvae of articulates resemble a trochophore.

Phylum Ectoprocta

Ectoprocta (ek'to-prok'ta) (Gr. *ektos,* outside, + *proktos,* anus) have long been called bryozoans (Gr. *bryon,* moss, + *zoōn,* animal), or moss animals, a term that originally included Entoprocta also.

Of the 4000 or so species of ectoprocts, few are more than 0.5 mm long; all are aquatic, both freshwater and marine, but they largely occur in shallow waters; and most, with very few exceptions, are colony builders. Ectoprocts, unlike most other phyla considered in this chapter, were abundant and widespread in the past and remain so today. They left a rich fossil record since the Ordovician era. Modern marine forms exploit all kinds of firm surfaces, such as shells, rock, large brown algae, mangrove roots, and ship bottoms. They are one of the most important groups of fouling organisms on boat hulls; they decrease efficiency of a hull passing through the water and make periodic scraping of the hull necessary.

Each member of a colony occupies a tiny chamber, called a **zoecium,** which is secreted by its epidermis (figure 13.6A). Each individual, or **zooid,** consists of a feeding **polypide** and a case-forming **cystid.** Polypides include a lophophore, digestive tract, muscles, and nerve centers. Cystids are the body wall of the animal, together with its secreted exoskeleton. An

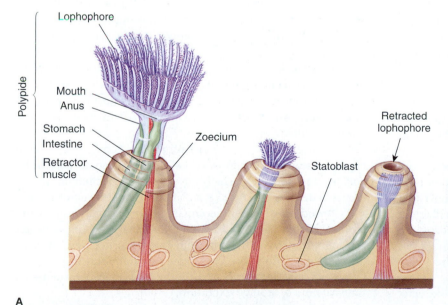
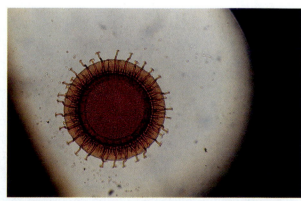

figure 13.6

A, Small portion of freshwater colony of *Plumatella* (phylum Ectoprocta), which grows on the underside of rocks. These tiny individuals disappear into their chitinous zoecia when disturbed. **B,** Statoblast of a freshwater ectoproct, *Cristatella.* Statoblasts are a kind of bud that survives over winter when a colony dies in the autumn. This one is about 1 mm in diameter and bears hooked spines.

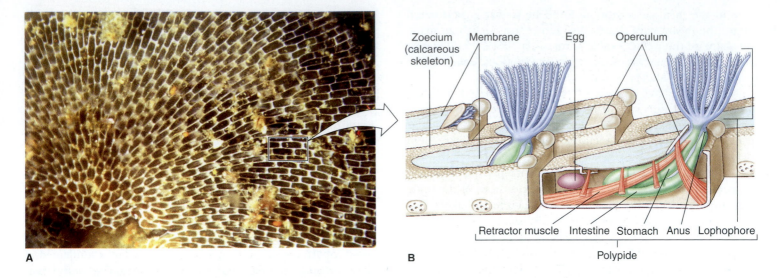

figure 13.7

A, Skeletal remains of a colony of *Membranipora,* a marine encrusting form of Ectoprocta. Each little oblong zoecium is the calcareous former home of a tiny ectoproct. **B,** In a living colony, each zoecium holds one polypide. To feed, each polypide lifts its operculum and extends its lophophore.

exoskeleton, or zoecium, may, according to the species, be gelatinous, chitinous, or stiffened with calcium and possibly also impregnated with sand. Its shape may be boxlike, vaselike, oval, or tubular.

Some colonies form limy encrustations on seaweed, shells, and rocks (figure 13.7); others form fuzzy or shrubby growths or erect, branching colonies that look like seaweed. Some ectoprocts might easily be mistaken for hydroids but can be distinguished under a microscope by presence of ciliated tentacles (figure 13.8). In some freshwater forms individuals are borne on finely branching stolons that form delicate tracings on the underside of rocks or plants. Other freshwater ectoprocts are embedded in large masses of gelatinous material. Although zooids are minute, colonies may be several centimeters in diameter; some encrusting colonies may be a meter or more in width, and erect forms may reach 30 cm or more in height.

A polypide lives a type of jack-in-the-box existence, popping up to feed and then quickly withdrawing into its little chamber, which often has a tiny trapdoor (operculum) that shuts to conceal its inhabitant. To extend their tentacular crown, certain muscles contract, which increases hydrostatic pressure within the body cavity and pushes the lophophore out. Other muscles contract to withdraw the crown to safety with great speed.

Lophophore ridges tend to be circular in marine ectoprocts and U-shaped in freshwater species. When feeding, an animal extends its lophophore, and its tentacles spread out to a funnel shape (figure 13.8). Cilia on the tentacles draw water into the funnel and out between tentacles. Food particles trapped in mucus in the funnel are drawn into the mouth, both by pumping action of their muscular pharynx and by action of cilia in their pharynx.

Respiratory, vascular, and excretory organs are absent. Gaseous exchange occurs through the body surface, and, since ectoprocts are small, coelomic fluid is adequate for internal transport. Coelomocytes engulf and store waste materials. There is a ganglionic mass and a nerve ring around the pharynx, but no sense organs are present. A septum divides the coelom into an anterior mesocoel in the lophophore and a larger posterior metacoel. A protocoel and epistome are present only in freshwater ectoprocts. Pores in the walls between adjoining zooids permit exchange of materials by way of coelomic fluid.

Feeding individuals dominate most colonies, but polymorphism also occurs. One type of modified zooid resembles a bird beak that snaps at small invading organisms that might foul a colony. Another type has a long bristle that sweeps away foreign particles.

Most ectoprocts are hermaphroditic. Some species shed eggs into seawater, but most brood their eggs, some within the coelom and some externally in a special zoecium in which embryos develop. Cleavage is radial but apparently determinate.

Brooding is often accompanied by degeneration of the lophophore and gut of adults, the remains of which contract into minute dark balls, or **brown bodies.** Later, new internal organs regenerate in old chambers. Brown bodies may remain passive or may be eliminated by the new digestive tract—an unusual kind of storage excretion.

Freshwater species reproduce both sexually and asexually. Asexual reproduction is by budding or by means of **statoblasts,** which are hard, resistant capsules containing a mass of germinative cells that form during the summer and fall (see figure 13.6B). When a colony dies in late autumn, statoblasts are released, and in spring they can give rise to new polypides and eventually to new colonies.

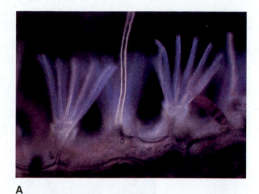

figure 13.8

A, Ciliated lophophore of *Electra pilosa,* a marine ectoproct. **B,** *Plumatella repens,* a freshwater ectoproct. It grows on the underside of rocks and vegetation in lakes, ponds, and streams.

Phylum Pentastomida

The wormlike Pentastomida (pen-ta-stom′i-da) (Gr. *pente,* five, + *stoma,* mouth) are parasites, 2.5 to 12 cm long, that are found in lungs and nasal passages of carnivorous vertebrates—most commonly in reptiles (figure 13.9). Some human infections have been found in Africa and Europe. Intermediate hosts are usually vertebrates that are eaten by a final host.

Pentastomids have four clawlike appendages at their anterior end, and their body is covered by a nonchitinous cuticle, which they periodically molt during juvenile stages. They lack organs for respiration, circulation, or excretion. Pentastomids show arthropod affinities; they seem closest to members of the crustacean subclass Branchiura in spermatozoan structure.

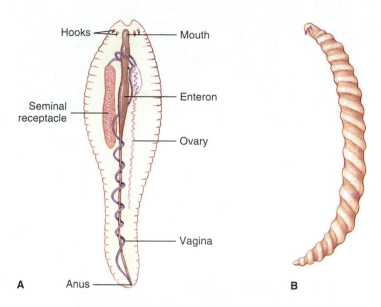

figure 13.9

Two pentastomids. **A,** *Linguatula,* found in the nasal passages of carnivorous mammals. A female is shown with some internal structures. **B,** Female *Armillifer,* a pentastomid with pronounced body rings. In parts of Africa and Asia, humans are parasitized by immature stages; adults (10 cm long or more) live in the lungs of snakes. Human infection may occur from eating snakes or from contaminated food and water.

Though some authorities consider them a subphylum of Arthropoda, their extensive modifications for parasitic life make their ancestry difficult to determine. A recent molecular study strongly supports their placement as highly derived crustaceans.

Phylum Onychophora

Members of phylum Onychophora (on-i-kof′o-ra) (Gr. *onyx,* claw, + *pherō,* to bear) are called "velvet worms" or "walking worms." There are about 70 species of caterpillar-like animals, 1.4 to 15 cm long, that live in rain forests and other tropical and semitropical leafy habitats.

The fossil record of onychophorans shows that they have changed little in their 500-million-year history. They were originally marine animals and were probably far more common than they are now. Today they are all terrestrial and are extremely retiring, coming out only at night or when the air is nearly saturated with moisture.

Onychophorans are covered by a soft cuticle, which contains chitin and protein. Their wormlike bodies are carried on 14 to 43 pairs of stumpy, unjointed legs, each ending with a flexible pad and two claws (figure 13.10). Their head bears a pair of flexible antennae with annelid-like eyes at the base.

They are air breathers, using a **tracheal system** that connects with pores scattered over their body. Their tracheal system, although similar to that of arthropods, probably evolved independently. Other arthropod-like characteristics include an open circulatory system with a tubular heart, a hemocoel for a body cavity, a large brain, and a cuticle that is molted. Annelid-

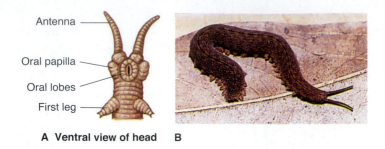

A **Ventral view of head**   B

figure 13.10

Peripatus, a caterpillar-like onychophoran that has characteristics in common with both annelids and arthropods. **A,** Ventral view of head. **B,** In natural habitat.

figure 13.11

Scanning electron micrograph of *Echiniscus maucci,* phylum Tardigrada. This species is 300 to 500 μm long. Unable to swim, it clings to moss or water plants with its claws, and if the environment dries up, it goes into a state of suspended animation and "sleeps away" the drought.

like characteristics include segmentally arranged nephridia, a muscular body wall, and pigment-cup ocelli.

Onychophorans are dioecious. In some species there is a placental attachment between mother and young, and young are born as juveniles (viviparous); others have young that develop in the uterus without attachment (ovoviviparous). Two Australian genera are oviparous and lay shell-covered eggs in moist places.

Phylum Tardigrada

Tardigrada (tar-di-gray′da) (L. *tardus,* slow, + *gradus,* step), or "water bears," are minute forms usually less than a millimeter in length. There are about 900 described species, some living in freshwater and marine habitats, but most species are terrestrial forms that live in water film that surrounds mosses and lichens.

Their body bears eight short, **unjointed legs,** each with claws (figure 13.11). Unable to swim, they creep awkwardly, clinging to the substrate with their claws. A pair of sharp stylets

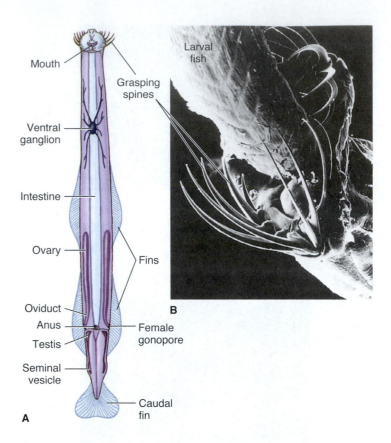

figure 13.12

Arrowworms. **A,** Internal structure of *Sagitta.* **B,** Scanning electron micrograph of a juvenile arrowworm, *Flaccisagitta hexaptera* (35 mm length), eating a larval fish.

and a sucking pharynx adapt them for piercing and sucking plant cells or small prey such as nematodes and rotifers.

A body covering of nonchitinous **cuticle** is molted several times during their life cycle. As in arthropods, muscle fibers are attached to the cuticular exoskeleton, and their body cavity is a hemocoel.

Their annelid-type nervous system is surprisingly complex, and in some species there is a pair of eyespots. Circulatory and respiratory organs are lacking.

One of the most intriguing features of terrestrial tardigrades is their capacity to enter a state of suspended animation, called **cryptobiosis,** during which metabolism is virtually imperceptible. Under gradual drying conditions they reduce water content of their body from 85% to only 3%, movement ceases, and their body becomes barrel shaped. In a cryptobiotic state tardigrades can withstand harsh environmental conditions: temperature extremes, ionizing radiation, oxygen deficiency, etc., and many survive for years. Activity resumes when moisture is again available.

Tardigrades may reproduce by parthenogenesis, or sexually. Females may deposit their eggs in the old cuticle as they molt or attach them to a substrate. Embryonic formation of the coelom is enterocoelous, a deuterostome characteristic. Nevertheless, molecular data and their numerous arthropod-like characteristics strongly suggest phylogenetic affinity with Arthropoda (see figure 13.13).

Phylum Chaetognatha: Arrowworms

Chaetognatha (ke-tog′na-tha) (Gr. *chaitē*, long flowing hair, + *gnathos,* jaw) is a group of less than 200 species of marine animals highly specialized for a planktonic existence. Their small, straight bodies resemble miniature torpedoes, or darts, ranging from 2.5 to 10 cm in length.

Arrowworms usually swim to the surface at night and descend during the day. Much of the time they drift passively, but they can dart forward in swift spurts, using their caudal fin and longitudinal muscles—a fact that no doubt contributes to their success as planktonic predators. Horizontal fins bordering the trunk function in flotation rather than in active swimming.

The body of arrowworms is unsegmented and is composed of head, trunk, and postanal tail (figure 13.12). They have teeth and chitinous spines on their head around their mouth. When an animal captures prey, its teeth and raptorial spines spread apart and then snap shut with startling speed. Arrowworms are voracious feeders, living on other planktonic forms, especially copepods, and even small fish (figure 13.12B).

Their body is covered with a thin cuticle. They have a complete digestive system, well-developed coelom, and nervous system with a nerve ring containing several ganglia. Vascular, respiratory, and excretory systems, however, are entirely lacking.

Arrowworms are hermaphroditic with either cross-fertilization or self-fertilization. Juveniles develop directly without metamorphosis. There is no true peritoneum lining their coelom. Cleavage has been described as radial, complete, and equal, but recent studies dispute this description, finding instead that cleavage planes in four-cell embryos are similar to those of crustaceans and nematodes. The phylogenetic significance of this description, in conjunction with other morphological and molecular characters, is discussed in the Phylogeny section.

Phylogeny

The evolutionary history of the phyla described in this chapter is the subject of intense discussion. Are they all protostomes, and if so, do they belong in either of the large protostome clades? Sipunculans and echiurans are similar to annelids in features of early embryological development, nervous system structure, and body-wall structure, but neither group is segmented. Recent analyses of molecular and morphological data place echiurans within Annelida, as a derived clade of polychaetes where segmentation was lost. Sipunculans share certain developmental features with molluscs, whereas other features and molecular data suggest phylogenetic affinity to annelids. Exact placement of sipunculans is unclear, but they do belong within Lophotrochozoa.

The phylogenetic position of the lophophorates has generated much controversy and debate. Sometimes they are considered protostomes with some deuterostome characters, and at other times deuterostomes with some protostome characters. Sequence analysis of the gene encoding small-subunit ribosomal RNA provides evidence that they are protostomes.

They appear clearly allied with Annelida and Mollusca and very close to Entoprocta within superphylum Lophotrochozoa. Their common possession of a lophophore has been considered a unique synapomorphy, but a detailed analysis of lophophore function[1] suggests that Ectoprocta does not form a monophyletic group with Phoronida and Brachiopoda. There is little doubt that phoronids and brachiopods are sister taxa, but analyses based on morphology place them in Deuterostomia: like some other deuterostomes, they have a three-part (trimerous, or tripartite) coelom. Dividing the lophophorate clade among protostomes and deuterostomes would mean that other shared features, such as a U-shaped digestive tract, metanephridia, and a tendency to secrete outer cases, are convergent. It would leave ectoprocts as the only lophophorate member of the Lophotrochozoa. The sharp contrast between outcomes using molecular data and those using morphological data indicates that much more phylogenetic work is needed.

The clade Panarthropoda was erected to encompass the phylum Arthropoda plus two allied phyla: Tardigrada and Onychophora. Although tardigrades have some similarities to rotifers in reproduction and cryptobiotic tendencies, as well as an unusual enterocoelic formation of mesoderm, other synapomorphies support their association with arthropods (figure 13.13). Sequence analysis supports placement of all three groups within Ecdysozoa.

The wormlike pentastomids (p. 267) were placed in Ecdysozoa near arthropods because their larval form resembles tardigrade larvae, their cuticle is molted, and there are other similarities in sperm morphology and larval appendages. Phylogenies based on sequences of ribosomal RNA genes indicate that pentastomids are closely related to crustaceans, in particular to members of subclass Branchiura (p. 234). A recent study of gene arrangements and base sequences of mitochondrial DNA confirms that placement; phylum Pentastomida actually belongs *within* Arthropoda (figure 13.13). These odd vertebrate parasites are highly derived crustaceans and should no longer be considered a phylum.

Recent discoveries of Cambrian fossil pentastomids and tardigrades and additional fossil onychophorans strongly suggest that these small phyla arose during the Cambrian explosion, just as did the major phyla. Because this period was long before terrestrial vertebrates evolved, the identity of hosts for Cambrian pentastomids remains enigmatic; some authors have suggested that they might have been conodonts (see p. 305).

Placement of chaetognaths on the metazoan tree is a very contentious issue. A chaetognath coelom forms by enterocoely, a deuterostome feature. Previously described cleavage patterns (radial, complete, and equal) also supported a deuterostome affinity, but new studies identified a cleavage pattern similar to that of crustaceans and nematodes. Chaetognaths and nematodes both lack circular muscles and have a similar arrangement of longitudinal muscles, in keeping with their placement in Ecdysozoa according to sequence-based phylogenies. However, molting has never been reported in chaetognaths. Chaetognaths

---

[1]Neilsen, C. 2002. Integ. and Comp. Biol., **42:**685–691.

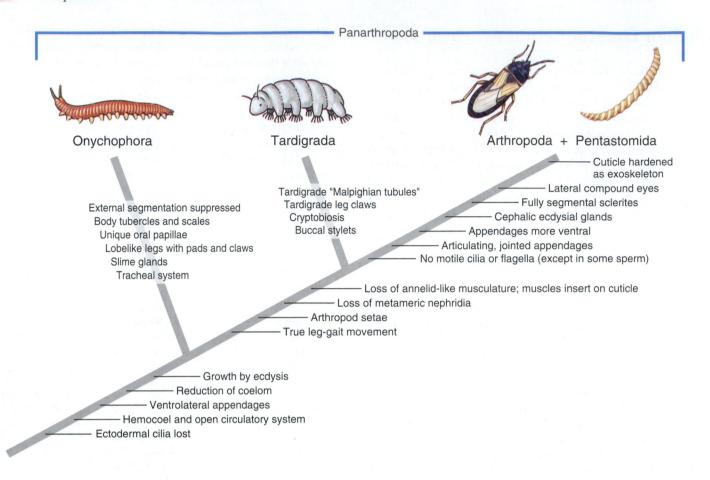

figure 13.13

Cladogram depicting hypothetical relationships of Onychophora and Tardigrada to arthropods. Onychophorans diverged from the arthropod line after development of such synapomorphies as hemocoel and growth by ecdysis. They share several primitive characters with annelids, such as metameric arrangement of nephridia, but molecular evidence indicates inclusion in Ecdysozoa, not Lophotrochozoa. Note that the tracheal system of onychophorans is not homologous to that of arthropods but represents a convergence. Other authors suggest that Tardigrada is the sister group of a clade consisting of Onychophora and Arthropoda. Phylum Pentastomida is no longer valid; pentastomids are highly derived crustaceans and belong within Phylum Arthropoda.

cannot be assigned unambiguously to either protostomes or deuterostomes. One recent phylogenetic study using *Hox* genes suggested that chaetognaths branched from the metazoan lineage before the protostome–deuterostome split.

Reconstructing the history of life is one of the most fascinating tasks undertaken by modern biologists. As we decipher the evolutionary branching patterns that led to what we now see as major metazoan phyla, some phyla are subsumed: Pentasomida belongs within Arthropoda; Pogonophora belongs within Annelida (see p. 209); Echiura may also be placed within Annelida. Other phyla can be united as superphyla or large clades: Panarthropoda unites Tardigrada with Onychophora and Arthropoda, all within the much larger clade Ecdysozoa; Ectoprocta belongs with Annelida and Mollusca in Lophotrochozoa. Despite this progress, the branching order for many of these taxa remains unclear, and the status of some groups, like lophophorates and chaetognaths, continues to puzzle systematists.

Summary

Some molecular evidence supports distribution of the 9 small coelomate phyla considered in this chapter between superphylum Lophotrochozoa (Sipuncula, Echiura, Ectoprocta, Phoronida and Brachiopoda), and superphylum Ecdysozoa (Pentastomida, Onychophora, Tardigrada, and Chaetognatha).

Sipunculans and echiurans are burrowing marine worms; neither shows metamerism.

Phoronida, Ectoprocta, and Brachiopoda all bear a lophophore, which is a crown of ciliated tentacles surrounding the mouth but not the anus and containing an extension of the mesocoel. They are also sessile as adults, have a U-shaped digestive tract, and have free-swimming larvae. The lophophore functions both as a respiratory and a feeding structure.

Phoronida are the least abundant of lophophorates. These wormlike animals live in tubes mostly in shallow coastal waters.

Brachiopods were a widely prevalent phylum in the Paleozoic era but have been declining since the early Mesozoic era. They have a dorsal and a ventral shell and usually attach to the substrate directly or by a pedicel.

Ectoprocta are abundant in marine habitats, living on a variety of submerged substrata, and a number of species are common in fresh water. Ectoprocts are colonial, and although each individual is quite small, colonies are commonly several centimeters or more in width or height.

Pentastomida have certain arthropod-like characteristics, but their specialized modifications as internal parasites complicate a determination of their phylogenetic relationships. Molecular and morphological evidence suggest a position within a crustacean subclass (Branchiura).

Onychophora are caterpillar-like animals that are metameric and show some annelid and some arthropod characteristics. Tardigrades are minute, mostly terrestrial animals that have a hemocoel, as do arthropods.

Sequence analysis of bases in the gene encoding small-subunit rRNA places both onychophorans and tardigrades in Ecdysozoa.

Chaetognaths are important predators on marine zooplankton. Based on developmental and morphological criteria, they have long been considered deuterostomes. But some molecular evidence indicates that chaetognaths are ecdysozoan protostomes.

Review Questions

1. Give the main distinguishing characteristics of the following, and tell where each lives: Sipuncula, Echiura, Pentastomida, Onychophora, Tardigrada, Chaetognatha.
2. What do the members of each of the aforementioned groups eat?
3. What evidence groups Sipuncula and Echiura phylogenetically with Annelida?
4. Some investigators regard Onychophora as a "missing link" between Annelida and Arthropoda. Give evidence for and against this hypothesis.
5. What is the survival value of cryptobiosis in tardigrades?
6. Some investigators consider lophophorates a clade within protostomes. What are some synapomorphies that distinguish this clade? What evidence suggests that lophophorates do not form a clade?
7. Define each of the following: lophophore, zoecium, zooid, polypide, cystid, brown bodies, statoblasts.
8. What coelomic compartments are found in lophophorates?
9. Brachiopods superficially resemble bivalve molluscs. How would you explain the difference to a layperson?
10. Lophophorates are sometimes placed in a phylogenetic position between protostomes and deuterostomes. How would you justify or oppose such placement?

Selected References
See also general references on page 415.

Conway Morris, S., B. L. Cohen, A. B. Gawthrop, and T. Cavalier-Smith. 1996. Lophophorate phylogeny. Science 272:282–283. *These authors urged caution in acceptance of the taxon Lophotrochozoa proposed by Halanych et al. (1995).*

Cutler, E. B. 1995. The Sipuncula. Their systematics, biology, and evolution. Ithaca, New York, Cornell University Press. *The author tried to "bring together everything known about" sipunculans.*

Eernisse, D. J., and K. J. Peterson. 2004. The history of animals, pp. 197–208. In J. Cracraft, and M. J. Donoghue, eds., Assembling the tree of life. New York, Oxford University Press. *A general overview of current metazoan phylogeny with discussion of the major clades and "problem" taxa, including chaetognaths.*

Garey, J. R., M. Krotec, D. R. Nelson, and J. Brooks. 1996. Molecular analysis supports a tardigrade-arthropod association. Invert. Biol. 115:79–88. *Relationship of tardigrades and arthropods based on morphological characters is supported by sequence analysis of the gene encoding small-subunit rRNA.*

Gould, S. J. 1995. Of tongue worms, velvet worms, and water bears. Natural History 104(1):6–15. *Intriguing essay on affinities of Pentastomida, Onychophora, and Tardigrada and how they, along with larger phyla, are products of the Cambrian explosion.*

Halanych, K. M. 1996. Testing hypotheses of chaetognath origins: long branches revealed by 18S ribosomal DNA. Syst. Biol. 45:223–246. *Analysis suggests chaetognaths are most closely related to nematodes.*

Halanych, K. M., J. D. Bachellar, A. M. A. Aguinaldo, S. M. Liva, D. M. Hillis, and J. A. Lake. 1995. Evidence from 18S ribosomal DNA that lophophorates are protostome animals. Science 267:1641–1643. *Despite much morphological and developmental evidence that lophophorates are basal deuterostomes, they clustered with annelids and molluscs in this analysis. The authors proposed Lophotrochozoa, defined as the last common ancestor of lophophorate taxa, annelids, and molluscs, and all descendants of that ancestor.*

Hoffman, P. F., and D. P. Schrag. 2000. Snowball Earth. Sci. Am. 282:68–75 (Jan.). *It appears that an extreme ice age prevailed on earth 600 million years ago, followed by an extremely warm period fueled by a brutal greenhouse effect. Did these events precipitate the Cambrian explosion?*

Lavrov, D. V., W. M. Brown, and J. L. Boore. 2004. Phylogenetic position of the Pentastomida and (pan)crustacean relationships. Proc. Royal Soc. Biol. Sciences B 271:537–544. *Mitochondrial gene arrangements and sequences indicate that pentastomids are modified crustaceans, and likely related to branchiurans.*

Menon, J., and A. J. Arp. 1998. Ultrastructural evidence of detoxification in the alimentary canal of *Urechis caupo*. Invert. Biol. **117:**307–317. *This curious echiuran has detoxification bodies in its gut cells and epithelial cells that allow it to live in a highly toxic sulfide environment.*

Neilsen, C. 2002. The phylogenetic position of Entoprocta, Ectoprocta, Phoronida, and Brachiopoda. Integ. and Comp. Biol. **42:**685–691. *Evidence is presented to support the idea that lophophorates do not form a monophyletic group, and that phoronids and brachiopods are deuterostomes.*

Papillon, D., Y. Perez, L. Fasano, Y. Le Parco, and X. Caubit. 2003. *Hox* gene survey in the chaetognath *Spadellacephaloptera:* evolutionary implications. Dev. Genes Evol. **213:**142–148. *Results of this study suggest that chaetognaths are basal bilaterians.*

Telford, M. J., and P. W. H. Holland. 1993. The phylogenetic affinities of the chaetognaths: a molecular analysis. Mol. Biol. Evol. **10:**660–676. *This article presents evidence that chaetognaths are protostomes.*

Wada, H. and N. Satoh. 1994. Details of the evolutionary history from invertebrates to vertebrates, as deduced from the sequence of 18S rDNA. Proc. Natl. Acad. Sci. **91:**1801–1804. *Evidence in support of placing chaetognaths with protostomes.*

Custom Website

The *Animal Diversity* Online Learning Center is a great place to check your understanding of chapter material. Visit www.mhhe.com/hickmanad4e for access to key terms, quizzes, and more! Further enhance your knowledge with Web links to chapter-related material.

Explore live links for these topics:

Classification and Phylogeny of Animals
Phylum Sipuncula
Phylum Echiura
The Marine Deep Sea Zone
Phylum Onychophora

Phylum Tardigrada
Phylum Phoronida
Phylum Ectoprocta
Phylum Brachiopoda

Echinoderms and Hemichordates

A Design To Puzzle the Zoologist

The distinguished American zoologist Libbie Hyman once described echinoderms as a "noble group especially designed to puzzle the zoologist." With a combination of characteristics that should delight the most avid reader of science fiction, echinoderms would seem to confirm Lord Byron's observation that

> 'Tis strange—but true;
> for truth is always strange;
> Stranger than fiction.

Despite the adaptive value of bilaterality for free-moving animals, and the merits of radial symmetry for sessile animals, echinoderms confounded the rules by becoming free moving but radial. That they evolved from a bilateral ancestor there can be no doubt, for their larvae are bilateral. They undergo a bizarre metamorphosis to a radial adult in which there is a 90° reorientation in body axis. Despite the limitations of radial symmetry, echinoderms are among the most abundant organisms in some marine habitats.

A compartment of the coelom has been transformed in echinoderms into a unique water-vascular system that uses hydraulic pressure to power a multitude of tiny tube feet used in food gathering and locomotion. Dermal ossicles may fuse together to invest echinoderms in armor or may be reduced to microscopic plates. Many echinoderms have miniature jawlike pincers (pedicellariae) scattered on their body surface, often stalked and some equipped with poison glands.

This constellation of characteristics is unique in the animal kingdom. It has both defined and limited evolutionary potential of the echinoderms. Despite the vast amount of research that has been devoted to them, we are still far from understanding many aspects of echinoderm biology.

The deuterostome phylum Hemichordata includes a small group of mostly wormlike, marine organisms that possess several intriguing characteristics. Most species of hemichordates share gill slits and a dorsal, hollow nerve cord with members of the Phylum Chordata. The importance of hemichordates for providing clues to the origin of chordates, a group that includes humans, was first recognized in the mid-nineteenth century and continues today.

A mass of sea stars (*Pisaster ochraceous*) above the waterline at low tide.

Echinodermata, along with Chordata and Hemichordata (acorn worms and pterobranchs) are deuterostomes. Typical deuterostome embryogenesis is shown only by some chordates such as amphioxus (p. 300), but these shared characters and molecular evidence support monophyly of Deuterostomia. Nonetheless, their evolutionary history has taken echinoderms to the point where they are very much unlike any other animal group.

Phylum Echinodermata

Echinoderms are marine forms and include sea stars (also called starfishes), brittle stars, sea urchins, sea cucumbers, and sea lilies. They have a combination of characteristics found in no other phylum: (1) an endoskeleton of plates or ossicles, (2) water-vascular system, (3) pedicellariae, (4) dermal branchiae, and (5) secondary radial or biradial symmetry. Their larvae are bilateral and undergo a metamorphosis to a radial adult. The water-vascular system and dermal ossicles have been particularly important in determining evolutionary potential and limitations of this phylum.

Ecological Relationships

Echinoderms are all marine; they have no ability to osmoregulate and so are rarely found in brackish waters. Virtually all benthic as adults, they are found in all oceans of the world and at all depths, from intertidal to abyssal regions.

Some sea stars (figure 14.1) are particle feeders, but many are predators, feeding particularly on sedentary or sessile prey. Brittle stars (see figure 14.9) are the most active echinoderms, moving by their arms; and they may be scavengers, browsers, or deposit or filter feeders. Some brittle stars are commensals with sponges. Compared to other echinoderms, sea cucumbers (see figure 14.15) are greatly extended in the oral-aboral axis and are oriented with that axis more or less parallel to their substrate and lying on one side. Most are suspension or deposit feeders. "Regular" sea urchins (see figure 14.11), which are radially symmetrical, prefer hard substrates and feed chiefly on algae or detritus. "Irregular" urchins (sand dollars and heart urchins) (see figure 14.12), which have become secondarily bilateral, are usually found on sand and feed on small particles. Sea lilies and feather stars (see figures 14.18 and 14.19) stretch their arms out and up like a flower's petals and feed on plankton and suspended particles.

Class Asteroidea: Sea Stars

Sea stars, often called starfishes, demonstrate basic features of echinoderm structure and function very well, and they are easily obtainable. We shall consider them first, then comment on major differences shown by other groups.

Sea stars are familiar along shorelines, where sometimes large numbers may aggregate on rocks. They also live on muddy or sandy bottoms and among coral reefs. They are often brightly colored and range in size from a centimeter in greatest diameter to about a meter across.

Position in the Animal Kingdom

1. Phylum Echinodermata (e-ki'no-der'ma-ta) (Gr. *echinos,* sea urchin, hedgehog, + *derma,* skin, + *ata,* characterized by) belongs to the **Deuterostomia** branch of the animal kingdom, the members of which are enterocoelous coelomates. The other phyla traditionally assigned to this group are Hemichordata and Chordata.

2. Primitively, deuterostomes have the following embryological features in common: anus developing from or near the blastopore, and mouth developing elsewhere; coelom budded from the archenteron (enterocoel); radial and regulative (indeterminate) cleavage; and endomesoderm (mesoderm derived from or with endoderm) from enterocoelic pouches.

Biological Contributions

There is one word that best describes echinoderms: strange. They have a unique constellation of characteristics found in no other phylum. Among the more striking features shown by echinoderms are these:

1. A system of channels composing the **water-vascular system,** derived from a coelomic compartment.

2. A **dermal endoskeleton** composed of calcareous ossicles.

3. A **hemal system,** whose function remains mysterious, also enclosed in a coelomic compartment.

4. Their **metamorphosis,** which changes a bilateral larva to a basic pentaradial adult.

A

C   D

figure 14.1

Some sea stars (class Asteroidea) from the Pacific. **A,** Pincushion star, *Culcita novaeguineae,* preys on coral polyps and also eats other small organisms and detritus. **B,** *Choriaster granulatus* scavenges dead animals on shallow Pacific reefs. **C,** *Tossia queenslandensis* from the Great Barrier Reef system browses encrusting organisms. **D,** Crown-of-thorns star, *Acanthaster planci,* a major coral predator (see note, p. 278), can have 10–20 arms.

Form and Function

External Features Reflecting their pentaradial symmetry, sea stars typically have five tapering arms (rays), but there may be more (figure 14.1D). These merge gradually with the central disc (figure 14.2). The mouth is on the oral surface; the side opposite the mouth is called the aboral surface.

Ambulacral (am-bu-la′kral) **grooves** (figure 14.2B) radiate out along the arms from the centrally located mouth on the under, or oral, side of the animal. **Tube feet (podia)** project from the grooves, which are bordered by movable spines. Viewed from the oral side, a large **radial nerve** can be seen in the center of each ambulacral groove (figure 14.3C), between the rows of tube feet. This nerve is very superficially located, covered only by thin epidermis. Under the nerve is an extension of the coelom and the radial canal of the water-vascular system (figure 14.3C). In all other classes of living echinoderms except crinoids, ossicles or other dermal tissue cover these structures; thus the ambulacral grooves in most asteroids and crinoids are *open,* and those of the other groups are *closed.*

The aboral surface is usually rough and spiny, although spines of many species are flattened, so that the surface appears smooth. Around the bases of spines in many sea stars are groups of minute pincerlike **pedicellariae** (ped-e-cell-ar′e-ee), bearing tiny jaws manipulated by muscles (figure 14.4). These jaws help keep the body surface free of debris, protect the papulae, and sometimes aid in food capture. **Papulae** (pap′u-lee) **(dermal branchiae** or **skin gills)** are soft, delicate projections of the coelomic cavity, covered only with epidermis and lined internally with peritoneum; they extend out through spaces between ossicles (figure 14.3C). Papulae greatly increase surface area available for respiratory gas exchange. Also on the aboral side are an inconspicuous **anus** and a circular **madreporite** (figure 14.3A), a calcareous sieve leading to the water-vascular system.

figure 14.2

External anatomy of asteroid.
A, Aboral view. **B,** Oral view.

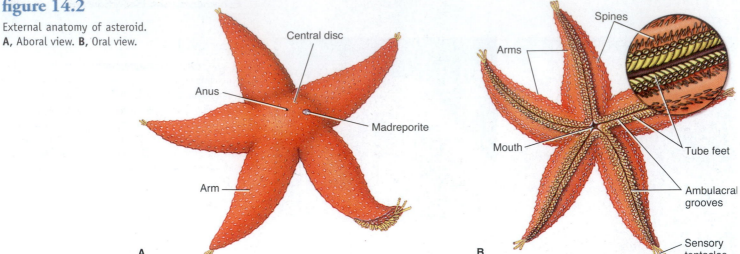

figure 14.3

A, Internal anatomy of a sea star. **B,** Water-vascular system. **C,** Cross section of arm at level of gonads, illustrating open ambulacral groove. Podia (tube feet) penetrate between ossicles. (Polian vesicles are not present in *Asterias*.)

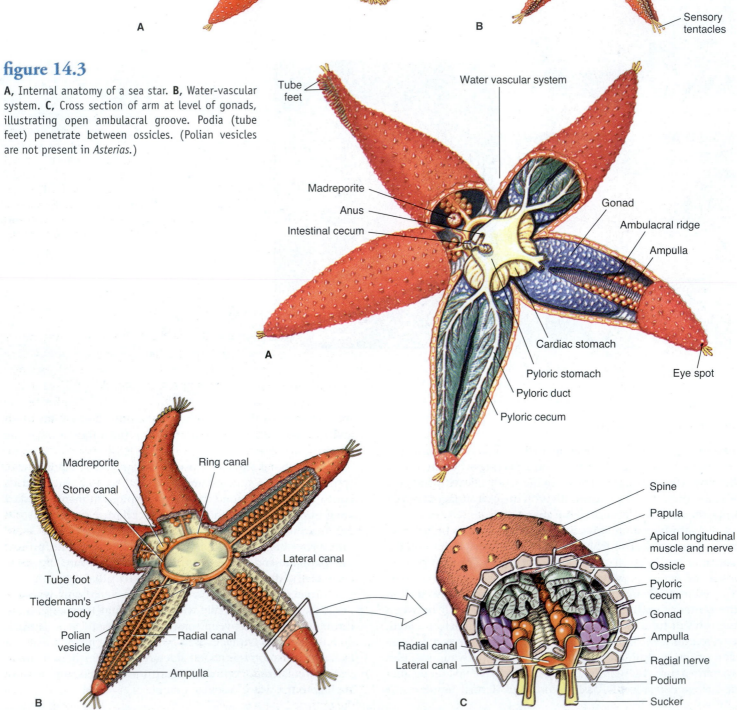

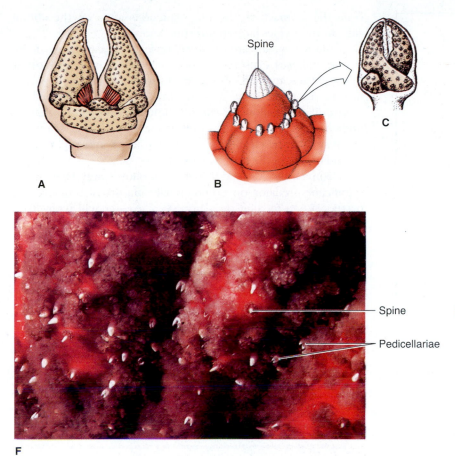

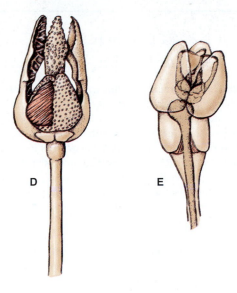

figure 14.4

Pedicellariae of sea stars and sea urchins. **A,** Forceps-type pedicellaria of *Asterias*. **B** and **C,** Scissors-type pedicellariae of *Asterias;* size relative to spine is shown in **B. D,** Tridactyl pedicellaria of *Strongylocentrotus,* cutaway showing muscle. **E,** Globiferous pedicellaria of *Strongylocentrotus.* **F,** Close-up view of the aboral surface of a sea star *Pycnopodia helianthoides.* Note the large pedicellariae, as well as groups of small pedicellariae around the spines. Many thin-walled papulae can be seen.

The function of the madreporite is still obscure. One suggestion is that it allows rapid adjustment of hydrostatic pressure within the water-vascular system in response to changes in external hydrostatic pressure resulting from depth changes, as in tidal fluctuations.

Endoskeleton and Coelom   Beneath the epidermis of sea stars is a mesodermal endoskeleton of small calcareous plates, or **ossicles,** bound together with connective tissue. From these ossicles project spines and tubercles that are responsible for their spiny surface. The ossicles are penetrated by a meshwork of spaces, usually filled with fibers and dermal cells. This internal meshwork structure is described as stereom (see figure 14.16) and is unique to echinoderms.

Coelomic compartments of larval echinoderms give rise to several structures in adults, one of which is a spacious body **coelom** filled with fluid. Coelomic fluid circulates around the body cavity and into the papulae, propelled by cilia on the peritoneal lining. Exchange of respiratory gases and excretion of nitrogenous waste, principally ammonia, occur by diffusion through the thin walls of the papulae and tube feet.

Water-Vascular System   The water-vascular system is another coelomic compartment and is unique to echinoderms. It is a system of canals and specialized tube feet that exploits hydraulic mechanisms to a greater degree than in any other animal group. In sea stars the primary functions of the water-vascular system are locomotion and food gathering in addition to respiration and excretion.

Structurally, the water-vascular system opens to the outside through small pores in the madreporite. The madreporite of asteroids is on the aboral surface (figure 14.2A), and leads into a **stone canal,** which descends toward a **ring canal** around the mouth (figure 14.3B). **Radial canals** diverge from the ring canal, one into the ambulacral groove of each ray. **Polian vesicles** are also attached to the ring canal of most asteroids (but not *Asterias*) and apparently serve as fluid reservoirs for the water-vascular system.

A series of small **lateral canals,** each with a one-way valve, connects the radial canal to the cylindrical podia (tube feet), along the sides of the ambulacral groove in each ray. Each podium is a hollow, muscular tube, the inner end of which is a muscular sac, or **ampulla,** which lies within the body coelom (figure 14.3), and the outer end of which usually bears a **sucker.** Ampullae of some species lack suckers. Podia pass to the outside between ossicles in the ambulacral groove.

characteristics
of Phylum Echinodermata

1. Body not segmented, adult with **pentaradial symmetry** characterized by five or more radiating areas
2. No head or brain; few specialized sensory organs
3. Nervous system with circumoral ring and radial nerves
4. **Endoskeleton** of **dermal calcareous ossicles** with **stereom** structure; covered by an epidermis (ciliated in most); **pedicellariae** (in some)
5. A **water-vascular system** of coelomic origin that extends from the body surface as a series of tentacle-like projections (**podia** or **tube feet**)
6. Locomotion by tube feet, which project from the **ambulacral areas,** or by movement of spines, or by movement of arms, which project from central disc of the body
7. Digestive system usually complete; axial or coiled; anus absent in ophiuroids
8. Coelom extensive, forming the perivisceral cavity and the cavity of the water-vascular system; coelom of enterocoelous type
9. Blood-vascular system (**hemal system**) much reduced, playing little, if any, role in circulation of body fluids, and surrounded by extensions of the coelom (perihemal sinuses)
10. Respiration by **papulae, tube feet, respiratory tree** (holothuroids), and **bursae** (ophiuroids)
11. Excretory organs absent
12. Sexes separate (except a few hermaphroditic); fertilization usually external
13. Development through **free-swimming, bilateral, larval stages** (some with direct development); metamorphosis to radial adult or subadult.

Locomotion by means of tube feet illustrates an interesting exploitation of hydraulic mechanisms by echinoderms. Valves in the lateral canals prevent backflow of fluid into the radial canals. Tube feet have in their walls connective tissue that maintains the cylinder at a relatively constant diameter. Contraction of muscles in an ampulla forces fluid into the tube foot, extending it. Conversely, contraction of longitudinal muscles in the tube foot retracts the podium, forcing fluid back into the ampulla. Contraction of muscles in one side of the tube foot bends the organ toward that side. Small muscles at the end of the tube foot can raise the middle of the disclike end, thus creating a suction-cup effect when the end is applied to a substrate. By combining mucous adhesion with suction, a single tube foot can exert a pull equal to 0.25 to 0.3 newtons. Coordinated action of all or many of the tube feet is sufficient to draw a sea star up a vertical surface or over rocks. On a soft surface, such as mud or sand, suckers are ineffective (numerous sand-dwelling species have no suckers), so the tube feet are employed as legs.

Feeding and Digestive System   The mouth leads into a two-part stomach located in the central disc (figure 14.3). In some species, the large, lower **cardiac stomach** can be everted. The smaller upper **pyloric stomach** connects with **digestive ceca** located in the arms. Digestion is largely extracellular, occurring in digestive ceca. A short **intestine** leads from the stomach to the inconspicuous anus on the aboral side. Some species lack an intestine and anus.

Many sea stars are relatively unselective carnivores, feeding on molluscs, crustaceans, polychaetes, echinoderms, other invertebrates, and sometimes small fish, but some show particular preferences. Some feed on brittle stars, sea urchins, or sand dollars, swallowing them whole and later regurgitating undigestible ossicles and spines. Some attack other sea stars, and if smaller than their prey, they may attack and begin eating an end of one of the prey's arms.

Some asteroids feed heavily on molluscs, and *Asterias* is a significant predator on commercially important mussels and oysters. When feeding on a bivalve, a sea star will wrap itself around its prey, attaching its podia to the valves, and then exert a steady pull, using its feet in relays. A force of some 12.75 newtons can thus be exerted. In half an hour or so the adductor muscles of the bivalve fatigue and relax. With a very small gap available, the star inserts its soft everted stomach into the space between the valves, wraps it around the soft parts of the bivalve, and secretes digestive enzymes to begin digestion. After feeding, the sea star withdraws its stomach by contraction of the stomach muscles and relaxation of body-wall muscles.

Some sea stars feed on small particles, either entirely or in addition to carnivorous feeding. Plankton or other organic particles coming in contact with a sea star's surface are carried by epidermal cilia to ambulacral grooves and then to the mouth.

Since 1963 there have been numerous reports of increasing numbers of crown-of-thorns starfish (*Acanthaster planci* [Gr. *akantha,* thorn, + *asteros,* star]) (see figure 14.1D), which damage large areas of coral reef in the Pacific Ocean. Crown-of-thorns stars feed on coral polyps, and they sometimes occur in large aggregations, or "herds." There is some evidence that outbreaks have occurred in the past, but an increase in frequency during the past 40 years suggests that some human activity may be affecting the starfish. Efforts to control the organism are very expensive and of questionable effectiveness. The controversy continues, especially in Australia, where it is exacerbated by extensive media coverage.

Hemal System   The so-called hemal system is not very well developed in asteroids, and in all echinoderms its function is unclear. It has little to do with circulation of body fluids. It is a system of tissue strands enclosing unlined channels and is itself enclosed in another coelomic compartment, the perihemal channels. Some research shows that absorbed nutrients appear in hemal fluids a few hours after feeding, suggesting that the hemal system plays a role in distributing digested nutrients.

Nervous and Sensory System   The nervous system in echinoderms comprises three subsystems, each formed by a nerve ring and radial nerves placed at different levels in the disc and arms. An epidermal nerve plexus, or nerve net, connects the systems. Sense organs include ocelli at the arm tips and sensory cells scattered all over the epidermis.

figure 14.5

Pacific sea star *Echinaster luzonicus* can reproduce itself by splitting across the disc, then regenerating missing arms. The one shown here has evidently regenerated six arms from the longer one at top left.

Reproductive System and Regeneration and Autotomy

Most sea stars have separate sexes. A pair of gonads lies in each interradial space (figure 14.3), and fertilization is external.

Echinoderms can regenerate lost parts. Sea star arms can regenerate readily, even if all are lost. Stars also use **autotomy** (the ability to detach part of their own bodies) and can cast off an injured arm near its base. An arm may take months to regenerate.

If a removed arm contains a part of the central disc (about one-fifth), the arm can regenerate a complete new sea star! In former times fishermen used to dispatch sea stars they collected from their oyster beds by chopping them in half with a hatchet—a worse than futile activity. Some sea stars reproduce asexually under normal conditions by cleaving the central disc, each part regenerating the rest of the disc and missing arms (figure 14.5).

Development

Some species brood their eggs, either under the oral side of the animal or in specialized aboral structures, and development is direct; but in most species embryonating eggs are free in the water and hatch to free-swimming larvae.

Early embryogenesis shows a typical primitive deuterostome pattern. Gastrulation is by invagination, and the anterior end of the archenteron pinches off to become the coelomic cavity, which expands in a U shape to fill the blastocoel. Each leg of the U, at the posterior end, constricts to become a separate vesicle, and these eventually give rise to the main coelomic compartments of the body (metacoels, called somatocoels in echinoderms) (figure 14.6). The anterior portions of the U undergo subdivision to form protocoels and mesocoels (called axocoels and hydrocoels in echinoderms). The left hydrocoel becomes the water-vascular system, and the left axocoel gives rise to the stone canal and perihemal channels. The right axocoel and hydrocoel disappear. The free-swimming larva, called a **bipinnaria** (figure 14.7A), has cilia arranged in bands, and these tracts extend onto the larval arms as development continues. The larva grows three adhesive arms and a sucker at its anterior end and is then called a **brachiolaria** (figure 14.7B). It attaches to the substratum with a temporary stalk and undergoes metamorphosis.

Metamorphosis involves a dramatic reorganization of a bilateral larva into a radial juvenile. The anteroposterior axis of the larva is lost. *What was the left side becomes the oral surface, and the larval right side becomes the aboral surface* (figure 14.6). Correspondingly, the larval mouth and anus disappear, and a new mouth and anus form on what were originally the left and right sides, respectively. As internal reorganization proceeds, short, stubby arms and the first podia appear. The animal then detaches from its stalk and begins life as a young sea star.

Sea Daisies

Strange little (less than 1 centimeter in diameter) disc-shaped animals, sometimes called sea daisies (figure 14.8), were discovered in 1986 in water more than 1000 meters deep off New Zealand. The two species in this group have no arms, and their tube feet are around the periphery of the disc, rather than along ambulacral areas. Their water-vascular system includes two concentric ring canals; the outer ring may represent radial canals because podia arise from it. A hydropore, homologous to the madreporite, connects the inner ring canal to the aboral surface. Although originally

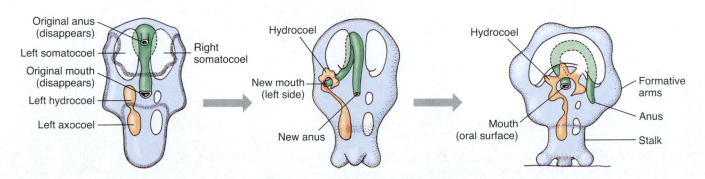

figure 14.6

Asteroid metamorphosis. The left somatocoel becomes the oral coelom, and the right somatocoel becomes the aboral coelom. The left hydrocoel becomes the water-vascular system and the left axocoel the stone canal and perihemal channels. The right axocoel and hydrocoel are lost.

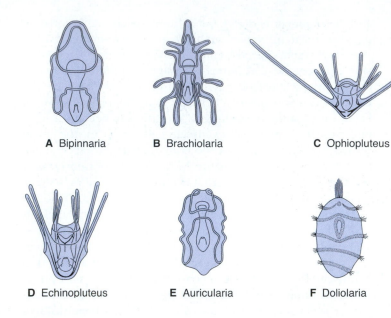

A Bipinnaria   **B** Brachiolaria   **C** Ophiopluteus

D Echinopluteus   **E** Auricularia   **F** Doliolaria

figure 14.7

Larvae of echinoderms. **A,** Bipinnaria of asteroids. **B,** Brachiolaria of asteroids. **C,** Ophiopluteus of ophiuroids. **D,** Echinopluteus of echinoids. **E,** Auricularia of holothuroids. **F,** Doliolaria of crinoids.

proposed as a class of echinoderms (Concentricycloidea), phylogenetic analysis of rDNA places them within Asteroidea.

Class Ophiuroidea: Brittle Stars and Basket Stars

Brittle stars are the largest group of echinoderms in numbers of species, with over 2000 extant species, and they are proba-bly the most abundant also. They abound in all types of benthic marine habitats, even carpeting the abyssal seafloor in many areas.

Apart from a typical possession of five arms, brittle stars are surprisingly different from asteroids. Arms of brittle stars are slender and sharply set off from their central disc (figure 14.9). They have no pedicellariae or papulae, and their ambulacral grooves are closed, covered with arm ossicles. Their tube feet are without suckers; they aid in feeding but are of limited use in locomotion. In contrast to asteroids, the madreporite of ophiuroids is located on the oral surface, on one of the oral-shield ossicles (figure 14.10).

Each jointed arm consists of a column of articulated ossi-cles connected by muscles and covered by plates. Locomotion is by arm movement.

Five movable plates surround the mouth, serving as **jaws** (figure 14.10). There is no anus. Their skin is leathery, with der-mal plates and spines arranged in characteristic patterns. Sur-face cilia are mostly lacking.

Visceral organs are all in the central disc, since the arms are too slender to contain them. The **stomach** is saclike and there is no intestine. Indigestible material is cast out of the mouth.

Five pairs of **bursae** (peculiar to ophiuroids) open toward the oral surface by **genital slits** at the bases of the arms. Water circulates in and out of these sacs for exchange of gases. On the coelomic wall of each bursa are small **gonads** that discharge into the bursa their ripe sex cells, which pass through the genital slits into the water for fertilization. Sexes are usually separate; a few ophiuroids are hermaphroditic. Ciliated bands of the larvae extend onto delicate, beautiful larval arms, like those of larval echinoids (figure 14.7C). During metamorphosis to the juvenile stage, there is no temporarily attached phase, as in asteroids.

figure 14.8

Xyloplax spp. (class Asteroidea) are peculiar little disc-shaped echinoderms. With their podia around the margin, they are the only echinoderms not having podia distributed along ambulacral areas.

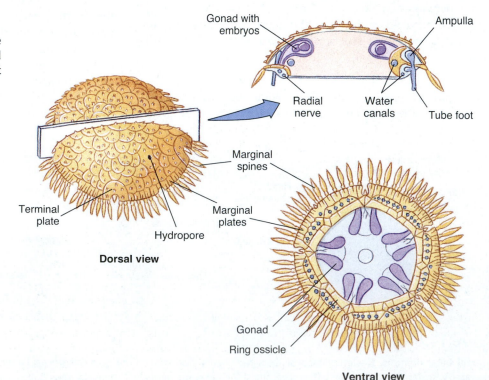

Gonad with embryos   Ampulla

Radial nerve   Water canals   Tube foot

Terminal plate   Marginal spines

Marginal plates

Hydropore

Dorsal view

Gonad

Ring ossicle

Ventral view

A

B

figure 14.9

A, Brittle star *Ophiothrix oerstedii* (class Ophiuroidea). Brittle stars do not use their tube feet for locomotion but can move rapidly (for an echinoderm) by means of their arms. **B,** Oral view of a basket star, *Gorgonocephalus eucnemis*. Basket stars extend their many-branched arms to filter feed, usually at night.

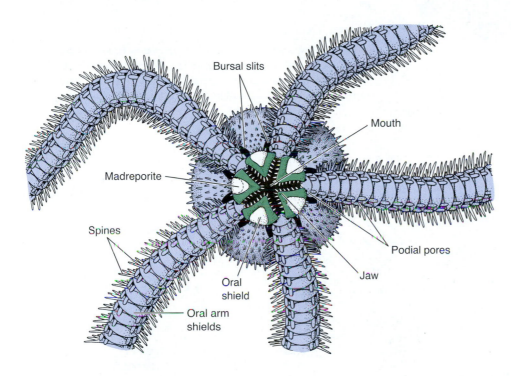

Bursal slits

Mouth

Madreporite

Spines

Podial pores

Jaw

Oral shield

Oral arm shields

figure 14.10

Oral view of spiny brittle star, *Ophiothrix*.

Water-vascular, nervous, and hemal systems are similar to those of sea stars.

Brittle stars tend to be secretive, living on hard substrates where no light penetrates. They are generally negatively phototropic and work themselves into small crevices between rocks, becoming more active at night. They are commonly fully exposed in the permanent darkness of a deep sea. Ophiuroids feed on a variety of small particles, either browsing food from the bottom or suspension feeding. Podia are important in transferring food to the mouth. Some brittle stars extend arms into the water and catch suspended particles in mucous strands between arm spines. Basket stars perch themselves on branched corals, extending their branched arms to capture plankton (figure 14.9B).

Regeneration and autotomy are even more pronounced in brittle stars than in sea stars. Many seem very fragile, releasing an arm or even part of the disc at the slightest provocation. Some can reproduce asexually by cleaving the disc; each progeny then regenerates missing parts.

Class Echinoidea: Sea Urchins, Sand Dollars, and Heart Urchins

Echinoids have a compact body enclosed in an endoskeletal **test,** or shell. Dermal ossicles, which have become closely fitting plates, form the test. Echinoids lack arms, but their tests reflect the typical five-part plan of echinoderms in their five ambulacral areas. Rather than extending from the oral surface to the tips of the arms, as in asteroids, the ambulacral areas follow contours of the test from the mouth around to the aboral side, ending at the area close to the anus **(periproct).** A majority of living species of sea urchins are termed "regular"; they are hemispherical in shape, radially symmetrical, and have medium to long spines (figure 14.11). Sand dollars and heart urchins (figure 14.12) are "irregular" because the orders to which they belong have become secondarily bilateral; their spines are usually very short. Regular urchins move by means of their tube feet, with some assistance from their spines, and irregular urchins move chiefly by their spines. Some echinoids are quite colorful.

figure 14.11

Purple sea urchin *Strongylocentrotus purpuratus* (class Echinoidea) is common along the Pacific Coast of North America where there is heavy wave action.

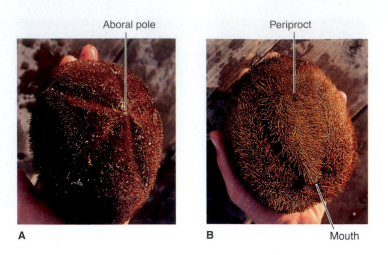

figure 14.12

An irregular echinoid *Meoma,* one of the largest heart urchins (test up to 18 cm). *Meoma* occurs in the West Indies and from the Gulf of California to the Galápagos Islands. **A,** Aboral view. **B,** Oral view. Note curved mouth at anterior end and periproct at posterior end.

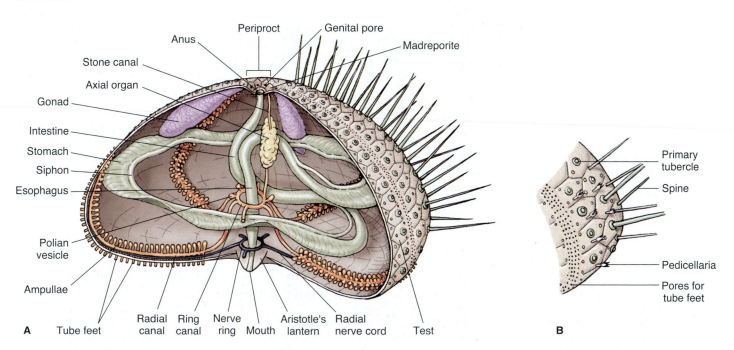

figure 14.13

A, Internal structure of the sea urchin; water-vascular system in tan. **B,** Detail of portion of endoskeleton.

Echinoids have wide distribution in all seas, from intertidal regions to deep oceans. Regular urchins often prefer rocky or hard substrates, whereas sand dollars and heart urchins like to burrow into a sandy substrate.

An echinoid test is a compact skeleton of 10 double rows of plates that bear movable, stiff spines (figure 14.13). The five pairs of ambulacral rows have pores (figure 14.13) through which long tube feet extend. Spines are moved by small muscles around their bases.

There are several kinds of pedicellariae, the most common of which have three jaws and are mounted on long stalks (see figure 14.4D, E). Pedicellariae of many species bear poison glands, and their toxin paralyzes small prey.

Five converging teeth surround the mouth of regular urchins and sand dollars. In some sea urchins branched **gills** (modified podia) encircle the region around the mouth, although these are of little importance in respiratory gas exchange. Anus, **genital openings,** and madreporite are aboral in the periproct region (figure 14.13). The mouth of sand dollars is located at about the center of the oral side, but their anus has shifted to the margin or even to the oral side of the disc, so that an anteroposterior axis and bilateral symmetry can be recognized. Bilateral symmetry is even more accentuated in heart urchins, with their anus near the posterior end on their oral side and their mouth moved away from the oral pole toward the anterior end (see figure 14.12).

Inside a sea urchin's test (figure 14.13) is a coiled digestive system and a complex chewing mechanism (in regular urchins and in sand dollars), called **Aristotle's lantern,** to which teeth are attached (figure 14.14). A ciliated siphon connects the esophagus to the intestine and enables water to

bypass the stomach to concentrate food for digestion in the intestine. Sea urchins eat algae and other organic material, and sand dollars collect fine particles on ciliated tracts.

The hemal and nervous systems are basically similar to those of asteroids. The ambulacral grooves are closed, and radial canals of the water-vascular system run just beneath the test, one in each ambulacral radius (see figure 14.13).

Sexes are separate, and both eggs and sperm are shed into the sea for external fertilization. Larvae (figure 14.7D) may live a planktonic existence for several months and then metamorphose quickly into young urchins. Sea urchins have been used extensively as models in studies of development.

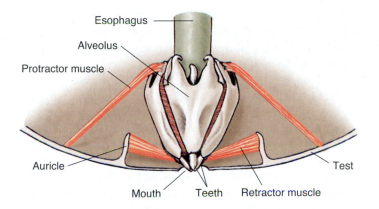

figure 14.14

Aristotle's lantern, a complex mechanism used by sea urchins for masticating their food. Five pairs of retractor muscles draw the lantern and teeth up into the test; five pairs of protractors push the lantern down and expose the teeth. Other muscles produce a variety of movements. Only major skeletal parts and muscles are shown in this diagram.

Class Holothuroidea: Sea Cucumbers

In a phylum characterized by odd animals, class Holothuroidea contains members that both structurally and physiologically are among the strangest. These animals have a remarkable resemblance to the vegetable after which they are named (figure 14.15). Compared to other echinoderms, holothurians are greatly elongated in their oral-aboral axis and have 10 to 30 prominent **oral tentacles,** modified tube feet, around the mouth. Despite a prominent anterior end, cephalization is absent. Ossicles (figure 14.16) are microscopic, numerous, and ornate in most species, but only comprise a modest portion of the body wall resulting in most species having leathery bodies.

Tube feet are typically used for movement in species that crawl over the surface of the sea bottom. Because holothurians typically lie on one side, their tube feet usually are well developed only in the three slender strips of ambulacra in contact with the substrate. Tube feet on the dorsal side are reduced and may serve a sensory role (figure 14.15A). Thus a secondary bilaterality is present, albeit of quite different origin from that of irregular urchins. Locomotion in burrowing species, which often lack

A

B

figure 14.15

Sea cucumber (class Holothuroidea). **A,** Common along the Pacific Coast of North America, *Parastichopus californicus* grows up to 50 cm in length. Its tube feet on the dorsal side are reduced to papillae and warts. **B,** *Bohadschia argus* expels its Cuverian tubules, organs attached to its respiratory tree, when it is disturbed. These sticky strands, containing a toxin, discourage potential predators.

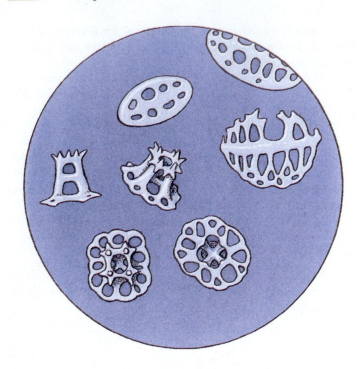

figure 14.16

Ossicles of sea cucumbers are usually microscopic bodies buried in the leathery dermis. They can be extracted from the tissue with commercial bleach and are important taxonomic characteristics. The ossicles shown here, called tables, buttons, and plates, are from the sea cucumber *Holothuria difficilis*. They illustrate the meshwork (stereom) structure observed in ossicles of all echinoderms at some stage in their development. (×250)

podia, primarily is accomplished by contraction of well-developed circular and longitudinal muscles in the body wall.

The coelomic cavity is spacious and filled with fluid, serving as a hydrostatis skeleton. A **respiratory tree** (figure 14.17) composed of two long, many-branched tubes empties into the **cloaca.** The respiratory tree is unique among echinoderms as a specialized, internal respiratory structure. Water is pumped in and out by both the muscular cloaca and the respiratory tree.

The hemal system is better developed in holothurians than in other echinoderms. The water-vascular system is peculiar in that the madreporite lies free in their coelom.

Sexes are separate, but some holothurians are hermaphroditic. Among echinoderms, only sea cucumbers have a single gonad, which is considered a primitive character. Fertilization is external.

Sea cucumbers are sluggish, moving partly by means of their ventral tube feet and partly by waves of contraction in the muscular body wall. More sedentary species trap suspended food particles in mucus of their outstretched oral tentacles; others are deposit feeders, collecting fine organic matter in their tentacles as they crawl. They stuff the tentacles into their pharynx, one by one, ingesting captured food. As deposit feeders, sea cucumbers are very successful, making up to 90% of the biomass on the surface of deep-sea floors.

Sea cucumbers have a peculiar power of what appears to be self-mutilation but in reality is a defense mechanism. When disturbed or subjected to unfavorable conditions, **Cuvierian tubules** may be discharged (figure 14.15B). These long, sticky, and sometimes toxic structures are attached to the respiratory tree and may entangle an enemy. Some species also discharge the digestive tract, respiratory tree, or gonads. All of these structures can be regenerated.

figure 14.17

Anatomy of a sea cucumber, *Sclerodactyla*. **A,** Internal; hemal system in red. **B,** External.

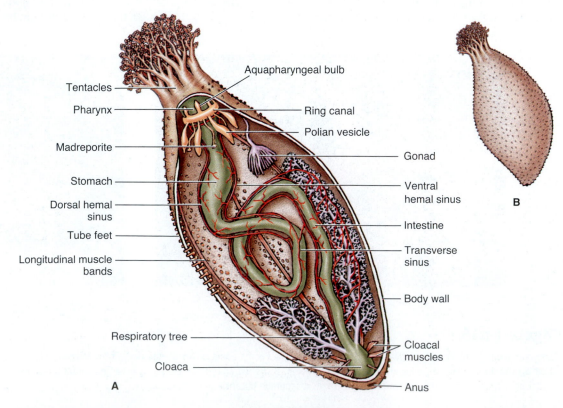

Tentacles
Pharynx
Madreporite
Stomach
Dorsal hemal sinus
Tube feet
Longitudinal muscle bands
Respiratory tree
Cloaca
Aquapharyngeal bulb
Ring canal
Polian vesicle
Gonad
Ventral hemal sinus
Intestine
Transverse sinus
Body wall
Cloacal muscles
Anus

A

B

Class Crinoidea: Sea Lilies and Feather Stars

Crinoids have several primitive characters. As fossil records reveal, crinoids were once far more numerous than now. They differ from other echinoderms by being attached during a substantial part of their lives. Many crinoids are deep-water forms, but feather stars may inhabit shallow waters, especially in Indo-Pacific and West Indian-Caribbean regions, where the largest numbers of species are found.

Their body disc has a leathery skin containing calcareous plates. Epidermis is poorly developed. Five flexible arms branch to form many more arms, each with many lateral **pinnules** arranged like barbs on a feather (figure 14.18). Sessile forms have a long, jointed **stalk** attached to the aboral side of the body (**calyx**) (figure 14.19). This stalk is composed of plates, appears jointed, and may bear a series of jointed, flexible appendages called **cirri**. Madreporite, spines, and pedicellariae are absent.

Their upper (oral) surface bears a mouth and anus. With the aid of tube feet and mucous nets, crinoids feed on small organisms caught in their ambulacral grooves. Their **ambulacral grooves** are open and ciliated and serve to carry food to the mouth. Tube feet in the form of tentacles are also found in the grooves. Their **water-vascular system** follows the basic echinoderm plan. Sense organs are scanty and simple.

The flower-shaped bodies of sea lilies are attached to the substratum by a stalk. During metamorphosis feather stars also become sessile and attached, but after several months they detach and become free moving. Although they may remain attached in the same location for long periods, they are capable of crawling and swimming short distances. They swim by alternate sweeping of their long, feathery arms.

figure 14.18

Comantheria briareus are crinoids found on Pacific coral reefs. They extend their arms into the water to catch food particles both during the day and at night.

Sexes are separate, and gonads are simple (figure 14.7F). Larvae swim freely for a time before they attach and metamorphose. Most living crinoids are from 15 to 30 cm long, but some fossil species had stalks 25 m in length.

Phylogeny and Adaptive Radiation

Phylogeny

Despite an extensive fossil record, there have been contesting hypotheses on echinoderm phylogeny. Based on the embryological evidence of bilateral larvae of echinoderms, there can be little doubt that their ancestors were bilateral and that their coelom had three pairs of spaces (trimeric). Some investigators have proposed that radial symmetry arose in a free-moving echinoderm ancestor and that sessile groups were derived several times independently from free-moving ancestors. However, this view does not consider the significance of radial symmetry as an adaptation for sessile existence. The more traditional view is that early echinoderms were sessile, became radial as an adaptation to that existence, and then gave rise to free-moving groups. Figure 14.20 is consistent with this hypothesis. It views evolution of endoskeletal plates with stereom structure and of external ciliary grooves for feeding as early echinoderm (or pre-echinoderm) developments. Extinct carpoids (Homalozoa,

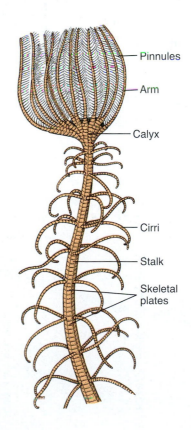

Pinnules

Arm

Calyx

Cirri

Stalk

Skeletal plates

figure 14.19

Sea lily (stalked crinoid) with portion of stalk. Modern crinoid stalks rarely exceed 60 cm, but fossil forms were as long as 20 m.

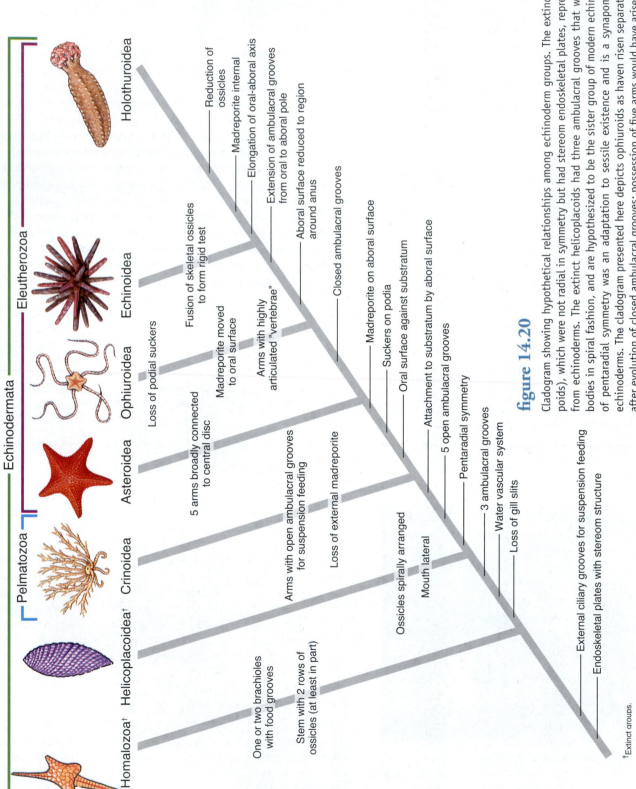

figure 14.20

Cladogram showing hypothetical relationships among echinoderm groups. The extinct Homalozoa (carpoids), which were not radial in symmetry but had stereom endoskeletal plates, represent an early split from echinoderms. The extinct helicoplacoids had three ambulacral grooves that wound around their bodies in spiral fashion, and are hypothesized to be the sister group of modern echinoderms. Evolution of pentaradial symmetry was an adaptation to sessile existence and is a synapomorphy for modern echinoderms. The cladogram presented here depicts ophiuroids as haven risen separately from asteroids, after evolution of closed ambulacral grooves; possession of five arms would have arisen independently in these groups. An alternate scenario, also well supported, unites Asteroidea and Ophiuroidea in a clade, with five arms being a synapomorphy and independent evolution of closed ambulacral grooves in ophiuroids and the common ancestor of echinoids and holothuroids.

Echinodermata

Pelmatozoa

Eleutherozoa

Homalozoa† Helicoplacoidea† Crinoidea Asteroidea Ophiuroidea Echinoidea Holothuroidea

Reduction of ossicles
Madreporite internal
Elongation of oral-aboral axis
Extension of ambulacral grooves from oral to aboral pole
Aboral surface reduced to region around anus
Fusion of skeletal ossicles to form rigid test
Closed ambulacral grooves
Madreporite moved to oral surface
Arms with highly articulated "vertebrae"
Madreporite on aboral surface
Loss of podial suckers
Suckers on podia
Oral surface against substratum
5 arms broadly connected to central disc
Attachment to substratum by aboral surface
5 open ambulacral grooves
Arms with open ambulacral grooves for suspension feeding
Loss of external madreporite
Pentaradial symmetry
Ossicles spirally arranged
3 ambulacral grooves
Water vascular system
Mouth lateral
Loss of gill slits
One or two brachioles with food grooves
Stem with 2 rows of ossicles (at least in part)
External ciliary grooves for suspension feeding
Endoskeletal plates with stereom structure

†Extinct groups.

classification of Phylum Echinodermata

There are about 7000 living and 20,000 extinct or fossil species of Echinodermata. The traditional classification placed all the free-moving forms that were oriented with oral side down in subphylum Eleutherozoa, containing most living species. The other subphylum, Pelmatozoa, contained mostly forms with stems and oral side up; most extinct classes and living Crinoidea belong to this group. Although alternative schemes have strong supporters, cladistic analysis provides evidence that the two traditional subphyla are monophyletic groups. This listing includes only groups with living members.

Subphylum Pelmatozoa (pel-ma'to-zo'a) (Gr. *pelmatos,* a stalk, + *zoōn,* animal). Body in form of cup or calyx, borne on aboral stalk during part or all of life; oral surface directed upward; open ambulacral grooves; madreporite absent; both mouth and anus on oral surface; several fossil classes plus living Crinoidea.

Class Crinoidea (krin-oy'de-a) (Gr. *krinon,* lily, + *eidos,* form, + *-ea,* characterized by): **sea lilies** and **feather stars.** Five arms branching at base and bearing pinnules; ciliated ambulacral grooves on oral surface with tentacle-like tube feet for food gathering; spines, madreporite, and pedicellariae absent. Examples: *Antedon, Comantheria* (see figure 14.18).

Subphylum Eleutherozoa (e-lu'ther-o-zo'a) (Gr. *eleutheros,* free, not bound, + *zoōn,* animal). Body form star-shaped, globular, discoidal, or cucumber shaped; oral surface directed toward substratum or oral-aboral axis parallel to substratum.

Class Asteroidea (as'ter-oy'de-a) (Gr. *aster,* star, + *eidos,* form, + *-ea,* characterized by): **sea stars (starfish).** Star-shaped, with arms not sharply demarcated from the central disc; ambulacral grooves open, with tube feet on oral side; tube feet often with suckers; anus and madreporite aboral; pedicellariae present. **Sea daisies,** included in this group, lack arms, have a disc-shaped body, and have a ring of suckerless podia near the body margin. Examples: *Asterias, Pisaster.*

Class Ophiuroidea (o'fe-u-roy'de-a) (Gr. *ophis,* snake, + *oura,* tail, + *eidos,* form, + *-ea,* characterized by): **brittle stars** and **basket stars.** Star shaped, with arms sharply demarcated from central disc; ambulacral grooves closed, covered by ossicles; tube feet without suckers and not used for locomotion; pedicellariae absent. Examples: *Ophiothrix* (see figure 14.9A), *Astrophyton* (see figure 14.9B).

Class Echinoidea (ek'i-noy'de-a) (Gr. *echinos,* sea urchin, hedgehog, + *eidos,* form, + *-ea* characterized by): **sea urchins, sea biscuits,** and **sand dollars.** More or less globular or disc-shaped, with no arms; compact skeleton or test with closely fitting plates; movable spines; ambulacral grooves closed; tube feet often with suckers; pedicellariae present. Examples: *Arbacia, Strongylocentrotus* (see figure 14.11), *Lytechinus, Meoma* (see figure 14.12).

Class Holothuroidea (hol'o-thu-roy'de-a) (Gr. *holothourion,* sea cucumber, + *eidos,* form, + *-ea,* characterized by): **sea cucumbers.** Cucumber-shaped, with no arms; spines absent; microscopic ossicles embedded in muscular body wall; anus present; ambulacral grooves closed; tube feet with suckers; circumoral tentacles (modified tube feet); pedicellariae absent; madreporite internal. Examples: *Sclerodactyla, Parastichopus* (see figure 14.15), *Cucumaria.*

figure 14.20) had stereom ossicles but were not radially symmetrical, and the status of their water-vascular system, if any, is uncertain. Some investigators regard carpoids as a separate subphylum of echinoderms (Homalozoa), and others consider them preechinoderms with affinities to chordates. Fossil helicoplacoids (figure 14.20) show evidence of three, true ambulacral grooves, and their mouth was on the side of their body.

Attachment to the substratum by the aboral surface would have selected for radial symmetry and the origin of Pelmatozoa. An ancestor that became free moving and applied its oral surface to the substratum would have given rise to Eleutherozoa. Phylogeny within Eleutherozoa is controversial. Most investigators agree that echinoids and holothuroids are related and form a clade, but opinions diverge on the relationships of ophiuroids and asteroids. Figure 14.20 illustrates the view that ophiuroids arose after closure of ambulacral grooves, but this scheme treats evolution of five ambulacral rays (arms) in ophiuroids and asteroids as independently evolved. Alternatively, if ophiuroids and asteroids form a clade, then closed ambulacral grooves must have evolved separately in ophiuroids and in the common ancestor of echinoids and holothuroids. Both hypotheses have support from molecular and morphological data.

Adaptive Radiation

Radiation of echinoderms has been determined by limitations and potentials of their most important characteristics: radial symmetry, water-vascular system, and dermal endoskeleton. If their ancestors had a brain and specialized sense organs, these were lost in the adoption of radial symmetry. Thus, it is not surprising that there are large numbers of creeping, benthic forms with filter-feeding, deposit-feeding, scavenging, and herbivorous habits, comparatively few predators, and very few pelagic species. In this light the relative success of asteroids as predators is impressive and probably attributable to the extent to which they have exploited the hydraulic mechanism of their tube feet.

Phylum Hemichordata

Hemichordata (hem'i-kor-da'ta) (Gr. *hemi,* half, + *chorda,* string, cord) are marine animals formerly considered a subphylum of chordates, based on their possession of gill slits, a rudimentary notochord, and a dorsal nerve cord. However, zoologists now agree that the so-called hemichordate "notochord" is really an

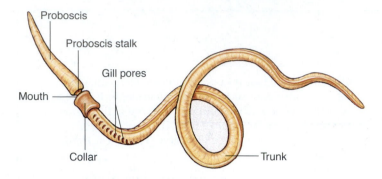

figure 14.21

External lateral view of an acorn worm, *Saccoglossus* (Hemichordata, class Enteropneusta).

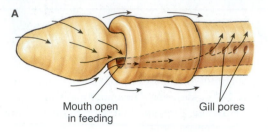

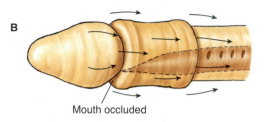

figure 14.22

Food currents of enteropneust hemichordate. **A,** Side view of acorn worm with mouth open, showing direction of currents created by cilia on proboscis and collar. Food particles are directed toward mouth and digestive tract. Rejected particles move toward outside of collar. Water leaves through gill pores. **B,** When mouth is occluded, all particles are rejected and passed onto the collar. Nonburrowing and some burrowing hemichordates use this feeding method.

evagination of their mouth cavity and not homologous with the chordate notochord, so hemichordates are considered a separate phylum. Hemichordates have the typical tricoelomate structure of deuterostomes.

Hemichordates are wormlike bottom dwellers, living usually in shallow waters. Some colonial species live in secreted tubes. Many are sedentary or sessile. They are widely distributed, but their secretive habits and fragile bodies make collecting them difficult.

Members of class Enteropneusta (acorn worms) range from 20 mm to 2.5 m in length. Members of class Pterobranchia (pterobranchs) are smaller, usually 1 to 7 mm, not including the stalk. About 75 species of enteropneusts and three small genera of pterobranchs are recognized.

Class Enteropneusta

Enteropneusts, or acorn worms (figure 14.21), are sluggish wormlike animals that live in burrows or under stones, usually in mud or sand flats of intertidal zones.

The mucus-covered body is divided into a tonguelike **proboscis,** a short **collar,** and a long **trunk** (protosome, mesosome, and metasome.)

Many acorn worms are **deposit feeders,** extracting the organic components of sediments. Others are **suspension feeders,** capturing plankton and detritus. The muscular proboscis is the active part of the animal, collecting small particles of food from mud and sand or water. Food particles captured in mucus on the proboscis are transported by ciliary action to the groove at the edge of the collar and then to the mouth (figure 14.22). The food-laden mucus continues

Position in the Animal Kingdom

1. Hemichordates belong to the deuterostome branch of the animal kingdom and are enterocoelous coelomates with radial cleavage.

2. A chordate plan of structure is suggested by gill slits and a restricted dorsal tubular nerve cord.
3. Similarity to echinoderms is seen in larval characteristics.

Biological Contributions

1. A **tubular dorsal nerve cord** in the collar zone may represent an early stage of the condition in chordates; a diffuse net of nerve cells is similar to the uncentralized, subepithelial plexus of echinoderms.

2. Cilia lining the **gill slits** create water flow permitting respiratory gas exchange in the pharynx.

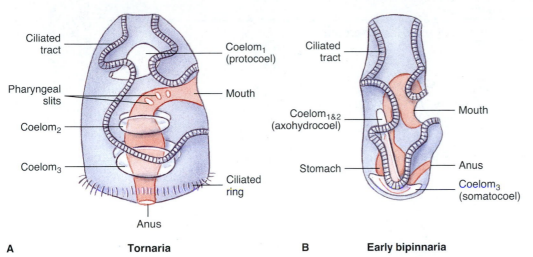

figure 14.23

Comparison of a hemichordate tornaria **(A)**, to an echinoderm bipinnaria **(B)**.

to be directed by cilia along the ventral part of the pharynx and esophagus to the intestine. A row of **gill pores** extends dorsolaterally on each side of the trunk just behind the collar (figure 14.21). These gill pores open from a series of densely ciliated gill chambers that in turn connect with a series of **gills slits** in the side of the pharynx. Much of the respiratory gas exchange occurs across the vascular epithelium of the pharynx. The gill slits allow removal of excess water from the pharynx, but they are not used as a filter-feeding structure.

In the posterior end of the proboscis is a small coelomic sac (protocoel) into which extends a **buccal diverticulum,** a slender, blindly ending pouch of the gut that was formerly considered a notochord. Above this buccal diverticulum is the heart, which receives blood carried forward from a dorsal blood vessel. Blood then flows into a network of blood sinuses called a **glomerulus,** which is believed to have an excretory function, then posteriorly through a ventral blood vessel, passing through extensive sinuses in the body wall and gut. Acorn worms have a ventral nerve cord and a larger dorsal nerve cord that is hollow in some species. Sexes are separate, and fertilization is external. In some species a ciliated **tornaria** larva develops that closely resembles the echinoderm bipinnaria (figure 14.23).

Class Pterobranchia

Pterobranchs are small colonial animals, usually 1–7 mm in length, although the stalk may be longer. They bear arms with tentacles containing extensions of the coelomic compartment of the mesosome, similar to a lophophore (figure 14.24). Food captured in mucus on the crown of tentacles is transported with ciliated grooves to the mouth. A dorsal, hollow nerve cord is lacking. Although the nervous system is similar to that of enteropneusts, one genus has a single pair of gill slits (figure 14.24). Some species are dioecious, others are monoecious, and asexual reproduction occurs by budding.

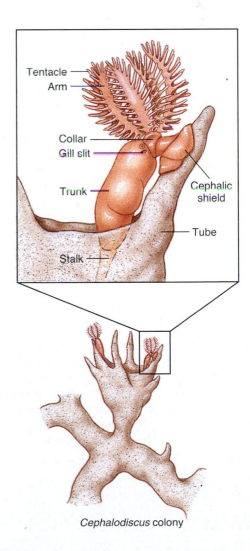

Cephalodiscus colony

figure 14.24

Cephalodiscus, a pterobranch hemichordate. These tiny (5 to 7 mm) forms live in tubes in which they can move freely. Ciliated tentacles and arms direct currents of food and water toward mouth.

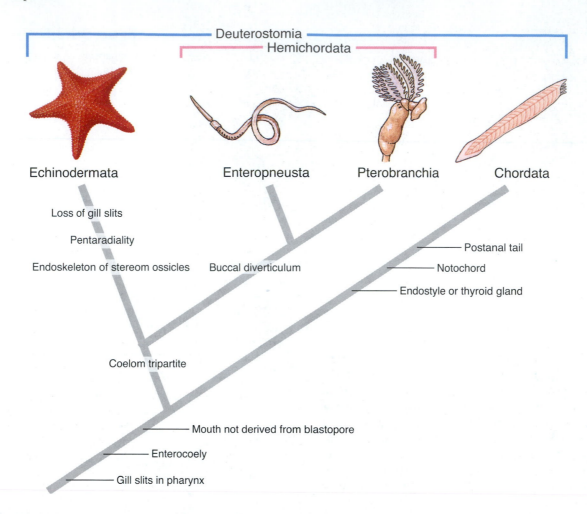

figure 14.25

Cladogram showing hypothetical relationships among deuterostome phyla. Based on this hypothesis, the ancestral deuterostome was a marine benthic form with gill slits, perhaps similar to enteropneusts. Gill slits apparently were present in stem echinoderms, but lost in modern echinoderms. Although this relationship is now widely accepted, some prefer an alternate hypothesis that hemichordates and chordates form a clade, based on shared gill slits and dorsal, hollow nerve cord.

Phylogeny

Which deuterostome phylum is the sister taxon to Hemichordata? Hemichordates share with chordates a dorsal hollow nerve cord and gill slits. However, early embryogenesis of hemichordates is remarkably like that of echinoderms, and early tornaria larvae are almost identical to bipinnaria larvae of asteroids. As a result, the relationship of Hemichordata to other deuterostome groups has been unclear. The more traditional hypothesis groups hemichordates with chordates as a clade, based on shared gill slits and dorsal nerve cord. Sequence analysis of 18S and 28S rDNA and a variation of *Hox* genes consistently place hemichordates with echinoderms (figure 14.25). Some morphological data support this relationship as well, including the similar, ciliated larvae and shared tricoelomic body. In addition, painstaking studies by Jefferies and his students described gill slits in basal echinoderms known as carpoids (figure 14.20), suggesting that gill slits are ancestral for deuterostomes.

Summary

Phyla Echinodermata, Chordata, and Hemichordata show characteristics of Deuterostomia. Echinoderms are an important marine group sharply distinguished from other ani-mals. They have a pentaradial symmetry but were derived from bilateral ancestors.

Sea stars (class Asteroidea) usually have five arms, and their arms merge gradually with a central disc. Like other echinoderms, they have no head and few specialized sensory organs. Their mouth is central on the under (oral) side of their body. They have an endoskeleton of dermal ossicles, open ambulacral grooves, respiratory papulae, and in many species, pedicellariae. Their water-vascular system is an elaborate hydraulic system derived from one of the coelomic cavities. Along the ambulacral areas, branches from the radial canals lead to the many tube feet, structures that are important in locomotion, food gathering, respiration, and excretion. Many sea stars are predators, whereas others feed on particulate organic matter. Sexes are separate, and reproductive systems are very simple. Bilateral, free-swimming larvae become attached, transform to radial juveniles, then detach and become motile sea stars.

Brittle stars (class Ophiuroidea) have slender arms that are sharply set off from their central disc; they have no pedicellariae or ampullae, and their ambulacral grooves are closed. Their madreporite is on the oral side. They crawl by means of their arms, and they can move more rapidly than other echinoderms.

In sea urchins (class Echinoidea), dermal ossicles have become closely fitting plates, and there are no arms. Some urchins (sand dollars and heart urchins) have evolved a return to bilateral symmetry.

Sea cucumbers (class Holothuroidea) have very small dermal ossicles and a soft body wall. They are greatly elongated in the oral-aboral axis and lie on their side. Tube feet around the mouth are modified as oral tentacles, with which they feed. They have an internal respiratory tree and can eject Cuvierian tubules for defense.

Sea lilies and feather stars (class Crinoidea) are the only group of living echinoderms, other than asteroids, with open ambulacral areas. They are mucociliary particle feeders and lie with their oral side upward.

Hemichordates are marine wormlike deposit or filter feeders that capture food using mucociliary action. Acorn worms capture food on their proboscis, whereas pterobranchs capture food on their tentacles. Hemichordates and echinoderms have in common a tripartite coelom and similar larvae and together form the probable sister group of chordates.

Review Questions

1. What constellation of characteristics possessed by echinoderms is found in no other phylum?
2. How do we know that echinoderms were derived from an ancestor with bilateral symmetry?
3. What is an ambulacral groove, and what is the difference between open and closed ambulacral grooves? Open ambulacra is considered the primitive condition, and closed ambulacra is considered derived. Can you suggest a reason why this is probably correct?
4. Trace or make a rough copy of figure 14.3B, without labels, and then from memory label parts of the water-vascular system of a sea star.
5. Name structures involved in following functions in sea stars, brittle stars, sea urchins, sea cucumbers, and crinoids, and briefly describe the action of each: respiration, feeding and digestion, excretion, reproduction.
6. Match these groups with *all* correct answers in the lettered column.
   ___ Crinoidea   ___ Asteroidea
   ___ Ophiuroidea   ___ Echinoidea
   ___ Holothuroidea
   a. Closed ambulacral grooves
   b. Oral surface generally upward
   c. With arms
   d. Without arms
   e. Approximately globular or disc shaped
   f. Elongated in oral-aboral axis
   g. With pedicellariae
   h. Madreporite internal
   i. Madreporite on oral plate
7. Define the following: pedicellariae, madreporite, respiratory tree, Aristotle's lantern, Cuvierian tubules.
8. Give three examples of how echinoderms are important to humans.
9. Distinguish Enteropneusta from Pterobranchia.
10. Hemichordata were once considered chordates. Why?
11. What evidence suggests that hemichordates and echinoderms form a monophyletic group?

Selected References

See also general references on page 415.

Baker, A. N., F. W. E. Row, and H. E. S. Clark. 1986. A new class of Echinodermata from New Zealand. Nature **321**:862–864. *Describes the strange class Concentricycloidea.*

Gilbert, S. F. 2003. Developmental biology, ed. 7. Sunderland, Massachusetts, Sinauer Associates. *Any modern text in developmental biology, such as this one, provides a multitude of examples in which studies on echinoderms have contributed (and continue to contribute) to our knowledge of development.*

Hendler, G., J. E. Miller, D. L. Pawson, and P. M. Kier. 1995. Sea stars, sea urchins, and allies: echinoderms of Florida and the Caribbean. Washington, Smithsonian Institution Press. *An excellent field guide for echinoderm identification.*

Hickman, C. P., Jr. 1998. A field guide to sea stars and other echinoderms of Galápagos. Lexington, Virginia, Sugar Spring Press. *Provides descriptions and clear photographs of members of classes Asteroidea, Ophiuroidea, Echinoidea, and Holothuroidea in the Galápagos Islands.*

Hughes, T. P. 1994. Catastrophes, phase shifts and large-scale degradation of a Caribbean coral reef. Science **265**:1547–1551. *Describes the sequence of events, including the die-off of sea urchins, leading to the destruction of the coral reefs around Jamaica.*

Janies, D. 2001. Phylogenetic relationships of extant echinoderm classes. Can. J. Zool. **79**:1232–1250. *This review indicates that Xyloplax (sea daisy) is an asteroid and that asteroids and ophiuroids form a monophyletic group.*

Lane, D. J. W. 1996. A crown-of-thorns outbreak in the eastern Indonesian Archipelago, February 1996. Coral Reefs **15**:209–210. *This is the first report of an outbreak of* Acanthaster planci *in Indonesia. It includes a good photograph of an aggregation of these sea stars.*

Mooi, R., and B. David. 1998. Evolution within a bizarre phylum: homologies of the first echinoderms. Amer. Zool. **38**:965–974. *The authors say, "The familiarity of a seastar or a sea urchin belies their overall weirdness." They describe the Extraxial/Axial Theory (EAT) of echinoderm skeletal homologies.*

Moran, P. J. 1990. *Acanthaster planci* (L.): biographical data. Coral Reefs **9**:95–96. *Presents a summary of essential biological data on* A. planci. *This entire issue of* Coral Reefs *is devoted to* A. planci.

Smith, A. B., K. J. Peterson, G. Wray, and D. T. J. Littlewood. 2004. From bilateral symmetry to pentaradiality: the phylogeny of hemichordates and echinoderms, pp. 365–383. *In:* Assembling the tree of life. J. Cracraft and M. J. Donoghue (eds.). New York, Oxford University Press. *Discusses evolution of echinoderms and hemichordates and summarizes evidence that they form a monophyletic group.*

Woodley, J. D., P. M. H. Gayle, and N. Judd. 1999. Sea-urchins exert top-down control of macroalgae on Jamaican coral reefs (2). Coral Reefs **18**:193. *In areas where* Tripneustes *(another sea urchin) have invaded fore reefs there is much less macroalgae, and such areas present a better chance that corals can recolonize.* Diadema *recovery has been slow.*

Wray, G. A., and R. A. Raff. 1998. Body builders of the sea. Nat. Hist. **107**:38–47. *Regulatory genes in bilateral animals have taken on new but analogous roles in radial echinoderms.*

Custom Website

The *Animal Diversity* Online Learning Center is a great place to check your understanding of chapter material. Visit www.mhhe.com/hickmanad4e for access to key terms, quizzes, and more! Further enhance your knowledge with Web links to chapter-related material.

Explore live links for these topics:

Classification and Phylogeny of Animals
Phylum Echinodermata
Class Asteroidea
Class Ophiuroidea
Class Echinoidea

Class Holothuroidea
Class Crinoidea
Class Concentricycloidea
Phylum Hemichordata

Vertebrate Beginnings:
The Chordates

It's a Long Way from Amphioxus

Along the more southern coasts of North America, half buried in sand on the seafloor, lives a small fishlike translucent animal quietly filtering organic particles from seawater. Inconspicuous, of no commercial value and largely unknown, this creature is nonetheless one of the famous animals of classical zoology. It is amphioxus, an animal that wonderfully exhibits the five distinctive hallmarks of the phylum Chordata—(1) dorsal, tubular nerve cord overlying (2) a supportive notochord, (3) pharyngeal slits and (4) endostyle for filter feeding, and (5) a postanal tail for propulsion—all wrapped up in one creature with textbook simplicity. Amphioxus is an animal that might have been designed by a zoologist for the classroom. During the nineteenth century, with interest in vertebrate ancestry running high, amphioxus was considered by many to resemble closely the direct ancestor of vertebrates. Its exalted position was later acknowledged by Philip Pope in a poem sung to the tune of "Tipperary." It ends with the refrain:

> It's a long way from amphioxus
> It's a long way to us,
> It's a long way from amphioxus
> To the meanest human cuss.
> Well, it's good-bye to fins and gill slits
> And it's welcome lungs and hair,
> It's a long, long way from amphioxus
> But we all came from there.

But amphioxus' place in the sun was not to endure. For one thing, amphioxus lacks one of the most important of vertebrate characteristics, a distinct head with special sense organs and the equipment for shifting to an active predatory mode of life. Absence of a head, together with several specialized features, suggests to zoologists today that amphioxus represents an early departure from the main line of chordate descent. It seems that we are a very long way indeed from amphioxus. Nevertheless, while amphioxus is denied the vertebrate ancestor award, it resembles the chordate condition immediately preceding the origin of vertebrates more closely than any other living animal we know.

Two amphioxus in feeding posture.

A nimals most familiar to most people belong to the phylum Chordata. Humans are members and share the characteristic from which the phylum derives its name—the **notochord** (Gr. *nōton,* back, + L. *chorda,* cord) (figure 15.1). All members of the phylum possess this structure, either restricted to early development or throughout life. The notochord is a rodlike, semirigid body of cells enclosed by a fibrous sheath, which extends, in most cases, the full length of the body just ventral to the central nervous system. Its primary purpose is to support and to stiffen the body, that is, to act as a skeletal axis.

The structural plan of chordates retains many of the features of nonchordate invertebrates, such as bilateral symmetry, anteroposterior axis, coelom tube-within-a-tube arrangement, metamerism, and cephalization. However, the exact phylogenetic position of chordates within the animal kingdom is unclear.

Two possible lines of descent have been proposed. Earlier speculations that focused on the arthropod-annelid-mollusc group (Protostomia branch) of the invertebrates have fallen from favor. Only members of the echinoderm-hemichordate assemblage (Deuterostomia branch) now deserve serious consideration as a chordate sister group. The chordates share with other deuterostomes several important characteristics: radial cleavage (p. 57), anus derived from the first embryonic opening (blastopore) and mouth derived from an opening of secondary origin, and a coelom primitively formed by fusion of enterocoelous pouches (except in vertebrates in which the coelom is basically schizocoelous, but independently derived). These common characteristics indicate a natural unity among the Deuterostomia.

Phylum Chordata shows a more fundamental unity of organs and organ systems than do other phyla. Ecologically, chordates are among the most adaptable of organic forms, able to occupy most kinds of habitat. They illustrate perhaps better than any other animal group the basic evolutionary processes of the origin of new structures, adaptive strategies, and adaptive radiation.

Traditional and Cladistic Classification of the Chordates

Traditional Linnean classification of chordates (p. 306) provides a simple and convenient way to indicate the taxa included in each major group. However, in cladistic usage, some traditional taxa, such as Agnatha and Reptilia, are no longer recognized. Such taxa do not satisfy the requirement of cladistics that only **monophyletic** groups are valid taxonomic entities, that is, groups that contain all known descendants of a single common ancestor. The reptiles, for example, are considered a **paraphyletic** grouping because this group does not contain all descendants of their most recent common ancestor. The common ancestor of reptiles as traditionally recognized is also the ancestor of birds and mammals. As shown in the cladogram (figure 15.3), reptiles, birds, and mammals compose a monophyletic group called Amniota, so named because all develop from an egg having special extraembryonic membranes, one of which is the amnion. Therefore according to cladistics, "reptiles" can be used only as a term of convenience to refer to amniotes that are not birds or mammals; there are no derived characters that unite the reptiles to the exclusion of birds and mammals. The reasons why nonmonophyletic groups are not used in cladistic taxonomy are explained in Chapter 4 (p. 76).

The phylogenetic tree of chordates (figure 15.2) and the cladogram of chordates (figure 15.3) provide different kinds of information (p. 76). The cladogram shows a nested hierarchy

Position in the Animal Kingdom

Phylum Chordata (kor-da′ta) (L. *chorda,* cord) belongs to the Deuterostomia branch of the animal kingdom that includes phyla Echinodermata and Hemichordata. From humble beginnings, chordates evolved a vertebrate body plan of enormous adaptability that always remains distinctive, while providing almost unlimited scope for specialization in life habit, form, and function.

Biological Contributions

1. The **endoskeleton** of vertebrates permits continuous growth and the attainment of large body size, and it provides an efficient framework for muscle attachment.
2. The **perforated pharynx** of protochordates originated as a suspension-feeding device and served as a framework for subsequent evolution of true internal gills with pharyngeal muscular pump, and jaws.
3. Adoption of a **predatory habit** by early vertebrates and the accompanying evolution of a **highly differentiated brain** and **paired special sense organs** contributed in large measure to the successful adaptive radiation of vertebrates.
4. **Paired appendages** that appeared in aquatic vertebrates were successfully adapted later as jointed limbs for efficient locomotion on land or as wings for flight.

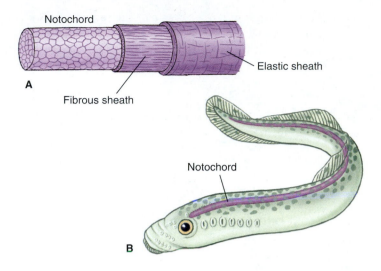

figure 15.1

A, Structure of the notochord and its surrounding sheaths. Cells of the notochord proper are thick walled, pressed together closely, and filled with semifluid. Stiffness is caused mainly by turgidity of fluid-filled cells and surrounding connective tissue sheaths. This primitive type of endoskeleton is characteristic of all chordates at some stage of the life cycle. The notochord provides longitudinal stiffening of the main body axis, a base for trunk muscles, and an axis around which the vertebral column develops. **B,** In hagfishes and lampreys it persists throughout life, but in other vertebrates it is largely replaced by vertebrae. In mammals slight remnants are found in nuclei pulposi of intervertebral discs. The method of notochord formation is different in various groups of vertebrates. In amphioxus it originates from the endoderm; in birds and mammals it arises as an anterior outgrowth of the embryonic primitive streak.

of taxa grouped by their sharing of derived characters. These characters may be morphological, physiological, embryological, behavioral, chromosomal, or molecular in nature. By contrast, the branches of a phylogenetic tree are intended to represent real lineages that occurred in the evolutionary past. Geological information regarding ages of lineages is added to the information from the cladogram to generate a phylogenetic tree for the same taxa.

In our treatment of chordates, we have retained the traditional Linnean classification (p. 306) because subfields of zoology are organized according to this scheme and because the alternative—thorough revision following cladistic principles—would require abandonment of familiar rankings. However, we have tried to use monophyletic taxa as much as possible, because such usage is necessary to reconstruct the evolution of morphological characters in chordates (see p. 76).

Several traditional divisions of phylum Chordata used in Linnean classifications are shown in table 15.1. A fundamental separation is Protochordata from the Vertebrata. Since the former lack a well-developed head, they also are called Acraniata. All vertebrates have a well-developed skull case enclosing the brain and are called Craniata. Some cladistic classifications exclude Myxini (hagfishes) from the group Vertebrata, because they lack vertebrae, but retain them in Craniata because they

characteristics
of Phylum Chordata

1. Bilateral symmetry; segmented body; three germ layers; well-developed coelom
2. **Notochord** (a skeletal rod) present at some stage in the life cycle
3. **Single, dorsal, tubular nerve cord;** anterior end of cord usually enlarged to form a brain
4. **Pharyngeal pouches** present at some stage in the life cycle; in aquatic chordates these develop into pharyngeal slits
5. **Endostyle** in floor of pharynx or a **thyroid gland** derived from the endostyle
6. **Postanal tail,** projecting beyond the anus at some stage but may or may not persist
7. Complete digestive system
8. **Segmentation,** if present, restricted to outer body wall, head, and tail and not extending into coelom

have a cranium. The vertebrates (craniates) may be variously subdivided into groups based on shared possession of characteristics. Two such subdivisions shown in table 15.1 are: (1) Agnatha, vertebrates lacking jaws (hagfishes and lampreys), and Gnathostomata, vertebrates having jaws (all other vertebrates) and (2) Amniota, vertebrates whose embryos develop within a fluid-filled sac, the amnion (reptiles, birds, and mammals), and Anamniota, vertebrates lacking this adaptation (fishes and amphibians). The Gnathostomata in turn can be subdivided into Pisces, jawed vertebrates with appendages, if any, in the form of fins; and Tetrapoda (Gr. *tetras,* four, + *podos,* foot), jawed vertebrates with appendages, if any, in the form of limbs. Note that several of these groupings are paraphyletic (Protochordata, Acraniata, Agnatha, Anamniota, Pisces) and consequently are not accepted in cladistic classifications. Accepted monophyletic taxa are shown at the top of the cladogram in figure 15.3 as a nested hierarchy of increasingly more inclusive groupings.

Five Chordate Hallmarks

Five distinctive characteristics that, taken together, set chordates apart from all other phyla are **notochord; single, dorsal, tubular nerve cord; pharyngeal pouches or slits; endostyle;** and **postanal tail.** These characteristics are always found at some embryonic stage, although they may be altered or may disappear in later stages of the life cycle.

Notochord

The notochord is a flexible, rodlike structure, extending the length of the body; it is the first part of the endoskeleton to appear in the embryo. The notochord is an axis for muscle

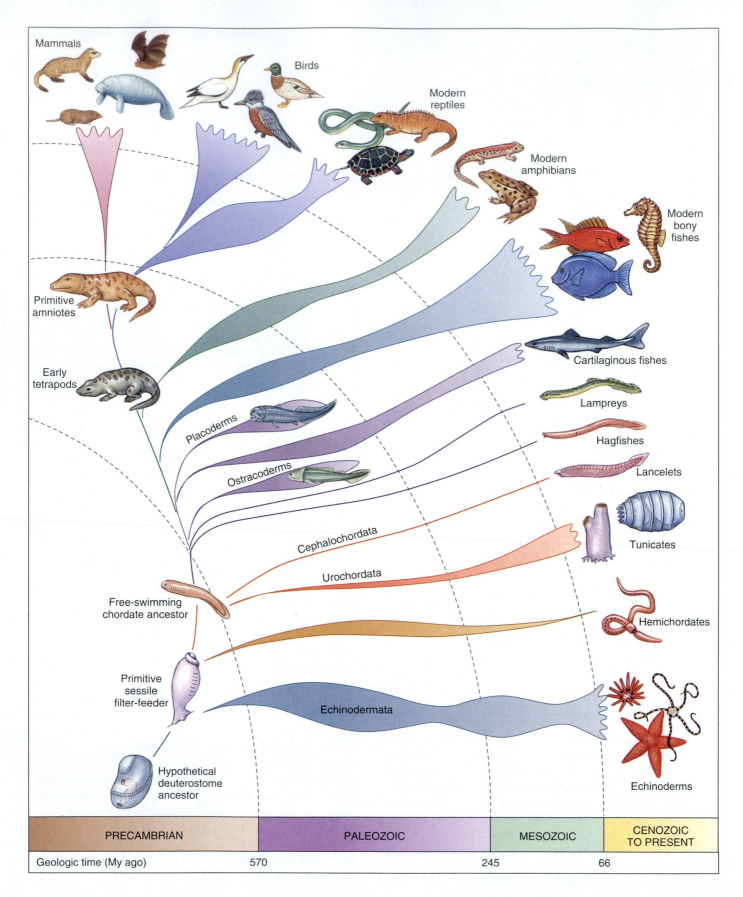

figure 15.2

Phylogenetic tree of the chordates, suggesting probable origin and relationships. Other schemes have been suggested and are possible. The relative abundance in numbers of species of each group through geological time, as indicated by the fossil record, is suggested by the bulging and thinning of that group's line of descent.

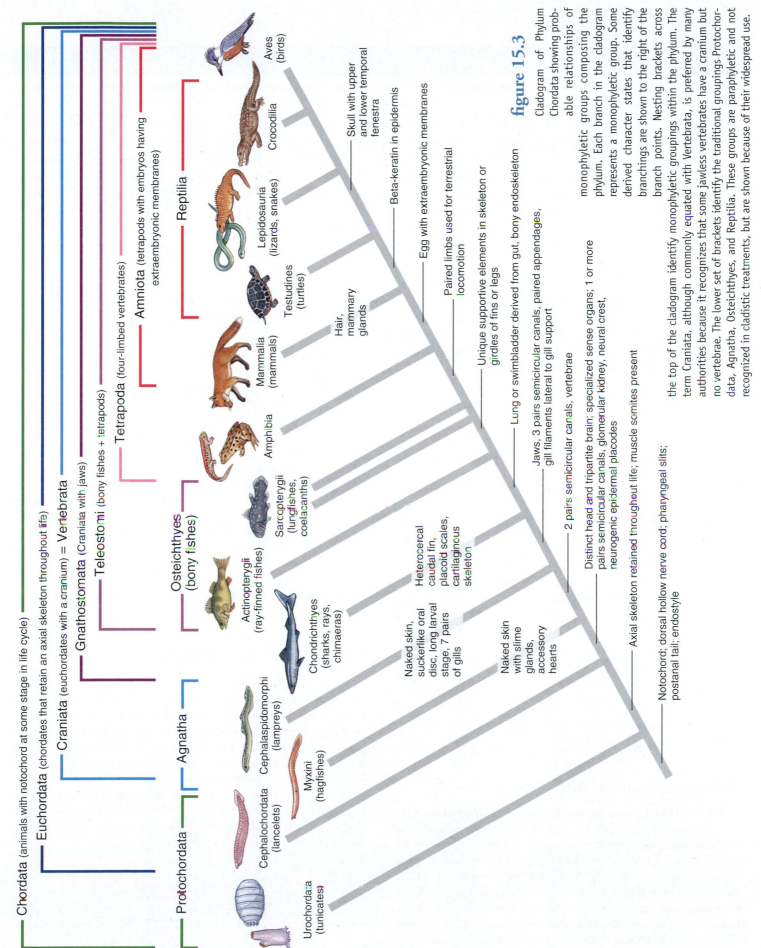

figure 15.3

Cladogram of Phylum Chordata showing probable relationships of monophyletic groups composing the phylum. Each branch in the cladogram represents a monophyletic group. Some derived character states that identify branchings are shown to the right of the branch points. Nesting brackets across the top of the cladogram identify monophyletic groupings within the phylum. The term Craniata, although commonly equated with Vertebrata, is preferred by many authorities because it recognizes that some jawless vertebrates have a cranium but no vertebrae. The lower set of brackets identify the traditional groupings Protochordata, Agnatha, Osteichthyes, and Reptilia. These groups are paraphyletic and not recognized in cladistic treatments, but are shown because of their widespread use.

Table 15.1   Traditional Divisions of the Phylum Chordata

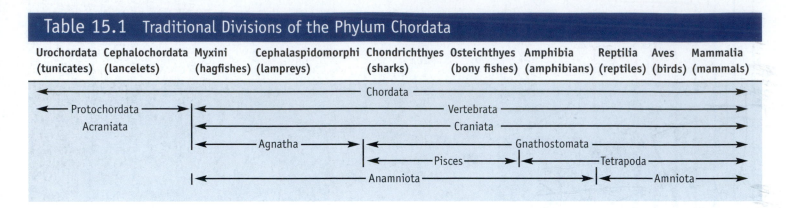

Urochordata (tunicates)	Cephalochordata (lancelets)	Myxini (hagfishes)	Cephalaspidomorphi (lampreys)	Chondrichthyes (sharks)	Osteichthyes (bony fishes)	Amphibia (amphibians)	Reptilia (reptiles)	Aves (birds)	Mammalia (mammals)

attachment, and because it can bend without shortening, it permits undulatory movements of the body. In most protochordates and in jawless vertebrates, the notochord persists throughout life (figure 15.1), but in all jawed vertebrates it is replaced, in whole or in part, by a series of cartilaginous or bony vertebrae.

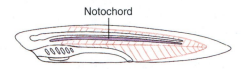

Notochord

Dorsal, Tubular Nerve Cord

In most invertebrate phyla that have a nerve cord, it is ventral to the alimentary canal and is solid, but in chordates the single cord is dorsal to the alimentary canal and notochord and is a tube (although the hollow center may be nearly obliterated during growth). In vertebrates the anterior end becomes enlarged to form a brain. The hollow cord is produced in embryos by infolding of ectodermal cells on the dorsal side of the body above the notochord. The nerve cord passes through the protective neural arches of the vertebrae, and the anterior brain is surrounded by a bony or cartilaginous cranium.

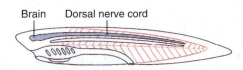

Brain   Dorsal nerve cord

Pharyngeal Pouches and Slits

Pharyngeal slits are openings that lead from the pharyngeal cavity to the outside. They are formed by the inpocketing of the outside ectoderm (pharyngeal grooves) and the evagination, or outpocketing, of the endodermal lining of the pharynx (pharyngeal pouches). In aquatic chordates, the two pockets break through the pharyngeal cavity where they meet to form the pharyngeal slit. In amniotes some pockets may not break through the pharyngeal cavity and only grooves are formed instead of slits. In tetrapod vertebrates pharyngeal pouches give

rise to several different structures, including the Eustachian tube, middle ear cavity, tonsils, and parathyroid glands.

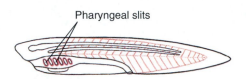

Pharyngeal slits

The perforated pharynx evolved as a filter-feeding apparatus and is used as such in protochordates. Water with suspended food particles is drawn by ciliary action through the mouth and flows out through pharyngeal slits, where food is trapped in mucus. In vertebrates, ciliary action is replaced by muscular pharyngeal contractions that drive water through the pharynx. The addition of a capillary network and thin gas-permeable walls in the pharyngeal arches led to the development of **internal gills,** completing the conversion of the pharynx from a filter-feeding apparatus in protochordates to a respiratory organ in aquatic vertebrates.

Endostyle or Thyroid Gland

Until recently the endostyle was not recognized as a chordate character. However, it or its derivative, the thyroid gland, is found in all chordates, but in no other animals. The endostyle, in the pharyngeal floor, secretes mucus that traps small food particles brought into the pharyngeal cavity. An endostyle is found in protochordates and lamprey larvae. Some cells in the endostyle secrete iodinated proteins. These cells are homologous with the iodinated-hormone-secreting thyroid gland of adult lampreys and the remainder of vertebrates. In other chordates, endostyle and perforated pharynx work together to create an efficient filter-feeding apparatus.

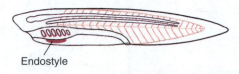

Endostyle

Postanal Tail

A postanal tail, together with somatic musculature and the stiffening notochord, provides the motility that larval tunicates and amphioxus need for their free-swimming existence. As a structure added to the body behind the anus, it clearly has evolved specifically for propulsion in water. Its efficiency is later increased in fishes with the addition of fins. A tail is evident in humans only as a vestige (the coccyx, a series of small vertebrae at the end of the spinal column) but most other mammals have a waggable tail as adults.

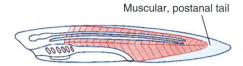

Muscular, postanal tail

Ancestry and Evolution

Since the mid-nineteenth century when Darwin's theory of common descent became the focal point for recognizing relationships among groups of living organisms, zoologists have debated the question of chordate origins. Zoologists at first speculated that chordates evolved within the protostome lineage (annelids and arthropods) but rejected such ideas when they realized that supposed morphological similarities had no developmental basis. Early in the twentieth century when further theorizing became rooted in developmental patterns of animals, it became apparent that chordates must have originated within the deuterostome branch of the animal kingdom. As explained in Chapter 3 (p. 61 and figure 3.18), Deuterostomia, a grouping that includes echinoderms, hemichordates, and chordates, has several important embryological features, as well as shared gene sequences, that clearly separate it from Protostomia and establish its monophyly. Thus deuterostomes are almost certainly a natural grouping of interrelated animals that have their common origin in ancient Precambrian seas. Somewhat later, at the base of the Cambrian period some 570 million years ago, the first distinctive chordates arose from a lineage related to echinoderms and hemichordates (figure 15.1; see also figure 14.25, p. 290).

Most early efforts to pin together invertebrate and chordate kinship are now recognized as based on similarities due to analogy rather than homology. Analogous structures are those that perform similar functions but have altogether different origins (such as wings of birds and butterflies). Homologous structures, on the other hand, share a common origin but may look quite different (at least superficially) and perform quite different functions. For example, all vertebrate forelimbs are homologous because they are derived from a pentadactyl limb of the same ancestor, even though they may be modified as differently as a human's arm and a bird's wing. Homologous structures share a genetic heritage; analogous structures do not. Obviously, only homologous similarities reveal common ancestry.

Subphylum Urochordata (Tunicata)

The urochordates ("tail-chordates"), more commonly called tunicates, include about 2000 species. They live in all seas from near shoreline to great depths. Most are sessile as adults, although some are free living. The name "tunicate" is suggested by the usually tough, nonliving **tunic,** or test, that surrounds the animal (figure 15.4). As adults, tunicates are highly specialized chordates, for in most species only the larval form, which resembles a microscopic tadpole, bears all the chordate hallmarks. During adult metamorphosis, the notochord (which, in larvae, is restricted to the tail, hence the group name Urochordata) and tail disappear, while the dorsal nerve cord becomes reduced to a single ganglion.

Urochordates are traditionally divided into three classes—**Ascidiacea** (Gr. *askiolion,* little bag, + *acea,* suffix), **Larvacea** (L. *larva,* ghost, + *acea,* suffix), and **Thaliacea** (Gr. *thalia,* luxuriance, + *acea,* suffix). Of these, members of Ascidiacea, commonly called ascidians, or sea squirts, are by far the most common and best known. Ascidians may be solitary, colonial, or compound. All but a few ascidian species are sessile, attached to rocks or other hard substances such as pilings or bottoms of ships. In some areas they are among the most abundant of intertidal animals.

Solitary (figure 15.4) or colonial ascidians are usually spherical or cylindrical forms, each bearing its own tunic. Lining the tunic is an inner membrane, the mantle. On the outside

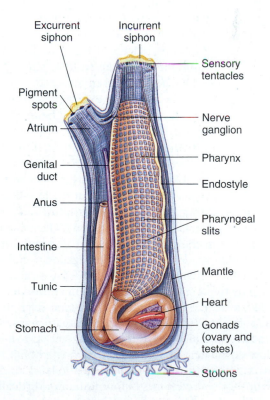

figure 15.4

Structure of a common tunicate, *Ciona* sp.

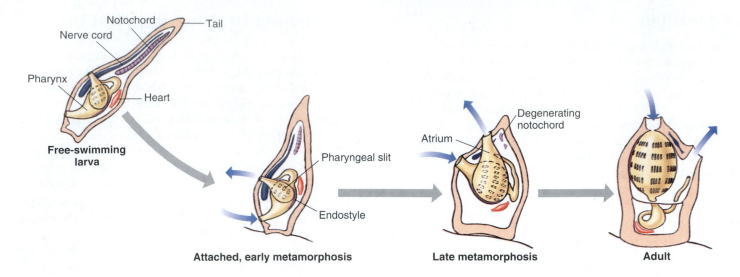

figure 15.5

Metamorphosis of a solitary ascidian from a free-swimming larval stage.

are two projections: **incurrent** and **excurrent siphons** (figure 15.4). Water enters the incurrent siphon and passes into the pharynx through the mouth. On the midventral side of the pharynx is a groove, the **endostyle,** which is ciliated and secretes mucus. As the mucous sheet is carried by cilia across the inner surface of the pharynx to the dorsal side, it sieves small food particles from water passing through slits in the wall of the pharynx. The mucus with its entrapped food is collected and passed posteriorly to the esophagus. The water, now largely cleared of food particles, is driven by cilia into the atrial cavity and finally out the excurrent siphon. Nutrients are absorbed in the midgut, and indigestable wastes are discharged form the anus, located near the excurrent siphon.

The circulatory system consists of a ventral **heart** near the stomach and two large vessels, one on either side of the heart. An odd feature found in no other chordate is that the heart drives the blood first in one direction for a few beats, then pauses, reverses, and drives the blood in the opposite direction. The excretory system is a type of nephridium near the intestine. The nervous system is restricted to a **nerve ganglion** and a few nerves that lie on the dorsal side of the pharynx. The animals are hermaphroditic. Germ cells are carried out the excurrent siphon into the surrounding water, where cross-fertilization occurs.

Of five chief characteristics of chordates, adult sea squirts have only two: pharyngeal gill slits and endostyle. However, their larvae reveal the secret of their true relationship. The tiny tadpole larva (figure 15.5) is an elongate, transparent form with a head and all five chordate characteristics: a notochord, hollow dorsal nerve cord, propulsive postanal tail, and a large pharynx with endostyle and gill slits. The larva does not feed but swims for several hours before fastening vertically by adhesive papillae to some solid object. It then metamorphoses to become a sessile adult.

The remaining two classes of the Urochordata— **Larvacea** and **Thaliacea**—are mostly small, transparent animals of the open sea (figure 15.6). Larvaceans are small, tadpolelike forms resembling the larval stage of ascidians. Thaliaceans are spindle shaped or cylindrical forms surrounded by delicate muscle bands. They are mostly carried by ocean currents and as such form a part of the plankton. Many are provided with luminous organs and emit a beautiful light at night.

Subphylum Cephalochordata

Cephalochordates are lancelets: slender, laterally compressed, translucent animals about 3 to 7 cm in length (figure 15.7) that inhabit sandy bottoms of coastal waters around the world. Lancelets originally bore the generic name *Amphioxus* (Gr. *amphi,* both ends, + *oxys,* sharp), later surrendered by priority to *Branchiostoma* (Gr. *branchia,* gills, + *stoma,* mouth).

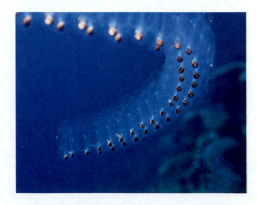

figure 15.6

Colonial thaliacean. The transparent individuals of this delicate, planktonic tunicate are grouped in a chain. Visible within each individual is an orange gonad, an opaque gut, and a long serrated gill bar.

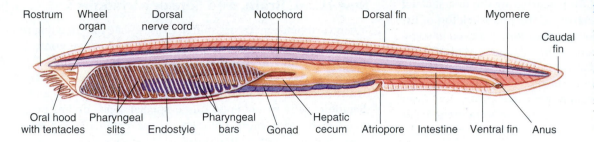

Rostrum   Wheel organ   Dorsal nerve cord   Notochord   Dorsal fin   Myomere   Caudal fin

Oral hood with tentacles   Pharyngeal slits   Endostyle   Pharyngeal bars   Gonad   Hepatic cecum   Atriopore   Intestine   Ventral fin   Anus

figure 15.7

Amphioxus. This interesting bottom-dwelling cephalochordate illustrates the five distinctive chordate characteristics (notochord, dorsal nerve cord, pharyngeal slits, endostyle, and postanal tail). The vertebrate ancestor probably had a similar body plan.

Amphioxus is still used, however, as a convenient common name for the approximately 29 species in this diminutive subphylum. Five species of amphioxus occur in North American coastal waters.

Amphioxus is especially interesting because it has the five distinctive characteristics of chordates in simple form. Water enters the mouth, driven by cilia in the buccal cavity, and then passes through numerous pharyngeal slits in the pharynx, where food is trapped in mucus, which is then moved by cilia into the intestine. Here the smallest food particles are separated from the mucus and passed into the **hepatic cecum,** where they are phagocytized and digested intracellularly. As in tunicates, filtered water passes first into an atrium, and then leaves the body by an **atriopore** (equivalent to the excurrent siphon of tunicates).

The closed circulatory system is complex for so simple a chordate. The flow pattern is remarkably similar to that of fishes, although there is no heart. Blood is pumped forward in the ventral aorta by peristaltic-like contractions of the vessel wall, and then passes upward through branchial arteries (aortic arches) in gill arches to the dorsal aorta. From here blood is distributed to body tissues by capillaries and then collected in veins, which return it to the ventral aorta. Their blood is colorless, lacking both erythrocytes and hemoglobin.

The nervous system is centered around a hollow nerve cord lying above the notochord. Sense organs are simple, unpaired bipolar receptors located in various parts of the body. The "brain" is a simple vesicle at the anterior end of the nerve cord.

Sexes are separate. Gametes are released in the atrium, then pass through the atriopore to the outside, where fertilization occurs. Larvae hatch soon after eggs are fertilized and gradually assume the shape of adults.

No other chordate shows the basic diagnostic chordate characteristics so well. In addition to the five chordate anatomical hallmarks, amphioxus possesses several structural features that resemble the vertebrate plan. Among these are a hepatic cecum, which secretes digestive enzymes, **segmented trunk musculature,** and the basic circulatory pattern of more advanced chordates.

Subphylum Vertebrata

The third subphylum of chordates is the large and diverse Vertebrata, the subject of the next five chapters of this book. This monophyletic group shares the basic chordate characteristics

with the other two subphyla, but in addition it reveals novel homologies that the others do not share. The alternative name of the subphylum, Craniata, more accurately describes the group because all have a cranium (bony or cartilaginous braincase) but some jawless fishes lack vertebrae.

Adaptations That Have Early Guided Vertebrate Evolution

The earliest vertebrates were substantially larger and considerably more active than the protochordates. Increased speed and mobility resulted from modifications of the skeleton and muscles. The higher activity level and size of vertebrates also required structures specialized in the location, capture, and digestion of food and adaptations designed to support a high metabolic rate.

Musculoskeletal Modifications

The endoskeleton of vertebrates permits almost unlimited body size, with much greater economy of building materials than the exoskeleton of arthropods. Some vertebrates have become the most massive organisms on earth. The endoskeleton forms an excellent jointed scaffolding for attachment of segmented muscles. The segmented body muscles (myomeres) changed from the V-shaped muscles of cephalochordates to the W-shaped muscles of vertebrates. This increased complexity of folding in the myomeres provides powerful control over an extended length of the body.

The endoskeleton probably was composed initially of cartilage that later gave way to bone. Cartilage, with its fast

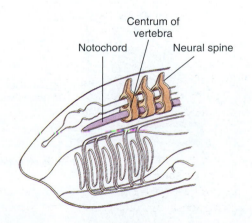

Centrum of vertebra

Notochord   Neural spine

growth and flexibility, is ideal for constructing the first skeletal framework of all vertebrate embryos. The endoskeleton of living hagfishes, lampreys, sharks, and their kin, and even in some "bony" fishes, such as sturgeons, is mostly composed of cartilage. Bone may have been adaptive in early vertebrates in several ways. The presence of bone in the skin of ostracoderms and other ancient fishes certainly provided protection from predators, although there are some more important benefits of bone. The structural strength of bone is superior to cartilage, making it ideal for muscle attachment in areas of high mechanical stress. One of the most interesting ideas is that the function associated with the origin of bone was for storage and regulation of minerals. Phosphorus and calcium are used for many physiological processes and are in particularly high demand in organisms with high metabolic rates.

We should note that most vertebrates possess an extensive exoskeleton, although it is highly modified in advanced forms. Some of the most primitive fishes, including the ostracoderms and placoderms, were partly covered in a bony dermal armor. This armor is modified as scales in later fishes. Most vertebrates are further protected with keratinized structures derived from the epidermis, such as reptilian scales, hair, feathers, claws, and horns.

Physiology Upgrade

Vertebrates have modifications to the digestive, respiratory, and circulatory systems to meet an increased metabolic demand. The perforated pharynx evolved as a filter-feeding device in early chordates, using cilia and mucus to move water and to trap small suspended food particles. In larger vertebrates, the addition of powerful muscles to the pharynx created a powerful pump for moving water. With the origin of highly vascularized gills, the function of the pharynx shifted primarily to gas exchange. Changes in the gut, including a shift from movement of food by ciliary action to muscular action and addition of accessory digestive glands, the liver and pancreas, were necessary to manage the increased amount of food ingested. A ventral three-chambered heart consisting of a sinus venosus, an atrium, and a ventricle, and erythrocytes with hemoglobin enhanced transportation of nutrients, gases, and other substances.

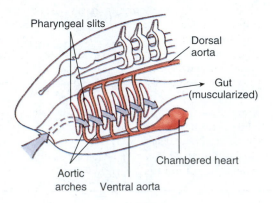

New Head, Brain, and Sensory Systems

When vertebrate ancestors shifted from sessile filter feeding to active predation, new sensory, motor, and integrative controls became essential for location and capture of larger prey. The anterior end of the nerve cord became enlarged as a **tripartite brain** (forebrain, midbrain, and hindbrain) protected by a cartilaginous or bony cranium. Paired special sense organs originated or become more complex. These included eyes with lenses and inverted retinas; pressure receptors, such as paired inner ears designed for equilibrium and sound reception; chemical receptors, including taste and exquisitely sensitive olfactory organs; lateral-line receptors for detecting water vibrations; and electroreceptors for detecting electrical currents that signal prey.

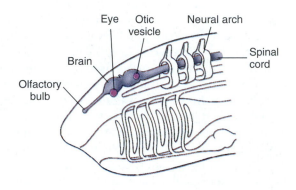

Neural Crest, Ectodermal Placodes, and *Hox* Genes

Development of the vertebrate head and special sense organs was largely the result of two embryonic innovations present only in vertebrates: the **neural crest** and **ectodermal placodes.** The neural crest, a population of ectodermal cells lying along the length of the embryonic neural tube, contributes to the formation of many different structures, among them most of the cranium, the pharyngeal skeleton, tooth dentine, Schwann cells, and some endocrine glands. The ectodermal placodes (Gr. *placo,* plate) are platelike ectodermal thickenings that appear on either side of the neural tube. These give rise to the olfactory epithelium, the lens of the eye, the inner-ear epithelium, some ganglia, some cranial nerves, lateral-line mechanoreceptors, and electroreceptors. Thus the vertebrate head, with its complex sensory structures located adjacent to the mouth (later equipped with prey-capturing jaws), apparently stemmed from the creation of completely new cell types.

Recent studies of the distribution of **homeobox** genes that control the body plan of chordate embryos suggest that the *Hox* genes were duplicated at about the time of the origin of vertebrates. One copy of *Hox* genes is found in amphioxus and other invertebrates, whereas living gnathostomes have four copies. Perhaps these additional copies of genes that control body plan provided genetic material free to evolve a more complex kind of animal.

The Search for the Ancestral Vertebrate Stock

Most of the early Paleozoic vertebrate fossils, the jawless ostracoderms (p. 304), share many novel features of organ system development with living vertebrates. These organ systems must have originated in an early vertebrate or invertebrate chordate lineage. Fossil invertebrate chordates are rare and known primarily from two fossil beds—the well-known middle-Cambrian Burgess Shales of Canada (p. 14) and the recently discovered early-Cambrian fossil beds of Chengjiang and Haikou, China. An ascidian tunicate and *Yunnanozoon,* a probable cephalochordate, are known from Chengjiang. Slightly better known is *Pikaia,* a ribbon-shaped, somewhat fishlike creature about 5 cm in length discovered in the Burgess Shales (figure 15.8). The presence of V-shaped myomeres and a notochord clearly identifies *Pikaia* as a chordate. The superficial resemblance of *Pikaia* to living amphioxus suggests it may be an early cephalochordate. *Haikouella lanceolata,* a small fishlike creature recently discovered from Haikou, possesses several characters that clearly identify it as a chordate, including notochord, pharynx, and dorsal nerve cord, but it also had characters, including pharyngeal muscles, paired eyes, and an enlarged brain, that are characteristic of vertebrates. Despite recent fossil discoveries of early chordates, most speculations regarding vertebrate ancestry have focused on the living protochordates, in part because they are much better known than the fossil forms.

Garstang's Hypothesis of Chordate Larval Evolution

The chordates have pursued two paths in their early evolution, one path leading to sedentary urochordates, the other to active, mobile cephalochordates and vertebrates. One hypothesis, proposed in 1928 by Walter Garstang of England, suggested that the chordate ancestral stock was derived by retaining into adulthood the larval form of sessile tunicate-like animals. The tadpole larvae of tunicates does indeed bear all the right attributes to qualify it as a possible vertebrate ancestral form: notochord, hollow dorsal nerve cord, pharyngeal slits, endostyle, and postanal tail. At some point, Garstang suggested, the tadpole failed to metamorphose into an adult tunicate, instead developing gonads and reproducing in the larval stage. With continued evolution, a new group of free-swimming animals appeared, the ancestors of cephalochordates and vertebrates (figure 15.9).

Garstang called this process **paedomorphosis** (Gr. *pais,* child, + *morphē,* form), a term that describes the evolutionary retention of juvenile or larval traits in adults. Garstang departed from previous thinking by suggesting that evolution may occur in larval stages of animals leading to adult vertebrate

---

characteristics
of Subphylum Vertebrata

1. Chief diagnostic features of chordates—**notochord, dorsal tubular nerve cord, pharyngeal pouches, endostyle** or **thyroid gland,** and **postanal tail**—all present at some stage of the life cycle

2. **Integument** basically of two divisions, an outer epidermis of stratified epithelium from ectoderm and an inner dermis of connective tissue from mesoderm

3. Distinctive cartilage or bone **endoskeleton** consisting of vertebral column (except in hagfishes, which lack vertebrae), and a head skeleton (cranium and pharyngeal skeleton) derived largely from **neural crest cells.**

4. **Muscular pharynx;** in fishes, pharyngeal pouches open to the outside as slits and bear gills; in tetrapods, pharyngeal pouches are sources of several glands

5. Complex, W-shaped muscle segments or **myomeres** to provide movement

6. Complete, **muscularized digestive tract** with distinct liver and pancreas

7. Circulatory system consisting of a **ventral heart** of multiple chambers; closed blood-vessel system of arteries, veins, and capillaries; blood fluid containing **erythrocytes** with **hemoglobin;** paired aortic arches connecting ventral and dorsal aortas and giving off branches to the gills among aquatic vertebrates; in terrestrial forms, aortic arches modified into pulmonary and systemic systems

8. Well-developed **coelom** divided into a pericardial cavity and a pleuroperitoneal cavity

9. Excretory system consisting of **paired, glomerular kidneys** provided with ducts to drain waste to the cloaca

10. Highly differentiated **tripartite brain;** 10 or 12 pairs of **cranial nerves; paired special sense organs** derived from **ectodermal placodes**

11. **Endocrine system** of ductless glands scattered throughout the body

12. Nearly always separate sexes; each sex containing gonads with ducts that discharge their products either into the cloaca or into special openings near the anus

13. Most vertebrates with two pairs of appendages supported by limb girdles and appendicular skeleton

---

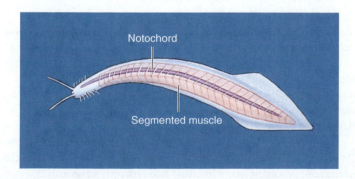

Notochord

Segmented muscle

figure 15.8

Pikaia, an early chordate, from the Burgess Shale of British Columbia, Canada

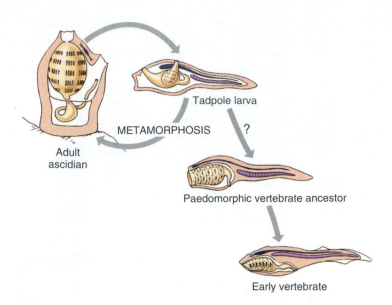

Tadpole larva

METAMORPHOSIS   ?

Adult
ascidian

Paedomorphic vertebrate ancestor

Early vertebrate

figure 15.9

Garstang's hypothesis of larval evolution. Adult tunicates live on the seafloor but reproduce through a free-swimming tadpole larva. More than 550 million years ago, some larvae began to reproduce in the swimming stage. These forms were ancestors of the first vertebrates.

characteristics. Paedomorphosis is a well-known phenomenon in several different animal groups (paedomorphosis in amphibians is described on p. 340). Furthermore, Garstang's hypothesis agrees with embryological evidence. Although long popular, Garstang's hypothesis has been challenged recently, because phylogenies generated from molecular data suggest that sessile ascidians represent a derived body form, and that free-swimming larvaceans are perhaps most similar in body form to the ancestral chordates.

Paedomorphosis, the displacement of ancestral larval or juvenile features into a descendant adult, can be produced by three different evolutionary-developmental processes: neoteny, progenesis, and post-displacement. In neoteny, the growth rate of body form is slowed so that the animal does not attain the ancestral adult form by the time it reaches reproductive maturity. Progenesis is precocious maturation of gonads in a larval (or juvenile) body that then stops growing and never attains an adult body form. In post-displacement, the onset of a developmental process is delayed relative to reproductive maturation, so that an ancestral adult form is not attained by the time of reproductive maturation. Neoteny, progenesis and post-displacement thus describe different ways in which paedomorphosis can happen. Zoologists use the inclusive term paedomorphosis to describe results of these evolutionary-developmental processes.

Position of Amphioxus

Zoologists consider the ceophalochordate amphioxus the closest living relative of vertebrates. Cephalochordates share several characters with vertebrates that are absent in urochordates, including segmented myomeres, dorsal and ventral aortas, and

branchial arches. However, amphioxus is unlike the most recent common ancestor of vertebrates because it lacks the tripartite brain, chambered heart, special sensory organs, muscular gut and pharynx, and neural crest tissue inferred to have been present in that ancestor. Despite these specializations, most zoologists consider amphioxus to have retained the primitive pattern of the immediate prevertebrate condition. Thus cephalochordates are probably the sister group of vertebrates (figure 15.3).

The Earliest Vertebrates

The earliest known vertebrate fossils, until recently, were armored jawless fishes called **ostracoderms** (os-trak'o-derm) (Gr. *ostrakon*, shell, + *derma*, skin) from late Cambrian and Ordovician deposits. In 1999 researchers described two fish-like 530-million-year-old vertebrates, *Myllokunmingia* and *Haikouichthys*, from the amazing Chengjiang deposits. These fossils push back the origin of vertebrates to at least the early Cambrian. Although they possess many typically vertebrate characters, such as W-shaped myomeres, a heart, a cranium, and fin rays, they lack evidence of mineralized tissues. Absence of mineralized tissue may explain the extreme rarity of vertebrate fossils prior to the late Cambrian.

The earliest ostracoderms were armored with bone in their dermis and usually lacked the paired fins so important to later fishes for stability (figure 15.10). The swimming movements of one of the early ostracoderms, the **heterostracans** (Gr. *heteros*, different, + *ostrakon*, shell), must have been imprecise, although sufficient to propel them along the ocean bottom where they searched for food. With fixed circular or slitlike mouth openings, they may have filtered small food particles from the water or ocean bottom. However, unlike the ciliary filter-feeding protochordates, ostracoderms sucked water into the pharynx by muscular pumping, an important innovation that suggests to some authorities that ostracoderms may have been mobile predators that fed on soft-bodied animals.

The term "ostracoderm" does not denote a natural evolutionary assemblage but instead is a descriptive term for several groups of heavily armored extinct jawless fishes.

During the Devonian period, the heterostracans underwent a major radiation, resulting in the appearance of numerous peculiar forms. Without ever evolving paired fins or jaws, these earliest vertebrates flourished for 150 million years until becoming extinct near the end of the Devonian period.

Coexisting with heterostracans throughout much of the Devonian period were **osteostracans** (Gr. *osteon*, bone + *ostrakon*, shell). Osteostracans had paired pectoral fins, an innovation that improved swimming efficiency by controlling yaw, pitch, and roll. A typical osteostracan is *Cephalaspis* (Gr. *kephalē*, head, + *aspis*, shield) (figure 15.10). A small animal, seldom exceeding 30 cm in length, it was covered with a heavy, dermal armor of bone, including a single-piece head

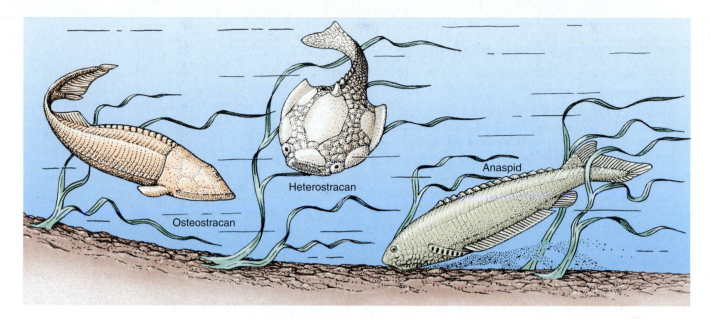

figure 15.10

Three ostracoderms, jawless fishes of Silurian and Devonian times. They are shown as they might have appeared while searching for food on the floor of a Devonian sea. All were probably suspension-feeders, but employed a strong pharyngeal pump to circulate water rather than the much more limiting mode of ciliary feeding used by their protovertebrate ancestors and by amphioxus today.

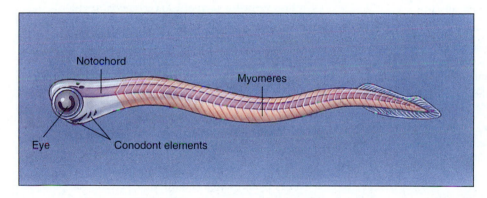

figure 15.11

Restoration of a living conodont. Conodonts superficially resembled amphioxus, but they possessed a much greater degree of encephalization (large paired eyes, possible auditory capsules, and cranium) and bonelike mineralized elements—all indicating that conodonts were chordates and probably vertebrates. Conodont elements are thought to be part of a food-handling apparatus.

shield. Examination of internal features of the braincase reveal a sophisticated nervous system and sense organs, similar to those of modern lampreys.

Another group of ostracoderms, the **anaspids** (figure 15.10), were more streamlined than other ostracoderms. These and other ostracoderms enjoyed an impressive radiation in the Silurian and Devonian periods. However, all ostracoderms became extinct by the end of the Devonian period.

For decades geologists have used strange, microscopic, toothlike fossils called **conodonts** (Gr. *kōnos*, cone, + *odontos*, tooth) to date Paleozoic marine sediments without having any idea what kind of creature originally possessed these elements. The discovery in the early 1980s of fossils of complete conodont animals has changed this situation. With their phosphatized toothlike elements, myomeres, cranium, notochord, and extrinsic eye muscles, conodonts clearly belong to the vertebrate clade (figure 15.11). Although their exact position in this clade is unclear, they are important in understanding the origin of vertebrates.

Early Jawed Vertebrates

All jawed vertebrates, whether extinct or living, are collectively called **gnathostomes** ("jaw mouth") in contrast to jawless vertebrates, the **agnathans** ("without jaw"). Gnathostomes are a monophyletic group because the presence of jaws is a derived character state shared by all jawed fishes and tetrapods. Agnathans, however, are defined principally by the absence of jaws, a character that is not unique to jawless fishes because jaws are lacking in vertebrate ancestors. Thus Agnatha is paraphyletic.

The origin of jaws was one of the most important events in vertebrate evolution. The utility of jaws is obvious: they allow predation on large and active forms of food not available to jawless vertebrates. Ample evidence suggests that jaws arose through modifications of the first or second of the serially repeated cartilaginous gill arches. But how did this mandibular arch change from a function of gill support and ventilation to one of feeding as jaws? Expansion of this arch

classification of Phylum Chordata

Phylum Chordata

Subphylum Urochordata (u'ro-kor-da'ta) (Gr. *oura*, tail, + L. *chorda*, cord, + *ata*, characterized by) **(Tunicata): tunicates.** Notochord and nerve cord in free-swimming larva only; ascidian adults sessile, encased in tunic. About 2000 species.

Subphylum Cephalochordata (sef'a-lo-kor-da'ta) (Gr. *kephalē*, head, + L. *chorda*, cord): **lancelets (amphioxus).** Notochord and nerve cord found along entire length of body and persist throughout life; fishlike in form. 29 species

Subphylum Vertebrata (ver'te-bra'ta) (L. *vertebratus*, backboned). **(Craniata): vertebrates.** Bony or cartilaginous cranium surrounding brain; notochord in embryonic stages, persisting in some fishes; usually with vertebrae; also may be divided into two groups (superclasses) according to presence of jaws.

    Superclass Agnatha (ag'na-tha) (Gr. *a*, without, + *gnathos*, jaw): **hagfishes, lampreys.** Without true jaws or paired appendages. A paraphyletic group.

        Class Myxini (mik-si'ny) (Gr. *myxa*, slime): **hagfishes.** Four pairs of tentacles around mouth; nasal sac with duct to pharynx; vertebrae absent. About 65 species.

        Class Cephalaspidomorphi (sef-a-lass'pe-do-morf'e) (Gr. *kephalē*, head, + *aspidos*, shield, *morphē*, form): **lampreys.** Buccal funnel with keratinized teeth, nasal sac not connected to pharynx; vertebrae present only as neural arches. 41 species.

    Class Gnathostomata (na'tho-sto'ma-ta) (Gr. *gnathos*, jaw + *stoma*, mouth): **jawed fishes, all tetrapods.** With jaws and (usually) paired appendages.

        Class Chondrichthyes (kon-drik'thee-eez) (Gr. *chondros*, cartilage, + *ichthys*, a fish): **sharks, skates, rays, chimaeras.** Cartilaginous skeleton, intestine with spiral valve; claspers present in males; no swim bladder. About 850 species.

Class Actinopterygii (ak'ti-nop-te-rij'ee-i) (Gr. *aktis*, ray, + *pteryx*, fin, wing): **ray-finned fishes.** Skeleton ossified; single gill opening covered by operculum; paired fins supported primarily by dermal rays; limb musculature within body; swim bladder mainly a hydrostatic organ, if present; atrium and ventricle not divided. About 27,000 species.

Class Sarcopterygii (sar-cop-te-rij'ee-i) (Gr. *sarkos*, flesh, + *pteryx*, fin, wing): **lobe-finned fishes.** Skeleton ossified; single gill opening covered by operculum; paired fins with sturdy internal skeleton and musculature within limb; diphycercal tail; intestine with spiral valve; usually with lunglike swim bladder; atrium and ventricle at least partly divided. 8 species. Not monophyletic unless tetrapods are included.

Class Amphibia (am-fib'e-a) (Gr. *amphi*, both or double, + *bios*, life): **amphibians.** Ectothermic tetrapods; respiration by lungs, gills, or skin; development through larval stage; skin moist, containing mucous glands, and lacking scales. About 5400 species.

Class Reptilia (rep-til'e-a) (L. *repere*, to creep): **reptiles.** Ectothermic tetrapods possessing lungs; embryo develops within shelled egg; no larval stage; skin dry, lacking mucous glands, and covered by epidermal scales. A paraphyletic group. About 7100 species.

Class Aves (ay'veez) (L. pl. of *avis*, bird): **birds.** Endothermic vertebrates with front limbs modified for flight; body covered with feathers; scales on feet. About 9900 species.

Class Mammalia (ma-may'lee-a) (L. *mamma*, breast): **mammals.** Endothermic vertebrates possessing mammary glands; body more or less covered with hair; well-developed neocerebrum; three middle ear bones. About 4800 species.

---

and evolution of new, associated muscles may have first assisted gill ventilation, perhaps to meet the increasing metabolic demands of early vertebrates. Once enlarged and equipped with extra muscles, the first pharyngeal arch could have easily been modified to serve as jaws (figure 15.12).

An additional feature characteristic of all gnathostomes is the presence of paired pectoral and pelvic appendages in the form of fins or limbs. These likely originated as stabilizers to check yaw, pitch, and roll generated during active swimming. The fin-fold hypothesis has been proposed to explain the origin of paired fins. According to this hypothesis, paired fins arose from paired continuous, ventrolateral folds or fin-forming zones. The addition of skeletal supports in the fins enhanced the fins' ability to provide stability during swimming. Evidence for this hypothesis is found in the paired flaps

of *Myllokunmingia, Haikouichthys,* and anapsids, and in the multiple paired fins of acanthodians. In one fish lineage the muscle and skeletal supports in the paired fins became strengthened, allowing them to become adapted for locomotion on land as limbs. The origin of jaws and paired appendages may be linked to a second *Hox* duplication, near the origin of the gnathostomes. The origins of jaws and paired fins were major innovations in vertebrate evolution, among the most important reasons for the subsequent major radiations of vertebrates that produced the modern fishes and all tetrapods, including you, the reader of this book.

Among the first jawed vertebrates were the heavily armored **placoderms** (plak'o-derm) (Gr. *plax*, plate, + *derma*, skin). They first appear in the fossil record in the early Devonian period (figure 15.13). Placoderms evolved a great variety of

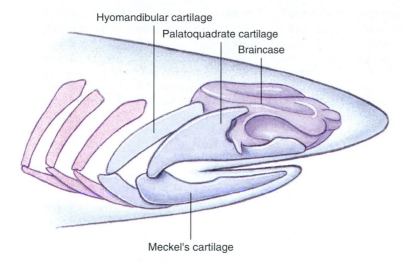

Hyomandibular cartilage
Palatoquadrate cartilage
Braincase

Meckel's cartilage

figure 15.12
How vertebrates got their jaw. The resemblance between jaws and gill supports of primitive fishes such as this Carboniferous shark suggests that the upper jaw (palatoquadrate) and lower jaw (Meckel's cartilage) evolved from structures that originally functioned as gill supports. The gill supports immediately behind the jaws are hinged like jaws and served to link the jaws to the braincase. Relics of this transformation are seen during the development of modern sharks.

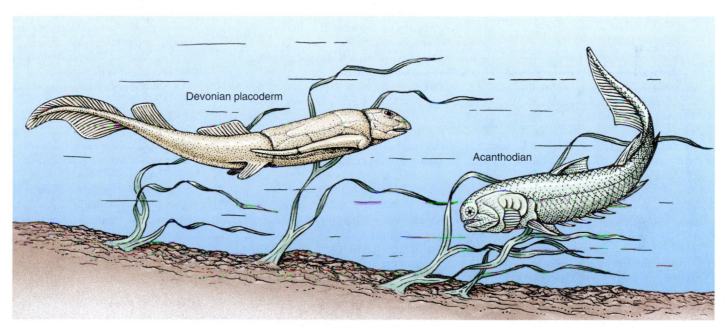

Devonian placoderm

Acanthodian

figure 15.13
Early jawed fishes of the Devonian period, 400 million years ago. Shown are a placoderm (*left*) and an acanthodian (*right*). Jaws and gill supports from which the jaws evolved develop from neural crest cells, a diagnostic character of vertebrates. Most placoderms were bottom dwellers that fed on benthic animals although some were active predators. Acanthodians carried less armor than placoderms and had a bony endoskeleton and prominent spines on paired fins. Most were marine but several species entered fresh water.

forms, some very large (one was 10 m in length!) and grotesque in appearance. They were armored fish covered with diamond-shaped scales or with large plates of bone. All became extinct by the end of the Paleozoic era and appear to have left no descendants. However, **acanthodians** (figure 15.13), a group of early jawed fishes characterized by fins with large spines, may have given rise to the great radiation of bony fishes that dominate the waters of the world today.

Summary

Phylum Chordata is named for the rodlike notochord that forms a stiffening body axis at some stage in the life cycle of every chordate. All chordates share five distinctive hallmarks that set them apart from all other phyla: notochord, dorsal tubular nerve cord, pharyngeal pouches, endostyle, and postanal tail. Two of the three chordate subphyla are invertebrates and lack a well-developed head. They are Urochordata (tunicates), most of which are sessile as adults, but all of which have a free-swimming larval stage; and Cephalochordata (lancelets), fishlike forms that include the famous amphioxus.

Chordates have evolutionary affinities to echinoderms or hemichordates, but the precise evolutionary origin of chordates is not yet known. Taken as a whole, chordates have a greater fundamental unity of organ systems and body plan than have many invertebrate phyla.

Subphylum Vertebrata includes the back-boned members of the animal kingdom (hagfishes actually lack vertebrae but are included with the Vertebrata by tradition because they share numerous homologies with vertebrates). As a group vertebrates are characterized by having a well-developed head and by their comparatively large size, high degree of motility, and distinctive body plan, which embodies several distinguishing features that permitted their exceptional adaptive radiation. Most important of these are living endoskeleton, which allows continuous growth and provides a sturdy framework for efficient muscle attach-ment and action; a muscular pharynx with slits (lost or greatly modified in higher vertebrates) with vastly increased respiratory efficiency; a muscularized gut and chambered heart for meeting higher metabolic demands; and an advanced nervous system with a distinct brain and paired sense organs. Evolution of jaws and paired appendages likely contributed to the incredible success of one group of vertebrates, the gnathostomes.

Review Questions

1. What characteristics are shared by the three deuterostome phyla that indicate a natural grouping of interrelated animals?
2. Explain how use of a cladistic classification for vertebrates results in important regroupings of traditional vertebrate taxa (refer to figure 15.3). Why are certain traditional groupings such as Reptilia and Agnatha not recognized in cladistic usage?
3. Name five hallmarks shared by all chordates, and explain the function of each.
4. In debating the question of chordate origins, zoologists eventually agreed that chordates must have evolved within the deuterostome assemblage rather than from a protostome group as earlier argued. What embryological evidences support this view?
5. Offer a description of an adult tunicate that would identify it as a chordate, yet distinguish it from any other chordate group.
6. Amphioxus long has been of interest to zoologists searching for a vertebrate ancestor. Explain why amphioxus captured such interest and why it no longer is considered to resemble closely the most recent common ancestor of all vertebrates.
7. Both sea squirts (urochordates) and lancelets (cephalochordates) are suspension-feeding organisms. Describe the suspension-feeding apparatus of a sea squirt and explain in what ways its mode of feeding is similar to, and different from, that of amphioxus.
8. Explain why it is necessary to know the life history of a tunicate to understand why tunicates are chordates.
9. List three groups of adaptations that guided vertebrate evolution, and explain how each has contributed to the success of vertebrates.
10. In 1928 Walter Garstang hypothesized that tunicates resemble the ancestral stock of the vertebrates. Explain this hypothesis.
11. Distinguish between ostracoderms and placoderms. What important evolutionary advances did each contribute to vertebrate evolution? What are conodonts?
12. Explain how we think the vertebrate jaw evolved.

Selected References

See also general references on page 415.

Alldredge, A. 1976. Appendicularians. Sci. Am. **235:**94–102 (July). *Describes the biology of larvaceans, which build delicate houses for trapping food.*

Bowler, P. J. 1996. Life's splendid drama: evolutionary biology and the reconstruction of life's ancestry 1860–1940. Chicago, University of Chicago Press. *Thorough and eloquent exploration of scientific debates over reconstruction of history of life on earth; chapter 4 treats theories of chordate and vertebrate origins.*

Carroll, R. L. 1997. Patterns and processes of vertebrate evolution. New York, Cambridge University Press. *A comprehensive analysis of the evolution-ary processes that have influenced large-scale changes in vertebrate evolution.*

Cohn, M. J. 2002. Lamprey *Hox* genes and the origin of jaws. Nature **416:**386–387. *Suppression of a Hox gene in the first pharyngeal arch may have led to the origin of jaws.*

Gans, C. 1989. Stages in the origin of vertebrates: analysis by means of scenarios. Biol. Rev. **64:**221–268. *Reviews the diagnostic characters of protochordates and ancestral vertebrates and presents a scenario for the protochordate-vertebrate transition.*

Gee, H. 1996. Before the backbone: views on the origin of the vertebrates. New York, Chapman & Hall. *Outstanding review of the many vertebrate origin hypotheses. Gee links much of the recent genetic, developmental, and molecular evidence in his discussion.*

Gould, S. J. 1989. Wonderful life: the Burgess Shale and the nature of history. New York, W. W. Norton & Company. *In this book describing the marvelous Cambrian fossils of the Burgess Shale, Gould "saves the best for last" by inserting an epilogue on Pikaia, the first known chordate.*

Gould, S. J. (ed.) 1993. The book of life. New York, W. W. Norton & Company. *A sweeping, handsomely illustrated view of (almost entirely) vertebrate life.*

Long, J. A. 1995. The rise of fishes: 500 million years of evolution. Baltimore, The

Johns Hopkins University Press. *An authoritative, liberally illustrated evolutionary history of fishes.*

Maisey, J. G. 1996. Discovering fossil fishes. New York, Henry Holt & Company. *Handsomely illustrated chronology of fish evolution with cladistic analysis of evolutionary relationships.*

Mallatt, J., and J.-Y. Chen. 2003. Fossil sister group of craniates: predicted and found.

J. Morph. **258**:1–31. *Reevaluation of Haikaouella lanceolata fossils reveals several features that suggest this is the sister group to craniates.*

Shimeld, S. M., and P. W. H. Holland. 2000. Vertebrate innovations. Proc. Natl. Acad. Sci. **97**:4449–4452. *Focuses on development characters, including neural crest, neural placodes, and Hox genes.*

Stokes, M. D., and N. D. Holland. 1998. The lancelet. Am. Sci. **86**(6):552–560. *Describes the historical role of amphioxus in early hypotheses of vertebrate ancestry and summarizes recent molecular data that has rekindled interest in amphioxus.*

Custom Website

The *Animal Diversity* Online Learning Center is a great place to check your understanding of chapter material. Visit www.mhhe.com/hickmanad4e for access to key terms, quizzes, and more! Further enhance your knowledge with Web links to chapter-related material.

Explore live links for these topics:

Classification and Phylogeny of Animals
General Chordate References
Subphylum Urochordata
Subphylum Cephalochordata
Subphylum Vertebrata

Systematics and Characteristics of the Craniates
Early Vertebrates

16

Fishes

Grey snappers (*Lutjanus griseus*) in the Florida Keys.

What Is a Fish?

In common (and especially older) usage, the term "fish" denotes a mixed assortment of water-dwelling animals. We speak of jellyfish, cuttlefish, starfish, crayfish, and shellfish, knowing full well that when we use the word "fish" in such combinations, we are not referring to a true fish. In earlier times, even biologists did not make such a distinction. Sixteenth-century natural historians classified seals, whales, amphibians, crocodiles, even hippopotamuses, as well as a host of aquatic invertebrates, as fish. Later biologists were more discriminating, eliminating first the invertebrates and then the amphibians, reptiles, and mammals from the narrowing concept of a fish. Today we recognize a fish as an aquatic vertebrate with gills, appendages, if present, in the form of fins, and usually a skin with scales of dermal origin. Even this modern concept of the term "fish" is used for convenience, not as a taxonomic unit. Fishes do not form a monophyletic group, because the ancestor of land vertebrates (tetrapods) is found within one group of fishes (the sarcopterygians). Thus fishes can be defined in an evolutionary sense as all vertebrates that are not tetrapods. Because fishes live in a habitat that is basically alien to humans, people have rarely appreciated the remarkable diversity of these vertebrates. Nevertheless, whether appreciated by humans or not, the world's fishes have enjoyed an effusive proliferation that has produced an estimated 28,000 living species—more than all other species of vertebrates combined—with adaptations that have fitted them to almost every conceivable aquatic environment. No other animal group threatens their domination of the seas.

The life of a fish is bound to its body form. Their mastery of river, lake, and ocean is revealed in the many ways that fishes have harmonized their design to the physical properties of their aquatic surroundings. Suspended in a medium that is 800 times more dense than air, a trout or pike can remain motionless, varying its neutral buoyancy by adding or removing air from its swim bladder. Or it may dart forward or at angles, using its fins as brakes and tilting rudders. With excellent organs for salt and water exchange, bony fishes can steady and finely tune their body fluid composition in their chosen freshwater or seawater environment. Their gills are the most effective respiratory devices in the animal kingdom for extracting oxygen from a medium that contains less than 1/20 as much oxygen as air. Fishes have excellent olfactory and visual senses and a unique lateral line system, which with its exquisite sensitivity to water currents and vibrations provides a "distance touch" in water. Thus in mastering the physical problems of their element, early fishes evolved a basic body plan and set of physiological strategies that both shaped and constrained the evolution of their descendants.

The use of "fishes" as the plural form of "fish" may sound odd to most people accustomed to using "fish" in both the singular and the plural. "Fish" refers to one or more individuals of the same species; "fishes" refers to more than one species.

Ancestry and Relationships of Major Groups of Fishes

Fishes are of ancient ancestry, having descended from an unknown free-swimming protochordate ancestor (hypotheses of chordate and vertebrate origins are discussed in Chapter 15). The earliest fishlike vertebrates were a paraphyletic assemblage of jawless **agnathan** fishes, the ostracoderms (see figure 15.10, p. 305). One group of ostracoderms gave rise to the jawed **gnathostomes** (see figure 15.13).

Agnathans, the least derived of the two groups, include along with the extinct ostracoderms the living **hagfishes** and **lampreys,** fishes adapted as scavengers or parasites. Although hagfishes have no vertebrae and lampreys have only rudimentary vertebrae, they nevertheless are included within the subphylum Vertebrata because they have a cranium and many other vertebrate homologies. The ancestry of hagfishes and lampreys is uncertain; they bear little resemblance to the extinct ostracoderms. Although hagfishes and the more derived lampreys superficially look much alike, they are in fact so different from each other that they have been assigned to separate classes by zoologists.

All remaining fishes have paired appendages and jaws and are included, along with tetrapods (land vertebrates), in the monophyletic group of gnathostomes. They appear in the fossil record in the late Silurian period with fully formed jaws, and no forms intermediate between agnathans and gnathostomes are known. By the Devonian period, the Age of Fishes, several distinct groups of jawed fishes were well represented. One of these, the placoderms (see figure 15.13, p. 307), became extinct in the following Carboniferous period, leaving no descendants. A second group, the **cartilaginous fishes** of class Chondrichthyes (sharks, rays, and chimaeras), lost the heavy dermal armor of the early jawed fishes and adopted cartilage for the endoskeleton. Most are active predators with a sharklike or raylike body form that has undergone only minor changes over the ages.

Of all gnathostomes, **bony fishes** (Osteichthyes) radiated most extensively and are the dominant fishes today (figure 16.1). We can recognize two distinct lineages of bony fishes. Of these two, by far the most diverse are the **ray-finned fishes** (class Actinopterygii), which radiated to form modern bony fishes. The other group, **lobe-finned fishes** (class Sarcopterygii), contains few living species but includes the sister group of the tetrapods. The lobe-finned fishes are represented today by **lungfishes** and **coelacanths**—meager remnants of important stocks that flourished in the Devonian period (figure 16.1). A classification of major fish taxa is on p. 331.

Living Jawless Fishes

Living jawless fishes are represented by approximately 106 species divided between two classes: Myxini (hagfishes) and Cephalaspidomorphi (lampreys) (figure 16.2). Members of each group lack jaws, internal ossification, scales, and paired fins, and both groups share porelike gill openings and an eellike body form. In other respects, however, the two groups are morphologically very different. Lampreys bear many derived morphological characters that place them phylogenetically much closer to jawed bony fishes than to hagfishes. Because of these differences, hagfishes and lampreys have been assigned to separate vertebrate classes, leaving the grouping "agnatha" as a paraphyletic assemblage of jawless fishes.

Hagfishes: Class Myxini

Hagfishes are an entirely marine group that feeds on dead or dying fishes, annelids, molluscs, and crustaceans. They are not parasitic like lampreys, but are scavengers and predators. There are 65 described species of hagfishes, of which the best known in North America are the Atlantic hagfish *Myxine glutinosa* (Gr. *myxa,* slime) (figure 16.3) and the Pacific hagfish *Eptatretus stouti* (NL, *ept*<Gr. *hepta,* seven, + *tretos,* perforated). Although almost completely blind, hagfishes are quickly attracted to food, especially dead or dying fish, by their keenly developed senses of smell and touch. After attaching itself to its prey by means of toothed, keratinized plates, the hagfish thrusts its tongue forward to rasp off pieces of tissue. For extra leverage, the hagfish often ties a knot in its tail, then passes the knot forward along its body until it is pressed securely against the side of its prey (figure 16.3).

Unlike any other vertebrate, hagfishes possess body fluids that are in osmotic equilibrium with seawater, as are most marine invertebrates. Hagfishes have several other anatomical

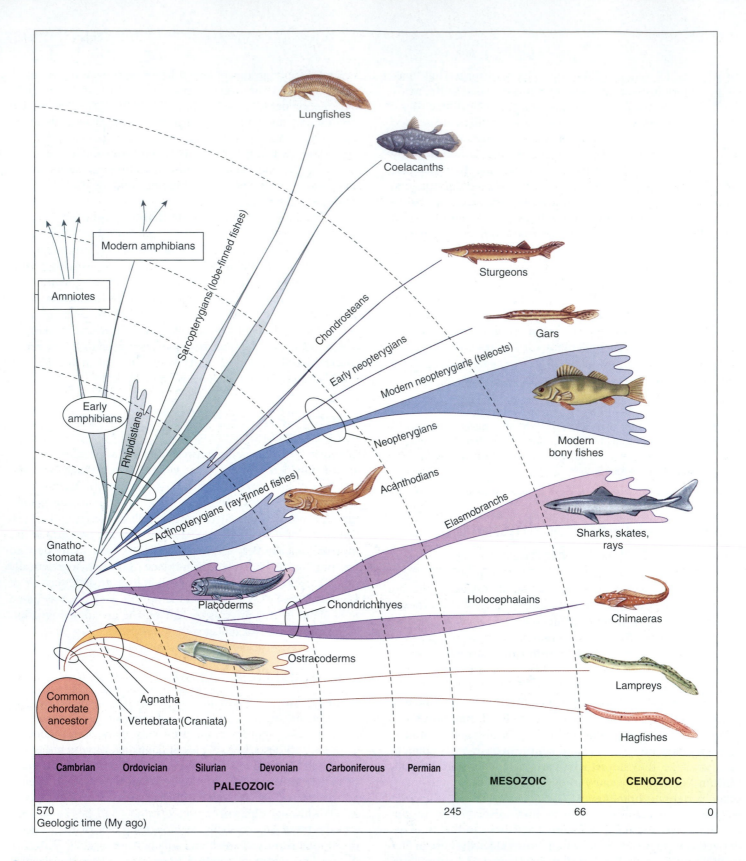

figure 16.1

Graphic representation of the family tree of fishes, showing evolution of major groups through geological time. Numerous lineages of extinct fishes are not shown. Widened areas in the lines of descent indicate periods of adaptive radiation and the relative number of species in each group. The lobe-finned fishes (sarcopterygians), for example, flourished in the Devonian period, but declined and are today represented by only four surviving genera (lungfishes and coelacanths). Homologies shared by the sarcopterygians and tetrapods suggest that they form a clade. Sharks and rays radiated during the Carboniferous period, declined in the Permian, then radiated again in the Mesozoic era. Johnny-come-latelies in fish evolution are the spectacularly diverse modern fishes, or teleosts, which include most living fishes.

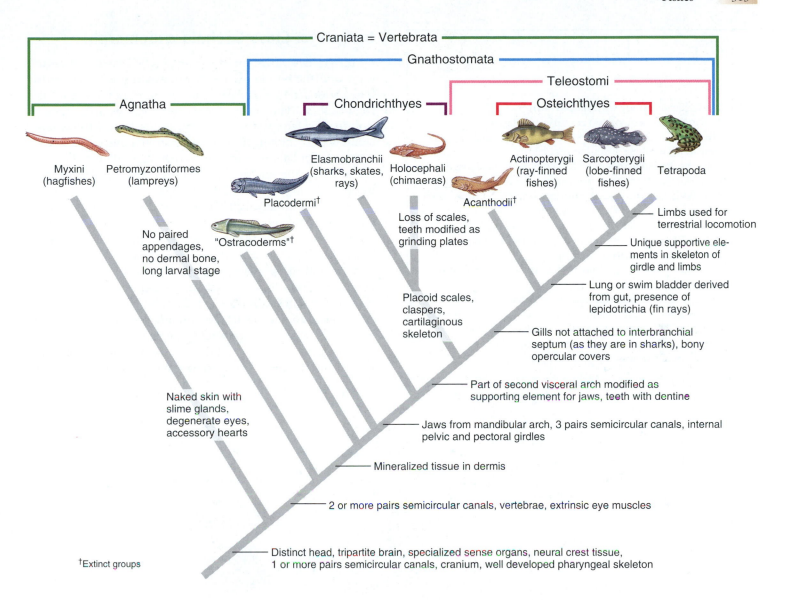

figure 16.2

Cladogram of the fishes, showing the probable relationships of major monophyletic fish taxa. Several alternative relationships have been proposed. Extinct groups are designated by a dagger (†). Some of the shared derived characters that mark the branchings are shown to the right of the branch points. The groups Agnatha and Osteichthyes, although paraphyletic structural grades considered undesirable in cladistic classification, are traditionally recognized in systematics because they share broad structural and functional patterns of organization.

and physiological peculiarities, including a low-pressure circulatory system served by three accessory hearts in addition to the main heart positioned behind the gills. Hagfishes are also renowned for their ability to generate enormous quantities of slime.

The reproductive biology of hagfishes remains largely a mystery. Both male and female gonads are found in each animal, but only one gonad becomes functional. Females produce small numbers of surprisingly large, yolky eggs 2 to 7 cm in diameter, depending on the species. There is no larval stage and growth is direct.

Although the strange features of hagfishes are of interest to biologists, hagfishes were not endeared to commercial fishermen. In earlier days of commercial fishing, mainly by gill nets and set lines, hagfishes often ate out the contents of captured fishes, leaving behind a useless sac of skin and bones. Hagfishes became a less important pest as large and efficient otter trawls came into use. Recently the commercial fishing industry "turned the tables" and began targeting hagfishes as a source of leather for golfbags and boots. Fishing pressure has been so intense that many species have greatly declined.

characteristics
of Jawless Fishes

1. Slender, eel-like body with **naked skin**
2. Median fins present, but no paired appendages
3. **Fibrous** and **cartilaginous skeleton;** notochord persistent; vertebrae absent or reduced
4. **Jaws absent;** biting mouth with keratinized plates in hagfishes; suckerlike oral disc with keratinized teeth in lampreys
5. Heart with sinus venosus, atrium and ventricle; **single circulation;** hagfishes with accessory hearts
6. Five to 16 pairs of gills in hagfishes; 7 pairs of gills in lampreys
7. **Pronephric** and **mesonephric kidneys** in hagfishes; **opisthonephric kidney** in lampreys
8. Dorsal nerve cord with **distinct brain;** 10 pairs of cranial nerves
9. Digestive system **without distinct stomach**
10. Sense organs of taste, smell, hearing; eyes poorly developed in hagfishes, but moderately well developed in adult lampreys; one pair (hagfishes) or two pairs (lampreys) of **semicircular canals**

Lampreys: Class Cephalaspidomorphi

Of the 41 described species of lampreys distributed around the world, the best known to North Americans is the destructive marine lamprey, *Petromyzon marinus,* of the Great Lakes (figure 16.4). The name *Petromyzon* (Gr. *petros,* stone, + *myzon,* sucking) refers to the lamprey's habit of grasping a stone with its mouth to hold position in a current. There are 22 species of lampreys in North America of which about half are parasitic; the rest are species that never feed after metamorphosis and die soon after spawning.

In North America all lampreys, marine as well as freshwater forms, spawn in the winter or spring in shallow gravel and sand in freshwater streams. Males begin nest building and are joined later by females. Using their oral discs to lift stones and pebbles and using vigorous body vibrations to sweep away light debris, they form an oval depression. As the female sheds eggs into the nest, they are fertilized by the male. The sticky eggs adhere to pebbles in the nest and soon become covered with sand. Adults die soon after spawning.

Eggs hatch in approximately 2 weeks, releasing small larvae (**ammocoetes**) (figure 16.5), which stay in the nest until

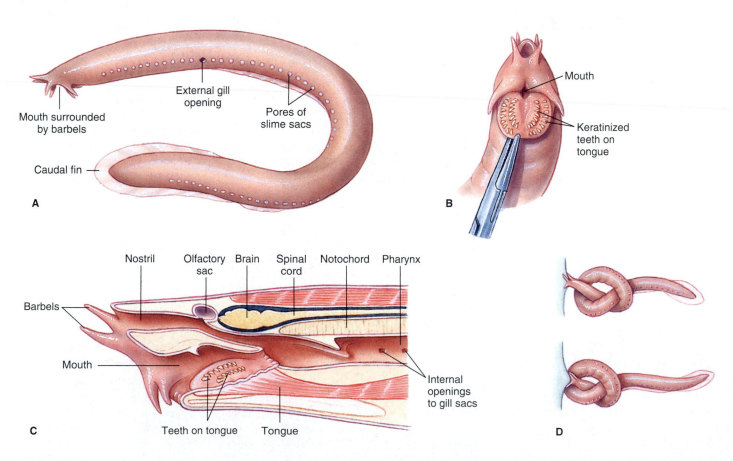

figure 16.3

The Atlantic hagfish *Myxine glutinosa* (class Myxini). **A,** External anatomy; **B,** Ventral view of head, showing keratinized teeth used to grasp food during feeding; **C,** Sagittal section of head region (note retracted position of rasping tongue and internal openings into a row of gill sacs); **D,** Hagfish knotting, showing how it obtains leverage to tear flesh from prey.

figure 16.4

Sea lamprey, *Petromyzon marinus,* feeding on the body fluids of a dying fish.

they are approximately 1 cm long; they then burrow into mud or sand and emerge at night to feed on small invertebrates, detritus, and other particulate matter in the water. The larval period lasts from 3 to 7 or more years before the larva rapidly metamorphoses into an adult.

Parasitic lampreys either migrate to the sea, if marine, or remain in fresh water, where they attach themselves by their suckerlike mouth to fish and with their sharp keratinized teeth rasp through flesh and suck body fluids. To promote the flow of blood, a lamprey injects an anticoagulant into the wound. When gorged, a lamprey releases its hold but leaves the fish with a wound that may prove fatal. Parasitic freshwater adults live a year or more before spawning and then die; marine forms may live longer.

Nonparasitic lampreys do not feed after emerging as adults, and their alimentary canal degenerates to a nonfunctional strand of tissue. Within a few months after spawning, they die.

The landlocked sea lamprey, *Petromyzon marinus* first entered the Great Lakes after the Welland Canal around Niagara Falls, a barrier to further western migration, was deepened between 1913 and 1918. Moving first through Lake Erie to Lakes Huron, Michigan, and Superior, sea lampreys, accompanied by overfishing, caused a total collapse of a multimillion dollar lake trout fishery in the early 1950s. Other less valuable fish species were attacked and destroyed in turn. After reaching a peak abundance in 1951 in Lakes Huron and Michigan and in 1961 in Lake Superior, sea lampreys began to decline, due in part to depletion of their food and in part to the effectiveness of control measures (mainly chemical larvicides placed in selected spawning streams). Lake trout, aided by a restocking program, are now recovering, but wounding rates are still high in some lakes.

figure 16.5

Life cycle of the "landlocked" form of the sea lamprey *Petromyzon marinus.*

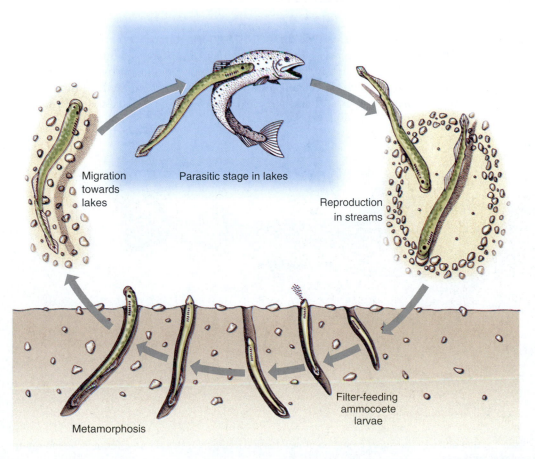

Migration towards lakes

Parasitic stage in lakes

Reproduction in streams

Metamorphosis

Filter-feeding ammocoete larvae

Cartilaginous Fishes: Class Chondrichthyes

There are nearly 850 living species in the class Chondrichthyes, an ancient, compact group. Although a much smaller and less diverse assemblage than bony fishes, their impressive combination of well-developed sense organs, powerful jaws and swimming musculature, and predaceous habits ensures them a secure and lasting niche in the aquatic community. One of their distinctive features is their cartilaginous skeleton. Although there is some limited calcification, bone is entirely absent throughout the class—a curious feature, since Chondrichthyes are derived from ancestors having well-developed bone.

Sharks and Rays: Subclass Elasmobranchii

Sharks, which include about 45% of the approximately 815 species in subclass Elasmobranchii, are typically predaceous fishes with five to seven pairs of gill slits and gills and (usually) a spiracle behind each eye. Larger sharks, such as the massive (but harmless) plankton-feeding whale shark, may reach 15 m in length, the largest of all fishes. Dogfish sharks, so widely used in zoological laboratories, rarely exceed 1 m.

Although to most people sharks have a sinister appearance and a fearsome reputation, they are at the same time among the most gracefully streamlined of all fishes (figure 16.6). Sharks are heavier than water and will sink if not swimming forward. The asymmetrical **heterocercal tail,** in which the vertebral column turns upward and extends into the dorsal lobe of the tail (see figure 16.13), provides lift and thrust as it sweeps to and fro in the water, and the broad head and flat pectoral fins act as planes to provide head lift.

Sharks are well equipped for their predatory life. Their tough leathery skin is covered with numerous dermal **placoid**

characteristics
of Sharks and Rays (Elasmobranchii)

1. **Body fusiform** (except rays) with a **heterocercal** caudal fin (figure 16.13)
2. **Mouth ventral** (figure 16.6); two olfactory sacs that do not connect to the mouth cavity; jaws present
3. Skin with **placoid scales** (figure 16.15) or naked; teeth **polyphyodont,** of modified placoid scales
4. **Endoskeleton entirely cartilaginous**
5. Digestive system with a J-shaped stomach and **intestine with spiral valve** (figure 16.7)
6. Circulatory system of several pairs of aortic arches; single circulation; heart with sinus venosus, atrium, ventricle, and conus arteriosus
7. Respiration by means of 5 to 7 pairs of gills leading to **exposed gill slits;** no operculum
8. No swim bladder or lung
9. Opisthonephric kidney and rectal gland (figure 16.7); blood isosmotic or slightly hyperosmotic to seawater; **high concentrations of urea and trimethylamine oxide in blood**
10. Brain of two olfactory lobes, two cerebral hemispheres, two optic lobes, a cerebellum, and a medulla oblongata; 10 pairs of cranial nerves; **three pairs of semicircular canals;** senses of smell, vision, vibration reception (lateral line system), and electroreception well developed
11. Separate sexes; oviparous, ovoviviparous, or viviparous; direct development; **internal fertilization**

scales (see figure 16.15) that are modified anteriorly to form replaceable rows of teeth in both jaws (figure 16.8). Placoid scales in fact consist of dentine enclosed by an enamel-like substance, and they very much resemble teeth of other vertebrates. Sharks have a keen sense of smell used to guide them to

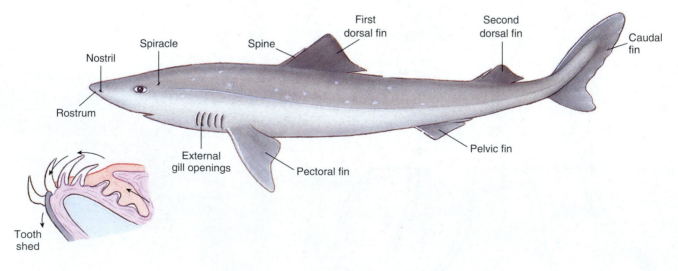

figure 16.6

Male spiny dogfish shark, *Squalus acanthias*. Inset: Section of lower jaw shows formation of new teeth developing inside the jaw. These move forward to replace lost teeth. Rate of replacement varies among species.

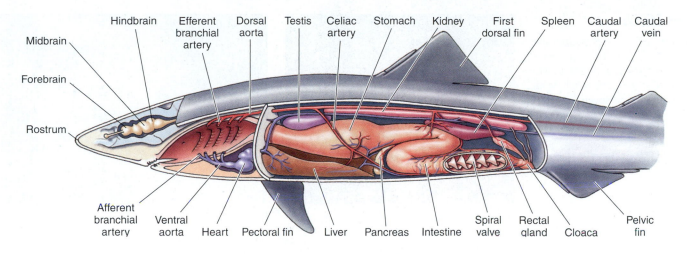

Hindbrain   Efferent   Dorsal   Testis   Celiac   Stomach   Kidney   First   Spleen   Caudal   Caudal
branchial   aorta   artery   dorsal fin   artery   vein
Midbrain   artery

Forebrain

Rostrum

Afferent   Ventral
branchial   aorta   Heart   Pectoral fin   Liver   Pancreas   Intestine   Spiral   Rectal   Cloaca   Pelvic
artery   valve   gland   fin

figure 16.7

Internal anatomy of a dogfish shark *Squalus acanthias*.

figure 16.8

Head of sand tiger shark *Carcharias* sp. Note the series of successional teeth. Also visible below the eye are ampullae of Lorenzini.

food. A well-developed **lateral line system** is used for detecting and locating objects and moving animals (predators, prey, and social partners). It is composed of a canal system extending along the side of the body and over the head (figure 16.9). Inside are special receptor organs (**neuromasts**) that are extremely sensitive to vibrations and currents in the water. At close range, a shark switches to vision as its primary method of tracking prey. Sharks can also detect and aim attacks at prey buried in the sand by sensing the bioelectric fields that surround all animals. The receptors, the **ampullary organs of Lorenzini,** are located on the shark's head.

All chondrichthyans have internal fertilization; sperm is introduced by a modified portion of the male's pelvic fin known as a **clasper** (figure 16.6). Many elasmobranchs, including some sharks and all skates, are **oviparous,** laying eggs soon after fertilization. Embryos are nourished for a long period—

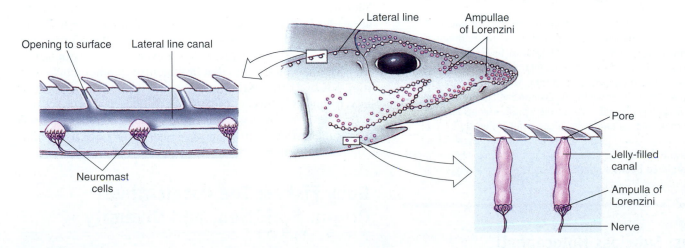

Opening to surface   Lateral line canal   Lateral line   Ampullae of Lorenzini

Neuromast cells

Pore

Jelly-filled canal

Ampulla of Lorenzini

Nerve

figure 16.9

Sensory canals and receptors in a shark. Ampullae of Lorenzini respond to weak electric fields, and possibly to temperature, water pressure, and salinity. Lateral line sensors, called neuromasts, are sensitive to disturbances in the water, enabling a shark to detect nearby objects by reflected waves in the water.

figure 16.10

Skates and rays are specialized for life on the sea floor. Both clearnose skates, *Raja eglanteria,* **A,** and southern stingrays, *Dasyatis americana,* **B,** are flattened dorsoventrally and move by undulations of winglike pectoral fins. This stingray, **B,** is followed by a pilot fish.

A

B

up to two years in one species—before hatching. Many sharks are **ovoviviparous,** retaining developing young in the uterus, where they are nourished by the contents of their yolk sac. Still others are **viviparous**—the embryos receive nourishment from the mother via a placenta or secretions from the uterus. Regardless of the initial amount of maternal support, all parental care ends once eggs are laid or young are born.

More than half of all elasmobranches are rays, a group that includes skates, stingrays, electric rays, and manta rays. Rays are distinguished by their dorsoventrally flattened bodies and enlarged pectoral fins, which they move in a wavelike fashion to propel themselves (figure 16.10). Gill openings are on the underside of the head, and the **spiracles** (on top of the head) are unusually large. Respiratory water enters through these spiracles to prevent clogging the gills, because the mouth is often buried in sand. Teeth are adapted for crushing prey—mainly molluscs, crustaceans, and an occasional small fish.

In stingrays, the slender and whiplike tail is armed with one or more saw-toothed spines that can inflict dangerous wounds. Electric rays have certain muscles on either side of the head, modified into powerful electric organs, which can give severe shocks to stun their prey.

The worldwide shark fishery is experiencing unprecedented pressure, driven by the high price of shark fins for shark-fin soup, an oriental delicacy (which may sell for as much as $50.00 per bowl). Coastal shark populations in general have declined so rapidly that "finning" has been outlawed in the United States; other countries, too, are setting quotas to protect threatened shark populations. Even in the Marine Resources Reserve of the Galápagos Islands, one of the world's exceptional wild places, tens of thousands of sharks have been killed illegally for the Asian shark-fin market. That illegal fishery continues at this writing. Contributing to the threatened collapse of shark fisheries worldwide is the long time required by most sharks to reach sexual maturity; some species take as long as 35 years.

Chimaeras: Subclass Holocephali

The approximately 35 species of chimaeras (ky-meer'uz; L. monster), distinguished by such suggestive names as ratfish (figure 16.11), rabbitfish, spookfish, and ghostfish, are rem-

figure 16.11

Spotted ratfish, *Hydrolagus collei,* of North American west coast. This species is one of the most handsome of chimaeras, which tend toward bizarre appearances.

nants of a group that diverged from the earliest shark lineage, which originated at least 360 million years ago (Devonian or Silurian periods of the Paleozoic). Fossil chimaeras first appeared in the Jurassic period, reached their zenith in the Cretaceous and early Tertiary periods (120 million to 50 million years ago), and have declined ever since. Anatomically, they have several features that link them to elasmobranchs, but they possess a suite of unique characters, too. Their food is a mixed diet of seaweed, molluscs, echinoderms, crustaceans, and fishes. Chimaeras are not commercial species and are seldom caught. Despite their bizarre shape, they are beautifully colored with a pearly iridescence.

Bony Fishes: The Osteichthyes
Origin, Evolution, and Diversity

In the early to middle Silurian, a lineage of fishes with bony endoskeletons gave rise to a clade of vertebrates that contains 96% of living fishes and all living tetrapods. Fishes of this clade have traditionally been termed "bony fishes" (Osteichthyes), because it was originally believed these were the only fishes with bony skeletons. Although we know now that bone occurs in many other early fishes (ostracoderms, placoderms, and acanthodians), bony fishes and their tetrapod relatives are united by the presence of endochondral bone (bone that replaces cartilage developmentally), presence of lungs or a swim bladder derived embryonically from the gut, and several cranial and dental characters. Because traditional usage of Osteichthyes does not describe a monophyletic (natural) group (figure 16.2), most recent classifications, including the one presented on page 331, do not recognize this term as a valid taxon. Rather, it is used as a term of convenience to describe vertebrates with endochondral bone that are conventionally termed "bony fishes."

Fossils of the earliest bony fishes show several structural similarities with acanthodians (p. 307 and figure 15.13), indicating that they likely descended from a unique common ancestor. By the middle of the Devonian bony fishes already had radiated extensively into two major clades, with adaptations that fitted them for every aquatic habitat except the most inhospitable. One of these clades, the **ray-finned fishes** (class Actinopterygii), includes modern bony fishes (figure 16.12), the most species-rich clade of living vertebrates. A second clade, the **lobe-finned fishes** (class Sarcopterygii), is represented today only by lungfishes and coelacanths (figures 16.19 and 16.20). The evolutionary history of this group is of particular interest because it gave rise to land vertebrates (tetrapods).

Several key adaptations contributed to their radiation. Bony fishes have an operculum over the gill composed of bony

characteristics
of Bony Fishes (Osteichthyes)

1. **Skeleton with bone of endochondral origin;** caudal fin heterocercal in primitive forms, **homocercal** in advanced forms (figure 16.13); skin with mucous glands and embedded dermal scales (figure 16.14); scales **ganoid** in primitive forms, scales **cycloid, ctenoid,** or absent in advanced forms (figure 16.15).
2. Paired and median **fins supported by dermal rays (lepidotrichia)**
3. Jaws present; teeth usually present with enamel or enameloid covering; spiral valves present in ancestral forms, absent in advanced forms
4. Respiration primarily by gills supported by arches and covered with an **operculum**
5. **Swim bladder** often present (figure 16.24), usually functioning in buoyancy, sometimes vascularized and functioning in respiration
6. Circulation consisting of a heart with a sinus venosus, atrium, undivided ventricle, and truncus arteriosus with **single circulation (actinopterygians)** or heart with a sinus venosus, two atria, partly divided ventricle, and conus arteriosus with **double circulation (sarcopterygians)**; erythrocytes nucleated
7. Nervous system of a brian with optic lobes, cerebellum, and small cerebrum; 10 pairs of cranial nerves; three pairs of semicircular canals
8. Excretory system of paired **opisthonephric kidneys;** sexes usually separate; fertilization usually external; larval forms may differ greatly from adults

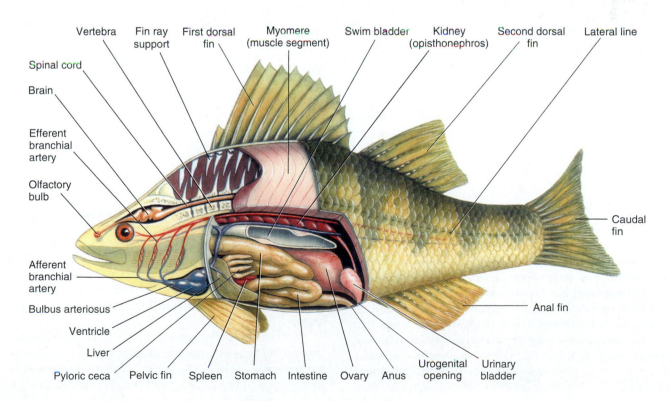

figure 16.12

Internal anatomy of a yellow perch, *Perca flavescens,* a freshwater teleost fish.

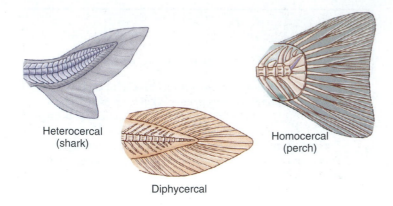

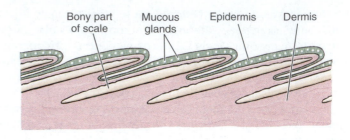

figure 16.14

Section through the skin of a bony fish, showing the overlapping scales (*yellow*). The scales lie in the dermis and are covered by epidermis.

figure 16.13

Types of caudal fins among fishes.

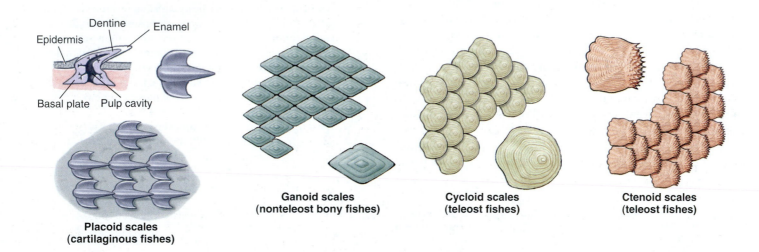

figure 16.15

Types of fish scales. Placoid scales are small, conical toothlike structures characteristic of Chondrichthyes. Diamond-shaped ganoid scales, present in gars and considered primitive for bony fishes, are composed of layers of silvery enamel (ganoin) on the upper surface and bone on the lower. Other bony fishes have either cycloid or ctenoid scales. These scales are thin and flexible and are arranged in overlapping rows.

plates and attached to a series of muscles. This feature increased respiratory efficiency because outward rotation of the operculum created a negative pressure so that water would be drawn across the gills, as well as pushed across by the mouth pump. The earliest bony fishes also had a gas-filled pouch branching from the esophagus that provided an additional means of gas exchange in oxygen-poor waters and an efficient means of achieving neutral buoyancy. In fishes that use these pouches primarily for gas exchange, the pouches are called **lungs,** while in fishes that use these pouches primarily for buoyancy, these pouches are called **swim bladders** (p. 324). Progressive specialization of jaw musculature and skeletal elements involved in feeding is another key feature of bony fish evolution.

Ray-Finned Fishes: Class Actinopterygii

Ray-finned fishes are an enormous assemblage containing all of our familiar bony fishes—more than 27,000 species. The group had its beginnings in Devonian freshwater lakes and streams. The ancestral forms were small, bony fishes, heavily armored with ganoid scales (figure 16.15), and heterocercal caudal fins (figure 16.13).

From these earliest ray-finned fishes, two major groups emerged. Those bearing the most primitive characteristics are **chondrosteans** (Gr. *chondros,* cartilage, + *osteon,* bone), a possibly paraphyletic assemblage, represented today by sturgeons, paddlefishes, and bichirs (figure 16.16). The bichir, *Polypterus* (Gr. *poly,* many, + *pteros,* winged), of African

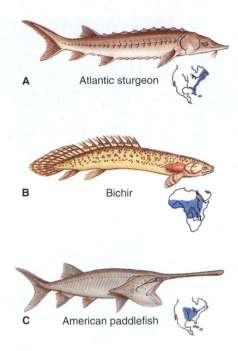

A

B

figure 16.16

Chondrostean ray-finned fishes of class Actinopterygii. **A,** Atlantic sturgeon, *Acipenser oxyrhynchus* (now uncommon) of Atlantic coastal rivers. **B,** Bichir, *Polypterus bichir,* of equatorial west Africa. It is a nocturnal predator. **C,** Paddlefish, *Polyodon spathula,* of the Mississippi River basin reaches a length of 2 m and a weight of 90 kg.

figure 16.17

Nonteleost neopterygian fishes. **A,** Bowfin *Amia calva.* **B,** Longnose gar *Lepisosteus osseus.* They frequent slow-moving streams and swamps of eastern North America where they may hang motionless in the water, ready to snatch passing fish.

waters is an interesting relic with a lunglike swim bladder and many other primitive characteristics; it resembles ancestral ray-finned fishes more than it does any other living fish.

The second major group to emerge from the early ray-finned stock were **neopterygians** (Gr. *neos,* new, + *pteryx,* fin). Neopterygians appeared in the late Permian and radiated extensively during the Mesozoic era. During the Mesozoic one lineage gave rise to a secondary radiation that led to modern bony fishes, the teleosts. The two surviving genera of non-teleost neopterygians are the bowfin *Amia* (Gr. tunalike fish) of shallow, weedy waters of the Great Lakes and Mississippi River basin, and gars *Lepisosteus* (Gr. *lepidos,* scale, + *osteon,* bone) of eastern and southern North America (figure 16.17). Gars are large, ambush predators that belie their lethargic appearance by suddenly dashing forward to grasp their prey with needle-sharp teeth.

The major clade of neopterygians are **teleosts** (Gr. *teleos,* complete, + *osteon,* bone), the modern bony fishes (figure 16.12). Teleost diversity is astounding, with about 27,000 described species, representing about 96% of all living fishes and almost half of all vertebrates (figure 16.18). They display an array of morphological form and size and occupy nearly every aquatic habitat on Earth, and may even make excursions onto land, as do mudskippers (figure 16.18B).

Several morphological trends in the teleost lineage allowed them to diversify into a truly incredible variety of habitats and forms. The heavy dermal armor of primitive ray-finned fishes was replaced by light, thin, flexible **cycloid** and **ctenoid scales** (figure 16.15). Increased mobility and speed that resulted from loss of heavy armor improved predatory avoidance and feeding efficiency. The symmetrical shape of the **homocercal** tail (figure 16.13) of most teleosts focused musculature contractions on the tail, resulting in greater speed. Various fins were elaborated into forms that permitted precise steering, camouflage, protection, attachment, or social

figure 16.18

Diversity among teleosts. **A,** Blue marlin, *Makaira nigricans,* one of the largest teleosts. **B,** Mudskippers, *Periopthalmus* sp., make extensive excursions on land to graze on algae and capture insects; they build nests in which the young hatch and are guarded by the mother. **C,** Protective coloration of the flamboyant lionfish, *Pterois* sp., advises caution; the dorsal spines are venomous. **D,** The sucking disc on the sharksucker, *Echeneis naucrates,* is a modification of the dorsal fin.

A

B

C

D

communication (figure 16.18). The teleost lineage demonstrated an increasingly fine control of gas resorption and secretion in the swim bladder, improving control of buoyancy. Changes in jaw suspension enabled the orobranchial cavity to expand rapidly, creating a highly sophisticated suction device. With so many innovations, teleosts have become the most diverse of fishes.

Lobe-Finned Fishes: Class Sarcopterygii

The ancestor of tetrapods is found within a group of extinct sarcopterygian fishes called **rhipidistians,** which included several lineages that flourished in shallow waters in the late Paleozoic (figure 17.1, p. 336). The evolution of tetrapods from rhipidistians is discussed in Chapter 17.

All early sarcopterygians had lungs as well as gills, and a tail of the **heterocercal** type. However, during the Paleozoic the orientation of the vertebral column changed so that the tail became symmetrical and **diphycercal** (figure 16.13). They had powerful jaws, heavy enameled scales, and strong, fleshy, paired lobed fins that may have been used like legs to scuttle along the bottom. The sarcopterygian clade today is represented by only eight fish species: six species of lungfishes and two species of coelacanths (figures 16.19 and 16.20).

Neoceratodus (Gr. *neos,* new, + *keratos,* horn, + *odes,* form), the living Australian lungfish, may attain a length of 1.5 m (figure 16.19). This lungfish is able to survive in stagnant, oxygen-poor water by coming to the surface and gulping air into its single lung, but it cannot live out of water. The South American lungfish, *Lepidosiren* (L. *lepidus,* pretty, + *siren,* Siren, mythical mermaid), and the African lungfish, *Protopterus*

(Gr. *protos,* first, + *pteron,* wing), can live out of water for long periods of time. *Protopterus* lives in African streams and ponds that are baked hard by the hot tropical sun during the dry season. The fish burrows down at the approach of the dry season and secretes a copious slime that mixes with mud to form a hard cocoon in which it remains dormant until the rains return.

Coelacanths also arose in the Devonian period, radiated somewhat, and reached their peak of diversity in the Mesozoic era. At the end of the Mesozoic era they nearly disappeared but left one remarkable surviving genus (figure 16.20). Because the last coelacanths were believed to have become extinct 70 million years ago, the astonishment of the scientific world may be imagined when the remains of a coelacanth were found on a trawl off the coast of South Africa in 1938. An intensive search to locate more specimens was successful off the coast of the Comoro Islands. There fishermen occasionally catch them at great depths with hand lines, providing specimens for research. This was believed to be the only population of *Latimeria* until 1998, when the scientific world was again surprised by the capture of a new species of coelacanth in Indonesia, 10,000 km from the Comoros!

The "modern" marine coelacanth is a descendant of the Devonian freshwater stock. The tail is diphycercal (figure 16.13) but possesses a small lobe between the upper and lower caudal lobes, producing a three-pronged structure (figure 16.20).

Coelacanths are a deep metallic blue with irregular white or brassy flecks, providing camouflage against the dark lava-cave reefs they inhabit. Young are born fully formed after hatching internally from eggs 9 cm in diameter—the largest among bony fishes.

Australian lungfish

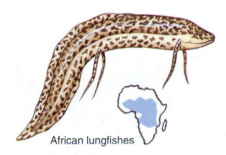
African lungfishes

South American lungfish

figure 16.19

Lungfishes are lobe-finned fishes of Sarcopterygii. The African lungfishes *Protopterus* are best adapted of the three genera for remaining dormant in mucous-lined cocoons breathing air during prolonged periods of drought.

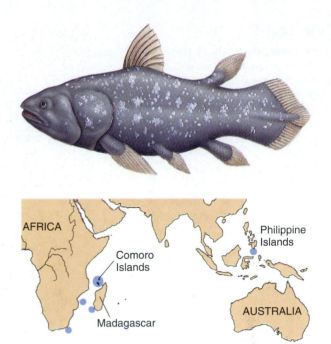

figure 16.20

The coelacanth genus *Latimeria* is a surviving marine relict of a group of lobe-finned fishes that flourished some 350 million years ago.

Measuring fish cruising speeds accurately is best done in a "fish wheel," a large ring-shaped channel filled with water that is turned at a speed equal and opposite to that of the fish. Much more difficult to measure are the sudden bursts of speed that most fish can make to capture prey or to avoid being captured. A hooked bluefin tuna was once "clocked" at 66 km per hour (41 mph); swordfish and marlin may be capable of incredible bursts of speed approaching, or even exceeding, 110 km per hour (68 mph). They can sustain such high speeds for no more than 1 to 5 seconds.

Structural and Functional Adaptations of Fishes

Locomotion in Water

To the human eye, some fishes appear capable of swimming at extremely high speeds. But our judgment is unconsciously tempered by our own experience that water is a highly resistant medium through which to move. Most fishes, such as a trout or a minnow, can swim maximally about 10 body lengths per second, obviously an impressive performance by human standards. Yet when these speeds are translated into kilometers per hour it means that a 30 cm (1 foot) trout can swim only about 10.4 km (6.5 miles) per hour. As a general rule, the larger the fish the faster it can swim.

The propulsive mechanism of a fish is its trunk and tail musculature. The axial, locomotory musculature is composed of zigzag bands, called **myomeres.** Muscle fibers in each myomere are relatively short and connect the tough connective tissue partitions that separate each myomere from the next. On the surface myomeres take the shape of a W lying on its side (figure 16.21) but internally the bands are complexly folded and nested so that the pull of each myomere extends over several vertebrae. This arrangement produces more power and finer control of movement since many myomeres are involved in bending a given segment of the body.

Understanding how fishes swim can be approached by studying the motion of a very flexible fish such as an eel (figure 16.21). The movement is serpentine, not unlike that of a snake, with waves of contraction moving backward along the body by alternate contraction of the myomeres on either side. The anterior end of the body bends less than the posterior end, so that each undulation increases in amplitude as it travels along the body. While undulations move backward, bending of the

figure 16.21

Trunk musculature of a teleost fish, partly dissected to show internal arrangement of the muscle bands (myomeres). The myomeres are folded into a complex, nested grouping, an arrangement that favors stronger and more controlled swimming.

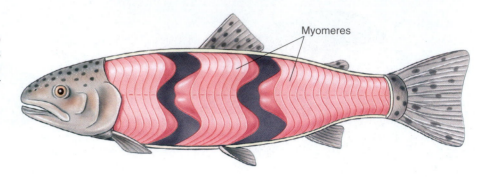

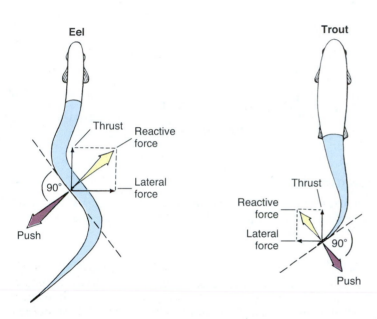

figure 16.22

Movements of swimming fishes, showing the forces developed by an eel-shaped and spindle-shaped fish.

Source: Pough, F. H.; Janis, Christine M; Heiser, John B., VERTEBRATE LIFE, 5th Edition, © 1999. Reprinted by permission of Pearson Education Inc., Upper Saddle River, NJ

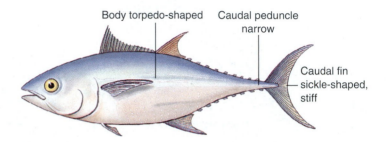

figure 16.23

Bluefin tuna, *Thunnus thynnus,* showing adaptations for fast swimming. Powerful trunk muscles pull on the slender caudal peduncle. Because the body does not bend, all of the thrust comes from beats of the stiff, sickle-shaped caudal fin.

body pushes laterally against the water, producing a **reactive force** that is directed forward, but at an angle. It can be analyzed as having two components: **thrust,** which is used to overcome drag and propels the fish forward, and **lateral force,** which tends to make the fish's head "yaw," or deviate from the course in the same direction as the tail. This side-to-side head movement is very obvious in a swimming eel or shark, but many fishes have a large, rigid head with enough surface resistance to minimize yaw.

Movement of an eel is reasonably efficient at low speed, but its body shape generates too much frictional drag for rapid swimming. Fishes that swim rapidly, such as trout, are less flexible and limit body undulations mostly to the caudal region (figure 16.22). Muscle force generated in the large anterior muscle mass is transferred through tendons to the relatively nonmuscular caudal peduncle and caudal fin where thrust is generated. This form of swimming reaches its highest development in tunas, whose bodies do not flex at all. Virtually all thrust is derived from powerful beats of the caudal fin (figure 16.23). Many fast oceanic fishes such as marlin, swordfish, amberjacks, and wahoo have swept-back caudal fins shaped much like a sickle. Such fins are the aquatic counterpart of the high-speed wings of the swiftest birds (p. 380). Swimming is the most economical form of animal locomotion, largely because aquatic animals are almost perfectly supported by their medium and need expend little energy to overcome the force of gravity.

Neutral Buoyancy and the Swim Bladder

All fishes are slightly heavier than water because their skeletons and other tissues contain heavy elements that are present only in trace amounts in natural waters. To keep from sinking, sharks must always keep moving forward in the water. The asymmetrical (heterocercal) tail of a shark provides the necessary tail lift as it sweeps to and fro in the water, and the broad head and flat pectoral fins (see figure 16.6) act as angled planes to provide head lift. Sharks also are aided in buoyancy by having very large livers containing a special fatty hydrocarbon called **squalene,** which has a density of only 0.86 grams per milliliter. The liver thus acts like a large sack of buoyant oil that helps to compensate for the shark's heavy body.

By far the most efficient flotation device is a gas-filled space. The **swim bladder** serves this purpose in bony fishes (figure 16.24). It arose from the paired lungs of primitive Devonian fishes. Swim bladders are present in most pelagic bony fishes but are absent from tunas, most abyssal fishes, and most bottom-dwellers, such as flounders and sculpins.

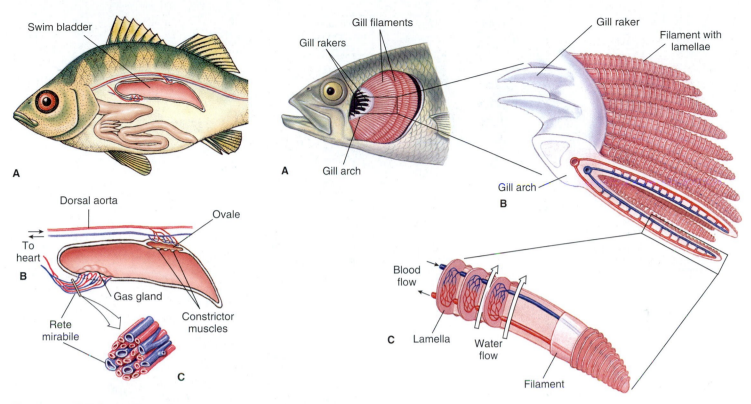

figure 16.24

A, Swim bladder of a teleost fish. The swim bladder lies in the coelom just beneath the vertebral column. **B,** Gas is secreted into the swim bladder by the gas gland. Gas from the blood is moved into the gas gland by the rete mirabile, a complex array of tightly-packed capillaries that act as a countercurrent multiplier to increase oxygen concentration. The arrangement of venous and arterial capillaries in the rete is shown in **C.** To release gas during ascent, a muscular valve opens, allowing gas to enter the ovale from which the gas is removed by diffusion into the blood.

figure 16.25

Gills of fish. Bony, protective flap covering the gills (operculum) has been removed, **A,** to reveal branchial chamber containing the gills. There are four gill arches on each side, each bearing numerous filaments. A portion of gill arch, **B,** shows gill rakers that project forward to strain food and debris, and gill filaments that project to the rear. A single gill filament, **C,** is dissected to show the blood capillaries within the platelike lamellae. Direction of water flow (*large arrows*) is opposite the direction of blood flow.

By adjusting the volume of gas in its swim bladder, a fish can achieve neutral buoyancy and remain suspended indefinitely at any depth with no muscular effort. If a fish swims to a greater depth, the greater pressure exerted by the surrounding water compresses the gas in the swim bladder, so that the fish becomes less buoyant and begins to sink. Gas must be added to the swim bladder to establish a new equilibrium buoyancy. When a fish swims upward, gas in the bladder expands because of the reduced surrounding water pressure, making the fish more buoyant. Unless gas is removed, the fish will continue to ascend with increasing speed as the swim bladder continues to expand.

Gases are removed from the swim bladder in two ways. Some fishes (trout, for example) have a **pneumatic duct** that connects the swim bladder to the esophagus, through which they may gulp or expel air. More advanced teleosts have lost the pneumatic duct and must exchange air in the swim bladder with the blood. Gas is absorbed by blood from a highly vascularized region of the swim bladder, the **ovale.** Gas is secreted into the swim bladder at the **gas gland.** The gas gland secretes lactic acid, causing oxygen to be released from hemoglobin. A remarkable network of blood capillaries, the **rete mirabile,** supplies the gas gland and acts as a countercurrent exchange system (figure 16.24). The rete allows oxygen to reach high concentrations in the gas gland, permitting it to diffuse into the swim bladder.

The amazing effectiveness of this device is exemplified by a fish living at a depth of 2400 m (8000 feet). To keep the bladder inflated at that depth, the gas inside (mostly oxygen) must have a pressure exceeding 240 atmospheres, which is much greater than the pressure in a fully charged steel gas cylinder. Yet the oxygen pressure in the fish's blood cannot exceed 0.2 atmosphere—in equilibrium with the oxygen pressure in the atmosphere at the sea surface.

Respiration

Fish gills are composed of thin filaments, each covered with a thin epidermal membrane that is folded repeatedly into platelike **lamellae** (figure 16.25). These are richly supplied with blood vessels. The gills are located inside a pharyngeal cavity and are covered with a movable flap, the **operculum.** This arrangement provides excellent protection to the delicate gill filaments,

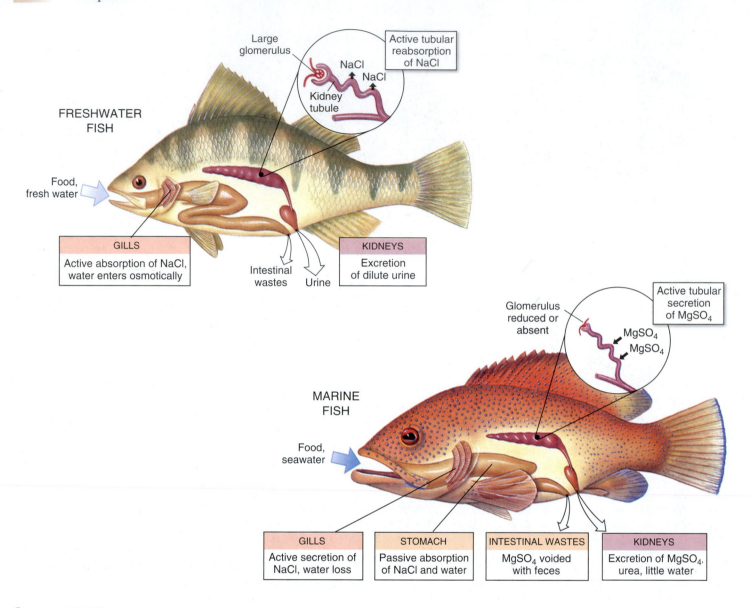

figure 16.26

Osmotic regulation in freshwater and marine bony fishes. A freshwater fish maintains osmotic and ionic balance in its dilute environment by actively absorbing sodium chloride across the gills (some salt is gained with food). To flush out excess water that constantly enters the body, the glomerular kidney produces a dilute urine by reabsorbing sodium chloride. A marine fish must drink seawater to replace water lost osmotically to its salty environment. Sodium chloride and water are absorbed from the stomach. Excess sodium chloride is actively transported outward by the gills. Divalent sea salts, mostly magnesium sulfate, are eliminated with feces and secreted by the tubular kidney.

streamlines the body, and makes possible a pumping system for moving water through the mouth, across the gills, and out the operculum. Instead of opercular flaps as in bony fishes, elasmobranchs have a series of **gill slits** (figure 16.6) out of which the water flows. In both elasmobranchs and bony fishes the branchial mechanism is arranged to pump water continuously and smoothly over the gills, although to an observer it appears that fish breathing is pulsatile.

The flow of water is opposite to the direction of blood flow (countercurrent flow), the best arrangement for extracting the greatest possible amount of oxygen from water. Some bony fishes can remove as much as 85% of the dissolved oxygen from water passing over their gills. Very active fishes, such

as herring and mackerel, can obtain sufficient water for their high oxygen demands only by swimming forward continuously to force water into their open mouth and across their gills. This process is called ram ventilation.

Osmotic Regulation

Fresh water is an extremely dilute medium with a salt concentration (0.001 to 0.005 gram moles per liter [M]) much below that of the blood of freshwater fishes (0.2 to 0.3 M). Water therefore tends to enter the bodies of freshwater fishes osmotically, and salt is lost by diffusion outward. Although the scaled and mucous-covered body surface is almost totally impermeable to water,

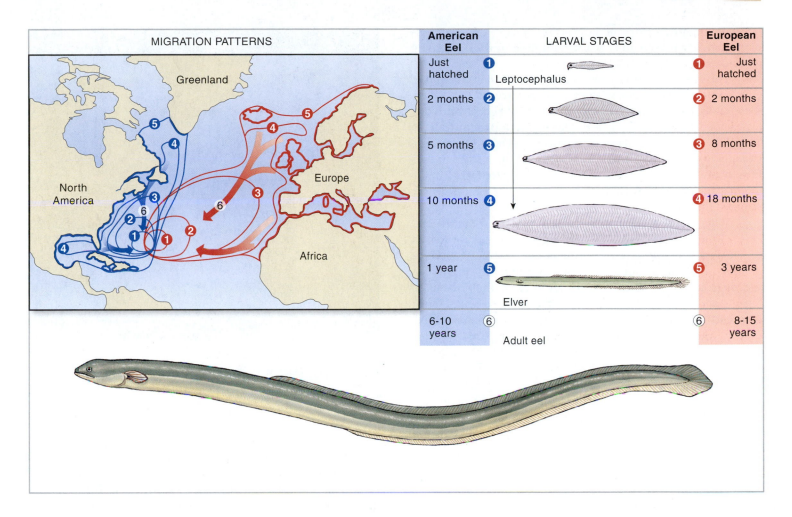

American Eel	LARVAL STAGES		European Eel
MIGRATION PATTERNS			
Just hatched ❶	Leptocephalus	❶	Just hatched
2 months ❷		❷	2 months
5 months ❸		❸	8 months
10 months ❹		❹	18 months
1 year ❺	Elver	❺	3 years
6–10 years ⑥	Adult eel	⑥	8–15 years

figure 16.27

Life histories of European eels, *Anguilla anguilla,* and American eels, *Anguilla rostrata.* Migration patterns of European species are shown in red. Migration patterns of American species are shown in blue. Boxed numbers refer to stages of development. Note that American eels complete their larval metamorphosis and sea journey in one year. It requires nearly three years for European eels to complete their much longer journey.

water gain and salt loss do occur across thin membranes of the gills. Freshwater fishes are **hyperosmotic regulators** with several defenses against these problems (figure 16.26). First, excess water is pumped out by the kidneys, which are capable of forming very dilute urine. Second, special **salt-absorbing cells** located in the gill epithelium actively move salt ions, principally sodium and chloride, from water to the blood. This, together with sale present in the fishes' food, replaces diffusive salt loss. These mechanisms are so efficient that a freshwater fish devotes only a small part of its total energy expenditure to keeping itself in osmotic balance.

Marine fishes are **hypoosmotic regulators** that encounter a completely different problem. Having a much lower blood salt concentration (0.3 to 0.4 M) than the seawater around them (about 1 M), they tend to lose water and gain salt. A marine teleost fish quite literally risks drying out, much like a desert mammal deprived of water. To compensate for water loss, a marine teleost drinks seawater (figure 16.26). Excess salt accompanying the seawater is expelled in two ways. Major sea salt ions (sodium, chloride, and potassium) are

carried by the blood to the gills, where they are secreted outward by special **salt-secretory cells.** The remaining sea salt ions, mostly magnesium, sulfate, and calcium, are voided with feces or excreted by the kidney.

Migration

Freshwater Eels

For centuries naturalists had been puzzled by the life history of freshwater eels, *Anguilla* (an-gwil′a) (L. eel), a common and commercially important species of coastal streams of the North Atlantic. Eels are **catadromous** (Gr. *kata,* down, + *dromos,* running), meaning that they spend most of their lives in fresh water but migrate to the sea to spawn. Each fall, people saw large numbers of eels swimming down rivers toward the sea, but no adults ever returned. Each spring countless numbers of young eels, called "elvers" (figure 16.27), each about the size of a wooden matchstick, appeared in coastal rivers and began swimming upstream. Beyond the assumption that eels must spawn somewhere at sea, the location of their breeding grounds was completely unknown.

The first clue was provided by two Italian scientists, Grassi and Calandruccio, who in 1896 discovered that elvers were advanced juvenile eels and that true larval eels were tiny leaf-shaped transparent creatures, known as **leptocephali.** In 1905 Johann Schmidt began a systematic study of eel biology, examining thousands of leptocephali caught in plankton nets from many areas of the Atlantic. By noting where larvae in different stages were captured, Schmidt and his colleagues reconstructed the spawning migration.

When adult eels leave the coastal streams of Europe and North America, they swim to the Sargasso Sea, a vast area of warm weather southeast of Bermuda (figure 16.27). Here, at depths of 300 m or more, the eels spawn and die. Minute larvae then begin an incredible journey back to the streams of Europe and North America. Because the Sargasso Sea is much closer to the American coastline, American eel larvae make their journey in only about eight months, compared to three years for European eel larvae. Males typically remain in brackish water of costal rivers, while females migrate as far as several hundred kilometers upstream. After 8 to 15 years of growth, females, now 1 m long, return to the ocean to join the smaller males in the journey back to the spawning grounds in the Sargasso Sea.

Recent enzyme electrophoretic analysis of eel larvae confirmed not only the existence of separate European and American species but also Schmidt's belief that the European and American eels spawn in partially overlapping areas of the Sargasso Sea.

Homing Salmon

The life history of salmon is nearly as remarkable as that of freshwater eels and certainly has received far more popular attention. Salmon are **anadromous** (Gr. *anadromous,* running upward), spending their adult lives at sea but returning to fresh water to spawn. Atlantic salmon (*Salmo salar*) and Pacific salmon (six species of the genus *Oncorhynchus* [on-ko-rink′us]) have this practice, but there are important differences among the seven species. Atlantic salmon (as well as the closely related steelhead trout) make upstream spawning runs year after year. The six Pacific salmon species (king, sockeye, silver, humpback, chum, and Japanese masu) each make a single spawning run (figures 16.28 and 16.29), after which they die.

The virtually infallible homing instinct of the Pacific species is legendary. After migrating downstream as a smolt (a juvenile stage, figure 16.29), a sockeye salmon ranges many hundreds of miles over the Pacific for nearly four years, grows to 2 to 5 kg in weight, and then returns almost unerringly to spawn in the headwaters of its parent stream. Some straying does occur and is an important means of increasing gene flow and populating new streams.

figure 16.28
Migrating Pacific sockeye salmon (*Oncorhynchus nerka*).

Experiments by A. D. Hasler and others show that homing salmon are guided upstream by the characteristic odor of their parent stream. When salmon finally reach the spawning beds of their parents (where they themselves were hatched), they spawn and die. The following spring, newly hatched fry transform into smolts before and during the downstream migration. At this time they are imprinted with the distinctive odor of the stream, which is apparently a mosaic of compounds released by the characteristic vegetation and soil in the watershed of the parent stream. They also seem to imprint on odors of other streams they pass while migrating downriver and use these odors in reverse sequence as a map during the upriver migration as returning adults.

Salmon runs along the Pacific coast of North America have been devastated by a lethal combination of spawning stream degradation by logging, pollution and, especially, by more than 50 hydroelectric dams, which obstruct upstream migration of adult salmon and kill downstream migrants as they pass through the dams' power-generating turbines. In addition, the chain of reservoirs behind the dams, which has converted the Columbia and Snake Rivers into a series of lakes, increases mortality of young salmon migrating downstream by slowing their passage to the sea. The result is that the annual run of wild salmon is today only about 3% of the 10 to 16 million fish that ascended the rivers 150 years ago. While recovery plans have been delayed by the power industry, environmental groups argue that in the long run losing the salmon will be more expensive to the regional economy than making the changes now that will allow salmon stocks to recover.

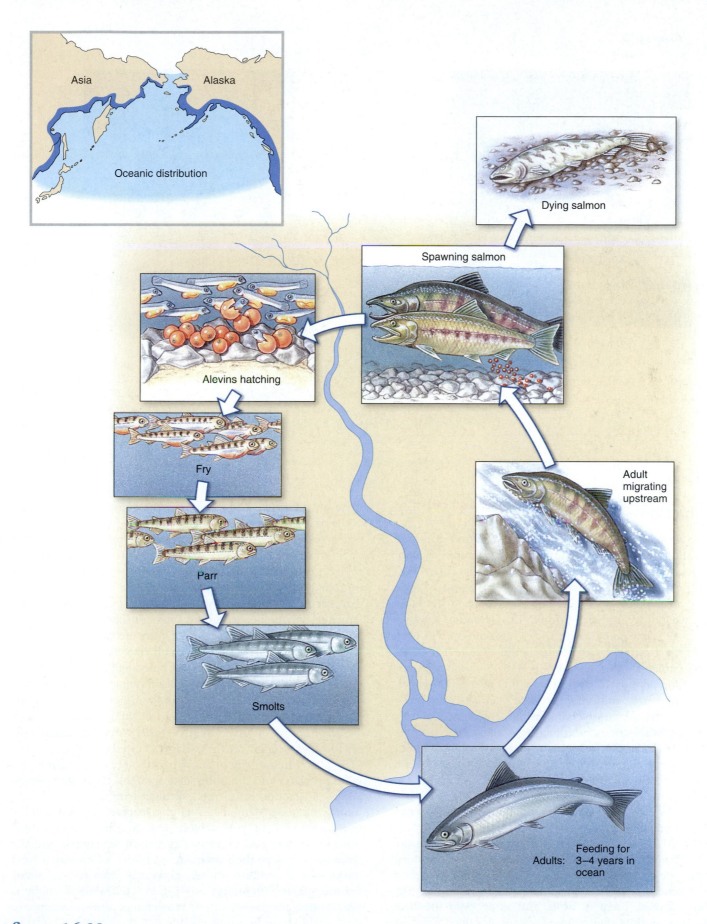

Asia

Alaska

Oceanic distribution

Dying salmon

Spawning salmon

Alevins hatching

Fry

Parr

Adult migrating upstream

Smolts

Adults: Feeding for 3–4 years in ocean

figure 16.29

Spawning Pacific salmon, *Oncorhynchus,* and development of eggs and young.

figure 16.30

Rainbow surfperch, *Hypsurus caryi,* giving birth. All of the West Coast surfperches (family Embiotocidae) are ovoviviparous.

figure 16.31

Male banded jawfish *Opistognathus macrognathus* orally brooding its eggs. The male retrieves the female's eggs and incubates them until they hatch. During brief periods when the jawfish is feeding, the eggs are left in the burrow.

Reproduction

In a group as diverse as fishes, it is no surprise to find extraordinary variations on the basic theme of sexual reproduction. Most fishes favor a simple theme: they are **dioecious,** with **external fertilization** and **external development** of their eggs and embryos (oviparity). However, as tropical fish enthusiasts are well aware, the ever-popular ovoviviparous guppies and mollies of home aquaria bear their young alive after development in the ovarian cavity of the mother (figure 16.30). As described earlier in this chapter (p. 318), some viviparous sharks develop a kind of placental attachment through which the young are nourished during gestation.

Oviparity is the most common mode of reproduction. Many marine fishes are extraordinarily profligate egg producers. Males and females aggregate in great schools and, without elaborate courtship behavior, release vast numbers of germ cells into the water to drift with currents. Large female cod may release 4 to 6 million eggs at a single spawning. Less than one in a million will survive the numerous perils of the ocean to reach reproductive maturity.

Unlike the minute, buoyant, transparent eggs of pelagic marine teleosts, those of many near-shore and bottom-dwelling (benthic) species are larger, typically yolky, nonbuoyant, and adhesive. Some bury their eggs, many attach them to vegetation, some deposit them in nests, and some even incubate them in their mouths (figure 16.31). Many benthic spawners guard their eggs. Intruders expecting an easy meal of eggs may be met with a vivid and often belligerent display by the guard, which is almost always the male.

Freshwater fishes almost invariably produce nonbuoyant eggs. Those, such as perch, that provide no parental care simply scatter their myriads of eggs among weeds or along the bottom. Freshwater fishes that do provide some form of egg care, such as bullhead catfish and some darters, produce fewer, larger eggs that enjoy a better chance for survival.

Elaborate preliminaries to mating are the rule for freshwater fishes. A female Pacific salmon, for example, performs a ritualized mating "dance" with her breeding partner after arriving at the spawning bed in a fast-flowing, gravel-bottomed stream (figure 16.29). She then turns on her side and scoops out a nest with her tail. As the eggs are laid by the female, they are fertilized by the male (figure 16.29). After the female covers the eggs with gravel, the exhausted fish dies.

Soon after an egg of an oviparous species is fertilized, it takes up water and the outer layer hardens. Cleavage follows, and the blastoderm is formed, sitting astride a relatively enormous yolk mass. Many fishes hatch as larvae, carrying a semitransparent sac of yolk which provides their food supply until the mouth and digestive tract have developed and the larvae can feed on their own. After a period of growth a larva undergoes a metamorphosis, especially dramatic in many marine species, including eels (figure 16.27). Body shape is refashioned, fin and color patterns change, and it becomes a juvenile bearing the unmistakable definitive body form of its species.

classification of Living Fishes

The following Linnean classification of major fish taxa follows that of Nelson (1994). The probable relationships of these traditional groupings together with the major extinct groups of fishes are shown in a cladogram in figure 16.2. Other schemes of classification have been proposed. Because of difficulty of determining relationships among the numerous living and fossil species, we can appreciate why fish classification has undergone, and will continue to undergo, continuous revision.

Phylum Chordata

    Subphylum Vertebrata (Craniata)

Superclass Agnatha (ag'na-tha) (Gr. *a*, not, + *gnathos*, jaw). No jaws; cartilaginous skeleton; paired fins absent; one or two semicircular canals; notochord persistent. Not a monophyletic group.

    Class Myxini (mik-sy'ny) (Gr. *myxa*, slime): **hagfishes.** Four pairs of tentacles around mouth; nasal sac with duct to pharynx; 5 to 16 pairs of gill pouches, accessory hearts and slime glands present; poorly developed eyes. Examples: *Myxine, Epaptretus;* about 65 species, marine.

    Class Cephalaspidomorphi (sef-a-lass'pe-do-morf'e) (Gr. *kephalē*, head, + *aspidos*, shield, + *morphē*, form): **lampreys.** Buccal funnel with keratinized teeth; nasal sac not connected to mouth; seven pairs of gill pouches, well-developed eyes. Examples: *Petromyzon, Lampetra;* 41 species, freshwater and anadromous.

Superclass Gnathostomata (na'tho-sto'ma-ta) (Gr. *gnathos*, jaw, + *stoma*, mouth). Jaws present; paired appendages present (secondarily lost in a few forms); three pairs of semicircular canals; notochord partly or completely replaced by centra.

    Class Chondrichthyes (kon-drik'thee-eez) (Gr. *chondros*, cartilage, + *ichthys*, fish): **cartilaginous fishes.** Cartilaginous skeleton; teeth not fused to jaws and usually replaced; no swim bladder; intestine with spiral valve; claspers present in males.

        Subclass Elasmobranchii (e-laz'mo-bran'kee'i) (Gr. *elasmos*, plated, + *branchia*, gills): **sharks, skates, and rays.** Placoid scales or derivatives (scutes and spines) usually present; five to seven gill arches and gill slits in separate clefts along pharynx; upper jaw not fused to cranium. Examples: *Squalus, Raja, Sphyrna;* about 815 species, mostly marine.

        Subclass Holocephali (hol'o-sef'a-li) (Gr. *holos*, entire, + *kephalē*, head): **chimaeras, ratfishes.** Scales absent; four gill slits covered by operculum; jaws with tooth plates; accessory clasping organ (tentaculum) in males; upper jaw fused to cranium. Examples: *Chimaera, Hydrolagus;* 31 species, marine.

    Class Actinopterygii (ak'ti-nop-te-rij'ee-i) (Gr. *aktis*, ray, + *pteryx*, fin, wing): **ray-finned fishes.** Skeleton ossified; single gill opening covered by operculum; paired fins supported primarily by dermal rays; limb musculature within body; swim bladder mainly a hydrostatic organ, if present; atrium and ventricle not divided; teeth with enameloid covering.

        Subclass Chondrostei (kon-dros'tee-i) (Gr. *chondros*, cartilage, + *osteon*, bone): **bichirs, paddlefishes, sturgeons.** Skeleton primarily cartilage; caudal fin heterocercal; scales ganoid, if present; spiral valve present; spiracle usually present; more fin rays than ray supports. Examples: *Polypterus, Polyodon, Acipenser;* 34 species, freshwater and anadromous.

        Subclass Neopterygii (nee'-op-te-rij'ee-i) (Gr. *neo*, new, + *pteryx*, fin, wing): **gars, bowfin, teleosts.** Skeleton primarily bone; caudal fin usually homocercal; scales cycloid, ctenoid, absent, or rarely, ganoid. Fin ray number equal to their supports in dorsal and anal fins. Examples: *Amia, Lepisosteus, Anguilla, Oncorhynchus, Perca;* about 27,000 species, nearly all aquatic habitats.

    Class Sarcopterygii (sar-cop-te-rij'ee-i) (Gr. *sarkos*, flesh, + *pteryx*, fin, wing): **lobe-finned fishes.** Skeleton ossified; single gill opening covered by operculum; paired fins with sturdy internal skeleton and musculature within limb; diphycercal tail; intestine with spiral valve; usually with lunglike swim bladder; atrium and ventricle at least partly divided; teeth with enamel covering. Examples: *Latimeria* (coelacanths); *Neoceratodus, Lepidosiren, Protopterus* (lungfishes); 8 species, marine and freshwater. Not monophyletic unless tetrapods are included.

Summary

Fishes are poikilothermic, gill-breathing aquatic vertebrates with fins for appendages. They include the oldest vertebrate groups, having originated from an unknown chordate ancestor in the Cambrian period or possibly earlier. Five classes of living fishes are recognized. The jawless hagfishes (class Myxini) and lampreys (class Cephalaspidomorphi), have an eel-like body form without paired fins, a cartilaginous skeleton, a notochord that persists throughout life, and a disclike mouth adapted for sucking and biting. All other vertebrates have jaws, a major development in vertebrate evolution.

Members of class Chondrichthyes (sharks, skates, rays, and chimaeras) have a cartilaginous skeleton (a degenerative feature), paired fins, excellent sensory organs, and an active, characteristically predaceous habit. Bony fishes may be divided into two classes. Lobe-finned fishes of the class Sarcopterygii, represented today by lungfishes and coelacanths, form a paraphyletic group if tetrapods are excluded, as is done in traditional classification. Terrestrial vertebrates arose from one lineage within this group. The second is ray-finned fishes (class Actinopterygii), a huge and diverse modern assemblage containing nearly all familiar freshwater and marine fishes.

Modern bony fishes (teleost fishes) have radiated into approximately 27,000 species that reveal an enormous diversity of adaptations, body form, behavior, and habitat preference. Most fishes swim by undulatory contractions of their body muscles, which generate thrust (propulsive force) and lateral force. Eel-like fishes oscillate the whole body, but in more rapid swimmers the undulations are limited to the caudal region or caudal fin alone.

Most pelagic bony fishes achieve neutral buoyancy in water using a gas-filled swim bladder, the most effective gas-secreting device known in the animal kingdom. Gills of fishes, having efficient countercurrent flow between water and blood, facilitate high rates of oxygen exchange. All fishes show well-developed osmotic and ionic regulation, achieved principally by kidneys and gills.

Many fishes are migratory, and some, such as freshwater eels and anadromous salmon, make remarkable migrations of great length and precision. Fishes reveal an extraordinary range of sexual reproductive strategies. Most fishes are oviparous, but ovoviviparous and viviparous fishes are not uncommon. Reproductive investment may be in large numbers of germ cells with low survival (many marine fishes) or in fewer germ cells with greater parental care for better survival (many freshwater fishes).

Review Questions

1. Provide a brief description of fishes citing characteristics that would distinguish them from all other animals.
2. What characteristics distinguish hagfishes and lampreys from all other fishes?
3. Describe feeding behavior in hagfishes and lampreys. How do they differ?
4. Describe the life cycle of the sea lamprey, *Petromyzon marinus,* and the history of its invasion of the Great Lakes.
5. In what ways are sharks well equipped for a predatory life habit?
6. The lateral line system has been described as a "distant touch" system for sharks. What function does the lateral line system serve? Where are the receptors located?
7. Explain how bony fishes differ from sharks and rays in the following systems or features: skeleton, scales, buoyancy, respiration, and reproduction.
8. Match the ray-finned fishes in the right column with the group to which each belongs in the left column:

   ___ Chondrosteans   a. Perch
   ___ Nonteleost   b. Sturgeon
       neopterygians   c. Gar
   ___ Teleosts   d. Salmon
                          e. Paddlefish
                          f. Bowfin

9. Although chondrosteans are today a relict group, they descended from one of two major clades that emerged from early ray-finned fishes of the Devonian period. Give examples of living chondrosteans. What does the term "Actinopterygii," the class to which the chondrosteans belong, literally mean (refer to the classification of living fishes on p. 331)?
10. List four characteristics of teleosts that contributed to their incredible evolutionary diversity.
11. Only eight species of lobe-finned fishes are alive today, remnants of a group that flourished in the Devonian period of the Paleozoic. What morphological characteristics distinguish lobe-finned fishes? What is the literal meaning of "Sarcopterygii," the class to which the lobe-finned fishes belong?
12. Give the geographical locations of the three surviving genera of lungfishes and explain how they differ in their ability to survive out of water.
13. Describe discovery of the living coelacanths. What is the evolutionary significance of the group to which they belong?
14. Compare the swimming movements of eels with those of trout, and explain why the latter are more efficient for rapid locomotion.

15. Sharks and bony fishes approach or achieve neutral buoyancy in different ways. Describe the methods evolved in each group. Why must a teleost fish adjust the gas volume in its swim bladder when it swims upward or downward? How is gas volume adjusted?
16. What is meant by "countercurrent flow" as it applies to fish gills?
17. Compare the osmotic problem and the mechanism of osmotic regulation in freshwater and marine bony fishes.
18. Describe the life cycle of European eels. How does the life cycle of American eels differ from that of the European?
19. How do adult Pacific salmon find their way back to their parent stream to spawn?
20. What mode of reproduction in fishes is described by each of the following terms: oviparous, ovoviviparous, viviparous?
21. Reproduction in marine pelagic fishes and in freshwater fishes is distinctively different. How and why do they differ?

Selected References

See also general references on page 415.

Bond, C. E. 1996. Biology of fishes, ed. 2. Fort Worth, Harcourt College Publishers. *A superior treatment of fish biology, anatomy, and genetics.*

Helfman, G. J., B. B. Collette, and D. E. Facey. 1997. The diversity of fishes. Malden, Massachusetts: Blackwell Science. *This delightful and information-packed textbook focuses on adaptation and diversity and*

is particularly strong in evolution, systematics, and history of fishes.

Horn, M. H., and R. N. Gibson. 1988. Intertidal fishes. Sci. Am. **258:**64–70 (Jan.). *Describes the special adaptations of* *fishes living in a demanding environment.*

Long, J. A. 1995. The rise of fishes: 500 million years of evolution. Baltimore, The Johns Hopkins University Press. *A lavishly illustrated evolutionary history of fishes.*

Martini, F. H. 1998. Secrets of the slime hag. Sci. Am. **279**:70–75 (Oct.). *Biology of the most basal living craniate.*

Nelson, J. S. 1994. Fishes of the world, ed. 3. New York, John Wiley & Sons, Inc. *Author-itative classification of all major groups of fishes.*

Paxton, J. R., and W. N. Eschmeyer. 1998. Encyclopedia of fishes, ed. 2. San Diego, Academic Press. *Excellent authoritative reference that focuses on diversity and is spectacularly illustrated.*

Springer, V. G., and J. P. Gold. 1989. Sharks in question. Washington, Smithsonian Institution Press. *Morphology, biology, and diversity of sharks, richly illustrated.*

Webb, P. W. 1984. Form and function in fish swimming. Sci. Am. **251**:72–82 (July). *Specializations of fish for swimming and analysis of thrust generation.*

Weinberg, S. 2000. A fish caught in time: the search for the coelacanth. London, Fourth Estate. *The exciting history of the coelacanth discoveries.*

Custom Website

The *Animal Diversity* Online Learning Center is a great place to check your understanding of chapter material. Visit www.mhhe.com/hickmanad4e for access to key terms, quizzes, and more! Further enhance your knowledge with web links to chapter-related material.

Explore live links for these topics:

Class Myxini
Class Cephalaspidomorphi
Subclass Elasmobranchii
Dissection Guides for Elasmobranchs
Vertebrate Laboratory Exercises
Class Osteichthyes

Primitive Bony Fish
Teleosts
Dissection Guides for Teleosts
Fisheries and Conservation Issues
    Concerning Teleosts
Fisheries
Impact of Humans on the Sea; Harvesting

17

The Early Tetrapods and Modern Amphibians

A frog tadpole undergoing metamorphosis.

Vertebrate Landfall

The chorus of frogs beside a pond on a spring evening heralds one of nature's dramatic events. Mating frogs produce masses of eggs, which soon hatch into limbless, gill-breathing tadpole larvae that feed and grow. Then a remarkable transformation occurs. Hindlimbs appear and gradually lengthen. The tail shortens. Larval teeth and gills are lost. Eyelids develop. The forelegs emerge. In a matter of weeks the aquatic tadpole has completed its metamorphosis to an adult frog.

The evolutionary transition from water to land occurred not in weeks but over millions of years. A lengthy series of alterations cumulatively fitted the vertebrate body plan for life on land. The origin of land vertebrates is no less a remarkable feat for this fact—a feat that is unlikely to happen again because well-established competitors make it impossible for a poorly adapted transitional form to gain a foothold.

Amphibians are the only living vertebrates that have a transition from water to land in both their ontogeny and phylogeny. Even after some 350 million years of evolution, amphibians remain quasiterrestrial, hovering between aquatic and land environments. This double life is expressed in their name. Even the amphibians best adapted for a terrestrial existence cannot stray far from moist conditions. Many, however, have developed ways to keep their eggs out of open water where the larvae would be exposed to enemies.

A daptation for life on land is a major theme of the remaining vertebrate groups. These animals form a monophyletic group called **tetrapods.** Amphibians and amniotes (including reptiles, birds, and mammals) represent the two major extant branches of tetrapod phylogeny. In this chapter, we review what is known about the origins of terrestrial vertebrates and discuss the amphibian branch in detail. We discuss the major amniote groups in Chapters 18 through 20.

Movement onto Land

Movement from water to land is perhaps the most dramatic event in animal evolution, because it involves invasion of a habitat that in many respects is more hazardous for life. Life originated in water. Animals are mostly water in composition, and all cellular activities occur in water. Nevertheless, organisms eventually invaded land, carrying their watery composition with them. Vascular plants, pulmonate snails, and tracheate arthropods made this transition much earlier than vertebrates, and winged insects were diversifying at approximately the same time that the earliest terrestrial vertebrates evolved. Although the invasion of land required modification of almost every system in the vertebrate body, aquatic and terrestrial vertebrates retain many basic structural and functional similarities. We see a transition between aquatic and terrestrial vertebrates most clearly today in the many living amphibians that make this transition during their own life histories.

Beyond the obvious difference in water content, there are several important physical differences that animals must accommodate when moving from water to land. These include (1) oxygen content, (2) density, (3) temperature regulation, and (4) habitat diversity. Oxygen is at least 20 times more abundant in air and it diffuses much more rapidly through air than through water. Consequently, terrestrial animals can obtain oxygen far more easily than aquatic ones once they possess appropriate adaptations, such as lungs. Air, however, has approximately 1000 times less buoyant density than water and is approximately 50 times less viscous. It therefore provides relatively little support against gravity, requiring terrestrial animals to develop strong limbs and to remodel their skeleton to achieve adequate structural support. Air fluctuates in temperature more readily than water does, and terrestrial environments therefore experience harsh and unpredictable cycles of freezing, thawing, drying, and flooding. Terrestrial animals require behavioral and physiological strategies to protect themselves from thermal extremes; one such strategy is homeothermy (regulated constant body temperature) of birds and mammals.

Despite its hazards, the terrestrial environment offers a great variety of habitats including coniferous, temperate, and tropical forests, grasslands, deserts, mountains, oceanic islands, and polar regions. Safe shelter for protection of vulnerable eggs and young may be found much more readily in many of these terrestrial habitats than in aquatic ones.

Early Evolution of Terrestrial Vertebrates

Devonian Origin of Tetrapods

The Devonian period, beginning some 400 million years ago, was a time of mild temperatures and alternating droughts and floods. During this period some primarily aquatic vertebrates evolved two features that would be important for permitting the subsequent evolution for life on land: lungs and limbs.

Devonian freshwater environments were unstable. During dry periods, many pools and streams evaporated, water became foul, and dissolved oxygen disappeared. Only those fishes able to acquire atmospheric oxygen survived such conditions. Gills were unsuitable because in air the filaments collapsed, dried, and quickly lost utility. Virtually all freshwater fishes surviving this period, including lobe-finned fishes and lungfishes (p. 322), had a kind of lung that developed as an outgrowth of the pharynx. It was relatively simple to enhance the efficiency of the air-filled cavity by improving its vascularity with a rich capillary network, and by supplying it with arterial blood from the last (sixth) pair of aortic arches. Oxygenated blood returned directly to the heart by a pulmonary vein to form a complete pulmonary circuit. Thus the **double circulation** characteristic of all tetrapods originated: a systemic circulation serving the body and a pulmonary circulation supplying the lungs.

Vertebrate limbs also arose during the Devonian period. Although fish fins at first appear very different from the jointed limbs of tetrapods, an examination of the bony elements of the paired fins of the lobe-finned fishes shows that they broadly resemble the homologous structures of amphibian limbs. In *Eusthenopteron,* a Devonian lobe-fin, we can recognize an upper arm bone (humerus) and two forearm bones (radius and ulna) as well as other elements that we can homologize with the wrist bones of tetrapods (figure 17.1). *Eusthenopteron* could paddle itself through the bottom mud of pools with its fins, but it could not walk upright because backward and forward movement of the fins was limited to about 20 to 25 degrees. *Acanthostega,* one of the earliest known Devonian tetrapods, had well-formed tetrapod limbs with clearly formed digits on both fore- and hindlimbs, but it was clearly an aquatically adapted form whose limbs were too weakly constructed for proper walking on land. *Ichthyostega,* however, with its fully developed shoulder girdle, bulky limb bones, well-developed muscles, and other adaptations for terrestrial life, must have been able to pull itself onto land, although it probably did not walk very well. Thus, the tetrapods evolved their limbs underwater and only later, for reasons unknown, invaded land.

Evidence points to lobe-finned fishes as the closest relatives of tetrapods; in cladistic terms they contain the sister group of tetrapods (figures 17.2 and 17.3). Both the lobe-finned fishes and early tetrapods such as *Acanthostega* and *Ichthyostega* shared several characteristics of the skull, teeth, and pectoral girdle. *Ichthyostega* (Gr. *ichthys,* fish, + *stegē,* roof, or covering, in reference to the roof of the skull

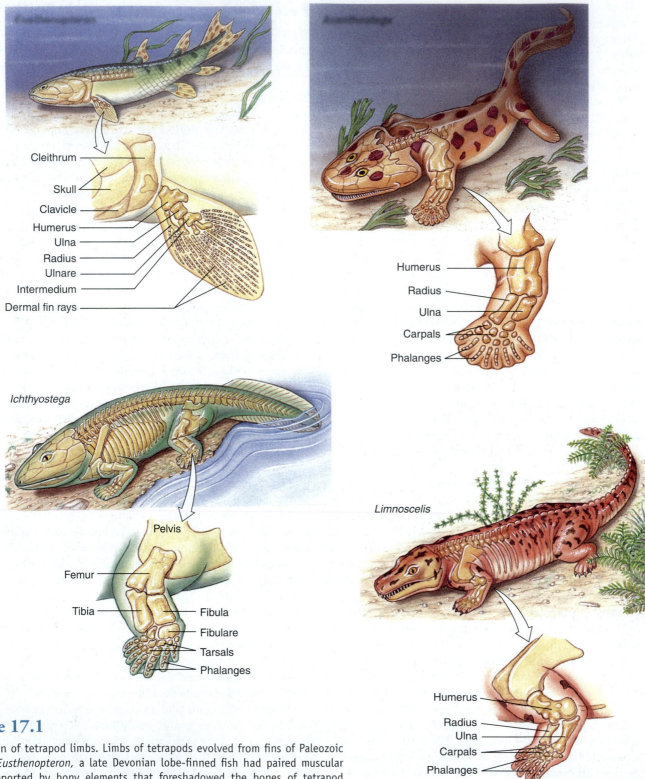

figure 17.1

Evolution of tetrapod limbs. Limbs of tetrapods evolved from fins of Paleozoic fishes. *Eusthenopteron,* a late Devonian lobe-finned fish had paired muscular fins supported by bony elements that foreshadowed the bones of tetrapod limbs. The anterior fin contained an upper arm bone (humerus), two forearm bones (radius and ulna), and smaller elements homologous to wrist bones of tetrapods. As typical of fishes, the pectoral girdle, consisting of the cleithrum, clavicle, and other bones, was firmly attached to the skull. In *Acanthostega,* one of the earliest known Devonian tetrapods (appearing about 360 million years BP), dermal fin rays of the anterior appendage were replaced by eight fully evolved fingers. *Acanthostega* was probably exclusively aquatic because its limbs were too weak for travel on land. *Ichthyostega,* a contemporary of *Acanthostega,* had fully formed tetrapod limbs and must have been able to walk on land. The hindlimb bore seven toes (the number of forelimb digits is unknown). *Limnoscelis,* an anthracosaur of the Carboniferous (about 300 million years BP) had five digits on both fore- and hindlimbs, the basic pentadactyl model that became the tetrapod standard.

(Sources: R. L. Carroll, *Vertebrate Paleontology and Evolution,* 1988, W. H. Freeman & Co., NY; M. I. Coates, and J. A. Clack, *Nature,* **347:**66–69, 1990; J. L. Edwards, *American Zoologist,* **29:**235–254, 1989; E. Jarvik, *Scientific Monthly,* **1955:**141–154, March 1955; and C. N. Zimmer, *Discover,* **16**(6):118–127, 1995.)

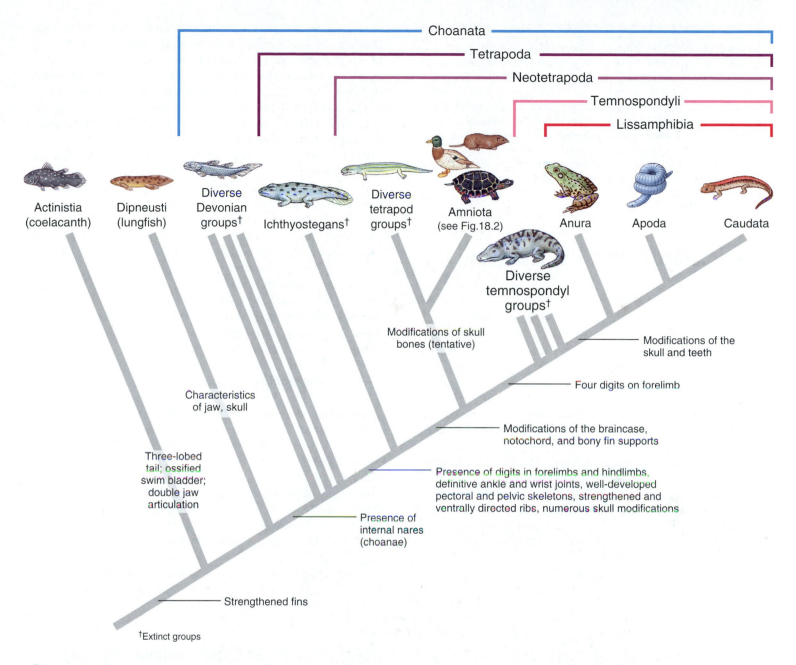

figure 17.2

Tentative cladogram of the Tetrapoda with emphasis on descent of the amphibians. Especially controversial are the relationships of major tetrapod groups (Amniota, Temnospondyli and diverse early tetrapod groups) and outgroups (Actinistia, Dipneusti, extinct Devonian groups). All aspects of this cladogram are controversial, however, including relationships of the Lissamphibia. The relationships shown for the three groups of Lissamphibia are based on molecular evidence. Extinct groups are marked with a dagger symbol (†).

(Source: From E. W. Gaffney in *Bulletin of the Carnegie Museum of Natural History*, **13**:92–105, 1979.)

which was shaped like that of a fish) is an early offshoot of tetrapod phylogeny. It possessed several adaptations, in addition to jointed limbs, that equipped it for life on land (figure 17.1): a stronger backbone and associated muscles to support the body in air, new muscles to elevate the head, strengthened shoulder and hip girdles, a protective rib cage, a more advanced ear structure for detecting airborne sounds, a foreshortening of the skull, and a lengthening of the snout that improved olfactory powers for detecting dilute airborne odors. Yet *Ichthyostega* still resembled aquatic forms in retaining a tail complete with fin rays and in having opercular (gill-covering) bones.

The bones of *Ichthyostega,* the most thoroughly studied of all early tetrapods, were first discovered on an east Greenland mountainside in 1897 by Swedish scientists looking for three explorers lost two years earlier during an ill-fated attempt to reach the North Pole by hot-air balloon. Later expeditions by Gunnar Säve-Söderberg uncovered skulls of *Ichthyostega* but Säve-Söderberg died, at age 38, before he could examine the skulls. After Swedish paleontologists returned to the Greenland site where they found the remainder of *Ichthyostega's* skeleton, Erik Jarvik, one of Säve-Söderberg's assistants, devoted his life's work to producing the detailed description of *Ichthyostega* available to us today.

Carboniferous Radiation of Tetrapods

The capricious Devonian period was followed by the Carboniferous period, characterized by a warm, wet climate during which mosses and large ferns grew in profusion on a swampy landscape. Tetrapods radiated quickly in this environment to produce a great variety of forms, feeding on the abundance of insects, insect larvae, and aquatic invertebrates available. Evolutionary relationships of early tetrapod groups are still controversial. We present a tentative cladogram (figure 17.2), which almost certainly will be revised as new data are collected. Several extinct lineages plus the **Lissamphibia,** which contains the modern amphibians, are placed in a group called **temnospondyls** (see figures 17.2 and 17.3). This group generally has only four digits on the forelimb rather than the five characteristic of most tetrapods.

Lissamphibians diversified during the Carboniferous to produce ancestors of the three major groups of amphibians alive today, **frogs** (Anura or Salientia), **salamanders** (Caudata or Urodela), and **caecilians** (Apoda or Gymnophiona). Amphibians improved their adaptations for living in water during this period. Their bodies became flatter for moving through shallow water. Early salamanders developed weak limbs, and their tail became better developed as a swimming organ. Even anurans (frogs and toads), which are now largely terrestrial as adults, developed specialized hindlimbs with webbed feet better suited for swimming than for movement on land. All amphibians use their porous skin as a primary or accessory breathing organ. This specialization was encouraged by swampy environments of the Carboniferous period but presented serious desiccation problems for life on land.

Modern Amphibians

The three living amphibian orders comprise more than 5400 species. Most share general adaptations for life on land, including skeletal strengthening and a shifting of special sense priorities from the ancestral lateral-line system to the senses of smell and hearing. Both the olfactory epithelium and the ear are restructured to improve sensitivities to airborne odors and sounds, respectively.

Nonetheless, most amphibians meet problems of independent life on land only halfway. In the ancestral life history of amphibians, eggs are aquatic and hatch to produce an aquatic larval form that uses gills for breathing. A metamorphosis follows in which gills are lost and lungs, which are present throughout larval life, are then activated for breathing air. Gaseous exchange across the skin also aids air breathing and is the primary means of respiration in many terrestrial salamanders. Many amphibians retain this general pattern but there are some important exceptions. Some salamanders lack a complete metamorphosis and retain a permanently aquatic, larval morphology throughout life. Some caecilians and other salamanders live entirely on land and lack the aquatic larval phase completely. Both alternatives are evolutionarily derived conditions. Some frogs also maintain a strictly terrestrial existence by eliminating the aquatic larval stage.

Even the most terrestrial amphibians remain dependent on very moist if not aquatic environments. Their skin is thin, and it requires moisture for protection against desiccation in air. An intact frog loses water nearly as rapidly as a skinless frog. Amphibians also require moderately cool environments. Being ectothermic, their body temperature is determined by and varies with their environment, greatly restricting where they can live. Cool and wet environments are especially important for reproduction. Eggs are not well protected from desiccation, and they must be shed directly into water or onto moist terrestrial surfaces. Completely terrestrial amphibians lay eggs under logs or rocks, in the moist forest floor, in flooded tree holes, in pockets on the mother's back, or in folds of the body wall. One species of Australian frog even broods its young in its stomach while interrupting digestion, and males of a South American species brood young in their vocal pouches (see figure 17.16, p. 346).

We now highlight special characteristics of the three major groups of amphibians. We expand coverage of some general amphibian features when discussing groups in which those features have been studied most extensively. For most features, this group is frogs.

Caecilians: Order Gymnophiona (Apoda)

The order Gymnophiona (jim'no-fy'o-na) (Gr. *gymnos,* naked, + *ophineos,* of a snake) contains approximately 160 species of elongate, limbless, burrowing creatures commonly called **caecilians** (figure 17.4). They occur in tropical forests of South America (their principal home), Africa, and Southeast Asia, and many species are aquatic. They possess a long, slender body, small scales in the skin of some, many vertebrae, long ribs, no limbs, and a terminal anus. Eyes are small, and most species are totally blind as adults. Their food consists mostly of worms and small invertebrates, which they find underground. Fertilization is internal, and the male is provided with a protrusible copulatory organ. The eggs are usually deposited in moist ground near water. In some species the eggs are carefully guarded in folds of the body during their development. Hatchling caecilians differ from adults in having a tail fin, an open gill slit, and external gills in some species. Viviparity also is common in

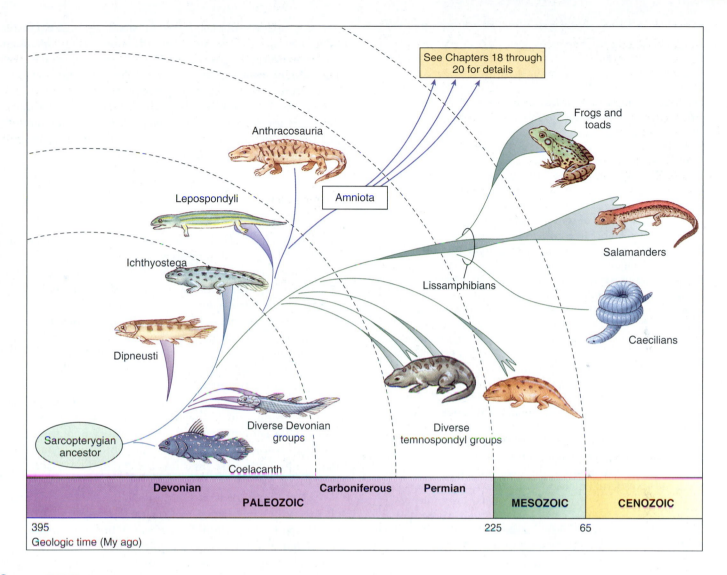

figure 17.3

Early tetrapod evolution and the descent of amphibians. The tetrapods share most recent common ancestry with the extinct Devonian groups (of living groups, the tetrapods are most closely related to the lungfishes). Amphibians share most recent common ancestry with the diverse temnospondyls of the Carboniferous and Permian periods of the Paleozoic, and Triassic period of the Mesozoic.

figure 17.4

Female caecilian coiled around eggs in her burrow.

Source: W. E. Duellmann and L. Trueb, *Biology of Amphibians,* 1986, McGraw-Hill Company, Inc., New York, NY.

some caecilians, with the embryos obtaining nourishment by eating the wall of the oviduct.

Salamanders: Order Urodela (Caudata)

Order Urodela (Gr. *oura,* tail + *delos,* evident) are tailed amphibians, approximately 500 species of salamanders. Salamanders occur in almost all northern temperate regions of the world, and they are abundant and diverse in North America. Salamanders are found also in tropical areas of Central America and northern South America. Salamanders are typically small; most of the common North American salamanders are less than 15 cm long. Some aquatic forms are considerably longer, and the Japanese giant salamander may exceed 1.5 m in length.

characteristics
of Modern Amphibians

1. Skeleton mostly bony, with varying numbers of vertebrae; ribs present in some, absent or fused to vertebrae in others
2. Body forms vary greatly from an elongated trunk with distinct head, neck, and tail to a compact, depressed body with fused head and trunk and no intervening neck
3. **Limbs usually four (tetrapod),** although some are limbless; forelimbs of some much smaller than hindlimbs, in others all limbs small and vestigial; webbed feet often present; no true nails or claws; **forelimb usually with four digits** but sometimes five and sometimes fewer
4. **Skin usually smooth and moist with many glands,** some of which may be poison glands; pigment cells (chromatophores) common, of considerable variety; no scales, except concealed dermal ones in some
5. Mouth usually large with small teeth in upper or both jaws; two nostrils open into anterior part of mouth cavity
6. Respiration by lungs (absent in some salamanders), skin, and gills in some, either separately or in combination; external gills in the larval form and may persist throughout life in some
7. **Heart with a sinus venosus,** two atria, one ventricle, a conus arteriosus, and a **double circulation through the heart;** skin abundantly supplied with blood vessels
8. Ectothermal
9. Excretory system of paired mesonephric or opisthonephric kidneys; urea main nitrogenous waste
10. Ten pairs of cranial nerves
11. Separate sexes; fertilization mostly internal in salamanders and caecilians, mostly external in frogs and toads; predominantly oviparous, some ovoviviparous or (rarely) viviparous; metamorphosis usually present; **moderately yolky eggs (mesolecithal) with jellylike membrane coverings**

Most salamanders have limbs set at right angles to their body, with forelimbs and hindlimbs of approximately equal size. In some aquatic and burrowing forms, limbs are rudimentary or absent.

Salamanders are carnivorous both as larvae and adults, preying on worms, small arthropods, and small molluscs. Most eat only moving animals. Like all amphibians, they are **ectothermic** and have a low metabolic rate.

Breeding Behavior

Some salamanders are either aquatic or terrestrial throughout their life cycle, but the ancestral life cycle is metamorphic, having aquatic larvae and terrestrial adults that live in moist places under stones and rotten logs. Eggs of most salamanders are fertilized internally; a female recovers in her vent a packet of sperm (**spermatophore**) deposited by a male on a leaf or stick (figure 17.5). Aquatic species lay their eggs in clusters or stringy masses in water. Their eggs hatch to produce an aquatic larva having external gills and a finlike tail. Completely terrestrial species deposit eggs in small, grapelike clusters under logs or in excavations in soft moist earth, and many species guard their eggs (figure 17.6). Terrestrial species have **direct development.** They bypass the larval stage and hatch as miniature versions of their parents. The most complex of salamander life cycles is observed in some American newts, whose aquatic larvae metamorphose to form terrestrial juveniles that later metamorphose again to produce secondarily aquatic, breeding adults (figure 17.7).

Respiration

At various stages of their life history, salamanders may have external gills, lungs, both, or neither of these structures. They also share the general amphibian condition of having extensive vascular nets in their skin that serve the respiratory exchange of oxygen and carbon dioxide. Salamanders that have an aquatic larval stage hatch with gills, but lose them later if a metamorphosis occurs. Several diverse lineages of salamanders have evolved permanently aquatic forms that fail to complete metamorphosis and retain their gills and finlike tail throughout life (figure 17.8). Lungs, the most widespread respiratory organ of terrestrial vertebrates, are present from birth in the salamanders that have them, and become the primary means of respiration following metamorphosis. Amphiumas, while having a completely aquatic life history, nonetheless lose their gills before adulthood and then breathe primarily by lungs. Amphiumas raise their nostrils above the surface of the water to get air.

Amphiumas provide an interesting contrast to many species of the large family Plethodontidae (figures 17.5, 17.6, and 17.9) that are entirely terrestrial but completely lack lungs. Efficiency of cutaneous respiration is increased by penetration of a capillary network into the epidermis or by thinning the epidermis over superficial dermal capillaries. Cutaneous respiration is supplemented by air pumped in and out of the mouth, where respiratory gases are exchanged across the vascularized membranes of the buccal (mouth) cavity (buccopharyngeal breathing). Lungless plethodontids probably originated in streams, where lungs would have been disadvantageous by providing too much buoyancy, and where water is so cool and well oxygenated that cutaneous respiration alone is sufficient for life. It is curious that the most completely terrestrial group of salamanders is one that lacks lungs.

Paedomorphosis

Whereas most salamanders complete their development by metamorphosis to the adult body form, some species reach sexual maturity while retaining their gills, aquatic lifestyle, and other larval characteristics. This condition illustrates **paedomorphosis** (Gr. "child form"), defined as the retention in adult descendants of features that were present only in preadult stages of their ancestors. Some characteristics of the ancestral adult morphology are consequently eliminated. Examples of such nonmetamorphic, permanently-gilled species are mud puppies of

figure 17.5

Courtship and sperm transfer in the pygmy salamander, *Desmognathus wrighti*. After judging the female's receptivity by the presence of her chin on his tail base, the male deposits a spermatophore on the ground, then moves forward a few paces. **A,** The white mass of the sperm atop a gelatinous base is visible at the level of the female's forelimb. The male moves ahead, the female following until the spermatophore is at the level of her vent. **B,** The female has recovered the sperm mass in her vent, while the male arches his tail, tilting the female upward and presumably facilitating recovery of the sperm mass. The female later uses sperm stored in her body to fertilize eggs internally before laying them.

figure 17.6

Female dusky salamander (*Desmognathus* sp.) attending eggs. Many salamanders care for their eggs, which includes rotating the eggs and protecting them from fungal infections and predation by various arthropods and other salamanders.

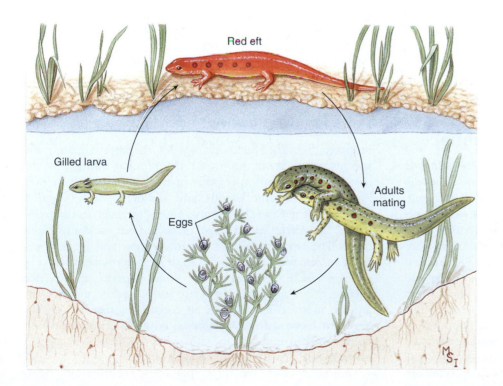

figure 17.7

Life history of the red-spotted newt, *Notophthalmus viridescens* of the family Salamandridae. In many habitats the aquatic larva metamorphoses into a brightly colored "red eft" stage, which remains on land from one to three years before transforming into a secondarily aquatic adult.

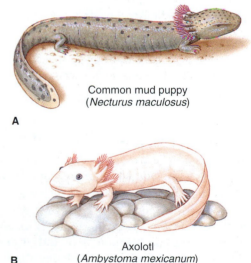

figure 17.8

Paedomorphosis in salamanders. **A,** The mud puppy *Necturus* sp. and **B,** the axolotl (*Ambystoma mexicanum*) are permanently gilled aquatic forms. Some other species of *Ambystoma* are facultatively paedomorphic; they may remain permanently gilled or, should their pond habitat evaporate, metamorphose to a terrestrial form that loses its gills and breathes by lungs. The axolotl shown is an albino form commonly used in laboratory experiments but uncommon in natural populations.

figure 17.9

Longtail salamander *Eurycea longicauda,* a common plethodontid salamander.

the genus *Necturus* (figure 17.8A), which live on bottoms of ponds and lakes; and the **axolotl** of Mexico (figure 17.8B). These species never metamorphose under any conditions.

There are other species of salamanders that reach sexual maturity with larval morphology but, unlike permanent larvae such as *Necturus,* may metamorphose to terrestrial forms under certain environmental conditions. We find good examples in *Ambystoma tigrinum* and related species from North America. Their typical habitat consists of small ponds that can disappear through evaporation in dry weather. When ponds evaporate the aquatic form metamorphoses to a terrestrial form, losing its gills and developing lungs. It then can travel across land in search of new sources of water in which to live and to reproduce.

Frogs and Toads: Order Anura (Salientia)

The more than 4840 species of frogs and toads that form the order Anura (Gr. *an,* without, + *oura,* tail) are for most people the most familiar amphibians. The Anura are an old group, known from the Jurassic period, 150 million years ago. Frogs and toads occupy a great variety of habitats, despite their aquatic mode of reproduction and water-permeable skin, which prevent them from wandering too far from sources of water, and their ectothermy, which bars them from polar and subarctic habitats. The name of the order, Anura, refers to an obvious group characteristic, the absence of tails in adults (although all pass through a tailed larval stage during development). Frogs and toads are specialized for jumping, as suggested by the alternative order name, Salientia, which means leaping.

We see in the appearance and life habit of their larvae further distinctions between the Anura and Caudata. Eggs of most frogs hatch into a tadpole ("polliwog"), having a long, finned tail, both internal and external gills, no legs, specialized mouthparts for herbivorous feeding (some tadpoles and all salamander larvae are carnivorous), and a highly specialized internal anatomy. They look and act altogether differently from adult frogs. Metamorphosis of a frog tadpole to an adult frog is thus a striking transformation. The permanently gilled larval condition never occurs in frogs and toads as it does in salamanders.

Frogs and toads are divided into 21 families. Best-known frog families in North America are Ranidae, which contains most of our familiar frogs (figure 17.10A), and Hylidae, the tree frogs (figure 17.10B). True toads, belonging to family Bufonidae, have short legs, stout bodies, and thick skins usually with prominent warts (figure 17.11). However, the term "toad" is used rather loosely to refer also to more or less terrestrial members of several other families.

The largest anuran is the West African *Conraua goliath,* which is more than 30 cm long from tip of nose to anus (figure 17.12). This giant eats animals as big as rats and ducks. The smallest frog recorded is *Phyllobates limbatus,* which is only approximately 1 cm long. This tiny frog, which can be more than covered by a dime, lives in Cuba. The largest American frog is the bullfrog, *Rana catesbeiana* (see figure 17.10A), which reaches a head and body length of 20 cm.

In addition to their importance in biomedical research and education, frogs have long served the epicurean frog-leg market. Some bullfrog species are in such heavy demand in Europe (especially France) and the United States—the worldwide harvest is an estimated 200 million bullfrogs (about 10,000 metric tons) annually—that their populations have fallen drastically from excessive exploitation. Most are Asian bullfrogs imported from India, Bangladesh, China, and Japan. With so many insect-eating frogs removed from the ecosystem, rice production is threatened from uncontrolled, flourishing insect populations. In the United States, attempts to raise bullfrogs in farms have not been successful, mainly because bullfrogs are voracious eating machines that normally require living prey, such as insects, crayfish, and other frogs.

Habitats and Distribution

Probably the most abundant frogs are the approximately 260 species of the genus *Rana* (Gr. frog), found throughout the temperate and tropical regions of the world except in New Zealand, the oceanic islands, and southern South America. They usually live near water, although some, such as the wood frog *R. sylvatica,* spend most of their time on damp forest floors. The larger bullfrogs, *R. catesbeiana,* and green frogs, *R. clamitans,* are nearly always in or near permanent water or swampy areas. Leopard frogs, *Rana pipiens* and related species, occur in nearly every state and Canadian province and are the most widespread of all North American frogs. The northern leopard frog, *R. pipiens,* is the species most commonly used in biology laboratories and for classical electrophysiological research.

Frog species are often patchy in distribution, being restricted to certain localities (for instance, to specific streams or pools) and absent or scarce in similar habitats elsewhere. Pickerel frogs (*R. palustris*) are especially noteworthy in this respect because they are abundant only in certain localized regions. Recent studies show that many populations of frogs worldwide may be suffering declines in numbers and becom-

A   **B**

figure 17.10

Two common North American frogs. **A**, Bullfrog, *Rana catesbeiana,* largest American frog and mainstay of the frog-leg epicurean market (family Ranidae). **B**, Green tree frog *Hyla cinerea,* a common inhabitant of swamps of the southeastern United States (family Hylidae). Note adhesive pads on the feet.

figure 17.11

American toad *Bufo americanus* (family Bufonidae). This principally nocturnal yet familiar amphibian feeds on large numbers of insect pests and on snails and earthworms. The warty skin contains numerous glands that produce a surprisingly poisonous milky fluid, providing the toad excellent protection from a variety of potential predators.

figure 17.12

Conraua (Gigantorana) goliath (family Ranidae) of West Africa, the world's largest frog. This specimen weighed 3.3 kg (approximately 7 1/2 pounds).

ing even more patchy than usual in their distributions. In most declining populations causes of decline are unknown.

Most larger frogs are solitary in their habits except during breeding season. During breeding periods most of them, especially males, are very noisy. Each male usually takes possession of a particular perch near water, where he may remain for hours or even days, trying to attract a female to that spot. At times frogs are mainly silent, and their presence is not detected until they are disturbed. When they enter the water, they dart swiftly to the bottom of the pool, where they kick the substrate to conceal themselves in a cloud of muddy water. In swimming, they hold the forelimbs near the body and kick backward with their webbed hindlimbs, which propel them forward. When they come to the surface to breathe, only the head and foreparts are exposed and, since they usually take advantage of any protective vegetation, they are difficult to see.

figure 17.13

African clawed frog, *Xenopus laevis.* The claws, an unusual feature in frogs, are on the hind feet. This frog has been introduced into California, where it is considered a serious pest.

figure 17.14

A male green frog, *Hyla cinerea,* clasps a larger female during the breeding season in a South Carolina swamp. Clasping (amplexus) is maintained until the female deposits her eggs, which are fertilized externally (outside the body). Like most tree frogs, these are capable of rapid and marked color changes; the male here, normally green, has darkened during amplexus.

Amphibian populations have declined in various parts of the world, whereas in other areas they thrive. No single explanation fits all declines, although loss of habitat predominates. In some populations, changes are simply random fluctuations caused by periodic droughts and other naturally occurring phenomena. Frog and toad eggs exposed on the surface of ponds are especially sensitive to the damaging action of ultraviolet radiation. Climatic changes that reduce water depth at oviposition sites increase ultraviolet exposure of embryos and make them more susceptible to fungal infection. Declines in population survival may be accompanied by an increased incidence of malformed individuals, such as frogs with extra limbs. Malformed limbs are often associated with infection by trematodes (p. 151).

Declines of some amphibian populations are caused by other amphibians. An exotic frog introduced into southern California thrives in its new American home. African clawed frogs, *Xenopus laevis* (figure 17.13), are voracious, aggressive, primarily aquatic frogs that rapidly displace native frogs and fish from waterways. This species was introduced into North America in the 1940s when it was used extensively in human pregnancy tests. Some hospitals dumped surplus frogs into nearby streams, where these prolific breeders became almost indestructible pests. Similar results occurred when giant toads, *Bufo marinus* (to 23 cm in length), were introduced to Queensland, Australia, and southern Florida to control agricultural pests. They rapidly spread, producing numerous ecological problems, including displacement of native anurans.

During winter months most frogs hibernate in soft mud under pools and streams. Their life processes reach a very low ebb during their hibernation period, and such energy as they need comes from glycogen and fat stored in their bodies during the spring and summer months. More terrestrial frogs, such as tree frogs, hibernate in humus of the forest floor. They tolerate low temperatures, and many actually survive prolonged freezing of all extracellular fluid, representing 35% of body water. Such frost-tolerant frogs prepare for winter by accumulating glucose and glycerol in body fluids, which protects tissues from ice-crystal formation.

Adult frogs have numerous enemies, such as snakes, aquatic birds, turtles, raccoons, and humans; fish prey upon tadpoles and only a few survive to maturity. Some defend themselves by feigning death. Most anurans can inflate their lungs so that they are difficult to swallow. When disturbed along the margin of a pond or brook, a frog often remains quite still. When it thinks it is detected it jumps, not always into the water where enemies may be lurking but into grassy cover on the bank. When held in the hand, a frog may cease its struggles for an instant to put its captor off guard and then leap violently, at the same time voiding its urine. A frog's best protection is its ability to leap and use of poison glands. Many frogs and toads in the tropics and subtropics are aggressive, jumping and biting at predators. Bullfrogs in captivity do not hesitate to snap at tormenters and can inflict painful bites.

Reproduction

Frogs and toads living in temperate regions breed, feed, and grow only during the warmer seasons of the year, usually in a predictable annual cycle. With warming spring temperatures, in combination with sufficient rainfall, males call vociferously to attract females. After a brief courtship, females enter the water and are clasped by the males in a process called **amplexus** (figure 17.14) in which eggs are fertilized externally (after leaving the female's body). As the female lays eggs, the male discharges seminal fluid containing sperm over the eggs to fertilize them. After fertilization, the jelly layers absorb water and swell (figure 17.15). Eggs are laid in large masses, often anchored to vegetation, then abandoned by the parents. The eggs begin development immediately.

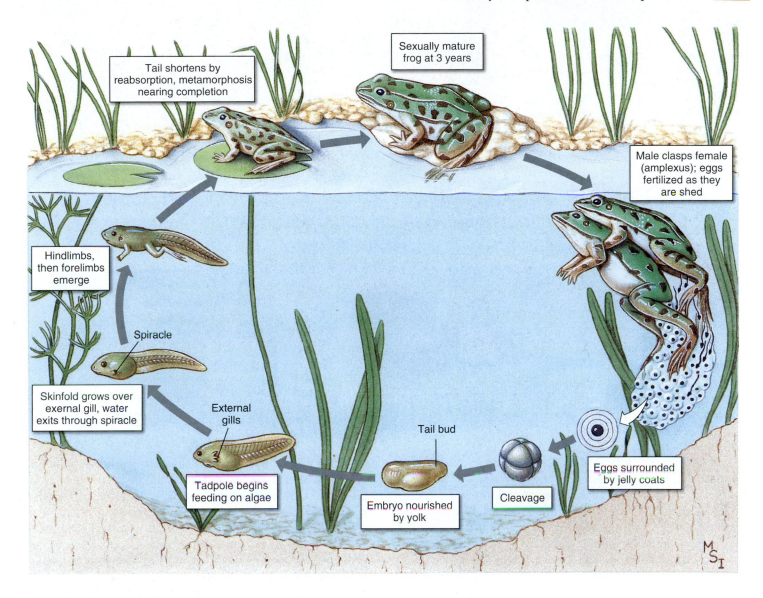

figure 17.15

Life cycle of a leopard frog.

classification of Class Amphibia

Order Gymnophiona (jim′no-fy′o-na) (Gr. *gymnos,* naked, + *ophioneos,* of a snake) **(Apoda): caecilians.** Body elongate; limbs and limb girdle absent; mesodermal scales present in skin of some; tail short or absent; 95 to 285 vertebrae; pantropical, 5 families, 33 genera, approximately 160 species

Order Urodela (yur′uh-dēl′uh) (Gr. *oura,* tail, + *delos,* evident) **(Caudata): salamanders.** Body with head, trunk, and tail; no scales; usually two pairs of equal limbs; 10 to 60 vertebrae;

predominantly Holarctic; 10 living families, 61 genera, approximately 500 species.

Order Anura (uh-nur′uh) (Gr. *an,* without, + *oura,* tail) **(Salientia): frogs, toads.** Head and trunk fused; no tail; no scales; two pairs of limbs; large mouth; lungs; 6 to 10 vertebrae including urostyle (coccyx); cosmopolitan, predominantly tropical; 29 families; 352 genera; approximately 4840 species.

figure 17.16

Unusual reproductive strategies of anurans. **A,** Female South American pygmy marsupial frog *Flectonotus pygmaeus* carries developing larvae in a dorsal pouch. **B,** Female Surinam frog carries eggs embedded in specialized brooding pouches on the dorsum; froglets emerge and swim away when development is complete. **C,** Male poison-dart frog *Phyllobates bicolor* carries tadpoles adhering to its back. **D,** Tadpoles of a male Darwin's frog *Rhinoderma darwinii* develop into froglets in its vocal pouch. When ready to emerge, a froglet crawls into the parent's mouth, which the parent opens to allow the froglet's escape.

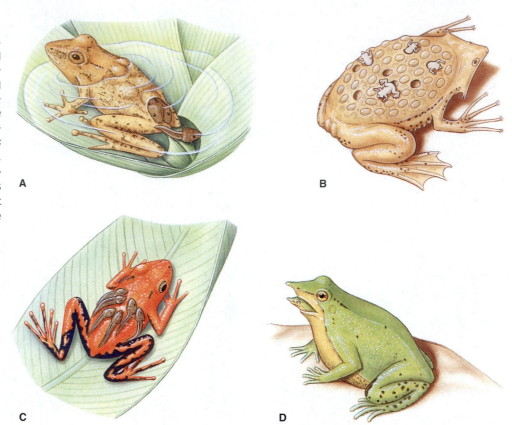

A   B   C   D

Within a few days the embryos have developed into tadpoles (figure 17.15) and hatch, often to face a precarious existence if the eggs were laid in a temporary pond or puddle. In such instances, tadpoles race against time to complete development before the habitat dries.

At hatching, a tadpole has a distinct head and body with a compressed tail. The mouth is on the ventral side of the head and has keratinized jaws for scraping vegetation from objects for food. Behind the mouth is a ventral adhesive disc for clinging to objects. In front of the mouth are two deep pits, which later become nostrils. Swellings occur on each side of the head, and these later become external gills. There are three pairs of external gills, which later transform into internal gills that become covered with a flap of skin (the operculum) on each side. On the right side the operculum completely fuses with the body wall, but on the left side a small opening, the **spiracle** (L. *spiraculum,* airhole) permits water to exit after entering the mouth and passing the internal gills. The hindlegs appear first, whereas the forelimbs are hidden for a time by folds of the operculum. The tail is absorbed, the intestine becomes much shorter, the mouth undergoes a transformation into the adult condition, lungs develop, and the gills are absorbed. Leopard frogs usually complete metamorphosis within three months; a bullfrog takes much longer to complete this process.

The reproductive mode just described, is typical of most temperate zone anurans, but only one of many reproductive patterns in tropical anurans. Some remarkable reproductive strategies of tropical anurans are illustrated in figure 17.16. Some species lay their eggs in foam masses that float on the surface of the water; some deposit their eggs on leaves overhanging ponds and streams into which the emerging tadpoles will drop; some lay their eggs in damp burrows; and others place their eggs in water trapped in tree cavities or in water-filled chambers of some bromeliads (epiphytic plants in the tropical forest canopy). While most frogs abandon their eggs, some, such as the tropical dendrobatids (a family that includes the poison-dart frogs), tend their eggs. When the tadpoles hatch, they squirm onto the parent's back to be carried for varying lengths of time (figure 17.16C). Marsupial frogs carry their developing eggs in a pouch on the back (figure 17.16A).

Although most frogs develop through a larval stage (the tadpole), many tropical frogs have evolved direct development. In direct development the tadpole stage is bypassed and the froglet that emerges is a miniature replica of the adult.

Summary

Amphibians are ectothermic, primitively quadrupedal vertebrates that have glandular skin and that breathe by lungs, gills, or skin. They form one of two major phylogenetic branches of living tetrapods, the other one being the amniotes. Living amphibians form three major evolutionary groups. Caecilians (order Gymnophiona) are a small tropical group of limbless, elongate forms. Salamanders (order Caudata) are tailed amphibians that have retained the generalized four-legged body plan of their Paleozoic ancestors. Frogs and toads (order Anura) are the largest group of modern amphibians; all are specialized for a jumping mode of locomotion.

Most amphibians have a biphasic life cycle that begins with an aquatic larva followed by metamorphosis to produce a terrestrial adult that returns to water to lay eggs. Some frogs, salamanders, and caecilians have evolved direct development that omits the aquatic larval stage and some caecilians have evolved viviparity. Salamanders are unique among amphibians in having evolved several permanently gilled species that retain a larval morphology throughout life, eliminating the terrestrial phase completely. The permanently gilled condition is obligate in some species, but others metamorphose to a terrestrial form only if the pond habitat dries.

Despite adaptations for terrestrial life, the adults and eggs of all amphibians require cool, moist environments if not actual pools or streams. Eggs and adult skin have no effective protection against very cold, hot, or dry conditions, greatly restricting the adaptive radiation of amphibians to environments with moderate temperatures and abundant water.

Review Questions

1. Compared with an aquatic habitat, a terrestrial habitat offers both advantages and problems for an animal making a transition from water to land. Summarize how these differences might have influenced the early evolution of tetrapods.
2. Describe the different modes of respiration used by amphibians. What paradox do the amphiumas and terrestrial plethodontids present regarding the association of lungs with life on land?
3. Evolution of the tetrapod limb was one of the most important advances in vertebrate history. Describe the inferred sequence of its evolution.
4. Compare the general life history patterns of salamanders with those of frogs. Which group shows a greater variety of evolutionary changes of the ancestral biphasic amphibian life cycle?
5. Give the literal meaning of the name Gymnophiona. What animals are in this amphibian order, what is their appearance, and where do they live?
6. What are the literal meanings of the order names Urodela and Anura? What major features distinguish members of these two orders from each other?
7. Describe the breeding behavior of a typical woodland salamander.
8. How is paedomorphosis important to evolutionary diversification of salamanders?
9. Briefly describe the reproductive behavior of frogs. In what important ways do frogs and salamanders differ in their reproduction?

Selected References

See also general references on page 415.

Clack, J. A. 2002. Gaining ground: the origin and evolution of tetrapods. Bloomington, Indiana, Indiana University Press. *An authoritative account of paleontological evidence regarding tetrapod origins.*

Conant, R., and J. T. Collins. 1998. A field guide to reptiles and amphibians: Eastern and Central North America. The Peterson field guide series. Boston, Houghton Mifflin Company. *Updated version of a popular field guide; color illustrations and distribution maps for all species.*

Duellman, W. E., and L. R. Trueb. 1994. Biology of amphibians. Baltimore, Johns Hopkins University Press. *Important comprehensive sourcebook of information on amphibians, extensively referenced and illustrated.*

Halliday, T. R., and K. Adler, eds. 2002. Firefly encyclopedia of reptiles and amphibians. Toronto, Canada, Firefly Books. *Excellent authoritative reference work with high-quality illustrations.*

Jamieson, B. G. M. (ed.). 2003. Reproductive biology and phylogeny of Anura. Enfield, New Hampshire, Science Publishers, Inc. *Provides detailed coverage of reproductive biology and early evolutionary diversification of frogs and toads.*

Lannoo, M. (ed.) 2005. Amphibian declines: the conservation status of United States species. Berkeley, California, University of California Press. *A survey of conservation status of American amphibians.*

Lewis, S. 1989. Cane toads: an unnatural history. New York, Dolphin/Doubleday. *Based on an amusing and informative film of the same title that describes the introduction of cane toads to Queensland, Australia and the unexpected consequences of their population explosion there. "If Monty Python teamed up with National Geographic, the result would be Cane Toads."*

Savage, J. M. 2002. The amphibians and reptiles of Costa Rica. Chicago, University of Chicago Press. *Costa Rica hosts diverse caecilians, frogs and salamanders. The Organization for Tropical Studies offers students a chance to study this amphibian fauna.*

Sever, D. M. (ed.). 2003. Reproductive biology and phylogeny of Urodela (Amphibia). Enfield, New Hampshire, Science Publishers, Inc. *A thorough review of reproductive biology and evolutionary relationships among salamanders.*

Stebbins, R. C., and N. W. Cohen. 1995. A natural history of amphibians. Princeton, New Jersey, Princeton University Press. *World-wide treatment of amphibian biology, emphasizing physiological adaptations, ecology, reproduction, behavior, and a concluding chapter on amphibian declines.*

Custom Website

The *Animal Diversity* Online Learning Center is a great place to check your understanding of chapter material. Visit www.mhhe.com/hickmanad4e for access to key terms, quizzes, and more! Further enhance your knowledge with web links to chapter-related material.

Explore live links for these topics:

Classification and Phylogeny of Animals
Class Amphibia

Order Gymnophiona
Order Caudata
Order Salienta
Dissection Guides for Amphibians
Conservation Issues Concerning Amphibians

Amniote Origins and Reptilian Groups

Hatching Komodo lizard, *Varanus komodoensis*.

Enclosing the Pond

Amphibians, with well-developed limbs, redesigned sensory and respiratory systems, and modifications of the postcranial skeleton for supporting the body in air, have made a notable conquest of land. But, with shell-less eggs and often gill-breathing larvae, their development remains hazardously tied to water. The lineage ancestral to reptile groups, birds, and mammals developed an egg that could be laid on land. This shelled egg, perhaps more than any other adaptation, unshackled the early reptiles from the aquatic environment by freeing the developmental process from dependence upon aquatic or very moist terrestrial environments. In fact, the "pond-dwelling" stages were not eliminated but enclosed within a series of extraembryonic membranes that provided complete support for embryonic development. One membrane, the amnion, encloses a fluid-filled cavity, the "pond," within which the developing embryo floats. Another membranous sac, the allantois, serves both as a respiratory surface and as a chamber for the storage of nitrogenous wastes. Enclosing these membranes is a third membrane, the chorion, through which oxygen and carbon dioxide freely pass. Finally, surrounding and protecting everything is a porous, parchmentlike or leathery shell.

With the last ties to aquatic reproduction severed, conquest of land by vertebrates was ensured. The Paleozoic tetrapods that developed this reproductive pattern were ancestors of a single, monophyletic assemblage called Amniota, named after the innermost of the three extraembryonic membranes, the amnion. Before the end of the Paleozoic era the amniotes had diverged into multiple lineages that gave rise to all reptilian groups, birds, and mammals.

embers of the paraphyletic class Reptilia (rep-til′e-a) (L. *repto,* to creep) include the first truly terrestrial vertebrates. With nearly 8000 species (approximately 340 species in the United States and Canada) occupying a great variety of aquatic and terrestrial habitats, they are diverse and abundant. Nevertheless, reptiles are perhaps remembered best for what they once were, rather than for what they are now. The Age of Reptiles, which lasted for more than 165 million years, saw the appearance of a great radiation of reptilian lineages into a bewildering array of terrestrial and aquatic forms. Among these were herbivorous and carnivorous dinosaurs, many of huge stature and awesome appearance, that dominated animal life on land—the ruling reptiles. Then, during a mass extinction at the end of the Mesozoic era, they suddenly declined. Among the few reptilian groups to emerge from the Mesozoic extinction are today's reptiles. One these groups includes the two species of tuataras (*Sphenodon*) of New Zealand, sole survivors of a group that otherwise disappeared 100 million years ago. But others, especially lizards and snakes, have radiated since the Mesozoic extinction into diverse and abundant groups (figure 18.1). Understanding the 300-million-year-old history of reptile life on earth has been complicated by widespread convergent and parallel evolution among many lineages and by large gaps in the fossil record.

Origin and Adaptive Radiation of Reptilian Groups

As mentioned in the prologue to this chapter, amniotes are a monophyletic group that originated in the late Paleozoic. Most paleontologists agree that amniotes arose from a group of amphibian-like tetrapods, the anthracosaurs, during the early Carboniferous period of the Paleozoic. By the late Carboniferous (approximately 300 million years ago), amniotes had separated into three major groups. The first group, the **anapsids** (Gr. *an,* without, + *apsis,* arch), characterized by a skull having no temporal opening behind the orbits (eye sockets), the skull behind the orbits being completely roofed with dermal bone (figure 18.2). This group is represented today only by turtles. Their morphology is an odd mix of ancestral and derived characters that has scarcely changed at all since turtles first appeared in the fossil record in the Triassic some 200 million years ago.

The second group, the **diapsids** (Gr. *di,* double, + *apsis,* arch), gave rise to all other reptilian groups and to birds (figure 18.1). The diapsid skull has two temporal openings: one pair located low on the cheeks, and a second pair positioned above the lower pair and separated from them by a bony arch (figure 18.2). These openings allow for additional jaw muscle attachment. Recesses in the temporal region of the anapsid turtle skull provide a similar function. Four subgroups of diapsids appeared. The **lepidosaurs** include all modern reptiles except turtles and crocodilians. The more derived **archosaurs** comprised dinosaurs and their relatives, and living crocodilians and birds. A third, smaller subgroup of diap-

sids, the **sauropterygians** includes several extinct aquatic groups, most conspicuous of which were the large, long-necked plesiosaurs. **Ichthyosaurs,** represented by extinct aquatic dolphinlike forms (figure 18.1), comprise the fourth subgroup.

The third group is the **synapsids** (Gr. *syn,* together, + *apsis,* arch), the mammals and extinct forms traditionally termed mammal-like reptiles. The synapsid skull has a single pair of temporal openings located low on the cheeks and bordered by a bony arch (figure 18.2). Synapsids were the first amniote group to diversify, giving rise first to pelycosaurs, later to therapsids, and finally to mammals (figure 18.1).

Changes in Traditional Classification of Reptilian Groups

With increasing use of cladistic methodology in zoology, and its insistence on hierarchical arrangement of monophyletic groups (see p. 82), important changes have been made in the traditional classification of reptiles. Class Reptilia is no longer recognized by cladists as a valid taxon because it is not monophyletic. As customarily defined, class Reptilia excludes birds

---

characteristics
of Class Reptilia

1. Body varied in shape, compact in some, elongated in others; **body covered with keratinized epidermal scales** with the addition sometimes of bony dermal plates; **integument with few glands**
2. **Two paired limbs, usually with five toes,** and adapted for climbing, running, or paddling; limbs absent or vestigial in snakes and some lizards and amphisbaenians
3. Skeleton well ossified; ribs with sternum (sternum absent in snakes) forming a complete thoracic basket; **skull with one occipital condyle** (rounded bony projection at the base of the skull)
4. Respiration by lungs; **no gills;** cloaca, pharynx, or skin used for respiration by some
5. Circulatory system functionally divided into **pulmonary and systemic circuits;** heart typically consisting of a sinus venosus, an atrium completely divided into two chambers, and a ventricle incompletely divided into three chambers; crocodilians with a sinus venosus, two atria, and two ventricles
6. **Ectothermic;** many thermoregulate behaviorally
7. **Metanephric kidney (paired); uric acid main nitrogenous waste**
8. Nervous system with the optic lobes on dorsal side of brain; **12 pairs of cranial nerves** in addition to nervus terminalis; enlarged cerebrum
9. Sexes separate; **fertilization internal**
10. **Eggs covered with calcareous or leathery shells; extraembryonic membranes (amnion, chorion,** and **allantois)** present during embryonic life; **no aquatic larval stages**

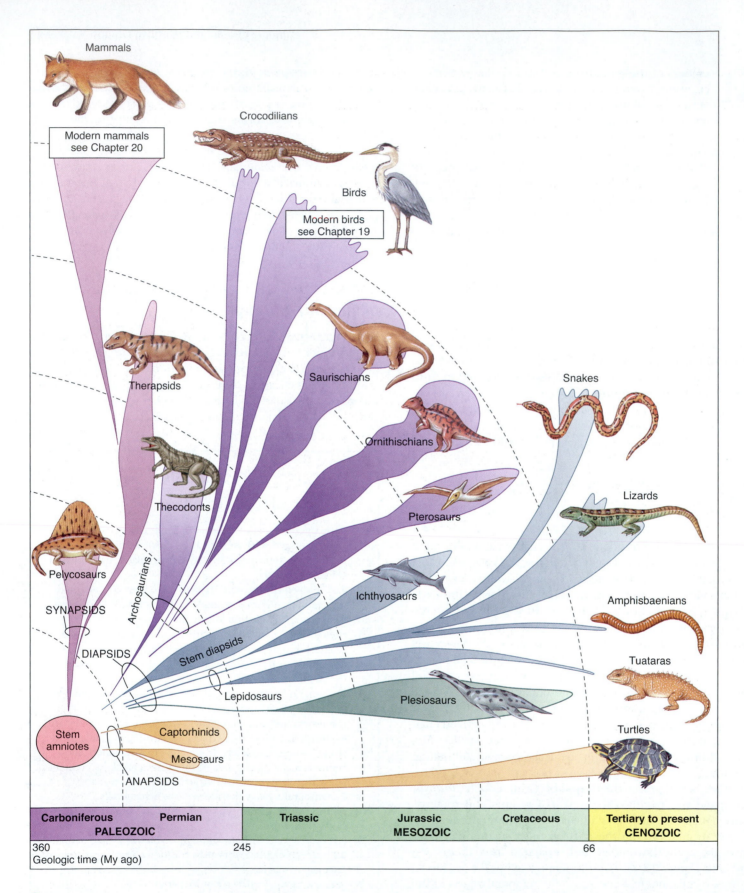

figure 18.1

Evolution of the amniotes. The evolutionary origin of amniotes occurred by the evolution of an amniotic egg that made reproduction on land possible, although this egg may well have developed before the earliest amniotes had ventured far on land. The amniote assemblage, which includes the reptiles, birds, and mammals, evolved from a lineage of small, lizardlike forms that retained the skull pattern of early tetrapods. First to diverge from the primitive stock was a lineage that evolved a skull pattern termed the synapsid condition. All other amniotes, including birds and all living reptiles except turtles, have a skull pattern known as diapsid. Turtles retained the primitive, anapsid skull pattern.

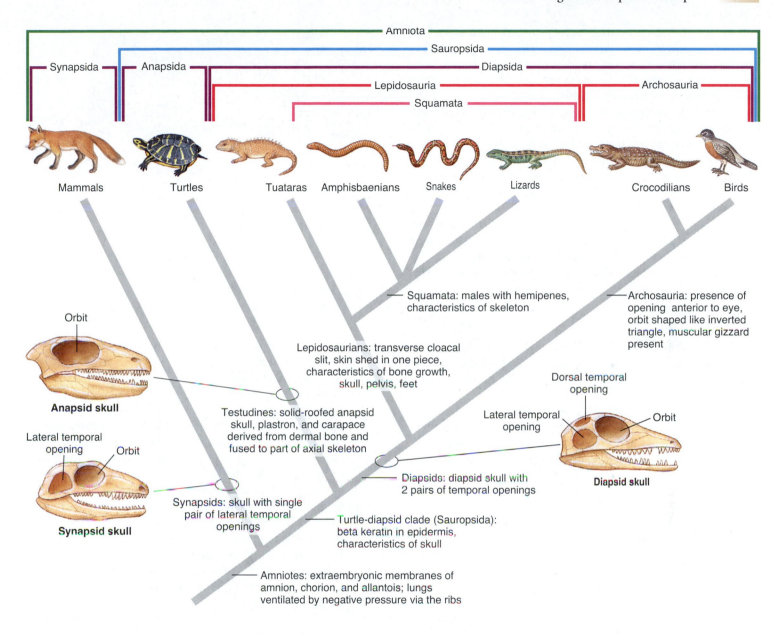

Squamata: males with hemipenes, characteristics of skeleton

Archosauria: presence of opening anterior to eye, orbit shaped like inverted triangle, muscular gizzard present

Lepidosaurians: transverse cloacal slit, skin shed in one piece, characteristics of bone growth, skull, pelvis, feet

Testudines: solid-roofed anapsid skull, plastron, and carapace derived from dermal bone and fused to part of axial skeleton

Anapsid skull

Orbit

Lateral temporal opening

Orbit

Synapsid skull

Dorsal temporal opening

Lateral temporal opening

Orbit

Diapsid skull

Synapsids: skull with single pair of lateral temporal openings

Diapsids: diapsid skull with 2 pairs of temporal openings

Turtle-diapsid clade (Sauropsida): beta keratin in epidermis, characteristics of skull

Amniotes: extraembryonic membranes of amnion, chorion, and allantois; lungs ventilated by negative pressure via the ribs

figure 18.2

Cladogram of the living Amniota showing monophyletic groups. Some shared derived characters (synapomorphies) of the groups are given. The skulls represent the ancestral condition of the three groups. Skulls of modern diapsids and synapsids are often highly modified by loss or fusion of skull bones, which obscures the ancestral condition. Representative skulls for anapsids are *Nyctiphruetus* of the upper Permian; for diapsids, *Youngina* of the upper Permian; for synapsids, *Aerosaurus,* a pelycosaur of the lower Permian. The relationships expressed in this cladogram are tentative and controversial. Relationships among lizards, snakes, and amphisbaenians are uncertain, but evidence suggests that both amphisbaenians and snakes evolved within the lizard clade.

(Source: F. H. Pough, J. B. Heiser, and W. N. McFarland, 1996, Vertebrate life, ed. 4. Upper Saddle River, NJ, Prentice Hall.)

which descended from the most recent common ancestor of reptiles. Consequently, reptiles are a **paraphyletic** group because they do not include all descendants of their most recent common ancestor (figure 18.1). Reptiles can be identified only as amniotes that are not birds.

An example of this problem is the shared ancestry of birds and crocodilians. Based solely on shared derived characteristics, crocodilians and birds are sister groups; they are more recently descended from a common ancestor than either is

from any other living reptilian lineage. In other words, birds and crocodilians belong to a monophyletic group apart from other living reptiles and, according to the rules of cladistics, should be assigned to a clade that separates them from the remaining reptiles. This clade is in fact recognized; it is the Archosauria (figures 18.1 and 18.2), a grouping that also includes the extinct dinosaurs. Therefore clade Reptilia includes birds and crocodilians, along with their sister group, the lepidosaurs (tuataras, lizards, snakes, and amphisbaenians), and turtles. The term "Reptilia" is thereby redefined to include birds in contrast to its traditional usage. However, evolutionary taxonomists argue that birds represent a novel adaptive zone and grade of organization whereas crocodilians remain within the traditionally recognized reptilian adaptive zone and grade. In this view, the morphological and ecological novelty of birds has been recognized by maintaining the traditional classification that places crocodilians in class Reptilia and birds in class Aves. Such conflicts of opinion between proponents of the two major competing schools of taxonomy (cladistics and evolutionary taxonomy) have had the healthy effect of forcing zoologists to reevaluate their views of amniote genealogy and how vertebrate classifications should represent genealogy and degree of divergence. In our treatment, "reptilian group," "reptile," and "reptilian" refer to members of four living monophyletic groups (turtles, crocodilians, squamates, tuataras) that are combined into the paraphyletic class Reptilia.

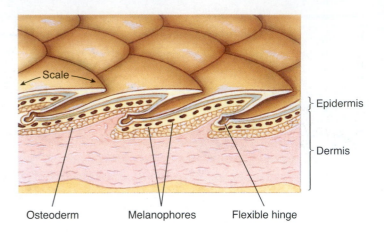

figure 18.3

Section of the skin of a reptile showing overlapping epidermal scales and bony osteoderms in the dermis.

Characteristics of Reptiles That Distinguish Them from Amphibians

1. **Reptiles have tough, dry, scaly skin offering protection against desiccation and physical injury.** The skin consists of a thin epidermis, shed periodically, and a much thicker, well-developed dermis (figure 18.3). The dermis is provided with **chromatophores,** the color-bearing cells that give many lizards and snakes their colorful hues. This layer, unfortunately for their bearers, is converted into alligator and snakeskin leather, so esteemed for expensive pocketbooks and shoes. Resistance to dessication is provided by hydrophobic lipids in the epidermis. The characteristic **scales** of reptiles, formed largely of beta keratin, provide protection against wear in terrestrial environments. Scales are derived mostly from the epidermis and thus are not homologous to fish scales, which are bony, dermal structures. In some reptiles, such as alligators, the scales remain throughout life, growing gradually to replace wear. In others, such as snakes and lizards, new scales grow beneath the old, which are shed at intervals. Turtles add new layers of keratin under the old layers of the platelike scutes, which are modified scales. In snakes the old skin (epidermis and scales) is turned inside out when discarded; lizards split out of the old skin leaving it mostly intact and right side out, or it may slough off in pieces.

2. **The amniotic egg of reptiles permits rapid development of large young in relatively dry environments.** Two membranes of amniotes, the **chorion** and **allantois** (figure 18.4), assist in exchange of oxygen and carbon dioxide with the environment, permitting relatively rapid development of large young. The **amnion** and shell provide support for the growing embryo and reduce the amount of water lost to the external environment. Although reptile eggs must still be maintained at relatively high humidity to avoid desiccation, they can be laid in drier environments than the eggs of even the most terrestrial amphibians.

3. **Reptilian jaws are efficiently designed for applying crushing or gripping force to prey.** The jaws of fish and amphibians are designed for quick closure, but once the prey is seized, little static force can be applied. In reptiles jaw muscles became larger, longer, and arranged for much better mechanical advantage.

4. **Reptiles have some form of copulatory organ, permitting internal fertilization.** Internal fertilization is obviously a requirement for a shelled egg, because sperm must reach the egg before the egg is enclosed. Sperm from paired testes are carried by the vasa deferentia to a copulatory organ, a penis or hemipenes, which is an evagination of the cloacal wall. Glandular walls of the female oviducts secrete albumin (source of amino acids, minerals, and water for the embryo) and shells for the large eggs.

5. **Reptiles have an efficient and flexible circulatory system and higher blood pressure than amphibians.** In all reptiles the right atrium, which receives unoxygenated blood from the body, is completely partitioned from the left atrium, which receives oxygenated blood from the lungs. Crocodilians have two completely separated ventricles as well (figure 18.5); in other reptiles the ventricle is incompletely separated. Even in reptiles with incomplete separation of ventricles, flow patterns within the heart prevent admixture of pulmonary (oxygenated) and systemic (unoxygenated) blood; all reptiles therefore have two functionally separate circulations. Incomplete separation between the right and left sides of the heart have the added benefit of permitting blood to bypass the lungs when pulmonary respiration is not occurring (for example, during diving or aestivation).

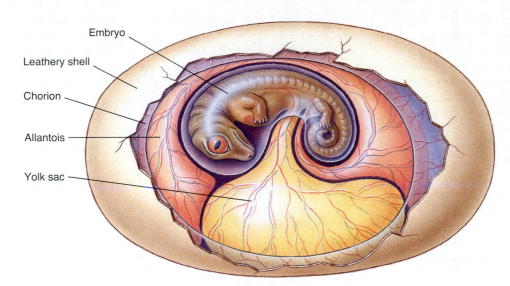

Embryo

Leathery shell

Chorion

Allantois

Yolk sac

figure 18.4

Amniotic egg. The embryo develops within the amnion and is cushioned by amniotic fluid. Food is provided by yolk from the yolk sac and metabolic wastes are deposited within the allantois. As development proceeds, the allantois fuses with the chorion, a membrane lying against the inner surface of the shell; both membranes are supplied with blood vessels that assist in the exchange of oxygen and carbon dioxide across the porous shell. Because this kind of egg is an enclosed, self-contained system, it is often called a "cleidoic" egg (Gr. *kleidoun,* to lock in).

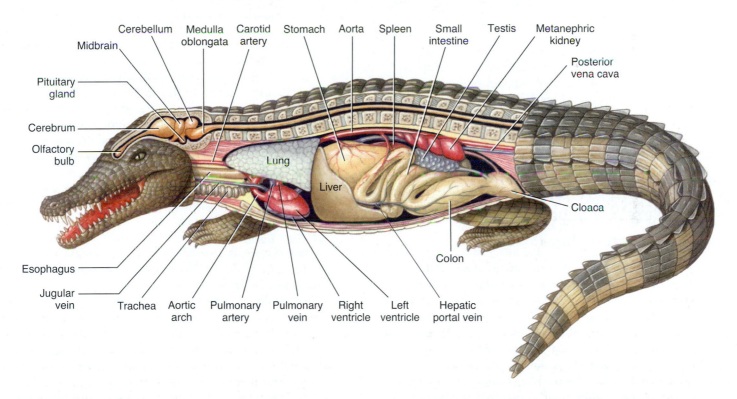

Cerebellum   Medulla   Carotid   Stomach   Aorta   Spleen   Small   Testis   Metanephric
Midbrain   oblongata artery   intestine   kidney

Pituitary gland

Cerebrum

Olfactory bulb

Posterior vena cava

Lung

Liver

Cloaca

Esophagus

Colon

Jugular vein   Trachea   Aortic   Pulmonary   Pulmonary   Right   Left   Hepatic
                         arch   artery   vein   ventricle ventricle portal vein

figure 18.5

Internal structure of a male crocodile.

6. **Reptilian lungs are better developed than those of amphibians.** Reptiles depend almost exclusively on lungs for gas exchange, supplemented by the pharynx, cloaca, or skin in some aquatic reptiles. Unlike amphibians, which *force* air into lungs with mouth muscles, reptiles *suck* air into lungs by enlarging the thoracic cavity, either by expanding the rib cage (snakes and lizards) or by movement of internal organs (turtles and crocodilians). Reptiles have no muscular diaphragm, a structure found only in mammals. Cutaneous respiration

(gas exchange across the skin), so important to amphibians, nearly has been abandoned by most reptiles.

7. **Reptiles have evolved efficient strategies for water conservation.** All amniotes have a metanephric kidney which is drained by its own passageway, the ureter. However, nephrons of the reptilian metanephros lack the specialized intermediate section of the tubule, the loop of Henle, that enables the kidney to concentrate solutes in urine. To remove salts from blood, many reptiles have located near the nose or eyes (in the tongue of saltwater

crocodiles) salt glands which secrete a salty fluid that is strongly hyperosmotic to the body fluids. Nitrogenous wastes are excreted by the kidney as uric acid, rather than urea or ammonia. Uric acid has a low solubility and precipitates out of solution readily, allowing water to be conserved; the urine of many reptiles is a semisolid suspension.

8. **All reptiles, except limbless members, have better body support than amphibians and more efficiently designed limbs for travel on land.** Nevertheless, most modern reptiles walk with their legs splayed outward and their belly close to the ground. Most dinosaurs, however, (and some modern lizards) walked on upright legs held beneath the body, the best arrangement for rapid movement and for support of body weight. Many dinosaurs walked on powerful hindlimbs alone.

9. **Reptilian nervous systems are considerably more complex than amphibian systems.** Although a reptile's brain is small, the cerebrum is larger relative to the rest of the brain. With the exception of hearing, sense organs in general are well developed. Jacobson's organ, a specialized olfactory chamber present in many tetrapods, is highly developed in lizards and snakes.

Characteristics and Natural History of Reptilian Orders

Anapsid Reptiles: Subclass Anapsida

Turtles: Order Testudines

Turtles descend from one of the earliest anapsid lineages, probably a group known as procolophonids of the late Permian, but turtles themselves do not appear in the fossil record until the Upper Triassic, some 200 million years ago. From the Triassic, turtles survived to the present with very little change to their early morphology. They are enclosed in shells consisting of a dorsal **carapace** (Fr., from Sp. *carapacho,* covering) and ventral **plastron** (Fr., breastplate). Clumsy and unlikely as they appear to be within their protective shells, they are nonetheless a varied and ecologically diverse group that seems able to adjust to human presence. The shell is so much a part of the animal that it is fused to thoracic vertebrae and ribs (figure 18.6). Like a medieval coat of armor, the shell offers protection for the head and appendages, which, in most turtles, can be retracted into it. Because the ribs are fused to the shell, the turtle cannot expand its chest to breathe. Instead, turtles employ certain abdominal and pectoral muscles as a "diaphragm." Air is drawn inward by contracting limb flank muscles to make the body cavity larger. Exhalation is also active: the shoulder girdle is drawn back into the shell, thus compressing the viscera and forcing air out of the lungs.

The terms "turtle," "tortoise," and "terrapin" are applied variously to different members of the turtle order. In North American usage, they are all correctly called turtles. The term "tortoise" is frequently given to land turtles, especially large forms. British usage of the terms is different: "tortoise" is the inclusive term, whereas "turtle" is applied only to the aquatic members.

Lacking teeth, turtle jaws are provided with tough (keratinized) plates for gripping food (figure 18.7). Sound perception is poor in turtles, and most turtles are mute (the biblical "voice of the turtle" refers to the turtledove, a bird). Compensating for poor hearing is a good sense of smell and color vision. Turtles are oviparous, and fertilization is internal. All turtles, even marine forms, bury their shelled, amniotic eggs in the ground. An odd feature of turtle reproduction is that in some turtle families, as in all crocodilians and some lizards, nest temperature determines sex of the hatchlings. In turtles, low temperatures during incubation produce males and high temperatures produce females.

Marine turtles, buoyed by their aquatic environment, can reach great size. Leatherbacks are largest, attaining a length of 2 m and weight of 725 kg. Green turtles, so named because of their greenish body fat, may exceed 360 kg, although most individuals of this economically valuable and heavily exploited species seldom live long enough to reach anything approaching this size. Some land tortoises weigh several hundred kilograms, such as the giant tortoises of the Galápagos Islands (figure 18.8) that so intrigued Darwin during his visit there in 1835. Most tortoises are rather slow moving; one hour of determined trudging carries a large Galápagos tortoise approximately 300 m. A low metabolism may explain the longevity of turtles, for some are believed to live more than 150 years.

figure 18.6

Skeleton and shell of a turtle, showing fusion of vertebrae and ribs with the carapace. The long and flexible neck allows the turtle to withdraw its head into its shell for protection.

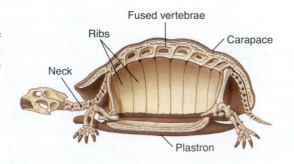

figure 18.7

Snapping turtle, *Chelydra serpentina,* showing the absence of teeth. Instead, the jaw edges are covered with a keratinized plate.

figure 18.8

Mating Galápagos tortoises, *Geochelone elaphantopus.* Males have a concave plastron that fits over the highly convex carapace of females, helping to provide stability during mating. Males utter a roaring sound during mating, the only time they are known to emit vocalizations.

Diapsid Reptiles: Subclass Diapsida

Diapsid reptiles, those having a skull with two pairs of temporal openings (see figure 18.2), are classified into three subgroups (superorders; see the Classification of Amniotes and Living Reptiles p. 362). Superorders with living representatives are Lepidosauria, containing lizards, snakes, worm lizards,

and *Sphenodon;* and Archosauria, containing crocodilians (and birds in cladistic taxonomy).

Lizards, Snakes, and Worm Lizards: Order Squamata

Squamates are the most recent and diverse products of diapsid evolution, comprising approximately 95% of all known living reptiles. Lizards appeared in the fossil record as early as the Jurassic, but they did not begin their radiation until the Cretaceous period of the Mesozoic when the dinosaurs were at the climax of their radiation. Snakes appeared during the late Jurassic period, probably from a group of lizards whose descendants include the Gila monster and monitor lizards. Two specializations in particular characterize snakes: extreme elongation of the body with accompanying displacement and rearrangement of internal organs; and specializations for eating large prey. Amphisbaenians (worm lizards), also likely evolved from a lizard group, and have structural specializations associated with a burrowing habit.

The diapsid skulls of squamates are modified from the ancestral diapsid condition by loss of dermal bone ventral and posterior to the lower temporal opening. This modification has allowed the evolution in most lizards and snakes of a **kinetic skull** having movable joints (figure 18.9). These joints, at the posterior of the quadrate, pterygoid, palate, and

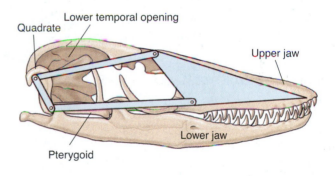

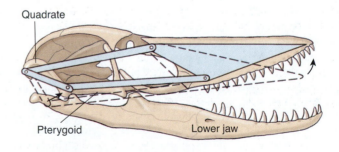

figure 18.9

Kinetic diapsid skull of a modern lizard (monitor lizard, *Varanus* sp.) showing the joints that allow the snout and upper jaw to move on the rest of the skull. The quadrate can move at its dorsal end and ventrally at both the lower jaw and the pterygoid. The front part of the braincase is also flexible, allowing the snout to be raised. Note that the lower temporal opening is very large with no lower border; this modification of the diapsid condition, common in modern lizards, provides space for expansion of large jaw muscles. The upper temporal opening lies dorsal and medial to the postorbital-squamosal arch and is not visible in this drawing.

The Mesozoic World of Dinosaurs

When, in 1842, the English anatomist Richard Owen coined the term *dinosaur* ("fearfully great lizard") to describe fossil Mesozoic reptiles of gigantic size, only three poorly known dinosaur genera were distinguished. But with new and marvelous fossil discoveries quickly following, by 1887 zoologists were able to distinguish two groups of dinosaurs based on differences in the structure of the pelvic girdles. The Saurischia ("lizard-hipped") had a simple, three-pronged pelvis with hip bones arranged much as they are in other reptiles. A large bladelike ilium is attached to one or two sacral vertebrae. The pubis and ischium extend ventrally and posteriorly, respectively, and all three bones meet at the hip socket. The Ornithischia ("bird-hipped") had a somewhat more complex pelvis. The ilium and ischium were arranged similarly in ornithischians and saurischians, but the ornithischian pubis was a narrow, rod-shaped bone with anteriorly and posteriorly directed processes lying alongside the ischium. Oddly, while the ornithischian pelvis, as the name suggests, was similar to that of birds, birds are of the saurischian lineage.

Dinosaurs and their living relatives, the birds, are archosaurs ("ruling lizards"), a group that includes thecodonts (early archosaurs restricted to the Triassic), crocodilians, and pterosaurs (refer to the classification of the reptiles on p. 362). As traditionally recognized, dinosaurs are a paraphyletic group because they do not include birds, which are descended from the theropod dinosaur lineage.

From among the various archosaurian radiations of the Triassic there emerged a thecodont lineage with limbs drawn under the body to provide an upright posture. This lineage gave rise to the earliest dinosaurs of the Late Triassic. In *Herrerasaurus,* a bipedal dinosaur from Argentina, we see one of the most distinctive characteristics of dinosaurs: walking upright on pillarlike legs, rather than on legs splayed outward as with modern amphibians and reptiles. This arrangement allowed legs to support the great weight of the body while providing an efficient and rapid stride.

Although their ancestry is unclear, two groups of saurischian dinosaurs have been distinguished by differences in feeding habits and locomotion: the carnivo-rous and bipedal theropods, and the herbivorous and quadrupedal sauropods *Coelophysis* was an early theropod with a body form typical of all theropods: powerful hindlegs with three-toed feet; long, heavy counterbalancing tail; slender, grasping forelimbs; flexible neck; and a large head with jaws armed with daggerlike teeth. Large predators such as *Allosaurus,* common during the Jurassic, were replaced by even more massively built carnivores of the Cretaceous, such as *Tyrannosaurus,* which reached a length of 14.5 m (47 ft), stood nearly 6 m high, and weighed more than 7200 kg (8 tons). Not all predatory saurischians were massive; several were swift and nimble, such as *Velociraptor* ("speedy predator") of the Upper Cretaceous.

Herbivorous saurischians, the quadrupedal sauropods, appeared in the Late Triassic. Although early sauropods were small- and medium-sized dinosaurs, those of the Jurassic and Cretaceous attained gigantic proportions, the largest terrestrial vertebrates ever to have lived. *Brachiosaurus* reached 25 m (82 ft) in length and may have weighed in excess of 30,000 kg (33 tons). Even larger sauropods have been discovered; *Argentinosaurus* was 40 m (132 ft) long and weighed at least 80,000 kg. With long necks and long front legs, the sauropods were the first vertebrates adapted to feed on trees. They reached their greatest diversity in the Jurassic and began to decline in overall abundance and diversity during the Cretaceous.

The second group of dinosaurs, the Ornithischia, were all herbivorous. Although more varied, even grotesque, in appearance than saurischians, ornithischians are united by several derived skeletal features that indicate common ancestry. The huge back-plated *Stegosaurus* of the Jurassic is a well known example of armored ornithischians which comprised two of the five major groups of ornithischians. Even more shielded with bony plates than stegosaurs were the heavily built ankylosaurs, "armored tanks" of the dinosaur world. As the Jurassic gave way to the Cretaceous, several groups of unarmored ornithischians appeared, although many bore impressive horns. The steady increase in ornithiscian diversity in the Cretaceous paralleled a concurrent gradual decline in giant sauropods, which had flourished in the Jurassic. *Triceratops* is representative of horned dinosaurs that were common in the Upper Cretaceous. Even more prominent in the Upper Cretaceous were the hadrosaurs, such as *Parasaurolophus,* which are believed to have lived in large herds. Many hadrosaurs had skulls elaborated with crests that probably functioned as vocal resonators to produce species-specific calls.

Recent discoveries have revealed that dinosaurs had complex parental care. Fossil nests of dinosaurs are known for several groups. In one case, a fossil adult of the small theropod *Oviraptor* was found with a nest of eggs. Originally it was believed that the adult was a predator on the eggs (*Oviraptor* means "egg seizer"). Later, an embryo in similar eggs was found and identified as *Oviraptor,* indicating that the adult was probably with its own eggs! Examination of baby *Maiasaura* (a hadrosaur) found in a nest revealed considerable wear on their teeth. This suggests the babies had remained in the nest and possibly were fed by adults during part of their early life. Although complex parental care is not present in most reptilian groups, it is present in nearly all members of the clade Archosauria (crocodilians, dinosaurs, and birds).

Sixty-five million years ago, the last Mesozoic dinosaurs became extinct, leaving birds and crocodilians as the only surviving archosaurs. The demise of the dinosaurs coincided with a large asteroid impact on the Yucatan peninsula that would have produced worldwide environmental upheaval. An alternate explanation, supported by many paleontologists, suggests extinctions were caused by changing climates and landforms at the close of the Cretaceous. However these hypotheses do not explain why dinosaurs became extinct, but many other vertebrate clades persisted. We continue to be fascinated by the awe-inspiring, often staggeringly large creatures that dominated the Mesozoic era for 165 million years—an incomprehensibly long period of time. Today, inspired by clues from fossils and footprints from a lost world, scientists continue to piece together the puzzle of how the various dinosaur groups arose, behaved, and diversified.

SAURISCHIANS

ORNITHISCHIANS

66 My ago

CRETACEOUS

Titanosaurus
12 m (40 ft)

Velociraptor
1.8 m (6 ft)

Parasaurolophus (duck-billed dinosaur)
10 m (33 ft)

Triceratops
9 m (30 ft)

144 My ago

JURASSIC

Brachiosaurus
25 m (82 ft)

Allosaurus
11 m (35 ft)

Stegosaurus
9 m (30 ft)

Ilium

Ischium

Pubis

208 My ago

TRIASSIC

Coelophysis
3 m (10 ft)

Ilium

Ischium

Pubis

Herrerasaurus 4 m (13 ft)
One of the oldest known
dinosaurs. Has characteristics
of both saurischians and
ornithischians.

245 My ago

figure 18.10

Tokay, *Gekko gecko,* of Southeast Asia has a true voice and is named after the strident repeated *to-kay, to-kay* call.

figure 18.11

A large male marine iguana, *Amblyrhynchus cristatus,* of the Galápagos Islands, feeding underwater on algae. This is the only marine lizard in the world. It has special salt-removing glands in the eye orbits and long claws that enable it to cling to the bottom while feeding on small red and green algae, its principal diet. It may dive to depths exceeding 10 m (33 feet) and remain submerged more than 30 minutes.

roof of the skull, allow the snout to be tilted upward. Specialized mobility of the skull enables squamates to seize and manipulate their prey. It also increases the effective closing force of the jaw musculature. The skull of snakes is even more kinetic than that of lizards. Such exceptional skull mobility is considered a major factor in diversification of lizards and snakes.

Lizards: Suborder Sauria   Lizards are an extremely diverse group, including terrestrial, burrowing, aquatic, arboreal, and aerial members. Among the more familiar groups in this varied suborder are **geckos** (figure 18.10), small, agile, mostly nocturnal forms with adhesive toe pads that enable them to walk upside down and on vertical surfaces; **iguanas,** often brightly colored New World lizards with ornamental crests, frills, and throat fans, and a group that includes the remarkable marine iguana of the Galápagos Islands (figure 18.11); **skinks,** with elongate bodies and reduced limbs; and **chameleons,** a group of arboreal lizards, mostly of Africa and Madagascar. Chameleons are entertaining creatures that catch insects with a sticky-tipped tongue that can be flicked accurately and rapidly to a distance greater than the length of their body (figure 18.12). The great majority of lizards have four limbs and relatively short bodies, but in many the limbs are degenerate, and a few such as the glass lizards (figure 18.13) are completely limbless.

Most lizards have movable eyelids, whereas a snake's eyes are permanently covered with a transparent cap. Lizards have keen vision for daylight (retinas rich in both cones and rods), although one group, the nocturnal geckos, has retinas composed entirely of rods for night vision. Most lizards have an external ear that snakes lack. However, as with other reptiles, hearing does not play an important role in the lives of most lizards. Geckos are exceptions because the males are strongly vocal (to announce territory and discourage the approach of other males), and they must, of course, hear their own vocalizations.

Many lizards have successfully invaded the world's hot and arid regions, aided by adaptations that make desert life possible. Lipids in their thick skin minimize water loss. Little

figure 18.12

A chameleon snares a dragonfly. After cautiously edging close to its target, the chameleon suddenly lunges forward, anchoring its tail and feet to the branch. A split second later, it launches its sticky-tipped, foot-long tongue to trap the prey. The eyes of this common European chameleon, *Chamaeleo chamaeleon,* are swiveled forward to provide binocular vision and excellent depth perception.

water is lost in their urine because they primarily excrete uric acid, as do other groups that are successful in arid habitats (birds, insects, and pulmonate snails). Some, such as the Gila monster of the southwestern United States deserts, store fat in their tails, which they use during drought to provide both energy and metabolic water (figure 18.14).

figure 18.13

A glass lizard, *Ophisaurus* sp., of the southeastern United States. This legless lizard feels stiff and brittle to the touch and has an extremely long, fragile tail that readily fractures when the animal is struck or seized. Most specimens, such as this one, have only a partly regenerated tip to replace a much longer tail previously lost. Glass lizards can be readily distinguished from snakes by the deep, flexible groove running along each side of the body. They feed on worms, insects, spiders, birds' eggs, and small reptiles.

figure 18.14

Gila monster, *Heloderma suspectum,* of southwestern United States desert regions and the related Mexican beaded lizard are the only venomous lizards known. These brightly colored, clumsy-looking lizards feed principally on birds' eggs, nesting birds, mammals, and insects. Unlike venomous snakes, the Gila monster secretes venom from glands in its lower jaw. The chewing bite is painful to humans but seldom fatal.

Lizards, like nearly all reptiles, are **ectothermic,** adjusting their body temperature by moving among different microclimates. Because cold climates provide few opportunities for ectotherms to raise their body temperature, there are relatively few reptile species in these habitats. However, because

figure 18.15

A worm lizard of suborder Amphisbaenia. Worm lizards are burrowing forms with a solidly constructed skull used as a digging tool. The species pictured, *Amphisbaena alba,* is widely distributed in South America.

ectotherms use considerably less energy than endotherms, reptiles are successful in ecosystems with low productivity and warm climates, such as tropical deserts and grasslands. Thus ectothermy is not an "inferior" characteristic of reptiles, but instead is a successful strategy for coping with specific environmental challenges.

Worm Lizards: Suborder Amphisbaenia The somewhat inappropriate common name "worm lizards" describes a group of highly specialized, burrowing forms that are neither worms nor true lizards but certainly are related to the latter. They have elongate, cylindrical bodies of nearly uniform diameter, and most lack any trace of external limbs (figure 18.15). With soft skin divided into numerous rings, and eyes and ears hidden under skin, amphisbaenians superficially resemble earthworms—a kind of structural convergence that often occurs when two very distantly related groups come to occupy similar habitats. Amphisbaenians have an extensive distribution in South America and tropical Africa; one species occurs in the United States.

Snakes: Suborder Serpentes Snakes are limbless and usually lack both pectoral and pelvic girdles (the latter persists as a vestige in pythons and boas). The numerous vertebrae of snakes, shorter and wider than those of most tetrapods, permit quick lateral undulations through grass and over rough terrain. Ribs increase rigidity of the vertebral column, providing more resistance to lateral stresses. Elevation of the neural spine gives the numerous muscles more leverage.

In addition to the highly kinetic skull that enables snakes to swallow prey several times their own diameter (figure 18.16), snakes differ from lizards in having no movable eyelids (snakes'

figure 18.16

The great mobility of the snake jaw and skeletal elements is evident in this snake, *Dasypeltis,* swallowing an egg.

figure 18.18

A blacktail rattlesnake, *Crotalus molossus,* flicks its tongue to smell its surroundings. Scent particles trapped on the tongue's surface are transferred to Jacobson's organs, olfactory organs in the roof of the mouth. Note the heat-sensitive pit organ between the nostril and eye.

figure 18.17

Parrot snake, *Leptophis ahaetulla.* The slender body of this Central American tree snake is an adaptation for sliding along branches without weighting them down.

eyes are permanently covered with upper and lower transparent eyelids fused together) and no external ears. Most snakes have relatively poor vision, but arboreal snakes possess excellent binocular vision, useful for tracking prey through branches where scent trails would be difficult to follow (figure 18.17). Snakes' internal ears are mainly sensitive to sounds in a limited range of low frequency (100 to 700 Hz). However, snakes are quite sensitive to vibrations conducted through the ground.

Nevertheless, most snakes employ chemical senses rather than vision or vibration detection to hunt their prey. In addition to the usual olfactory areas in the nose, which are not well developed, snakes have a pair of pitlike **Jacobson's**

organs in the roof of the mouth. These organs are lined with an olfactory epithelium and are richly innervated. The **forked tongue,** flicked through the air, picks up scent molecules (figure 18.18); the tongue is then drawn past Jacobson's organs. Information is transmitted from Jacobson's organs to the brain, where scents are identified.

Most snakes capture their prey by grabbing it with their mouth and swallowing it when it is still alive. Swallowing a struggling, kicking, biting animal is dangerous, so most snakes that swallow prey alive specialize on small prey, such as worms, insects, fish, frogs, and, less frequently, small mammals. Many of these snakes are quite fast and locate prey by actively foraging. Snakes that first kill their prey by constriction (figure 18.19) often specialize on larger, often mammalian, prey. The largest constrictors are able to kill and swallow prey as large as deer, leopards, and crocodilians. However, the muscle rearrangements that permit constricting reduce the speed at which constrictors can travel. As a result, most constrictors tend to ambush prey.

Other snakes kill their prey before swallowing by injecting it with venom. Less than 20% of all snakes are venomous, although venomous species outnumber nonvenomous species by 4 to 1 in Australia. Venomous snakes are usually divided into five families, based in part on type of fangs. Vipers (family Viperidae) have large, movable, tubular fangs at the front of their mouth (figure 18.20). This family includes American pit vipers (figure 18.18), and Old World true vipers, which lack facial heat-sensing pits. A second family of venomous snakes (family Elapidae) has short, permanently erect fangs in the front of the mouth and includes cobras (figure 18.21), mambas, coral snakes, and kraits. Smaller groups include the highly venomous sea snakes (family Hydrophiidae) and the poorly known **fossorial** mole vipers (family Atractaspididae). The large family

figure 18.19

Nonvenomous African house snake, *Boaedon fuluginosus,* constricting a mouse before swallowing it.

figure 18.21

Spectacled, or Indian, cobra, *Naja naja*. Cobras erect the front part of the body when startled and as a threat display. Although the cobra's strike range is limited, all cobras are dangerous because of the extreme toxicity of the venom.

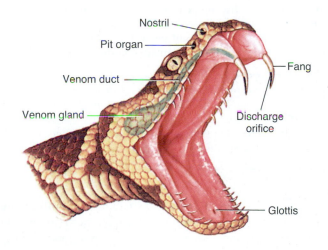

figure 18.20

Head of rattlesnake showing the venom apparatus. The venom gland, a modified salivary gland, is connected by a duct to the hollow fang.

Colubridae, which contains most familiar (and nonvenomous) snakes, does include a few venomous species, including the rear-fanged African boomslang and the twig snake, both of which are responsible for some human fatalities.

Snakes of subfamily Crotalinae within family Viperidae are called **pit vipers** because they possess special heat-sensitive **pit organs** on their heads, located between their nostrils and eyes (figures 18.18, 18.20, and 18.22). All of the best-known North American venomous snakes are pit vipers, such as rattlesnakes, water moccasins, and copperheads. The pits are supplied with a dense packing of free nerve endings from the fifth cranial nerve. They are exceedingly sensitive to radiant energy (long-wave infrared) and can distinguish temperature differences smaller

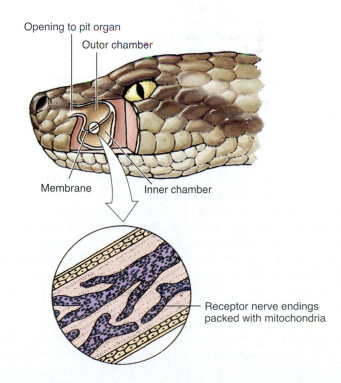

figure 18.22

Pit organ of rattlesnake, a pit viper. Cutaway shows location of a deep membrane that divides the pit into inner and outer chambers. Heat-sensitive nerve endings are concentrated in the membrane.

classification of Living Reptilian Groups

Subclass Anapsida (a-nap′se-duh) (Gr. *an*, without, + *apsis*, arch): **Anapsids.** Amniotes having a skull with no temporal opening.

Order Testudines (tes-tu′din-eez) (L. *testudo*, tortoise) **(Chelonia): turtles.** Body in a bony case of dorsal carapace and ventral plastron; jaws with keratinized beaks instead of teeth; vertebrae and ribs fused to overlying carapace; tongue not extensible; neck usually retractable; approximately 300 species.

Subclass Diapsida (di-ap′se-duh) (Gr. *di*, double, + *apsis*, arch): **Diapsids.** Amniotes having a skull with two pairs of temporal openings.

Superorder Lepidosauria (lep-i-do-sor′ee-uh) (Gr. *lepidos*, scale, + *sauros*, lizard). Diapsid lineage appearing in the Triassic; characterized by sprawling posture; no bipedal specializations; diapsid skull often modified by loss of one or both temporal arches; skin shed in one piece.

Order Squamata (squa-ma′ta) (L. *squamatus*, scaly, + *ata*, characterized by): **Snakes, lizards,** and **amphisbaenians.** Skin of keratinized epidermal scales or plates, which is shed; quadrate movable; skull kinetic (except worm lizards); vertebrae usually concave in front; paired copulatory organs.

Suborder Lacertilia (lay-sur-till′ee-uh) (L. *lacerta*, lizard) **(Sauria): lizards.** Body slender, usually with four limbs; rami of lower jaw fused; eyelids movable; external ear present; this paraphyletic suborder contains approximately 4600 species.

Suborder Amphisbaenia (am′fis-bee′nee-a) (L. *amphis*, double, + *baina*, to walk): **worm lizards.** Body elongate and of nearly uniform diameter; no legs (except one genus with short front legs); skull bones interlocked for burrowing (not kinetic); limb girdles vestigial; eyes hidden beneath skin; only one lung; approximately 160 species.

Suborder Serpentes (sur-pen′teez) (L. *serpere*, to creep): **snakes.** Body elongate; limbs, ear openings, and middle ear absent; mandibles joined anteriorly by ligaments; eyelids fused into transparent spectacle; tongue forked and protrusible; left lung reduced or absent; approximately 2900 species.

Order Sphenodonta (sfen′o-don′tuh) (Gr. *sphēnos*, wedge, + *odontos*, tooth): **tuataras.** Primitive diapsid skull; vertebrae biconcave; quadrate immovable; median parietal eye present. *Sphenodon* only extant genus (two species).

Superorder Archosauria (ark′o-sor′ee-uh) (Gr. *archōn*, ruling, + *sauros*, lizard). Advanced diapsids, mostly terrestrial, but some specialized for flight; gizzard present; includes birds (covered in the next chapter) and dinosaurs of the Mesozoic.

Order Crocodilia (croc′o-dil′ee-uh) (L. *crocodilus*, crocodile): **crocodilians.** Skull elongate and massive; nares terminal; secondary palate present; four-chambered heart; vertebrae usually concave in front; forelimbs usually of five digits; hindlimbs of four digits; quadrate immovable; advanced social behavior; 23 species.

---

than 0.003°C from a radiating surface. Pit vipers use their pits to track warm-blooded prey and to aim strikes, which they can make as effectively in total darkness as in daylight.

All vipers have a pair of teeth, modified as fangs, on the maxillary bones. The fangs lie in a membranous sheath when the mouth is closed. When a viper strikes, a special muscle and bone lever system erects the fangs as the mouth opens (figure 18.20). Fangs are driven into prey by the thrust, and venom is injected into the wound through a canal in the fangs. A viper immediately releases its prey after the bite and follows it until it is paralyzed or dies. Then the snake swallows the prey whole. Approximately 8000 bites but only 5–10 deaths from pit vipers are reported each year in the United States.

Even the saliva of all harmless snakes possesses limited toxic qualities, and it is logical that there was a natural selection for this toxic tendency as snakes evolved. Snake venoms have traditionally been divided into two types. The **neurotoxic** type acts mainly on the nervous systems, affecting the optic nerves (causing blindness) or the phrenic nerve of the diaphragm (causing paralysis of respiration). The **hemorrhagin** type destroys red blood cells and blood vessels and produces extensive hemorrhaging of blood into tissue spaces. In fact, most snake venoms are complex mixtures of various frac-

tions that attack different organs in specific ways. Although sea snakes and the Australian tiger snake have perhaps the most toxic of snake venoms, several larger snakes are more dangerous. The aggressive king cobra, which may exceed 5.5 m in length, is the largest and probably the most dangerous of all venomous snakes. It is estimated that, worldwide, 50,000–60,000 people die from snakebite each year. Most of the deaths occur in India, Pakistan, Myanmar, and nearby countries where poorly shod people frequently come into contact with venomous snakes and frequently do not get immediate medical attention following snakebite. The snakes primarily responsible for deaths in these areas are Russell's viper, the saw-scaled viper, and several species of cobras.

Most snakes are **oviparous** (L. *ovum*, egg, + *parere*, to bring forth) species that lay their shelled, elliptical eggs beneath rotten logs, under rocks, or in holes dug in the ground. Most of the remainder, including all American pit vipers except the tropical bushmaster, are **ovoviviparous** (L. *ovum*, egg, + *vivus*, living, + *parere*, to bring forth), giving birth to well-formed young. Very few snakes are **viviparous** (L. *vivus*, living, + *parere*, to bring forth); in these snakes a primitive placenta forms, permitting the exchange of materials between the embryonic and maternal bloodstreams.

Tuataras: Order Sphenodonta

The order Sphenodonta is represented by two living species of the genus *Sphenodon* (Gr. *sphenos*, wedge, + *odontos*, tooth) of New Zealand (figure 18.23). Tuataras are sole survivors of the sphenodontid lineage that radiated modestly during the early Mesozoic era but declined toward the end of the Mesozoic. Tuataras were once widespread throughout the two main islands of New Zealand but are now restricted to small islets of Cook Strait and off the northeast coast of North Island.

Tuataras are lizardlike forms 66 cm long or less that live in burrows often shared with petrels. They are slow-growing animals with a long life; one is recorded to have lived 77 years.

Tuataras captured the interest of zoologists because of numerous features that are almost identical to those of lizard-like diapsids 200 million years old. These features include a diapsid skull with two temporal openings bounded by complete arches, and a well-developed median parietal "third eye." *Sphenodon* represents one of the slowest rates of morphological evolution known among vertebrates.

Crocodiles and Alligators: Order Crocodilia

Modern crocodilians and birds are the only surviving representatives of the archosaurian lineage that gave rise to the great Mesozoic radiation of dinosaurs and their kin. Crocodilians differ little in structural details from crocodilians of the early Mesozoic. Having remained mostly unchanged for nearly 200 million years, crocodilians face an uncertain future in a world dominated by humans.

All crocodilians have an elongate, robust, well-reinforced skull and massive jaw musculature arranged to provide a wide gape and rapid, powerful closure. Teeth are set in sockets, a type of dentition typical of all extinct archosaurs as well as the earliest birds. Another adaptation, found in no other vertebrate except mammals, is a complete secondary palate. This innovation allows crocodilians to breathe when the mouth is filled with water or food (or both).

Estuarine crocodiles (*Crocodylus porosus*), found in southern Asia, and Nile crocodiles (*C. niloticus*) (figure 18.24A) grow to great size (adults weighing 1000 kg have been reported) and

A

B

figure 18.23

Tuatara, *Sphenodon* sp., a living representative of order Sphenodonta. This "living fossil" reptile has, on top of the head, a well-developed parietal "eye" with retina, lens, and nervous connections to the brain. Although covered with scales, this third eye is sensitive to light. The parietal eye may have been an important sense organ in early reptiles. Tuataras are found today only on certain islands off the coastline of New Zealand.

figure 18.24

Crocodilians. **A,** Nile crocodile, *Crocodylus niloticus*, basking. The lower jaw tooth fits *outside* the slender upper jaw; alligators lack this feature. **B,** American alligator, *Alligator mississipiensis*, an increasingly noticeable resident of rivers, bayous, and swamps of the southeastern United States.

are swift and aggressive. Crocodiles are known to attack animals as large as cattle, deer, and people. Alligators (figure 18.24B) are less aggressive than these crocodiles and certainly far less dangerous to people. They are unusual among reptiles in being able to make definite vocalizations. Male alligators emit loud bellows in the mating season. In the United States, *Alligator mississipiensis* is the only species of alligator; *Crocodylus acutus,* restricted to extreme southern Florida, is the only species of crocodile.

Crocodilians are oviparous. Usually from 20 to 50 eggs are laid in a mass of dead vegetation or buried in sand and guarded by the mother. The mother hears vocalizations from hatching young and responds by opening the nest to allow the hatchlings to escape. As with many turtles and some lizards, the incubation temperature of the eggs determines the sex ratio of the offspring. However, unlike turtles (p. 354), low nest temperatures produce only females, whereas high nest temperatures produce only males.

Summary

Reptiles and other amniotes diverged phylogenetically from a group of early tetrapods during the late Paleozoic era, over 300 million years ago. Their success as terrestrial vertebrates is attributed in large part to evolution of the amniotic egg, which, with its three extraembryonic membranes, provided support for full embryonic development within the protection of a shell. Thus reptiles could lay their eggs on land. Reptiles are also distinguished from amphibians by their dry, scaly skin that limits water loss; more powerful jaws; internal fertilization; and advanced circulatory, respiratory, excretory, and nervous systems. Like amphibians, reptiles are ectotherms, but most exercise considerable behavioral control over their body temperature.

Before the end of the Paleozoic era, amniotes began a radiation that separated into three groups: anapsids, which gave rise to the turtles; synapsids, a lineage that led to modern mammals; and the diapsid lineage, which led to all other reptiles and to birds. The great burst of reptilian radiation during the Mesozoic era produced a worldwide fauna of great diversity.

Turtles (order Testudines) with their distinctive shells have changed little in design since the Triassic period. Turtles are a small group of long-lived terrestrial, semiaquatic, aquatic, and marine species. They lack teeth. All are oviparous and all, including marine forms, bury their eggs.

Lizards, snakes, and worm lizards (order Squamata) form 95% of all living reptiles. Lizards (suborder Lacertilia) are a diversified and successful group adapted for walking, running, climbing, swimming, and burrowing. They are distinguished from snakes by typically having two pairs of limbs (some species are limbless), fused lower jaw halves, movable eyelids, and external ears. Many lizards are well adapted for survival under hot, arid desert conditions.

Worm lizards (suborder Amphisbaenia) are a small tropical group of limbless squamates highly adapted for burrowing.

Snakes (suborder Serpentes), in addition to being entirely limbless, are characterized by elongate bodies and a highly kinetic skull that permits them to swallow whole prey that may be much larger than the snake's own diameter. Most snakes rely on chemical senses, especially Jacobson's organs, to hunt prey, rather than on weakly developed visual and auditory senses. Pit vipers have unique infrared-sensing organs for tracking warm-bodied prey. Some snakes are venomous.

Tuataras of New Zealand (order Sphenodonta) represent a relict genus and sole survivor of a group that otherwise disappeared 100 million years ago. They bear several features that are almost identical to those of Mesozoic fossil diapsids.

Crocodiles and alligators (order Crocodilia) are the only living reptilian representatives of the archosaurian lineage that gave rise to the extinct dinosaurs and living birds. Crocodilians have several adaptations for a carnivorous, semiaquatic life, including a massive skull with powerful jaws, and a secondary palate. They have the most complex social behavior of any living reptile.

Review Questions

1. What were the three major amniote radiations of the Mesozoic and from which major group do birds and mammals descend? How could you distinguish the skulls characteristic of these different radiations?
2. What changes in egg design allowed reptiles to lay eggs on land? Why is the egg often called an "amniotic" egg? What are the "amniotes"?
3. Why are reptiles considered a paraphyletic rather than a monophyletic group? How have cladistic taxonomists revised the content of this taxon to make it monophyletic?
4. Describe ways in which reptiles are more functionally or structurally suited for terrestriality than amphibians.
5. What are the main characteristics of reptilian skin and how would you distinguish reptilian skin from frog skin?
6. Describe the principal structural features of turtles that would distinguish them from any other reptilian order.
7. How might nest temperature affect egg development in turtles? In crocodilians?
8. What is meant by a "kinetic" skull and what benefit does it confer? How are snakes able to eat such large prey?
9. In what ways are the special senses of snakes similar to those of lizards, and in what ways have they evolved for specialized feeding strategies?

10. What are amphisbaenians? What morphological adaptations do they have for burrowing?

11. Distinguish ornithiscian and saurischian dinosaurs, based on their hip anatomy. Was a dinosaur's parental care more like a lizard's or a crocodilian's?

12. How do crocodilians breathe when their mouths are full of food?

13. What is the function of Jacobson's organ of snakes and lizards?

14. What is the function of the "pit" of pit vipers?

15. What is the difference in structure or location of fangs of a rattlesnake, a cobra, and an African boomslang?

16. Most snakes are oviparous, but some are ovoviviparous or viviparous. What do these terms mean and what would you have to know to be able to assign a particular snake to one of these reproductive modes?

17. Why are tuataras (*Sphenodon*) of special interest to biologists? Where would you have to go to see one in its natural habitat?

18. From which diapsid lineage have crocodilians descended? What other major fossil and living vertebrate groups belong to this same lineage? In what structural and behavioral ways do crocodilians differ from other living reptiles?

Selected References

See also general references on page 415.

Alexander, R. M. 1991. How dinosaurs ran. Sci. Am. **264**:130–136 (April). *By applying the techniques of modern physics and engineering, a zoologist calculates that the large dinosaurs walked slowly but were capable of a quick run; none required the buoyancy of water for support.*

Alvarez, W., and F. Asaro. 1990. An extraterrestrial impact. Sci. Am. **263**:78–84 (Oct.). *This article and an accompanying article by V. E. Courtillot, "A volcanic eruption," present opposing interpretations of the cause of the Cretaceous mass extinction that led to the demise of the dinosaurs.*

Cogger, H. G., and R. G. Zweifel, eds. 1998. Reptiles and amphibians. San Diego, Academic Press. *This comprehensive, up-to-date, and lavishly illustrated volume was written by some of the best-known herpetologists in the field.*

Crews, D. 1994. Animal sexuality. Sci. Am. **270**:108–114 (Jan.) *The reproductive strategies of reptiles, including nongenetic sex determination, provide insights into the origins and functions of sexuality.*

Greene, H. W. 1997. Snakes: The evolution of mystery in nature. Berkeley, University of California Press. *Beautiful photographs accompany a well-written volume for the scientist or novice.*

Halliday, T. R., and K. Adler, eds. 1986. The encyclopedia of reptiles and amphibians. New York, Facts on File, Inc. *Comprehensive and beautifully illustrated treatment of the reptilian groups with helpful introductory sections on origins and characteristics.*

Lohmann, K. J. 1992. How sea turtles navigate. Sci. Am. **266**:100–106 (Jan.). *Recent evidence suggests that sea turtles use the earth's magnetic field and the direction of ocean waves to navigate back to their natal beaches to nest.*

Mattison, C. 1995. The encyclopedia of snakes. New York, Facts on File, Inc. *Generously illustrated book treating evolution, physiology, behavior, and classification of snakes.*

Norman, D. 1991. Dinosaur! New York, Prentice-Hall. *Highly readable account of the life and evolution of dinosaurs with fine illustrations.*

Paul, G. 2000. The Scientific American book of dinosaurs. New York, St. Martin's Press. *Essays emphasizing functional morphology, behavior, evolution, and extinction of dinosaurs.*

Pianka, E. R., and L. J. Vitt. 2003. Lizards: Windows to the evolution of diversity. Berkeley, University of California Press. *Hundreds of color photographs highlight a discussion of behavior and evolution of lizards.*

Pough, F. H., R. M. Andrews, J. E. Cadle, M. L. Crump, A. H. Savitzky, and K. D. Wells. 2001. Herpetology. Upper Saddle River, New Jersey, Prentice Hall. *A comprehensive textbook treating diversity, physiology, behavior, ecology, and conservation of reptiles and amphibians.*

Custom Website

The *Animal Diversity* Online Learning Center is a great place to check your understanding of chapter material. Visit www.mhhe.com/hickmanad4e for access to key terms, quizzes, and more! Further enhance your knowledge with web links to chapter-related material.

Explore live links for these topics:

Class Reptilia
Order Testudines
Marine Turtles
Order Crocodilia
Suborder Lacertilia

Suborder Serpentes
Suborder Sphenodonta
Superorder Archosauria and Related Mesozoic Reptiles
Dissection Guides for Reptiles
Conservation Issues Concerning Reptiles

19

Birds

Flock of dunlins, *Calidris alpina,* in flight.

Long Trip to a Summer Home

Perhaps it was ordained that birds, having mastered flight, often use this power to make the long and arduous seasonal migrations that have captured human wonder and curiosity. For the advantages of migration are many. Moving between southern wintering regions and northern summer breeding regions with long summer days and an abundance of insects provides parents with ample food to rear their young. Predators of birds are not so abundant in the far North, and a brief once-a-year appearance of vulnerable young birds does not encourage buildup of predator populations. Migration also vastly increases the amount of space available for breeding and reduces aggressive territorial behavior. Finally, migration favors homeostasis—the balancing of physiological processes that maintains internal stability—by allowing birds to avoid climatic extremes.

Still, the wonder of the migratory pageant remains, and there is much yet to learn about its mechanisms. What times migration, and what determines that each bird shall store sufficient fuel for the journey? How did the sometimes difficult migratory routes originate, and what cues do birds use in navigation? And what was the origin of this instinctive force to follow the retreat of winter northward? It is instinct that drives the migratory waves in spring and fall, instinctive blind obedience that carries most birds successfully to their northern nests, while countless others fail and die, winnowed by the ever-challenging environment.

Of the vertebrates, birds of class Aves (ay'veez) (L. pl. of *avis,* bird) are the most noticeable, the most melodious, and many think the most beautiful. With more than 9900 species distributed over nearly the entire earth, birds far outnumber all other vertebrates except fishes. Birds occur in forests and deserts, in mountains and prairies, and on all oceans. Four species are known to have visited the North Pole, and one, a skua, was seen at the South Pole. Some birds live in total darkness in caves, finding their way by echolocation, and others dive to depths greater than 45 m to prey on aquatic life.

The single unique feature that distinguishes birds from other living animals is their feathers. If an animal has feathers, it is a bird; if it lacks feathers, it is not a bird. No other vertebrate group bears such an easily recognizable and foolproof identification tag.

There is great uniformity of structure among birds. Despite approximately 150 million years of evolution, during which they proliferated and adapted to specialized ways of life, we have no difficulty recognizing a bird as a bird. In addition to feathers, all birds have forelimbs modified into wings (although not always used for flight); all have hindlimbs adapted for walking, swimming, or perching; all have keratinized beaks; and all lay eggs. The reason for this great structural and functional uniformity is that birds evolved into flying machines. This fact greatly restricts morphological diversity, so much more evident in other vertebrate classes. For example, birds do not begin to approach the morphological diversity seen in their endothermic evolutionary peers, the mammals, a group that includes forms as dissimilar as whales, porcupines, bats, and giraffes.

A bird's entire anatomy is designed around flight. An airborne life for a large vertebrate is a highly demanding evolutionary challenge. A bird must, of course, have wings for support and propulsion. Bones must be light and hollow yet serve as a rigid airframe. The respiratory system must be highly efficient to meet the intense metabolic demands of flight and serve also as a thermoregulatory device to maintain a constant body temperature. A bird must have a rapid and efficient digestive system to process an energy-rich diet; it must have a high metabolic rate; and it must have a high-pressure circulatory system. Above all, birds must have a finely tuned nervous system and acute senses, especially superb vision, to handle complex demands of headfirst, high-velocity flight.

Origin and Relationships

Approximately 147 million years ago, a flying animal drowned and settled to the bottom of a shallow marine lagoon in what is now Bavaria, Germany. It was rapidly covered with a fine silt and eventually fossilized. There it remained until discovered in 1861 by a workman splitting slate in a limestone quarry. The fossil was approximately the size of a crow, with a skull not unlike that of modern birds except that the beaklike jaws bore small bony teeth set in sockets like those of dinosaurs (figure 19.1). The skeleton

characteristics
of Class Aves

1. Body usually spindle shaped, with four divisions: head, neck, trunk, and tail; **neck disproportionately long** for balancing and food gathering
2. Limbs paired; **forelimbs usually modified for flying;** posterior pair variously adapted for perching, walking, and swimming; foot with four toes (two or three toes in some)
3. Epidermal **covering of feathers** and **leg scales;** thin integument of epidermis and dermis; no sweat glands; oil or preen gland at base of tail; **pinna of ear rudimentary**
4. **Fully ossified skeleton with air cavities;** skull bones fused with **one occipital condyle;** each jaw covered with a keratinized sheath, forming a **beak; no teeth;** ribs with strengthening, uncinate processes; **tail not elongate;** sternum usually well developed with keel; **single bone in middle ear**
5. Nervous system well developed, with 12 pairs of cranial nerves and brain with **large cerebellum and optic lobes**
6. Circulatory system of four-chambered heart with two atria and two ventricles; completely **separate pulmonary and systemic circuits; right aortic arch persisting;** nucleated red blood cells
7. **Endothermic**
8. Respiration by slightly expansible lungs, with thin **air sacs** among the visceral organs and skeleton; **syrinx (voice box)** near junction of trachea and bronchi
9. Excretory system of metanephric kidney; ureters open into cloaca; **no bladder;** semisolid urine; uric acid main nitrogenous waste
10. Sexes separate; testes paired, with the vas deferens opening into the cloaca; **females with left ovary and oviduct only;** copulatory organ (penis) in ducks, geese, paleognathids, and a few others
11. Fertilization internal; **amniotic eggs with much yolk and hard calcareous shells;** embryonic membranes in egg during development; **incubation external;** young active at hatching **(precocial)** or helpless and naked **(altricial);** sex determination by females (females heterogametic)

was decidedly reptilian with a long bony tail, clawed fingers, and abdominal ribs. It might have been classified as a theropod dinosaur except that it carried an unmistakable imprint of **feathers,** those marvels of biological engineering that only birds possess. *Archaeopteryx lithographica* (ar-kee-op'ter-iks lith-o-graf'e-ka, Gr., meaning "ancient wing inscribed in stone"), as the fossil was named, was an especially fortunate discovery because it demonstrated beyond reasonable doubt the phylogenetic relatedness of birds and theropod dinosaurs.

Zoologists had long recognized the similarity of birds and reptiles because of their many shared morphological, developmental, and physiological homologies. The distinguished English zoologist Thomas Henry Huxley was so impressed with these affinities that he called birds "glorified reptiles" and classified them with a group of dinosaurs called theropods that displayed several birdlike characteristics

A   **B**

figure 19.1

Archaeopteryx, a 147-million-year-old relative of modern birds. **A,** Cast of the second and most nearly perfect fossil of *Archaeopteryx,* which was discovered in a Bavarian stone quarry. Seven specimens of *Archaeopteryx* have been discovered, the most recent one in 1992. **B,** Reconstruction of *Archaeopteryx.*

(figures 19.2 and 19.3). Theropod dinosaurs share many derived characters with birds, the most obvious of which is the elongate, mobile, S-shaped neck.

Dromeosaurs, a group of theropods that includes *Velociraptor,* share many additional derived characters with birds, including a furcula (fused clavicles) and lunate wrist bones that permit swiveling motions used in flight (figure 19.3). Additional evidence linking birds to dromeosaurs comes from recently described fossils from late Jurassic and early Cretaceous deposits in Liaoning Province, China. These spectacular dromeosaur-like fossils include some with filaments, such as *Sinosauropteryx,* and some with feathers, such as *Protarchaeopteryx* and *Caudipteryx.* These feathered dinosaurs could not fly, however, as they had short forelimbs and symmetrical vaned feathers (the flight feathers of modern flying birds are asymmetrical). Clearly these filaments and feathers were not used for flight; perhaps they were used for thermoregulation or were colorful and used in social displays. Other fossils from Spain and Argentina of birds more derived than *Archaeopteryx* document the development of the keeled sternum and alula (see figure 19.5), loss of teeth, and fusion of bones characteristic of modern birds. A phylogenetic approach to classification would group birds with theropod dinosaurs. With this view, dinosaurs are not extinct—they are with us today as birds!

Living birds (Neonithes) are divided into two groups: (1) **Paleognathae** (Gr. *palaios,* ancient, + *gnathos,* jaw), the large, flightless ostrichlike birds and kiwis, often called **ratite** birds, which have a flat sternum with poorly developed pectoral muscles; and (2) **Neognathae** (Gr. *neos,* new, + *gnathos,* jaw), flying birds that have a keeled sternum on which powerful flight muscles insert. This division assumes that flightless ratites form a monophyletic group. However, evidence supporting this grouping is weak, and relationships of ratites to other birds are controversial. Ostrichlike paleognathids clearly have descended from flying ancestors. Furthermore, not all neognathous birds can fly and many of them even lack keels. Flightlessness has appeared independently among many groups of birds; the fossil record reveals flightless wrens, pigeons, parrots, cranes, ducks, auks, and even a flightless owl. Penguins are flightless although they use their wings to "fly" through water (see figure 4.6, p. 79). Flightlessness has evolved almost always on islands where few terrestrial predators are found. Flightless birds living on continents today are the large paleognathids (ostrich, rhea, cassowary, emu), that can run fast enough to escape predators. Ostriches can run 70 km (42 miles) per hour, and claims of speeds of 96 km (60 miles) per hour have been made. The evolution of flightless birds is discussed on p. 18 and in figure 1.17.

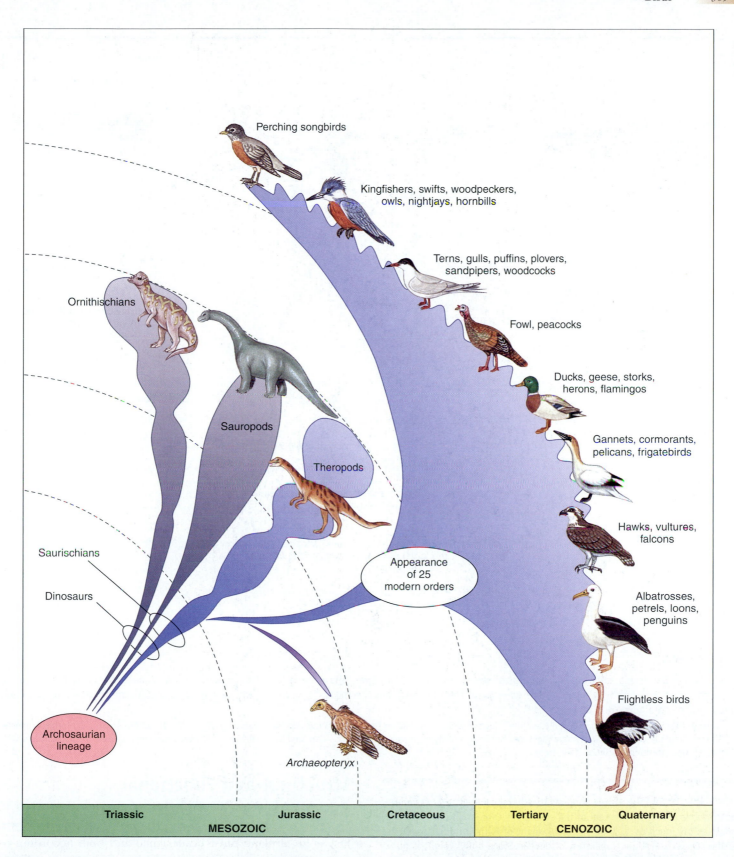

Perching songbirds

Kingfishers, swifts, woodpeckers,
owls, nightjays, hornbills

Terns, gulls, puffins, plovers,
sandpipers, woodcocks

Ornithischians

Fowl, peacocks

Ducks, geese, storks,
herons, flamingos

Sauropods

Gannets, cormorants,
pelicans, frigatebirds

Theropods

Appearance
of 25
modern orders

Hawks, vultures,
falcons

Saurischians

Albatrosses,
petrels, loons,
penguins

Dinosaurs

Archosaurian
lineage

Flightless birds

Archaeopteryx

Triassic	Jurassic	Cretaceous	Tertiary	Quaternary
MESOZOIC			CENOZOIC	

figure 19.2

Evolution of modern birds. Of 25 living bird orders, 9 of the largest are shown. The earliest known bird, *Archaeopteryx,* lived in the Upper Jurassic, about 147 million years ago. *Archaeopteryx* uniquely shares many specialized aspects of its skeleton with the smaller theropod dinosaurs and is considered to have evolved within the theropod lineage. Evolution of modern bird orders occurred rapidly during the Cretaceous and early Tertiary periods.

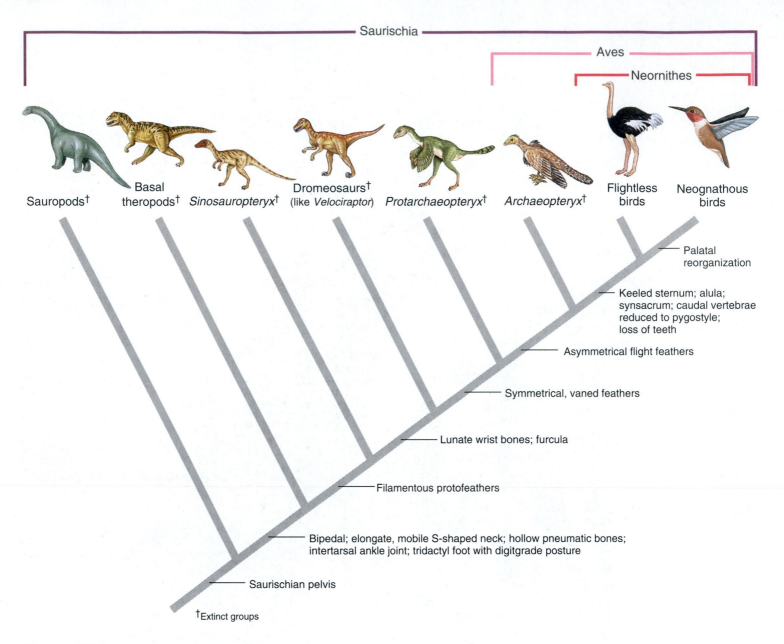

figure 19.3

Cladogram of Saurischia, showing relationships of several taxa to modern birds. Shown are a few of the shared derived characters, mostly those relating to flight, that were used to construct the genealogy. The ornithischians are the sister group to the saurischians, and all are members of the clade Archosauria (see figure 18.1, p. 350).

Sources: J. Gauthier, "Saurischian monopoly and the origin of birds" in K. Padian *The origin of birds and the evolution of flight. No. 18, 1986,* Memoirs California Academy of Science; and J.M.V. Rayner, "Vertebrate flight and the origins of flying vertebrates" in K.C. Allen, and D.E.G. Briggs, 1989, *Evolution and the fossil record,* Smithsonian Institute Press, Washington, D.C.

The bodies of flightless birds are dramatically redesigned to remove all restrictions of flight. The keel of the sternum is lost, and heavy flight muscles (as much as 17% of the body weight of flying birds), as well as other specialized flight apparatus, disappear. Since body weight is no longer a restriction, flightless birds tend to become large. Several extinct flightless birds were enormous: the giant moas of New Zealand weighed more than 225 kg (500 pounds) and elephantbirds of Madagascar, the largest birds that ever lived, probably weighed nearly 450 kg (about 1000 pounds) and stood nearly 2 m tall.

Structural and Functional Adaptations for Flight

Just as an airplane must be designed and built according to rigid aerodynamic specifications if it is to fly, so too must birds meet stringent structural requirements if they are to stay airborne. All the special adaptations found in flying birds contribute to two things: more power and less weight. Flight by humans became possible when they developed an internal combustion engine and learned how to reduce the weight-to-power ratio to a critical point. Birds accomplished flight millions of years ago, but birds must do much more than fly. They must feed themselves and convert food into high-energy fuel; they must escape predators; they must be able to repair their own injuries; they must be able to air-condition themselves when overheated and heat themselves when too cool; and, most important of all, they must reproduce themselves.

Feathers

Feathers are very lightweight, yet possess remarkable toughness and tensile strength. Most typical of bird feathers are **contour feathers,** vaned feathers that cover and streamline the bird's body. A contour feather consists of a hollow **quill,** or calamus, emerging from a skin follicle, and a **shaft,** or rachis, which is a continuation of the quill and bears numerous **barbs** (figure 19.4). Barbs are arranged in closely parallel fashion and spread diagonally outward from both sides of the central shaft to form a flat, expansive, webbed surface, the **vane.** There may be several hundred barbs in a vane.

If we examine a feather with a microscope, each barb appears to be a miniature replica of the feather with numerous parallel filaments called **barbules** set in each side of the barb and spreading laterally from it. There may be 600 barbules on each side of a barb, adding up to more than 1 million barbules for the feather. The barbules of one barb overlap the barbules of a neighboring barb in a herringbone pattern and are held together with great tenacity by tiny hooks. Should two adjoining barbs become separated—and considerable force is needed to pull the vane apart—they are instantly zipped together again by drawing the feather through the fingertips. Birds do this preening with their bill.

Like a reptile's scale to which it is homologous, a feather develops from an epidermal elevation overlying a nourishing dermal core. However, rather than flattening like a scale, a feather bud rolls into a cylinder and sinks into the follicle from which it is growing. During growth, pigments (lipochromes and melanin) are added to epidermal cells. As the feather enlarges and nears the end of its growth, the soft rachis and barbs are transformed into hard structures by deposition of keratin. The protective sheath splits apart, allowing the end of the feather to protrude and the barbs to unfold.

When fully grown, a feather, like mammalian hair, is a dead structure. Shedding, or molting, of feathers is a highly orderly process, with feathers discarded gradually to avoid appearance of bare spots. Flight and tail feathers are lost in exact pairs, one from each side, so that balance is maintained. Replacements emerge before the next pair is lost, and most birds can continue to fly unimpaired during the molting period; however, many water birds (ducks, geese, loons, and others) lose all their primary feathers at once and are grounded during the molt. Many prepare for molting by moving to isolated bodies of water where they can find food and more easily escape enemies. Nearly all birds molt at least once a year, usually in late summer after nesting season.

figure 19.4

Contour feather. Inset enlargement of the vane shows the minute hooks on the barbules that cross-link loosely to form a continuous surface of vane.

Skeleton

A major structural requirement for flight is a light, yet sturdy, skeleton (figure 19.5A). As compared with the earliest known bird, *Archaeopteryx* (figure 19.5B), bones of modern birds are phenomenally light, delicate, and laced with air cavities. Such **pneumatized** bones (figure 19.6) are nevertheless strong. The skeleton of a frigate bird with a 2.1 m (7-foot) wingspan weighs only 114 grams (4 ounces), less than the weight of all its feathers.

As archosaurs, birds evolved from ancestors with diapsid skulls (p. 349). However, skulls of modern birds are so specialized that it is difficult to see any trace of the original diapsid condition. Bird skulls are built lightly and mostly fused into one piece. A pigeon skull weighs only 0.21% of its body weight; by comparison a rat's skull weighs 1.25% of its body weight. The braincase and orbits are large in bird skulls to accommodate a bulging brain and large eyes needed for quick motor coordination and superior vision.

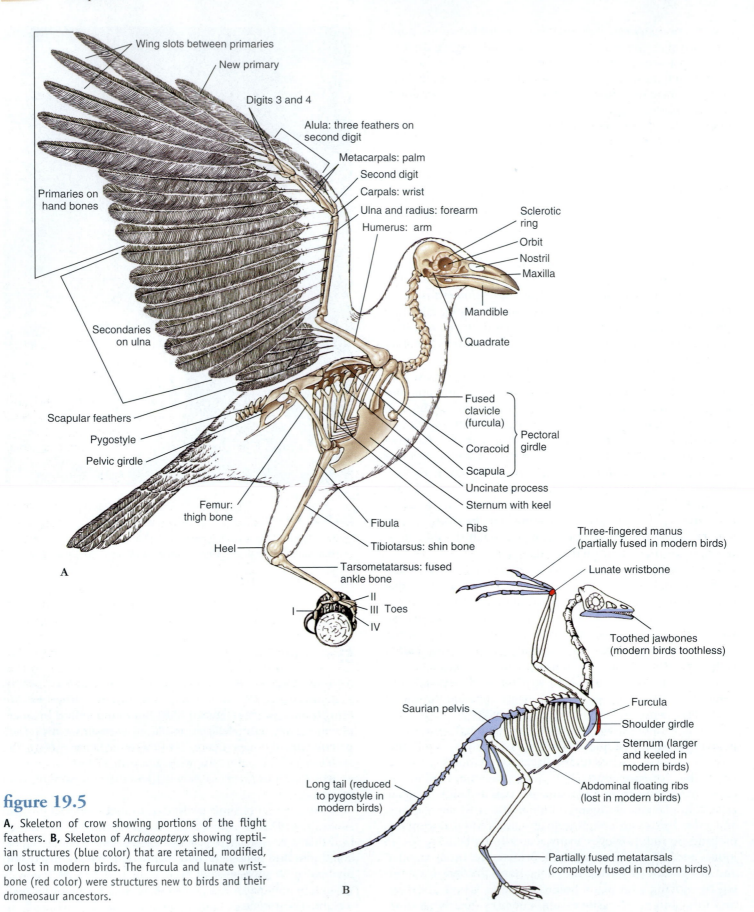

Wing slots between primaries

New primary

Digits 3 and 4

Alula: three feathers on second digit

Metacarpals: palm

Second digit

Carpals: wrist

Ulna and radius: forearm

Humerus: arm

Primaries on hand bones

Sclerotic ring

Orbit

Nostril

Maxilla

Mandible

Quadrate

Secondaries on ulna

Scapular feathers

Pygostyle

Pelvic girdle

Fused clavicle (furcula)

Coracoid

Scapula

Pectoral girdle

Uncinate process

Sternum with keel

Femur: thigh bone

Fibula

Ribs

Heel

Tibiotarsus: shin bone

Tarsometatarsus: fused ankle bone

II

I   III Toes

IV

A

Three-fingered manus (partially fused in modern birds)

Lunate wristbone

Toothed jawbones (modern birds toothless)

Saurian pelvis

Furcula

Shoulder girdle

Sternum (larger and keeled in modern birds)

Abdominal floating ribs (lost in modern birds)

Long tail (reduced to pygostyle in modern birds)

Partially fused metatarsals (completely fused in modern birds)

B

figure 19.5

A, Skeleton of crow showing portions of the flight feathers. **B,** Skeleton of *Archaeopteryx* showing reptilian structures (blue color) that are retained, modified, or lost in modern birds. The furcula and lunate wristbone (red color) were structures new to birds and their dromeosaur ancestors.

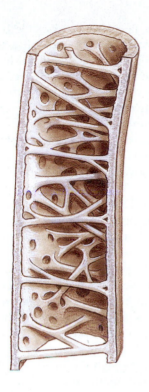

figure 19.6

Hollow wing bone of a songbird showing stiffening struts and air spaces that replace bone marrow. Such "pneumatized" bones are remarkably light and strong.

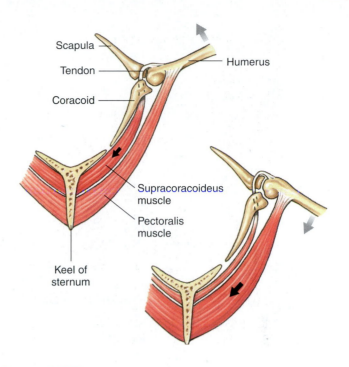

figure 19.7

Flight muscles of a bird are arranged to keep the center of gravity low in the body. Both major flight muscles are anchored on the sternum keel. Contraction of the pectoralis muscle pulls the wing downward. Then, as the pectoralis relaxes, the supracoracoideus muscle contracts and, acting as a pulley system, pulls the wing upward.

In *Archaeopteryx,* both jaws contained teeth set in sockets, an archosaurian characteristic. Modern birds are completely toothless, having instead a horny (keratinous) beak molded around the bony jaws. The mandible is a complex of several bones hinged to provide a double-jointed action which permits the mouth to gape widely. Most birds have kinetic skulls (kinetic skulls of lizards are described on p. 355) with a flexible attachment between upper jaw and skull. This attachment allows the upper jaw to move slightly, thus increasing the gape.

The most distinctive feature of the vertebral column is its rigidity. Most vertebrae except the **cervicals** (neck vertebrae) are fused together and with the pelvic girdle form a stiff but light framework to support legs and provide rigidity for flight. To assist in this rigidity, ribs are mostly fused with vertebrae, pectoral girdle, and sternum and braced against each other with uncinate processes (figure 19.5A). Except in flightless birds, the sternum bears a large, thin keel that provides an attachment for powerful flight muscles. Fused clavicles form an elastic furcula that apparently stores energy as it flexes during wing beats. The asymmetrical flight feathers and distinct furcula indicate that *Archaeopteryx* was capable of flight, but the small sternum, offering relatively little area for flight muscle attachment, suggests it was not a strong flier (figure 19.5B).

Bones of the forelimbs are highly modified for flight. They are reduced in number, and several are fused together. Despite these alterations, bird wings are clearly a rearrangement of the basic vertebrate tetrapod limb from which they

arose (see figure 17.1, p. 336), and all the elements—arm, forearm, wrist, and fingers—are represented in modified form (see figure 19.5A).

Muscular System

Locomotor muscles of wings are relatively massive to meet demands of flight. The largest of these is the **pectoralis,** which depresses the wings in flight. Its antagonist is the **supracoracoideus** muscle, which raises the wing (figure 19.7). Surprisingly, perhaps, this latter muscle is not located on the backbone (anyone who has been served the back of a chicken knows that it offers little meat) but is positioned under the pectoralis on the breast. It is attached by a tendon to the upper side of the humerus of the wing so that it pulls from below by an ingenious "rope-and-pulley" arrangement. Both pectoralis and supracoracoideus are anchored to the keel of the sternum. Positioning the main muscle mass low in the body improved aerodynamic stability.

From the main leg muscle mass located in the thigh, thin but strong tendons extend downward through sleevelike sheaths to the toes. Consequently the feet are nearly devoid of muscles, explaining the thin, delicate appearance of a bird's leg. This arrangement places the main muscle mass near a bird's center of gravity and at the same time allows great agility to the slender, lightweight feet. Because the feet are composed mostly of bone, tendon, and tough, scaly skin, they are highly

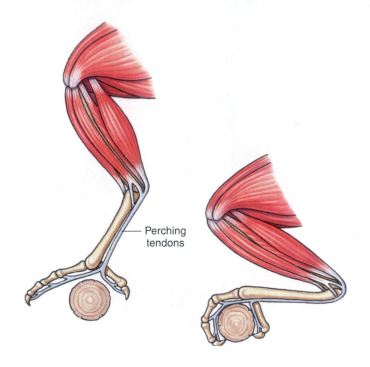

Perching
tendons

figure 19.8

Perching mechanism of a bird. When a bird settles on a branch, tendons automatically tighten, closing the toes around the perch.

resistant to damage from freezing. When a bird perches on a branch, an ingenious toe-locking mechanism (figure 19.8) is activated, which prevents the bird from falling off its perch when asleep. The same mechanism causes the talons of a hawk or owl automatically to sink deeply into its prey as the legs bend under the impact of the strike. The powerful grip of a bird of prey was described by L. Brown.[1]

> When an eagle grips in earnest, one's hand becomes numb, and it is quite impossible to tear it free, or to loosen the grip of the eagle's toes with the other hand. One just has to wait until the bird relents, and while waiting one has ample time to realize that an animal such as a rabbit would be quickly paralyzed, unable to draw breath, and perhaps pierced through and through by the talons in such a clutch.

Digestive System

Birds process an energy-rich diet rapidly and thoroughly with efficient digestive equipment. A shrike can digest a mouse in 3 hours, and berries pass completely through the digestive tract of a thrush in just 30 minutes. Although many animal foods find their way into diets of birds, insects comprise by far the

[1]Brown, L. 1970, Eagles, New York, Arco Publishing.

largest component. Because birds lack teeth, foods that require grinding are reduced in the gizzard. Salivary glands are poorly developed and mainly secrete mucus for lubricating food and the slender, horn-covered **tongue.** There are few taste buds, although all birds can taste to some extent. From the short **pharynx** a relatively long, muscular, elastic **esophagus** extends to the **stomach.** In many birds there is an enlargement **(crop)** at the lower end of the esophagus, which serves as a storage chamber.

In pigeons, doves, and some parrots the crop not only stores food but, during nesting season, produces "milk" by breakdown of epithelial cells of the crop lining. For the first few days after hatching, the helpless young are fed regurgitated crop milk by both parents. Crop milk is especially rich in fat and protein.

The stomach proper consists of a **proventriculus,** which secretes gastric juice, and a muscular **gizzard,** a region specialized for grinding food. To assist grinding food, grain-eating birds swallow gritty objects or pebbles, which lodge in the gizzard. Certain birds of prey, such as owls, form pellets of indigestible materials, mainly bones and fur, in the proventriculus and eject them through the mouth. At the junction of the intestine with the rectum there are paired **ceca;** these are well developed in herbivorous birds in which they serve as fermentation chambers. The terminal part of the digestive system is the **cloaca,** which also receives genital ducts and ureters.

Beaks of birds are strongly adapted to specialized food habits—from generalized types, such as strong, pointed beaks of crows and ravens, to grotesque, highly specialized ones in flamingos, pelicans, and avocets (figure 19.9). The beak of a woodpecker is a straight, hard chisel-like device. Anchored to a tree trunk with its tail serving as a brace, a woodpecker delivers powerful, rapid blows to excavate nest cavities or expose burrows of wood-boring insects. It then uses its long, flexible, barbed tongue to seek insects in their galleries. A woodpecker's skull is especially thick to absorb shock.

Circulatory System

The general plan of circulation in birds is not greatly different from that of mammals, although it evolved independently. Their four-chambered heart is large with strong ventricular walls; thus birds share with mammals a complete separation of respiratory and systemic circulations. Their heartbeat is extremely fast, and as in mammals there is an inverse relationship between heart rate and body weight. For example, a turkey has a heart rate at rest of approximately 93 beats per minute, a chicken has a rate of 250 beats per minute, and a black-capped chickadee has a heart rate of 500 beats per minute when asleep, which may increase to a phenomenal 1000 beats per minute during exercise. Blood pressure in birds is roughly equivalent to that in mammals of similar size. Birds' blood contains **nucleated biconvex erythrocytes.** (Mammals, the only other endothermic vertebrates, have biconcave erythrocytes without nuclei that are somewhat smaller than those of birds.) **Phagocytes,** or mobile ameboid cells of

Raven
Generalized bill

Cardinal
Seed cracker

Flamingo
Mud sifter

American avocet
Worm burrow probe

Pelican
Dip net

Parrot
Nut cracker

Eagle
Meat tearer

Anhinga
Fish spear

figure 19.9

Some bills of birds showing variety of adaptations.

Respiratory System

The respiratory system of birds differs radically from lungs of reptiles and mammals and is marvelously adapted for meeting high metabolic demands of flight. In birds the finest branches of the bronchi, rather than ending in saclike alveoli as in mammals, are tubelike **parabronchi** through which air flows continuously. Also unique is the extensive system of nine interconnecting **air sacs** that are located in pairs in the thorax and abdomen and even extend by tiny tubes into the centers of the long bones (figure 19.10A). Air sacs are connected to the lungs in such a way that perhaps 75% of the inspired air bypasses the lungs and flows directly into the posterior air sacs, which serve as reservoirs for fresh air. On expiration, this oxygenated air is passed through the lungs and collected in the anterior air sacs. From there it flows directly to the outside. Thus, it takes two respiratory cycles for a single breath of air to pass through the respiratory system (figure 19.10B). The advantage of such a system is that an almost continuous stream of oxygenated air is passed through a system of richly vascularized parabronchi.

blood, are particularly efficient in birds in repairing wounds and destroying microbes.

The remarkable efficiency of a bird's respiratory system is emphasized by bar-headed geese that routinely migrate over the Himalayan mountains and have been sighted flying over Mt. Everest (8848 meters or 29,141 feet) under conditions that are severely hypoxic to humans. They reach altitudes of 9000 meters in less than a day, without the acclimatization that is absolutely essential for humans even to approach the upper reaches of Mt. Everest.

Excretory System

The relatively large paired metanephric kidneys are composed of many thousands of **nephrons,** each consisting of a renal corpuscle and a nephric tubule. As in other vertebrates, urine is formed by glomerular filtration followed by selective modification of the filtrate in the tubule.

Birds, like reptiles, excrete their nitrogenous wastes as uric acid rather than urea. In shelled eggs, all excretory products must remain within the eggshell with the growing embryo. Uric acid crystallizes from solution and can be stored harmlessly within the eggshell. Because of uric acid's low solubility, a bird can excrete 1 g of uric acid in only 1.5 to 3 ml of water, whereas a mammal may require 60 ml of water to

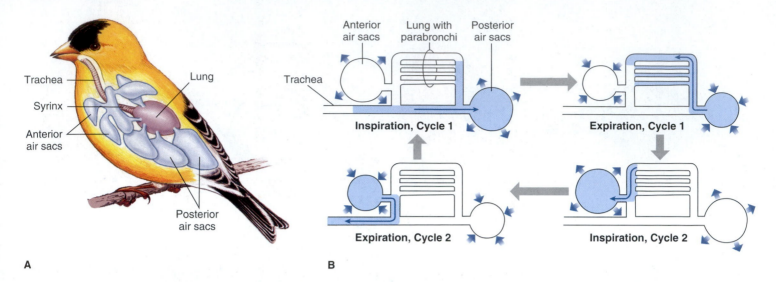

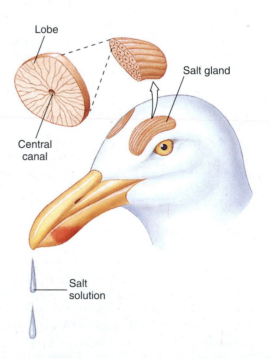

figure 19.10

Respiratory system of a bird. **A,** Lungs and air sacs. One side of the bilateral air sac system is shown. **B,** Movement of a single volume of air through a bird's respiratory system. Two full respiratory cycles are required to move air through the system.

excrete 1 g of urea. Uric acid is combined with fecal material in the cloaca. Excess water is reabsorbed in the cloaca, resulting in formation of a white paste. Thus, despite having kidneys that are less effective in true concentrative ability than mammalian kidneys, birds can form urine containing uric acid nearly 3000 times more concentrated than in their blood. Even the most effective mammalian kidneys, those of certain desert rodents, can excrete urea only about 25 times plasma concentration.

However, bird kidneys are much less efficient than mammalian kidneys in removal of salts, such as sodium, potassium, and chloride. Marine birds (also marine turtles) have evolved a unique method for excreting large loads of salt eaten with their food and in seawater they drink. Seawater contains approximately 3% salt and is three times saltier than a bird's body fluids. Because a bird's kidney cannot concentrate salt in urine above approximately 0.3%, excess salt is removed from their blood by special **salt glands,** one located above each eye (figure 19.11) . These glands are capable of excreting a highly concentrated solution of sodium chloride—up to twice the concentration of seawater. The salt solution runs out the internal or external nostrils, giving gulls, petrels, and other sea birds a perpetual runny nose.

Nervous and Sensory System

The design of a bird's nervous and sensory system reflects the complex problems of flight and a highly visible existence, in which it must gather food, mate, defend territory, incubate and rear young, and correctly distinguish friend from foe. Their brain has well-developed **cerebral hemispheres, cerebellum,** and **optic lobes.** The **cerebral cortex**—chief coordinating center of a mammalian brain—is thin, unfissured, and poorly developed in birds. But the core of the cerebrum, the **dorsal ventricular**

figure 19.11

Salt glands of a marine bird (gull). One salt gland is located above each eye. Each gland consists of several lobes arranged in parallel. One lobe is shown in cross section, much enlarged. Salt is secreted into many radially arranged tubules, then flows into a central canal that leads into the nose.

ridge, has enlarged into the principal integrative center, controlling such activities as eating, singing, flying, and all complex instinctive reproductive activities. Relatively intelligent birds, such as crows and parrots, have larger cerebral hemispheres than do less intelligent birds, such as chickens and pigeons. The

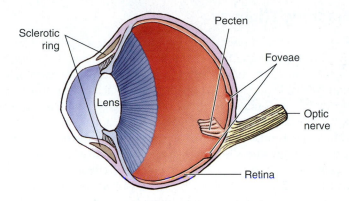

figure 19.12

A hawk eye has all the structural components of a mammalian eye, plus a peculiar pleated structure, or pecten, believed to provide nourishment to the retina. The extraordinarily keen vision of hawks is attributed to the extreme density of cone cells in the foveae: 1.5 million per fovea compared to 0.2 million for humans.

cerebellum is a crucial coordinating center where muscle-position sense, equilibrium sense, and visual cues are assembled and used to coordinate movement and balance. The **optic lobes,** laterally bulging structures of the midbrain, form a visual apparatus comparable to the visual cortex of mammals.

Except in flightless birds, ducks, and vultures, smell and taste are poorly developed in birds. This deficiency, however, is more than compensated by good hearing and superb vision, the keenest in the animal kingdom. The organ of hearing, the **cochlea,** is much shorter than the coiled mammalian cochlea, yet birds can hear roughly the same range of sound frequencies as humans. Actually a bird's ear far surpasses our capacity to distinguish differences in intensities and to respond to rapid fluctuations in pitch.

A bird's eye resembles that of other vertebrates in gross structure but is relatively larger, less spherical, and almost immobile; instead of turning their eyes, birds turn their heads with their long flexible necks to scan the visual field. The light-sensitive **retina** (figure 19.12) is generously equipped with rods (for dim light vision) and cones (for color vision). Cones predominate in day birds, and rods are more numerous in nocturnal birds. A distinctive feature of a bird's eye is the **pecten,** a highly vascularized organ attached to the retina and jutting into the vitreous humor (figure 19.12). The pecten is thought to provide nutrients and oxygen to the eye. It may do more, but its function remains largely a mystery.

The **fovea,** or region of keenest vision on the retina, is in a deep pit (in birds of prey and some others), which makes it necessary for the bird to focus exactly on the subject. Many birds moreover have two sensitive spots (foveae) on the retina (figure 19.12), the central one for sharp monocular views and the posterior one for binocular vision. The visual acuity of a hawk is about eight times that of humans (enabling a hawk to see clearly a crouching rabbit 2 km away), and an owl's ability to see in dim light is more than 10 times that of a human.

Many birds can see at ultraviolet wavelengths, enabling them to view environmental features inaccessible to us but accessible to insects (such as flowers with ultraviolet-reflecting "nectar guides" that attract pollinating insects). Several species of ducks, hummingbirds, kingfishers, and passerines (songbirds) can see at near ultraviolet (UV) wavelengths down to 370 nm (human eyes filter out ultraviolet light below 400 nm). For what purpose do birds use their UV-sensitivity? Some, such as hummingbirds, may be attracted to nectar-guiding flowers, like insects. But, for others, the benefit derived from UV-sensitivity is a matter of conjecture.

Flight

What prompted evolution of flight in birds, the ability to rise free of earthbound concerns, as almost every human has dreamed of doing? The air was a relatively unexploited habitat stocked with flying insects for food. Flight also offered escape from terrestrial predators and opportunity to travel rapidly and widely to establish new breeding areas and to benefit from year-round favorable climate by migrating north and south with the seasons.

Bird Wing as a Lift Device

A bird's wing is an airfoil subject to recognized laws of aerodynamics. It is streamlined in cross section, with a slightly concave lower surface (**cambered**) and with small, tight-fitting feathers where the leading edge meets the air (figure 19.13). Air slips smoothly over the wing, creating lift with minimum drag. Some lift is produced by positive pressure against the undersurface of the wing. But on the upper side, where the airstream must travel farther and faster over a convex surface, negative pressure is created that provides more than two-thirds of the total lift.

The lift-to-drag ratio of an airfoil is determined by the angle of tilt (angle of attack) and airspeed (figure 19.13). A wing carrying a given load can pass through the air at high speed and small angle of attack or at low speed and larger angle of attack. As speed decreases, lift can be increased by increasing the angle of attack, but drag forces also increase. Finally a point is reached at which the angle of attack becomes too steep; turbulence appears on the upper surface, lift is destroyed, and stalling occurs. Stalling can be delayed or prevented by placing a **wing slot** along the leading edge so that a layer of rapidly moving air is directed across the upper wing surface. Wing slots were and still are used in aircraft traveling at a low speed. In birds, two kinds of wing slots have developed: (1) the **alula,** or group of small feathers on the thumb (see figures 19.5A and 19.16), which provides a mid-wing slot, and (2) **slotting between the primary feathers,** which provides a wing-tip slot. In a number of songbirds, these together provide stall-preventing slots for nearly the entire outer (and aerodynamically more important) half of the wing.

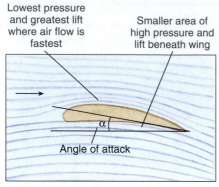

Air flow around wing

Lowest pressure and greatest lift where air flow is fastest

Smaller area of high pressure and lift beneath wing

Angle of attack

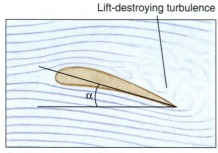

Stalling at low speed

Lift-destroying turbulence

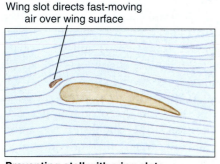

Preventing stall with wing slots

Wing slot directs fast-moving air over wing surface

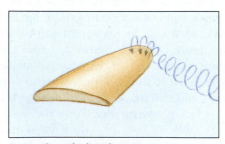

Formation of wing tip vortex

figure 19.13

Air patterns formed by the airfoil, or wing, moving from right to left. At low speed the angle of attack (α) must increase to maintain lift but this increases the threat of stalling. The upper figures show how low-speed stalling can be prevented with wing slots. Wing tip vortex (*bottom*), a turbulence that tends to develop at high speeds, reduces flight efficiency. The effect is reduced in wings that sweep back and taper to a tip.

Flapping Flight

Two forces are required for flapping flight: a vertical *lifting* force to support the bird's weight, and a horizontal *thrusting* force to move the bird forward against resistive forces of friction. Thrust is provided mainly by primary feathers at the wing tips, while secondary feathers of the highly cambered inner wing, which do not move so far or so fast, act as an airfoil, providing mainly lift. Greatest power is applied on the downstroke. The primary feathers bend upward and twist to a steep angle of attack, biting into the air like a propeller (figure 19.14). The entire wing (and the bird's body) is pulled forward. On the upstroke, the primary feathers bend in the opposite direction so that their upper surfaces twist into a positive angle of attack to produce thrust, just as the lower surfaces did on the downstroke. A powered upstroke is essential for hovering flight, as in hummingbirds (figure 19.15), and is important for fast, steep takeoffs by small birds with elliptical wings.

Basic Forms of Bird Wings

Bird wings vary in size and form because successful exploitation of different habitats has imposed special aerodynamic requirements. Four types of bird wings are easily recognized.

Elliptical Wings

Birds such as sparrows, warblers, doves, woodpeckers, and magpies (figure 19.16A) that must maneuver in forested habitats, have elliptical wings. This type has a **low aspect ratio** (ratio of length to average width). Wings of the highly maneuverable British Spitfire fighter plane of World War II fame conformed closely to the outline of a sparrow's wing. Elliptical wings are slotted with an alula and between the primary feathers; this slotting helps prevent stalling during sharp turns, low-speed flight, and frequent landing and takeoff. Each separated primary feather behaves as a narrow wing with a high angle of attack, providing high lift at low speed. High maneuverability of elliptical wings is exemplified by the tiny chickadee, which, if frightened, can change course within 0.03 second.

High-Speed Wings

Birds that feed during flight, such as swallows, hummingbirds, and swifts, or that make long migrations, such as plovers, sandpipers, terns and gulls (figure 19.16B), have wings that sweep back and taper to a slender tip. They are rather flat in section, have a moderately high aspect ratio, and lack wing-tip slotting characteristic of elliptical wings. Sweepback and wide separation of wing tips reduce "tip vortex" (see figure 19.13, bottom panel), a drag-creating turbulence that tends to develop at wing tips at faster speeds. This type of wing is aerodynamically efficient for high-speed flight but cannot easily keep a bird airborne at low speeds. The fastest birds, such as sandpipers, clocked at 175 km (109 miles) per hour, belong to this group.

figure 19.14

In normal flapping flight of strong fliers like ducks, the wings sweep downward and forward fully extended. Thrust is provided by primary feathers at the wing tips. To begin the upbeat, the wing is bent, bringing it upward and backward. The wing then extends, ready for the next downbeat.

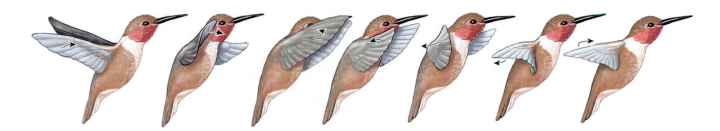

figure 19.15

The secret of a hummingbird's ability to change direction instantly, or hang motionless in the air while sipping nectar from a flower, lies in its wing structure. The wing is nearly rigid, but hinged at the shoulder by a swivel joint and powered by a supracoracoideus muscle that is unusually large for the bird's size. When hovering the wing moves in a sculling motion. The leading edge of the wing moves forward on the forward stroke, then swivels nearly 180 degrees at the shoulder to move backward on the backstroke. The effect is to provide lift without propulsion on *both* forward and backstrokes.

Soaring Wings

Oceanic soaring birds have **high aspect ratio** wings resembling those of sailplanes. This group includes albatrosses, frigate birds, and gannets (figure 19.16C). Such long, narrow wings lack wing slots and are adapted for high speed, high lift, and dynamic soaring. They have the highest aerodynamic efficiency of all wings but are less maneuverable than the wide, slotted wings of land soarers. Dynamic soarers have learned to exploit the highly reliable sea winds, using adjacent air currents of different velocities.

High-Lift Wings

Vultures, hawks, eagles, owls, and ospreys (figure 19.16D)—predators that carry heavy loads—have wings with slotting, alulas, and pronounced camber, all of which promote high lift at low speed. Many of these birds are land soarers, with broad, slotted wings that provide the sensitive response and maneuverability required for static soaring in the capricious air currents over land.

Migration and Navigation

We described advantages of migration in the prologue to this chapter. Not all birds migrate, of course, but most North American and European species do, and the biannual journeys of some are truly extraordinary undertakings. Migration is both the greatest adventure and the greatest risk in the life of a migratory bird.

Migration Routes

Most migratory birds have well-established routes trending north and south. Since most birds (and other animals) breed in the Northern Hemisphere, where most of the earth's landmass is concentrated, most birds migrate south in the northern winter and north in the northern summer to nest. Of the 4000 or more species of migrant birds (a little less than half the total bird species), most breed in the more northern latitudes of the hemisphere; the percentage of migrants breeding in Canada is far higher than the percentage of migrants breeding in Mexico,

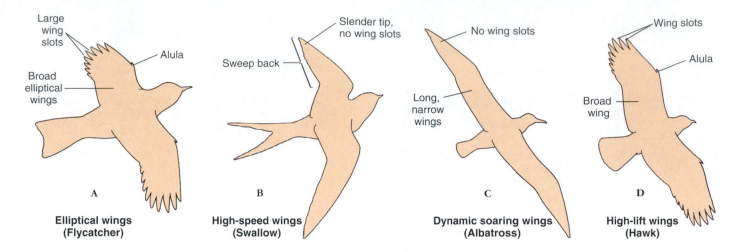

Large
wing
slots

Alula

Broad
elliptical
wings

A

**Elliptical wings
(Flycatcher)**

Slender tip,
no wing slots

Sweep back

B

**High-speed wings
(Swallow)**

No wing slots

Long,
narrow
wings

C

**Dynamic soaring wings
(Albatross)**

Wing slots

Alula

Broad
wing

D

**High-lift wings
(Hawk)**

figure 19.16

Four basic forms of bird wings.

for example. Some use different routes in the fall and spring (figure 19.17). Some, especially certain aquatic species, complete their migratory routes in a very short time. Others, however, make a leisurely trip, often stopping along the way to feed. Some warblers are known to take 50 to 60 days to migrate from their winter quarters in Central America to their summer breeding areas in Canada.

Some species are known for their long-distance migrations. Arctic terns, greatest globe spanners of all, breed north of the Arctic Circle during the northern summer then migrate to Antarctic regions for the northern winter. This species is also known to take a circuitous route in migrations from North America, passing over to the coastlines of Europe and Africa and then to winter quarters, a trip that may exceed 18,000 km (11,200 miles).

Many small songbirds, such as warblers, vireos, thrushes, flycatchers, and sparrows, also make great migratory treks (figure 19.17). Migratory birds that nest in Europe or Central Asia spend the northern winter in Africa.

Stimulus for Migration

Humans have known for centuries that onset of reproductive cycles of birds is closely related to season. Only relatively recently, however, has it been shown that lengthening days of late winter and early spring stimulate development of gonads and accumulation of fat—both important internal changes that predispose birds to migrate northward. Increasing day length stimulates the anterior lobe of the pituitary into activity. Release of pituitary gonadotropic hormone in turn sets in motion a complex series of physiological and behavioral changes, resulting in gonadal growth, fat deposition, migration, courtship and mating behavior, and care of the young.

Direction Finding in Migration

Numerous experiments suggest that most birds navigate chiefly by sight. Birds recognize topographical landmarks and follow familiar migratory routes—a behavior assisted by flock migration, during which navigational resources and experience of older birds can be pooled. In addition to visual navigation, birds use a variety of orientation cues at their disposal. Birds have a highly accurate sense of time. Recent work adds credence to an old, much debated hypothesis that birds can detect and navigate by the earth's magnetic field. These navigational abilities are primarily instinctive, although they may require calibration with existing landmarks and may improve with experience.

In the early 1970s W. T. Keeton showed that the flight bearings of homing pigeons were significantly disturbed by magnets attached to the birds' heads, or by minor fluctuations in the geomagnetic field. But until recently the nature and position of a magnetic receptor in pigeons remained a mystery. Deposits of a magnetic substance called magnetite (Fe_3O_4) have been discovered in the neck musculature of pigeons and migratory white-crowned sparrows. If this material is coupled to sensitive muscle receptors, as has been proposed, the structure could serve as a magnetic compass that would enable birds to detect and orient their migrations to the earth's magnetic field.

Experiments by German ornithologists G. Kramer and E. Sauer and American ornithologist S. Emlen demonstrated convincingly that birds can navigate by celestial cues: the sun by day and the stars by night. Using special circular cages, Kramer concluded that birds maintain compass direction by referring to the sun, regardless of the time of day (figure 19.18). This process is called **sun-azimuth orientation** (*azimuth,* com-

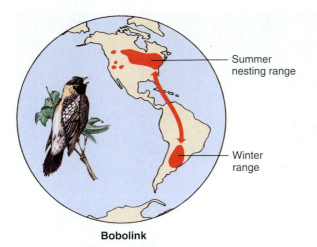

Bobolink

American golden plover

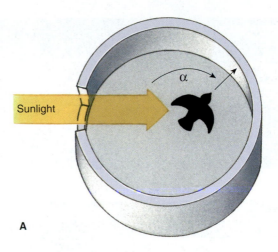

A

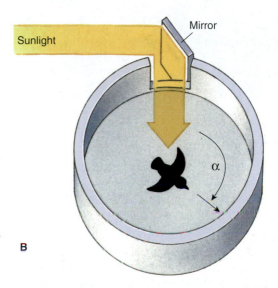

B

figure 19.17

Migrations of bobolinks, *Dolichonyx oryzivorus,* and American golden plovers, *Pluvialis dominica.* Bobolinks commute 22,500 km (14,000 miles) each year between nesting sites in North America and their range in Argentina, where they spend the northern winters, a phenomenal feat for such a small bird. Although the breeding range has extended to colonies in western areas, these birds take no shortcuts but adhere to the ancestral seaboard route. American golden plovers fly a loop migration, striking out across the Atlantic in their southward autumnal migration but returning in the spring by way of Central America and the Mississippi Valley because ecological conditions are more favorable at that time.

figure 19.18

Gustav Kramer's experiments with sun-compass navigation in starlings. **A,** In a windowed, circular cage, the bird fluttered to align itself in the direction it would normally follow if it were free. **B,** When the true angle of the sun is deflected with a mirror, the bird maintains the same relative position to the sun. This shows that these birds use the sun as a compass. The bird navigates correctly throughout the day, changing its orientation to the sun as the sun moves across the sky.

pass bearing of the sun). Sauer's and Emlen's ingenious planetarium experiments also strongly suggest that some birds, probably many, are able to detect and navigate by the North Star axis around which the constellations appear to rotate.

Some remarkable feats of bird navigation still defy rational explanation. Most birds undoubtedly use a combination of environmental and innate cues to migrate. Migration is a rigorous undertaking. The target is often small, and natural selection relentlessly prunes off individuals making errors in migration, leaving only the best navigators to propagate the species.

Social Behavior and Reproduction

The adage says "birds of a feather flock together," and many birds are indeed highly social creatures. Especially during the breeding season, sea birds gather, often in enormous colonies, to nest and rear young. Land birds, with some conspicuous exceptions, such as starlings and rooks, tend to be less gregarious than sea birds during breeding and to seek isolation for rearing their brood. But these same species that covet separation from their kind during breeding may aggregate for migration or feeding.

A

B

figure 19.19

Cooperative feeding behavior by white pelicans, *Pelecanus onocrotalus.*
A, Pelicans form a horseshoe to drive fish together. **B,** Then they plunge simultaneously to scoop fish in their huge bills. These photographs were taken 2 seconds apart.

figure 19.20

Copulation in waved albatrosses, *Diomeda irrorata.* In most bird species males lack a penis. A male copulates by standing on the back of a female, pressing his cloaca against that of the female, and passing sperm to the female.

Togetherness offers advantages: mutual protection from enemies, greater ease in finding mates, less opportunity for individual straying during migration, and mass huddling for protection against low night temperatures during migration. Certain species, such as pelicans (figure 19.19), may use highly organized cooperative behavior to feed. At no time are the highly organized social interactions of birds more evident than during the breeding season, as they stake out territorial claims, select mates, build nests, incubate and hatch their eggs, and rear their young.

Reproductive System

During most of the year the **testes** of males are tiny bean-shaped bodes. But during the breeding season they enlarge greatly, to as much as 300 times their nonbreeding size. Because males of most species lack a penis, copulation is a matter of bringing cloacal surfaces into contact, usually while the male stands on the back of the female (figure 19.20). Some swifts copulate in flight.

In females of most birds, only the **left ovary and oviduct** develop (figure 19.21); those on the right dwindle to vestigial structures (loss of one ovary is another adaptation of birds for reducing weight). Eggs discharged from the ovary are picked up by the oviduct, which runs posteriorly to the cloaca.

While eggs are passing down the oviduct, **albumin,** or egg white, from special glands is added to them; farther down the oviduct, shell membrane, shell, and shell pigments are secreted around the egg. Fertilization occurs in the upper oviduct several hours before layers of albumin, shell membranes, and shell are added. Sperm remain alive in the female oviduct for many days after a single mating.

Mating Systems

Two types of mating systems are **monogamy,** in which an individual has only one mate, and **polygamy,** in which an individual has more than one mate during a breeding period. Monogamy is rare in most animal groups, but it is common in birds; more than 90% are monogamous. In a few bird species such as swans and geese, partners are chosen for life and often remain together throughout the year. Seasonal monogamy is more common; as the great majority of migrant birds pair during the breeding season but lead independent lives the rest of the year.

One reason monogamy is much more common among birds than among mammals is that male and female birds are equally adept at most aspects of parental care. Male mammals do not gestate the young and do not lactate and thus can provide little help in caring for the young. Female and male birds can alternate care of the nest and young, which permits one parent to be at the nest at all times. For some species, a female remains on the nest for months at a time but is brought food by the male. This constant attendance to the nest may be particularly important in species that would experience high loss

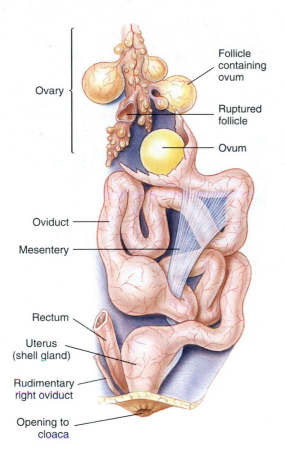

figure 19.21

Reproductive system of a female bird. For most birds, only the left ovary and reproductive tract are functional. Structures on the right dwindle to vestiges.

Labels: Ovary, Follicle containing ovum, Ruptured follicle, Ovum, Oviduct, Mesentery, Rectum, Uterus (shell gland), Rudimentary right oviduct, Opening to cloaca

figure 19.22

Dominant male sage grouse, *Centrocercus urophasianus,* surrounded by several hens that have been attracted by his "booming" display.

(figure 19.22). There is nothing of value in a lek to the female except the male, and all he can offer are his genes, for only females care for the young. Usually there are a dominant male and several subordinate males in a lek. Competition among males for females is intense, but females appear to choose the dominant male for mating because, presumably, social rank correlates with genetic quality.

Nesting and Care of Young

To produce offspring, all birds lay eggs that must be incubated by one or both parents. Most duties of incubation fall on females, although in many instances both parents share the task, and occasionally only males incubate the eggs.

Most birds build some form of nest in which to rear their young. Some birds simply lay eggs on bare ground or rocks, making no pretense of nest building. Others build elaborate nests such as the pendant nests constructed by orioles, the delicate lichen-covered mud nests of hummingbirds (figure 19.23) and flycatchers, the chimney-shaped mud nests of cliff swallows, the floating nests of rednecked grebes, and the huge sand and vegetation mounds of Australian brush turkeys. Most birds take considerable pains to conceal their nests from enemies. Nest parasites such as brown-headed cowbirds and European cuckoos build no nests at all but simply lay their eggs in nests of birds smaller than themselves. When their eggs hatch, the foster parents care for the cowbird young which outcompete the host's own hatchlings.

Newly hatched birds are of two types: **precocial** and **altricial.** Precocial young, such as quail, fowl, ducks, and most water birds, are covered with down when hatched and can run or swim as soon as their plumage is dry (figure 19.24). Altricial young, on the other hand, are naked and helpless at birth and remain in the nest for a week or more. Young of both types require care from parents for some time after hatching. Parents of altricial species must carry food to their young almost constantly, for most young birds will eat more than their weight each day. Some birds are not easily categorized as precocial or

of eggs or young to predators or rival birds if a nest were left unguarded. For many bird species, the high demands on a male to care for the young or his mate preclude the establishment of nests with additional females.

Although most birds have a monogamous mating system, either member of a pair may mate with an individual that is not the partner. Recent DNA analyses have shown that most passerine species frequently are "unfaithful," engaging in extra-pair copulations. As a result, nests of many of these monogamous species contain a sizeable portion (30% or more) of young with fathers other than the attendant male. Why do individuals engage in extra-pair copulations? By mating with an individual of better genetic quality, fitness of the offspring can be improved. Also, mating with multiple partners increases the genetic variation of the offspring. Males are able to father more offspring with additional partners, and receive an additional benefit in that the mate of the extra-pair female he has mated provides parental care to his offspring!

The most common form of polygamy in birds is **polygyny** ("many females"), in which a male mates with more than one female. In many species of grouse, males gather in a collective display ground, or **lek,** which is divided into individual territories, each vigorously defended by a displaying male

figure 19.23

Anna's hummingbird, *Calypte anna,* feeding its young in its nest of plant down and spider webs and decorated on the outside with lichens. A female builds a nest, incubates two pea-sized eggs, and rears the young with no assistance from a male. Anna's hummingbird is a common resident of California. It is the only hummingbird to overwinter in the United States.

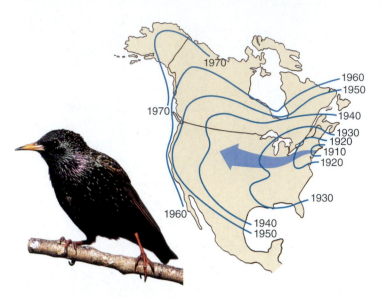

figure 19.25

Colonization of North America by starlings, *Sturnus vulgaris,* after the introduction of 120 birds into Central Park in New York City in 1890. There are now perhaps 200 million starlings in the United States alone, testimony to the great reproductive potential of birds. Starlings are omnivorous, eating mostly insects in spring and summer and shifting to wild fruits in the fall.

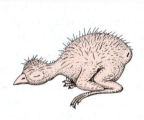 

Altricial
One-day-old meadowlark

Precocial
One-day-old ruffed grouse

figure 19.24

Comparison of 1-day-old altricial and precocial young. The altricial meadowlark (*left*) is born nearly naked, blind, and helpless. The precocial ruffed grouse (*right*) is covered with down, alert, strong legged, and able to feed itself.

altricial because their young are intermediate in development at birth. For example, gulls and terns are born covered with down and with eyes open, but are unable to leave the nest for some time.

Although it may seem that precocial young have all the advantages, with their greater ability to find food and escape predation, altricial birds have some advantages of their own. Because altricial birds lay relatively small eggs with minimal yolk supplies, the mother has a small investment in the eggs and can easily replace eggs lost to predation or extreme

weather conditions. Altricial young also grow faster, perhaps due to the higher growth potential of immature tissue.

Humans and Bird Populations

Occasionally activities of people may cause spectacular changes in bird distribution. Both starlings (figure 19.25) and house sparrows have been accidentally or deliberately introduced into numerous countries, to become the two most abundant bird species on earth, with the exception of domestic fowl.

Humans also are responsible for the extinction of many bird species. More than 80 species of birds have, since 1695, followed the last dodo to extinction. Many were victims of changes in their habitat or competition with introduced species. Overhunting contributed to extinction of some species, among them passenger pigeons, which only a century ago darkened skies over North America in incredible numbers estimated in the billions (figure 19.26).

Today, game bird hunting is a well-managed renewable resource in the United States and Canada, and while hunters kill millions of game birds each year, none of the 74 bird species legally hunted are endangered. Hunting interests, by acquiring large areas of wetlands for migratory bird refuges and sanctuaries, have contributed to the recovery of both game and nongame birds.

figure 19.26

Sport-shooting of passenger pigeons in Louisiana during the nineteenth century. Relentless sport and market hunting before establishment of state and federal hunting regulations, in addition to clearing of the hardwood forests that served as nesting habitats, eventually dropped the population too low to sustain colonial breeding. The last passenger pigeon died in captivity in 1914.

Of particular concern is the recent sharp decline of songbirds in the United States and southern Canada. Amateur bird-watchers and ornithologists have recorded that many songbird species that were abundant as recently as 40 years ago are now suddenly scarce. There are several reasons for the decline. Intensification of agriculture, permitted by the use of herbicides, pesticides, and fertilizers, has deprived ground-nesting birds of fields that were left fallow before use of these agents. Excessive fragmentation of forests throughout much of the United States has increased exposure of nests of forest-dwelling species to nest predators such as blue jays, raccoons, and opossums, and to nest parasites such as brown-headed cowbirds. House cats also kill millions of small birds every year. From a study of radio-collared farm cats in Wisconsin, researchers estimated that in that state alone, cats may kill 19 million songbirds in a single year.

The rapid loss of tropical forests—approximately 170,000 square kilometers each year, an area about the size of the state of Washington—is depriving some 250 species of songbird migrants of their wintering homes. Of all long-term threats facing songbird populations, tropical deforestation is the most serious and most intractable to change.

Lead poisoning of waterfowl is a side effect of hunting. Before long-delayed federal regulations went into effect in 1991, requiring use of nonlead shot for all inland and coastal waterfowl hunting, shotguns scattered more than 3000 tons of lead each year in the United States alone. When waterfowl eat the pellets (which they mistake for seeds), the pellets are ground and eroded in their gizzards, facilitating absorption of lead into their blood. Lead poisoning paralyzes or weakens birds, leading to death by starvation. Today, birds are still dying from ingesting lead shot that has accumulated over the years.

Some birds, such as robins, house sparrows, and starlings, can accommodate these changes, and may even thrive on them. But for most birds the changes are adverse. Terborgh (1992) warns that unless we take leadership in managing our natural resources wisely we soon could be facing "the silent spring" that Rachel Carson envisioned in 1962.

classification of Class Aves

Class Aves contains more than 9900 species distributed among 25 orders of living birds and a few fossil birds. There is little consensus for the relationships among bird orders, and the monophyly of several bird orders is in question. Until recently classifications have relied on shared, derived morphological characters. An alternative taxonomy has been proposed that is based on DNA hybridization studies by Sibley and Ahlquist (1990). This study suggests a number of surprising relationships, including placement of penguins, loons, grebes, albatrosses, and birds of prey in order Ciconiiformes, which traditionally only contained the herons and relatives. Relationships implied by this biochemical study have not been tested with other phylogenetic studies using molecular techniques. Questions persist about the validity of the "molecular clock" which is critical for DNA hybridization studies. Given this uncertainty, we present a traditional classification, primarily based on morphological characters.

Class Aves (L. *avis*, bird)

Subclass Archaeornithes (Gr. *archaios*, ancient, + *ornis*, bird). Birds of the late Jurassic and early Cretaceous bearing many primitive characteristics. *Archaeopteryx.*

Subclass Neornithes (Gr. *neos*, new, + *ornis*, bird). Extinct and living birds with well-developed sternum and usually with keel; tail reduced; metacarpals and some carpals fused together. Cretaceous to Recent.

Superorder Paleognathae (Gr. *palaios*, ancient, + *gnathos*, jaw). Modern birds with primitive archosaurian palate. Ratites (with unkeeled sternum) and tinamous (with keeled sternum).

Order Struthioniformes (stroo′thi-on-i-for′meez) (L. *struthio*, ostrich, + *forma*, form): **ostrich, rheas, cassowaries, emus, kiwis.** Fifteen species of flightless birds of Africa, South America, Australia, New Guinea, and New Zealand. The ostrich, *Struthio camelus* (figure 19.27), of Africa, is the largest of living birds, with some specimens being 2.4 m tall and weighing 135 kg. The feet are provided with only two toes of unequal size covered with pads, which enable the birds to travel rapidly over sandy ground. The kiwis, about the size of domestic fowl, are unusual in having only the merest vestige of a wing.

Order Tinamiformes (tin-am′i-for′meez) (N.L. *Tinamus*, type genus, + form): **tinamous.** Ground-dwelling, grouselike birds of Central and South America. About 47 species.

Superorder Neognathae (Gr. *neos*, new, + *gnathos*, jaw). Modern birds with flexible palate.

Order Sphenisciformes (sfe-nis′i-for′meez) (Gr. *Sphēiskos*, dim. of *sphen*, wedge, from the shortness of the wings, + form): **penguins.** Web-footed marine swimmers of southern seas from Antarctica north to the Galápagos Islands. Although penguins are carinate birds, they use their wings as paddles for swimming rather than for flight. About 17 species.

Order Gaviiformes (gay′vee-i-for′meez) (L. *gavia*, bird, probably sea mew, + form): **loons.** The five species of loons are remarkable swimmers and divers with short legs and heavy bodies. They live exclusively on fish and small aquatic forms. The familiar great northern diver, *Gavia immer,* is found mainly in northern waters of North America and Eurasia.

Order Podicipediformes (pod′i-si-ped′i-for′meez) (L. *podex*, rump; *pes, pedis*, foot): **grebes.** These are short-legged divers with lobate-webbed toes. The pied-billed grebe, *Podilymbus podiceps,* is a familiar example of this order. Grebes are most common in old ponds where they build their raftlike floating nests. Twenty-one species, worldwide distribution.

Order Procellariiformes (pro-sel-lar′ee-i-for′meez) (L. *procella*, tempest, + form): **albatrosses, petrels, fulmars, shearwaters.** All are marine birds with hooked beak and tubular nostrils. In wingspan (more than 3.6 m in some), albatrosses are the largest of flying birds. About 115 species, worldwide distribution.

Order Pelecaniformes (pele-can-i-for′meez) (Gr. *pelekan*, pelican, + form): **pelicans, cormorants, gannets, boobies, and others.** These are colonial fish-eaters with throat pouch and all four toes of each foot included within the web. About 65 species, worldwide distribution, especially in the tropics.

Order Ciconiiformes (si-ko′nee-i-for′meez) (L. *ciconia*, stork, + form): **herons, bitterns, storks, ibises, spoonbills, flamingos** (figure 19.28) **vultures.** These are long-necked, long-legged, mostly colonial waders and vultures. A familiar eastern North American representative is the great blue heron, *Ardea herodias,* which frequents marshes and ponds. About 120 species, worldwide distribution.

Order Falconiformes (fal′ko-ni-for′meez) (L. *falco*, falcon, + form): **eagles, hawks, falcons, condors, buzzards.** Diurnal birds of prey. All are strong fliers with keen vision and sharp, curved talons. About 310 species, worldwide distribution.

figure 19.27

Ostrich, *Struthio camelus,* of Africa, the largest of all living birds. Order Struthioniformes.

Order Anseriformes (an′ser-i-for′meez) (L. *anser,* goose + form): **swans, geese, ducks.** Members of this order have broad bills with filtering ridges at their margins, a foot web restricted to the front toes, and a long breastbone with a low keel. About 160 species, worldwide distribution.

Order Galliformes (gal′li-for′meez) (L. *gallus,* cock, + form): **quail, grouse, pheasants, ptarmigan, turkeys, domestic fowl.** Chickenlike ground-nesting herbivores with strong beaks and heavy feet. The bobwhite quail, *Colinus virginianus,* occurs across the eastern half of the United States. The ruffed grouse, *Bonasa umbellus,* is found in about the same region, but in woods instead of the open pastures and grain fields, that the bobwhite frequents. About 290 species, worldwide distribution.

Order Gruiformes (groo′i-for′meez) (L. *grus,* crane, + form): **cranes, rails, coots, gallinules.** Mostly prairie and marsh breeders. About 215 species, worldwide distribution.

Order Charadriiformes (ka-rad′ree-i-for′meez) (N.L. *Charadrius,* genus of plovers, + form): **gulls** (figure 19.29), **oyster catchers, plovers, sandpipers, terns, woodcocks, turnstones, lapwings, snipe, avocets, phalaropes, skuas, skimmers, auks, puffins.** All are shorebirds. They are strong fliers and are usually colonial. About 330 species, worldwide distribution.

Order Columbiformes (ko-lum′bi-for′meez) (L. *columba,* dove, + form): **pigeons, doves.** All have short necks, short legs, and a short, slender bill. The flightless dodo, *Raphus cucullatus,* of the Mauritius Islands became extinct in 1681. About 320 species, worldwide distribution.

Order Psittaciformes (sit′ta-si-for′meez) (L. *psittacus,* parrot, + form): **parrots, parakeets.** Birds with hinged and movable upper beak, fleshy tongue. About 370 species, pantropical distribution.

Order Musophagiformes (myu′-so-fa-ji-for′meez) (L. *musa,* banana, + Gr. *phagō,* to eat + form): **turacos.** Medium to large birds of dense forest or forest edge with a conspicuous patch of crimson on the spread wing. Bill brightly colored, wings short and rounded. Twenty-three species restricted to Africa.

Order Cuculiformes (ku-koo′li-for′meez) (L. *cuculus,* cuckoo, + form): **cuckoos, roadrunners.** European cuckoos, *Cuculus canorus,* lay their eggs in nests of smaller birds, which rear the young cuckoos. American cuckoos, black billed and yellow billed, usually rear their own young. About 150 species, worldwide distribution.

Order Strigiformes (strij′i-for′meez) (L. *strix,* screech owl, + form): **owls.** Nocturnal predators with large eyes, powerful beaks and feet, and silent flight. About 185 species, worldwide distribution.

Order Caprimulgiformes (kap′ri-mul′ji-for′meez) (L. *caprimulgus,* goatsucker, + form): **goatsuckers, nighthawks, whippoorwills.** Night and twilight feeders with small, weak legs and wide mouths fringed with bristles. Whippoorwills, *Antrostomus vociferus,* are common in woods of the eastern states, and nighthawks, *Chordeiles minor,* are often seen and heard in the evening flying around city buildings. About 115 species, worldwide distribution.

Order Apodiformes (up-pod′i-for′meez) (Gr. *apous,* footless, + form): **swifts, hummingbirds.** These are small birds with short legs and rapid wingbeat. The familiar chimney swift, *Chaetura pelagia,* fastens its nest in chimneys by means of saliva. A swift found in China builds a nest of saliva that is used by Chinese people for soup making. Most species of hummingbirds are found in the tropics, but there are 14 species in the

figure 19.28
Greater flamingos, *Phoenicopterus ruber,* on an alkaline lake in East Africa. Order Ciconiiformes.

figure 19.29
Laughing gulls, *Larus atricilla,* in flight. Order Charadriiformes.

United States, of which only one, the ruby-throated hummingbird, is found in the eastern part of the country. About 435 species, worldwide distribution.

Order Coliiformes (ka-lee′i-for′meez) (Gr. *kolios,* green woodpecker, + form): **mousebirds.** Small crested birds of uncertain relationship. Six species restricted to southern Africa.

Order Trogoniformes (tro-gon′i-for′meez) (Gr. *trōgon,* gnawing, + form): **trogons.** Richly colored, long-tailed birds. About 40 species, pantropical distribution.

Order Coraciiformes (ka-ray′see-i-for′meez or kor′uh-sigh′uh-for′meez) (N.L. *coracii* from Gr. *korakias,* a kind of raven, + form): **kingfishers, hornbills, and others.** Birds with strong, prominent bills that nest in cavities. In the eastern half of the United States, belted kingfishers, *Megaceryle alcyon,* are common along most waterways of any size. About 220 species, worldwide distribution.

Order Piciformes (pis′i-for′meez) (L. *picus,* woodpecker, + form): **woodpeckers, toucans, puffbirds, honeyguides.** Birds with highly specialized bills and having two toes extending forward and two backward. All nest in cavities. There are many species of woodpeckers in North America, most common of which are flickers and downy, hairy, red-bellied, redheaded, and yellow-bellied woodpeckers. Largest is the pileated woodpecker, which is usually found in deep and remote woods. About 410 species, worldwide distribution.

Order Passeriformes (pas′er-i-for′meez) (L. *passer,* sparrow, + form): **perching songbirds** (figure 19.30).

This is the largest order of birds, containing 56 families and 60% of all birds. Most have a highly developed syrinx. Their feet are adapted for perching on thin stems and twigs. The young are altricial. To this order belong many birds with beautiful songs such as thrushes, warblers, mockingbird, meadowlark, and hosts of others. Others of this order, such as swallows, magpie, starling, crows, raven, jays, nuthatches, and creepers, have no songs worthy of the name. More than 5900 species, worldwide distribution.

figure 19.30

Ground finch, *Geospiza fuliginosa,* one of the famous Darwin's finches of the Galápagos Islands. Order Passeriformes.

Summary

The more than 9900 species of living birds are egg-laying, endothermic vertebrates with feathers and having forelimbs modified as wings. Birds are closest phylogenetically to theropods, a group of Mesozoic dinosaurs with several birdlike characteristics. The oldest known fossil bird, *Archaeopteryx* from the Jurassic period of the Mesozoic era, had numerous reptilian characteristics and was almost identical to certain theropod dinosaurs except that it had feathers. It is probably the sister taxon of modern birds.

Adaptations of birds for flight are of two basic kinds: those reducing body weight and those promoting more power for flight. Feathers, the hallmark of birds, are complex derivatives of reptilian scales and combine lightness with strength, water repellency, and high insulative value. Body weight is further reduced by elimination of some bones, fusion of others (to provide rigidity for flight), and presence in many bones of hollow, air-filled spaces. The light, keratinized bill, replacing the heavy jaws and teeth of reptiles, serves as both hand and mouth for all birds and is variously adapted for different feeding habits.

Adaptations that provide power for flight include a high metabolic rate and body temperature coupled with an energy-rich diet; a highly efficient respiratory system consisting of a system of air sacs arranged to provide a constant, one-way flow of air through the lungs; powerful flight and leg muscles arranged to place muscle weight near the bird's center of gravity; and an efficient, high-pressure circulation.

Birds have keen eyesight, good hearing, and superb coordination for flight. Their metanephric kidneys produce uric acid as the principal nitrogenous waste.

Birds fly by applying the same aerodynamic principles as an airplane and using similar equipment: wings for lift, support, and propulsion; a tail for steering and landing control, and wing slots for control at low flight speed. Flightlessness in birds is unusual but has evolved independently in several bird orders, usually on islands where terrestrial predators are absent; all are derived from flying ancestors.

Bird migration refers to regular movements between summer nesting places and wintering regions. Spring migration to the north, where more food is available for nestlings, enhances reproductive success. Many cues are used for finding direction during migration, including innate sense of direction and ability to navigate by the sun, stars, or earth's magnetic field.

The highly developed social behavior of birds is manifested in vivid courtship displays, mate selection, territorial behavior, and incubation of eggs and care of the young.

Review Questions

1. Explain the significance of the discovery of *Archaeopteryx*. Why did this fossil demonstrate beyond reasonable doubt that birds share an ancestor with some reptilian groups?
2. The special adaptations of birds contribute to two essentials for flight: more power and less weight. Explain how each of the following contributes to one or both of these two essentials: feathers, skeleton, muscle distribution, digestive system, circulatory system, respiratory system, excretory system, reproductive system.
3. How do marine birds rid themselves of excess salt?
4. In what ways are a bird's ears and eyes specialized for demands of flight?
5. Explain how a bird wing is designed to provide lift. What design features help to prevent stalling at low flight speeds?
6. Describe four basic forms of bird wings. How does wing shape correlate with bird size and nature of flight (whether powered or soaring)?
7. What are advantages of seasonal migration for birds?
8. Describe different navigational resources birds may use in long-distance migration.
9. What are some advantages of social aggregation among birds?
10. More than 90% of all bird species are monogamous. Explain why monogamy is much more common among birds than among mammals.
11. Briefly describe an example of polygyny among birds.
12. Define precocial and altricial as they relate to birds.
13. Offer some examples of how human activities have affected bird populations.

Selected References

See also general references on page 415.

Ackerman, J. 1998. Dinosaurs take wing. National Geographic **194**(1):74–99. *Beautifully illustrated synopsis of dinosaur-to-bird evolution.*

Bennett, P. M., and I. F. F. Owens. 2002. Evolutionary ecology of birds: Life histories, mating systems, and extinction. Oxford, UK, Oxford University Press. *A phylogenetic approach to understanding how natural and sexual selection have led to the incredible diversity of bird mating systems.*

Brooke, M., and T. Birkhead, eds. 1991. The Cambridge encyclopedia of ornithology. New York, Cambridge University Press. *Comprehensive, richly illustrated treatment that includes a survey of all modern bird orders.*

Elphick, J. ed. 1995. The atlas of bird migration: tracing the great journeys of the world's birds. New York, Random House. *Lavishly illustrated collection of maps of birds' breeding and wintering areas, migration routes, and many facts about each bird's migration journey.*

Emlen, S. T. 1975. The stellar-orientation system of a migratory bird. Sci. Am. **233**:102–111 (Aug.). *Describes fascinating research with indigo buntings, revealing their ability to navigate by the center of celestial rotation at night.*

Feduccia, A. 1996. The origin and evolution of birds. New Haven, Yale University Press. *An updated successor to the author's* The Age of Birds *(1980) but more comprehensive; rich source of information on evolutionary relationships of birds.*

Norbert, U. M. 1990. Vertebrate flight. New York, Springer-Verlag. *Detailed review of mechanics, physiology, morphology, ecology, and evolution of flight. Covers bats as well as birds.*

Proctor, N. S., and P. J. Lynch. 1993. Manual of ornithology: avian structure and function. New Haven, Connecticut, Yale University Press.

Sibley, C. G., and J. E. Ahlquist. 1990. Phylogeny and classification of birds: a study in molecular evolution. New Haven, Yale University Press. *A comprehensive application of DNA annealing experiments to the problem of resolving avian phylogeny.*

Terborgh, J. 1992. Why American songbirds are vanishing. Sci. Am. **266**:98–104 (May). *The number of songbirds in the United States has been dropping sharply. The author suggests reasons why.*

Waldvogel, J. A. 1990. The bird's eye view. Am. Sci. **78**:342–353 (July–Aug.). *Birds possess visual abilities unmatched by humans. So how can we know what they really see?*

Wellnhofer, P. 1990. *Archaeopteryx*. Sci. Am. **262**:70–77 (May). *Description of perhaps the most important fossil ever discovered.*

Custom Website

The *Animal Diversity* Online Learning Center is a great place to check your understanding of chapter material. Visit www.mhhe.com/hickmanad4e for access to key terms, quizzes, and more! Further enhance your knowledge with web links to chapter-related material.

Explore live links for these topics:

Classification and Phylogeny of Animals
Class Aves

Marine Birds
Dissection Guides for Birds
Conservation Issues Concerning Birds
Movement of Populations

20

Mammals

Juvenile grizzly bear, *Ursus arctos horribilis*.

The Tell-Tale Hair

If Fuzzy Wuzzy, the bear that had no hair (according to the children's rhyme), was truly hairless, he could not have been a mammal or a bear. For hair is as much an unmistakable characteristic of mammals as feathers are of birds. If an animal has hair it is a mammal; if it lacks hair it must be something else. It is true that many aquatic mammals are nearly hairless (whales, for example) but hair can usually be found (with a bit of searching) at least in vestigial form somewhere on the body of an adult. Unlike feathers, which evolved from converted reptilian scales, mammalian hair is a completely new epidermal structure. Mammals use their hair for protective coloration and concealment, for waterproofing and buoyancy, and for behavioral signaling; they have turned hairs into sensitive vibrissae on their snouts and into prickly quills. Perhaps most important, mammals use their hair for thermal insulation, which allows them to enjoy the great advantages of homeothermy. Warm-blooded animals in most climates and at sunless times benefit from this natural and controllable protective insulation.

Hair, of course, is only one of several features that together characterize a mammal and help us to understand the mammalian evolutionary achievement. Among these are a highly developed placenta for feeding an embryo; mammary glands for nourishing the newborn; specialized teeth and jaws for processing diverse foods; and a surpassingly advanced nervous system that far exceeds in performance that of any other animal group. It is doubtful, however, that even with this winning combination of adaptations, mammals could have triumphed as they have without their hair.

M ammals, with their highly developed nervous system and numerous adaptations, occupy almost every environment on earth that supports life. Although not a large group (about 4800 species as compared with more than 9900 species of birds, approximately 28,000 species of fishes, and 800,000 species of insects), class Mammalia (mam-may'lee-a) (L. *mamma,* breast) is among the most biologically differentiated groups in the animal kingdom. Mammals are exceedingly diverse in size, shape, form, and function. They range in size from the recently discovered Kitti's hognosed bat, weighing only 1.5 g, to blue whales, exceeding 130 metric tons.

Yet, despite their adaptability and in some instances because of it, mammals have been influenced by humans more than any other group of animals. We have domesticated numerous mammals for food and clothing, as beasts of burden, and as pets. We use millions of mammals each year in biomedical research. We have introduced alien mammals into new habitats, occasionally with benign results but more frequently with unexpected disaster. Although history provides us with numerous warnings, we continue to overcrop valuable wild stocks of mammals. The whale industry has threatened itself with total collapse by exterminating its own resource—a classic example of self-destruction in the modern world, in which competing segments of an industry are intent only on reaping all they can today as though tomorrow's supply were of no concern whatever. In some cases destruction of a valuable mammalian resource has been deliberate, such as the officially sanctioned (and tragically successful) policy during the Indian wars of exterminating bison to drive the Plains Indians into starvation. Although commercial hunting has declined, the ever-increasing human population with accompanying destruction of wild habitats has harassed and disfigured the mammalian fauna.

We are becoming increasingly aware that our presence on this planet as the most powerful product of organic evolution makes us responsible for our natural environment. Since our welfare has been and continues to be closely related to that of the other mammals, it is clearly in our interest to preserve the natural environment of which all mammals, ourselves included, are a part. We need to remember that nature can do without humans but humans cannot exist without nature.

Origin and Evolution of Mammals

The evolutionary descent of mammals from their earliest amniote ancestors is perhaps the most fully documented transition in vertebrate history. From the fossil record, we can trace the derivation over 150 million years of endothermic, furry mammals from their small, ectothermic, hairless ancestors. The structure of the skull roof permits us to identify three major groups of amniotes that diverged in the Carboniferous period of the Paleozoic era, **synapsids, anapsids,** and **diapsids** (figure 20.1; see also p. 349). The synapsid group, which includes mammals and their closest ancestors, has a pair of openings in the skull roof associated with attachment of jaw muscles. Synapsids were the first amniote group to radiate widely into terrestrial habitats. The anapsid group is characterized by solid skulls and includes turtles and their ancestors (figure 20.1A

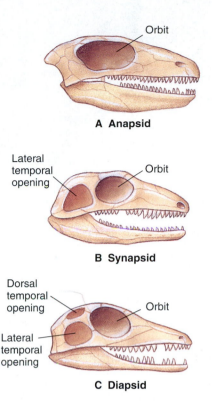

figure 20.1

Skulls of early amniotes, showing the pattern of temporal openings that distinguish the three groups.

and p. 354). Diapsids have two pairs of openings in their skull roof (figure 20.1C; see also figure 18.2, p. 351) and this group contains dinosaurs, lizards, snakes, crocodilians, birds, and their ancestors.

The earliest synapsids radiated extensively into diverse herbivorous and carnivorous forms often collectively called **pelycosaurs** (figures 20.2 and 20.3). Pelycosaurs share a general outward resemblance to lizards, but this resemblance is misleading. Pelycosaurs are not closely related to lizards, which are diapsids, nor are they a monophyletic group. From one group of early carnivorous synapsids arose the **therapsids** (figure 20.3), the only synapsid group to survive beyond the Paleozoic. With therapsids we see for the first time an efficient erect gait with upright limbs positioned beneath the body. Since stability was reduced by raising the animal from the ground, the muscular coordination center of the brain, the cerebellum, assumed an expanded role. Changes in the morphology of the skull and jaw-closing muscles associated with increased feeding efficiency began with the early therapsids. The therapsids radiated into numerous herbivorous and carnivorous forms but most disappeared during a great extinction at the end of the Permian.

One therapsid group to survive into the Mesozoic era was the **cynodonts.** Cynodonts evolved several features that supported a high metabolic rate: increased and specialized jaw musculature, permitting a stronger bite; several skeletal changes, supporting greater agility; **heterodont** teeth, permitting better

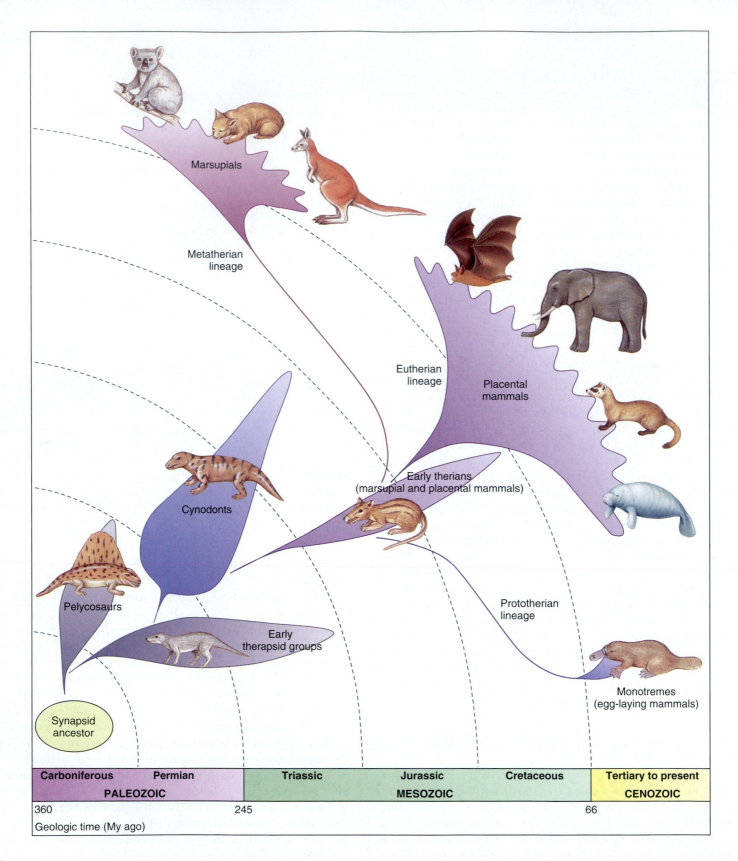

figure 20.2

Evolution of major groups of synapsids. The synapsid lineage, characterized by lateral temporal openings in the skull, began with pelycosaurs, early mammal-like amniotes of the Permian. Pelycosaurs radiated extensively and evolved changes in jaws, teeth, and body form that presaged several mammalian characteristics. These trends continued in their successors, the therapsids, especially in cynodonts. One lineage of cynodonts gave rise in the Triassic to therians (marsupial and placental mammals). Fossil evidence, as currently interpreted, indicates that all three groups of living mammals—monotremes, marsupials, and placentals—are derived from the same cynodont lineage. The great radiation of modern placental orders occurred during the Cretaceous and Tertiary periods.

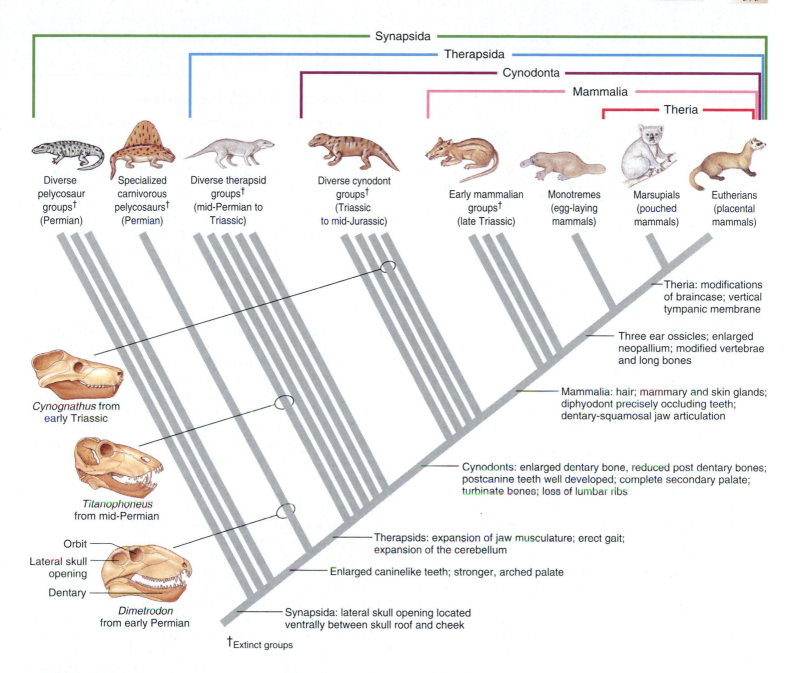

figure 20.3

Abbreviated cladogram of synapsids emphasizing origins of important characteristics of mammals (shown to the right of the cladogram). Extinct groups are indicated by a dagger. The skulls show a progressive increase in size of the dentary relative to other bones in the lower jaw.

*Sources: J. Gauthier, A. G. Kluge, and T. Rowe, "Amniote phylogeny and the importance of fossils" in Cladistics **4**:105–209 (1988); R. L. Carroll, Vertebrate Paleontology and Evolution, W. H. Freeman, New York, 1988; and F. H. Pough, C. M. Janis, J. B. Heiser, Vertebrate Life, 7th edition, Prentice Hall, New Jersey, 2005.*

food processing; **turbinate bones** in the nasal cavity, aiding retention of body heat (figure 20.4); and a secondary bony palate (figure 20.4), enabling an animal to breathe while holding prey or chewing food. The secondary palate would be important to subsequent mammalian evolution by permitting the young to breathe while suckling. Cynodonts had limbs oriented vertically under the body, rather than sprawled out to the side, as in modern lizards and in primitive pelycosaurs, which better supported the weight of the body and permitted more efficient locomotion. Long bones became more slender and developed

bony processes at the joints for firmer muscle attachment. Loss of lumbar ribs in cynodonts is correlated with the evolution of a diaphragm and also may have provided greater dorsoventral flexibility of the spinal column.

The earliest mammals of the late Triassic period were small mouse- or shrew-sized animals with enlarged crania, jaws redesigned for shearing action, and a new type of dentition, called **diphyodont,** in which teeth are replaced only once (deciduous and permanent teeth). This event contrasts with the primitive amniote pattern of continual tooth replacement throughout life

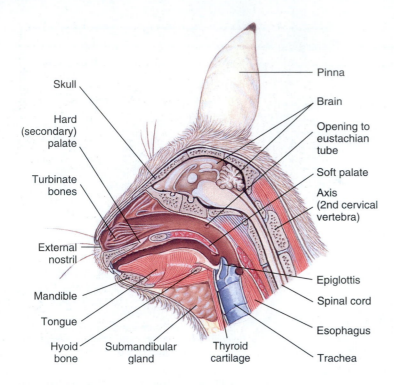

figure 20.4

Sagittal section through the head of a rabbit. Hard and soft palates together form the secondary palate, a roof that separates mouth and nasal cavities, and a characteristic of all mammals.

(polyphodont teeth). Two bones, the articular and quadrate, which previously served as the jaw joint, were reduced in size and relocated in the middle ear, becoming the malleus and incus, respectively. The earliest mammals were almost certainly endothermic, although their body temperature would have been rather lower than modern placental mammals. Hair was essential for insulation, and the presence of hair implies that sebaceous and sweat glands must have evolved at this time to lubricate the hair and promote heat loss. The fossil record is silent on the appearance of mammary glands, but they must have evolved before the end of the Triassic.

Oddly, early mammals of the mid-Triassic, having developed nearly all novel attributes of modern mammals, had to wait for another 150 million years before they could achieve their great diversity. While dinosaurs became diverse and abundant, all nonmammalian synapsid groups became extinct. But mammals survived, first as shrewlike, probably nocturnal, creatures. Then, beginning in the Cretaceous period, but especially during the Eocene epoch that began about 54 million years ago, modern mammals began to diversify rapidly. The great Cenozoic radiation of mammals is partly attributed to numerous habitats vacated by the extinction of many amniote groups at the end of the Cretaceous. Mammalian radiation was almost certainly promoted by the facts that mammals were agile, endothermic, intelligent, adaptable, and gave birth to living young, which they protected and nourished from their own milk supply, thus dispensing with vulnerable eggs laid in nests.

Class Mammalia includes 26 orders: one order containing **monotremes,** seven orders of **marsupials,** and 18 orders of placentals. A complete classification is on pp. 409–411.

Structural and Functional Adaptations of Mammals

Integument and Its Derivatives

Mammalian skin and especially its modifications distinguish mammals as a group. As the interface between an animal and its environment, skin is strongly molded by an animal's way of life. In general, skin is thicker in mammals than in other classes of vertebrates, although as in all vertebrates it is composed of **epidermis** and **dermis.** Among mammals the dermis becomes much thicker than the epidermis. The epidermis is thinner where it is well protected by hair, but in places that are subject to much contact and use, such as the palms or soles, its outer layers become thick and cornified with keratin.

Hair

Hair is especially characteristic of mammals, although humans are not very hairy creatures and, in whales, hair is reduced to only a few sensory bristles on their snout. A hair grows from a hair follicle that, although an epidermal structure, is sunk into the dermis of the skin (figure 20.5). A hair grows continuously by rapid proliferation of cells in a follicle. As a hair shaft is pushed upward, new cells are carried away from their source of nourishment and die, becoming filled with the same dense type of fibrous protein, called **keratin,** that constitutes nails, claws, hooves, and feathers.

Mammals characteristically have two kinds of hair forming their **pelage** (fur coat): (1) dense and soft **underhair** for insulation and (2) coarse and longer **guard hair** for protection against wear and to provide coloration. Underhair traps a layer of insulating air. In aquatic mammals, such as fur seals, otters, and beavers, it is so dense that it is almost impossible to wet. In water, guard hairs become wet and mat down, forming a protective blanket over the underhair (figure 20.6).

When a hair reaches a certain length, it stops growing. Normally it remains in its follicle until a new growth starts, whereupon it falls out. In most mammals there are periodic molts of the entire coat. In humans, hair is shed and replaced throughout life (although balding males confirm that replacement is not assured!).

A hair is more than a strand of keratin. It consists of three layers: the medulla or pith in the center of the hair, the cortex with pigment granules next to the medulla, and the outer cuticle composed of imbricated scales. The hair of different mammals shows a considerable range of structure. It may be deficient in cortex, such as the brittle hair of deer, or it may be deficient in medulla, such as the hollow, air-filled hairs of a wolverine. Hairs of rabbits and some others are scaled to interlock when pressed together. Curly hair, such as that of sheep, grows from curved follicles.

In the simplest cases, such as foxes and seals, the coat is shed every summer. Most mammals have two annual molts, one in spring and one in fall. Summer coats are always much

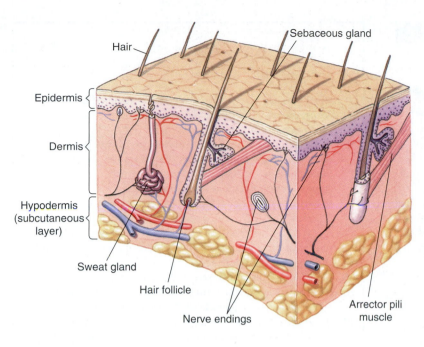

Hair

Sebaceous gland

Epidermis

Dermis

Hypodermis (subcutaneous layer)

Sweat gland

Hair follicle

Nerve endings

Arrector pili muscle

figure 20.5

Structure of human skin (epidermis and dermis) and hypodermis, showing hair and glands.

figure 20.6

American beaver, *Castor canadensis,* gnawing on an aspen tree. This second largest rodent (the South American capybara is larger) has a heavy waterproof pelage consisting of long, tough guard hairs overlying the thick, silky underhair so valued in the fur trade. Order Rodentia, family Castoridae.

thinner than winter coats and in some mammals may be a different color. Several northern mustelid carnivores, for example, weasels, have white winter coats and brown-colored summer coats. It was once believed that the white inner pelage of arctic animals conserved body heat by reducing radiation loss; in fact, dark and white pelages radiate heat equally well. Winter white pelage of arctic animals is simply camouflage in a land of snow. The varying hare of North America has three annual molts: the white winter coat is replaced by a brownish gray summer coat, and this is replaced in autumn by a grayer coat, which is soon shed to reveal the winter white coat beneath (figure 20.7).

Outside the Arctic, most mammals wear somber colors that are protective. Often the species is marked with "salt-and-pepper" coloration or a disruptive pattern that helps make it inconspicuous in its natural surroundings. Examples are spots of leopards and fawns and stripes of tigers. Skunks advertise their presence with conspicuous warning coloration.

The hair of mammals has become modified to serve many purposes. Bristles of hogs, spines of porcupines and their kin, and vibrissae on the snouts of most mammals are examples. **Vibrissae,** commonly called "whiskers," are really sensory hairs that provide a tactile sense to many mammals. The slightest movement of a vibrissa generates impulses in sensory nerve endings that travel to special sensory areas in the brain. Vibrissae are especially long in nocturnal and burrowing animals.

Porcupines, hedgehogs, echidnas, and a few other mammals have developed an effective and dangerous spiny armor. When cornered, the common North American porcupine turns its back toward its attacker and lashes out with its barbed tail. The lightly attached quills break off at their bases

A

B

figure 20.7

Snowshoe, or varying, hare, *Lepus americanus* in **A**, brown summer coat and, **B**, white winter coat. In winter, extra hair growth on the hind feet broadens the animal's support in snow. Snowshoe hares are common residents of the taiga and are an important food for lynxes, foxes, and other carnivores. Population fluctuations of hares and their predators are closely related. Order Lagomorpha, family Leporidae.

figure 20.8

Dogs are frequent victims of the porcupine's impressive armor. Unless removed (usually by a veterinarian) the quills will continue to work their way deeper in the flesh causing great distress and may lead to the victim's death.

when they enter the skin and, aided by backward-pointing hooks on the tips, work deeply into tissues. Dogs are frequent victims (figure 20.8) but fishers, wolverines, and bobcats are able to flip the porcupine onto its back to expose vulnerable underparts.

Horns and Antlers

Several kinds of horns or hornlike structures are found in mammals. **True horns,** found in members of the family Bovidae (for example, sheep and cattle), are sheaths of keratinized epidermis that embrace a core of bone arising from the skull. True horns are not shed, usually are not branched (although they may be greatly curved), grow continuously, and are found in both sexes. Horns may be absent from pronghorn antelope females but, if present, are shorter than those of the male.

Antlers of the deer family Cervidae are branched and composed of solid bone when mature. During their annual spring growth, antlers develop beneath a covering of highly vascular soft skin called **velvet** (figure 20.9). When growth of antlers is complete just before the breeding season, blood vessels constrict and the stag tears off the velvet by rubbing its antlers against trees. Antlers are shed after the breeding season. New buds appear a few months later to herald the next set of antlers. For several years each new pair of antlers is larger and more elaborate than the previous set. Annual growth of antlers places a strain on mineral metabolism, since during the growing season an older moose or elk must accumulate 50 or more pounds of calcium salts from its vegetable diet.

Horns of the pronghorn antelope (family Antilocapridae) are similar to true horns of bovids except the keratinized portion is forked and shed annually. Giraffe horns are similar to antlers but retain their integumentary covering and are not

characteristics
of Mammals

1. **Body mostly covered with hair,** but reduced in some
2. **Integument** with **sweat, scent, sebaceous,** and **mammary glands**
3. Skull with **two occipital condyles; turbinate bones** in nasal cavity; jaw joint between the squamosal and dentary bones; middle ear with **three ossicles** (malleus, incus, stapes); **seven cervical vertebrae** (except some xenarthrans [edentates] and manatees); **pelvic bones fused**
4. Mouth with **diphyodont teeth** (milk, or deciduous, teeth replaced by a permanent set of teeth); teeth **heterodont** in most (varying in structure and function); lower jaw a **single enlarged bone (dentary)**
5. Movable eyelids and **fleshy external ears (pinnae)**
6. Circulatory system of a four-chambered heart, **persistent left aorta,** and **nonnucleated, biconcave red blood cells**
7. Respiratory system of lungs with alveoli, and voice box (larynx); **secondary palate** (anterior bony palate and posterior continuation of soft tissue, the soft palate) separates air and food passages (figure 20.4); **muscular diaphragm** for air exchange separates thoracic and abdominal cavities
8. Excretory system of metanephric kidneys and ureters that usually open into a bladder
9. Brain highly developed, especially **cerebral cortex** (a highly folded superficial layer of the cerebrum); 12 pairs of cranial nerves
10. Endothermic and homeothermic
11. Cloaca present only in monotremes (present, but shallow in marsupials)
12. Separate sexes; copulatory organ is a penis; testes usually in scrotum; female reproductive tracts usually partly fused; sex determined by chromosomes (male is heterogametic)
13. Internal fertilization; **embryos develop in a uterus** with **placental attachment** (placenta absent in monotremes); **fetal membranes (amnion, chorion, allantois)**
14. Young nourished by **milk from mammary glands**

shed. Rhinoceros horn consists of hairlike keratinized filaments that arise from dermal papillae cemented together, but they are not attached to the skull.

An escalating trade in rhinoceros products—especially rhinoceros horn—during the last three decades, is pushing Asian and African rhinos to the brink of extinction. Rhinoceros horn is valued in China as an agent for reducing fever, and for treating heart, liver, and skin diseases; and in North India as an aphrodisiac. Such supposed medicinal values are totally without pharmacological basis. The principal use of rhinoceros horns, however, is to fashion handles for daggers in the Middle East. Because of their phallic shape, rhinoceros horn daggers are traditional gifts at puberty rites. Between 1969 and 1977, horns from 8000 slaughtered rhinos were imported into North Yemen alone.

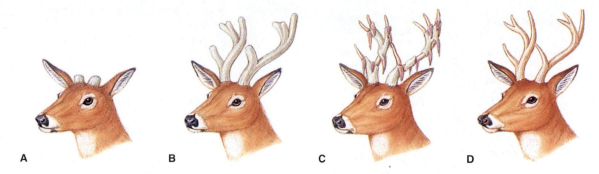

figure 20.9

Annual growth of buck deer antlers. **A,** Antlers begin growth in late spring, stimulated by pituitary gonadotropins. **B,** The bone grows very rapidly until halted by a rapid rise in testosterone production by the testes. **C,** The skin (velvet) dies and sloughs off. **D,** Testosterone levels peak during the fall breeding season. The antlers are shed in January as testosterone levels subside.

Glands

Of all vertebrates, mammals have the greatest variety of integumentary glands. Most fall into one of four classes: sweat, scent, sebaceous, and mammary. All are derivatives of epidermis.

Sweat glands are tubular, highly coiled glands that occur over much of the body surface in most mammals. They are not present in other vertebrates. There are two kinds of sweat glands: eccrine and apocrine (see figure 20.5). **Eccrine glands** secrete a watery fluid that, if evaporated on the skin's surface, draws heat away from the skin and cools it. Eccrine glands occur in hairless regions, especially foot pads, in most mammals, although in horses and most primates they are scattered over the body. **Apocrine glands** are larger than eccrine glands and have longer and more convoluted ducts. Their secretory coil is in the dermis and extends deep into the hypodermis. They always open into a hair follicle or where a hair once was. Apocrine glands develop near sexual puberty and are restricted (in the human species) to the axillae (armpits), mons pubis, breasts, prepuce, scrotum, and external auditory canals. In contrast to the watery secretions of eccrine glands, apocrine secretions are milky fluids, whitish or yellow in color, that dry on skin to form a film. Apocrine glands are not involved in heat regulation. Their activity is correlated with certain aspects of the reproductive cycle.

Scent glands are present in nearly all mammals. Their location and functions vary greatly. They are used for communication with members of the same species, for marking territorial boundaries, for warning, or for defense. Scent-producing glands are located variously in orbital, metatarsal, and interdigital regions. The most odoriferous of all glands are those of skunks, which open by ducts into the anus; their secretions can be discharged forcefully for 2 to 3 m. During mating season many mammals produce strong scents for attracting the opposite sex. Humans also are endowed with scent glands. However we dislike our own scent, a concern that has stimulated a lucrative deodorant industry to produce an endless output of soaps and odor-masking concoctions.

Sebaceous glands are intimately associated with hair follicles (figure 20.5), although some are free and open directly onto the surface. The cellular lining of a gland is discharged in the secretory process and must be renewed for further secretion. These gland cells become distended with a fatty accumulation, then die, and are expelled as a greasy mixture called **sebum** into the hair follicle. Called a "polite fat" because it does not turn rancid, it serves as a dressing to keep skin and hair pliable and glossy. Most mammals have sebaceous glands over their entire body; in humans they are most numerous in the scalp and on the face.

Mammary glands, for which mammals are named, occur on all female mammals and in a rudimentary form on all male mammals. They develop by thickening of the epidermis to form a milk line along each side of the abdomen in the embryo. On certain parts of these lines the mammae appear while the intervening parts of the ridge disappear. Mammary glands increase in size at maturity, becoming considerably larger during pregnancy and subsequent nursing of young. In human females, adipose tissue begins to accumulate around mammary glands at puberty to form the breast. In most mammals, milk is secreted from mammary glands via nipples or teats, but monotremes lack nipples and simply secrete milk into a depression on the mother's belly where it is lapped by the young.

Food and Feeding

Mammals exploit an enormous variety of food sources; some mammals require highly specialized diets, whereas others are opportunistic feeders that thrive on diversified diets. Food habits and physical structure are thus inextricably linked. A mammal's adaptations for attack and defense and its specializations for finding, capturing, chewing, swallowing, and digesting food all determine a mammal's shape and habits.

Teeth, perhaps more than any other single physical characteristic, reveal the life habit of a mammal (figure 20.10). All

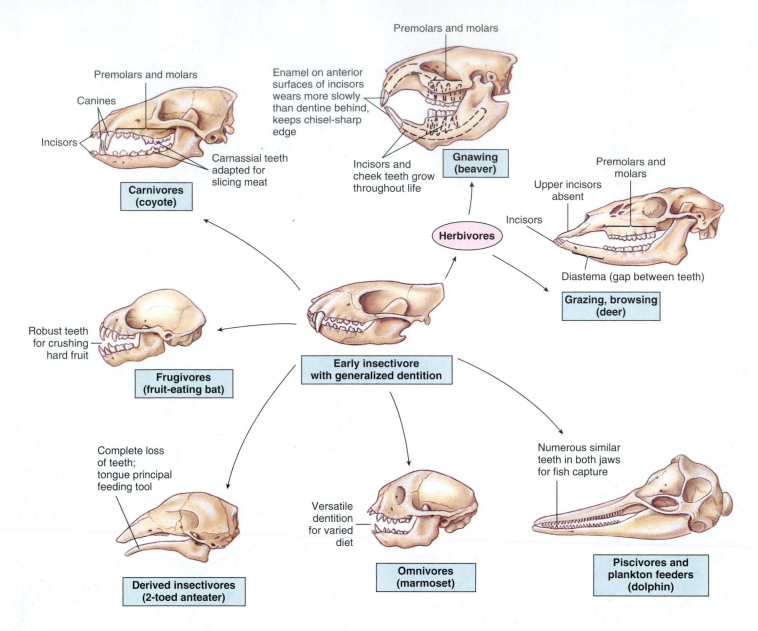

figure 20.10

Feeding specializations of major trophic groups of eutherian mammals. Early eutherians were insectivores; all other types are descended from them.

mammals have teeth (with few exceptions) and their modifications are correlated with what the mammal eats.

As mammals evolved during the Mesozoic, major changes occurred in teeth and jaws. Unlike the uniform **homodont** dentition of the first synopsids, mammalian teeth became differentiated to perform specialized functions such as cutting, seizing, gnawing, tearing, grinding, and chewing. Teeth differentiated in this manner are called **heterodont.** Mammalian dentition is differentiated into four types: **incisors,** with simple crowns and sharp edges, used mainly for snipping or biting; **canines,** with long conical crowns, specialized for piercing; **premolars** and **molars,** with compressed crowns and one or more cusps, suited for shearing, slicing, crushing, or grinding. The primitive tooth formula, which expresses the number of each tooth type in one-half of the upper and lower jaw, was I 3/3, C 1/1, PM 4/4, M 3/3. Members of order Insectivora, some

omnivores, and carnivores come closest to this primitive pattern (figure 20.10).

Most mammals grow just two sets of teeth: a temporary set, called **deciduous,** or **milk,** teeth, which is replaced by a permanent set when the skull has grown large enough to accommodate a full set. Only incisors, canines, and premolars are deciduous; molars are never replaced and a single permanent set must last a lifetime.

Feeding Specializations

The feeding, or trophic, apparatus of a mammal—teeth and jaws, tongue, and alimentary canal—are adapted to its particular feeding habits. Mammals are customarily divided among four basic trophic categories—insectivores, carnivores, omnivores, and herbivores—but many other feeding specializations have

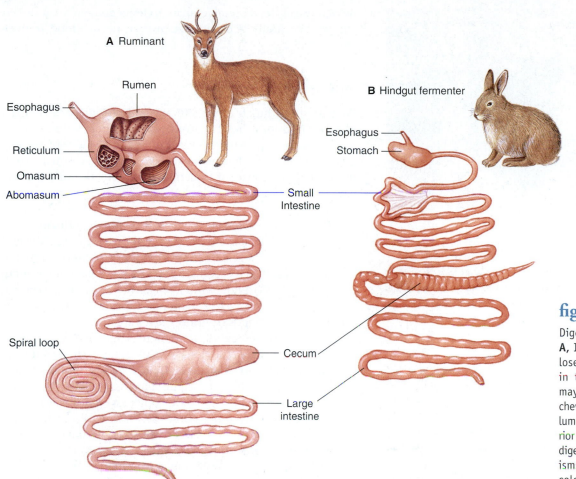

A Ruminant

Esophagus

Rumen

Reticulum

Omasum

Abomasum

Spiral loop

Small Intestine

B Hindgut fermenter

Esophagus

Stomach

Cecum

Large intestine

figure 20.11

Digestive tracts of two mammals. **A,** In ruminants, digestion of cellulose by microorganisms takes place in the rumen. Food in the rumen may be returned to the mouth for chewing or sent directly to the reticulum, omasum, abomasum, and posterior portions of the gut. **B,** In rabbits, digestion of cellulose by microorganisms takes place in the cecum and colon.

evolved in mammals, as in other living organisms, and feeding habits of many mammals defy exact classification. Principal feeding specializations of mammals are shown in figure 20.10.

Insectivorous mammals, such as shrews, moles, anteaters, and bats, feed on a variety of small invertebrates, such as worms and grubs, as well as insects. Insectivorous mammals have teeth with pointed cusps, permitting them to puncture the exoskeleton or skin of prey.

Herbivorous mammals that feed on grasses and other vegetation form two main groups: (1) **browsers** and **grazers,** such as ungulates (hooved mammals including horses, deer, antelope, cattle, sheep, and goats); and (2) **gnawers,** such as rodents, and rabbits and hares. In herbivores, canines are absent or reduced in size, whereas molars, which are adapted for grinding, are broad and usually high-crowned. Rodents have chisel-sharp incisors that grow throughout life and must be worn away to keep pace with their continual growth (figure 20.10).

Herbivores have a number of adaptations for dealing with their fibrous diet of plant food. **Cellulose,** the structural carbohydrate of plants, is composed of long chains of glucose molecules and therefore is a potentially nutritious food resource. However, no vertebrates synthesize cellulose-splitting enzymes **(cellulases).** Instead, herbivorous vertebrates

harbor anaerobic bacteria and protozoans that produce cellulase in fermentation chambers in their gut. Simple carbohydrates, proteins, and lipids produced by the microorganisms can be absorbed by the host animal, and the host can digest the microorganisms themselves.

Fermentation in some herbivores, such as horses, zebras, rabbits, elephants, some primates, and many rodents; takes place primarily in the colon and in a spacious side pocket, or diverticulum, called a **cecum** (figure 20.11). Although some absorption takes place in the colon and cecum, most fermentation occurs after the primary absorptive area (the small intestine), and many nutrients are lost in the feces. Rabbits and many rodents eat their fecal pellets **(coprophagy),** giving food a second pass through the gut to extract additional nutrients.

Ruminants (cattle, bison, buffalo, goats, antelopes, sheep, deer, giraffes, and okapis) have a huge **four-chambered stomach** (figure 20.11). As a ruminant feeds, grass passes down the esophagus to the **rumen,** where it is broken down by bacteria and protozoa and then formed into small balls of cud. At its leisure the ruminant returns a cud to its mouth where the cud is deliberately chewed at length to crush the fiber. Swallowed again, the food returns to the rumen where the cellulolytic bacteria and protozoa continue fermentation. The pulp passes to the **reticulum,** then to the **omasum,** where water, soluble

figure 20.12

Lionesses, *Panthera leo*, eating a wildebeest. Lions stalk prey and then charge suddenly to surprise the victim. They lack stamina for a long chase. Lions gorge themselves with the kill, then sleep and rest for periods as long as one week before eating again. Order Carnivora, family Felidae.

food, and microbial products are absorbed. The remainder proceeds to the **abomasum** ("true" acid stomach) and small intestine, where proteolytic enzymes are secreted and normal digestion occurs. Perhaps because ruminants are particularly good at extracting nutrients from forage, they are the primary large herbivores in ecosystems with a shortage of food, such as tundras and deserts.

Herbivores generally have large, long digestive tracts and must eat a considerable amount of plant food to survive. A large African elephant weighing 6 tons must consume 135 to 150 kg (300 to 400 pounds) of rough fodder each day to obtain sufficient nourishment for life.

Carnivorous mammals feed mainly on herbivores. This group includes foxes, dogs, weasels, wolverines, fishers, and cats. Carnivores are well-equipped with biting and piercing teeth and powerful clawed limbs for killing their prey. Since their protein diet is more easily digested than is the woody food of herbivores, their digestive tract is shorter and the cecum small or absent. Carnivores organize their feeding into discrete meals rather than feeding continuously (as do most herbivores) and therefore have much more leisure time (figure 20.12).

Note that the terms "insectivores" and "carnivores" have two different uses in mammals: to describe diet and to denote specific taxonomic orders of mammals. For example, not all carnivores belong to the order Carnivora (many marsupials, and cetaceans are carnivorous) and not all members of the order Carnivora are carnivorous. Many are opportunistic feeders and some, such as pandas, are strict vegetarians.

Omnivorous mammals use both plants and animals for food. Examples are pigs, raccoons, many rodents, bears, and most primates (including humans). Many carnivorous forms also eat fruits, berries, and grasses when hard pressed. Foxes,

which usually feed on mice, small rodents, and birds, eat frozen apples, beechnuts, and corn when their normal food sources are scarce.

Migration

Migration is a much more difficult undertaking for mammals than for birds or fishes, because terrestrial locomotion is more energetically expensive than swimming or flying. Not surprisingly, few terrestrial mammals make regular seasonal migrations, preferring instead to center their activities in a defined and limited home range. Nevertheless, there are some striking examples of terrestrial mammalian migrations. More migrators are found in North America than on any other continent.

An example is the barren-ground caribou of Canada and Alaska, which undertakes direct and purposeful mass migrations spanning 160 to 1100 km (100 to 700 miles) twice annually (figure 20.13). From winter ranges in boreal forests (taiga), they migrate rapidly in late winter and spring to calving ranges on the barren grounds (tundra). Calves are born in mid-June. As summer progresses, caribou are increasingly harassed by warble and nostril flies that bore into their flesh, by mosquitoes that drink their blood (estimated at a liter per caribou each week during the height of the mosquito season), and by wolves that prey on their calves. They move southward in July and August, feeding little along the way. In September they reach the taiga and feed there almost continuously on low ground vegetation. Mating (rut) occurs in October.

The longest mammalian migrations are made by oceanic seals and whales. One of the most remarkable migrations is that of northern fur seals, which breed on the Pribilof Islands approximately 300 km (185 miles) off the coast of Alaska and north of the Aleutian Islands. From wintering grounds off southern California females journey as much as 2800 km (1740 miles) across open ocean, arriving in spring at the Pribilofs where they congregate in enormous numbers (figure 20.14). Young are born within a few hours or days after arrival of the cows. Then bulls, having already arrived and established territories, collect harems of cows, which they guard with vigilance. After calves have been nursed for approximately three months, cows and juveniles leave for their long migration southward. Bulls do not follow but remain in the Gulf of Alaska during winter.

Caribou have suffered a drastic decline in numbers since the nineteenth century when there were several million of them. By 1958 less than 200,000 remained in Canada. Their decline has been attributed to several factors, including habitat alteration from exploration and development in the north, but especially to excessive hunting. For example the western arctic herd in Alaska exceeded 250,000 caribou in 1970. Following five years of heavy unregulated hunting, a 1976 census revealed only about 65,000 animals left. After restricting hunting, the herd had increased to 340,000 by 1988 and 490,000 by 2003. However, this recovery is threatened by proposed expansion of oil extraction in several wildlife refuges, including the Arctic National Wildlife Refuge.

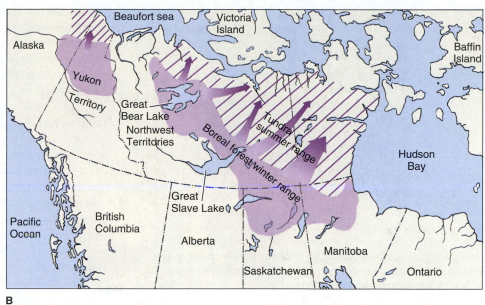

figure 20.13

Barren-ground caribou, *Rangifer tarandus,* of Canada and Alaska. **A,** Adult male caribou in autumn pelage and antlers in velvet. **B,** Summer and winter ranges of some major caribou herds in Canada and Alaska (other herds not shown occur on Baffin Island and in western and central Alaska). The principal spring migration routes are indicated by arrows; routes vary considerably from year to year. The same species is known as reindeer in Europe. Order Artiodactyla, family Cervidae.

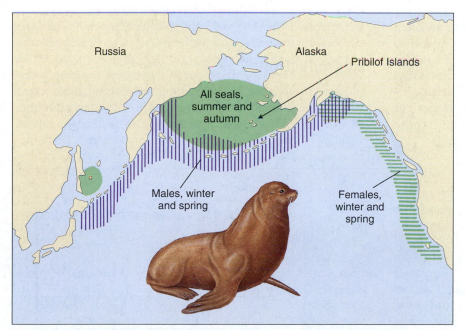

figure 20.14

Annual migrations of northern fur seals, *Callorhinus ursinus,* showing the separate wintering grounds of males and females. Both males and females of the large Pribilof population migrate in early summer to the Pribilof Islands, where females give birth to their pups and then mate with males. Order Carnivora, family Otariidae.

Although we might expect bats, the only winged mammals, to use their gift of flight to migrate, few of them do. Most spend winters in hibernation. Four species of American bats that migrate spend their summers in northern or western states and their winters in the southern United States or Mexico.

Flight and Echolocation

Many mammals scamper about in trees with amazing agility; some can glide from tree to tree, and one group, bats, is capable of full flight. Gliding and flying evolved independently in

figure 20.15

Northern flying squirrel, *Glaucomys sabrinus,* gliding in for a landing. Area of undersurface is nearly trebled when gliding skin is spread. Glides of 40 to 50 m are possible. Good maneuverability during flight is achieved by adjusting the position of the gliding skin with special muscles. Flying squirrels are nocturnal and have superb night vision. Order Rodentia, family Sciuridae.

several groups of mammals, including marsupials, rodents, flying lemurs, and bats. Flying squirrels (figure 20.15) actually glide rather than fly, using the gliding skin (patagium) that extends from the sides of the body.

Bats are nocturnal and thus hold a niche unoccupied by most birds. Their achievement is attributed to two features: flight and capacity to navigate by echolocation. Together these adaptations enable bats to fly and avoid obstacles in absolute darkness, to locate and catch insects with precision, and to find their way deep into caves (a habitat largely unexploited by other mammals and birds) where they sleep during the daytime hours.

When in flight, bats emit short pulses 5 to 10 msec in duration in a narrow directed beam from the mouth or nose

(figure 20.16). Each pulse is frequency modulated; it is highest at the beginning, up to 100,000 Hz (hertz, cycles per second), and sweeps down to perhaps 30,000 Hz at the end. Sounds of this frequency are ultrasonic to human ears, which have an upper limit of about 20,000 Hz. When bats are searching for prey, they produce about 10 pulses per second. If prey is detected, the rate increases rapidly up to 200 pulses per second in the final phase of approach and capture. Pulses are spaced so that the echo of each is received before the next pulse is emitted, an adaptation that prevents jamming. Since transmission-to-reception time decreases as a bat approaches an object, the bat can increase pulse frequency to obtain more information about the object. Pulse length is also shortened as it nears an object. It is interesting that some prey of bats, certain nocturnal moths for example, have evolved ultrasonic detectors used to detect and avoid approaching bats.

External ears of bats are large, like hearing trumpets, and shaped variously in different species. Less is known about the inner ear of bats, but it obviously is capable of receiving the ultrasonic sounds emitted. Biologists believe bat navigation is so refined that a bat builds a mental image of its surroundings from echo scanning that approaches the resolution of a visual image from eyes of diurnal animals.

Some bats, including the approximately 170 species of Old World fruit bats, lack echolocation abilities. Even so, most remain nocturnal, using large eyes and olfaction to find their meals of fruits, flowers, and nectar. Vampire bats are provided with razor-sharp incisors used to shave away the epidermis of their prey, exposing underlying capillaries. After infusing an anticoagulant to keep blood flowing, the bat laps up its meal and stores it in a specially modified stomach.

Reproduction

Most mammals have definite mating seasons, usually in winter or spring and timed to coincide with the most favorable time of year for rearing young after birth. Many male mammals are

figure 20.16

Echolocation of an insect by the little brown bat *Myotis lucifugus.* Frequency modulated pulses are directed in a narrow beam from the bat's mouth. As the bat nears its prey, it emits shorter, lower signals at a faster rate. Order Chiroptera, family Vespertilionidae.

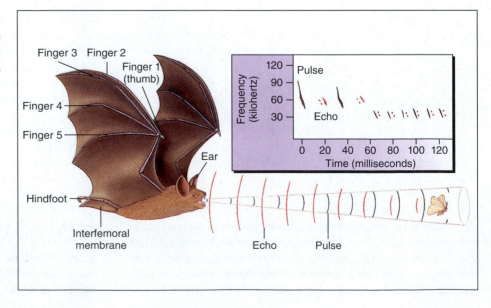

figure 20.17

African lions, *Panthera leo,* mating. Lions breed at any season, although predominantly in spring and summer. During the short period a female is receptive, she may mate repeatedly. Three or four cubs are born after gestation of 100 days. Once the mother introduces cubs into the pride, they are treated with affection by both adult males and females. Cubs go through an 18- to 24-month apprenticeship learning how to hunt and then are frequently driven from the pride to manage themselves. Order Carnivora, family Felidae.

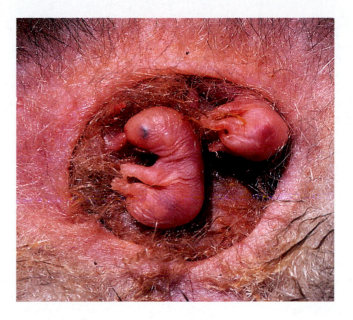

figure 20.18

Opossums, *Didelphis marsupialis,* 15 days old, fastened to teats in mother's pouch. When born after a gestation period of only 12 days, they are the size of honey bees. They remain attached to the nipples for 50 to 60 days. Order Marsupialia, family Didelphidae.

capable of fertile copulation at any time, but female fertility is restricted to a time during a periodic cycle, known as the **estrous cycle.** Females only copulate with males during a relatively brief period known as **estrus,** or heat (figure 20.17).

There are three different patterns of reproduction in mammals. One pattern is represented by egg-laying (oviparous) mammals, the **monotremes.** The duck-billed platypus has one breeding season each year. Embryos develop for 10–12 days in the uterus, where they are nourished by yolk supplies, deposited prior to ovulation, and secretions from the mother. A thin leathery shell is secreted around the embryos before the eggs are laid. The platypus lays its eggs in a burrow, where they hatch at a relatively underdeveloped state after about 12 days. After hatching, the young feed on milk produced by the mother's mammary glands. Because monotremes have no nipples, the young lap milk secreted onto the belly fur of the mother.

Marsupials are pouched, viviparous mammals that exhibit a second pattern of reproduction. Although only eutherians are called "placental mammals," marsupials do have a primitive type of yolk sac placenta. An embryo (blastocyst) of a marsupial is at first encapsulated by shell membranes and floats free for several days in uterine fluid. After "hatching" from the shell membranes, the embryo does not implant, or "take root" in the uterus as it would in eutherians, but it does erode a shallow depression in the uterine wall in which it lies and absorbs nutrient secretions from the mucosa by way of the vascularized yolk sac. Gestation (the intrauterine period of

development) is brief in marsupials, and therefore all marsupials give birth to tiny young that are effectively still embryos, both anatomically and physiologically (figure 20.18). However, early birth is followed by a prolonged interval of lactation and parental care (figure 20.19).

The third pattern of reproduction is that of viviparous **placental mammals,** eutherians. In placentals, the reproductive investment is in prolonged gestation, unlike marsupials in which the reproductive investment is in prolonged lactation (figure 20.19). Embryos remain in the uterus, nourished by food primarily supplied through a chorioallantoic type of placenta, an intimate connection between mother and young. Length of gestation is longer in placentals than marsupials, and in large mammals it is much longer. For example, mice have a gestation period of 21 days; rabbits and hares, 30 to 36 days; cats and dogs, 60 days; cattle, 280 days; and elephants, 22 months (the longest). But there are important exceptions (nature seldom offers perfect correlations). Baleen whales, the largest mammals, carry their young for only 12 months, while bats, no larger than mice, have gestation periods of 4 to 5 months. The condition of the young at birth also varies. An antelope bears its young well furred, eyes open, and able to run. Newborn mice, however, are blind, naked, and helpless. We all know how long it takes a human baby to gain its footing. Human growth is in fact slower than that of any other mammal, and this is one of the distinctive attributes that sets us apart from other mammals.

figure 20.19

Comparison of gestation and lactation periods between matched pairs of ecologically similar species of marsupial and placental mammals. The graph shows that marsupials have shorter intervals of gestation and much longer intervals of lactation than in similar species of placentals.

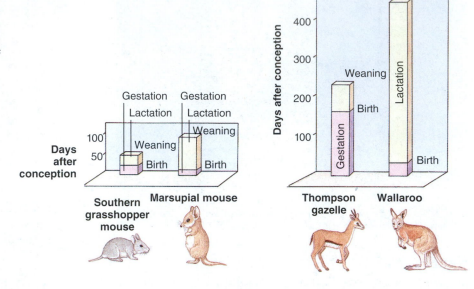

A curious phenomenon that lengthens the gestation period of many mammals is delayed implantation. The blastocyst remains dormant while its implantation in the uterine wall is postponed for periods of a few weeks to several months. For many mammals (for example, bears, seals, weasels, badgers, bats, and many deer) delayed implantation is a device for extending gestation so that the young are born at a time of year that is best for their survival.

Mammalian Populations

A population of animals includes all members of a species that share a particular space and potentially interbreed. All mammals (like other organisms) live in ecological communities, each composed of numerous populations of different animal and plant species. Each species is affected by the activities of other species and by other changes, especially climatic, that occur. Thus populations are always changing in size. Populations of small mammals are lowest before breeding season and greatest just after addition of new members. Beyond these expected changes in population size, mammalian populations may fluctuate from other causes.

Irregular fluctuations are commonly produced by variations in climate, such as unusually cold, hot, or dry weather, or by natural catastrophes, such as fires, hailstorms, and hurricanes. These are **density-independent** causes because they affect a population whether it is crowded or dispersed. However, the most spectacular fluctuations are **density dependent;** that is, they correlate with population crowding. These extreme limits to growth are discussed in Chapter 2 (p. 41).

Cycles of abundance are common among many rodent species. Among the best known examples are mass migrations of Scandinavian and arctic North American lemmings following population peaks. Lemmings (figure 20.20) breed all year,

The renowned fecundity of small rodents, and the effect of removing natural predators from rodent populations, is felicitously expressed in this excerpt from Thornton Burgess's "Portrait of a Meadow Mouse."

> He's fecund to the nth degree
> In fact this really seems to be
> His one and only honest claim
> To anything approaching fame.
> In just twelve months, should all survive,
> A million mice would be alive—
> His progeny. And this, 'tis clear,
> Is quite a record for a year.
> Quite unsuspected, night and day
> They eat the grass that would be hay.
> On any meadow, in a year,
> The loss is several tons, I fear.
> Yet man, with prejudice for guide,
> The checks that nature doth provide
> Destroys. The meadow mouse survives
> And on stupidity he thrives.

figure 20.20

Collared lemming, *Dicrostonyx* sp., a small rodent of the far north. Populations of lemmings fluctuate widely. Order Rodentia, family Muridae.

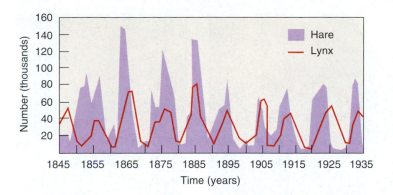

figure 20.21

Changes in population size of varying hare and lynx in Canada as indicated by pelts received by the Hudson's Bay Company. The abundance of lynx (predator) follows that of the hare (prey).

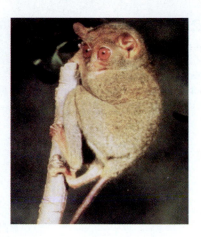

figure 20.22

A prosimian, the Mindanao tarsier, *Tarsius syrichta carbonarius,* of Mindanao Island in the Philippines. Order Primates, family Tarsiidae.

although more in summer than in winter. The gestation period is only 21 days; young born at the beginning of summer are weaned in 14 days and are capable of reproducing by the end of summer. At the peak of their population density, having devastated vegetation by tunneling and grazing, lemmings begin long, mass migrations to find new undamaged habitats for food and space. They swim across streams and small lakes as they go but cannot distinguish these from large lakes, rivers, and the sea, in which they drown. Since lemmings are the main diet of many carnivorous mammals and birds, any change in lemming population density affects all their predators as well.

Varying hares (snowshoe rabbits) of North America show 10-year cycles in abundance. The well-known fecundity of rabbits enables them to produce litters of three or four young as many as five times per year. The density may increase to 4000 hares competing for food in each square mile of northern forest. Predators (owls, minks, foxes, and especially lynxes) also increase (figure 20.21). Then the population crashes pre-

cipitously for reasons that have long been a puzzle to scientists. Rabbits die in great numbers, not from lack of food or from an epidemic disease (as was once believed) but evidently from some density-dependent psychogenic cause. As crowding increases, hares become more aggressive, show signs of fear and defense, and stop breeding. The entire population reveals symptoms of pituitary-adrenal gland exhaustion, an endocrine imbalance called "shock disease," which results in death. These dramatic crashes are not well understood. Whatever the causes, population crashes that follow superabundance, although harsh, permit vegetation to recover, providing survivors with a much better chance for successful breeding.

Human Evolution

Darwin devoted an entire book, *The Descent of Man and Selection in Relation to Sex,* largely to human evolution. The idea that humans share common descent with apes and other animals was repugnant to the Victorian world, which responded with predictable outrage (figure 1.16, p. 18). When Darwin's views were first debated, few human fossils had been unearthed, but the current accumulation of fossil evidence has strongly vindicated Darwin's hypothesis that humans descend from primate ancestors. All primates share certain significant characteristics: grasping fingers on all four limbs, flat fingernails instead of claws, and forward-pointing eyes with binocular vision and excellent depth perception. The following synopsis will highlight probable relationships among major primate groups.

The earliest primate was probably a small, nocturnal animal similar in appearance to tree shrews. This ancestral primate stock split into two major lineages, one of which gave rise to **prosimians,** which include lemurs, tarsiers (figure 20.22), and lorises, and the other to **simians,** which include monkeys (figure 20.23) and apes (figure 20.24). Prosimians and many simians are arboreal (tree-dwellers), which is probably the ancestral lifestyle for both groups. Arboreality probably selected

In his book *The Arctic* (1974. Montreal, Infacor, Ltd.), Canadian naturalist Fred Bruemmer describes the growth of lemming populations in arctic Canada: "After a population crash one sees few signs of lemmings; there may be only one to every 10 acres. The next year, they are evidently numerous; their runways snake beneath the tundra vegetation, and frequent piles of rice-sized droppings indicate the lemmings fare well. The third year one sees them everywhere. The fourth year, usually the peak year of their cycle, the populations explode. Now more than 150 lemmings may inhabit each acre of land and they honeycomb it with as many as 4000 burrows. Males meet frequently and fight instantly. Males pursue females and male after a brief but ardent courtship. Everywhere one hears the squeak and chitter of the excited, irritable, crowded animals. At such times they may spill over the land in manic migrations."

figure 20.23

Monkeys. **A,** Red-howler monkeys, *Alovatta senicu-lus,* order Primates, family Cebidae, an example of New World monkeys. **B,** The olive baboon, *Papio homadryas,* order Primates, family Cercopitheci-dae, an example of Old World monkeys.

A

B

figure 20.24

The gorilla, *Gorilla gorilla,* order Primates, family Hominidae, an anthropoid ape.

for increased intelligence. Flexible limbs are essential for active animals moving through trees. Grasping hands and feet, in contrast to the clawed feet of squirrels and other rodents, enable primates to grip limbs, hang from branches, seize food and manipulate it, and, most significantly, use tools. Highly developed sense organs, especially good vision, and proper coordination of limb and finger muscles are essential for an active

arboreal life. Of course, sense organs are no better than the brain processing sensory information. Precise timing, judgment of distance, and alertness require a large cerebral cortex.

The earliest simian fossils appeared in Africa some 40 million years ago. Many of these primates became day-active rather than nocturnal, making vision the dominant special sense, now enhanced by color vision. We recognize three major simian clades. These are (1) New World monkeys of South America (ceboids), including howler monkeys (figure 20.23A), spider monkeys, and tamarins; (2) the Old World monkeys (cercopithecoids), including baboons (figure 20.23B), mandrills, and colobus monkeys; and (3) anthropoid apes (figure 20.24). In addition to their geographic separation, Old World monkeys differ from New World monkeys in lacking a grasping tail while having close-set nostrils, better opposable, grasping thumbs, and more advanced teeth. Apes first appear in 25-million-year-old fossils. At this time woodland savannas were arising in Africa, Europe, and North America. Perhaps motivated by the greater abundance of food on the ground, these apes left the trees and became largely terrestrial. Because of the benefits of standing upright (better view of predators, freeing of hands for using tools, defense, caring for young, and gathering food), emerging hominids gradually evolved upright posture.

Evidence of the earliest humans of this period is sparse. Yet in 2001 the desert sands of Chad yielded one of the most astonishing and important discoveries of modern paleontology, a remarkably complete skull of a hominid, *Sahelanthropus tchadensis* (Sahel hominid of Chad), dated at nearly 7 million years ago (figure 20.25). Although its brain is no larger than that of a chimp (between 320 and 380 cm^3), its relatively small canine teeth, massive brow ridges on a short face, and mouth and jaw that protrude less than in most apes confirm that the skull truly is phylogenetically closer to humans than chimpanzees. Until this skull was discovered, the earliest fossil placed closer to humans than chimpanzees was

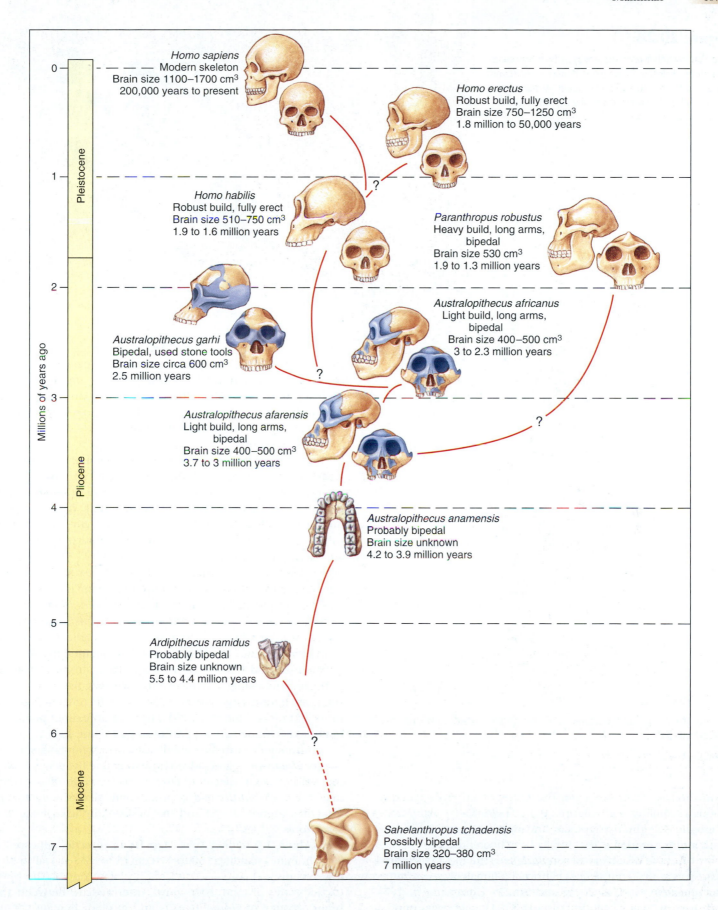

figure 20.25

Human skulls, showing several of the best-known human fossils preceding modern humans (*Homo sapiens*).

figure 20.26

Lucy (*Australopithecus afarensis*), the most nearly complete skeleton of an early human ever found. Lucy is dated at 3.0 million years old. A nearly complete skull of *A. afarensis* was discovered in 1994.

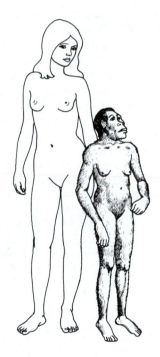

figure 20.27

A reconstruction of the appearance of Lucy (*right*) compared with a modern human (*left*).

Ardipithecus ramidus, from the sands of Ethiopia, dated at about 4.4 million years (figure 20.25). *Ardipithecus ramidus* is a mosaic of primitive apelike and derived humanlike traits, with indirect evidence that it may have been bipedal.

Until the discovery of *S. tchadensis,* the most celebrated fossil was a 40% complete skeleton of a female *Australopithecus afarensis* (figures 20.26 and 20.27). Unearthed in 1974 and named "Lucy" by its discoverer, Donald Johanson, *A afarensis* was a short, bipedal hominid with face and brain size

resembling those of a chimpanzee. Numerous fossils, dating from 3.7 to 3.0 million years ago, of this species have now been discovered. In 1995 *Australopithecus anamensis* was discovered in the Rift Valley of Kenya. Many researchers believe that this species, which lived between 4.2 and 3.9 million years ago, is an intermediate between *A. ramidus* and *A. afarensis* (Lucy).

In the last two decades there has been an explosion of australopithecine fossil finds, with eight putative species requiring interpretation. Most of these species are known as gracile australopithecines because of their relatively light build, especially in the skull and teeth (although all were more robust than modern humans). The gracile australopithecines are generally considered direct ancestors of populations of early *Homo* and, by extension, direct ancestors of modern humans. In 1998 a partial skull, dated at 2.5 million years ago, was discovered in Ethiopia. Named *Australopithecus garhi,* this species appears to descend from *A. afarensis* and possibly an ancestor for *Homo.* Coexisting with the earliest *Homo* was a different lineage of robust australopithecines that existed between 2.5 and 1.2 million years ago. These "robust" australopithecines, including *Paranthropus robustus,* had heavy jaws, skull crests, and large back molars used for chewing coarse roots and tubers (figure 20.25). They are an extinct branch in hominid evolution and not part of our own lineage.

Although researchers are divided over who the first members of *Homo* were, and indeed how to define the genus *Homo,* the earliest known species of *Homo* was *Homo habilis* ("handy man"). *Homo habilis* had a larger brain than the australopithecines (figure 20.25) and unquestionably used bone and stone tools.

About 1.5 million years ago *Homo erectus* appeared, a large hominid standing 150 to 170 cm (5 to 5.5 feet) tall, with a low but distinct forehead and strong brow ridges. The brain capacity was around 1000 mm^3, intermediate between the brain capacity of *Homo habilis* and modern humans (figure 20.25). *Homo erectus* was a social species living in tribes of 20

classification of Living Mammalian Orders[1]

All modern mammals are placed in two subclasses, Prototheria, containing the monotremes, and Theria, containing marsupials and placentals. Of the 26 recognized mammalian orders, three of the smaller marsupial orders and seven of the smaller placental orders are omitted.

Class Mammalia

Subclass Prototheria (pro′to-thir′ee-a) (Gr. *prōtos,* first, + *thēr,* wild animal). Cretaceous and early Cenozoic mammals. Extinct except for egg-laying monotremes.

Infraclass Ornithodelphia (or′ni-tho-del′fee-a) (Gr. *ornis,* bird, + *delphys,* womb). Monotreme mammals.

Order Monotremata (mon′o-tre′ma-tah) (Gr. *monos,* single, + *trēma,* hole): **egg-laying (oviparous) mammals: duck-billed platypus, echidnas.** Three species in this order from Australia, Tasmania, and New Guinea. The most noted member of order is the duck-billed platypus (*Ornithorhynchus anatinus*). Spiny anteaters, or echidnas (*Tachyglossus*), have a long, narrow snout adapted for feeding on ants, their chief food.

Subclass Theria (thir′ee-a) (Gr. *thēr,* wild animal).

Infraclass Metatheria (met′a-thir′e-a) (Gr. *meta,* after, + *thēr,* wild animal). Marsupial mammals.

Order Didelphimorphia (dy′del-fi-mor′fee-a) (Gr. *di,* two, + *delphi,* uterus, + *morph,* form): **American opossums.** These mammals, like other marsupials, are characterized by an abdominal pouch, or marsupium, in which they rear their young. Most species occur in Central and South America, but one species, the Virginia opossum, is widespread in North America; 66 species.

Order Dasyuormorphia (das-ee-yur′o-mor′fee-a) (Gr. *dasy,* hairy, + *uro,* tail, + *morph,* form): **Australian carnivorous mammals.** In addition to a number of larger carnivores, it includes a number of marsupial "mice," all of which are carnivorous. Confined to Australia, Tasmania, and New Guinea; 64 species.

Order Peramelemorphia (per′a-mel-e-mor′fee-a) (Gr. *per,* pouch, + *mel,* badger, + *morph,* form): **Bandicoots.** Like placentals, members of this group have a chorioallantoic placenta and a high rate of reproduction for marsupials. Confined to Australia, Tasmania, and New Guinea; 22 species.

Order Diprotodontia (dy′pro-to-don′-tee-a) (Gr. *di,* two, + *pro,* front, + *odont,* tooth): **Koala, wombats, possums, wallabies, kangaroos.** Diverse marsupial group containing some of the largest and most familiar marsupials. Present in Australia, Tasmania, New Guinea, and many islands of the East Indies; 131 species.

Infraclass Eutheria (yu-thir′e-a) (Gr. *eu,* true, + *thēr,* wild animal). Viviparous placental mammals.

Order Insectivora (in-sec-tiv′o-ra) (L. *insectum,* an insect, + *vorare,* to devour): **insect-eating mammals: shrews** (figure 20.28), **hedgehogs, tenrecs, moles.** Small, sharp-snouted animals with primitive characters that feed principally on insects; 440 species.

Order Chiroptera (ky-rop′ter-a) (Gr. *cheir,* hand, + *pteron,* wing): **bats.** Flying mammals with forelimbs modified into wings; use of echolocation by most bats; most nocturnal; second largest mammalian order, exceeded in species numbers only by order Rodentia; 977 species.

Order Primates (pry-may′teez) (L. *prima,* first): **prosimians, monkeys, apes, humans.** First in the animal kingdom in brain development with especially large cerebral hemispheres; mostly arboreal; large eyes with binocular vision; usually with grasping hands; five digits (usually provided with flat nails) on both forelimbs and hindlimbs; group singularly lacking in claws, scales, horns, and hooves; two suborders; 279 species.

Suborder Strepsirhini (strep′suh-ry-nee) (Gr. *strepsō,* to turn, twist, + *rhinos,* nose): **lemurs, aye-ayes, lorises, pottos, bush babies.** Seven families of arboreal primates, formerly called prosimians, concentrated on Madagascar, but with species in Africa, Southeast Asia, and Malay peninsula. With comma-shaped nostrils, long nonprehensile tail, and second toe provided with a claw; 49 species.

figure 20.28

Shorttail shrew, *Blarina brevicauda,* eating a grasshopper. This tiny but fierce mammal, with a prodigious appetite for insects, mice, snails, and worms, spends most of its time underground and so is seldom seen by humans. Shrews are believed to resemble the insectivorous ancestors of placental mammals. Order Insectivora, family Soricidae.

---

[1]Based on Nowak, R. M. 1999. *Walker's Mammals of the world,* ed. 6. Baltimore, The Johns Hopkins University Press.

Suborder Haplorhini (hap′lo-ry-nee) (Gr. *haploos,* single, simple + *rhinos,* nose): **tarsiers, marmosets, New and Old World monkeys, gibbons, gorilla, chimpanzees, orangutans, humans.** Six families, four of which were formerly called Anthropoidea. Haplorhine primates have dry, hairy noses, ringed nostrils, and differences in uterine anatomy, placental development, and skull morphology that distinguish them from strepsirhine primates; 230 species.

Order Xenarthra (ze-nar′thra) (Gr. *xenos,* intrusive, + *arthron,* joint) (formerly Edentata [L. *edentatus,* toothless]): **anteaters, armadillos, sloths.** Either toothless (anteaters) or with simple peglike teeth (sloths and armadillos); restricted to South and Central America with the nine-banded armadillo in the southern United States; 29 species.

Order Lagomorpha (lag′o-mor′fa) (Gr. *lagos,* hare; + *morphē,* form): **rabbits, hares, pikas** (figure 20.29). Dentition resembling that of rodents but with four upper incisors rather than two as in rodents; 81 species.

Order Rodentia (ro-den′che-a) (L. *rodere,* to gnaw): **gnawing mammals: squirrels** (figure 20.30), **rats, woodchucks.** Most numerous of all mammals both in numbers and species; dentition with two upper and two lower chisel-like incisors that grow continually and are adapted for gnawing; 2052 species.

Order Carnivora (car-niv′o-ra) (L. *caro,* flesh, + *vorare,* to devour): **flesh-eating mammals: dogs, wolves, cats, bears** (figure 20.31), **weasels, seals, sea lions, walruses.** All with predatory habits (except giant pandas); teeth especially adapted for tearing flesh; in most, canines used for killing prey; worldwide except in Australian and Antarctic regions; 280 species.

Order Proboscidea (pro′ba-sid′e-a) (Gr. *proboskis,* elephant's trunk, from *pro,* before, + *boskein,* to feed): **proboscis mammals: elephants.** Largest of living land

figure 20.30

Eastern gray squirrel, *Sciurus carolinensis.* This common resident of Eastern towns and hardwood forests serves as an important reforestation agent by planting numerous nuts that sprout into trees. Order Rodentia, family Sciuridae.

figure 20.29

A pika, *Ochotona princeps,* atop a rockslide in Alaska. This little rat-sized mammal does not hibernate but prepares for winter by storing dried grasses beneath boulders. Order Lagomorpha, family Ochotonidae.

figure 20.31

Grizzly bear, *Ursus arctos horribilis,* of Alaska. Grizzlies, once common in the lower 48 states, are now confined largely to northern wilderness areas. Order Carnivora, family Ursidae.

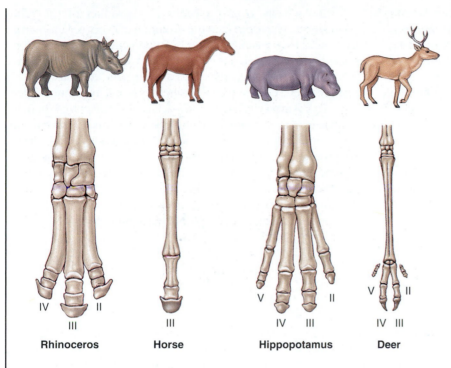

figure 20.32

Odd-toed and even-toed ungulates. Rhinoceroses and horses (order Perissodactyla) are odd-toed; hippopotamuses and deer (order Artiodactyla) are even-toed. Lighter, faster mammals run on only one or two toes.

| Rhinoceros | Horse | Hippopotamus | Deer |

animals, two upper incisors elongated as tusks, molar teeth well developed; two extant species: Indian elephant, with relatively small ears, and African elephant, with large ears.

Order Perissodactyla(pe-ris′so-dak′ti-la) (Gr. *perissos,* odd, + *dactylos,* toe): **odd-toed hoofed mammals: horses, asses, zebras, tapirs, rhinoceroses.** Mammals with an odd number (one or three) of toes and with well-developed hooves (figure 20.32); all herbivorous; both Perissodactyla and Artiodactyla often called **ungulates,** or hoofed mammals, with teeth adapted for chewing; 17 species.

Order Artiodactyla (ar′te-o-dak′ti-la) (Gr. *artios,* even, + *daktylos,* toe): **even-toed hoofed mammals: swine, camels, deer and their allies, hippopotamuses, antelopes, cattle, sheep, goats.** Each toe sheathed in a cornified hoof; most have two toes, although hippopotamuses and some others have four (figure 20.32); many, such as cattle, deer, and sheep, with horns or antlers; many are ruminants, that is, herbivores with partitioned stomachs; 221 species.

Order Cetacea (see-tay′she-a) (L. *cetus,* whale): **whales** (figure 20.33), **dolphins, porpoises.** Anterior limbs of cetaceans modified into broad flippers; posterior limbs absent; nostrils represented by a single or double blowhole on top of the head; hair limited to a few hairs on muzzle, no skin glands except mammary and those of eye; no external ear; 78 species.

figure 20.33

Humpback whale, *Megaptera novaeangliae,* breaching. Among the most acrobatic of whales, humpbacks appear to breach to stun fish schools or to communicate information to other herd members. Order Cetacea, family Balaenopteridae.

to 50 individuals. *Homo erectus* had a successful and complex culture and became widespread throughout the tropical and temperate Old World. Their brain shape, revealed by casts of skulls, suggests that limited speech was possible, and charcoal deposits indicate that they controlled and made use of fire. Yet another amazing hominid find was announced in 2004, with the discovery of *Homo floresiensis,* a species only 1 m tall, from the island of Flores, Indonesia. This species apparently arose from *H. erectus* and became extinct only about 13,000 years ago.

After the disappearance of *Homo erectus* about 50,000 years ago, subsequent human evolution and establishment of the lineage leading to *Homo sapiens* ("wise man") threaded a complex course. From among the many early subcultures of this lineage, Neandertals, *Homo neanderthalensis,* emerged about 150,000 years ago. With a brain capacity well within the range of modern humans, Neandertals were proficient hunters and tool users. They dominated the Old World in the late Pleistocene epoch. About 30,000 years ago the Neandertals were replaced and quite possibly exterminated by modern humans, tall people with a culture very different from that of Neandertals. Implement crafting developed rapidly, and human culture became enriched with aesthetics, artistry, and sophisticated language.

Biologically, *Homo sapiens* is a product of the same processes that have directed the evolution of every organism from the time of life's origin. Mutation, isolation, genetic drift, and natural selection have operated for us as they have for other animals. Yet we have what no other animal has, a nongenetic cultural evolution that provides a constant feedback between past and future experience. Our symbolic languages, capacities for conceptual thought, knowledge of our history, and abilities to manipulate our environment emerge from this nongenetic cultural endowment. Finally, we owe much of our cultural and intellectual achievements to our arboreal ancestry which gave us binocular vision, superb visuotactile discrimination, and manipulative skills in use of our hands. If the horse (with one toe instead of five fingers) had human mental capacity, could it have accomplished what humans have?

Summary

Mammals are endothermic and homeothermic vertebrates whose bodies are insulated by hair and who nurse their young with milk. The approximately 4800 species of mammals are descended from the synapsid lineage of amniotes that arose during the Carboniferous period of the Paleozoic era. Their evolution can be traced from pelycosaurs of the Permian period to therapsids of the late Permian and Triassic periods of the Mesozoic era. One group of therapsids, the cynodonts, gave rise during the Triassic to the true mammals. Mammalian evolution was accompanied by the appearance of many important derived characteristics, among these enlarged brain with greater sensory integration, high metabolic rate, endothermy, heterodont teeth, and many changes in the skeleton that supported a more active life. Mammals diversified rapidly during the Tertiary period of the Cenozoic era.

Mammals are named for the glandular milk-secreting organs of females (rudimentary in males), a unique adaptation which, combined with prolonged parental care, buffers infants from demands of foraging for themselves and eases the transition to adulthood. Hair, the integumentary outgrowth that covers most mammals, serves variously for mechanical protection, thermal insulation, protective coloration, and waterproofing. Mammalian skin is rich in glands: sweat glands that function in evaporative cooling, scent glands used in social interactions, and sebaceous glands that secrete lubricating skin oil. All placental mammals have deciduous teeth that are replaced by permanent teeth (diphyodont dentition). Four groups of teeth—incisors, canines, premolars, and molars—may be highly modified in different mammals for specialized feeding tasks, or they may be absent.

Food habits of mammals strongly influence their body form and physiology. Insectivores have pointed teeth for piercing the exoskeleton of insects and other small invertebrates. Herbivorous mammals have specialized teeth for grinding cellulose and silica-rich plants and have specialized regions of the gut for harboring bacteria that can break down cellulose. Carnivorous mammals have adaptations, including specialized jaw muscles and teeth, for killing and processing their prey, mainly herbivorous mammals. Omnivores feed on both plant and animal foods and have a variety of tooth types.

Some marine, terrestrial, and aerial mammals migrate; some migrations, such as those of fur seals and caribou, are extensive. Migrations are usually made toward climatic conditions favorable for finding food, mating, or rearing young.

Mammals with true flight, the bats, are mainly nocturnal and thus avoid direct competition with birds. Most employ ultrasonic echolocation to navigate and feed in darkness.

Living mammals with the most primitive characters are egg-laying monotremes of the Australian region. After hatching, young are nourished with their mother's milk. All other mammals are viviparous. Embryos of marsupials have brief gestation periods, are born underdeveloped, and complete their early growth in the mother's pouch, nourished by milk. The remaining mammals are eutherians, mammals that develop an advanced placental attachment between mother and embryos through which embryos are nourished for a prolonged period.

Mammal populations fluctuate from both density-dependent and density-independent causes and some mammals, particularly rodents, may experience extreme cycles of abundance in population density. The unqualified success of mammals as a group cannot be attributed to greater organ system perfection, but rather to their impressive overall adaptability—the capacity to fit more perfectly in total organization to environmental conditions and thus exploit virtually every habitat on earth.

Darwinian evolutionary principles give us great insight into our own origins. Humans are primates, a mammalian group that descends from a shrewlike ancestor. The common ancestor of all modern primates was arboreal and had grasping fingers and forward-facing eyes capable of binocular vision. Primates radiated over the last 80 million years to form two major lines of descent: prosimians (lemurs, lorises, and tarsiers) and simians (monkeys, apes, and humans).

Humans appeared in Africa about 7 million years ago, and gave rise to several species of australopithecines, which persisted for more than 3 million years. Australopithecines gave rise to, and coexisted with, *Homo habilis,* the first user of stone tools. *Homo erectus* appeared about 1.8 million years ago and was eventually replaced by modern humans, *Homo sapiens.*

Review Questions

1. Describe the evolution of mammals, tracing their synapsid lineage from early amniote ancestors to true mammals. How would you distinguish the skull of a synapsid from that of diapsid?

2. Describe structural and functional adaptations that appeared in early amniotes that foreshadowed the mammalian body plan. Which mammalian attributes were especially important to successful radiation of mammals?

3. Hair is thought to have evolved in therapsids as an adaptation for insulation, but modern mammals have adapted hair for several other purposes. Describe these.

4. What is distinctive about each of the following: horns of bovids, antlers of the deer, and horns of rhinos? Describe the growth cycle of antlers.

5. Describe the location and principal function(s) of each of the following skin glands: sweat glands (eccrine and apocrine), scent glands, sebaceous glands, and mammary glands.

6. Define the terms "diphyodont" and "heterodont" and explain how both terms apply to mammalian dentition.

7. Describe the food habits of each of the following groups: insectivores, herbivores, carnivores, and omnivores. Give the common names of some mammals belonging to each group.

8. Most herbivorous mammals depend on cellulose as their main energy source, yet no mammal synthesizes cellulose-splitting enzymes. How are the digestive tracts of mammals specialized for symbiotic digestion of cellulose?

9. How does fermentation differ between horses and cattle?

10. Describe the annual migrations of barren-ground caribou and fur seals.

11. Explain what is distinctive about the life habit and mode of navigation in bats.

12. Describe and distinguish patterns of reproduction in monotremes, marsupials, and placental mammals. What aspects of mammalian reproduction are present in *all* mammals but in no other vertebrates?

13. What is the difference between density-dependent and density-independent causes of fluctuations in size of mammalian populations?

14. Describe the hare-lynx population cycle, considered a classic example of a prey-predator relationship (figure 20.21). From your examination of the cycle, formulate a hypothesis to explain the oscillations.

15. What anatomical characteristics set primates apart from other mammals?

16. What role does the fossil named "Lucy" play in reconstruction of human evolutionary history?

17. In what ways do the genera *Australopithecus* and *Homo,* which coexisted for at least 2 million years, differ?

18. When did the different species of *Homo* appear and how did they differ socially?

19. What major attributes make the human position in animal evolution unique?

Selected References

See also general references on page 415.

Feldhamer, G. A., L. C. Drickamer, S. H. Vessey, and J. F. Merritt. 2003. Mammalogy: adaptation, diversity, and ecology, ed. 2. Dubuque, Iowa, WCB/McGraw-Hill. *Modern well-illustrated textbook.*

Grzimek's encyclopedia of mammals. 1990. vol. 1–5. New York, McGraw-Hill Publishing Company. *Valuable source of information on all mammalian orders.*

Macdonald, D., ed. 1984. The encyclopedia of mammals. New York, Facts on File Publications. *Coverage of all mammalian orders and families, enhanced with fine photographs and color artwork.*

Nowak, R. M. 1999. Walker's mammals of the world, ed. 6. Baltimore, The Johns Hopkins University Press. *The definitive illustrated reference work on mammals, with descriptions of all extant and recently extinct species.*

Pilbeam, D., R. D. Martin, and D. Jones. 1994. Cambridge encyclopedia of human evolution. New York, Cambridge University Press. *Comprehensive and informative encyclopedia written for nonspecialists. Highly readable and highly recommended.*

Preston-Mafham, R., and K. Preston-Mafham. 1992. Primates of the world. New York, Facts on File Publications. *A small "primer" with high quality photographs and serviceable descriptions.*

Rice, J. A., ed. 1994. The marvelous mammalian parade. Natural History **103**(4):39–91. *A special multi-authored section on mammalian evolution.*

Rismiller, P. D., and R. S. Seymour. 1991. The echidna. Sci. Am. **294**:96–103 (Feb.). *Recent studies of this fascinating monotreme have revealed many secrets of its natural history and reproduction.*

Stringer, C. B. 1990. The emergence of modern humans. Sci. Am. **263**:98–104 (Dec.). *A review of the geographical origins of modern humans.*

Suga, N. 1990. Biosonar and neural computation in bats. Sci. Am. **262**:60–68 (June). *How the bat nervous system processes echolocation signals.*

Custom Website

The *Animal Diversity* Online Learning Center is a great place to check your understanding of chapter material. Visit www.mhhe.com/hickmanad4e for access to key terms, quizzes, and more! Further enhance your knowledge with web links to chapter-related material.

Explore live links for these topics:

Classification and Phylogeny of Animals
Class Mammalia
Marine Mammals
Cat Dissections
Rat Dissections

Fetal Pig Dissections
Vertebrate Laboratory Exercises
Conservation Issues Concerning Mammals

General References

These references pertain to groups covered in more than one chapter. They include a number of very valuable field manuals that aid in identification, as well as general texts.

Barrington, E. J. W. 1979. Invertebrate structure and function. ed. 2. New York, John Wiley & Sons, Inc. *Excellent account of function in major invertebrate groups.*

Brusca, R. C., and G. J. Brusca. 2003. Invertebrates, ed. 2. Sunderland, Massachusetts, Sinauer Associates, Inc. *Invertebrate text organized around the bauplan ("body plan") concept—structural range, architectural limits, and functional aspects of a design—for each phylum. Includes cladistic analysis of phylogeny for most groups.*

Cracraft, J., and M. J. Donoghue (eds.). 2004. Assembling the tree of life. New York, Oxford University Press. *This volume summarizes much of the current work on the Tree of Life project, an attempt to uncover the evolutionary relationships for all living taxa. The early chapters of the book discuss the relevance of the Tree of Life project for society. The overview and introductory sections are very readable, and these are paired with results of recent phylogenetic analyses for all domains of life.*

Fotheringham, N. 1980. Beachcomber's guide to Gulf Coast marine life. Houston, Texas, Gulf Publishing Company. *Coverage arranged by habitats. No keys, but common forms that occur near shore can be identified.*

Gosner, K. L. 1979. A field guide to the Atlantic seashore: invertebrates and seaweeds of the Atlantic coast from the Bay of Fundy to Cape Hatteras. The Peterson Field Guide Series. Boston, Houghton Mifflin Company. *A helpful aid for students of invertebrates found along the northeastern coast of the United States.*

Halstead, B. W. 1992. Dangerous aquatic animals of the world: a color atlas: with prevention, first aid, and emergency treatment procedures. Princeton, New Jersey, Darwin Press. *Excellent coverage of threats to humans by dangerous aquatic animals and how to avoid fateful encounters.*

Humann, P., and N. DeLoach. 2002. Reef creature identification. Florida, Caribbean, Bahamas. Jacksonville, Florida, New World Publications. *Excellent field guide to aid identification of Atlantic reef invertebrates except corals.*

Hyman, L. H. 1940–1967. The invertebrates, 6 vols. New York, McGraw-Hill Book Company. *Informative discussions on the phylogenies of most invertebrates are treated in this outstanding series of monographs. Volume I contains a discussion of the colonial hypothesis of the origin of metazoa, and volume 2 contains a discussion of the origin of bilateral animals, body cavities, and metamerism.*

Kaplan, E. H. 1988. A field guide to southeastern and Caribbean seashores: Cape Hatteras to the Gulf Coast, Florida, and the Caribbean. A. Peterson Field Guide Series. Boston, Houghton Mifflin Company. *More than just a field guide, this comprehensive book is filled with information on the biology of seashore animals: complements Gosner's field guide which covers animals north of Cape Hatteras.*

Kardong, K. V. 2006. Vertebrates: comparative anatomy, function, evolution, ed. 4. New York, McGraw-Hill. *Evolution of vertebrate structure.*

Kozloff, E. N. 1996. Marine invertebrates of the Pacific Northwest. Seattle, University of Washington Press. *Contains keys for many marine groups.*

Lane, R. P., and H. W. Crosskey. 1993. Medically important insects and arachnids. London, Chapman and Hall. *The most up-to-date medical entomology text available.*

Meglitsch, P. A., and F. R. Schram. 1991. Invertebrate zoology, ed. 3. New York, Oxford University Press. *Schram's thorough revision of the older Meglitsch text. Includes cladistic treatments.*

Morris, R. H., D. P. Abbott, and E. C. Haderlie. 1980. Intertidal invertebrates of California. Stanford, Stanford University Press. *An essential reference on the most important invertebrates of the intertidal zone in California. Contains 900 color photographs.*

Nielsen, C. 2001. Animal evolution: interrelationships of the living phyla. New York, Oxford University Press. *Cladistic analysis of morphology is used to develop sister-group relationships of the living Metazoa. An advanced but essential reference.*

Pechenik, J. A. 2005. Biology of the invertebrates, ed. 5. Dubuque, McGraw-Hill Companies, Inc. *Up-to-date text, including recognition of superphyla Ecdysozoa and Lophotrochozoa.*

Pennak, R. W. 1989. Freshwater invertebrates of the United States, ed. 3. New York, John Wiley & Sons, Inc. *Contains keys for identification of freshwater invertebrates with brief accounts of each group. Indispensable for freshwater biologists.*

Pough, F. H., J. B. Heiser, and W. N. McFarland. 2005. Vertebrate life, ed. 7. Upper Saddle River, New Jersey, Prentice Hall. *Vertebrate morphology, physiology, ecology, and behavior cast in a cladistic framework.*

Ricketts, E. F., J. Calvin, and J. W. Hedgpeth (revised by D. W. Phillips). 1985. Between Pacific tides, ed. 5. Stanford, Stanford University Press. *A revision of a classic work in marine biology. It stresses the habits and habitats of the Pacific coast invertebrates, and the illustrations are revealing. It includes an excellent, annotated systematic index and bibliography.*

Roberts, L. S., and J. Janovy, Jr. 2005. Gerald D. Schmidt and Larry S. Roberts' Foundations of parasitology, ed. 7. Dubuque, McGraw-Hill Publishers. *Highly readable and up-to-date account of parasitic protozoa, worms and arthropods.*

Ruppert, E. E., and R. D. Barnes. 1994. Invertebrate zoology, ed. 6. Philadelphia, Saunders College Publishing. *Authoritative, detailed coverage of invertebrate phyla.*

Smith, D. L. and K. B. Johnson. 1996. A guide to marine coastal plankton and marine invertebrate larvae. Dubuque, Iowa,

Kendall/Hunt Publishing Company. *Valuable manual for identification of marine plankton, which is usually not covered in most field guides.*

Tudge, C. 2000. The variety of life; a survey and celebration of all the creatures that have ever lived. New York, Oxford University Press. *This well-written book might be called a* celebration of systematics, the discipline, as the author says, that introduces us to the creatures themselves. A fine introduction that explains cladistics is followed by chapters that describe . . . almost everything!*

Valentine, J. V. 2004. On the origin of phyla. Chicago, Illinois, University of Chicago Press. *A thorough coverage of paleontological, morphological,* and molecular evidence on relationships among the animal phyla and the evolutionary origins of animal body plans.*

Willmer, P., G. Stone, and I. A. Johnston. 2005. Environmental physiology of animals. Malden, Massachusetts, Blackwell Publishers. *A comparative treatment of the physiological ecology of animals.*

Glossary

This glossary lists definitions, pronunciations, and derivations of the most important recurrent technical terms, units, and names (excluding taxa) used in the text.

A

abiotic (ā′bī-äd′ik) (Gr. *a*, without, + *biōtos*, life, livable) Characterized by the absence of life.

abomasum (ab′ō-mā′səm) (L. *ab*, from, + *omasum*, paunch) Fourth and last chamber of the stomach of ruminant mammals.

aboral (ab-o′rəl) (L. *ab*, from, + *os*, mouth) A region of an animal opposite the mouth.

acanthodians (a′kan-thō′dē-əns) (Gr. *akantha*, prickly, thorny) A group of jawed fishes, characterized by large spines in their fins, from Lower Silurian to Lower Permian.

acclimatization (ə-klī′mə-də-zā-shən) (L, *ad*, to, + Gr. *klima*, climate) Gradual physiological adjustment of an organism in response to relatively long-lasting environmental changes.

aciculum (ə-sik′ū-ləm) (L. *acicula*, dm. of *acus*, a point) Supporting structure in notopodium and neuropodium of some polychaetes.

acoelomate (a-sēl′ə-māt′) (Gr. *a*, not, + *koilōma*, cavity) Without a coelom, as in flatworms and proboscis worms.

acontium (ə-kän′chē-əm), pl. **acontia** (Gr. *akontion*, dart) Threadlike structure bearing nematocysts located on mesentery of sea anemone.

adaptation (L. *adaptatus*, fitted) An anatomical structure, physiological process, or

behavioral trait that evolved by natural selection and improves an organism's ability to survive and leave descendants.

adaptive radiation Evolutionary diversification that produces numerous ecologically disparate lineages from a single ancestral one, especially when this diversification occurs within a short interval of geological time.

adaptive zone A characteristic reaction and mutual relationship between environment and organism ("way of life") demonstrated by a group of evolutionarily related organisms.

adductor (ə-duk′tər) (L. *ad*, to, + *ducere*, to lead) A muscle that draws a part toward a median axis, or a muscle that draws the two valves of a mollusc shell together.

adipose (ad′ə-pōs) (L. *adeps*, fat) Fatty tissue; fatty.

aerobic (a-rō′bik) (Gr. *aēr*, air, + *bios*, life) Oxygen-dependent form of respiration.

afferent (af′ə-rənt) (L. *ad*, to, + *ferre*, to bear) Adjective meaning leading or bearing toward some organ, for example, nerves conducting impulses toward the brain or blood vessels carrying blood toward an organ; opposed to efferent.

age structure An accounting of the ages of individuals in a population at a particular time and place.

agnathan (ag′nə-thən (Gr. *a*, without, + *gnathos*, jaw) A jawless fish of the paraphyletic superclass Agnatha of the phylum Chordata.

alate (ā′lāt) (L. *alatus*, wing) Winged.

albumin (al-bū′mən) (L. *albuman*, white of egg) Any of a large class of simple proteins that are important constituents of vertebrate blood plasma and tissue fluids and also present in milk, whites of eggs, and other animal substances.

allele (ə-lēl′) (Gr. *allēlōn*, of one another) Alternative forms of genes coding for the

same trait; situated at the same locus in homologous chromosomes.

allelic frequency The proportion of all copies of a particular gene in a population **gene pool** represented by a particular **allele.**

allopatric (Gr. *allos*, other, + *patra*, native land) In separate and mutually exclusive geographical regions.

allopatric speciation The hypothesis that new species are formed by dividing an ancestral species into geographically isolated subpopulations that evolve **reproductive barriers** between them through independent evolutionary divergence from their common ancestor.

altricial (al-tri′shəl) (L. *altrices*, nourishers) Referring to young animals (especially birds) having the young hatched in an immature dependent condition.

alula (al′yə-lə) (L. dim of *ala*, wing) The first digit or thumb of a bird's wing, much reduced in size.

alveolate (al-vē′ə-lət) (L. dim. Of *alveus*, cavity, hollow). Any member of a protozoan clade having alveolar sacs; includes ciliates, apicomplexans, and dinoflagellates.

alveoli (al-vē′ə-lī) Pockets or spaces bounded by membrane or epithelium.

ambulacra (am′byə-lak′rə) (L. *ambulare*, to walk) In echinoderms, radiating grooves where podia of water-vascular system characteristically project outside the organism.

ameboid (ə-mē′boid) (Gr. *amoibē*, change, + *oid*, like) Ameba-like in putting forth pseudopodia.

amebozoa (ə-mē′bō-zō′ə) (Gr. *amoibē*, change, + *zōon*, animal). A protozoan clade containing slime molds and amebas with lobose pseudopodia.

amensalism (ā-men′səl-iz′əm) An asymmetric competitive interaction between two species in an ecological community in which only one of the species is affected.

bat/āpe/ärmadillo/herring/fēmale/finch/līce/ crocodile/crōw/duck/ūnicorn/tūna/ə indicates unaccented vowel sound "uh" as in mammal, fishes, cardinal, heron, vulture/stress as in bi-ol′o-gy, bi′o-log′i-cal

amictic (ə-mik′tic) (Gr. *a*, without + *miktos*, mixed or blended) Pertaining to female rotifers, which produce only diploid eggs that cannot be fertilized, or to the eggs produced by such females. Compare with mictic.

ammocoetes (am-ə-sēd′ēz) (Gr. *ammos*, sand, + *koitē*, bed) Larvae of any of various species of lampreys.

amnion (am′nē-än) (Gr. *amnion*, membrane around the fetus) The innermost of the extraembryonic membranes forming a fluid-filled sac around the embryo in amniotes.

amniote (am′nē-ōt) Having an amnion; as a noun, an animal that develops an amnion in embryonic life; refers collectively to reptiles, birds, and mammals. Adj., **amniotic.**

amplexus (am-plek′səs) (L. embrace) The copulatory embrace of frogs or toads.

ampulla (am-pūl′ə) (L. flask) Membranous vesicle; dilation at one end of each semicircular canal containing sensory epithelium; muscular vesicle above the tube foot in water-vascular system of echinoderms.

anadromous (an-ad′rə-məs) (Gr. *anadromos*, running upward) Refers to fishes that migrate up streams from the sea to spawn.

anaerobic (an′ə-rō′bik) (Gr. *an*, not + *aēr*, air, + *bios*, life) Not dependent on oxygen for respiration.

analogy (L. *analogous*, ratio) Similarity of function but not of origin.

anapsids (ə-nap′səds) (Gr. *an*, without, + *apsis*, arch) Amniotes in which the skull lacks temporal openings, with turtles the only living repesentatives.

ancestral character state The condition of a taxonomic character inferred to have been present in the most recent common ancestor of a taxonomic group being studied cladistically.

androgenic gland (an′drō-jen′ək) (Gr. *anēr*, male, + *gennaein*, to produce) Gland in Crustacea that causes development of male characteristics.

annulus (an′yəl-əs) (L. ring) Any ringlike structure, such as superficial rings on leeches.

antenna (L. sail yard) A sensory appendage on the head of arthropods, or the second pair of the two such pairs of structures in crustaceans.

antennal gland Excretory gland of Crustacea located in the antennal metamere.

bat/āpe/ärmadillo/herring/fēmale/finch/līce/crocodile/crōw/duck/ūnicorn/tūna/ə indicates unaccented vowel sound "uh" as in mammal, fishes, cardinal, heron, vulture/stress as in bi-ol′o-gy, bi′o-log′i-cal

anterior (L. comparative of *ante*, before) The head of an organism or (as the adjective) toward that end.

anthropoid (an′thrə-poyd) (Gr. *anthrōpos*, man, + *eidos*, form) Resembling humans; especially the great apes.

aperture (ap′ər-chər) (L. *apertura* from *aperire*, to uncover) An opening; the opening into the first whorl of a gastropod shell.

apical (ā′pə-kl) (L. *apex*, tip) Pertaining to the tip or apex.

apical complex A certain combination of organelles found in the protozoan phylum Apicomplexa.

apocrine (ap′ə-krən) (Gr. *apo*, away, + *krinein*, to separate) Applies to a type of mammalian sweat gland that produces a viscous secretion by breaking off a part of the cytoplasm of secreting cells.

apopyle (ap′ə-pīl) (Gr. *apo*, away from, + *pylē*, gate) In sponges, opening of the radial canal into the spongocoel.

arboreal (är-bōr′ē-al) (L. *arbor*, tree) Living in trees.

archaeocyte (ar′kē-ə-sīt) (Gr. *archaios*, beginning + *kytos*, hollow vessel) Ameboid cells of varied function in sponges.

archenteron (ärk-en′tə-rän) (Gr. *archē*, beginning + *eneron*, gut) The main cavity of an embryo in the gastrula stage; it is lined with endoderm and represents the future digestive cavity.

archosaur (är′kə-sor) (Gr. *archōn*, ruling, + *sauros*, lizard) A clade of diapsid vertebrates that includes the living crocodiles and birds and the extinct pterosaurs and dinosaurs.

Aristotle's lantern Masticating apparatus of some sea urchins.

artiodactyl (är′ti-o-dak′təl) (Gr. *artios*, even, + *daktylos*, toe) One of an order of mammals with two or four digits on each foot.

asconoid (Gr. *askos*, bladder) Simplest form of sponges, with canals leading directly from the outside to the interior.

asexual Without distinct sexual organs; not involving formation of gametes.

asymmetric competition See **amensalism**.

atoke (ā′tōk) (Gr. *a*, without, + *tokos*, offspring) Anterior, nonreproductive part of a marine polychaete, as distinct from the posterior, reproductive part (epitoke) during the breeding season.

atrium (ā′trē-əm) (L. *atrium*, vestibule) One of the chambers of the heart; also, the tympanic cavity of the ear, also, the large cavity containing the pharynx in tunicates and cephalochordates.

auricle (aw′ri-kl) (L. *auricula*, dim. of *auris*, ear) One of the less muscular chambers of the heart; atrium; the external ear, or pinna; any earlike lobe or process.

auricularia (ə-rik′u-lar′ē-ə) (L. *auricula*, a small ear) A type of larva found in Holothuroidea.

autogamy (aw-täg′ə-me) (Gr. *autos*, self, + *gamos*, marriage) Condition in which the gametic nuclei produced by meiosis fuse within the same organism that produced them to restore the diploid number.

autotomy (aw-tät′ə-me) (Gr. *autos*, self, + *tomos*, a cutting) Detachment of a part of the body by the organism itself.

autotroph (aw′tō-trōf) (Gr. *autos*, self, + *trophos*, feeder) An organism that makes its organic nutrients from inorganic raw materials.

autotrophic nutrition (Gr. *autos*, self, + *trophia*, denoting nutrition) Nutrition characterized by the ability to use simple inorganic substances for the synthesis of more complex organic compounds, as in green plants and some bacteria.

axial (L. *axis*, axle) Relating to the axis, or stem; on or along the axis.

axocoel (aks′ə-sēl) (Gr. *axon*, axle + *koilos*, hollow) Anterior coelomic compartment in echinoderms; corresponds to protocoel.

axolotl (ak′sə-lot′l) (Nahuatl *atl*, water, + *xolotl*, doll, servant, spirit) Salamanders of the species *Ambystoma mexicanum*, which do not metamorphose, and retain aquatic larval characteristics throughout adulthood.

axon (ak′sän) (Gr. *axōn*) Elongate extension of a neuron that conducts impulses away from the cell body and toward the synaptic terminals.

axoneme (aks′ə-nēm) (L. *axis*, axle, + Gr. *nēma*, thread) The microtubules in a cilium or flagellum, usually arranged as a circlet of nine pairs; enclosing one central pair; also, the microtubules of an axopodium.

axopodium (ak′sə-pō′dē-əm) (Gr. *axon*, an axis, + *podion*, small foot) Long, slender, more or less permanent pseudopodium found in certain amebas. (Also **axopod.**)

axostyle (aks′ō-stīl) Tubelike organelle in some flagellate protozoa, extending from the area of the kinetosomes to the posterior end, where it often protrudes.

B

basal body Also known as kinetosome and blepharoplast, a cylinder of nine triplets of microtubules found basal to a flagellum or cilium; same structure as a centriole.

basal disc Aboral attachment site on a cnidarian polyp.

basis, basipodite (bā′səs, bā-si′pə-dīt) (Gr. *basis,* base, + *pous, podos,* foot) The distal or second joint of the protopod of a crustacean appendage.

benthos (ben′thäs) (Gr. depth of the sea) Organisms that live along the bottom of the seas and lakes. Adj., **benthic.** Also, the bottom itself.

Bilateria (bī′lə-tir′ē-ə) (L. *bi-,* two, + *latus,* side) Bilaterally symmetrical animals.

binary fission A mode of asexual reproduction in which the animal splits into two approximately equal offspring.

binomial nomenclature The Linnean system of naming species in which the first word is the name of the genus (first letter capitalized) and the second word is the specific epithet (uncapitalized), usually an adjective modifying the name of the genus. Both of these words are written in italics.

biogenetic law A statement postulating a characteristic relationship between **ontogeny** and **phylogeny.** Examples include Haeckel's law of **recapitulation** and Von Baer's law that general characteristics (those shared by many species) appear earlier in ontogeny than more restricted ones; neither of these statements is universally true.

biogeochemical cycle A description of the flow of elementary matter, such as carbon or phosphorus, through the component parts of an ecosystem and its **abiotic** environment, including the amount of an element present at the various stages of a **food web.**

biological species concept A reproductive community of populations (reproductively isolated from others) that occupies a specific niche in nature.

bioluminescence Method of light production by living organisms in which usually certain proteins (luciferins), in the presence of oxygen and an enzyme (luciferase), are converted to oxyluciferins with the liberation of light.

biomass (Gr. *bios,* life, + *maza,* lump or mass). The weight of total living organisms or of a species population per unit of area.

biome (bī′ōm), (Gr. *bios,* life, + *ōma,* abstract group suffix) Complex of plant and animal communities characterized by climatic and soil conditions; the largest ecological unit.

biosphere (Gr. *bios,* life + *sphaira,* globe) That part of earth containing living organisms.

biotic (bī-äd′ik) (Gr. *biōtos,* life, livable) Of or relating to life.

bipinnaria (L. *bi,* double, + *pinna,* wing, + *aria,* like or connected with) Free-swimming, ciliated, bilateral larva of the asteroid echinoderms; develops into the brachiolaria larva.

biradial symmetry A type of radial symmetry in which only two planes passing through the oral-aboral axis yield mirror images because some structure is paired.

biramous (bī-rām′əs) (L. *bi,* double, + *ramus,* a branch) Adjective describing appendages with two distinct branches, contrasted with uniramous, unbranched.

blastocoel (blas′tə-sēl) (Gr. *blastos,* germ, + *koilos,* hollow) Cavity of the blastula.

blastomere (Gr. *blastos,* germ, + *meros,* part) An early cleavage cell.

blastopore (Gr. *blastos,* germ, + *poros,* passage, pore) External opening of the archenteron in the gastrula.

blastula (Gr. *blastos,* germ, + *ula,* dim) Early embryological state of many animals; consists of a hollow mass of cells.

blepharoplast (blə-fä′rə-plast) (Gr. *blepharon,* eyelid, + *plastos,* formed) See **basal body.**

book gill Respiratory structure of aquatic chelicerates (Arthropoda) where many thin blood-filled gills are layered like the pages of a book. Gas exchange occurs as seawater passes between each pair of gills.

book lung Respiratory structure of terrestrial chelicerates (Arthropoda) where many thin-walled air pockets extend into a blood-filled chamber in the abdomen.

BP Before the present.

brachial (brak′ē-əl) (L. *brachium,* forearm) Referring to the arm.

branchial (brank′ē-əl) (Gr. *branchia,* gills) Referring to gills.

brown bodies Remnants of the lophophore and digestive tract of a degenerating adult ectoproct left behind in the chamber as a new lophophore and digestive tract are formed.

buccal (buk′əl) (L. *bucca,* cheek) Referring to the mouth cavity.

budding Reproduction in which the offspring arises as an outgrowth from the parent and is initially smaller than the parent. Failure of the offspring to separate from the parent leads to colony formation.

bursa pl. **bursae** (M.L. *bursa,* pouch, purse made of skin) A saclike cavity. In ophiuroid echinoderms, pouches opening at bases of arms and functioning in respiration and reproduction (genitorespiratory bursae).

C

calyx (kā′-liks) (L. bud cup of a flower) Any of various cup-shaped zoological structures.

capitulum (ka-pi′chə-ləm) (L. small head) Term applied to small, headlike structures of various organisms, including projection from body of ticks and mites carrying mouthparts.

carapace (kar′ə-pās) (F. from Sp. *carapacho,* shell) Shieldlike plate covering the cephalothorax of certain crustaceans; dorsal part of the shell of a turtle.

carinate (kar′ə-nāt) (L. *carina,* keel) Having a keel, in particular the flying birds with a keeled sternum for the insertion of flight muscles.

carnivore (kar′nə-vōr′) (L. *carnivorus,* flesh eating) One of the mammals of the order Carnivora. Also, any organism that eats animals. Adj., **carnivorous.**

carrying capacity The maximum number of individuals that can persist under specified environmental conditions.

cartilage (L. *cartilago;* akin to L. *cratis,* wickerwork) A translucent elastic tissue that forms most of the skeleton of embryos, very young vertebrates, and adult cartilaginous fishes, such as sharks and rays; in higher forms much of it is converted into bone.

caste (kast) (L. *castus,* pure, separated) One of the polymorphic forms within an insect society, each caste having its specific duties, as queen, worker, soldier, and others.

catadromous (kə-tad′rə-məs) (Gr. *kata,* down, + *dromos,* a running) Refers to fishes that migrate from fresh water to the ocean to spawn.

catastrophic species selection Differential survival among species during a time of mass extinction based on character variation that permits some species but not others to withstand severe environmental disturbances, such as those caused by an asteroid impact.

caudal (käd′l) (L. *cauda,* tail) Constituting, belonging to, or relating to a tail.

cecum (sē′kəm) (L. *caecus,* blind) A blind pouch at the beginning of the large intestine; any similar pouch.

cellulase (sel′ū-lās) (L. *cella,* small room) An enzyme that cleaves cellulose; only synthesized by bacteria and some protists.

cellulose (sel′ū-lōs) (L. *cella,* small room) Chief polysaccharide constituent of the cell wall of green plants and some fungi;

an insoluble carbohydrate $(C_6H_{10}O_5)_n$ that is converted into glucose by hydrolysis.

centriole (sen′trē-ōl) (Gr. *kentron,* center of a circle, + L. *ola,* small) A minute cytoplasmic organelle usually found in the centrosome and considered to be the active division center of the animal cell; organizes spindle fibers during mitosis and meiosis. Same structure as basal body or kinetosome.

cephalization (sef′ə-li-zā-shən) (Gr. *kephalē,* head) The evolutionary process by which specialization, particularly of the sensory organs and appendages, became localized in the head end of animals.

cephalothorax (sef′ə-lä-thō′raks) (Gr. *kephalē,* head, + thorax) A body division found in many Arachnida and higher Crustacea in which the head is fused with some or all of the thoracic segments.

cerata (sə-ra′tə) (Gr. *keras,* a horn, bow) Dorsal processes on some nudibranchs for gaseous exchange.

cercaria (ser-kar′ē-ə) (Gr. *kerkos,* tail, + L. *aria,* like or connected with) Tadpolelike juveniles of trematodes (flukes).

cervical (sər′və-kəl) (L. *cervix,* neck) Relating to a neck.

character (kar′ik-tər). A component of phenotype (including specific molecular, morphological, behavioral, or other features) used by systematists to diagnose species or higher taxa, or to evaluate phylogenetic relationships among different species or higher taxa, or relationships among populations within a species.

character displacement Differences in morphology or behavior within a species caused by competition with another species; characteristics typical of one species differ according to whether the other species is present or absent from a local community.

chela (kēl′ə) (Gr. *chēlē,* claw) Pincerlike claw.

chelicera (kə-lis′ə-rə), pl. **chelicerae** (Gr. *chēlē,* claw, + *keras,* horn) One of a pair of the most anterior head appendages on members of subphylum Chelicerata.

chelipeds (kēl′ə-peds) (Gr. *chēlē, claw,* + L. *pes,* foot) Pincerlike first pair of legs in most decapod crustaceans; specialized for seizing and crushing.

chemoautotroph (ke-mō-aw′tō-trōf) (Gr. *chemeia,* transmutation, + *autos,* self, + *trophos,* feeder) An organism utilizing

inorganic compounds as a source of energy.

chitin (kī′tən) (Fr. *chitine,* from Gr. *chitōn,* tunic) A hard substance that forms part of the cuticle of arthropods and is found sparingly in certain other invertebrates; a nitrogenous polysaccharide insoluble in water, alcohol, dilute acids, and digestive juices of most animals.

chloragogen cells (klōr′ə-gog-ən) (Gr. *chlōros,* light green, + *agōgos,* a leading, a guide) Modified peritoneal cells, greenish or brownish, clustered around the digestive tract of certain annelids; apparently they aid in elimination of nitrogenous wastes and in food transport.

chlorophyll (klō′rə-fil) (Gr. *chlōros,* light green, + *phyllōn,* leaf) Green pigment found in plants and in some animals necessary for photosynthesis.

chloroplast (Gr. *chlōros,* light green, + *plastos,* molded) A plastid containing chlorophyll and usually other pigments, found in cytoplasm of plant cells.

choanoblast (kō-an′ə-blast) (Gr. *choanē,* funnel + *blastos,* germ) One of several cellular elements within the syncytial tissue of a hexactinellid sponge.

choanocyte (kō-an′ə-sīt) (Gr. *choanē,* funnel, + *kytos,* hollow vessel) One of the flagellate collar cells that line cavities and canals of sponges.

choanoflagellate (kō-an′ə-fla-jel′āt) Any member of a protozoan clade having a single flagellum surrounded by a column of microvilli; some form colonies, and all are included within the larger clade of opisthokonts.

chorion (kō′rē-on) (Gr. *chorion,* skin) The outer of the double membrane that surrounds the embryo of reptiles, birds, and mammals; in mammals it contributes to the placenta.

chromatophore (krō-mat′ə-fōr) (Gr. *chrōma,* color, + *herein,* to bear) Pigment cell, usually in the dermis, in which usually the pigment can be dispersed or concentrated.

chromosomal theory of inheritance The well-established principle, initially proposed by Sutton and Boveri in 1903–1904, that nuclear chromosomes are the physical bearers of genetic material in eukaryotic organisms. It is the foundation for modern evolutionary genetics.

chrysalis (kris′ə-lis) (L. from Gr. *chrysos,* gold) The pupal stage of a butterfly.

cilium (sil′ē-əm), pl. **cilia** (L. eyelid) A hairlike, vibratile organelle process found on many animal cells. Cilia may be used in moving particles along the cell surface or, in ciliate protozoans, for locomotion.

cirrus (sir′əs) (L. curl) A hairlike tuft on an insect appendage; locomotor organelle of fused cilia; male copulatory organ of some invertebrates.

clade (klād) (Gr. *klados,* branch) A taxon or other group consisting of a particular ancestral lineage and all of its descendants, forming a distinct branch on a phylogenetic tree.

cladistics (klad-is′təks) (Gr. *klados,* branch, sprout) A system of arranging taxa by analysis of primitive and derived characteristics so that the arrangement will reflect phylogenetic relationships.

cladogram (klād′ə-gram) (Gr. *klados,* branch, + *gramma,* letter) A branching diagram showing the pattern of sharing of evolutionarily derived characters among species or higher taxa.

clasper Digitiform projection on the medial side of pelvic fins of male chondrichthians and some placoderms; used as an intromittent organ to transfer sperm to female reproductive tract

clitellum (klī-tel′əm) (L. *clitellae,* pack-saddle) Thickened saddlelike portion of certain midbody segments of many oligochaetes and leeches.

cloaca (klō-ā′kə) (L. sewer) Posterior chamber of digestive tract in many vertebrates, receiving feces and urogenital products. In certain invertebrates, a terminal portion of digestive tract that serves also as respiratory, excretory, or reproductive duct.

cloning (klō′ning) Production of genetically identical organisms by asexual reproduction.

cnida (nī′də) (Gr. *knidē* nettle). Stinging or adhesive organelles formed within cnidocytes in phylum Cnidaria; nematocysts are a common type.

cnidoblast (nī′də-blast) (Gr. *knidē,* nettle, + Gr. *blastos,* germ) A cnidocyte is called a cnidoblast during the time when a cnida is forming within it.

cnidocil (nī′dō-sil) (Gr. *knidē,* nettle, + L. *cilium,* hair) Triggerlike spine on nematocyst.

cnidocyte (nī′dō-sīt) (Gr. *knidē,* nettle, + *kytos,* hollow vessel) Modified interstitial cell that holds the cnida.

coccidian (kok-sid′ē-ən) (Gr. *kokkis,* kernel, grain) Intracellular protozoan parasite belonging to a class within phylum Apicomplexa; the organism causing malaria is an example.

cochlea (kōk′lē-ə) (L. snail, from Gr. *kochlos,* a shellfish) A tubular cavity of the inner ear containing the essential organs of hearing; occurs in crocodiles, birds, and mammals; spirally coiled in mammals.

cocoon (kə-kün′) (Fr. *cocon,* shell) Protective covering of a resting or developmental stage, sometimes used to refer to both the covering and its contents; for example, the cocoon of a moth or the protective covering for the developing embryos in some annelids.

coelenteron (sē-len′tər-on) (Gr. *koilos,* hollow, + *enteron,* intestine) Internal cavity of a cnidarian; gastrovascular cavity; archenteron.

coelom (sē′lōm) (Gr. *koilōma,* cavity) The body cavity in triploblastic animals, lined with mesodermal peritoneum.

coelomoduct (sē-lō′mə-dukt) (Gr. *koilos,* hollow, + *ductus,* a leading) A duct that carries gametes or excretory products (or both) from the coelom to the exterior.

cohort (kō′hort) All organisms of a population born within a specified interval of time.

collagen (käl′ə-jən) (Gr. *kolla,* glue, + *genos,* descent) A tough, fibrous protein occurring in vertebrates as the chief constituent of collagenous connective tissue; also occurs in invertebrates, for example, the cuticle of nematodes.

collar bodies Extensions of choanoblasts bearing flagellated collars in hexactinellid sponges.

collar cells Cells having a single flagellum surrounded by a ring of microvilli. Sponge choanocytes are collar cells, as are choanoflagellates, but collar cells also occur outside of these taxa.

collencyte (käl′ən-sīt) (Gr. *kolla,* glue, + *kytos,* hollow vessel) A type of cell in sponges that secretes fibrillar collagen.

colloblast (käl′ə-blast) (Gr. *kolla,* glue, + *blastos,* germ) A glue-secreting cell on the tentacles of ctenophores.

comb plate One of the plates of fused cilia that are arranged in rows for ctenophore locomotion.

commensalism (kə-men′səl-iz′əm) (L. *com,* together with, + *mensa,* table) A relationship in which one individual lives close to or on another and benefits, and the host is unaffected; often symbiotic.

common descent Darwin's theory that all forms of life are derived from a shared ancestral population through a branching of evolutionary lineages.

community (L. *communitas,* community, fellowship) An assemblage of organisms that are associated in a common environment and interact with each other in a self-sustaining and self-regulating relation.

comparative biochemistry Studies of the structures of biological macromolecules, especially proteins and nucleic acids, and their variation within and among related species to reveal homologies of macromolecular structure.

comparative cytology Studies of the structures of chromosomes within and among related species to reveal homologies of chromosomal structure.

comparative method Use of patterns of similarity and dissimilarity among species or populations to infer their phylogenetic relationships; use of phylogeny to examine evolutionary processes and history.

comparative morphology Studies of organismal form and its variation within and among related species to reveal homologies of organismal characters.

competition Some degree of overlap in ecological niches of two populations in the same community, such that both depend on the same food source, shelter, or other resources, and negatively affect each other's survival.

competitive exclusion An ecological principle stating that two species whose niches are very similar cannot coexist indefinitely in the same community; one species is driven to extinction by competition between them.

conjugation (kon′jū-gā′shən) (L. *conjugare,* to yoke together) Temporary union of two ciliate protozoa while they are exchanging chromatin material and undergoing nuclear phenomena resulting in binary fission. Also, formation of cytoplasmic bridges between bacteria for transfer of plasmids.

conodont (kōn′ə-dänt) (Gr. *kōnos,* cone, + *odontos,* tooth) Toothlike element from a Paleozoic animal now believed to have been an early marine chordate.

conspecific (L. *com,* together, + species) Of the same species.

consumer (kən-sū′-mer) An organism whose energy and matter are acquired by eating other organisms, which may be **primary producers, herbivores,** or **carnivores.**

contractile vacuole A clear fluid-filled cell vacuole in protozoa and a few lower metazoa; collects water and releases it to the outside in a cyclical manner, for osmoregulation and some excretion.

control That part of a scientific experiment to which the experimental variable is not applied but similar to the experimental group in all other respects.

coprophagy (kə-prä′fə-jē) (Gr. *kopros,* dung, + *phagein,* to eat) Feeding on dung or excrement as a normal behavior among animals; reingestion of feces.

copulation (Fr. from L. *copulare,* to couple) Sexual union to facilitate the reception of sperm by the female.

corneum (kor′nē-əm) (L. *corneus,* horny) Epithelial layer of dead, keratinized cells. Stratum corneum.

corona (kə-rō′nə) (L. crown) Head or upper portion of a structure; ciliated disc on anterior end of rotifers.

corpora allata (kor′pə-rə əl-la′tə) (L. *corpus,* body, + *allatum,* aided) Endocrine glands in insects that produce juvenile hormone.

cortex (kor′teks) (L. bark) The outer layer of a structure.

coxa, coxopodite (kox′ə, kəx-ä′pə-dīt) (L. *coxa,* hip, + Gr. *pous, podos,* foot) The proximal joint of an insect or arachnid leg; in crustaceans, the proximal joint of the protopod.

Cretaceous extinction A **mass extinction** that occurred 65 million years ago in which 76% of existing species, including all dinosaurs, became extinct, marking the end of the Mesozoic era.

crop A region of the esophagus specialized for storing food.

cryptobiotic (Gr. *kryptos,* hidden, + *bioticus,* pertaining to life) Living in concealment; refers to insects and other animals that live in secluded situations, such as underground or in wood; also tardigrades and some nematodes, rotifers, and others that survive harsh environmental conditions by assuming for a time a state of very low metabolism.

crystalline style A rod containing digestive enzymes present in the stomach of a bivalve.

ctenoid scales (ten′oyd) (Gr. *kteis, ktenos,* comb) Thin, overlapping dermal scales of teleost fishes; exposed posterior margins have fine, toothlike spines.

cuticle (kū′ti-kəl) (L. *cutis,* skin) A protective, noncellular, organic layer secreted by the external epithelium (hypodermis) of many invertebrates. In vertebrates the term refers to the epidermis or outer skin.

Cuvierian tubules (*Cuvier,* 19th-century French comparative vertebrate anatomist) Sticky, often toxic elongate internal organs of holothurians expelled to entangle potential predators; these can be regenerated.

cyanobacteria (sī-an′ō-bak-ter′ē-ə) (Gr. *kyanos,* a dark-blue substance, + *bakterion,* dim. of *baktron,* a staff) Photosynthetic prokaryotes, also called blue-green algae, cyanophytes.

cycloid scales (sī′kloyd) (Gr. *kyklos,* circle) Thin, overlaping dermal scales of teleost fishes; posterior margins are smooth.

cynodonts (sin′ə-dänts) (Gr. *kynodōn,* canine tooth) A group of mammal-like carnivorous synapsids of the Upper Permian and Triassic.

cyst (sist) (Gr. *kystis,* a bladder, pouch) A resistant, quiescent stage of an organism, usually with a secreted wall.

cysticercus (sis′tə-ser′kəs) (Gr. *kystis,* bladder, + *kerkos,* tail) A type of juvenile tapeworm in which an invaginated and introverted scolex is contained in a fluid-filled bladder.

cystid (sis′tid) (Gr. *kystis,* bladder) In an ectoproct, the dead secreted outer parts plus the adherent underlying living layers.

cytopharynx (Gr. *kytos,* hollow vessel, + *pharynx,* throat) Short tubular gullet in ciliate protozoa.

cytoplasm (si′tə-plasm) (Gr. *kytos,* hollow vessel, + *plasma,* mold) The living matter of the cell, excluding the nucleus.

cytoproct (si′tə-prokt) (Gr. *kytos,* hollow vessel, + *prōktos,* anus) Site on a protozoan where indigestible matter is expelled.

cytopyge (si′tə-pīj) (Gr. *kytos,* hollow vessel, + *pyge,* rump or buttocks) In some protozoa, localized site for expulsion of wastes.

cytostome (si′tə-stōm) (Gr. *kytos,* hollow vessel, + *stoma,* mouth) The cell mouth in many protozoa.

D

dactylozooid (dak-til′ə-zō-id) (Gr. *dakos,* bite, sting, + *tylos,* knob, + *zōon,* animal) A polyp of a colonial hydroid specialized for defense or killing food.

Darwinism Theory of evolution emphasizing common descent of all living organisms, gradual change, multiplication of species and natural selection.

data sing. **datum** (Gr. *dateomai,* to divide, cut in pieces) The results in a scientific experiment, or descriptive observations, upon which a conclusion is based.

deciduous (də-sij′ü-wəs) (L. *deciere,* to fall off) Shed at the end of a growing period.

decomposer (dē′-kəm-pō′zər) A **consumer** that breaks organic matter into soluble components available to plants at the base of the food web; most are bacteria or fungi.

deduction (L. *deductus,* led apart, split, separated) Reasoning from the general to the particular, from given premises to their necessary conclusion.

definitive host The host in which sexual reproduction of a symbiont occurs; if no sexual reproduction, then the host in which the symbiont becomes mature and reproduces; contrast intermediate host.

deme (dēm) (Gr. populace) A local population of closely related animals.

demography (də-mäg′rə-fē) (Gr. *demos,* people, + *graphy*) The properties of the rate of growth and the age structure of populations.

density-dependent Refers to biotic environmental factors, such as predators and parasites, whose effects on a population vary according to the number of organisms in the population.

density-independent Refers to abiotic environmental factors, such as fires, floods, and temperature changes, whose effects on a population are unaffected by the number of organisms in the population.

derived character state A condition of a taxonomic character inferred by cladistic analysis to have arisen within a taxon being examined cladistically rather than having been inherited from the most recent common ancestor of all members of the taxon.

dermal (Gr. *derma,* skin) Pertaining to the skin; cutaneous.

dermal ostia (Gr. *derma,* skin; L. *ostium,* door) Incurrent pores in a sponge.

dermis The inner, sensitive mesodermal layer of skin; corium.

determinate cleavage The type of cleavage, usually spiral, in which the fate of the blastomeres is determined very early in development; mosaic cleavage.

detritus (də-trī′tus) (L. that which is rubbed or worn away) Any fine particulate debris of organic or inorganic origin.

Deuterostomia (dū′də-rō-stō′mē-ə) (Gr. *deuteros,* second, secondary, + *stoma,* mouth) A group of higher phyla in which cleavage is indeterminate and primitively radial. The endomesoderm is enterocoelous, and the mouth is derived away from the blastopore. Includes Echinodermata, Chordata, and Hemichordata. Compare with **Protostomia.**

diapause (dī′ə-pawz) (Gr. *diapausis,* pause) A period of arrested development in the life cycle of insects and certain other animals in which physiological activity is very low and the animal is highly resistant to unfavorable external conditions.

diapsids (dī-ap′səds) (Gr. *di,* two, + *apsis,* arch) Amniotes in which the skull bears two pairs of temporal openings; includes reptiles (except turtles) and birds.

dictyosome (dik′tē-ə-sōm) (Gr. *diction,* to throw, + *sōma,* body). A part of the secretory system of endoplasmic reticulum in protozoans; also called Golgi bodies.

diffusion (L. *diffusion,* dispersion) The movement of particles or molecules from area of high concentration of the particles or molecules to area of lower concentration.

digitigrade (di′jəd-ə-grād) (L. *digitus,* finger, toe, + *gradus,* step, degree) Walking on the digits with the posterior part of the foot raised; compare **plantigrade.**

dimorphism (dī-mor′fizm) (Gr. *di,* two, + *morphē,* form) Existence within a species of two distinct forms according to color, sex, size, organ structure, or behavior. Occurrence of two kinds of zooids in a colonial organism.

dinoflagellate (dī-nō-fla′jə-lāt) (Gr. *dinos,* whirling, + L. *flagellum,* a whip). Any member of a protozoan clade having two flagella, one in the equatorial region of the body and other trailing; cells naked or with a test of cellulose plates.

dioecious (dī-esh′əs) (Gr. *di,* two, + *oikos,* house) Having male and female organs in separate individuals.

diphycercal (dif′i-sər′kəl) (Gr. *diphyēs,* twofold, + *kerkos,* tail) A tail that tapers to a point, as in lungfishes; vertebral column extends to tip without upturning.

diphyodont (dī-fī′ə-dänt) (Gr. *diphyēs,* twofold, + *odous,* tooth) Having deciduous and permanent sets of teeth successively.

diploblastic (di′plə-blas′tək) (Gr. *diploos,* double, + *blastos,* bud) Organism with two germ layers, endoderm and ectoderm.

diploid (dip′loid) (Gr. *diploos,* double, + *eidos,* form) Having the somatic (double, or 2n) numbers of chromosomes or twice the number characteristic of a gamete of a given species.

diplomonad (dip′lō-mō′nad) (Gr. *diploos,* double, + *monos,* lone, single). Any member of a protozoan clade having four kinetosomes and lacking mitochondria.

direct development A postnasal ontogeny featuring primarily growth in size rather than a major change in body shape or organs present; a life history that lacks **metamorphosis.**

directional selection Natural selection that favors one extreme value of a continuously varying trait and disfavors other values.

disruptive selection Natural selection that favors simultaneously two different extreme values of a continuously varying trait but disfavors intermediate values.

distal (dis´təl) Farther from the center of the body than a reference point.

domain (dō-mān) An informal taxonomic rank above the Linnean kingdom; Archaea, Bacteria, and Eucarya are ranked as domains.

dorsal (dor´səl) (L. *dorsum,* back) Toward the back, or upper surface, of an animal.

double circulation A blood-transport system having a distinct pulmonary circuit of blood vessels separate from the circuit of blood vessels serving the remainder of the body.

dual-gland adhesive organ Organs in the epidermis of most turbellarians, with three cell types: viscid and releasing gland cells and anchor cells.

E

eccrine (ek´rən) (Gr. *ek,* out of, + *krinein,* to separate) Applies to a type of mammalian sweat gland that produces a watery secretion.

ecdysiotropin (ek-dē-zē-ə-trō´pən) (Gr. *ekdysis,* to strip off, escape, + *tropos,* a turn, change) Hormone secreted in brain of insects that stimulates prothoracic gland to secrete molting hormone. Prothoracicotropic hormone; brain hormone.

ecdysis (ek´də-sis) (Gr. *ekdysis,* to strip off, escape) Shedding of outer cuticular layer; molting, as in insects or crustaceans.

ecdysone (ek-di´sōn) (Gr. *ekdysis,* to strip off) Molting hormone of arthropods, stimulates growth and ecdysis, produced by prothoracic glands in insects and Y organs in crustaceans.

ecdysozoan protostome (ek´dī-sō-zō´ən prō´tə-stōm) (Gr. *ekdysis,* to strip off, escape, + *zōon,* animal. Gr. *protos,* first, + *stoma* mouth). Any member of a clade within Protostomia whose members shed the cuticle as they grow; includes arthropods, nematodes and several smaller phyla.

ecological pyramid A quantitative measurement of a **food web** in terms of amount of **biomass,** numbers of **organisms,** or energy at each of the different **trophic** levels present (**producers, herbivores,** first-level **carnivores,** higher-level carnivores).

ecology (Gr. *oikos,* house, + *logos,* discourse) Part of biology that concerns the relationship between organisms and their environment.

ecosystem (ek´ō-sis-təm) (eco[logy] from Gr. *oikos,* house, + *system*) An ecological unit consisting of both the biotic communities and the nonliving (abiotic) environment, which interact to produce a stable system.

ectoderm (ek´tō-derm) (Gr. *ektos,* outside, + *derma,* skin) Outer layer of cells of an early embryo (gastrula stage); one of the germ layers, also sometimes used to include tissues derived from ectoderm.

ectognathous (ek´tə-nā´thəs) (Gr. *ektos,* outside, without, + *gnathos,* jaw) A derived character shared by most insects; mandibles and maxillae not in pouches.

ectolecithal (ek´tō-les´ə-thəl) (Gr. *ektos,* ouside, + *ekithos,* yolk) Yolk for nutrition of the embryo contributed by cells that are separate from the egg cell and are combined with the zygote by envelopment within the eggshell.

ectoparasite (ek´tō-par´ə-sīt) A parasite that resides on the outside surface of its host organism; contrasts with **endoparasite.**

ectoplasm (ek´tō-pla-zm) (Gr. *ektos,* outside, + *plasma,* form) The cortex of a cell or that part of cytoplasm just under the cell surface; contrasts with **endoplasm.**

ectothermic (ek´tō-therm´ic) (Gr. *ektos,* outside, + *thermē,* heat) Having a variable body temperature derived from heat acquired from the environment; contrasts with **endothermic.**

efferent (ef´ə-rənt) (L. *ex,* out, + *ferre,* to bear) Leading or conveying away from some organ, for example, nerve impulses conducted away from the brain, or blood conveyed away from an organ; contrasts with **afferent.**

egestion (ē-jes´chən) (L. *egestus,* to discharge) Act of dispelling indigestible or waste matter from the body by any normal route.

elephantiasis (el-ə-fən-ti´ə-səs) Disfiguring condition caused by chronic infection with filarial worms *Wuchereria bancrofti* and *Brugia malayi.*

Eltonian pyramid An ecological **pyramid** showing numbers of organisms at each of the **trophic** levels.

embryogenesis em´brē-ō-jen´ə-səs (Gr. *embryon,* embryo, + *genesis,* origin) The origin and development of the embryo; embryogeny.

emigrate (L. *emigrare,* to move out) To move *from* one area to another to establish residence.

encystment Process of cyst formation.

endemic (en-dem´ik) (Gr. *en,* in, + *demos,* populace) Peculiar to a certain region or country; native to a restricted area; not introduced.

endoderm (en´də-dərm) (Gr. *endon,* within, + *derma,* skin) Innermost germ layer of an embryo, forming the primitive gut; also may refer to tissues derived from endoderm.

endognathous (en´də-nā-thəs) (Gr. *endon,* within, + *gnathous,* jaw) An ancestral character in insects, found in orders Diplura, Collembola, and Protura, in which the mandibles and maxillae are located in pouches.

endolecithal (en´də-les´ə-thəl) (Gr. *endon,* within, + *lekithos,* yolk) Yolk for nutrition of the embryo incorporated into the egg cell itself.

endoparasite (en´dōpar´ə-sīt) A pararsite that resides inside the body of its host organism; contrasts with **ectoparasite.**

endoplasm (en´də-pla-zm) (Gr. *endon,* within, + *plasma,* mold or form) The portion of cytoplasm that immediately surrounds the nucleus.

endopod, endopodite (en´də-päd, en-dop´ə-dīt) (Gr. *endon,* within, + *pous, podos,* foot) Medial branch of a biramous crustacean appendage.

endoskeleton (Gr. *endon,* within, + *skeletos,* hard) A skeleton or supporting framework within the living tissues of an organism; contrasts with **exoskeleton.**

endostyle (en´də-stīl) (Gr. *endon,* within, + *stylos,* a pillar) Ciliated groove(s) in the floor of the pharynx of tunicates, cephalochordates, and larval lampreys, used for accumulating and moving food particles to the stomach.

endosymbiosis A symbiosis in which the symbiont lives inside its host; origin of eukaryotes, in which one prokaryote (symbiont) came to live inside another prokaryote (host), and symbionts eventually became organelles, such as mitochondria, of the host.

endothermic (en´də-therm´ik) (Gr. *endon,* within, + *thermē,* heat) Having a body temperature determined by heat derived from an animal's own oxidative metabolism; contrasts with **ectothermic.**

energy budget An economic analysis of the energy used by an organism, partitioned into **gross productivity, net productivity,** and **respiration.**

enterocoel (en´tər-ō-sēl´) (Gr. *enteron,* gut, + *koilos,* hollow) A type of coelom formed by the outpouching of a mesodermal sac from the endoderm of the primitive gut.

enterocoelomate (en´ter-ō-sēl´ō-māt) (Gr. *enteron,* gut, + *koilōma,* cavity, + Engl. *ate,* state of) An animal having an enterocoel, such as an echinoderm or a vertebrate.

enterocoelous mesoderm formation Embryonic formation of mesoderm by a pouchlike outfolding from the archenteron, which then expands and obliterates the blastocoel, thus forming a large cavity, the coelom, lined with mesoderm.

enteron (en′tə-rän) (Gr. intestine) The digestive cavity.

entomology (en′tə-mol′ə-jē) (Gr. *entoma,* an insect, + *logos,* discourse) Study of insects.

ephyra (ef′ə-rə) (Gr. *Ephyra,* Greek city) Juvenile medusa that is budded from a strobilating polyp in class Scyphozoa, phylum Cnidaria.

epidermis (ep′ə-dər′məs) (Gr. *epi,* on, upon, + *derma,* skin) The outer, nonvascular layer of skin of ectodermal origin; in invertebrates, a single layer of ectodermal epithelium.

epipod, epipodite (ep′ə-päd, e-pip′ə-dīt) (Gr. *epi,* on, upon, + *pous, podos,* foot) A lateral process on the protopod of a crustacean appendage, often modified as a gill.

epistome (ep′i-stōm) (Gr. *epi,* on, upon, + *stoma,* mouth) Flap over the mouth in some lophophorates bearing the protocoel.

epithelium (ep′i′thē′lē-um) (Gr. *epi,* on, upon, + *thēlē,* nipple) A cellular tissue covering a free surface or lining a tube or cavity.

epitoke (ep′i-tōk) (Gr. *epitokos,* fruitful) Posterior part of a marine polychaete when swollen with developing gonads during the breeding season; contrast with **atoke.**

erythrocyte (ə-rith′rō-sīt) (Gr. *erythros,* red, + *kytos,* hollow vessel) Red blood cell; has hemoglobin to carry oxygen from lungs or gills to tissues; during formation in mammals, erythrocytes lose their nuclei, those of other vertebrates retain the nuclei.

estrous cycle Periodic episodes of estrus, or "heat," when females of most mammalian species become sexually receptive.

estrus (es′trəs) (L. *oestrus,* gadfly, frenzy) The period of heat, or rut, especially of the female during ovulation of the eggs. Associated with maximum sexual receptivity.

eukaryotic, eucaryotic (ū′ka-rē-ot′ik) (Gr. *eu,* good, true, + *karyon,* nut, kernel) Organisms whose cells characteristically contain a membrane-bound nucleus or nuclei; contrasts with **prokaryotic.**

eumetazoan (ü-met-ə-zō′ən) (Gr. *eu,* good, true, + *meta,* after, + *zōon,* animal). Any multicellular animal with distinct germ layers that form true tissues; animals beyond the cellular grade of organization.

euryhaline (ū-rə-hā′līn) (Gr. *eurys,* broad, + *hals,* salt) Able to tolerate wide ranges of saltwater concentrations.

euryphagous (yə-rif′ə-gəs) (Gr. *eurys,* broad, + *phagein,* to eat) Eating a large variety of foods.

eurytopic (ū-rə-täp′ik) (Gr. *eurys,* broad, + *topos,* place) Refers to an organism with a wide environmental range.

eutely (u′te-lē) (Gr. *euteia,* thrift) Condition of a body composed of a constant number of cells or nuclei in all adult members of a species, as in rotifers, acanthocephalans, and nematodes.

evagination (ē-vaj-ə-nā′shən) (L. *e,* out, + *vagina,* sheath) An outpocketing from a hollow structure.

evolution (L. *evolvere,* to unfold) Organic evolution encompasses all changes in the characteristics and diversity of life on earth throughout its history.

evolutionary sciences Empirical investigation of ultimate causes in biology using the comparative method.

evolutionary species concept A single lineage of ancestral-descendant populations that maintains its identity from other such lineages and has its own evolutionary tendencies and historical fate; differs from the biological species concept by explicitly including a time dimension and including asexual lineages.

evolutionary taxonomy A system of classification, formalized by George Gaylord Simpson, that groups species into Linnean higher taxa representing a hierarchy of distinct adaptive zones; such taxa may be monophyletic or paraphyletic but not polyphyletic.

exopod, exopodite (ex′ə-päd, ex-äp′ə-dīt) (Gr. *exō,* outside, + *pous, podos,* foot) Lateral branch of a biramous crustacean appendage.

exoskeleton (ek′sō-skel′ə-tən) (Gr. *exō,* outside, + *skeletos,* hard) A supporting structure secreted by ectoderm or epidermis; external, not enveloped by living tissue, as opposed to endoskeleton; in vertebrates, a supporting structure formed within the integument.

experiment (L. *experiri,* to try) A trial made to support or to disprove a hypothesis.

experimental method A general procedure for testing hypotheses by predicting how a biological system will respond to a disturbance, making the disturbance under controlled conditions, and then comparing the observed results with the predicted ones.

experimental sciences Empirical investigation of proximate causes in biology using the **experimental method.**

extrusome (eks′trə-sōm) (L. *extrusus,* driven out, + *soma,* body). Any membrane-bound organelle used to extrude something from a protozoan cell.

extrinsic factor An environmental variable that influences the biological properties of a population, such as observed number of individuals or rate of growth.

F

filipodium (fi′li-pō′dē-əm) (L. *filum,* thread, + Gr. *pous, podos,* a foot) A type of pseudopodium that is very slender and may branch but does not rejoin to form a mesh.

filter feeding Any feeding process by which particulate food is filtered from water in which it is suspended.

fission (L. *fissio,* a splitting) Asexual reproduction by a division of the body into two or more parts.

fitness Degree of adjustment and suitability for a particular environment. Genetic fitness is the relative contribution of a genotype to the next generation; organisms with high genetic fitness are those favored by natural selection.

flagellum (flə-jel′əm) pl. **flagella** (L. a whip) Whiplike organelle of locomotion.

flame cell Specialized hollow excretory or osmoregulatory structure of one or several small cells containing a tuft of flagella (the "flame") and situated at the end of a minute tubule; connected tubules ultimately open to the outside. See **protonephridium.**

fluke (O.E. *flōc,* flatfish) A member of class Trematoda or class Monogenera. Also, certain of the flatfishes (order Pleuronectiformes).

food vacuole A digestive organelle in the cell.

food web An analysis relating species in an ecological community according to how they acquire nutrition, such as fixing atmospheric carbon (**producers**), consuming producers (**herbivores**), consuming herbivores (first-level **carnivores**), or consuming carnivores (higher-level carnivores).

foraminiferan (fə-ra-mə-nif′-ə-rən) (L. *foramin,* hole, perforation, + *fero,* to bear). Granuloreticulosean amebas bearing a test with many openings.

fossil (fos′əl) Any remains or impression of an organism from a past geological age that has been preserved by natural processes, usually by mineralization in the earth's crust.

fossorial (fä-sōr′ē-əl) (L. *fossor,* digger) Adapted for digging.

fouling Contamination of feeding or respiratory areas of an organism by excrement, sediment, or other matter. Also, accumulation of sessile marine organisms on the hull of a boat or ship so as to impede its progress through the water.

fovea (fō′vē-ə) (L. a small pit) A small pit or depression; especially the fovea centralis, a small rodless pit in the retina of some vertebrates, a point of acute vision.

frontal plane A plane parallel to the main axis of the body and at right angles to the sagittal plane.

fundamental niche A variety of roles potentially performed by an organism or population in an ecological community; limits on such roles are set by the intrinsic biological attributes of an organism or population. See also **niche** and **realized niche.**

funnel The tube from which a jet of water exits the mantle cavity of a cephalopod mollusc.

G

gamete (ga′mēt, gə-mēt′) (Gr. *gamos,* marriage) A mature haploid sex cell; usually male and female gametes can be distinguished. An egg or a sperm.

gametic meiosis Meiosis that occurs during formation of the gametes, as in humans and other metazoa.

gametocyte (gə-mēt′ə-sīt) (Gr. *gametēs,* spouse, + *kytos,* hollow vessel) The mother cell of a gamete, an immature gamete.

ganglion (gang′lē-ən) pl. **ganglia** (Gr. little tumor) An aggregation of nerve tissue containing nerve cells.

ganoid scales (ga′noyd) (Gr. *ganos,* brightness) Thick, bony, rhombic scales of some bony fishes; not overlapping.

gastrocoel (gas′trō-sēl) (Gr. *gastēr,* stomach, + *koilos,* hollow). An embryonic cavity forming in gastrulation that becomes the adult gut; also called an archenteron.

gastrodermis (gas′tro-dər′mis) (Gr. *gastēr,* stomach, + *derma,* skin) Lining of the digestive cavity of cnidarians.

gastrovascular cavity (Gr. *gastēr,* stomach, + L. *vasculum,* small vessel) Body cavity in certain lower invertebrates that functions in both digestion and circulation and has a single opening serving as both mouth and anus.

gastrozooid (gas′trō-zō-id) (Gr. *gastēr,* stomach, + *zoon,* animal). An embryonic cav-

ity forming in gastrulation that becomes the adult gut; also called an archenteron.

gastrula (gas′trə-lə) (Gr. *gastēr,* stomach, + L. *ula,* dim) Embryonic stage, usually cap- or sac-shaped, with walls of two layers of cells surrounding a cavity (archenteron) with one opening (blastopore).

gemmule (je′mūl) (L. *gemma,* bud, + *ula,* dim.) Asexual, cystlike reproductive unit in freshwater sponges; formed in summer or autumn and capable of overwintering.

gene (Gr. *genos,* descent) The part of a chromosome that is the hereditary determiner and is transmitted from one generation to another. Specifically, a gene is a nucleic acid sequence (usually DNA) that encodes a functional polypeptide or RNA sequence.

gene pool A collection of all the alleles of all the genes in a population.

genetic drift Change in gene frequencies by chance processes in the evolutionary process of animals. In small populations one allele may drift to fixation, becoming the only representative of that gene locus.

genotype (jēn′ə-tīp) (Gr. *genos,* offspring, + *typos,* form) The genetic constitution, expressed and latent, of an organism; the total set of genes present in the cells of an organism; contrasts with **phenotype.**

genus (jē′nus) pl. **genera** (L. race) A group of related species with taxonomic rank between family and species.

germinative zone The site immediately following the scolex on the body of a mature tapeworm where new proglottids are produced.

germ layer In the animal embryo, one of three basic layers (ectoderm, endoderm, mesoderm) from which the various organs and tissues arise in the multicellular animal.

germovitellarium (jer′mə-vit-əl-ar′ē-əm) (L. *germen,* a bud, offshoot, + *vitellus,* yolk) Closely associated ovary (germarium) and yolk-producing structure (vitellarium) in rotifers.

germ plasm The germ cells of an organism, as distinct from the somatoplasm; the hereditary material (genes) of the germ cells.

gestation (je-stā′shən) (L. *gestare,* to bear) The period in which offspring are carried in the uterus.

glochidium (glō-kid′ē-əm) (Gr. *glochis,* point, + *idion,* dimin. suffix) Bivalved larval stage of freshwater mussels.

glycogen (glī′kə-jən) (Gr. *glykys,* sweet, + *genes,* produced) A polysaccharide constituting the principal form in which carbohydrate is stored in animals; animal starch.

gnathobase (nath′ə-bās′) (Gr. *gnathos,* jaw, base) A median basic process on certain appendages in some arthropods, usually for biting or crushing food.

gnathostomes (nath′ə-stōmz) (Gr. *gnathos,* jaw, + *stoma,* mouth) Vertebrates with jaws.

gonad (gō′nad) (N.L. *gonas,* a primary sex organ) An organ that produces gametes (ovary in the female and testis in the male).

gonangium (gō-nan′jē-əm) (N.L. *gonas,* primary sex organ, + *angeion,* dimin. of vessel) Reproductive zooid of hydroid colony (Cnidaria).

gonoduct (Gr. *gonos,* seed, progeny, + duct) Duct leading from a gonad to the exterior.

gonophore (gon′ə-for) (Gr. *gonos,* seed, progeny, + *phoros,* bearer). Sexual reproductive structure developing from reduced medusae in some hydrozoans; it may be retained on the colony or released.

gonopore (gän′ə-pōr) (Gr. *gonos,* seed, progeny, + *poros,* an opening) A genital pore found in many invertebrates.

grade (L. *gradus,* step) A level of organismal complexity or adaptive zone characteristic of a group of evolutionarily related organisms.

gradualism (graj′ə-wal-iz′əm) A component of Darwin's evolutionary theory postulating that evolution occurs by the temporal accumulation of small, incremental changes by populations, usually across very long periods of geological time; it opposes claims that evolution can occur by large, discontinuous or macromutational changes.

granuloreticulosan (gran′yə-lō-ru-tik-yə-lō′sən) (L. *granulus,* small grain, + *reticulum,* a net). Any member of a protozoan clade having branched and netlike pseudopodia; includes foraminiferans.

green gland Excretory gland of certain Crustacea; the antennal gland.

gregarine (gre-ga-rin′) (L. *gregarious,* belonging to a herd or flock) Protozoan parasites belonging to class Gregarinea within phylum Apicomplexa; these organisms infect guts or body cavities of invertebrates.

gross productivity A measurement of the total energy assimilated by an organism.

ground substance The matrix in which connective tissue fibers are embedded.

growth rate The proportion by which a population changes in numbers of individuals at a given time by reproduction and possibly immigration.

guild (gild) (M.E. *gilde,* payment, tribute) Species of a local community that partition resources through character displacement to avoid niche overlap and competition, such as Galápagos finch communities whose component species differ in beak size for specializing on different-size seeds.

gynecophoric canal (gī′nə-kə-fōr′ik) (Gr. *gynē,* woman, + *pherein,* to carry) Groove in male schistosomes (certain trematodes) that carries the female.

H

habitat (L. *habitare,* to dwell) The place where an organism normally lives or where individuals of a population live.

halter (hal′tər) pl. **halteres** (hal-ti′rēz) (Gr. leap) In Diptera, small club-shaped structure on each side of the metathorax representing the hind wings; thought to be sense organs for balancing; also called balancer.

haplodiploidy (Gr. *haploos,* single, + *diploos,* double, + *eidos,* form) Reproduction in which haploid males are produced parthenogenetically and diploid females develop from fertilized eggs.

haploid (Gr. *haploos,* single) The reduced, or n, number of chromosomes, typical of gametes, as opposed to the diploid, or 2n, number found in somatic cells. In certain groups, some mature organisms have a haploid number of chromosomes.

Hardy-Weinberg equilibrium Mathematical demonstration that the Mendelian hereditary process does not change the populational frequencies of alleles or genotypes across generations, and that change in allelic or genotypic frequencies requires factors such as natural selection, genetic drift in finite populations, recurring mutation, migration of individuals among populations, and nonrandom mating.

hemal system (hē′məl) (Gr. *haima,* blood) System of small vessels in echinoderms; function unknown.

hemimetabolous (he′mē-mə-ta′bə-ləs) (Gr. *hēmi,* half, + *metabolē,* change) Refers to gradual metamorphosis during development of insects, without a pupal stage.

hemocoel (hē′mə-sēl) (Gr. *haima,* blood, + *koilos,* hollow) Main body cavity of arthropods; may be subdivided into sinuses, through which blood flows.

hemoglobin (Gr. *haima,* blood, + L. *globulus,* globule) An iron-containing respiratory pigment occurring in vertebrate red blood cells and in blood plasma of many invertebrates; a compound of an iron porphyrin heme and a protein globin.

hemolymph (hē′mə-limf) (Gr. *haima,* blood + L. *lympha,* water) Fluid in the coelom or hemocoel of some invertebrates that functions as the blood and lymph of vertebrates.

herbivore ([h]erb′ə-vōr′) (L. *herba,* green crop, + *vorare,* to devour) Any organism subsisting on plants. Adj., **herbivorous.**

hermaphrodite (hər-maf′rə-dīt) (Gr. *hermaphroditos,* containing both sexes; from Greek mythology. Hermaphroditos, son of Hermes and Aphrodite) An organism with both male and female functional reproductive organs. **Hermaphroditism** may refer to an aberration in unisexual animals; **monoecism** implies that this is the normal condition for the species.

heterocercal (het′ər-o-sər′kəl) (Gr. *heteros,* different, + *kerkos,* tail) In some fishes, a tail with the upper lobe larger than the lower, and the end of the vertebral column somewhat upturned in the upper lobe, as in sharks.

heterochrony (hed′ə-rō-krōn-ē) (Gr. *heteros,* different, + *chronos,* time) Evolutionary change in the relative time of appearance or rate of development of characteristics from ancestor to descendant.

heterodont (hed′ə-ro-dänt) (Gr. *heteros,* different, + *odous,* tooth) Having teeth differentiated into incisors, canines, and molars for different purposes.

heterolobosea (hed′ə-rō-lo-bō′sē-ə) (Gr. *heteros,* other, different, + *lobos,* lobe). A protozoan clade in which most members can assume both ameboid and flagellate forms.

heterotroph (hed′ə-rō-trōf) (Gr. *heteros,* different, + *trophos,* feeder) An organism that obtains both organic and inorganic raw materials from the environment in order to live; includes most animals and those plants that do not have photosynthesis.

heterozygous Refers to an organism in which homologous chromosomes contain different allelic forms (often dominant and recessive) of a gene; derived from a zygote formed by union of gametes of dissimilar allelic constitution.

hexamerous (hek-sam′ər-əs) (Gr. *hex,* six, + *meros,* part) Six parts, specifically, symmetry based on six or multiples thereof.

hibernation (L. *hibernus,* wintry) Condition, especially of mammals, of passing the win-

ter in a torpid state in which the body temperature drops nearly to freezing and the metabolism drops close to zero.

hierarchical system A scheme arranging organisms into a series of taxa of increasing inclusiveness, as illustrated by Linnean classification.

histology (hi-stäl′-ə-jē) (Gr. *histos,* web, tissue, + *logos,* discourse) The study of the microscopic anatomy of tissues.

holometabolous (hō′lō-mə-ta′bə-ləs) (Gr. *holo,* complete, + *metabolē,* change) Complete metamorphosis during development.

holophytic nutrition (hä-lō-fit′ik) (Gr. *holo,* whole, + *phyt,* plant) Occurs in green plants and certain protozoa and involves synthesis of carbohydrates from carbon dioxide and water in the presence of light, chlorophyll, and certain enzymes.

holozoic nutrition (hä-lo-zō′ik) (Gr. *holo,* whole, + *zoikos,* of animals) Type of nutrition involving ingestion of liquid or solid organic food particles.

homeobox (hō′mē-ō-box) (Gr. *homolos,* like, resembling, + L. *buxus,* boxtree [used in the sense of enclosed, contained]). A highly conserved 180-base-pair sequence found in homeotic genes, regulatory sequences of protein-coding genes that regulate development.

homeotic genes (hō-mē-ät′ik) (Gr. *homolos,* like, resembling). Genes, identified through mutations, that give developmental identity to specific body segments.

homeothermic (hō′mē-ō-thər′mik) (Gr. *homeo,* alike, + *thermē,* heat) Having a nearly uniform body temperature, regulated independent of the environmental temperature; "warm-blooded."

home range The area over which an animal ranges in its activities. Unlike territories, home ranges are not defended.

hominid (häm′ə-nid) (L. *homo, hominis,* man) A member of the family Hominidae. This taxon formerly contained only one living species, *Homo sapiens,* and its closest fossil relatives, but now is expanded to include chimpanzees, gorillas, and orangutans.

homocercal (hō′mə-ser′kəl) (Gr. *homos,* same, common, + *kerkos,* tail) A tail with the upper and lower lobes symmetrical and the vertebral column ending near the middle of the base, as in most teleost fishes.

homodont (hō′mō-dänt) (Gr. *homos,* same, + *odous,* tooth) Having all teeth similar in form.

homology (hō-mäl′ə-jē) (Gr. *homologos,* agreeing) Similarity of parts or organs of different organisms caused by evolutionary derivation from a corresponding part or organ in a remote ancestor, and usually having a similar embryonic origin. May also refer to a matching pair of chromosomes. Serial homology is the correspondence in the same individual of repeated structures having the same origin and development, such as the appendages of arthropods. Adj., **homologous.**

homonoid (häm′ə-noyd) Relating to the Hominoidea, a superfamily of primates to which the great apes and humans are assigned.

homoplasy (hō′mə-plā′sē) Phenotypic similarity among characteristics of different species or populations (including molecular, morphological, behavioral, or other features) that does not accurately represent patterns of common evolutionary descent (= nonhomologous similarity); it is produced by evolutionary parallelism, convergence and/or reversal, and is revealed by incongruence among different characters on a cladogram or phylogenetic tree.

hyaline (hī′ə-lən) (Gr. *hyalos,* glass) Adj., glassy, translucent. Noun, a clear, glassy structureless material occurring in, for example, cartilage, vitreous bodies, mucin, and glycogen.

hydatid cyst (hī-da′təd) (Gr. *hydatis,* watery vesicle) A type of cyst formed by juveniles of certain tapeworms (*Echinococcus*) in their vertebrate hosts.

hydranth (hī′dranth) (Gr. *hydōr,* water, + *anthos,* flower) Nutritive zooid of hydroid colony.

hydrocoel (hī′drə-sēl) (Gr. *hydōr,* water, + *koilos,* hollow) Second or middle coelomic compartment in echinoderms; left hydrocoel gives rise to water vascular system.

hydrocoral (Gr. *hydōr,* water, + *korallion,* coral) Certain members of the cnidarian class Hydrozoa that secrete calcium carbonate, resembling true corals.

hydrogenosomes (hī-drə-jen′ə-sōmz) Small organelles in certain anaerobic protozoa that produce molecular hydrogen as an end product of energy metabolism.

hydroid The polyp form of a cnidarian as distinguished from the medusa form. Any cnidarian of the class Hydrozoa, order Hydroida.

hydrostatic skeleton A mass of fluid or plastic parenchyma enclosed within a muscular wall to provide the support necessary for antagonistic muscle action; for example, parenchyma in acoelomates and perivisceral fluids in pseudocoelomates serve as hydrostatic skeletons.

hydrothermal vent A submarine hot spring; seawater seeping through the sea bottom is heated by magma and expelled back into the sea through a hydrothermal vent.

hyperosmotic (Gr. *hyper,* over, + *ōsmos,* impulse). Refers to a solution that contains a greater concentration of dissolved particles than another solution to which it is compared; gains water through a selectively permeable membrane from a solution containing fewer particles; contrasts with **hypoosmotic.**

hyperparasitism A parasite itself parasitized by another parasite.

hypodermis (hī′pə-dər′mis) (Gr. *hypo,* under, + L. *dermis,* skin) The cellular layer lying beneath and secreting the cuticle of annelids, arthropods, and certain other invertebrates.

hypoosmotic (Gr. *hypo,* under, + *ōsmos,* impulse). Refers to a solution that contains a lesser concentration of dissolved particles than another solution to which it is compared; loses water through a selectively permeable membrane from a solution containing more particles; contrasts with **hyperosmotic.**

hypostome (hī′pə-stōm) (Gr. *hypo,* under, + *stoma,* mouth) Name applied to a structure in various invertebrates (such as mites and ticks), located at posterior or ventral area of mouth; elevation supporting mouth of hydrozoan.

hypothesis (Gr. *hypothesis,* foundation, supposition) A statement or proposition that can be tested by observation or experiment.

hypothetico-deductive method The central procedure of scientific inquiry in which a postulate is advanced to explain a natural phenomenon and then is subjected to observational or experimental testing that potentially could reject the postulate.

I

immediate cause See **proximate cause.**

inbreeding The tendency among members of a population to mate preferentially with close relatives.

indeterminate cleavage A type of embryonic development in which the fate of the blastomeres is not determined very early as to tissues or organs, for example, in echinoderms and vertebrates.

indigenous (in-dij′ə-nəs) (L. *indigna,* native) Pertains to organisms that are native to a particular region; not introduced.

induction (L. *inducere, inductum,* to lead) Reasoning from the particular to the general; that is, deriving a general statement (hypothesis) based on individual observations. In embryology, the alteration of cell fates as the result of interaction with neighboring cells.

infraciliature (in′frə-sil′ē-ə-chər) (L. *infra,* below, + *cilia,* eyelashes) The organelles just below the cilia in ciliate protozoa.

inheritance of acquired characteristics The discredited Lamarckian notion that organisms, by striving to meet the demands of their environments, obtain new adaptations and pass them by heredity to their offspring.

instar (inz′tär) (L. form) Stage in the life of an insect or other arthropod between molts.

integument (in-teg′ū-mənt) (L. *integumentum,* covering) An external covering or enveloping layer.

intermediary meiosis Meiosis that occurs neither during gamete formation nor immediately after zygote formation, resulting in both haploid and diploid generations, such as in foraminiferan protozoa.

intermediate host A host in which some development of a symbiont occurs, but in which maturation and sexual reproduction do not occur (contrasts with **definitive host).**

interstitial (in′tər-sti′shəl) (L. *inter,* among, + *sistere,* to stand) Situated in the interstices or spaces between structures such as cells, organs, or grains of sand.

intracellular (in-trə-sel′yə-lər) (L. *intra,* inside, + *cellula,* chamber) Occurring within a body cell or within body cells.

intrinsic growth rate Exponential growth rate of a population, the difference between the density-independent components of the birth and death rates of a natural population with stable age distribution.

intrinsic rate of increase See **intrinsic growth rate.**

introvert (L. *intro,* inward, + *vertere,* to turn) The anterior narrow portion that can be withdrawn (introverted) into the trunk of a sipunculid worm.

iteroparity (i′tər-o-pā′ri-tē′) A life history in which individual organisms of a population normally reproduce more than one time before dying; contrasts with **semelparity.**

J

Jacobson's organ (*Jacobson,* 19th-century Danish surgeon and anatomist) A chemosensory organ in the roof of the mouth of many terrestrial vertebrates; odors are transferred to this organ, also known as the vomeronasal organ, by the tongue.

juvenile hormone Hormone produced by the corpora allata of insects; among its effects are maintenance of larval or nymphal characteristics during development.

K

keratin (ker'ə-tən) (Gr. *kera,* horn, + *in,* suffix of proteins) A scleroprotein found in epidermal tissues and modified into hard structures such as horns, hair, and nails.

keystone species A species (typically a predator) whose removal leads to reduced species diversity within the community.

kinetoplast (kī-nēt'ə-plast) (Gr. *kinētos,* moving, + *plastos,* molded, formed). Cellular organelle that functions in association with a kinetosome at the base of a flagellum; presumed to be derived from a mitochondrion.

kinetosome (kin-et'ə-sōm) (Gr. *kinētos,* moving, + *sōma,* body) The self-duplicating granule at the base of the flagellum or cilium; similar to centriole, also called basal body or blepharoplast.

L

labium (lā'bē-əm) (L. a lip) The lower lip of the insect formed by fusion of the second pair of maxillae.

labrum (lā'brəm) (L. a lip) The upper lip of insects and crustaceans situated above or in front of the mandibles; also refers to the outer lip of a gastropod shell.

labyrinthodont (lab'ə-rin'thə-dänt) (Gr. *labyrinthos,* labyrinth, + *odous, odontos,* tooth) A group of Paleozoic amphibians containing the temnospondyls and the anthracosaurs.

lacunar system A netlike set of circulatory canals filled with fluid in an acanthocephalan.

Lamarckism Hypothesis, as expounded by Jean-Baptiste de Lamarck, of evolution by acquisition during an organism's lifetime of characteristics that are transmitted to offspring.

lamella (lə-mel'ə) (L. dim. of *lamina,* plate) One of the two plates forming a gill in a bivalve mollusc. One of the thin layers of bone laid concentrically around an osteon

bat/āpe/ärmadillo/herring/fēmale/finch/līce/
crocodile/crōw/duck/ūnicorn/tūna/ə indicates
unaccented vowel sound "uh" as in mammal,
fishes, cardinal, heron, vulture/stress as in
bi-ol'o-gy, bi'o-log'i-cal

(Haversian) canal. Any thin, platelike structure.

larva (lar'və) pl. **larvae** (L. a ghost) An immature stage that is quite different from the adult.

lateral (L. *latus,* the side, flank) Of or pertaining to the side of an animal; a *bilateral* animal has two sides.

lateral line system Sensory organ that detects water vibrations; consisting of neuromast organs in canals and grooves on the head and sides of the body of fishes and some amphibians.

lek (lek) (Sw. play, game) An area where animals assemble for communal courtship display and mating.

lemniscus (lem-nis'kəs) (L. ribbon) One of a pair of internal projections of the epidermis from the neck region of Acanthocephala, which functions in fluid control in the protrusion and invagination of the proboscis.

lepidosaurs (lep'ə-dō-sors) (L. *lepidos,* scale, + *sauros,* lizard) A lineage of diapsid reptiles that appeared in the Permian and that includes the modern snakes, lizards, amphisbaenids, and tuataras.

leptocephalus (lep'tə-sef'ə-ləs) pl. **leptocephali** (Gr. *leptos,* thin, + *kephalē,* head) Transparent, ribbonlike migratory larva of the European or American eel.

limiting resource A particular source of nutrition, energy, or living space whose scarcity is causally associated with a population having fewer individuals than otherwise expected in a particular environment.

lobopodium (lō'bə-pō'dē-əm) (Gr. *lobos,* lobe, + *pous, podos,* foot) Blunt, lobelike pseudopodium.

lobosea (lə-bō'sē-ə) (Gr. *lobos,* lobe). A protozoan clade comprising amebas with lobopodia.

lophophore (lōf'ə-fōr) (Gr. *lophos,* crest, + *phoros,* bearing) Tentacle-bearing ridge or arm within which is an extension of the coelomic cavity in lophophorate animals (ectoprocts, brachiopods, and phoronids).

lophotrochozoan protostome (lō'fō-trō'kō-zō'ən) (Gr. *lophos,* crest, + *trochos,* wheel, + *zōon,* animal). Any member of a clade within Protostomia whose members generally possess either a trochophore larva or a lophophore; examples are annelids, molluscs, and ectoprocts.

lorica (lor'ə-kə) (L. *lorica,* corselet) A secreted, protective covering, as in phylum Loricifera.

lymph (limf) (L. *lympha,* water) The interstitial (intercellular) fluid in the body, also the fluid in the lymphatic space.

M

macroevolution (L. *makros,* long, large, + *evolvere,* to unfold) Evolutionary change on a grand scale, encompassing the origin of novel designs, evolutionary trends, adaptive radiation, and mass extinction.

macrogamete (mak'rə-gam'ēt) (Gr. *makros,* long, large, + *gamos,* marriage) The larger of the two gamete types in a heterogametic organism, considered the female gamete.

macronucleus (ma'krō-nū'klē-əs) (Gr. *makros,* long, large, + *nucleus,* kernel) The larger of the two kinds of nuclei in ciliate protozoa; controls all cell function except reproduction.

madreporite (ma'drə-pōr'īt) (Fr. *madrépore,* reef-building coral, + *ite,* suffix for some body parts) Sievelike structure that is the intake of the water-vascular system of echinoderms.

malacostracan (mal'ə-käs'trə-kən) (Gr. *malako,* soft, + *ostracon,* shell) Any member of the crustacean subclass Malacostraca, which includes both aquatic and terrestrial forms of crabs, lobsters, shrimps, pillbugs, sand fleas, and others.

malaria (mə-lar'ē-ə) (It. *malaria,* bad air) A disease marked by periodic chills, fever, anemia, and other symptoms, caused by *Plasmodium* spp.

Malpighian tubules (Mal-pig'ē-ən) (Marcello Malpighi, Italian anatomist, 1628–94) Blind tubules opening into the hindgut of nearly all insects and some myriapods and arachnids and functioning primarily as excretory organs.

mandible (L. *mandibula,* jaw) One of the lower jaw bones in vertebrates; one of the head appendages in arthropods.

mantle Soft extension of the body wall in certain invertebrates, for example, brachiopods and molluscs, which usually secretes a shell; thin body wall of tunicates.

manubrium (mə-nü'brē-əm) (L. handle) The portion projecting from the oral side of a jellyfish medusa, bearing the mouth; oral cone; presternum or anterior part of sternum; handlelike part of malleus of ear.

marsupial (mär-sü'pē-əl) (Gr. *marsypion,* little pouch) One of the pouched mammals of the subclass Metatheria.

mass extinction A relatively short interval of geological time in which a large portion (75%–95%) of existing species or higher taxa are eliminated nearly simultaneously.

mastax (mas'təx) (Gr. jaws) Pharyngeal mill of rotifers.

matrix (mā′triks) (L. *mater,* mother) The extracellular substance of a tissue, or that part of a tissue into which an organ or process is set.

maxilla (mak-sil′ə) (L. dim. of *mala,* jaw) One of the upper jawbones in vertebrates; one of the head appendages in arthropods.

maxilliped (mak-sil′ə-ped) (L. *maxilla,* jaw, + *pes,* foot) One of the pairs of head appendages located just posterior to the maxilla in crustaceans; a thoracic appendage that has become incorporated into the feeding mouthparts.

medial (mē′dē-əl) Situated, or occurring, in the middle.

medulla (mə-dül′ə) (L. marrow) The inner portion of an organ in contrast to the cortex or outer portion. Also, hindbrain.

medusa (mə-dü′-sə) (Gr. mythology, female monster with snake-entwined hair) A jellyfish, or the free-swimming stage that reproduces sexually in the life cycle of cnidarians.

Mehlis' gland (me′ləs) Glands of uncertain function surrounding the junction of yolk duct, oviduct, and uterus in trematodes and cestodes.

meiosis (mī-ō′səs) (Gr. from *meioun,* to make small) The nuclear changes by means of which the chromosomes are reduced from the diploid to the haploid number; in animals, usually occurs in the last two divisions in the formation of the mature egg or sperm.

melanin (mel′ə-nin) (Gr. *melas,* black) Black or dark-brown pigment found in plant or animal structures.

membranelle A tiny membrane-like structure, may be formed by fused cilia.

merozoite (me′rə-zō′īt) (Gr. *meros,* part, + *zōon,* animal) A very small trophozoite at the stage just after cytokinesis has been completed in multiple fission of a protozoan.

mesenchyme (me′zn-kīm) (Gr. *mesos,* middle, + *enchyma,* infusion) Embryonic connective tissue; irregular or amebocytic cells often embedded in gelatinous matrix.

mesocoel (mez′ō-sēl) (Gr. *mesos,* middle, + *koilos,* hollow) Middle body coelomic compartment in some deuterostomes; anterior in lophophorates, corresponds to hydrocoel in echinoderms.

mesoderm (me′zə-dərm) (Gr. *mesos,* middle, + *derma,* skin) The third germ layer, formed in the gastrula between the ectoderm and endoderm; gives rise to connective tissues, muscle, urogenital and vascular systems, and the peritoneum.

mesoglea (mez′ō-glē′ə) (Gr. *mesos,* middle, + *glia,* glue) The layer of jellylike or cement material between the epidermis and gastrodermis in cnidarians and ctenophores.

mesohyl (me′zō-hil) (Gr. *mesos,* middle, + *hyle,* a wood) Gelatinous matrix surrounding sponge cells; mesoglea, mesenchyme.

mesonephros (me-zō-nef′rōs) (Gr. *mesos,* middle, + *nephros,* kidney) The middle of three pairs of embryonic renal organs in vertebrates. Functional kidney of embryonic amniotes; its collecting duct is a Wolffian duct. Adj., **mesonephric.**

mesosome (mez′ə-sōm) (Gr. *mesos,* middle, + *sōma,* body) The portion of the body in lophophorates and some deuterostomes that contains the mesocoel.

metacercaria (me′tə-sər-ka′rē-ə) (Gr. *meta,* after, + *kerkos,* tail, + L. *aria,* connected with) Fluke juvenile (cercaria) that has lost its tail and has become encysted.

metacoel (met′ə-sēl) (Gr. *meta,* after, + *koilos,* hollow) Posterior coelomic compartment in some deuterostomes and lophophorates; corresponds to somatocoel in echinoderms.

metamere (met′ə-mēr) (Gr. *meta,* after, + *meros,* part) A repeated body unit along the longitudinal axis of an animal, a somite, or segment.

metamerism (mə-ta′-mə-ri′zəm) (Gr. *meta,* between, after, + *meros,* part) Condition of being composed of serially repeated parts (metameres); serial segmentation.

metamorphosis (Gr. *meta,* after, + *morphē,* form, + *osis,* state of) Sharp change in form during postembryonic development, for example, tadpole to frog or larval insect to adult.

metanephridium (me′tə-nə-fri′di-əm) (Gr. *meta,* after, + *nephros,* kidney) A type of tubular nephridium with the inner open end draining the coelom and the outer open end discharging to the exterior.

metapopulation dynamics The structure of a large population that comprises numerous semi-autonomous subpopulations, termed demes, with some limited movement of individuals among demes. Demes of a metapopulation are often geographically distinct.

metasome (met′ə-som) (Gr. *meta,* after, behind, + *sōma,* body) The portion of the body in lophophorates and some deuterostomes that contains the metacoel.

metazoa (met-ə-zō′ə) (Gr. *meta,* after, + *zōon,* animal) Multicellular animals.

microevolution (mī′krō-ev-ə-lü-shən) (L. *mikros,* small, + *evolvere,* to unfold) A change in the gene pool of a population across generations.

microfilariae (mīk′rə-fil-ar′ē-ē) (Gr. *mikros,* small, + L. *filum,* a thread) Partially developed juveniles borne alive by filarial worms (phylum Nematoda).

microgamete (mīk′rə-ga′-mēt) (Gr. *mikros,* small, + *gamos,* marriage) The smaller of the two gamete types in a heterogametic organism, considered the male gamete.

micron (′m) (mī′-krän) (Gr. neuter of *mikros,* small) One-thousandth of a millimeter; about 1/25,000 of an inch. Now largely replaced by micrometer (′m).

microneme (mī′krə-nēm) (Gr. *mikros,* small, + *nēma,* thread) One type of structure forming the apical complex in the Phylum Apicomplexa, slender and elongate, leading to the anterior and thought to function in host cell penetration.

micronucleus A small nucleus found in ciliate protozoa; controls the reproductive functions of these organisms.

microsporidian (mī′krō-spo-rid′ē-ən) (Gr. *micros,* small, + *spora,* seed, + *idion,* dim. Suffix). Any member of a protozoan clade comprising intracellular parasites with a distinctive spore morphology.

microthrix See **microvillus.**

microtubule (Gr. *mikros,* small, + L. *tubule,* pipe) A long, tubular cytoskeletal element with an outside diameter of 20 to 27 nm. Microtubules influence cell shape and play important roles during cell division.

microvillus (Gr. *mikros,* small, + L. *villus,* shaggy hair) Narrow, cylindrical cytoplasmic projection from epithelial cells; microvilli form the brush border of several types of epithelial cells. Also, microvilli with unusual structure cover the surface of cestode tegument (also called **microthrix** [pl. **microtriches**]).

mictic (mik′tik) (Gr. *miktos,* mixed or blended) Pertaining to haploid egg of rotifers of the females that lay such eggs.

mimic (mim′ik) (Gr. *mimicus,* imitator) A species whose morphological or behavioral characteristics copy those of another species because those characteristics deter shared predators.

miracidium (mīr′ə-sid′ē-əm) (Gr. *meirakid-ion,* youthful person) A minute ciliated larval stage in the life of flukes.

mitochondrion (mīd′ə-kän′drē-ən) (Gr. *mitos,* a thread, + *chondrion,* dim. of *chondros,* corn, grain) An organelle in the cell in which aerobic metabolism occurs.

mitosis (mī-tō′səs) (Gr. *mitos,* thread, + *osis,* state of) Nuclear division in which there is an equal qualitative and quantitative division of the chromosomal material between the two resulting nuclei; ordinary cell division (indirect).

model (mod'l) (Fr. *modèle*, pattern) A species whose morphological or behavioral characteristics are copied by another species because those characteristics deter shared predators.

modular (mäj'ə-lər) Describes the structure of a colony of genetically identical organisms that are physically associated and produced asexually by cloning.

molting Shedding of the outer cuticular layer; see **ecdysis.**

monoecious (mə-nē'shəs) (Gr. *monos*, single, + *oikos*, house) Having both male and female gonads in the same organism; hermaphroditic.

monogamy (mə-näg'ə-mē) (Gr. *monos*, single, + *gamos*, marriage) The condition of having a single mate at any one time. Adj. **monogamous.**

monophyly (män'ō-fī'lē) (Gr. *monos*, single, + *phyle*, tribe) The condition that a taxon or other group of organisms contains the most recent common ancestor of the group and all of its descendants. Adj., **monophyletic.**

monotreme (mä'nō-trēm) (Gr. *monos*, single, + *trēma* hole) Egg-laying mammal of the order Monotremata.

morphogenesis (mor'fə-je'nə-səs) (Gr. *morphē*, form, + *genesis*, origin) Development of the architectural features of organisms; formation and differentiation of tissues and organs.

morphology (Gr. *morphē*, form, + *logos*, discourse) The science of structure. Includes cytology, the study of cell structure; histology, the study of tissue structure; and anatomy, the study of gross structure.

mosaic cleavage Embryonic development characterized by independent differentiation of each part of the embryo; determinate cleavage.

mucus (mū'kəs) (L. *mucus*, nasal mucus) Viscid, slippery secretion rich in mucins produced by secretory cells such as those in mucous membranes. Adj., **mucous.**

multiple fission A mode of asexual reproduction in some protistans in which the nuclei divide more than once before cytokinesis occurs.

multiplication of species The Darwinian theory that the evolutionary process generates new species through a branching

of evolutionary lineages derived from an ancestral species.

mutation (mū-tā'shən) (L. *mutare*, to change) A stable and abrupt change of a gene; the heritable modification of a character.

mutualism (mū'chə-wə-li'zəm) (L. *mutuus*, lent, borrowed, reciprocal) A type of interaction in which two different species derive benefit from their association and in which the association is necessary to both; often symbiotic.

myocyte (mī'ə-sīt) (Gr. *mys*, muscle, + *kytos*, hollow vessel) Contractile cell (pinacocyte) in sponges.

myofibril (Gr. *mys*, muscle, + L. dim. of *fibra*, fiber) A contractile filament within muscle or muscle fiber.

myomere (mī'ə-mer) (Gr. *mys*, muscle, + *meros*, part) A muscle segment of successive segmental trunk musculature.

myotome (mī'ə-tōm) (Gr. *mys*, muscle, + *tomos*, cutting) A voluntary muscle segment in cephalochordates and vertebrates; that part of a somite destined to form muscles; the muscle group innervated by a single spinal nerve.

N

nacre (nā'kər) (F. mother-of-pearl) Innermost lustrous layer of mollusc shell, secreted by mantle epithelium. Adj., **nacreous.**

nares (na'rēz), sing. **naris** (L. nostrils) Openings into the nasal cavity, both internally and externally, in the head of a vertebrate.

natural selection The interactions between organismal character variation and the environment that cause differences in rates of survival and reproduction among varying organisms in a population; leads to evolutionary change if variation is heritable.

nauplius (naw'plē-əs) (L. a kind of shellfish) A free-swimming microscopic larval stage of certain crustaceans, with three pairs of appendages (antennules, antennae, and mandibles) and a median eye. Characteristic of ostracods, copepods, barnacles, and some others.

nekton (nek'tən) (Gr. neuter of. *nēktos*, swimming) Term for actively swimming organisms, essentially independent of wave and current action. Compare with **plankton.**

nematocyst (ne-mad'ə-sist') (Gr. *nēma*, thread, + *kystis*, bladder) Stinging organelle of cnidarians.

neo-Darwinism (nē'ō'där'wə-niz'əm) A modified version of Darwin's evolutionary theory that eliminates elements of the

Lamarckian inheritance of acquired characteristics and pangenesis that were present in Darwin's formulation; this theory originated with August Weissmann in the late nineteenth century and, after incorporating Mendelian genetic principles, has become the currently favored version of Darwinian evolutionary theory.

neopterygian (nē-äp'tə-rij'ē-ən) (Gr. *neos*, new, + *pteryx*, fin) Any of a large group of bony fishes that includes most modern species.

nephridium (nə-frid'ē-əm) (Gr. *nephridios*, of the kidney) One of the segmentally arranged, paired excretory tubules of many invertebrates, notably the annelids. In a broad sense, any tubule specialized for excretion and/or osmoregulation; with an external opening and with or without an internal opening.

nephron (ne'frän) (Gr. *nephros*, kidney) Functional unit of kidney structure of vertebrates, consisting of Bowman's capsule, an enclosed glomerulus, and the attached uriniferous tubule.

nephrostome (nef'rə-stōm) (Gr. *nephros*, kidney, + *stoma*, mouth) Ciliated, funnel-shaped opening of a nephridium.

nested hierarchy A pattern in which species are ordered into a series of increasingly more inclusive clades according to the taxonomic distribution of synapomorphies.

net productivity The energy stored by an organism, equal to the energy assimilated (**gross productivity**) minus the energy used for metabolic maintenance (**respiration**).

neural crest Populations of ectodermally derived embryonic cells that differentiate into many skeletal, neural, and sensory structures; unique to vertebrates.

neuroglia (nü-räg'lē-ə) (Gr. *neuron*, nerve, + *glia*, glue) Tissue supporting and filling the spaces between the nerve cells of the central nervous system.

neuromast (Gr. *neuron*, sinew, nerve, + *mastos*, knoll) Cluster of sense cells on or near the surface of a fish or amphibian that is sensitive to vibratory stimuli and water current.

neuron (Gr. nerve) A nerve cell.

neuropodium (nü'rə-pō'dē-əm) (Gr. *neuron*, nerve, + *pous*, *podos*, foot) Lobe of parapodium nearer the ventral side in polychaete annelids.

neurosecretory cell (nü'rō-sə-krēd'ə-rē) Any cell (neuron) of the nervous system that produces a hormone.

niche The role of an organism in an ecological community; its unique way of life and its relationship to other biotic and abiotic factors.

niche overlap A comparison of two species quantifying the proportion of each species' resources that are utilized also by the other species.

notochord (nōd′ə-kord′) (Gr. *nōtos*, back, + *chorda*, cord) An elongated cellular cord, enclosed in a sheath, which forms the primitive axial skeleton of chordate embryos, adult cephalochordates, and jawless vertebrates.

notopodium (nō′tə-pō′dē-əm) (Gr. *nōtos*, back, + *pous, podos,* foot) Lobe of parapodium nearer the dorsal side in polychaete annelids.

nucleolus (nü-klē′ə-ləs) (dim. of L. *nucleus,* kernel) A deeply staining body within the nucleus of a cell and containing RNA; nucleoli are specialized portions of certain chromosomes that carry multiple copies of the information to synthesize ribosomal RNA.

nucleoplasm (nü′klē-ə-pla′zəm) (L. *nucleus,* kernel, + Gr. *plasma,* mold) Protoplasm of nucleus, as distinguished from cytoplasm.

nucleus (nü′klē-əs) (L. *nucleus,* a little nut, the kernel) The organelle in eukaryotes that contains the chromatin and which is bounded by a double membrane (nuclear envelope).

nurse cells Single cells or layers of cells surrounding or adjacent to other cells or structures for which the nurse cells provide nutrient or other molecules (for example, for insect oocytes or *Trichinella* spp. juveniles).

nymph (L. *nympha,* nymph, bride) An immature stage (following hatching) of a hemimetabolous insect that lacks a pupal stage.

O

ocellus (ō-sel′əs) (L. dim. of *oculus,* eye) A simple eye or eyespot in many types of invertebrates.

octomerous (ok-tom′ər-əs) (Gr. *oct,* eight, + *meros,* part) Eight parts, specifically, symmetry based on eight.

odontophore (ō-don′tə-for′) (Gr. *odous,* tooth, + *pherein,* to carry) Tooth-bearing organ in molluscs, including the radula, radular sac, muscles, and cartilages.

omasum (ō-mā′səm) (L. paunch) The third compartment of the stomach of a ruminant mammal.

ommatidium (ä′mə-tid′ē-əm) (Gr. *omma,* eye, + *idium,* small) One of the optical units of the compound eye of arthropods.

omnivore (äm′nə-vōr) (L. *omnis,* all, + *vorare,* to devour) An animal that uses a variety of animal and plant material in its diet.

oncosphere (än′kō-sfiər) (Gr. *onkinos,* a hook, + *sphaira,* ball) Rounded larva common to all cestodes; bears hooks.

ontogeny (än-tä′jə-nē) (Gr. *ontos,* being, + *geneia,* act of being born, from *genēs,* born) The course of development of an individual from egg to senescence.

oocyst (ō′ə-sist) (Gr. *ōion,* egg, + *kystis,* bladder) Cyst formed around zygote of malaria and related organisms.

oocyte (ō′ə-sīt) (Gr. *ōion,* egg, + *kytos,* hollow) Stage in formation of ovum, just preceding first meiotic division (primary oocyte) or just following first meiotic division (secondary oocyte).

ookinete (ō-ə-kī′nēt) (Gr. *ōion,* egg, + *kinein,* to move) The motile zygote of malarial parasites.

operculum (ō-per′kū-ləm) (L. cover) The gill cover in body fishes; keratinized plate in some snails.

opisthaptor (ō′pəs-thap′tər) (Gr. *opisthen,* behind, + *haptein,* to fasten) Posterior attachment organ of a monogenetic trematode.

opisthokont (ō-pis′thō-kont) (G. *opisthen,* behind, + *kontos,* a pole). Any member of a eukaryotic clade comprising fungi, microsporidians, choanoflagellates, and metazoans; if present, flagellated cells possess a single posterior flagellum.

oral disc The end of a cnidarian polyp bearing the mouth.

oral lobe A flaplike extension of the mouth of a scyphozoan medusa that aids in feeding.

organelle (Gr. *organon,* tool, organ, + L. *ella,* dimin. suffix) Specialized part of a cell; literally, a small organ that performs functions analogous to organs of multicellular animals.

organism (or′gə-niz′-əm) A biological individual composed of one or more cells, tissues, and/or organs whose parts are interdependent in producing a collective physiological system. Organisms of the same species may form **populations.**

osculum (os′kū-ləm) (L. *osculum,* a little mouth) Excurrent opening in a sponge.

osmoregulation Maintenance of proper internal salt and water concentrations in a cell or in the body of a living organism, active regulation of internal osmotic pressure.

osmosis (oz-mō′sis) (Gr. *ōsmos,* act of pushing, impulse) The flow of solvent (usually water) through a semipermeable membrane.

osmotroph (oz′mə-trōf) (Gr. *ōsmos,* a thrusting, impulse, + *trophē,* to eat) A heterotrophic organism that absorbs dissolved nutrients.

osphradium (äs-frā′dē-əm) (Gr. *osphradion,* small bouquet, dim., + of *osphra,* smell) A sense organ in aquatic snails and bivalves that tests incoming water.

ossicles (L. *ossiculum,* small bone) Small separate pieces of echinoderm endoskeleton. Also, tiny bones of the middle ear of vertebrates.

ostium (L. door) Opening.

otolith (ōd′ə-lith′) (Gr. *ous, otos,* ear, + *lithos,* stone) Calcerous concretions in the membranous labyrinth of the inner ear of lower vertebrates or in the auditory organ of certain invertebrates.

outgroup In phylogenetic systematic studies, a species or group of species closely related to but not included within a taxon whose phylogeny is being studied, and used to polarize variation of characters and to root the phylogenetic tree.

outgroup comparison A method for determining the polarity of a character in cladistic analysis of a taxonomic group. Character states found within the group being studied are judged ancestral if they occur also in related taxa outside the study group (= outgroups); character states that occur only within the taxon being studied but not in outgroups are judged to have been derived evolutionarily within the group being studied.

oviger (ō′vi-jər) (L. *ovum,* egg, + *gerere,* to bear) Leg that carries eggs in pycnogonids.

oviparity (ō′və-pa′rəd-ē) (L. *ovum,* egg, + *parere,* to bring forth) Reproduction in which eggs are released by the female; development of offspring occurs outside the maternal body. Adj., **oviparous** (ō-vip′ə-rəs).

ovipositor (ō′ve-päz′əd-ər) (L. *ovum,* egg, + *positor,* builder, placer, + *or,* suffix denoting agent or doer) In many female insects a structure at the posterior end of the abdomen for laying eggs.

ovoviviparity (ō′vo-vī-və-par′ə-dē) (L. *ovum,* egg, + *vivere,* to live, + *parere,* to bring forth) Reproduction in which eggs develop within the maternal body without additional nourishment from the parent and hatch within the parent or immediately after laying. Adj., **ovoviviparous** (ō-vo-vī-vip′ə-rəs).

ovum (L. *ovum,* egg) Mature female germ cell (egg).

P

paedomorphosis (pē-dō-mor′fə-səs) (Gr. *pais,* child, + *morphē,* form) Displacement of ancestral juvenile features to later stages of the ontogeny of descendants.

pangenesis (pan-jen′ə-sis) (Gr. *pan,* all, + *genesis,* descent) Darwin's hypothesis that hereditary characteristics are carried by individual body cells that produce particles collecting in the germ cells.

papilla (pə-pil′ə) pl. **papillae** (L. nipple) A small nipplelike projection. A vascular process that nourishes the root of a hair, feather, or developing tooth.

papula (pa′pū-lə) pl. **papulae** (L. pimple) Respiratory processes on skin of sea stars; also, pustules on skin.

parabasal bodies Cellular organelles similar to Golgi bodies, presumed to function as part of the secretory system in endoplasmic reticulum.

parabasalid (pa′rə-bə′sa-lid) (Gr. *para,* beside, + *basis,* body). Any member of protozoan clade having a flagellum and parabasal bodies.

parabronchi (par-ə-brong′kī) (Gr. *para,* beside, + *bronchos,* windpipe) Fine airconduction pathways of the bird lung.

paraphyly (par′ə-fī′lē) (Gr. *para,* before, + *phyle,* tribe) The condition that a taxon or other group of organisms contains the most recent common ancestor of all members of the group but excludes some descendants of that ancestor. Adj., **paraphyletic.**

parapodium (pa′rə-pō′dē-əm) (Gr. *para,* beside, + *pous, podos,* foot) One of the paired lateral processes on each side of most segments in polychaete annelids; variously modified for locomotion, respiration, or feeding.

parasite (par′ə-sīt) An organism that lives physically on or in and at the expense of another organism.

parasitism (par′ə-sit′iz-əm) (Gr. *parasitos,* from *para,* beside, + *sitos,* food) The condition of an organism living in or on another organism (host) at whose expense the parasite is maintained; destructive symbiosis.

parasitoid An organism that is a typical parasite early in its development but that finally kills the host during or at the completion of development; used in reference to many insect parasites of other insects.

parenchyma (pə-ren′kə-mə) (Gr. anything poured in beside) In simpler animals, a spongy mass of vacuolated mesenchyme cells filling spaces between viscera, muscles, or epithelia; in some, the cells are cell bodies of muscle cells. Also, the specialized tissue of an organ as distinguished from the supporting connective tissue.

parenchymula (pa′rən-kīm′yə-lə) (Gr. *para,* beside, + *enchyma,* infusion) Flagellated, solid-bodied larva of some sponges.

parietal (pä-rī′-ə-təl) (L. *paries,* wall) Something next to, or forming part of, a wall of a structure.

parthenogenesis (pär′thə-nō-gen′ə-sis) (Gr. *parthenos,* virgin, + L. from Gr. *genesis,* origin) Unisexual reproduction involving the production of young by females not fertilized by males; common in rotifers, cladocerans, aphids, bees, ants, and wasps. A parthenogenetic egg may be diploid or haploid.

pecten (L. comb) Any of several types of comblike structures on various organisms, for example, a pigmented, vascular, and comblike process that projects into the vitreous humor from the retina at a point of entrance of the optic nerve in the eyes of all birds and many reptiles.

pectoral (pek′tə-rəl) (L. *pectoralis,* from *pectus,* the breast) Of or pertaining to the breast or chest; to the pectoral girdle; or to a pair of keratinized shields of the plastron of certain turtles.

pedalium (pə-dal′ē-əm) (Gr. *pedalion,* a prop, rudder) The flattened, bladelike base of a tentacle or group of tentacles in the cnidarian class Cubozoa.

pedal laceration Asexual reproduction found in sea anemones, a form of fission.

pedicel (ped′ə-sel) (L. *pediculus,* little foot) A small or short stalk or stem. In insects, the second segment of an antenna or the waist of an ant.

pedicellaria (ped′ə-sə-lar′ē-ə) (L. *pediculus,* little foot, + *aria,* like or connected with) One of many minute pincerlike organs on the surface of certain echinoderms.

pedipalps (ped′ə-palps′) (L. *pes, pedis,* foot, + *palpus,* stroking, caress) Second pair of appendages of arachnids.

peduncle (pē-dun′kəl) (L. *pedunculus,* dim. of *pes,* foot) A stalk. Also, a band of white matter joining different parts of the brain.

pelage (pel′ij) (Fr. fur) Hairy covering of mammals.

pelagic (pə-laj′ik) (Gr. *pelagos,* the open sea) Pertaining to the open ocean.

pellicle (pel′ə-kəl) (L. *pellicula,* dim. of *pelis,* skin) Thin, translucent, secreted envelope covering many protozoa.

pen A flattened flexible internal support in a squid; a remnant of the ancestral shell.

pentadactyl (pen-tə-dak′təl) (Gr. *pente,* five, + *daktylos,* finger) With five digits, or five fingerlike parts, to the hand or foot.

perennibranchiate (pə-ran′ə-brank′ē-āt) (L. *perennis,* throughout the year, + Gr. *branchia,* gills) Having permanent gills, relating especially to certain paedomorphic salamanders.

periostracum (pe-rē-äs′trə-kəm) (Gr. *peri,* around, + *ostrakon,* shell) Outer keratinized layer of a mollusc shell.

peripheral (pə-ri′fər-əl) (Gr. *peripherein,* to move around) Structure or location distant from center, near outer boundaries.

periproct (per′ə-präkt) (Gr. *peri,* around, + *prōktos,* anus) Region of aboral plates around the anus of echinoids.

perisarc (per′ə-särk) (Gr. *peri,* around, + *sarx,* flesh) Sheath covering the stalk and branches of a hydroid.

perissodactyl (pə-aris′ə-dak′təl) (Gr. *perissos,* odd, + *daktylos,* finger, toe) Pertaining to an order of ungulate mammals with an odd number of digits.

peristomium (per′ə-stō′mē-əm) (Gr. *peri,* around, + *stoma,* mouth) One of two parts forming the annelid head; it bears the mouth.

peritoneum (per′ə-tə-nē′əm) (Gr. *peritonaios,* stretched around) The membrane that lines the coelom and covers the coelomic viscera.

Permian extinction A mass extinction that occurred 245 million years ago in which 96% of existing species became extinct, marking the end of the Paleozoic era.

perpetual change The most basic theory of evolution, that the living world is neither constant nor cycling, but is always undergoing irreversible modification through time.

phagocyte (fag′ə-sīt) (Gr. *phagein,* to eat, + *kytos,* hollow vessel) Any cell that engulfs and devours microorganisms or other particles.

phagocytosis (fag′ə-sī-tō′səs) (Gr. *phagein,* to eat, + *kytos,* hollow vessel) The engulfment of a particle by a phagocyte or a protozoan.

phagosome (fa′gə-sōm) (Gr. *phagein,* to eat, + *sōma,* body) Membrane-bound vessel in cytoplasm containing food material engulfed by phagocytosis.

phagotroph (fag′ə-trōf) (Gr. *phagein*, to eat, + *trophē*, food) A heterotrophic organism that ingests solid particles for food.

pharynx (far′inks) pl. **pharynges** (Gr. *pharynx*, gullet) The part of the digestive tract between the mouth cavity and the esophagus that, in vertebrates, is common to both digestive and respiratory tracts. In cephalochordates the gill slits open from it.

phenetic taxonomy (fə-ne′tik) (Gr. *phaneros*, visible, evident) Refers to the use of a criterion of overall similarity to classify organisms into taxa; contrasts with classification based explicitly on a reconstruction of phylogeny.

phenotype (fē′nə-tīp) (Gr. *phainein*, to show) The visible or expressed characteristics of an organism, controlled by the genotype, but not all genes in the genotype are expressed.

pheromone (fer′ə-mōn) (Gr. *pherein*, to carry, + *hormōn*, exciting, stirring up) Chemical substance released by one organism that influences the behavior or physiological processes of another organism.

photoautotroph (fōd-ə-aw′tō-trōf) (Gr. *phōtos*, light, + *autos*, self, + *trophos*, feeder) An organism requiring light as a source of energy for making organic nutrients from inorganic raw materials.

photosynthesis (fōt-ō-sin′thə-sis) (Gr. *phōs*, light, + *synthesis*, action or putting together). The synthesis of carbohydrates from carbon dioxide and water in chlorophyll-containing cells exposed to light.

phototaxis (fōd′ō-tak′sis) (Gr. *phōtos*, light, + *taxis*, arranging, order) A taxis in which light is the orienting stimulus. An involuntary tendency for an organism to turn toward (positive) or away from (negative) light.

phototrophs (fōt′ō-trōfs) (Gr. *phōs*, *phōtos*, light, + *trophē*, nouishment) Organisms capable of using CO_2 in the presence of light as a source of metabolic energy.

phyletic gradualism A model of evolution in which morphological evolutionary change is continuous and incremental and occurs mainly within unbranched species or lineages over long periods of geological time; contrasts with **punctuated equilibrium.**

phylogenetic species concept An irreducible (basal) cluster of organisms, diagnosably distinct from other such clusters, and within which there is a parental pattern of ancestry and descent.

phylogenetic systematics See **cladistics.**

phylogenetic tree A branching diagram whose branches represent evolutionary lineages and depicts the common descent of species or higher taxa.

phylogeny (fī′läj′ə-nē) (Gr. *phylon*, tribe, race, + *geneia*, origin) The origin and diversification of any taxon, or the evolutionary history of its origin and diversification, usually presented as a dendrogram.

phylum (fī′ləm) pl. **phyla** (N.L. from Gr. *phylon*, race, tribe) A chief category, between kingdom and class, of taxonomic classifications into which are grouped organisms of common descent that share a fundamental pattern of organization.

physiology (L. *physiologia*, natural science) A branch of biology covering the organic processes and phenomena of an organism or any of its parts or a particular bodily process.

phytoflagellates (fī-tə-fla′jə-lāts) Members of the former class Phytomastigophorea, plantlike flagellates.

pinacocyte (pin′ə-kō-sīt′) (Gr. *pinax*, tablet, + *kytos*, hollow vessel) Flattened cells comprising dermal epithelium in sponges.

pinna (pin′ə) (L. feather, sharp point) The external ear. Also a feather, wing, or fin or similar part.

pinocytosis (pin′ō-sī-tō′sis, pīn′ō-sī-tō′sis) (Gr. *pinein*, to drink, + *kytos*, hollow vessel, + *osis*, condition) Acquiring fluid by a cell; cell drinking.

placenta (plə-sen′tə) (L. flat cake) The vascular structure, embryonic and maternal, through which the embryo and fetus are nourished while in the uterus.

placoderms (plak′ə-dərmz) (Gr. *plax*, plate, + *derma*, skin) A group of heavily armored jawed fishes of the Lower Devonian to Lower Carboniferous.

placoid scale (pla′koyd) (Gr. *plax, plakos*, tablet, plate) Type of scale found in cartilaginous fishes, with basal plate of dentin embedded in the skin and a backward-pointing spine tipped with enamel.

plankton (plank′tən) (Gr. neuter of *planktos*, wandering) The passively floating animal and plant life of a body of water; contrasts with **nekton.**

plantigrade (plan′tə-grād′) (L. *planta*, sole, + *gradus*, step, degree) Pertaining to animals that walk on the whole surface of the foot (for example, humans and bears); contrasts with **digitigrade.**

planula (plan′yə-lə) (N.L. dim. from L. *planus*, flat) Free-swimming, ciliated larval type of cnidarians; usually flattened and ovoid, with an outer layer of ectodermal cells and an inner mass of endodermal cells.

planuloid ancestor (plan′yə-loid) (L. *planus*, flat, + Gr. *eidos*, form) Hypothetical form representing ancestor of Cnidaria and Platyhelminthes.

plasma membrane (plaz′mə) (Gr. *plasma*, a form, mold) A living, external, limiting, protoplasmic structure that functions to regulate exchange of nutrients across the cell surface.

plastron (plast′trən) (Fr. *plastron*, breast plate) Ventral body shield of turtles; structure in corresponding position in certain arthropods; thin film of gas retained by epicuticle hairs of aquatic insects.

pleura (plü′rə) (Gr. side, rib) The membrane that lines each half of the thorax and covers the lungs.

podium (pō′dē-əm) (Gr. *pous, podos*, foot) A footlike structure, for example, the tube foot of echinoderms.

poikilothermic (poi-ki′lə-thər′mik) (Gr. *poikilos*, variable, + thermal) Pertaining to animals whose body temperature is variable and fluctuates with that of the environment; cold-blooded; contrasts with **ectothermic.**

polarity (Gr. *polos*, axis) In systematics, the ordering of alternative states of a taxonomic character from ancestral to successively derived conditions in an evolutionary transformation series. In developmental biology, the tendency for the axis of an ovum to orient corresponding to the axis of the mother. Also, condition of having opposite poles; differential distribution of gradation along an axis.

Polian vesicles (pō′le-ən) (From G. S. Poli, 1746–1825, Italian naturalist) Vesicles opening into ring canal in most asteroids and holothuroids.

polyandry (pol′y-an′drē) (Gr. *polys*, many, + *anēr*, man) Condition of having more than one male mate at one time.

polygamy (pə-lig′ə-mē) (Gr. *polys*, many, + *gamos*, marriage) Condition of having more than one mate at one time.

polygyny (pə-lij′ə-nē) (Gr. *polys*, many + *gynē*, woman) Condition of having more than one female mate at one time.

polymorphism (pä′lē-mor′fi-zəm) (Gr. *polys*, many, + *morphē*, form) The presence in a species of more than one structural type of individual.

polyp (päl′əp) (Fr. *polype*, octopus, from L. *polypus*, many footed) The sessile stage in the life cycle of cnidarians.

polyphyletic (pä′lē-fī-led′-ik) (Gr. *polys*, many, + *phylon*, tribe) Derived from more than one ancestral source; contrasts with monophyletic and paraphyletic.

polyphyly (pä′lē-fī′lē) (Gr. *polys*, full + *phylon*, tribe) The condition that a taxon or other group of organisms does not contain the most recent common ancestor of all members of the group, implying that it has multiple evolutionary origins; such groups are not valid as formal taxa and are recognized as such only through error.

polyphyodont (pä-lē-fī′ə-dänt) (Gr. *polyphyes,* manifold, + *odous,* tooth) Having several sets of teeth in succession.

polypide (pä′lē-pīd) (L. *polypus,* polyp) An individual or zooid in a colony, specifically in ectroprocts, which has a lophophore, digestive tract, muscles, and nerve centers.

population (L. *populus,* people) A group of organisms of the same species inhabiting a specific geographical locality.

porocyte (pō′rə-sīt) (Gr. *porus,* passage, pore, + *kytos,* hollow vessel) Type of cell in asconoid sponges through which water enters the spongocoel.

portal system (L. *porta,* gate) System of large veins beginning and ending with a bed of capillaries; for example, hepatic portal and renal portal system in vertebrates.

positive assortative mating A tendency of an individual to mate preferentially with others whose phenotypes are similar to its own.

posterior (L. latter) Situated at or toward the rear of the body; in bilateral forms, the end of the main body axis opposite the head.

preadaptation The possession of a trait that coincidentally predisposes an organism for survival in an environment different from those encountered in its evolutionary history.

precocial (prē-kō′shəl) (L. *praecoquere,* to ripen beforehand) Referring (especially) to birds whose young are covered with down and are able to walk when newly hatched.

predaceous, predacious (prē-dā′shəs) (L. *praedator,* a plunderer, *praeda,* prey) Living by killing and consuming other animals; predatory.

predation (prə-dā′shən) An interaction between species in an ecological community in which members of one species (prey) serve as food for another species (**predator**).

predator (pred′ə-tər) (L. *praedator,* a plunderer, *praeda,* prey) An organism that preys on other organisms for its food.

prehensile (prē-hen′səl) (L. *prehendere,* to seize) Adapted for grasping.

primary producer A species whose members begin **productivity** by acquiring energy and matter from **abiotic** sources, such as plants that synthesize sugars from water and carbon dioxide using solar energy (see **photosynthesis**).

primate (prī-māt) (L. *primus,* first) Any mammal of the order Primates, which includes the tarsiers, lemurs, marmosets, monkeys, apes, and humans.

primitive (L. *primus,* first) Primordial; ancient; little evolved; characteristics closely approximating those possessed by early ancestral types.

proboscis (prō-bäs′əs) (Gr. *pro,* before, + *boskein,* feed) A snout or trunk. Also, tubular sucking or feeding organ with the mouth at the end as in planarians, leeches, and insects. Also, the sensory and defensive organ at the anterior end of certain invertebrates.

producers (L. *producere,* to bring forth) Organisms, such as plants, able to produce their own food from inorganic substances.

production In ecology, the energy accumulated by an organism that becomes incorporated into new biomass.

productivity (prō′dak-tiv′-ət-ē) A property of a biological system measured by the amount of energy and/or materials that it incorporates.

proglottid (prō-gläd′əd) (Gr. *proglōttis,* tongue tip, from *pro,* before, + *glōtta,* tongue, + *id,* suffix) Portion of a tapeworm containing a set of reproductive organs; usually corresponds to a segment.

prokaryotic, procaryotic (pro-kar′ē-ät′ik) (Gr. *pro,* before, + *karyon,* kernel, nut) Not having a membrane-bound nucleus or nuclei. Prokaryotic cells characterize bacteria and cyanobacteria.

pronephros (prō-nef′rəs) (Gr. *pro,* before, + *nephros,* kidney) Most anterior of three pairs of embryonic renal organs of vertebrates; functional only in adult hagfishes and larval fishes and amphibians; vestigial in mammalian embryos. Adj., **pronephric.**

prosimian (prō-sim′ē-ən) (Gr. *pro,* before, + L. *simia,* ape) Any member of a group of arboreal primates including lemurs, tarsiers, and lorises but excluding monkeys, apes, and humans.

prosopyle (präs′-ə-pīl) (Gr. *prosō,* forward, + *pylē,* gate) Connections between the incurrent and radial canals in some sponges.

prostomium (prō-stō′mē-əm) (Gr. *pro,* before, + *stoma,* mouth) In most annelids

and some molluscs, that part of the head located in front of the mouth.

protein (prō′tēn, prō′tē-ən) (Gr. *protein,* from *proteios,* primary) A macromolecule of carbon, hydrogen, oxygen, and nitrogen and usually containing sulfur; composed of chains of amino acids joined by peptide bonds; present in all cells.

prothoracic glands Glands in the prothorax of insects that secrete the hormone ecdysone.

prothoracicotropic hormone See **ecdysiotropin.**

protist (prō′-tist) (Gr. *prōtos,* first) A member of the kingdom Protista, generally considered to include the protozoa and eukaryotic algae.

protocoel (prōd′ə-sēl) (Gr. *prōtos,* first, + *koilos,* hollow) The anterior coelomic compartment in some deuterostomes, corresponds to the axocoel in echinoderms.

protocooperation A mutually beneficial interaction between organisms in which the interaction is not physiologically necessary to the survival of either.

protonephridium (prōd′ə-nə-frid′ē-əm) (Gr. *prōtos,* first, + *nephros,* kidney) Primitive osmoregulatory or excretory organ consisting of a tubule terminating internally with flame bulb or solenocyte; the unit of a flame bulb system.

proton pump Active transport of hydrogen ions (protons) across an inner mitochondrial membrane during cellular respiration.

protopod, protopodite (prōd′ə-päd, prō-top′ə-dīt) (Gr. *prōtos,* first, + *pous, podos,* foot) Basal portion of crustacean appendage, containing coxa and basis.

protostome (prōd′ə-stō′m) (Gr. *protos,* first, + *stoma,* mouth) A member of the group Protostomia. Protostome taxa have recently been divided into ecdysozoan protostomes and lophotrochozoan protostomes.

Protostomia (prōd′ə-stō′mē-ə) (Gr. *prōtos,* first, + *stoma,* mouth) A group of phyla in which cleavage is determinate, the coelom (in coelomate forms) is formed by proliferation of mesodermal bands (schizocoelic formation), the mesoderm is formed from a particular blastomere (called 4d), and the mouth is derived from or near the blastopore in most members. However, arthropods and related taxa have a unique cleavage pattern and do not form mesoderm from the 4d cell. Includes the Annelida, Arthropoda, Mollusca, and a number of minor phyla. Contrasts with **Deuterostomia.**

proventriculus (pro′ven-trik′ū-ləs) (L. *pro,* before, + *ventriculum,* ventricle) In birds the glandular stomach between the crop and gizzard. In insects, a muscular dilation of foregut armed internally with chitinous teeth.

proximal (L. *proximus*, nearest) Situated toward or near the point of attachment; opposite of distal, distant.

proximate cause (L. *proximus*, nearest, + *causa*) The factors that underlie the functioning of a biological system at a particular place and time, including those responsible for metabolic, physiological, and behavioral functions at the molecular, cellular, organismal, and population levels. Immediate cause.

pseudocoel (sü′də-sēl) (Gr. *pseudēs*, false, + *koilos*, hollow) A body cavity not lined with peritoneum and not a part of the blood or digestive systems, embryonically derived from the blastocoel.

pseudopodium (sü′də-pō′dē-əm) (Gr. *pseudēs*, false, + *podion*, small foot, + *eidos*, form) A temporary cytoplasmic protrusion extended out from a protozoan or ameboid cell and serving for locomotion or for engulfing food.

punctuated equilibrium A model of evolution in which morphological evolutionary change is discontinuous, being associated primarily with discrete, geologically instantaneous events of speciation leading to phylogenetic branching; morphological evolutionary stasis characterizes species between episodes of speciation; contrasts with **phyletic gradualism.**

pupa (pū′pə) (L. girl, doll, puppet) Inactive quiescent state of the holometabolous insects. It follows the larval stages and precedes the adult stage.

pygidium (pə-jid′ē-əm) (Gr. *pyge*, rump, buttocks, + *idion*, dim. Ending) Posterior region of a segmented animal bearing the anus.

Q

queen In entomology, a reproductive female in a colony of social insects such as bees, ants, and termites, distinguished from workers, nonreproductive females, and soldiers.

R

radial canals Canals along the ambulacra radiating from the ring canal of echinoderms; also choanocyte-lined canals in syconoid sponges.

radial cleavage Embryonic development in which early cleavage planes are symmetri-

cal to the polar axis, each blastomere of one tier lying directly above the corresponding blastomere of the next layer; indeterminate cleavage.

radial symmetry A morphological condition in which the parts of an animal are arranged concentrically around an oral-aboral axis, and more than one imaginary plane through this axis yields halves that are mirror images of each other.

Radiata (rā′dē-ä′tə) (L. *radius*, ray) Phyla showing radial symmetry, specifically Cnidaria and Ctenophora.

radiolarian (rā′dē-ō-la′rē-ən) (L. *radius*, ray, spoke of a wheel, + *Lar*, tutelary god house and field). Amebas with actinopodia and silica tests.

radiole (rā′dē-ōl) (L. *radiolus*, dim. of *radius*, ray, spoke of a wheel) Structure extending from head of some sedentary polychaetes used in feeding on suspended particles.

radula (ra′jə-lə) (L. scraper) Rasping tongue found in most molluscs.

ramicristate (rä-mi-kris′tāt) (L. *ramus*, branch, + *cristatus*, crested). Any member of a protozoan clade having branched tubular cristae in mitochondria; typically ameboid forms, naked or testate, including true slime molds.

ratite (ra′tīt) (L. *ratis*, raft) Having an unkeeled sternum; contrasts with **carinate.**

realized niche The role actually performed by an organism or population in its ecological community at a particular time and place as constrained by both its intrinsic biological attributes and particular environmental conditions. See also **niche** and **fundamental niche.**

recapitulation Summarizing or repeating; hypothesis that an individual repeats its phylogenetic history in its development.

redia (rē′dē-ə) pl. **rediae** (rē′dē-ē) (from Francesco Redi, 1626–97, Italian biologist) A larval stage in the life cycle of flukes; it is produced by a sporocyst larva, and in turn gives rise to many cercariae.

regulative cleavage See **radial cleavage.**

regulative development Progressive determination and restriction of initially totipotent embryonic material.

reproductive barrier (L. re, + *producere*, to lead forward; M.F. *barriere*, bar) The factors that prevent one sexually propagating population from interbreeding and exchanging genes with another population.

resource (rē′so(ə)rs) An available source of nutrition, energy or space in which to live.

respiration (L. *respiratio*, breathing). Gaseous interchange between an organ-

ism and its surrounding medium. In the cell, the release of energy by the oxidation of food molecules.

rete mirabile (rē′tē mə-rab′ə-lē) (L. wonderful net) A network of small blood vessels so arranged that the incoming blood runs countercurrent to the outgoing blood and thus makes possible efficient exchange between the two bloodstreams. Such a mechanism serves to maintain the high concentration of gases in the fish swim bladder.

reticulopodia (rə-tik′ū-lə-pō′dē-ə) (L. *reticulum*, dim. of *rete*, net, + *podos, pous*, foot) Pseudopodia that branch and rejoin extensively.

reticulum (rə-tik′yə-ləm) (L. *rete*, dim. *reticulum*, a net) Second stomach of ruminants; a netlike structure.

retortamonad (rə-tort′ə-mō′nad) (L. *retro*, bend backward, + *monas*, single). Any member of a protozoan clade composed of certain heterotrophic flagellates.

rhabdite (rab′dīt) (Gr. *rhabdos*, rod) Rodlike structures in the cells of the epidermis or underlying parenchyma in certain turbellarians. They are discharged in mucous secretions.

rheoreceptor (rē′ō-rē-cep′tor) (Gr. *rheos*, a flowing, + L. *receptus*, accept) A sensory organ of aquatic animals that responds to water current.

rhinophore (rī′nə-fōr) (Gr. *rhis*, nose, + *pherein*, to carry) Chemoreceptive tentacles in some molluscs (opisthobranch gastropods).

rhipidistian (rip-ə-dis′tē-ən) (Gr. *rhipis*, fan, + *histion*, sail, web) Member of a group of Paleozoic lobe-finned fishes.

rhizopodia (rī′zə-pō′dē-ə) (Gr. *rhiza*, root, + *podos*, foot). Branched filamentous pseudopodia made by some amebas.

rhopalium (rō-pā′lē-əm) (N.L. from Gr. *rhopalon*, a club) One of the marginal, club-shaped sense organs of certain jellyfishes; tentaculocyst.

rhoptries (rōp′trēz) (Gr. *thopalon*, club, + *tryō*, to rub, wear out) Club-shaped bodies in Apicomplexa forming one of the structures of the apical complex; open at anterior and apparently functioning in penetration of host cell.

rhynchocoel (rink′ō-sēl) (Gr. *rhynchos*, snout, + *koilos*, hollow) In nemertines, the dorsal tubular cavity that contains the inverted proboscis. It has no opening to the outside.

rostrum (räs′trəm) (L. ship's beak) A snoutlike projection on the head.

rumen (rü′mən) (L. cud) The large first compartment of the stomach of ruminant mammals. Fermentation of cellulose by microorganisms occurs in it.

S

ruminant (rüm′ə-nənt) (L. *ruminare*, to chew the cud) Cud-chewing artiodactyl mammals with a complex four-chambered stomach.

sagittal (saj′ə-dəl) (L. *sagitta*, arrow) Pertaining to the median anteroposterior plane that divides a bilaterally symmetrical organism into right and left halves.

sagittal plane (saj′i-dəl) (L. *saggita*, arrow) Pertaining to the median anteroposterior plane that divides a bilaterally symmetrical organism into right and left halves.

saprophagous (sə-präf′ə-gəs) (Gr. *sapros*, rotten, + *phagos*, from *phagein*, to eat) Feeding on decaying matter; saprobic; saprozoic.

saprophyte (sap′rə-fīt) (Gr. *sapros*, rotten, + *phyton*, plant) A plant living on dead or decaying organic matter.

saprozoic nutrition (sap′rə-zō′ik) (Gr. *sapros*, rotten + *zōon*, animal) Animal nutrition by absorption of dissolved salts and simple organic nutrients from surrounding medium; also refers to feeding on decaying matter.

sauropterygians (so-räp′tə-rij′ē-əns) (Gr. *sauros*, lizard, + *pteryginos*, winged) Mesozoic marine reptiles.

scalids (skā-lədz) (Gr. *skalis*, hoe, mattock) Recurved spines on the head of kinorhynchs.

schistosomiasis (shis′tō-sō-mī′-ə-sis) (Gr. *schistos*, divided, + *soma*, body, + *iasis*, a diseased condition) Infection with blood flukes of the genus *Schistosoma*.

schizocoel (skiz′ə-sēl) (Gr. *schizo*, from *schizein*, to split, + *koilos*, hollow) A coelom formed by the splitting of embryonic mesoderm. Noun, **schizocoelomate,** an animal with a schizocoel, such as an arthropod or mollusc. Adj., **schizocoelous.**

schizocoelous mesoderm formation (skiz′ō-sē-ləs) Embryonic formation of the mesoderm as cords of cells between ectoderm and endoderm; splitting of these cords results in the coelomic space.

schizogony (skə-zä′gə-ne) (Gr. *schizein*, to split, + *gonos*, seed) Multiple asexual fission.

sclerite (skle′rīt) (Gr. *skleros*, hard) A hard chitinous or calcareous plate or spicule;

one of the plates forming the exoskeleton of arthropods, especially insects.

scleroblast (skler′ə-blast) (Gr. *skleros*, hard, + *blastos*, germ) An amebocyte specialized to secrete a spicule, found in sponges.

sclerocyte (skler′ə-sīt) (Gr. *skleros*, hard, + *kytos*, hollow vessel) An amebocyte in sponges that secretes spicules.

sclerotin (skler′-ə-tən) (Gr. *sklērotēs*, hardness) Insoluble, tanned protein permeating the cuticle of arthropods.

sclerotization (skle′rə-tə-zā′shən) Process of hardening of the cuticle of arthropods by the formation of stabilizing cross linkages between peptide chains of adjacent protein molecules.

scolex (skō′leks) (Gr. *skōlex*, worm, grub) The holdfast, or so-called head, of a tapeworm; bears suckers and, in some, hooks; posterior to it new proglottids are differentiated.

scyphistoma (sī-fis′tə-mə) (Gr. *skyphos*, cup, + *stoma*, mouth) A stage in the development of scyphozoan jellyfishes just after the larva becomes attached; the polyp form of a scyphozoan.

sebaceous (sə-bāsh′əs) (L. *sebaceus*, made of tallow) A type of mammalian epidermal gland that produces a fatty substance.

sebum (s-e′bəm) (L. grease, tallow) Oily secretion of the sebaceous glands of the skin.

sedentary (sed′ən-ter-ē) Stationary, sitting, inactive; staying in one place.

segmentation Division of the body into discrete segments or metameres; also called **metamerism.**

semelparity (se′məl-pā′ri-tē′) A life history in which individual organisms of a population normally reproduce only one time before dying, although numerous offspring may be produced at the time of reproduction; contrasts with **iteroparity.**

sensillum pl. **sensilla** (sin-si′ləm) (L. *sensus*, sense) A small sense organ, especially in arthropods.

septum pl. **septa** (L. fence) A wall between two cavities.

serial homology See **homology.**

serosa (sə-rō′sə) (N.L. from L. *serum*, serum) The outer embryonic membrane of birds and reptiles; chorion. Also, the peritoneal lining of the body cavity.

serous (sir′əs) (L. *serum*, serum) Watery, resembling serum; applied to glands, tissue, cells, fluid.

serum (sir′əm) (L. whey, serum) The liquid that separates from the blood after coagulation; blood plasma from which fibrinogen has been removed. Also, the clear portion of a biological fluid separated from its particular elements.

sessile (ses′əl) (L. *sessilis*, low, dwarf) Attached at the base; fixed to one spot.

seta (sēd′ə), pl. **setae** (sē′tē) (L. *bristle*) A needlelike chitinous structure of the integument of annelids, arthropods, and others.

sex ratio An accounting of the proportion of males versus females in a population at a particular time and place.

sexual selection Charles Darwin's theory that there exists a struggle among males for mates and that characteristics favorable for mating may prevail through reproductive success even if they are not advantageous in the struggle for survival.

siliceous (sə-li′shəs) (L. *silex*, flint) Containing silica.

simian (sim′ē-ən) (L. *simia*, ape) Pertaining to monkeys or apes.

sink deme A subpopulation (deme) whose members are drawn disproportionately from other subpopulations of the same species (see **metapopulation dynamics**); for example, a deme occupying an environmentally unstable area whose members are periodically destroyed by climatic changes and then replenished by colonists from other demes when favorable conditions are restored.

sinus (sī′nəs) (L. curve) A cavity or space in tissues or in bone.

siphon (sī′-fən) A tube for directing water flow.

siphonoglyph (sī′fan′ə-glif′) (Gr. *siphōn*, reed, tube, siphon, + *glyphē*, carving) Ciliated furrow in the gullet of sea anemones.

siphuncle (sī′fun-kəl) (L. *siphunculus*, small tube) Cord of tissue running through the shell of a nautiloid, connecting all chambers with the animal's body.

sister taxon (=**"sister group"**) The relationship between a pair of species or higher taxa that are each other's closest phylogenetic relatives.

solenia (sə-len′ē-ə) (Gr. *solen*, pipe) Channels through the coenenchyme connecting the polyps in an octocorallian colony (phylum Cnidaria).

soma (sō′mə) (Gr. body) The whole of an organism except the germ cells (germ plasm).

somatic (sō-mat′ik) (Gr. *sōma*, body) Refers to the body, for example, somatic cells in contrast to germ cells.

somatocoel (sə-mat′ə-sēl) (Gr. *sōma*, the body, + *koilos*, hollow) Posterior coelomic compartment of echinoderms; left somatocoel gives rise to oral coelom, and right somatocoel becomes aboral coelom.

somite (sō′mīt) (Gr. *sōma*, body) One of the blocklike masses of mesoderm arranged segmentally (metamerically) in a longitudinal series beside the digestive tube of the embryo; metamere.

sorting Differential survival and reproduction among varying individuals; often confused with natural selection, which is one possible cause of sorting.

source deme A stable subpopulation (deme) that serves differentially as a source of colonists for establishing, joining or replacing other such subpopulations of the same species (see **metapopulation dynamics**); for example, a deme inhabiting an environmentally stable area whose members routinely establish transitory populations in environmentally unstable nearby areas.

speciation (spē'sē-ā'shən) (L. *species,* kind) The evolutionary process or event by which new species arise.

species (spē'shez, spē'sēz) sing. and pl. (L. particular kind) A group of interbreeding individuals of common ancestry that are reproductively isolated from all other such groups; a taxonomic unit ranking below a genus and designated by a binomen consisting of its genus and the species name.

species diversity The number of different **species** that coexist at a given time and place to form an ecological **community.**

species epithet The second, uncapitalized word in the binomial name of a species. It is usually an adjective modifying the first word, which identifies the genus into which the species is placed.

species selection Differential rates of speciation and/or extinction among varying evolutionary lineages caused by interactions among species-level characteristics and the environment.

spermatheca (spər'mə-thē'kə) (Gr. *sperma,* seed, + *thēkē,* case) A sac in the female reproductive organs for the reception and storage of sperm.

spermatophore (spər-mad'ə-fōr') (Gr. *sperma, spermatos,* seed, + *pherein,* to bear) Capsule or packet enclosing sperm, produced by males of several invertebrate groups and a few vertebrates.

spicule (spi'kyul) (L. dim. of *spica,* point) One of the minute calcareous or siliceous skeletal bodies found in sponges, radiolarians, soft corals, and sea cucumbers.

spiracle (spi'rə-kəl) (L. *spiraculum,* from *spirare,* to breathe) External opening of a trachea in arthropods. One of a pair of openings on the head of elasmobranchs for passage of water. Exhalant aperture of tadpole gill chamber.

spiral cleavage A type of early embryonic cleavage in which cleavage planes are diagonal to the polar axis and unequal cells are produced by the alternate clockwise and counterclockwise cleavage around the axis of polarity; determinate cleavage.

spongin (spun'jin) (L. *spongia,* sponge) Fibrous, collagenous material forming the skeletal network of some sponges.

spongocoel (spun'jō-sēl) (Gr. *spongos,* sponge, + *koilos,* hollow) Central cavity in sponges.

spongocyte (spun'jō-sīt) (Gr. *spongos,* sponge, + *kytos,* hollow vessel) A cell in sponges that secretes spongin.

sporocyst (spō'rə-sist) (Gr. *sporos,* seed, + *kystis,* pouch) A larval stage in the life cycle of flukes; it originates from a miracidium.

sporogony (spor-äg'ə-nē) (Gr. *sporos,* seed, + *gonos,* birth) Multiple fission to produce sporozoites after zygote formation.

sporozoite (spō'rə-zō'īt) (Gr. *sporos,* seed, + *zōon,* animal, + *ite,* suffix for body part) A stage in the life history of many sporozoan protozoa; released from oocysts.

stabilizing selection Natural selection that favors average values of a continuously varying trait and disfavors extreme values.

statoblast (stad'ə-blast) (Gr. *statos,* standing, fixed, + *blastos,* germ) Biconvex capsule containing germinative cells and produced by most freshwater ectoprocts by asexual budding. Under favorable conditions it germinates to give rise to a new zooid.

statocyst (Gr. *statos,* standing, + *kystis,* bladder) Sense organs of equilibrium; a fluid-filled cellular cyst containing one or more granules (statoliths) used to sense direction of gravity.

stenohaline (sten-ə-ha'līn, -lən) (Gr. *stenos,* narrow, + *hals,* salt) Pertaining to aquatic organisms that have restricted tolerance to changes in environmental saltwater concentration.

stenophagous (stə-näf'ə-gəs) (Gr. *stenos,* narrow, + *phagein,* to eat) Eating few kinds of foods.

stereom (ster'ē-ōm) (Gr. *stereos,* solid, hard, firm) Meshwork structure of endoskeletal ossicles of echinoderms.

sternum (ster'nəm) (L. breastbone) Ventral plate of an arthropod body segment; breastbone of vertebrates.

stigma (Gr. *stigma,* mark, tattoo mark) Eyespot in certain protozoa. Spiracle of certain terrestrial arthropods.

stolon (stō'lən) (L. *stolō, stolonis,* a shoot, or sucker of a plant) A rootlike extension of the body wall giving rise to buds that may develop into new zooids, thus forming a compound animal in which the zooids remain united by the stolon. Found in some colonial anthozoans, hydrozoans, ectoprocts, and ascidians.

stoma (stō'mə) (Gr. mouth) A mouthlike opening.

strobila (strō'bə-lə) (Gr. *strobilē,* lint plug like a pine cone [*strobilos*]) A scyphozoan jellyfish polyp with a stack of ephyrae atop it. Also, the chain of proglottids of a tapeworm.

strobilation (strō'bi-lā'shən) (Gr. *strobilos,* a pine cone) Repeated linear budding of individuals, as in scyphozoan ephyrae (phylum Cnidaria), or sets of reproductive organs in tapeworms (phylum Platyhelminthes).

stroma (strō'mə) (Gr. *stroma,* bedding) Supporting connective tissue framework of an animal organ; filmy framework of red blood corpuscles and certain cells.

subnivean (səb-ni'vē-ən) (L. *sub,* under, below, + *nivis,* snow) Applied to environments beneath snow, in which snow insulates against a colder atmospheric temperature.

survivorship (sər-vī'vər-ship) The proportion of individuals of a cohort or population that persist from one point in their life history, such as birth, to another one, such as reproductive maturity or a specified age.

suspension feeder Aquatic organisms that collect suspended food particles from the surrounding water; particles may be filtered or taken by other methods.

swim bladder Gas-filled sac of many bony fishes used in buoyancy and, in some cases, respiratory gas exchange.

sycon (sī'kon) (Gr. *sykon,* fig) A type of canal system in certain sponges. Sometimes called syconoid.

symbiosis (sim-bī-ōs'əs, sim'bē-ōs'əs) (Gr. *syn,* with, + *bios,* life) The living together of two different species in an intimate relationship. Symbiont always benefits; host may benefit, may be unaffected, or may be harmed (mutualism, commensalism, and parasitism).

synapomorphy (sin-ap'ə-mor'fē) (Gr. *syn,* together with, + *apo,* of, + *morphē,* form) Shared, evolutionarily derived character states that are used to recover patterns of common descent among two or more species.

synapsids (si-nap'sədz) (Gr. *synapsis,* contact, union) An amniote lineage comprising the mammals and their immediate ancestors, having a skull with a single pair of temporal openings.

syncytium (sin-sish′ē-əm) (Gr. *syn*, with, + *kytos*, hollow vessel) A mass of protoplasm containing many nuclei and not divided into cells.

syngamy (sin′gə-mē) (Gr. *syn*, with, + *gamos*, marriage) Fertilization of one gamete with another individual gamete to form a zygote, found in most animals with sexual reproduction.

syrinx (sir′inks) (Gr. shepherd's pipe) The vocal organ of birds located at the base of the trachea.

systematics (sis-tə-mad′iks) Science of classification and reconstruction of phylogeny.

T

tactile (tak′til) (L. *tactilis,* able to be touched, from *tangere,* to touch) Pertaining to touch.

taenidia (tə′nid′ē-ə) (Gr. *tainia,* ribbon) Spiral thickenings of the cuticle that support tracheae (phylum Arthropoda).

tagma pl. **tagmata** (Gr. *tagma,* arrangement, order, row) A compound body section of an arthropod resulting from embryonic fusion of two or more segments; for example, head, thorax, abdomen.

tagmatization, tagmosis Organization of the arthropod body into tagmata.

taxon (tak′son) pl. **taxa** (Gr. *taxis,* order, arrangement) Any taxonomic group or entity.

taxonomy (tak-sän′ə-mē) (Gr. *taxis,* order, arrangement, + *nomas,* law) Study of the principles of scientific classification; systematic ordering and naming of organisms.

tegument (teg′ū-ment) (L. *tegumentum,* from tegere, to cover) An integument; specifically external covering in cestodes and trematodes, formerly considered a cuticle.

teleology (tel′ē-äl′ə-jē) (Gr. *telos,* end, + L. *logia,* study of, from Gr. *logos,* word) The philosophical view that natural events are goal-directed and are preordained, as opposed to the scientific view of mechanical determinism.

teleost (tē′lē-ost) A clade of advanced ray-finned fishes characterized by a homocercal caudal fin.

telson (tel′sən) (Gr. extremity) Posterior projection of the last body segment in many crustaceans.

tergum (ter′gəm) (L. back) Dorsal part of an arthropod body segment.

territory (L. *territorium,* from *terra,* earth) A restricted area preempted by an animal or pair of animals, usually for breeding purposes, and guarded from other individuals of the same species.

test (L. *testa,* shell) A shell or hardened outer covering.

tetrapods (te′trə-päds) (Gr. *tetras,* four, + *pous, podos,* foot) Four-limbed vertebrates; the group includes amphibians, reptiles, birds, and mammals.

theory A scientific hypothesis or set of related hypotheses that offer very powerful explanations for a wide variety of related phenomena and serve to organize scientific investigation of those phenomena.

therapsid (thə-rap′sid) (Gr. *theraps,* an attendant) Extinct Mesozoic synapsid amniotes reptiles from which true mammals evolved.

thoracic (thō-ra′sək) (L. *thōrax,* chest) Pertaining to the thorax or chest.

Tiedemann's bodies (tēd′ə-mənz) (from F. Tiedemann, German anatomist) Four or five pairs of pouchlike bodies attached to the ring canal of sea stars, apparently functioning in production of coelomocytes.

tissue (ti′shü) (M.E. *tissu,* tissue) An aggregation of cells and cell products organized to perform a common function.

tornaria (tor-na′re-ə) (Gr. *tornos,* compass, circle, wheel) A hemichordate (acorn worm) larva that closely resembles bipinnaria larvae of sea stars.

torsion (L. *torquere,* to twist) A twisting phenomenon in gastropod development that alters the position of the visceral and pallial organs by 180 degrees.

toxicyst (tox′i-sist) (Gr. *toxikon,* poison, + *kystis,* bladder) Structures possessed by predatory ciliate protozoa, which on stimulation expel a poison to subdue the prey.

trabecular reticulum (trə-bek′yə-lər rə-tik′yə-ləm) An extensive syncytial tissue in hexactinellid sponges, bearing choanoblasts and collar bodies, forming flagellated chambers.

trachea (trā′kē-ə) (M.L. windpipe) The windpipe. Also, any of the air tubes of insects.

tracheal system (trāk′ē-əl) (L. *trachia,* windpipe) A network of thin-walled tubes that branch throughout the entire body of terrestrial insects; used for respiration.

tracheole (trāk′ē-ōl) (L. *trachia,* windpipe) Fine branches of the tracheal system, filled with fluid, but not shed at ecdysis.

transverse plane (L. *transversus,* across) A plane or section that lies or passes across a body or structure dividing it into cephalic and caudal pieces.

trend A directional change in the characteristic features or patterns of diversity in a group of organisms when viewed over long periods of evolutionary time in the fossil record.

trichinosis (trik-ən-o′səs) Disease caused by infection with the nematode *Trichinella* spp.

trichocyst (trik′ə-sist) (Gr. *thrix,* hair, + *kystis,* bladder) Saclike protrusible organelle in the ectoplasm of ciliates, which discharges as a threadlike weapon of defense.

triploblastic (trip′lō-blas′tik) (Gr. *triploos,* triple, + *blastos,* germ) Pertaining to metazoa in which the embryo has three primary germ layers—ectoderm, mesoderm, and endoderm.

trochophore (trōk′ə-fōr) (Gr. *trochos,* wheel, + *pherein,* to bear) A free-swimming ciliated marine larva characteristic of most molluscs and certain ectoprocts, brachiopods, and marine worms; an ovoid or pyriform body with preoral circlet of cilia and sometimes a secondary circlet behind the mouth.

trophallaxis (trōf′ə-lak′səs) (Gr. *trophē,* food, + *allaxis,* barter, exchange) Exchange of food between young and adults, especially certain social insects.

trophic (trō′fək) (Gr. *trophē,* food) Pertaining to nutrition.

trophic level Position of a species in a **food web,** such as producer, herbivore, first-level carnivore, or higher-level carnivore.

trophosome (trōf′ə-sōm) (Gr. *trophē,* food, + *soma,* body) Organ in pogonophorans bearing mutualistic bacteria; derived from midgut.

trophozoite (trōf′ə-zō′īt) (Gr. *trophē,* food, + *zōon,* animal) Adult stage in the life cycle of a protozoan in which it is actively absorbing nourishment.

tube feet (podia) Numerous small, muscular, fluid-filled tubes projecting from an echinoderm; part of water-vascular system; used in locomotion, clinging, food handling, and respiration.

tubulin (tü′bū-lən) (L. *tubulus,* small tube, *in,* belonging to) Globular protein forming the hollow cylinder of microtubules.

tunic (L. *tunica,* tunic, coat) In tunicates, a cuticular, cellulose-containing covering of the body secreted by the underlying body wall.

turbinates (tər′bin-āts) (L. *turbin,* whirling) Highly convoluted bones covered in mucous membrane in the nasal cavity of endotherms that serve to reduce heat and water lost during respiration.

typhlosole (tif′lə-sōl′) (Gr. *typhlos*, blind, + *sōlēn*, channel, pipe) A longitudinal fold projecting into the intestine in certain invertebrates such as the earthworm.

typological species concept The discredited, pre-Darwinian notion that species are classes defined by the presence of fixed, unchanging characters (= "essence") shared by all members.

U

ultimate cause (L. *ultimatus*, last, + *causa*) The evolutionary factors responsible for the origin, state of being, or role of a biological system.

umbo (um′bō) pl. **umbones** (əm-bō′nēz) (L. boss of a shield) One of the prominences on either side of the hinge region in a bivalve mollusc shell. Also, the "beak" of a brachiopod shell.

undulating membrane A membranous structure on a protozoan associated with a flagellum; on other protozoa may be formed from fused cilia.

ungulate (un′gū-lət) (L. *ungula*, hoof) Hooved. Noun, any hooved mammal.

uniformitarianism (ū′nə-fōr′mə-ter′ē-ə-niz′əm) Methodological assumptions that the laws of chemistry and physics have remained constant throughout the history of the earth, and that past geological events occurred by processes that can be observed today.

unitary (ū′nə-ter′ē) Describes the structure of a population in which reproduction is strictly sexual and each organism is genetically distinct from others.

uniramous (ū′nə-rām′əs) (L. *unus*, one, + *ramus*, a branch) Adjective describing unbranched appendages (phylum Arthropoda).

uropod (ū′rə-pod) (Gr. *oura*, tail, + *pous, podos*, foot) Posteriormost appendage of many crustaceans.

V

vacuole (vak′yə-wōl) (L. *vacuus*, empty, + Fr. *ole;* dimin. suffix) A membrane-bounded, fluid-filled space in a cell.

valve (L. *valva*, leaf of a double door) One of the two shells of a typical bivalve mollusc or brachiopod.

variation (L. *varius*, various) Differences among individuals of a group or species that cannot be ascribed to age, sex, or position in the life cycle.

velarium (və-la′rē-əm) (L. *velum*, veil, covering). Shelf-like extension of the subumbrella edge in cubozoans (phylum Cnidaria).

veliger (vēl′ə-jər) (L. *velum*, veil, covering) Larval form of certain molluscs; develops from the trochophore and has the beginning of a foot, mantle, and shell.

velum (vē′ləm) (L. veil, covering) A membrane on the subumbrellar surface of jellyfishes of class Hydrozoa. Also, a ciliated swimming organ of the veliger larva.

ventral (ven′trəl) (L. *venter,* belly) Situated on the lower or abdominal surface.

vestige (ves′tij) (L. *vestigium*, footprint) A rudimentary organ that may have been well developed in some ancestor or in the embryo.

vibrissa (vī-bris′ə), pl. **vibrissae** (L. nostril-hair) Stiff hairs that grow from the nostrils or other parts of the face of many mammals and that serve as tactile organs; "whiskers."

villus (vil′əs) pl. **villi** (L. tuft of hair) A small fingerlike, vascular process on the wall of the small intestine. Also one of the branching, vascular processes on the embryonic portion of the placenta.

viscera (vis′ər-ə) (L. pl. of *viscus*, internal organ) Internal organs in the body cavity.

visceral (vis′ər-əl) Pertaining to viscera.

viviparity (vī′və-par′ə-dē) (L. *vivus*, alive, + *parere,* to bring forth) Reproduction in which eggs develop within the female body, which supplies nutritional aid as in therian mammals, many reptiles, and some fishes; offspring are born as juveniles. Adj., **viviparous** (vī-vip′ə-rəs).

W

water-vascular system System of fluid-filled closed tubes and ducts peculiar to echinoderms; used to move tentacles and tube feet that serve variously for clinging, food handling, locomotion, and respiration.

weir (wēr) (Old English *wer,* a fence placed in a stream to catch fish). Interlocking extensions of a flame cell and a collecting tubule cell in some protonephridia.

X

X-organ Neurosecretory organ in eyestalk of crustaceans that secretes molt-inhibiting hormone.

Y

Y-organ Gland in the antennal or maxillary segment of some crustaceans that secretes molting hormone.

Z

zoecium, zooecium (zō-ē′shē-əm) (Gr. *zōon*, animal, + *oikos*, house) Cuticular sheath or shell of Ectoprocta.

zoochlorella (zō′ə-klōr-el′ə) (Gr. *zōon*, animal, + *Chlorella*) Any of various minute green algae (usually *Chlorella*) that live symbiotically within the cytoplasm of some protozoa and other invertebrates.

zooflagellates (zō′ə-fla′jə-lāts) Members of the former Zoomastigophorea, animal-like flagellates (former phylum Sarcomastigophora).

zooid (zō-oid) (Gr. *zōon*, animal) An individual member of a colony of animals, such as colonial cnidarians and ectoprocts.

zooxanthella (zō′ə-zan-thəl′ə) (Gr. *zōon*, animal, + *xanthos*, yellow) A minute dinoflagellate alga living in the tissues of many types of marine invertebrates.

zygote (Gr. *zygōtos*, yoked) A fertilized egg.

zygotic meiosis Meiosis that occurs within the first few divisions after zygote formation; thus all stages in the life cycle other than the zygote are haploid.

Credits

Line Art

Chapter 1

Figure 1.5: Source: A. Moorehead, *Darwin and the Beagle,* 1969, Harper & Row, New York, NY. Figure 1.14: Source: J. J. Sepkoski, Jr., *Paleobiology,* 7:36–53, 1981. Figure 1.15: From Peter H. Raven and George B. Johnson, *Biology,* 4th edition. Copyright © 1996 The McGraw-Hill Companies. All rights reserved. Figure 1.17: Source: J. Cracraft in *IBIS,* 116:294–521, 1974. Figure 1.20: Source: P. R. Grant, "Speciation and Adaptive Radiation of Darwin's Finches" in *American Scientist,* 69:653–663, 1981. Figure 1.21a: From Peter H. Raven and George B. Johnson, *Biology,* 4th edition. Copyright © 1996 The McGraw-Hill Companies. All Rights Reserved. Figure 1.25c: Source: P. M. Brakefield, "Industrial Melanism: Do We Have the Answers?" in *Trends in Ecology and Evolution,* 2:117–122, 1987. Figure 1.26: Source: A. E. Mourant, *The Distribution of Human Blood,* 1954, Ryerson Press, Toronto, Ontario, Canada. Figure 1.30: Source: D. M. Raup and J. J. Sepkoski, Jr., "Mass Extinctions in the Marine Fossil Record" in *Science,* 215:1502–1504, 1982.

Chapter 2

Figure 2.16: Source: Data from R. L. Smith, *Biology and field biology,* 3d ed. Harper and Row. New York, 1980; and E. P. Odum *Foundations of Ecology,* 3d ed. W. B. Saunders. Philadelphia, 1971.

Chapter 3

Figure 3.17: Source: J. T. Bonner, *The Evolution of Complexity,* 1988, Princeton University Press. Figure 3.18: Source: C. R. Taylor, et al., "Scaling of Energetic Costs of Running to Body Size Animals" in *American Journal of Physiology,* 219(4): 1104–1107, October 1970. Figure 3.16: From Kent M. Van De Graaff and Stuart Ira Fox, *Concepts of Human Anatomy & Physiology,* 4th edition. Copyright © 1995 McGraw-Hill Company, Inc., Dubuque, IA. All rights reserved. Reprinted by permission.

Chapter 5

Figure 5.2: Source: J. Lasman in *Journal of Parasitology,* 24:244–248, 1977. Figure 5.3a: From Peter H. Raven and George B. Johnson, *Biology,* 4th edition. Copyright © 1996 The McGraw-Hill Companies. All rights reserved.

Chapter 8

Figure 8.6: Source: Based on drawing by L. T. Threadgold from Larry S. Roberts and J. Janovy, Jr., *Foundations of Parasitology,* 5th edition, 1996, McGraw-Hill Company, Inc., Dubuque, IA. All rights reserved. Reprinted by permission. Figure 8.12: Source: J. F. Mueller and H. J. Van Cleave, *Roosevelt Wildlife Annals,* 1932. Figure 8.14: Source: D. J. Morseth, *Journal of Parasitology,* 53:492–500, 1967. Figure 8.20: Source: W. E. Sterrer, "Systematics and Evolution within the Gnathostomulida" in *Systematic Zoology,* 21:151, 1972. Figure 8.21: Source: Modified from D. R. Brooks. The phylogeny of the Cercomeria (Platyhelminthes: Rhabdocoela) and general evolutionary principles. Journal of Parasitology 75:606–616, 1989.

Chapter 9

Figure 9.5: Source: C. Con, "Kamptozoa" in H. G. Bronn, ed., *Klassen und Ordnungen des Tier-Reichs,* Vol. 4, Part 2, 1936, Akademische Verlagsgesselschaft, Leipzig. Figure 9.16: Source: R. M. Kristensen, "Loricifera, a New Phylum with Aschelminthes Characters from the Meiobenthos" in *Zeitsch. Zool Syst. Evol.,* 21:163, 1983.

Chapter 11

Figure 11.7: Source: P. Fauvel, "Annelides Polychetes. Reproduction" in P. P. Grasse, ed., *Traite de Zoologie,* Vol. 5, Part 1, 1959, Masson et Cie, Paris; modified from W. M. Woodworth, 1907.

Chapter 12

Figure 12.3: From *Synopsis and Classification of Living Organisms* edited by S. P. Parker, 1982, Volume 2. Copyright © 1982 McGraw-Hill Company, Inc., New York, NY. All rights reserved. Reprinted by permission. Figure 12.44: From Peter H. Raven and George B. Johnson, *Biology,* 4th edition. Copyright © 1996 McGraw-Hill Company, Inc., Dubuque, IA. All rights reserved. Reprinted by permission.

Chapter 14

Figure 14.19: Source: A. N. Baker, et al., "A New Class of Echinodermata from New Zealand" in *Nature,* 321:862–864, 1986. Figure 14.22: Source: W. D. Russell-Hunger, *A Biology of Higher Invertebrates,* 1969, Macmillan Publishing Company, New York, NY.

Chapter 15

Lyrics: "IT'S A LONG WAY FROM AMPHIOXUS," © Alpha Music Inc. All rights reserved. Used by permission. Figure 15.11: Source: R. J. Aldridge, et al., "The Anatomy of Conodonts" in *Phil. Trans. Roy. Soc. London B,* 340:405–421, 1993. Figure 15.12: Source: R. Zangerl and M. E. Williams in *Paleontology,* 18:333–341, 1975.

Chapter 16

Figure 16.20: From Peter Castro and Michael E. Huber, *Marine Biology.* Copyright © 1992 Mosby-Year Book. Reprinted by permission of Times Mirror Higher Education Group, Inc., Dubuque, Iowa. All rights reserved.

Chapter 17

Figure 17.1: Sources: R. L. Carroll, *Vertebraste Pale-ontology and Evolution,* 1988, W. H. Freeman & Co., NY; M. I. Coates, and J. A. Clack, *Nature,* 347:66–69, 1990; J. L. Edwards, *American Zoologist,* 29:235–254, 1989; E. Jarvik, *Scientific Monthly,* 1955: 141–154, March 1955; and C. N. Zimmer, *Discover,* 16(6): 118–127, 1995. Figure 17.2: Source: From E. W. Gaffney in Bulletin of the Carnegie Museum of Natural History, 13:92–105, 1979. Figure 17.4: Source: W. E. Duellmann and L. Trueb, *Biology of Amphibians,* 1986, McGraw-Hill Company, Inc., New York, NY.

Chapter 18

Figure 18.9: Source: After R. M. Alexander, *The Chordates.* Cambridge University Press, England, 1975.

Chapter 20

Figure 20.16: Source: N. Suga, "Biosonar and Neural Computation in Bats" in *Scientific America,* 262:60–68, June 1990. Figure 20.19: Source: J. A. Lillegraven, et al., "The Origin of Eutherian Mammals" in *Biological Journal of Linnean Society,* 32:281–336, 1987.

Photographs

Chapter 1

1.1a: © William Ober; 1.1b: Cleveland P. Hickman, Jr.; 1.1c: Courtesy Duke University Marine Laboratory; 1.1d,e: Cleveland P. Hickman, Jr.; 1.2a: Courtesy American Museum of Natural History Neg. 32662; 1.2b: © 2001 The Natural History Museum, London; pg. 6: Courtesy Foundation for Biomedical Research; 1.3, 1.4: © 2001 The Natural History Museum, London; 1.6a: BAL Portrait of Charles Darwin, 1840 by George Richmond (1809–96) Downe House, Downe, Kent, UK/Bridgeman Art Library, London/ New York; 1.6b: ©Stock Montage; 1.7, 1.8: Cleveland P. Hickman, Jr.; pg. 11: © The Natural History Museum, London; 1.10a: © A.J. Copley/Visuals Unlimited; 1.10b: © G.O. Poinar/Oregon State University, Corvallis; 1.10c: © Ken Lucas; 1:10c: © Roberta Hess Poinar; 1.11a: Boehm Photography; 1.12: Cleveland P. Hickman, Jr.; 1.16: Courtesy of The Library of Congress; 1.21b: Cleveland P. Hickman, Jr.; 1.22: Courtesy of Storrs Agricultural Experiment Station, University of Connecticut at Storrs; 1.25a,b: © Michael Tweedie/Photo Researchers, Inc.; 1.27: © Timothy W. Ransom/Biological Photo Service; 1.28: © S. Karsemann/Photo Researchers, Inc.; 1.31: Courtesy Natural Resources, Canada

Chapter 2

Opener: Cleveland P. Hickman, Jr.; 2.3 left, right: Courtesy Carl Gans; 2.08: © Noble Proctor/Photo Researchers, Inc.; 2.9 top, bottom: Cleveland P. Hickman, Jr.; 2.13a, top: © Patti Murray/Animals Animals/Earth Scenes; 2.13a, bottom: © Bill Beatty/Animals Animals/Earth Scenes; 2.13b, top, bottom, 2.13c: © James L. Castner; pg. 46: © D. Foster/WHOI/Visuals Unlimited

Chapter 3

Opener: Larry S. Roberts; 3.12a-c, 3.13 (top, bottom), 3.14 top left, top right, bottom right: © E. Reschke; 3.14 bottom left: Cleveland P. Hickman, Jr.; 3.15: © E. Reschke

Chapter 4

Opener: Cleveland P. Hickman, Jr.; 4.1: Courtesy of The Library of Congress; 4.2: Courtesy American Museum of Natural History Neg. 334101; 4.6a: © OSF/M.Cole/Animals Animals/Earth Scenes; 4.6b: ©OSF/D. Allen/Animals Animals/Earth Scenes; 4.8: Courtesy of George W. Byers/University of Kansas

Chapter 5

Opener: © Michael Abbey/Visuals Unlimited; 5.2: Courtesy of Jerry Y. Niederkorn, Ph.D., UT Southwestern Medical Center, Dallas, Texas; 5.3b: Courtesy Dr. Ian R. Gibbons; 5.4 all: © M. Abbey/Visuals Unlimited; 5.6: © L. Evans Roth/Biological Photo Service; 5.21a: Courtesy Gustaf M. Hallegraeff; 5.21b: ©A.M. Siegelman/Visuals Unlimited; 5.22: Courtesy of J. and M. Cachon, from Lee, J.J., Hunter, S.H., and Boves, E.C., (editors) 1985, "An Illustrated Guide to the Protozoa," Society of Protozoologists; Permission by Edna Kaneshiro

Chapter 6

Opener, 6.2, 6.4, 6.12a-c: Larry S. Roberts

Chapter 7

Opener: Larry S. Roberts; 7.1a: © William Ober; 7.6: © R. Harbo; 7.7, 7.8: ©CABISCO/Phototake; 7.10: © Daniel W. Gotshall; 7.12: © OSF/Peter Parks/Animals Animals/Earth Scenes; 7.13a, b: Larry S. Roberts; 7.14, 7.15: ©Rick Harbo; 7.17: Larry S. Roberts; 7.19: Daniel W. Gotshall; 7.20a: © Jeff Rotman Photography; 7.20b, 7.22, 7.23a-c, 7.25, 7.27: Larry S. Roberts; 7.29a: © Jeff Rotman Photography; 7.29b: © Kjell Sandved/Butterfly Alphabet

Chapter 8

Opener: Larry S. Roberts; 8.2: © James L. Castner; 8.3: © CABISCO/Phototake; 8.9a, 8.10: Larry S. Roberts; 8.14: © CABISCO/Visuals Unlimited; 8.16: Larry S. Roberts; 8.17: © Stan Elems/Visuals Unlimited; 8.19: Cleveland P. Hickman, Jr.

Chapter 9

Opener: Courtesy D. Despommier/From H. Zaiman A Pictorial Presentation of Parasites; 9.1 © John Walsh/Photo Researchers, Inc.; 9.3 © Robert Calentine/Visuals Unlimited; pg. 166: SEM image kindly supplied by Dr. Peter Funch, University of Aarhus; 9.4 © Larry Stepanowicz/Visuals Unlimited; 9.5: Image courtesy of Dr. Kerstin Wasson; 9.8a: Courtesy of Francas M. Hickman; 9.8b: Larry S. Roberts; 9.9: Photo by E. Pike/ From A Pictorial Presentation of Parasites; 9.10: H. Zaiman/ From A Pictorial Presentation of Parasites; 9.11a: © R. Calentine/Visuals Unlimited; 9.11b: Larry S. Roberts; 9.12: H. Zaiman/ From A Pictorial Presentation of Parasites; 9.13: Larry S. Roberts; 9.14: © David Scharf/Photo Researchers, Inc.

Chapter 10

Opener: Larry S. Roberts; 10.1a: © Kjell B. Sanvad/Visuals Unlimited, Inc.; 10.1b,c: © Rick Harbo; 10.1d: © Daniel W. Gotshall; 10.1e: Larry S. Roberts; 10.5: © Dr. Thurston C. Lacalli, Biology Department, University of Victoria; 10.6: © Kjell Sandved/Visuals Unlimited, Inc.; 10.8, 10.14a,b: © Daniel Gotshall/Visuals Unlimited, Inc.; 10.15a, b: © Alex Kerstitch/ Sea of Cortez Enterprises; 10.16a, b: Cleveland P. Hickman, Jr.; 10.18a: © Rick Harbo; 10.18b: © William Ober; 10.19, 10.20a: Larry S. Roberts; 10.20b: Cleveland P. Hickman, Jr.; 10.21, 10.24a, b, 10.25: Larry S. Roberts; 10.26, 10.28: © Rick Harbo; 10.29b: Courtesy Richard J. Neves; 10.30: Dave Fleetham/Tom Stack & Associates; 10.31: Courtesy M. Butschler, Vancouver Public Aquarium; 10.33: Larry S. Roberts

Chapter 11

Opener: Photo Gear# CRAB 02.TIF; 11.3a: © William Ober; 11.3b: Larry S. Roberts; 11.5: Cleveland P. Hickman, Jr.; 11.6: Larry S. Roberts; 11.15: © G.L. Twiest/ Visuals Unlimited; 11.17: © Tomothy Branning; 11.19: Cleveland P. Hickman, Jr.

Chapter 12

Opener: Larry S. Roberts; 12.1a, b: © A.J. Copley/ Visuals Unlimited; 12.5a: © James L. Castner; 12.5b, 12.6, 12.7: © J.H. Gerard/Nature Press; 12.8: © Todd Zimmerman/Natural History Museum of Los Angeles County; 12.9a, b: © James L. Castner; 12.10a: © Todd Zimmerman/Natural History Museum of Los Angeles County; 12.10b: © James L. Caster; 12.11: © John H. Gerard/Nature Press; 12.12: Larry S. Roberts; 12.13: © D. S. Snyder/ Visuals Unlimited; 12.22: © CABISCO/Phototake; 12.26a, b: © Rick Habro; 12.27a: Cleveland P. Hickman, Jr.; 12.28: Larry S. Roberts; 12.29a: © Rick Harbo; 12.29b, c: © Kjell Sandved/Butterfly Alphabet; 12.31a: Cleveland P. Hickman, Jr.; 12.31b, c: © Rick Harbo; 12.31d, e: Larry S. Roberts; 12.32, 12.33, 12.35: © James L. Castner; 12.36a, b: © Ron West/Nature Photography; 12.37: © Dwight R.

Kuhn; 12.40: © John D. Cunningham/Visuals Unlimited; 12.41: Courtesy Jay Georgi; 12.42: James L. Castner; 12.43a: Cleveland P. Hickman, Jr.; 12.43b: © J.H. Gerard/Nature Press; 12.45c: © James L. Castner; 12.47, 12.49a: Cleveland P. Hickman, Jr.; 12.49b: © J.H. Gerard/Nature Press; 12.49c: © CABISCO/Phototake; 12.50a, b: © James L. Castner; 12.51: © J.H. Gerard/Nature Press; 12.52: James E. Lloyd; 12.53: © James L. Castner; 12.54: © K. Lorensen/Andromedia Productions; 12.55a: © J.H. Gerard/Nature Press; 12.55b: James L. Castner; 12.56: Larry S. Roberts; 12.57a-c: © James L. Castner; 12.58a: © Leonard Lee Rue, III; 12.58b: © James L. Castner; 12.58c: © J.H. Gerard/Nature Press; 12.59, 12.60, 12.61b, 12.62, 12.63: © James L. Castner

Chapter 13

Opener: Larry S. Roberts; 13.1b: © Cleveland P. Hickman, Jr.; 13.6b: Larry S. Roberts; 13.7a: ©William Ober; 13.8a, b: © Robert Brons/Biological Photo Service; 13.10b: © James L. Castner; 13.11: Courtesy Diane R. Nelson; 13.12b: From Theusen, E.V. and R. Bertl, 1987. Canadian Journal of Zoology, 65:181-87/NRC Canada

Chapter 14

Opener: © Peter Gier/Visuals Unlimted; 14.1a-c: Larry S. Roberts; 14.1d: © Godrey Merlin; 14.4f: © Rick Harbo; 14.5: Larry S. Roberts; 14.8a: ©William Ober; 14.8b, 14.10: © Rick Harbo; 14.11a, b: Larry S. Roberts; 14.14a: © Rick Harbo; 14.14b, 14.17: Larry S. Roberts

Chapter 15

Opener: Lancelets©Heather Angel/Biophoto; 15.6: Larry S. Roberts

Chapter 16

Opener: Larry S. Roberts; 16.4: © Berthoule-Scott/Jacana/Photo Researchers; 16.8: © Jeff Rotman Photography; 16.10a: © William Ober; 6.10b: © Jeff Rotman Photography; 16.17a, b: © John G. Shedd Aquarium 1997/Patrice Ceisel; 16.18a: © James D. Watt/Animals Animals/Earth Scenes; 16.18b: © Francois Odendaal/Biological Photo Service; 16.18c:© Jeff Rotman Photography; 16.18d: © Fred McConnaughey/Photo Researchers, Inc.; 16.28: © William Troyer/Visuals Unlimited; 16.30: © Daniel W. Gotshall; 16.31: © Frederick R. McConnaughey/Photo Researchers, Inc.

Chapter 17

Opener: © Gary Mezaros/Visuals Unlimited; 17.5a, b: Courtesy L. Houck; 17.9: Cleveland P. Hickman, Jr.; 17.10a: © Ken Lucas; 17.10b, 17.11: Cleveland P. Hickman, Jr.; 17.12: Courtesy American Museum of Natural History Neg. 125617; 17.13, 17.14: Cleveland P. Hickman, Jr.

Chapter 18

Opener: Courtesy of Ron Magill/Miami Metrozoo; 18.7, 18.8: Cleveland P. Hickman, Jr.; 18.10: © John Mitchell/Photo Researchers, Inc.; 18.11: Cleveland P. Hickman, Jr.; 18.12: © John D. Cunningham/Visuals Unlimited; 18.13, 18.14: © Leonard Lee Rue, III; 18.15: © Paul Freed/Animals Animals/Earth Scenes; 18.16: © Austin J. Stevens/Animals Animals/Earth Scenes; 18.17: Cleveland P. Hickman, Jr.; 18.18: © Joe McDonald/Visuals Unlimited; 18.19: Cleveland P. Hickman, Jr.; 18.21: © Renee Lynn; 18.23: © Carmela Leszczynski/Animals Animals/Earth Scenes; 18.24a, b: Cleveland P. Hickman, Jr.

Chapter 19

Opener: PhotoDisc Vol 6 #6321; 19.1a: Courtesy American Museum of Natural History Neg; 125065; 19.19a, b: © Leonard Lee Rue, III; 19.20: Cleveland P. Hickman, Jr.; 19.22: © John Gerlach/Visuals Unlimited; 19.23: Richard R. Hansen/Photo Researchers, Inc.; 19.27, 19.28, 19.29, 19.30: Cleveland P. Hickman, Jr.

Chapter 20

Opener: © Jeff Lepore/Photo Researchers; 20.6: © Leonard Lee Rue, III; 20.7a: © PhotoDisc; 20.7b: © Corbis; 20.8: Courtesy Robert E. Treat; 20.12: © Leonard Lee Rue, III; 20.13a: Cleveland P. Hickman, Jr.; 20.15: © S. Maslowski/ Visuals Unlimited; 20.17: © Kjell Sandved/Visuals Unlimited; 20.18: © Leonard Lee Rue, III; 20.20: © G. Herben/ Visuals Unlimited; 20.22, 22.23a: Courtesy of Zoological Society of San Diego; 20.23b: Cleveland P. Hickman, Jr.; 20.24: © Milton H. Tierney, Jr./Visuals Unlimited; 20.26: © John Reader; 20.28: © John Gerlach/Visuals Unlimited; 20.29, 20.30, 20.31: Cleveland P. Hickman, Jr.; 20.33: © William Ober

Index

T

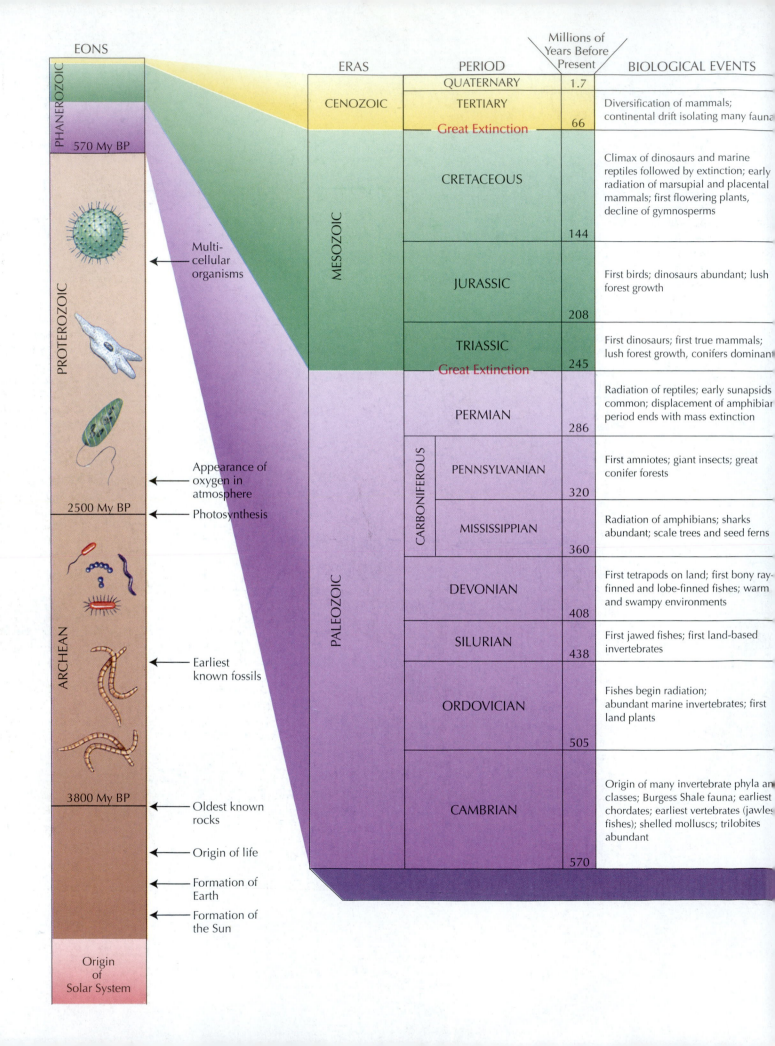

EONS		ERAS	PERIOD	Millions of Years Before Present	BIOLOGICAL EVENTS
PHANEROZOIC		CENOZOIC	QUATERNARY	1.7	
			TERTIARY		Diversification of mammals; continental drift isolating many fauna
570 My BP			Great Extinction	66	
		MESOZOIC	CRETACEOUS		Climax of dinosaurs and marine reptiles followed by extinction; early radiation of marsupial and placental mammals; first flowering plants, decline of gymnosperms
				144	
			JURASSIC		First birds; dinosaurs abundant; lush forest growth
				208	
			TRIASSIC		First dinosaurs; first true mammals; lush forest growth, conifers dominant
			Great Extinction	245	
PROTEROZOIC		PALEOZOIC	PERMIAN		Radiation of reptiles; early sunapsids common; displacement of amphibian period ends with mass extinction
				286	
			CARBONIFEROUS — PENNSYLVANIAN		First amniotes; giant insects; great conifer forests
2500 My BP				320	
			CARBONIFEROUS — MISSISSIPPIAN		Radiation of amphibians; sharks abundant; scale trees and seed ferns
				360	
			DEVONIAN		First tetrapods on land; first bony ray-finned and lobe-finned fishes; warm and swampy environments
				408	
ARCHEAN			SILURIAN		First jawed fishes; first land-based invertebrates
				438	
			ORDOVICIAN		Fishes begin radiation; abundant marine invertebrates; first land plants
				505	
3800 My BP			CAMBRIAN		Origin of many invertebrate phyla and classes; Burgess Shale fauna; earliest chordates; earliest vertebrates (jawless fishes); shelled molluscs; trilobites abundant
				570	

Multi-cellular organisms

Appearance of oxygen in atmosphere

Photosynthesis

Earliest known fossils

Oldest known rocks

Origin of life

Formation of Earth

Formation of the Sun

Origin of Solar System